T4-AJQ-037

AIP CONFERENCE PROCEEDINGS 251

SUPERCONDUCTIVITY AND ITS APPLICATIONS

 BUFFALO, NY 1991

EDITORS:

Y. H. KAO, A. E. KALOYEROS, & H. S. KWOK

STATE UNIVERSITY OF NEW YORK, BUFFALO

American Institute of Physics New York

L.C. Catalog Card No. 92-52726
ISBN 1-56396-016-8
DOE CONF-9109102

Printed in the United States of America.

CONTENTS

CHEMICAL VAPOR DEPOSITION

BULK PROCESSING AND PROPERTIES

FUNDAMENTAL PROPERTIES

APPLICATIONS

PREFACE

The papers collected in this volume were presented at the Fifth Annual Conference on Superconductivity and Applications sponsored by the New York State Institute on Superconductivity. This year's meeting was held at the Hyatt Regency Hotel in Buffalo, New York on September 24–26, 1991. Over 200 scientists from 8 countries participated in technical and poster sessions covering the full breadth of high-temperature superconductivity research.

Since the first meeting five years ago, the Conference on Superconductivity and Applications has chronicled the continued progress in thin film and bulk processing techniques. The emphasis of the annual meeting is on those fundamental and applied studies that will promote the development of practical high-temperature superconducting materials. Again this year, the Conference was proud to present the latest work in this area.

We wish to express our gratitude to the plenary and invited speakers for their cogent and illuminating presentations. We are particularly indebted to the other members of the Program Committee, P. R. Aron, P. Coppens, M. Gurvitch, J. R. Kwo, K. W. Lay, M. J. Naughton, and D. T. Shaw, for their organization of the individual symposia. The efforts of the Conference Chair, Dr. R. S. Hamilton, his assistant, Ms. B. Routhier, and Ms. P. Barnett, must also be acknowledged.

Y. H. Kao
A. E. Kaloyeros
H. S. Kwok

PLENARY REVIEWS

EPITAXIAL GROWTH

T. H. Geballe
Department of Applied Physics
Stanford University
Stanford, CA 94305

Epitaxial growth offers unique advantages for the deposition of superconducting structures. It brings an additional thermodynamic variable, interface energy, under experimental control. It makes possible the formation of new phases and metastable compositions with enhanced properties. In heterostructures it makes possible pin-hole-free insulating layers, diffusion-barrier buffer layers, seed layers for controlling orientation, and model superlattices for studying dimensional crossover and other types of phenomena. Thus, improving the understanding of epitaxial growth has many virtues.

The use of epitaxial growth to obtain enhanced superconducting properties was employed in the growth of Nb_3Ge using polycrystalline Nb_3Ir and Nb_3Rh (1). The latter are stable low T_c A15 structures which have the correct lattice constant for stoichiometric Nb_3Ge. The latter forms from the melt as a Nb-rich superconductor with a T_c of 6 K which has the A15 structure with an enlarged lattice constant. Growth on Nb_3Ir produces T_c's above 20 K. Another interesting use of epitaxy, this time employing homogeneous epitaxy in the NbAl system (that is, depositing a more nearly stoichiometric metastable composition of Nb_{3+x} Al with a slightly smaller lattice constant on a previously deposited layer with a stable composition and larger lattice constant) resulted in a film with an enhanced T_c. Tunneling experiments then were used to show that the enhanced T_c composition was able to maintain itself for up to 80 angstroms (2).

Fortunately oxides have a greater propensity for epitaxial growth than metals. This can be understood in terms of the strong and directional bonds which oxides make in contrast to metallic bonds. Qualitatively it is easy to see that the optimization of the bond energy associated with coincident sites across an interface will be important in controlling orientation. This was demonstrated soon after the discovery of high temperature superconductivity in the cuprates when highly oriented films of $YBa_2Cu_3O_7$ were grown on single crystal $SrTiO_3$ substrates at IBM and at Stanford which had very high critical currents (3). Since then epitaxial relationships have been found in many situations which are under experimental control. This makes it possible to obtain model systems and devices of increased complexity which are optimized for particular purposes.

Perhaps the most advanced heterostructure to date is the integrated squid magnetometer recently reported by Lee, Char, Colclough and Zarhachuk (4) which was pioneered by the work of Chaudhari, Dimos, Mannhart and others at IBM (5). They used bicrystals of $SrTiO_3$ of varying bicrystal angle as substrates to establish a rather general relationship between critical current and angle. There is a sharp fall off of critical current above 10 degrees which offers another experimental possibility for controlling the coupling. Recent work from Sweden shows that the fall off with angle is exponential (6). It appears also that there are a few specific angles which favor a large number of coincident sites and enhanced coupling. A 90 degree twist boundary has recently been identified in which the coupling is as strong as in unperturbed CuO_2 planes (7).

Buffer layers can be used to grow high T_c epitaxial YBCO films on Si, sapphire, and on other substrates. Currently the best structural and electrical films on Si are obtained using yttrium stabilized zirconia, (YSZ) or Y_2O_3/(YSZ) buffer layers. An interesting secondary effect occurs presumably because oxygen diffuses through the YSZ during the growth after the epitaxial relationship has been established (8). As a consequence an intermediate layer of amorphous SiO_x separates the expitaxial YSZ and silicon substrates. Most of the epitaxial growth of high T_c films has been with c-axis normal. The highly conducting CuO_2 planes are in the plane of the films and the short ξ coherence length (c < 3Å) is perpendicular. It is desirable, in order to study proximity effects and Josephson coupling in sandwich junctions and superlattices, to use a-axis films where the longer $\xi_{ab} \sim 15$Å is normal to the plane of the film. This has been borne out by initial tunneling studies (9). Excellent quality a-axis films can be grown on $LaAlO_3$ and $SrTiO_3$. X-ray studies show that the films are twin-free with small grains and 90° tilt boundaries. Rocking curves and their back scattering channeling yields compare favorably with single crystals. Superlattices between YBCO and PrBCO can be grown with little or no deterioration of the crystal properties (10). The half-width of superlattice peaks show that structural coherence is maintained for over 40 interfaces (900Å) which is the limit of the x-ray diffractometer used.

Transport (I-V) measurements on superlattices with the field perpendicular to the CuO_2 planes show crossover behavior for 24/24, 48/24 and 48/48 superlattices where the numbers refer, respectively, to the YBCO and PrBCO layers in angstroms and where the unit cell is nominally 4Å (11). The crossover from field-independent to field-dependent behavior is governed by the PBCO thickness. At present there is some ambiguity in interpreting the current-voltage characteristics of YBCO/PBCO/YBCO sandwich junctions. Evidence has been reported for a very long proximity coherence length (12) which would seem to be contrary to the interpretation of the superlattice results. For YBCO layering <24Å the conduction in the planes takes place in strips that are <24Å. Since the CuO_2 planes are weakly coupled to each other in the presence of a magnetic field the system becomes one dimensional and as expected no superconducting transition is observed.

Perhaps the most perfect c-axis multilayers have been grown layer by layer using 2201 (T_c <10K) and 2212 (T_c 85-75 K) structures of the BSCCO system at Varian by Bozovic, Eckstein, et. al. (13). In the BSCCO system there is better lattice matching and less charge transfer than in the corresponding c-axis YBCO-PrBCO system. There is no appreciable change in the resistively measured transition with superlattice modulation over an impressive range down to a single CuO_2-Ca-CuO_2 sequence of the 2212 structure separated by as much as 60 angstroms of the 2201 compound. Further studies are in progress in order to obtain quantitative information.

From the examples cited here, which are representative, but not comprehensive and do not give credit to the many who have made valuable contributions, it should be clear that there is much to be gained by developing an even deeper understanding of the thermodynamics and kinetics involved in epitaxial growth. We are no where near the limit of the benefits that should accrue from finer control of surface mobility during growth, for example. The use of epitaxy to stabilize new and unknown phases which cannot be reached by other methods has hardly been exploited. There is much more to be done.

REFERENCES

(1) A. H. Dayem, T. H. Geballe, A. B. Hallak and G. W. Hull, Jr., Appl. Phys. Lett. **30** 541 (1977).

(2) J. Kwo and T. H. Geballe, Phys. Rev. B **23**, 3230 (1981).

(3) P. Chaudhari, R. H. Koch et. al., Phys. Rev. Lett. **58**, 2684 (1987); B. Oh, M. Naito S. Arnason et. al., Appl. Phys. Lett. **51**, 852 (1987).

(4) Lee et. al., Appl. Phys. Lett. in press

(5) D. Dimos, P. Chaudhari, F. K. Mannhart, and LeGoves, Phys. Rev. Lett. **61**, 219 (1988).

(6) Z. G. Ivanov, P. A. Nilsson, D. Winkler, J. A. Alarco, G. Drorossou and T. Claeson, Extended Abstract 3rd Inst. Superconducting Electronics Conf., Strathclyde, 1991.

(7) C. B. Eom, A. F. Marshall, Y. Suzuki, B. Boyer, R.F.W. Pease and T. H. Geballe, Nature, in press.

(8) D. K. Fork, D. B. Fenner, A. Barrera, J. M. Phillips, T. H. Geballe, G.A.N. Connell and J. B. Boyce, IEEE Transactions on Applied Superconductivity, Vol. 1, March 1991.

(9) M. Lee, D. Lew, C. B. Eom, T. H. Geballe, and M. R. Beasley, Appl. Phys. Lett. **58** (11), 1990.

(10) E. C. Eom, A. F. Marshall, J.-M. Triscone, B. Wilkens, S. S. Laderman and T. H. Geballe, Science, **251**, 780 (1991).

(11) Triscone, C.B., Eom, Y. Suzuki and T. H. Geballe, Journal of Less Common Metals, in press.

(12) J. B. Barner, C. T. Rogers, A. Inam, R. Ramesh, and S. Bersey, Submitted to Appl. Phys. Lett.

(13) I. Bozovic, J. Eckstein et. al., in press.

MODELS OF CURRENT DENSITY IN BISMUTH-BASED HIGH-TEMPERATURE SUPERCONDUCTING TAPES

A. P. Malozemoff
American Superconductor Corporation
149 Grove St., Watertown MA 02172 USA

ABSTRACT

The discovery of a critical current density persisting out to high fields violates early expectations of weak-link behavior in polycrystalline $YBa_2Cu_3O_7$ and Bi-based superconductors. Explanations in terms of non-weak-linked portions of grain boundaries or the brick-wall model are assessed.

INTRODUCTION

One of the most important recent developments in high temperature superconductivity was the discovery, by Heine et al.,[1] of critical current density J_c in tape-shaped wires of bismuth-based high-temperature superconductors extending out to applied magnetic fields as high as 30 T. The original results on the Bi-2212 composition ($Bi_2Sr_2CaCu_2O_x$) have been extended, particularly by Sato et al.[2] to the Bi-2223 composition.

Typically, J_c of these tapes at 4.2 K is maximum at zero field, shows a first drop at fields of 0.1 to 1 T in the tape plane, and has a plateau at higher fields.[1-3] As temperature is increased, the absolute values drop off gradually, but the high-field plateau persists up to at least 20 K. At higher temperatures, J_c drops off more strongly with field, and by 77 K in Bi-2223, the field limit for finite J_c is only slightly larger than 1 T. A similar behavior occurs for field perpendicular to the tape plane, with only a modest anisotropy at low temperature but a strong anisotropy at 77 K, where the field limit is only 0.1 T in Bi-2223.

The absolute values of J_c are impressive, reaching levels, at least in short samples, which compete favorably with the conventional low-temperature superconductors NbTi and Nb_3Sn at 4.2 K. The record values for Bi-2212, prepared by Kase et al.[3] using a doctor-blade technique, are almost 200,000 A/cm2 in the high-field plateau. Sato et al.[2] have attained 100,000 A/cm2 at high fields in Bi-2223 tapes prepared by an oxide-powder-in-tube method with multiple pressing and heat-treat cycles. This progress has permitted the fabrication of magnet coils of steadily increasing strength, already exceeding 1 T at 4.2 K.[2,4,5]

Several other features of the data are of interest for the discussion below. Measurements with field parallel to the current flow (the "force-free" configuration) give J_c essentially identical to that with field perpendicular to the current flow and in the plane of the tape, at least at 77 K.[2] This is surprising in the context of conventional models where the Lorentz force **JxB** on the vortices is expected to play a significant role in determining the current density. Another interesting observation is of hysteresis in the measured current density for increasing and decreasing fields.[2]

These successes have opened up serious interest in applications of high-temperature superconducting wires. At the same time, the question arises as to the mechanism controlling J_c of these tapes. This paper addresses the latter issue and reviews recent work in the field. Surprisingly, there has been relatively little attention in the literature so far to the mechanism of J_c in these tapes.

The context for this problem is set by the work of Dimos et al.[6] on controlled grain boundaries in $YBa_2Cu_3O_7$ thin films. They found a steady decrease of the grain-boundary critical current as a function of the angle of the grain boundary. This work confirmed the hypothesis, first enunciated on a theoretical basis by Deutscher and Mueller,[7] that planar structures like grain boundaries can form weak links in the short-coherence-length high-temperature superconductors.

Weak links are expected to follow Josephson behavior, which usually implies not only a reduced critical current but also a rapid falloff in field. The predicted dependence of net current density j across a uniform rectangular Josephson junction of length L, on magnetic field H in the plane of the junction is

$$j(H) = j_0(\phi_0/\pi H\Lambda L)\sin(\pi H\Lambda L/\phi_0) \quad , \qquad (1)$$

where j_0 is the local maximum current density of the junction, ϕ_0 is the flux quantum and $\Lambda = \Lambda_0$

$$\Lambda_0 = 2\lambda_L + d \, , \tag{2}$$

with λ_L the London penetration depth of the superconductors flanking the Josephson junction and d the thickness of the insulating layer forming the junction. Because of its resemblance to the well-known diffraction pattern of optics, Eq. 1 is often called the Josephson Fraunhofer diffraction pattern.

Peterson and Ekin[8] have shown that a distribution of junction sizes leads to a net falloff like the envelope of Eq. 1, namely as 1/H. With a typical grain size L of 2 μm and Λ_0 of order 3000 Angstroms, the characteristic falloff field, just as the first zero of Eq. 1, occurs at a field of only 33 Oe. This weak-link mechanism is widely believed to explain the poor field-performance of ceramic $YBa_2Cu_3O_7$ superconductors.[9]

THE NON-WEAK-LINKED COMPONENT MODEL

Thus the discovery of a high-field plateau in the Bi-tapes violated conventional wisdom. The simplest hypothesis, made by Tenbrink et al.[1] and Jin et al.,[10] was that the grain boundaries in the Bi materials behaved differently from those in $YBa_2Cu_3O_7$. In particular, the grain-boundaries of the Bi-materials were assumed to be "non-weak-linked" or "strongly linked," that is, they behaved as if there was no Josephson junction at all. The high-field plateau was called the "non-weak-linked component" of the current density.

A puzzling feature was that bulk, unoriented Bi-ceramics showed as rapid dropoff with field as $YBa_2Cu_3O_7$, suggesting that under some circumstances they could also have weak-linked grain boundaries.[10,11] At the same time, evidence for a small non-weak-linked component in oriented - and even some unoriented - $YBa_2Cu_3O_7$ ceramics and bicrystals appeared in the work of Ekin et al.[11] and Babcock et al.[12] In particular, Ekin et al.[11] showed a commonality of behavior in three families of high temperature superconductors: $YBa_2Cu_3O_7$ as well as the bismuth and thallium families: all three showed both weak- and strong-linked components.

These results led Ekin et al.[11] to propose that the grain boundaries could be subdivided into weak- and strong-linked sections. The weak-linked sections could account for the low-field peak in the J_c data, since this component would be expected to be suppressed rapidly by field according to Eq. 1. Since the high-field J_c data in polycrystalline $YBa_2Cu_3O_7$ resembled the data in crystals, it was natural to suppose that the size of the non-weak-linked component was gated by the extent of the strongly-linked regions in the grain boundary. Furthermore, in aligned polycrystalline material, these data showed the same anisotropy as crystals, with J_c at 76 K persisting out to 30 T when the field was oriented along the ab-planes. Grain alignment was assumed to greatly enhance the fractional area of non-weak-linked material; this rationalized why grain-aligned material showed the enhanced high-field J_c. Nevertheless, at low temperatures, the non-weak-linked component in the best polycrystalline $YBa_2Cu_3O_7$ remains much smaller than in the best polycrystalline Bi-tapes.

Larbalestier et al.[12] studied a series of bicrystals of $YBa_2Cu_3O_7$ with different misorientations. They found that even at the same misorientation angle, J_c could exhibit either weak-linked Josephson junction character, or strong-linked "flux-pinning" character. In one case they induced a transition from one character to the other simply by oxygenating the boundary. The reason that the bicrystals appeared to give different results from the thin-film technique of Dimos et al.[6] might be that the thin-film grain boundary orientations were imposed by the substrate orientation, while the bicrystals, grown from a flux, might find optimal orientations with special low-energy grain boundaries of potentially very different physical properties.[12]

However, microstructural examination has so far not revealed direct evidence for the structures which might cause the strong- and weak-linked components, except in low angle grain boundaries, where regular arrays of grain boundary dislocations have been detected.[6,12] Given the similarity of the superconducting planes of $YBa_2Cu_3O_7$ and Bi-materials, it is also hard to understand why grain boundaries with tilt or twist of the a or b-axes might be different in these two systems. Finally, in this picture, it is hard to understand the independence of J_c with field orientation in the plane.

One difference between $YBa_2Cu_3O_7$ and Bi-materials lies in the c-axis twist boundary between platelet-shaped grains, as pointed out by Lay:[13] In the Bi-materials, the twist occurs in the middle of a BiO double layer, so that the distance between superconducting CuO_2 layers is about the same as in the bulk of the grain. But in $YBa_2Cu_3O_7$, c-axis twist boundaries tend to occur at the chain layer, with the insertion of an extra chain layer. This makes the distance between neighboring CuO_2

layers flanking the twist boundary substantially larger than in the bulk of the grain. In well-textured Bi-tapes, with the platelet planes lying in the plane of the tape and the c-axes perpendicular to the tape, these twist boundaries are aligned orthogonal to the principal axis of current flow, and so it is initially not obvious why these boundaries could affect the net J_c.

THE BRICK-WALL MODEL

An alternative model to explain Jc(B) in Bi-tapes has been proposed by Bulaevsky et al.[14] based on a "brick-wall model" discussed by Mannhart and Tsuei,[15] Clarke et al.[16] and Malozemoff.[17] The model, illustrated schematically in Fig. 1, is suggested directly by the microstructure of the tapes,[2] with platelet-like grains lying in the plane of the tape, with random a-b orientations but c-axes aligned perpendicular to the tape plane. The basic idea is that the current weaves through a first large-area c-axis twist boundary, then through the bulk of a grain, and finally through a second large-area c-axis twist boundary, so as to bypass the small a-axis or b-axis tilt boundaries in the direct path of the current.

The explanation of how such a devious current path can lead to substantial high-field J_c rests on three insights:[14] The first is that large aspect-ratio grains, oriented along the current direction, can enhance the effective grain-boundary area and so increase the net current density of the tape, as first pointed out in this context by Mannhart and Tsuei.[15]

In particular, let us consider the simple brick-wall geometry of Fig. 1 where each grain is rectangular with length 2L (and overlap length L with neighboring grains) and thickness D, and where $L >> D$. Let us also assume that the pinning inside the grains is sufficiently strong not to limit J_c. This means that the bottleneck for current flow is at the grain boundaries. If the current had to flow through the narrow "head-on" a- or b-axis twist and tilt boundaries, it would be fully limited by the low current density of those boundaries.

But if the current flows through the large-area c-axis twist boundaries, the net current density of the wire will be enhanced over that of the grain boundary by a ratio L/D, which is proportional to the aspect ratio of the grain. This mechanism gives a natural explanation of why Jc of the Bi-tapes can be so much larger than that of $YBa_2Cu_3O_7$: Bi-materials tend to grow with more highly aspected grains, with an almost mica-like morphology. The difference in the c-axis twist boundaries, mentioned above, may also contribute: These boundaries in $YBa_2Cu_3O_7$ are likely to have much lower current densities since Josephson tunneling currents depend exponentially on the separation between superconducting planes and that separation is larger in $YBa_2Cu_3O_7$.

The second insight is that the Josephson Fraunhofer diffraction formula of Eq. 1 is modified when the magnetic field exceeds the lower critical field of the grains. In this range, Abrikosov vortices penetrate the bulk of the grains, and, in a simple physical picture, they limit the region in which screening currents can flow around the junction. As the field increases, the vortices crowd closer to the junction, reducing Λ from its value in Eq. 2. In the limit of high field, Bulaevsky et al.[14] show that Λ approaches

$$\Lambda = \lambda_{ab}(8\pi M(H)/H)^{1/2} + d \quad (3)$$

where λ_{ab} is the London penetration depth for in-plane currents, and M(H) is the bulk superconductor magnetization. At the highest fields, Λ approaches d because the currents are confined to flow at the surface of the insulating portion of the Josephson junction.

The effect of this change in Λ is to expand the field-period of oscillation of the Fraunhofer diffraction pattern significantly, but it is still insufficient to explain the extraordinary field-range for the J_c plateau observed in experiment.

The third insight suggested by Bulaevsky et al.[14] to explain the high-field J_c is that the Josephson junction has a random variation in the local coupling strength, characterized by an average zero-field current density j_0, a mean square current-density deviation δj^2 and a correlation radius r_0. Such a variation could arise from random strains, from the incommensurate distortion often observed in the Bi-materials, or from the random location of vortices neighboring the junction. Yanson[18] first treated this problem for one-dimensional disorder and found that a plateau appears in the high-field portion of the Fraunhofer diffraction pattern, with magnitude of order $(\delta j^2/N)^{1/2}$, where $N=r_0/L$ is the number of random independent sections of the junction. A simple extension of this result is to the more realistic two-dimensional case with $N=LW/r_0^2$, where W is now the width of the grain in the tape plane perpendicular to its axis.

In this model, it can be shown[14,18] that the plateau starts at a first characteristic field $H_1=\phi_0/L\Lambda$ corresponding to the first zero in the diffraction pattern of the ideal uniform junction, but now with Λ given by Eq. 3. It ends at a second characteristic field $H_2=\phi_0/r_0\Lambda$. This is just the field at which the first zero would occur for a junction of length r_0, and so this model in some respect resembles a junction divided into sections of length r_0. Since both r_0 and Λ can be quite small, this result can explain the very large field-range of J_c. In fact, the end of the plateau has not been observed so far at 4.2 K, except in one study[19] of oriented sintered $DyBa_2Cu_3O_7$, where a fairly abrupt dropoff was observed for B//ab around 25 T. For the Bi-materials, the maximum measured field of 30 T sets a lower limit for the value of H_2. In the case where r_0 is determined by randomness in the positions of Abrikosov vortices with a spacing $(\phi_0/B)^{1/2}$, and where Λ approaches d as in Eq. 3, H_2 becomes simply ϕ_0/d^2, which, with d=10 Angstroms, gives the phenomenally high field of 2000 T.

Putting these results together, with $\Lambda = d$ and using the maximum plausible mean-square deviation $\delta j^2=(j_0)^2/2$, Bulaevsky et al.[14] found simple formulas describing the various features of $J_c(H)$, including the zero-field J_{c0}, the plateau value J_{c1}, the first drop-off field H_1 and the high-field dropoff H_2:

$$J_{c0} = j_0 L/D \,, \tag{4}$$

$$H_1 = \phi_0/Ld \,, \tag{5}$$

$$J_{c1} = (j_0 r_0/D)(L/2W)^{1/2} \,, \tag{6}$$

$$H_2 = \phi_0/r_0 d \,. \tag{7}$$

Comparing to experiment, they found an inconsistency in the experimentally observed ratios $J_{c0}/2J_{c1}$ and H_2/H_1, both of which should scale as L/r_0. They hypothesized that the experimental J_{c0} was reduced by trapped flux and self-field effects, as suggested by observed J_c hysteresis mentioned above and as explained by Svistunov et al.[20] Bulaevsky et al.[14] found that the other experimental parameters of, say, Sato et al.'s Bi-2223 data[2] at 4.2 K could be accounted for with a grain size 2L of 4 µm, a junction inhomogeneity radius r_0=1000 Angstroms and a grain thickness of 0.1 µm. These are plausible values.

This model also explains in a natural way the independence of J_c on orientation of the field in the tape plane, provided the grains are equiaxed (W=2L) in the tape plane. This is because the current bottleneck is at the c-axis twist boundaries where the current flows normal to the tape plane, as first pointed out by Sato et al.[2]

However, many issues still need to be resolved. This model does not explain the interesting similarity of current densities in wires and single-crystals, pointed out recently by Schmitt et al.[21] Nor does it treat, at least so far, the complex temperature dependence. The low anisotropy for fields perpendicular to the tape axis is also puzzling; perhaps pinning in the grains causes significant distortions of the Abrikosov vortices, causing sections to lie parallel to the junction.[14] Svistunov et al.[22] reported a weak dependence of critical current on hydrostatic pressure in unoriented pressed pellets of Bi-2223. The effect could be accounted for simply by a T_c change, and so it appears to rule out a weak-link mechanism for the critical current. This important experiment needs to be repeated on high-current-density tapes. Most of all, controlled study of single c-axis twist boundaries and direct confirmation of the twisting current paths is needed. Finally, it must be recognized that there is an interesting similarity between this and the non-weak-linked-component model: In both there are strong inhomogeneities in the grain-boundaries, and the structural origin of these inhomogeneities needs to be identified.

Clearly, there is a need for more detailed scientific study of this complex problem, and one can expect accelerating progress as the tapes themselves improve in quality.

The author has benefited from many valuable discussions with L. Bulaevsky, J. R. Clem, P. Chaudhari, D. R. Clarke, J. Ekin, T. Sheahan, D. Larbalestier and M. Tinkham.

REFERENCES

1. K. Heine, J. Tenbrink and M. Thoener, Appl. Phys. Lett. 55, 2441 (1989); J. Tenbrink, K. Heine and H. Krauth, Cryogenics 30, 422 (1990).
2. K. Sato, T. Hikata, H. Mukai, T. Masuda, M. Ueyama, H. Hitotsuyanagi, T. Mitsui and M. Kawashima, in Advances in Superconductivity II, ed. T. Ishiguro and K. Kajimura (Springer Verlag, Tokyo 1990), p. 335; T. Hikata, M. Ueyama, H. Mukai

and K. Sato, Cryogenics 30, 924 (1990); K. Sato, T. Hikata, H. Mukai, M. Ueyama, N. Shibuta, T. Kato, T. Masuda, M. Nagata, K. Iwata and T. Mitsui, IEEE Trans. Magn. 27,1231 (1991).
3. J. Kase, K. Togano, D. R. Dieterich, N. Irisawa, T. Morimoto and H. Maeda, Japan J. Appl. Phys. 29, L1096 (1990).
4. K. Togano, H. Kumakura, K. Kadowaki, H. Kitaguchi, H. Maeda, J. Kase, J. Shimoyama and K. Nomura, Proc. ICMC'91, Huntsville, Alabama, June 11-14, 1991.
5. H. Kitamura, T. Hasegawa, H. Takeshita, K. Yamamoto, T. Murase and H. Ogiwara, Cryogenics Society of Japan, Makuhari, May 14-16, 1991, Abstract p. 231.
6. D. Dimos, P. Chaudhari and J. Mannhart, Phys. Rev. B41, 4038 (1990).
7. G. Deutscher and K. A. Mueller, Phys. Rev. Lett. 59, 1160 (1987).
8. R. L. Peterson and J. W. Ekin, Phys. Rev. B 37, 9848 (1988).
9. A. P. Malozemoff, in Physical Properties of High Temperature Superconductors, Vol. 1, ed. D. Ginsberg (World Scientific, Singapore 1989), p. 71.
10. S. Jin, R. B. van Dover, T. H. Tiefel, J. E. Graebner and N. D. Spencer, Appl. Phys. Lett. 58, 868 (1991); S. Jin, T. H. Tiefel, R. B. van Dover, J. E. Graebner, T. Siegrist and G. W. Kammlott, Appl. Phys. Lett., to be published.
11. J. W. Ekin, T. M. Larson, A. M. Hermann, Z. Z. Sheng, K. Togano and H. Kumakura, Physica C 160, 489 (1989); J. W. Ekin, H. R. Hart, Jr., and A. R. Gaddipati, J. Appl. Phys. 68, 2285 (1990).
12. S. E. Babcock, X. Y. Cai, D. L. Kaiser and D. C. Larbalestier, Nature 347, 6289 (1990); D. C. Larbalestier, S. E. Babcock, X. Y. Cai, M. B. Field, Y. Gao, N. F. Heinig, D. L. Kaiser, K. Merkle, L. K. Williams and N. Zhang, Physica C, to be published; Y. Feng, K. E. Hautenen, Y. E. High, D. C. Larbalestier, R. Ray II, E. E. Hellstrom and S. E. Babcock, Physica C, to be published.
13. K. Lay, AIP Conf. Proc., ed. Y.-H. Kao et al. (AIP, New York 1991), p. 119.
14. L. N. Bulaevskii, J. R. Clem, L. I. Glazman and A. P. Malozemoff, Phys. Rev. B, Rapid Commun., to be published.
15. J. Mannhart and C. Tsuei, Z. Physik B77, 53 (1989).
16. D. R. Clarke, T. M. Shaw and D. Dimos, J. Am. Ceram. Soc. 72, 1103 (1989).
17. A. P. Malozemoff, in High Temperature Superconducting Compounds II, ed. S. H. Whang et al. (TMS Publications, Warrendale PA 1990), p. 3.
18. I. K. Yanson, Sov. Phys. JETP 31, 800 (1970).
19. D. P. Hampshire, J. Seuntjens, L. D. Cooley and D. C. Larbalestier,

Appl. Phys. Lett. 53, 814 (1988).
20. V. M. Svistunov, A. I. D'yachenko and V. Yu. Tarenkov, Physica C, to be published.
21. P. Schmitt, P. Kummeth, L. Schultz and G. Saemann-Ischenko, Phys. Rev. Lett. 67, 267 (1991).
22. V. M. Svistunov, A. I. D'yachenko and V. Yu. Tarenkov, Preprint DonPhTI-91-12, Donetsk Physico-Technical Institute of the Ukrainian Academy of Sciences, 340114 Donetsk USSR.

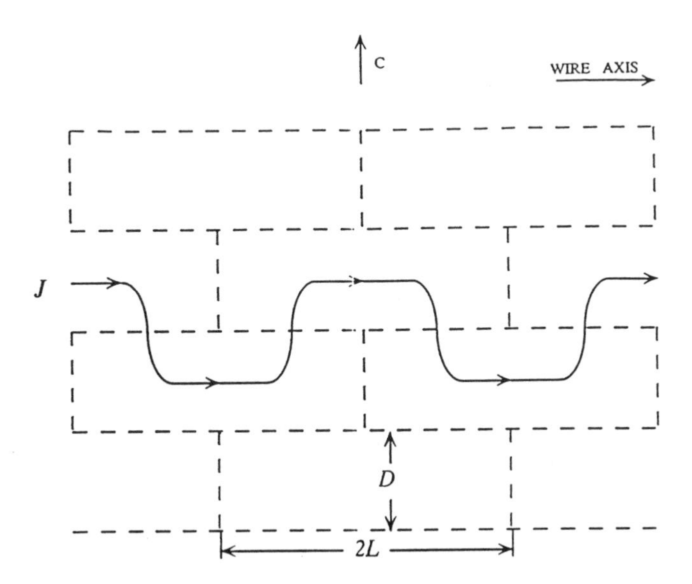

Fig. 1. Brick-wall model of current flow in Bi-based high-temperature superconducting tapes. Each brick, representing a grain with length 2L and thickness D, has a random a-b orientation but a common c-axis perpendicular to the tape plane. The current weaves across the c-axis twist boundaries as shown.

CRITICAL CURRENT, GRAIN BOUNDARIES, AND SQUIDS

P. Chaudhari

IBM Research Division, T. J. Watson Research Center,
Yorktown Heights, NY 10598, USA

ABSTRACT

This note is a summary of the talk presented at the meeting. It describes our current understanding of the weak link nature of grain boundaries in the high temperature cuprate superconductors. An hypothesis is advanced which suggests how large critical currents in polycrystalline wires can be attainable.

SUMMARY

This paper is concerned with grain boundaries in the high temperature cuprate superconductors. The grain boundaries can be detrimental when these materials are used as wires, but offer a unique opportunity to produce Superconducting Quantum Interference Devices (SQUIDs). Both aspects stem from the observation that large angle grain boundaries in these materials are Josephson coupled weak links. This limits the critical current density in wires and, in the case of SQUIDs, provides naturally occurring Josephson junctions. In this paper I shall describe our current understanding of the nature of the grain boundary that leads to Josephson coupling. We shall use this understanding to suggest what microstructure might be optimal for producing wires. We shall also touch on the behavior of SQUIDs using grain boundary junctions.

In terms of detail, this written version of the paper is divided into two parts. Most of the information is published and this part is therefore presented with great brevity. The new thought, concerning the behavior of grain boundaries in large magnetic fields such as those encountered in wires, is presented in greater detail.

Experiments carried out on single grain boundaries between two crystals show that the boundaries are weak links.[1] For grain boundaries formed between two c-axis films deposited on [100] $SrTiO_3$ bicrystals, the weak link shows flux flow behavior for small misorientation angles and Josephson coupled weak link when the misorientation angle is large.[2] This weak link behavior is observed in the Nd, Y, Bi, and Tl superconductors.[3] The critical angle at which this switchover in behavior occurs depends upon the superconducting system.

The weak link behavior cannot be explained in terms of the symmetry (p- or d-wave) of the wave function.[4] Nor can it be explained in terms of large concentration variations.[5,6] The role of grain boundary dislocations as the possible cause of weak link behavior was discussed as soon as the observation of decreasing critical current with increasing misorientation angle was made.[2] However, the precise mechanism was never delineated. It has been suggested that hydrostatic strain might reduce the order parameter. However, the similarity in decrease of the critical current with increasing misorientation angle of both tilt and twist boundaries and the presence of an oscillatory strain field associated with a dislocation array are inconsistent with a simple hydrostatic strain model.[4] The other possibility, associated with dislocations, is to assume that the core of the dislocation suppresses the order parameter. This model has the feature that for small angles of misorientation pseudo one-dimensional flux flow along the grain boundary is expected. As the misorientation angle increased and the spacing between the dislocations approaches the coherence length, the order parameter never recovers its full value along the boundary. When this order parameter is less than that imposed by the adjoining grain, Josephson coupling dominates.[7] Qualitatively, the behavior of the order parameter as a function of the misorientation angle, θ, is expected to decrease with increasing misoriention. Both Josephson and intrinsic boundary coupling decrease with increasing angle. However, the intrinsic part decreases more rapidly and eventually Josephson coupling dominates. When $\theta < \theta_c$, flux flow behavior dominates and for $\theta > \theta_c$ Josephson coupling.

The anticipated behavior of the critical current as a function of magnetic field can be described as follows, when $\theta > \theta_c$. There is a large initial drop arising from suppression of the Josephson current followed by a slow decrease which is the flux flow component. It is this component which is desirable for wires and it is the Josephson component which is desirable for SQUIDs.

The microstructure required for wires is a combination of controlling the c-axis texture and the relative misorientation between grains once the c-axes of the grains have been aligned parallel to each other. We note that the average texture in the plane could be random. It is only the nearest neighbor orientation that controls the extent of the critical current dominated by flux flow. These considerations are in addition to the geometrical shape arguments.[8]

Experimental evidence that the grain boundary has a superconducting gap, albeit suppressed, is, at best, indirect. Direct experimental evidence would be very desirable and a possible approach is to use a low temperature STM. These experiments are underway. The indirect experimental confirmation comes from the following observations. Firstly, the critical current of wires is found to have a magnetic field dependence of the type sketched in Figure 2. Secondly, as the temperature is increased, the slowly varying component of the critical current with magnetic field decreases more rapidly than the zero field value, suggestive of a lower gap. Thirdly, the behavior of the grain boundary junction is consistent with models that assume it is a dirty metal rather than a semiconductor or insulator. The junction resistance, R_n, is independent of temperature;[2] the temperature dependence of the critical current is quadratic with reduced temperature;[4,9,10] and the I_cR_n product is a function of the junction resistance.[11-13] The value of resistivity, associated with the boundary region of approximately 20Å width, is of the order of $10^{-1}\Omega - cm$. Assuming that the scattering rate is substantially larger or alternatively the mean free path is close to atomic dimensions, we estimate the number of charge carriers in the boundary region to be within an order of magnitude of the bulk value. Experimental evidence that the scattering rate is enhanced in the grain boundary region comes from 1/f noise measurements and ozone annealing experiments.[14]

Experimental evidence for the Josephson dominated critical current is decisive. For large angle grain boundaries we observe all of the signatures generally associated with Josephson coupling. Perhaps the most direct evidence is the observation of SQUID oscillations in single grain boundary junctions.[15] These junctions have been used to produce the lowest noise dc SQUIDs operating at 77 K. More recently, these grain boundary junctions have been used to produce sub-micron SQUIDs[16], and, using lithographic techniques, a fully integrated SQUID.[17] In another development, a SQUID response has also been obtained from patterned superconducting films deposited on zirconia bicrystals.[18]

In summary, we have narrowed the range of possible mechanisms responsible for the weak link behavior of grain boundaries in cuprate superconductors. We believe that a description of the boundary in terms of a dirty metal with a suppressed gap is consistent with experimental information. The presence of Josephson coupling can be usefully exploited to produce superconducting devices, such as SQUIDs.

REFERENCES

1. P. Chaudhari, J. Mannhart, D. Dimos, C.C. Tsuei, C.C. Chi, M.M. Oprysko, and M. Scheuermann, Phys. Rev. Lett. **60**, 1653 (1988).
2. D. Dimos, P. Chaudhari, J. Mannhart, and F. LeGoues, Phys. Rev. Lett. **61**, 219 (1988); J. Mannhart. P. Chaudhari, D. Dimos, C.C. Tsuei, and T.R. McGuire, Phys. Rev. Lett.**61**, 2746 (1988); D. Dimos, P. Chaudhari, and J. Mannhart, Phys. Rev. B **41**, 4038 (1990).
3. P. Chaudhari, A. Gupta, M. Kawasaki, J. Lacey, W. Lee, and E. Sarnelli, to be published.
4. P. Chaudhari, in M^{2}S-HTSC III Proceedings of the Third International Conference on Materials and Mechanisms of Superconducvity High-Temperature Superconductors, (Kanazawa, Japan, July 1991), to be published in Physica C.
5. M. Chisholm and S.J. Pennycook, Nature 351, 47 (1991).

6. D. Larbalestier, Phys. Today **44**, 74 (1991).
7. P. Chaudhari, in Proceedings of the Toshiba International School of Superconductivity, (Kyoto, Japan, July 1991), to be published in Solid-State Sciences.
8. A.P. Malozemoff, this proceeding.
9. G. Deutscher, IBM J. Res. Develop. **33**, 293 (1989).
10. R. Gross, P. Chaudhari, D. Dimos, A. Gupta, and G. Koren, Phys. Rev. Lett. **64**, 228 (1990).
11. R. Gross, P. Chaudhari, M. Kawasaki, and A. Gupta, Phys. Rev. B **42**, 10735 (1990).
12. S.E. Russek D.K. Lathrop, R.A. Burhmann, and D.H. Shin, Appl. Phys. Lett. **57**, 1155 (1990).
13. G. Deutscher and P. Chaudhari, Phys. Rev. B, in print.
14. M. Kawasaki, P. Chaudhari, and A. Gupta, Phys. Rev. Lett., submitted for publication.
15. R. Gross, P. Chaudhari, M. Kawasaki, M.B. Ketchen, and A. Gupta, Appl. Phys. Lett. **57**, 727 (1990). See also Physica C **170**, 315 (1990).
16. M. Kawasaki, P. Chaudhari, T. Newman, and A. Gupta, Appl. Phys. Lett., June (1991).
17. L.P. Lee, K. Char, M.S. Colclough, G. Zaharchuk, J.J. Kingston, A.H. Miklich, F.C. Wellstood, and John Clarke, to be published.
18. Z.G. Ivanov, P.A. Nilsson, D. Winkler, J.A. Alarco, G. Brorsson, T. Claeson, E.A. Stepantsov, and A. Ya. Tzalenchuk, to be published in Proceedings of the SQUID '91 Meeting, (Berlin, Germany, 1991).

GROWTH AND SUPERCONDUCTIVITY OF SINGLE ONE UNIT-CELL $YBa_2Cu_3O_{7-x}$ LAYER

Y. Bando, T. Terashima, K. Shimura and T. Sato
Institute for Chemical Research, Kyoto University, Uji, Kyoto-fu 611, Japan

Y. Matsuda and S. Komiyama
Department of Pure and Applied Science, University of Tokyo, 381 Komaba, Meguro-ku, Tokyo 153, Japan

K. Kamigaki and H. Terauchi
Faculty of Science, Kwansei-Gakuin University, Nishinomiya 662, Japan

ABSTRACT

Structure and superconductivity of ultrathin films as thin as a unit-cell size have been investigated. Oscillations in the reflection high energy electron diffraction (RHEED) specular intensity during growth of the superconducting oxides on $SrTiO_3$ indicate that the growth occurs in monomolecular layer by monomolecular layer mode. The film thickness can be exactly controlled by monitoring the RHEED intensity. The isolated ultrathin $YBa_2Cu_3O_{7-x}$ (YBCO) films sandwiched in between $PrBa_2Cu_3O_{7-x}$ (PrBCO) layers were prepared using the growth interruption technique by oscillations monitoring. A one unit-cell YBCO layer has a superconducting transition with the onset temperature at 70K and the zero resistance at 30K. The overlayer of PrBCO was required to provide the underlying one unit-cell YBCO layer with hole carriers sufficient to induce superconductivity. The reduction of the onset temperature is found to be mainly due to the decrease of the hole carriers but not due to the absence of the interlayer coupling. The resistive transitions for the YBCO layers with the thickness below 5 unit-cells and also for the YBCO (1 unit-cell) / PrBCO (1 unit-cell) superlattice are magnetic field independent up to 10T when the field is applied parallel to the CuO_2 layers. In perpendicular magnetic fields, the activation energy of flux flow for the isolated ultrathin layers is nearly proportional to the number of unit-cells. The activation energy for the YBCO (1 unit-cell) / PrBCO (1 unit-cell) superlattice is comparable to the value of the isolated layers, suggesting that the flux lines along c-axis are limited to each one unit-cell layer of YBCO.

INTRODUCTION

Recently, superlattices consisting of superconducting YBCO and nonsuperconducting PrBCO have been intensively studied in order to solve the following problems.

1. Are interlayer coupling between the CuO_2 bilayers in adjacent unit-cells necessary for superconductivity to occur?
2. What role does the interlayer coupling play in determining T_c?

Triscone et al[1]. reported that a coupling between the CuO_2 bilayers was necessary for occurrence of superconductivity. Contrarily, Li et al[2]. and Lowndes et al[3]. have shown that superconductivity can occur in one unit-cell YBCO layer and that a coupling between CuO_2 bilayers is needed to achieve a T_c of 90K. Interests are furthermore focussed on the transition characteristics[4-7] and flux flow[8] related to the reduced dimensionality. Particularly, when a magnetic field was applied parallel to the c-axis, Brunner et al. found very interesting fact that, for thick enough PrBCO layers in the superlattices, the activation energy of flux motion was proportional to the YBCO thickness.

Although utilization of artificial superlattice is effective to address the superconducting properties of the ultrathin layers, interlayer coupling effect and interconnection between YBCO layers through PrBCO layers may not be completely eliminated. Comparison of the characteristics of isolated ultrathin films with those of the superlattices is desirable for solving the problems.

However, there have never been means of measurements for the thickness of a ultrathin YBCO layer sandwiched between PrBCO layers with the same structure. Recently, we have found definite intensity oscillations in reflection high energy electron diffraction (RHEED) caused by a two-dimensional unit-cell by unit-cell growth during the epitaxial growth of YBCO on $SrTiO_3$[9]. The increase in the thickness during an oscillation period precisely corresponds to the one unit-cell height in the c direction. The monitor of the RHEED oscillations makes it possible to estimate exactly in-situ the film thickness.

There are many problems to be solved to realize a superconducting YBCO film as thin as one unit-cell. One of the most crucial problem is the lattice distortion of YBCO due to the lattice mismatch. The $SrTiO_3$ substrate give the ultrathin YBCO layers with no orthorhombicity (a=b) and with the same in-plane lattice constant as the substrate which exhibit nonsuperconductivity in the thickness below 4 unit-cells[10]. The two unit-cells YBCO layer on MgO exhibiting superconductivity has no orthorhombicity but the same in-plane lattice constant as that of the bulk[11]. The one unit-cell layer of YBCO on MgO exhibits nonsuperconductivity and large lattice distortion[12]. Additionally, the surface films grown on MgO is rough and therefore the film thickness may be not uniform. The problem on superconductivity of one unit-cell layer of YBCO remains unsolved. Therefore, we have used a PrBCO buffer layer on $SrTiO_3$ as a substrate, since PrBCO has the same 1:2:3 structure with lattice constants close to those of YBCO (lattice mismatch≤1.5%). In the present paper, we discuss superconductivity in a few unit-cells thick layers of YBCO, in the isolated state and in the superlattices, and behavior of the resistive transition in magnetic fields.

EXPERIMENTALS

The films were grown on $SrTiO_3$ by reactive coevaporation. Detail growth conditions were described elsewhere[9,13]. In order to control the film thickness, the specular intensity in RHEED during growth was measured using photomultiplier through optical fiber. The acceleration voltage of the incident electron beam for RHEED observation was 20kV. X-ray diffraction measurements were carried out using a conventional double-axis diffractometer. We took the $[001]_s$ traces with a symmetric scattering mode and the $[100]_s$ (or $[010]_s$) traces with an asymmetric mode. The $[001]_s$ trace provides information on the lattice along the growth direction and $[100]_s$ trace gives the in-plane lattice properties[10]. Resistivity was measured by a usual four prove method. The typical distance between electrodes was 1mm. The current density was less than $10A/cm^2$.

RESULTS AND DISCUSSIONS

1. Fabrication and structural examination of ultrathin films and superlattice

Figure 1 shows RHEED oscillations for the sequence of growth of the 6 unit-cells buffer layer of PrBCO on $SrTiO_3$, the one unit-cell of YBCO and the overlayer of PrBCO by interruption at every one unit-cell growth[6]. The specular intensity decreased to form the minimum, then increased and had a maximum peak. As one period (peak

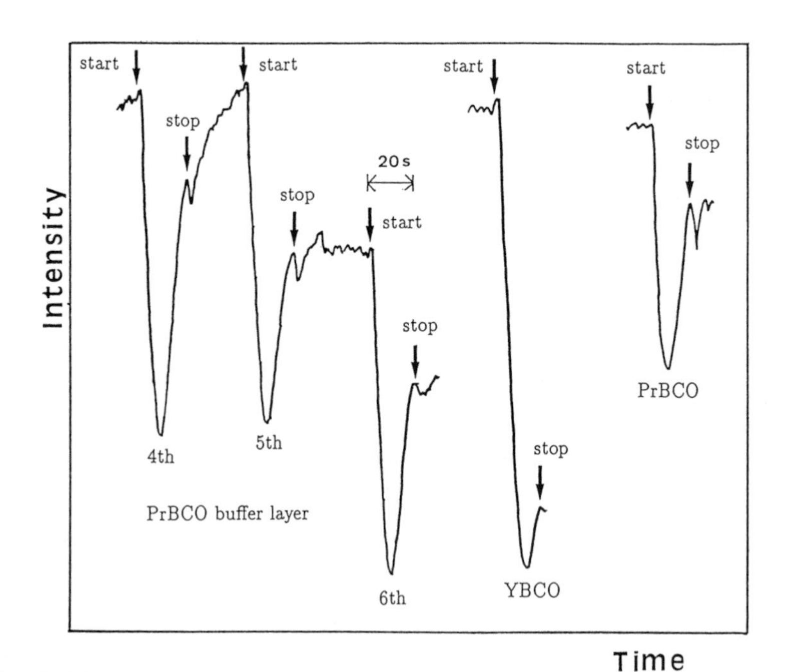

Fig. 1. RHEED intensity oscillations during epitaxial growth of the PrBCO buffer layer (6 unit-cells), the YBCO one unit-cell layer, and the PrBCO overlayer.

to peak) corresponds to the growth of a one unit-cell thick layer, the growth of each one unit-cell layer can be controlled by monitoring RHEED intensity. After each growth interruption, we took 20 sec before the next growth was started, in order to smooth the surface by allowing surface adatoms to migrate. The rocking curve through (004) was measured in order to evaluate the crystallinity of PrBCO film (230Å thick) grown on $SrTiO_3$ by the growth interruption technique of every one unit-cell. The mosaic spread is very narrow, as the full width at half maximum of the rocking curve is 0.09° and comparable to that of the substrate of $SrTiO_3$ (0.08°).

Cross-sectional transmission electron microscopic (TEM) image of PrBCO (4 unit-cells) / YBCO (1 unit-cell) / PrBCO (6 unit-cells) / $SrTiO_3$ (100) trilayer film was observed[14]. The fringe contrasts of a unit-cell in the image are parallel to the surface without large imperfections, and the numbers of the unit-cells are 11, being the same as calculated in RHEED oscillations. This indicates that the layer thickness is exactly controlled in a unit-cell scale. In order to examine the uniformity of one unit-cell layer grown by the reactive coevaporation, the structure of the bilayer superlattice [YBCO (1 unit-cell) / PrBCO (1 unit-cell)] × 10 grown using RHEED intensity monitoring was investigated. Generally, if the layer-by-layer growth is not achieved enough, the interface roughness would introduce very serious breaking of superlattice coherency. It will appear as broad superlattice peaks in X-ray spectrum. Figure 2 shows the X-ray scattering spectrum scanned along the direction of c-axis. The strong and sharp satellite peaks at the middle between the fundamental peaks (00*l*) are caused by the superlattice modulation. The full width (0.4°) at the half maximum of the superlattice peaks is entirely the same as that of the fundamental peaks. The oscillatory peaks between the fundamental and satellite peaks are attributed to the finite layer thickness of the film and called Laue oscillations. The narrow width of the satellite peaks and

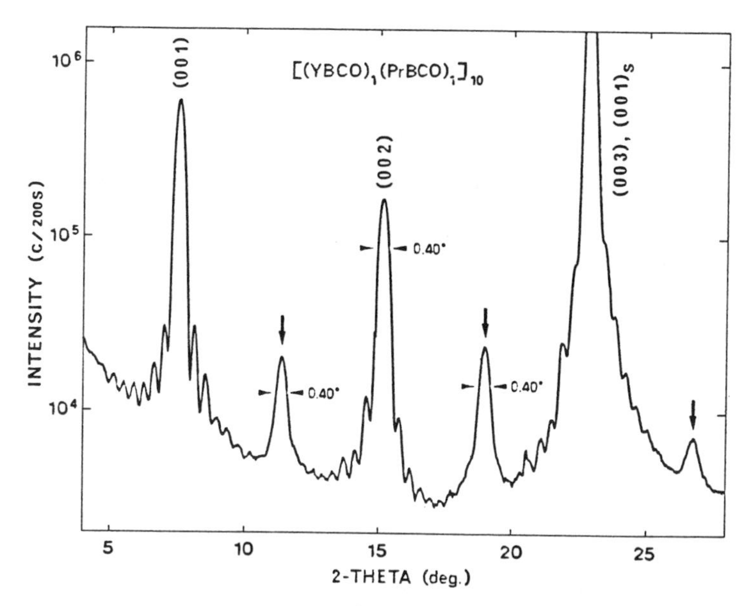

Fig. 2. X-ray diffraction profile scanned along the direction normal to the crystal surface in [YBCO (1 unit-cell) PrBCO / (1 unit-cell)]×10. Satellite peaks are indicated by arrows.

the sharp Laue oscillations indicate that there is almost no fluctuation of compositional modulation and no intermixing of Y and Pr due to the interface roughness. The result of X-ray diffraction study demonstrated that exactly unit-cell controlled growth was established in the reactive coevaporation with the RHEED oscillations. It is clear that the ultrathin layers of YBCO on the PrBCO buffer layer are uniform in thickness and continuous enough to investigate superconductivity for one unit-cell layer.

2. Superconductivity of ultrathin films

Figure 3 depicts resistive transition curves for PrBCO (6 unit-cells) / YBCO (n unit-cell, n=1, 2, 3, 5, 10) / PrBCO (6 unit-cells) / $SrTiO_3$ (100) grown by the reactive coevaporation using RHEED oscillation monitoring[6]. Even the layer of n=1 is found to exhibit superconductivity with the transition onset T_c^{on} at 70K and the zero resistance T_{co} at 30K. The T_{co} increases rapidly and the transition becomes sharp, as the n increases. The T_c^{on} for n=1, 2, 3, 5 layers of YBCO are ~70, 86, 88, 90K and does not change drastically with decreasing n. Hall coefficient measurements show the carrier number for n=1 is smaller than the value of bulk sample by a factor of 1.3~1.7. It is suggested that the reduction of T_c^{on} is due to the reduction of hole carrier concentrations rather than due to the absence of the interlayer coupling. The reduction of the hole carrier concentration in the n=1 may come from sharing holes doped from Cu-O chain layer between the unit-cells of PrBCO and YBCO.

The feature of the resistive curve for n=1 is a significantly broad transition. A pure two dimensional superconductor would be described by the Kosterlitz-Thouless (KT) theory of phase transition. In the superconductor, vortices and antivortices are induced by a thermal fluctuation even in the absence of magnetic field, and are bounded in pairs below T_{KT}. If the temperature at which resistivity goes to zero is T_{KT}, an effective thickness of the superconduction layer would be about 2Å[15]. This

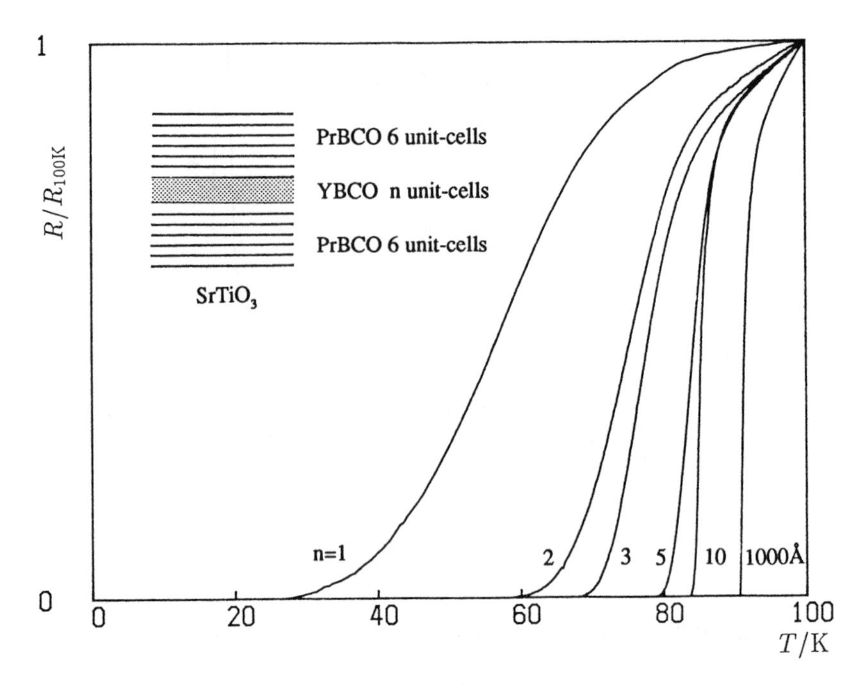

Fig. 3. Normalized resistance vs. temperature curves for n unit-cells thick layers (n=1,2,3,5,10) of YBCO sandwiched in between the 6 unit-cells layers of PrBCO and for a 1000Å thick YBCO film.

thickness seems to be rather small. Slight local randomness due to the intermixing of PrBCO layer in the one unit-cell layer of YBCO would give rise to a phase breaking effect in KT transition and lead to the reduction of T_{KT}.

3. Conditions of one unit-cell layer of YBCO needed for superconductivity to occur

We discuss superconductivity of the ultrathin layer of YBCO, taking effects of the adjacent PrBCO layers into account. The underlying PrBCO layer, indeed, plays a role of a buffer layer having a better lattice matching with YBCO than $SrTiO_3$. At first, we discuss the lattice modification introduced by lattice mismatches among YBCO, PrBCO and $SrTiO_3$. Figure 4 shows the resistive transition for the 2 unit-cell layers of YBCO located between PrBCO layers with various thicknesses. The resistive transition occurs at the same temperature except that for the sample (a) which is thinner than other samples. An asymmetric X-ray scattering spectrum along [100] direction for the multilayer film (d) (total thickness is 281Å) is found to exhibit the orthorhombic structure with a=3.879Å and b=3.917Å which are quite similar to those of the PrBCO lattice[16]. However, a multilayer film (a) (total thickness is 105Å) is the mixture of the orthorhombic structure with the same lattice constant as PrBCO and the tetragonal-like structure with the lattice constant similar to that of $SrTiO_3$. The component part of the tetragonal-like structure which may be somewhat more than that of the orthorhombic structure reduces the transition temperature.

Although the sample of n=2 (total thickness is 164Å) in Fig. 3 contains the component of tetragonal-like structure which is less than that of the orthorhombic structure, its transition temperature is entirely similar to that of the sample (d). The

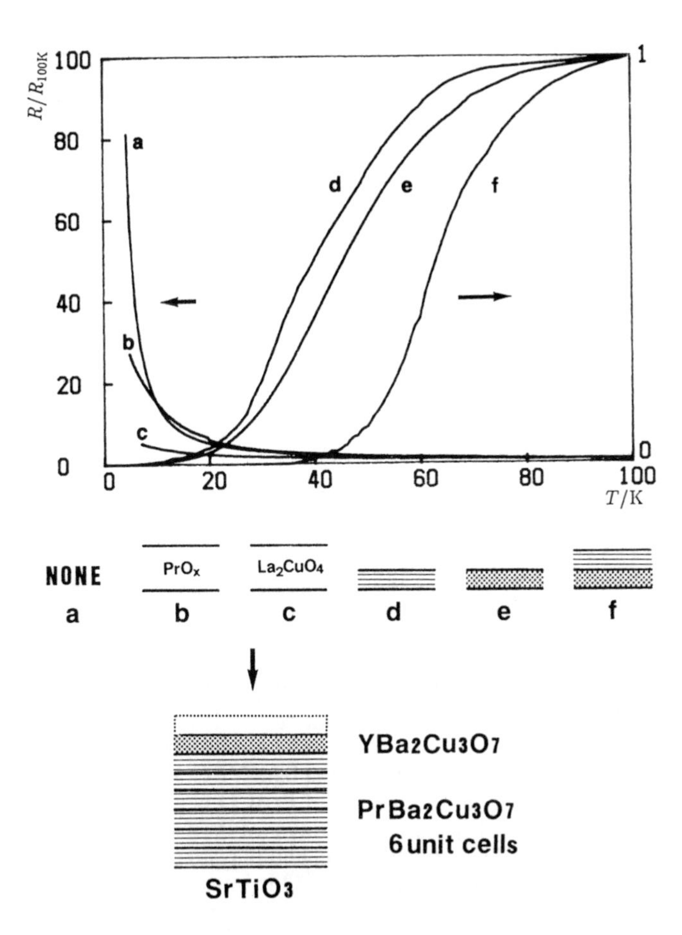

Fig. 4. Normalized resistance vs temperature curves showing effect of the overlayer. The overlayers are nothing (a), PrO_x (b), La_2CuO_4 (c), PrBCO (1 unit-cell) (d), YBCO (1 unit-cell) (e), PrBCO (1 unit-cell) / YBCO (1 unit-cell) (f).

tetragonality caused by lattice matching with the substrate of $SrTiO_3$ turns to the orthorhombicity with increase of the total thickness of the multilayer, being the same as the tendency appearing in the structures of ultrathin layers of YBCO on $SrTiO_3$[10]. An enough thickness of the PrBCO layers interposing the YBCO layer is needed for the intrinsic resistive transition of the YBCO layer to be achieved. The problem about the effect of the biaxixial tension in the orthorhombic YBCO layer on T_c remains unsolved but will not be so significant.

Next, we discuss the role of the overlayer of PrBCO on superconductivity of the YBCO layer[6]. Figure 5 shows the resistive curves against temperature for the one unit-cell layers of YBCO covered with various oxide layers. The bare one unit-cell layer of YBCO (a) shows nonsuperconductivity. Furthermore, when insulating PrO_x (b) and epitaxial La_2CuO_4 layers (c) are grown, the films are still nonsuperconductive. The film (d) covered with a one unit-cell layer of PrBCO turns to superconductor with

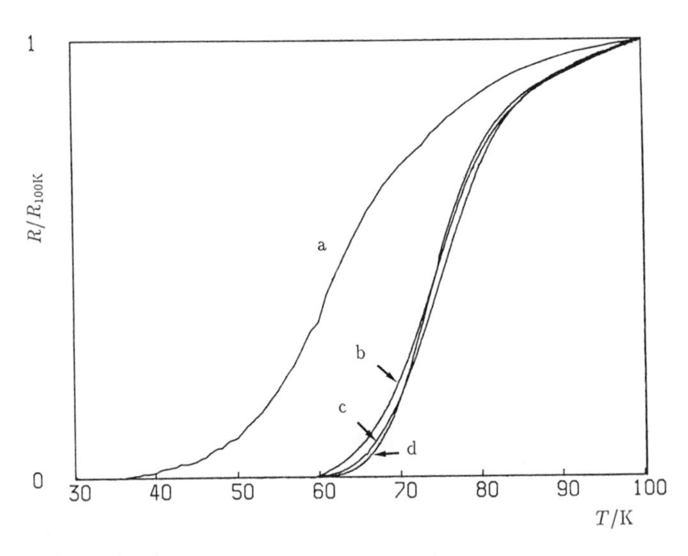

Fig. 5. Normalized resistance vs. temperature curves for the 2 unit-cells layer of YBCO sandwiched in between various thick PrBCO layers on $SrTiO_3$.
(a) PrBCO (1 unit-cell) / YBCO (2 unit-cells) / PrBCO (6 unit-cells)
(b) PrBCO (1 unit-cell) / YBCO (2 unit-cells) / PrBCO (11 unit-cells)
(c) PrBCO (6 unit-cells) / YBCO (2 unit-cells) / PrBCO (6 unit-cells)
(d) PrBCO (11 unit-cells) / YBCO (2 unit-cells) / PrBCO (11 unit-cells)

the transition temperature similar to that of the bare YBCO layer of two unit-cells (e). The T_c increases when a one unit-cell layer of PrBCO is grown on the two unit-cells layer of YBCO. It is noteworthy that the overlayer of PrBCO (d) makes the underlying one unit-cell layer of YBCO superconductive, and plays the same role as the overlayer of YBCO (e). The top unit-cell layer of YBCO is intrinsically nonsuperconductive. Generally, it is believed that superconductivity would occur in CuO_2 bilayer interposing the Y layer, for which hole carriers are supplied by CuO_x chain layer. The necessity of an overlayer of 1:2:3 compound for occurrence of superconductivity means that the both up and down BaO-CuO_x-BaO blocks for the CuO_2 bilayer is needed to supply sufficient hole carriers. The similarity of transition temperature between (d) and (e) in the figure implies that the top unit-cell layer of YBCO acts only as a hole dopant for the underlying YBCO layer and does not show superconductivity itself. Since the two dimensional layer growth occurs in the unit-cell, its terminating layer should always be definite. Although we can't specify the terminating layer, CuO_2 layer on the Y layer is most probable for it and CuO_x chain is next. In either case, the above explanation for the role of the overlayer is available.

4. Superconductivity in the magnetic field

The energy dissipation of high T_c superconductor in the magnetic field exhibits the strong anisotropy, when the magnetic field is applied parallel and perpendicular to the CuO_2 plane. An enhanced anisotropy effect should be expected as a result of reducing the dimensionality of the film. Figure 6 shows resistive transitions in magnetic fields parallel to CuO_2 layer with a precision of 0.1° for n=10, 5 and 3, where the transport current and the magnetic field are perpendicular to each other[4]. In parallel configuration the resistive transition for n=1 and 3 is independent of the

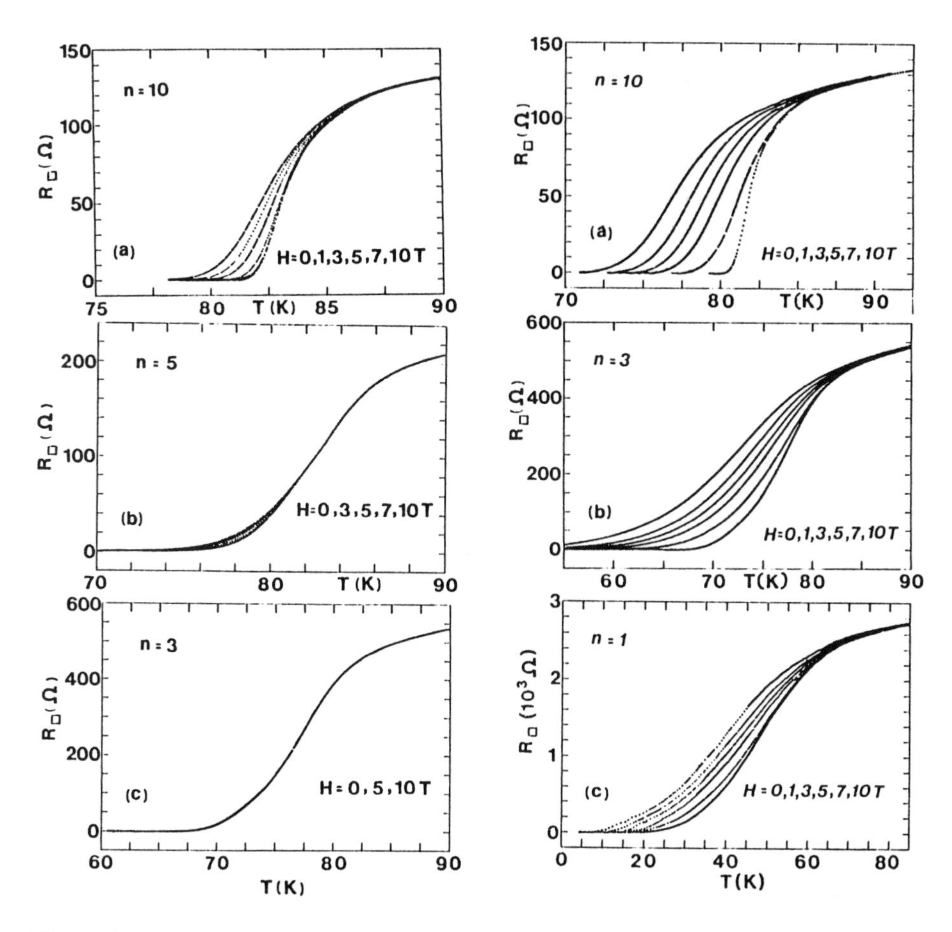

Fig. 6. Sheet resistance in magnetic fields parallel to CuO_2 layer for n=10, 5 and 3.

Fig. 7. Sheet resistance in magnetic fields perpendicular to CuO_2 layer for n=10, 3 and 1.

magnetic field up to 10T. The magnetic field is assumed to penetrate uniformly without the formation of vortices for the ultrathin film. The ultrathin layer may have a very high H_{c1}, because the vortex creation is not energetically favorable.

On the other hand, the magnetic fields perpendicular to the CuO_2 layer make the resistive transition broader, as shown in Fig. 7. Recently, we found that the activation energies estimated from the slope of the Arrhenius plots are proportional to the film thickness[17]. This suggests that the flux line along c-axis is limited to the layer thickness of YBCO, because activation energy may be proportional to the correlation length of flux line along c-axis, as found by Brunner et al[8].

The resistive transition for the 1 unit / 1 unit superlattice in the parallel configuration exhibits no broadening except of slight broadening in the tail[6]. In the perpendicular configuration, the activation energies for the 1 unit / 1 unit superlattice are about one third of values for the 3 unit-cells layer. Brunner et al.[8] found that the activation energy increased when the PrBCO separation thickness decreased from 48Å to 24Å. In our superlattice sample, no increase of activation energy even in the PrBCO separation thickness of 12Å in the superlattice is observed, suggesting that the flux lines in the superlattice are limited to each one unit-cell layer of YBCO. This indicates that the separation thickness of 12Å is sufficient to suppress the coupling between each group of CuO_2 bilayers to produce relatively rigid flux lines.

CONCLUSION

High quality samples of isolated ultrathin films of YBCO sandwiched in between PrBCO and YBCO (1 unit) / PrBCO (1 unit) superlattices were obtained by reactive coevaporation using RHEED intensity monitoring. An isolated one unit-cell layer of YBCO sandwiched between PrBCO layers has a superconducting transition with the onset temperature at 70K and the zero resistance at 30K. The overlayer of PrBCO was required to provide the underlying one unit-cell YBCO layer with hole carriers sufficient to introduce superconductivity. The reduction of the onset temperature is found to be mainly due to the decrease of the hole carriers but not due to the absence of the interlayer coupling. When the magnetic field is applied parallel to the CuO_2 layer, we find that the resistive transition is field independent for the YBCO layer with the thickness below 5 unit-cells and also for the YBCO (1 unit-cell) / PrBCO(1 unit-cell) superlattice. In perpendicular field, the activation energy of flux motion for the isolated ultrathin layers is nearly proportional to the number of unit-cell of YBCO. The activation energy for the superlattice is comparable to the value of the isolated one unit-cell layers, indicating that the flux lines along c-axis are limited to each one unit-cell layer of YBCO.

REFERENCES

1. J.-M. Triscone, Ø. Fischer, O. Brunner, L. Antognazza, A.D. Kent and M.G. Karkut, Phys. Rev. Lett. 64, 804 (1990).
2. Q. Li, X.X. Xi, X.D. Wu, A. Inam, S. Vadlamannati, W.L. McLean, T. Venkatesan, R. Ramesh, D.M. Hwang, J.A. Martinez and L. Nazar, Phys. Rev. Lett. 64, 3086 (1990).
3. D.H. Lowndes, D.P. Norton and J.D. Budai, Phys. Rev. Lett. 65, 1160 (1990).
4. Y. Matsuda, A. Fujiyama, S. Komiyama, T. Terashima, K. Shimizu and Y. Bando, to be published in Proc. ISSP-PCOS '91, Tokyo, Japan, Jan. 16-18 (1991).
5. S. Vadlamannati, Q. Li, T. Venkatesan, W.L. McLean and P. Lindenfeld, to be published in Phys. Rev. B. Rapid Comm.
6. T. Terashima, K. Shimura, Y. Bando, Y. Matsuda, A. Fujiyama and S. Komiyama, Phys. Rev. Lett. 67, 1362 (1991).
7. D.P. Norton, D.H. Lowndes, S.J. Pennycook and J.D. Budai, Phys. Rev. Lett. 67, 1358 (1991).
8. O. Brunner, L. Antognazza, J.-M. Triscone, L. Miéville, and Ø. Fischer, Phys. Rev. Lett. 67, 1354 (1991).
9. T. Terashima, Y. Bando, K. Iijima, K. Yamamoto, K. Hirata, K. Hayashi, K. Kamigaki and H. Terauchi, Phys. Rev. Lett. 65, 2684 (1990).
10. K. Kamigaki, H. Terauchi, T. Terashima, Y. Bando, K. Iijima, K. Yamamoto, K. Hirata, K. Hayashi, I. Nakagawa and Y. Tomii, J. Appl. Phys. 69, 3653 (1991).
11. Y. Bando, T. Terashima, K. Iijima, K. Yamamoto, K. Hirata, K. Hayashi, K. Kamigaki and H. Terauchi, Proc. XIIth Int. Cong. for Electron Microscopy,

Seattle, Aug. 12-18, 1990 (Eds. L.D. Peachey, D.B. Williams) Material Sciences 4, 2 (1990).
12. T. Terashima, K. Iijima, K. Yamamoto, K. Hirata, Y. Bando and T. Takada, Jpn. J. Appl. Phys., 28, L987 (1989).
13. T. Terashima, K. Iijima, K. Yamamoto, Y. Bando and H. Mazaki, Jpn. J. Appl. Phys. 27, L91 (1988).
14. Y. Bando, T. Terashima, K. Shimura, T. Sato, Y. Matsuda, S. Komiyama, K. Kamigaki and H. Terauchi, to appear in Physica C, International Confernce on Materials and Mechanisms of Superconductivity, High-Temperature Superconductors (M^2S-HTSC III) Kanazawa, Japan, July 22-26, (1991).
15. B.I. Halperin and D.R. Nelson, J. Low. Temp. Phys. 36, 599 (1979).
16. K. Kamigaki, H. Terauchi, T. Terashima, K. Shimura and Y. Bando, to be submitted to Appl. Phys. Lett.
17. Y. Matsuda, A. Fujiyama,. S. Komiyama, T. Terashima, K. Shimura and Y. Bando, submitted to Phys. Rev. Lett.

THIN FILM AND MULTILAYER HETEROSTRUCTURES

PROPERTIES OF EPITAXIAL $YBa_2Cu_3O_{7-\delta}$-BASED SUPERCONDUCTING SUPERLATTICES

David P. Norton, Douglas H. Lowndes, Z. -Y. Zheng,* Ron Feenstra, and Shen Zhu**
**Department of Physics and Astronomy
The University of Tennessee, Knoxville, TN 37996
*Health and Safety Research Division
Solid State Division, Oak Ridge National Laboratory, P. O. Box 2008,
Oak Ridge, Tennessee 37831-6056

ABSTRACT

Epitaxial $YBa_2Cu_3O_{7-\delta}$-based superconducting superlattices have been fabricated using pulsed laser deposition in which *c*-axis oriented $YBa_2Cu_3O_{7-\delta}$ layers as thin as one unit cell thick are separated by relatively thick $PrBa_2Cu_3O_{7-\delta}$-based barrier layers. The superlattice T_c (R = 0) decreases rapidly with increasing barrier layer thickness, but then saturates at a finite T_c for $YBa_2Cu_3O_{7-\delta}$ layers as thin as a single *c*-axis unit cell. The superconducting properties of $YBa_2Cu_3O_{7-\delta}$-based superlattices are shown to depend strongly on the electronic properties of the barrier layers. The resistive transition width decreases significantly as the hole carrier density in the barrier layers is increased. However, T_c (onset) does not change, contrary to predictions of hole filling models. Theoretical analyses suggest that the broadening of the resistive transition for the thinnest $YBa_2Cu_3O_{7-\delta}$ layers is most likely due to a crossover to 2D resistive behavior involving thermally-generated vortices. Scanning tunneling microscopy reveals that epitaxial $YBa_2Cu_3O_{7-\delta}$ thin films grow unit cell-by-unit cell, by a terraced-island growth mode. This terraced microstructure explains the steps found in ultrathin $YBa_2Cu_3O_{7-\delta}$ layers in these superlattices. These steps may act as superconducting weak links, providing support for 2D Josephson-coupled-array models of superconducting superlattices.

INTRODUCTION

High-temperature superconductivity is associated with layered, quasi-two-dimensional (2D) crystal structures and with carriers moving in CuO_2 planes. Because the *c*-axis superconducting coherence length is very short (e.g., $\xi_c \sim 3$–6Å, vs a lattice constant $c \sim 11.7$ Å in $YBa_2Cu_3O_{7-x}$), questions arise whether isolated single-cell-thick layers of these materials are superconducting and, if so, how their superconductivity is affected by their extreme anisotropy (including possible reduced dimensionality) and by residual interlayer coupling or other interactions. Several groups recently have reported on the electrical transport properties of high temperature superconducting/semiconducting superlattices, including $YBa_2Cu_3O_{7-\delta}/PrBa_2Cu_3O_{7-\delta}$,[1–4] $YBa_2Cu_3O_x/Nd_{1.83}Ce_{0.17}CuO_\delta$,[5] and $Bi_2Sr_2Ca_{0.85}Y_{0.15}Cu_2O_8/Bi_2Sr_2Ca_{0.5}Y_{0.5}Cu_2O_8$[6] structures. The superconducting properties of $YBa_2Cu_3O_{7-\delta}/PrBa_2Cu_3O_{7-\delta}$ (YBCO/PrBCO) superlattices are sensitive functions of both the superconducting (YBCO) and the barrier (PrBCO) layer thicknesses, d. T_c decreases as the YBCO layer thickness decreases or as the PrBCO layer thickness increases, but for all YBCO layer thicknesses, including layers one unit cell thick, the superconducting transition temperature saturates at nonzero values; e.g., for YBCO layer thicknesses of one, two, and three unit cells isolated in a relatively thick PrBCO matrix, the zero resistance transition temperature, T_{c0}, is ~19 K, ~54 K, and ~70 K, respectively.[3] Thus, there appears to be "coupling" between thin YBCO layers that are separated by PrBCO layers only a few unit cells thick. This is evident as, for a given YBCO layer thickness, T_c does not become independent of PrBCO layer thickness until $d_{PrBCO} > 5$ nm. In addition, the widths of the superconducting transitions are large, with $\Delta T_c \sim 37$ K for YBCO layers one unit cell thick isolated in a PrBCO matrix.

Recently, we have determined that the superconducting transition of YBCO-based superlattice structures depend on the electronic properties of the barrier layers, focusing specifically on the mobile hole concentration (resistivity) of the barrier layers.[7] Superlattice structures were fabricated using three different barrier layer materials, namely $PrBa_2Cu_3O_{7-\delta}$ (PrBCO), $Pr_{0.7}Y_{0.3}Ba_2Cu_3O_{7-\delta}$ (PrYBCO), and $Pr_{0.5}Ca_{0.5}Ba_2Cu_3O_{7-\delta}$ (PrCaBCO). By systematically varying the hole concentration of the barrier layers used to isolate ultrathin YBCO layers, we find that the superconducting transition width depends on the carrier density in the barrier layer, while T_c(onset) does not.

Several explanations have been proposed for the resistive transitions observed in these superlattice structures, including proximity effect,[8] localization effects,[9] and hole-filling.[10] Rasolt, Edis, and Tesanovic recently suggested[11] that these YBCO/PrBCO superlattices provide a 3D system in which the interlayer (c-axis) coupling can be weakened to zero, as indicated by the saturation of T_c values that occurs for large PrBCO thicknesses.[3] Thus, a crossover to 2D behavior occurs, accompanied by characteristic 2D dissipation.[11] Minnhagen and Olsson also find that the resistive transition data for superlattices containing isolated YBCO layers (one or two cells thick) are well explained by the 2D Ginzburg-Landau Coulomb Gas model.[12] Finally, Ariosa and Beck have suggested[13] that each superconducting CuO_2 bilayer in a superlattice structure may be modeled as a 2D array of ultrasmall Josephson junctions. Thus, several recent papers suggest that reduced dimensionality plays an important role in the resistive behavior of these structures.

RESULTS AND DISCUSSION

The c-axis perpendicular superlattice structures were fabricated using in situ pulsed-laser deposition, as has been described elsewhere.[3,7] We

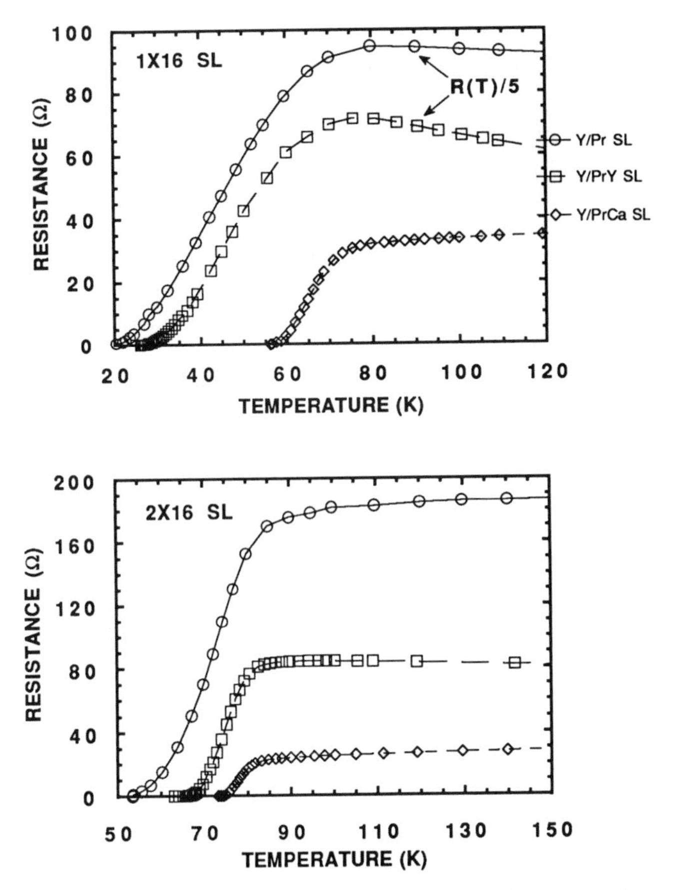

Fig. 1. R(T) for 1 × 16 and 2 × 16 YBCO/PrBCO (O), YBCO/PrYBCO (□), and YBCO/PrCaBCO (◊) superlattice structures.

describe them using the nomenclature "$N \times M$", where N and M are the numbers of YBCO and barrier layer unit cells per superlattice period, respectively. Most of the structures consist of 30 superlattice periods. Of the three barrier layer materials utilized, PrBCO is the least conductive with the lowest mobile hole concentration. PrYBCO is more conductive, but still demonstrates a divergent resistivity at low temperatures. The properties of the PrCaBCO thin films, in which divalent Ca is added to introduce holes into an otherwise semiconducting compound, are nearly metallic with very little temperature dependence in the resistivity. A description of the superconducting properties of the PrCaBCO thin film system has been reported elsewhere.[14]

Figure 1 shows the R(T) behavior for 1 × 16 and 2 × 16 superlattices with either PrBCO, PrYBCO, or PrCaBCO utilized as the barrier layer material. A systematic dependence of the resistive transitions on the carrier density of the barrier layers is observed. The most interesting effect is a significant increase in T_{c0} with increasing barrier layer carrier density. For the 1 x 16 superlattices, T_{c0} increases from ~20 K for the YBCO/PrBCO structure to > 50 K for the YBCO/PrCaBCO structure. However, note that T_c(onset) [defined as the initial inflection of R(T)] is not significantly influenced by the hole concentration in the barrier layers. This is more clearly seen in Fig. 2 where T_c(onset) and T_{c0} are plotted as functions of barrier layer thickness. Although T_{c0} increases (transition width decreases) significantly as the barrier layer carrier density is increased, T_c (onset) is insensitive to the barrier layer composition, depending only on the thicknesses of the YBCO and barrier layers.

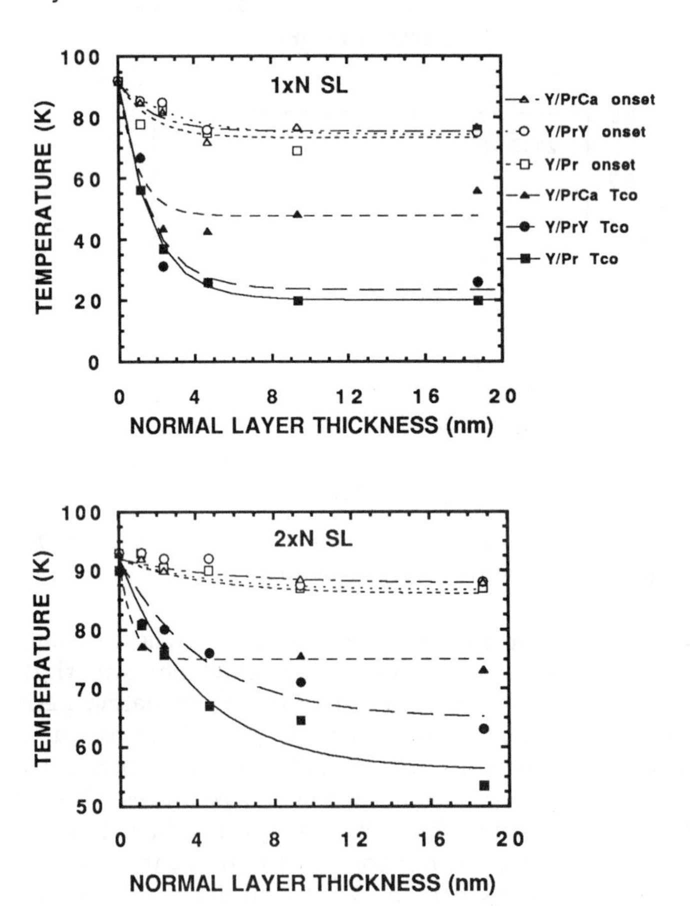

Fig. 2. T_c(onset) (open symbols) and T_{c0} (filled symbols) as a function of normal layer thickness for 1 × N and 2 × N YBCO/PrBCO (□,■), YBCO/PrYBCO (O, ●), and YBCO/PrCaBCO (Δ,▲) superlattice structures.

It recently was proposed that electron transfer from the PrBCO layers into the YBCO layers (resulting in hole-filling in the YBCO layers) can explain the depression of T_c for YBCO/PrBCO superlattices.[10] Hole-filling was previously used to explain the suppression of T_c as Pr is substituted into superconducting "123" oxide materials. In these alloyed systems, the mixed valence of the Pr leads to a reduction of mobile hole density on the CuO_2 planes, with a subsequent reduction in T_c.[15, 16] In YBCO/PrBCO superlattice structures, ultra-thin (perhaps 1 unit cell thick) YBCO layers with a high mobile hole density are layered alternately with a low carrier density material (PrBCO) of comparable thickness. The possibility of electron transfer from the PrBCO to the YBCO therefore must be considered. If there is a significant reduction of the mobile hole density in the YBCO layers, then a reduction in the transition temperature, T_c (onset), should result. Conversely, by adding holes to the barrier layers (for instance, by doping the PrBCO with divalent Ca), the transfer of holes from the YBCO layers into the barrier layers should be reduced, and T_c (onset) should increase.

Thus, within the hole-filling model, it is difficult to explain why there is no significant change in T_c (onset) as the carrier density in the barrier layers is increased, as seen in Figs. 1 and 2. The hole-filling model predicts that an increase in the hole density in the barrier layers should lead to an increase in T_c (onset), with little effect on the transition width. One does not expect the transition width to be a strong function of the YBCO carrier density. Recent experiments in which the mobile hole concentration was varied directly, by removing oxygen from YBCO thin films and YBCO/PrBCO superlattice structures, showed that T_{c0} and T_c (onset) shift together as the mobile hole concentration is decreased, with little or no additional broadening of the transition.[7] Figure 3 shows the R(*T*)/R(100 K) behavior of a 2 × 8 $YBa_2Cu_3O_{7-\delta}/PrBa_2Cu_3O_{7-\delta}$ and a 2 × 4 $YBa_2Cu_3O_{7-\delta}/Pr_{0.5}Y_{0.5}Ba_2Cu_3O_{7-\delta}$ superlattice structure after partial removal of oxygen by means of low pressure oxygen annealing. As oxygen is removed, no significant broadening of the superconducting transition region is observed, while T_{c0} and T_c (onset) both shift to lower temperatures. Thus, the insensitivity of T_c (onset) to the barrier layer carrier density provides evidence against a simple change in the hole carrier density as the explanation for the depression of T_c in YBCO/PrBCO superlattices.

A reduction of T_c (onset) and/or an increased transition width has also been observed for conventional low T_c superconducting systems as the superconducting layer thickness is reduced.[18-21] In these materials, the reduction of T_c (onset) has been attributed to a suppression of the superconducting pair wave function amplitude, whereas the broadening involves a lack of phase coherence. The reduction in T_c (onset) for low T_c,

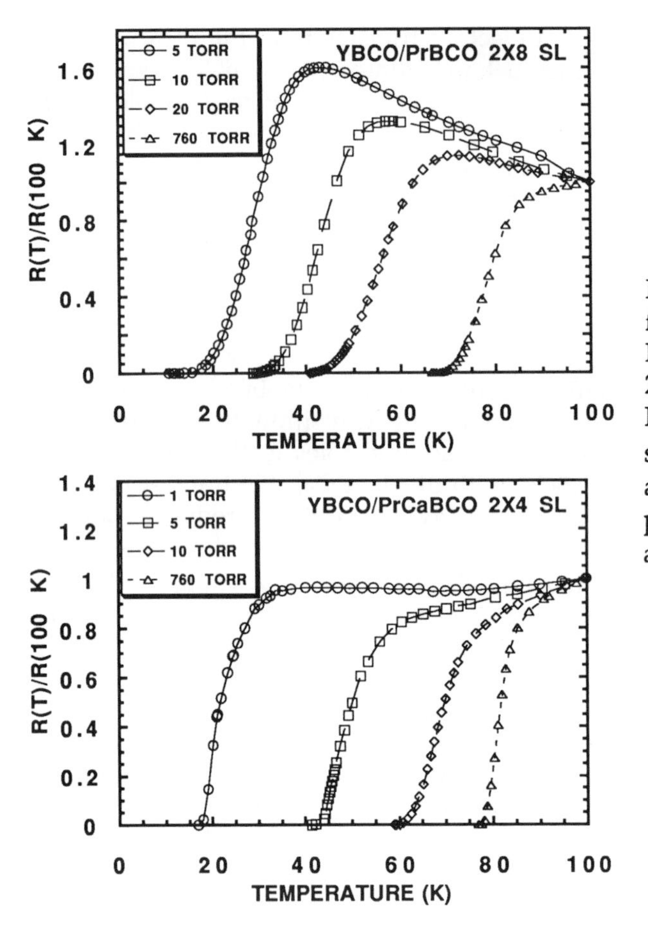

Fig. 3. R(*T*) R (100 K) for 2×8 $YBa_2Cu_3O_{7-\delta}$/ $PrBa_2Cu_3O_{7-\delta}$ and 2 × 4 $YBa_2Cu_3O_{7-\delta}$/ $Pr_{0.5}Ca_{0.5}Ba_2Cu_3O_{7-x}$ superlattice structures after various low pressure oxygen anneals.

ultrathin films has been explained in terms of the proximity effect,[8] in terms of localization in disordered 2D systems,[9] and in terms of the boundary conditions for the order parameter within the Ginzburg-Landau free energy expression.[22] Predictions based on proximity effect and order parameter boundary conditions agree qualitatively with the superconducting properties of YBCO/PrBCO superlattices. However, the length scale over which the YBCO layers become "decoupled", as the PrBCO barrier layer thickness is increased, is somewhat larger than the normal metal electronic coherence length in the semiconducting PrBCO. (This coherence length is expected to be less than the c-axis coherence length in metallic YBCO.) If localization in a disordered, 2D system, is responsible for the reduction of T_c (onset), the resistivity of the superconducting layer should increase as T_c (onset) decreases, in agreement with our observations. For the YBCO/PrBCO superlattice system, the 1×16 (ρ_{YBCO}(100 K) = 400 $\mu\Omega$-cm, T_c (onset) = 76 K) and the 2×16 (ρ_{YBCO}(100 K) = 300 $\mu\Omega$-cm, T_c (onset) = 87 K) structures follow this trend.

Although some depression of T_c (onset) occurs as a function of layer thickness in the YBCO-based superlattice structures (Fig. 2), a more prominent feature is the broadening of the superconducting transition. In general, the transition width represents a lack of long-range phase coherence of the superconducting order parameter. Transition width broadening has also been observed in ultrathin films of low-temperature superconductors, and has been attributed to the presence of a 2-D array of weakly in-plane-coupled Josephson junctions and possibly to 2-D vortex-antivortex pair unbinding.[23-26] Rasolt et al.[11] and Minnhagen and Olsson[12] have pointed out that a crossover to 2D resistive behavior occurs for sufficiently thin YBCO layers, as the PrBCO thickness is increased. When crossover occurs, characteristic 2D dissipation processes should be observed, specifically resistance in the 2D superconducting state due to free vortices that are produced by thermal unbinding of vortex-antivortex pairs.[27–30] A low temperature Kosterlitz-Thouless (KT) transition also may occur, if a 3D superconducting transition caused by residual interplane coupling does not intervene first[12] (for intermediate PrBCO thicknesses).

One of the characteristic "signatures" of a 2D conductor is that in the temperature range well below the mean-field (bulk, or BCS) superconducting transition temperature, T_{cB}, but just above the Kosterlitz-Thouless temperature, T_c, where thermal unbinding of the vortex-antivortex pairs begins to produce free vortices,[27,28] the thermally generated vortices produce a nonzero resistivity with temperature dependence[29,30]

$$R(T)/R_N \sim \xi_+(T)^{-2} = \xi_{ab}(T_c)^{-2} \exp[-2b'(\tau_c/\tau)^{1/2}]. \qquad (1)$$

In Eq. (1) R_N is the normal-state resistance, $\xi_+(T)$ is the phase correlation length for the superconducting order parameter,[29,30] $\xi_{ab}(T)$ is the Ginzburg-Landau correlation length within the *a-b* plane, $\tau_c = (T_{cB} - T_c)/T_c$, $\tau = (T - T_c)/T_c$, and b' is a constant of order unity. From Eq. (1) it follows that log R(T) should scale as $\tau^{-1/2}$.

Figure 4 shows the result of plotting log $R_n(T)$ vs $\tau^{-1/2}$, where $R_n(T) = R(T)/R(100\text{ K})$, for the M(YBCO) $\times$ 16(PrBCO) superlattice structures.[3,31] These are the structures with the thickest PrBCO layers, for which the saturation of T_c suggests that a crossover to 2D resistive behavior has occurred. Good agreement with the expected scaling behavior is observed, and from these plots we extracted least-squares best-fit values for the *KT* transition temperature of T_c = 14, 44, 70, and 86.5 K, for the structures with 1-, 2-, 3-, and 8-cell-thick YBCO layers, respectively. We should note,

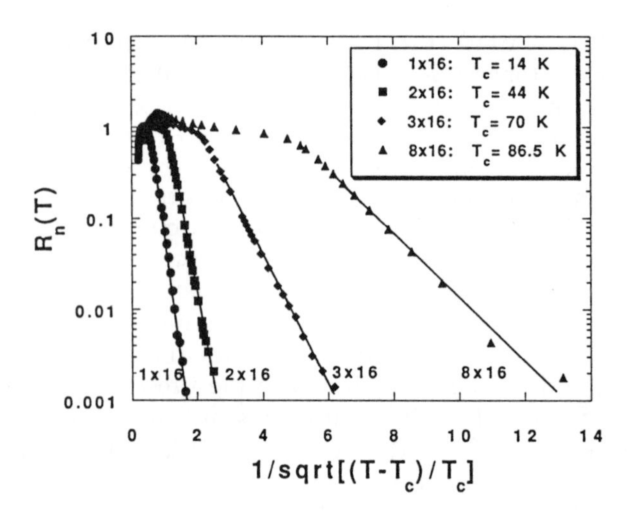

Fig. 4. Superconducting transitions for M(YBCO)×16(PrBCO) superlattices with M = 1, 2, 3, and 8 c-axis unit cells, plotted as log $R_n(T)$ vs $\tau^{-1/2}$, where $R_n(T)$ is the normalized resistance $R(T)/R(100\text{ K})$ and $\tau = (T - T_c)/T_c$. The T_c values determined by best least-squares fits to the data are given in the text.

however, that the R(T) data are outside the region ($T_c < T \ll T_{cB}$) where Eq. (1) is strictly applicable.[29]

Similar broadening of the superconducting transition, associated with a Kosterlitz-Thouless transition, is predicted for 2D films with a granular nature, consisting of an array of superconducting islands connected by weak links.[24,32] As the temperature is decreased, the individual islands become superconducting at T_c (onset). However, long-range phase coherence is established by Josephson coupling between islands only at a lower temperature, leading to a broad transition. Although the YBCO-based superlattices are high quality, fully epitaxial structures, some evidence supporting a weak link description has recently been provided by Z-contrast transmission electron microscopy (TEM) images.[7,33] These images indicate that the YBCO layers are not perfectly flat relative to the lattice structure; i.e., the YBCO layer shifts up (or down) by one c-axis unit cell increment as one progresses parallel to the a-b planes. These one-cell-thick "kinks" in the YBCO layers are due to steps on the growing film surface. Their significance for current flow in the YBCO layers is that conduction along the c-axis will be necessary at the kinks, in order for a continuous conducting path to be established. These kinks should influence the transport properties of these superlattice structures, contributing to the high resistivity seen in the YBCO layers, and introducing regions of weakened superconductivity. Since the boundaries of these "kinks" are defined by the barrier layers, the properties of the weak links, in particular the perturbation of phase coherence across the region of weakened superconductivity, should be influenced by the electronic properties of the barriers.

We have also investigated the surface microstructure, determined by scanning tunneling microscopy (STM), in epitaxial YBCO thin films grown by pulsed-laser deposition.[34] STM images were obtained using a Nanoscope I STM. No special treatment of the film surface was required, and all of the images were obtained in air at room temperature. The samples were stored in a desiccator prior to being scanned. Figure 5 shows a STM image of a *c*-axis oriented epitaxial YBCO thin film grown at T_{sub} ~680°C on (100) $SrTiO_3$. Well-defined islands are clearly evident, with each island composed of stacks of terraces. The terrace step heights are multiples of the *c*-axis unit cell height (1.17 nm). The island-like appearance seen in Fig. 5 was common to *c*-axis oriented films, with the most symmetric, well-defined, and largest grains obtained at higher growth temperatures. The terrace width, defined as the distance between terrace steps, is ~10 nm, which is the same order of magnitude as the distance between "kinks" in the YBCO layers of YBCO/PrBCO superlattices, as observed by cross-section Z-contrast TEM.[7,33]

The surface microstructure observed in the *c*-axis oriented thin films may have important implications concerning the transport properties of YBCO/PrBCO superlattice structures. The terraced "roughness" that is present on the growing surface (Fig. 5) will be translated into the individual layers in a multilayered structure. Based on the observed surface

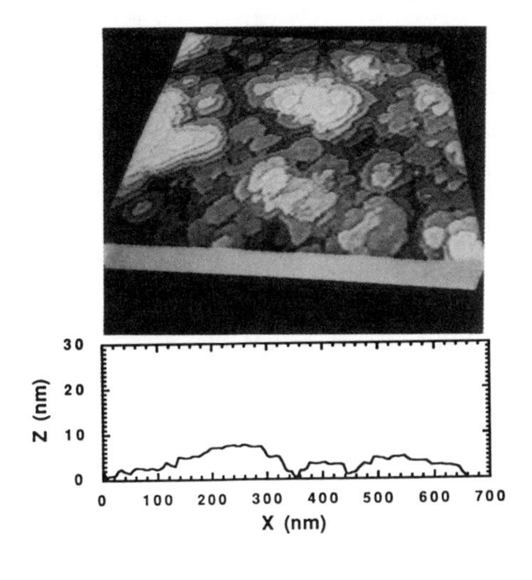

Fig. 5. STM image (upper) and line-scan profile (lower) of a *c*-axis oriented YBCO epitaxial thin film grown on (100) $SrTiO_3$ at T_{sub} ~680°C. The dimensions of the STM image are 670×670 nm^2.

microstructure,[34] together with Z-contrast TEM images of the superlattice structures,[7,33] it appears that a complete theory of the transport properties of YBCO/PrBCO superlattice structures must include the effects of these "kinks," and that perhaps one should regard the YBCO layers as a 2D array of superconducting terraces connected by weak links.

In summary, we have found that T_c (onset) is relatively insensitive to the hole carrier density in the barrier layers of YBCO-based superlattices, suggesting that hole-filling is not a major contributor to the depression of T_c (onset) or T_{c0}, although it cannot be completely dismissed. T_c (onset) appears to depend intrinsically on the YBCO layer thickness, possibly determined by localization effects. On the other hand, the values of T_{c0} (and the transition widths) measured for these superlattice structures are not intrinsic to YBCO layers of a given thickness, but are highly dependent on the boundary conditions and the barrier layer material. Theoretical analyses show that the systematic depression of T_c and the broadening of the resistive transition for the thinnest YBCO layers may be related to a crossover from 3D to 2D behavior.[11,12] The resistance in the region of the broadened superconducting transitions scales with temperature as expected for the (flux flow) resistance produced by thermally generated 2D vortices[29,30], or for a 2D array of superconducting weak links.[32] The terraced-island morphology, as revealed by STM, implies that ultrathin YBCO layers in superlattice structures should contain a high density of "steps" or "kinks". These should play a significant role as superconducting weak links in broadening the resistive transition in such multilayered structures.[7,31]

ACKNOWLEDGEMENTS

We would like to thank P. H. Fleming for assistance with sample characterization. This research was sponsored by the Division of Materials Sciences, U.S. Department of Energy under contract DE-AC05-84OR21400 with Martin Marietta Energy Systems, Inc.

REFERENCES

1. J. -M. Triscone et al., *Phys. Rev. Lett.* **64,** 804 (1990).
2. Q. Li. et al., *Phys. Rev. Lett.* **64,** 3086 (1990).
3. D. H. Lowndes, D. P. Norton, and J. D. Budai, *Phys. Rev. Lett.* **65,** 1160 (1990).
4. C. B. Eom et al., *Science* **251,** 780 (1991).
5. A. Gupta et al., *Phys. Rev. Lett.* **64,** 3191 (1990).
6. M. Kanai, T. Kawai, and S. Kawai, *Appl. Phys. Lett.* **57,** 198 (1990).
7. D. P. Norton, D. H. Lowndes, S. J. Pennycook, and J. D. Budai, *Phys. Rev. Lett* **67,** 1358 (1991).

7. D. P. Norton, D. H. Lowndes, S. J. Pennycook, and J. D. Budai, *Phys. Rev. Lett* **67**, 1358 (1991).
8. N. R. Werthamer, *Phys. Rev.* **132**, 2440 (1963).
9. S. Maekawa and H. Fukuyama, *J. Phys. Soc. Jpn.* **51**, 1380 (1981).
10. R. F. Wood, *Phys. Rev. Lett.* **66**, 829 (1991).
11. M. Rasolt, T. Edis, and Z. Tesanovic, *Phys. Rev. Lett.* **66**, 2927 (1991).
12. P. Minnhagen and P. Olsson, submitted to *Physical Review Letters.*
13. D. Ariosa and H. Beck, *Phys. Rev. B* **43**, 344 (1991).
14. D. P. Norton et al., *Phys. Rev. Lett.* **66**, 1537 (1991).
15. M. E. Lopez-Morales et al., *Phys. Rev. B* **41**, 6655 (1990).
16. J. J. Neumeier et al., *Phys. Rev. Lett.* **63**, 2516 (1989).
17. D. P. Norton and R. Feenstra, unpublished data.
18. H. K. Wong et al., *J. Low Temp. Phys.* **63**, 307 (1986).
19. C. S. L. Chun et al., *Phys. Rev. B* **29**, 4915 (1984).
20. S. T. Ruggiero, T. W. Barbee, and M. R. Beasley, *Phys. Rev. Lett.* **45**, 1299 (1980).
21. B. J. Jin et al., *J. Appl. Phys.* **57**, 2543 (1985).
22. J. Simonin, *Phys. Rev. B* **33**, 7830 (1986).
23. S. A. Wolf et al., *Phys. Rev. Lett.* **47**, 1071 (1981).
24. A. E. White, R. C. Dynes, and J. P. Garno, *Phys. Rev. B* **33**, 3549 (1986).
25. A. F. Hebard and A. T. Fiory, *Phys. Rev. Lett.* **44**, 291 (1980).
26. P. A. Bancel and K. E. Gray, *Phys. Rev. Lett.* **46**, 148 (1981).
27. J. M. Kosterlitz and D. J. Thouless, *J. Phys. C* **6**, 1181 (1972); J. M. Kosterlitz and D. J. Thouless, *Prog. Low Temp. Phys.* **7**, 373 (1978).
28. J. M. Kosterlitz, *J. Phys. C* **7**, 1046 (1974).
29. B. I. Halperin and D. R. Nelson, *J. Low Temp. Phys.* **36**, 599 (1979).
30. V. Ambegaokar, B. I. Halperin, D. R. Nelson, and E. D. Siggia, *Phys. Rev. B* **21**, 1806 (1980).
31. D. H. Lowndes and D. P. Norton, "Kosterlitz-Thouless-Like Behavior Over Extended Ranges of Temperature and Layer Thickness in Crystalline $YBa_2Cu_3O_{7-\delta}$ /$PrBa_2Cu_3O_{7-\delta}$ Superlattices," *University of Miami Workshop on Electronic Structure and Mechanisms for High-Temperature Superconductivity* (Plenum Publishing Co., 1991, in press).
32. C. J. Lobb, D. W. Abraham, and M. Tinkham, *Phys. Rev. B* **27**, 150 (1983).
33. S. J. Pennycook, M. F. Chisholm, D. E. Jesson, D. P. Norton, D. H. Lowndes, R. Feenstra, and H. R. Kerchner, *Phys. Rev. Lett.* **67**, 765 (1991).
34. D. P. Norton, D. H. Lowndes, X.-Y. Zheng, Shen Zhu, and R. J. Warmack, *Phys. Rev. B: Rapid Communications* (in press).

CORRELATION OF STRUCTURAL AND SUPERCONDUCTING PROPERTIES OF $Ba_2YCu_3O_{7-\delta}$ THIN FILMS

Julia M. Phillips and M. P. Siegal
AT&T Bell Laboratories, Murray Hill, NJ 07974

ABSTRACT

We review our results on the optimization of the production parameters which has led to the fabrication of very high quality thin films of superconducting $Ba_2YCu_3O_{7-\delta}$ by co-evaporation of Y, Cu, and BaF_2 followed by a two stage anneal. The significance of critical annealing and deposition parameters is discussed. We also discuss our work correlating some of the superconducting properties of these films with their structural features.

INTRODUCTION

Recent work on the growth of epitaxial $Ba_2YCu_3O_{7-\delta}$ (BYCO) thin films has centered largely on various methods of producing films which are superconducting when they are removed from the growth chamber (*in situ* films). While a number of papers have discussed the advantages of such approaches over techniques which require the films to be annealed after removal from the growth apparatus (*ex situ* techniques),[1] high quality *ex situ* films also have appeal. This attractiveness originates from the relative simplicity of the *ex situ* growth process, its demonstrated ability to produce uniform large area films on one or both sides of a substrate, and its compatibility with lift-off processing. Of the *ex situ* growth techniques, the so-called BaF_2 process,[2] which involves co-evaporation of Y, Cu, and BaF_2 followed by a two-stage anneal, has probably received the most attention due to a large extent to the stability of the non-superconducting precursor films.[3]

We have previously shown that it is possible to produce films with excellent structural and electrical properties using the BaF_2 process.[4] We have also demonstrated the ability to alter the film preparation conditions to study correlations between various aspects of the film structure and its superconducting properties.[5] This work has revealed a number of similarities in the properties of BaF_2 process films and *in situ* films, but a number of important differences have been noted as well. Chief among these are the extreme importance of stoichiometric control in the preparation of BaF_2 process films,[6] the deviation of the behavior of the penetration depth from BCS-like behavior under certain preparation conditions,[5] and the weak flux pinning in the films even in the absence of weak links.[3,5] These findings suggest that there are important differences in the preparation conditions for films grown by the two classes of techniques which have ramifications for their superconducting properties.

In this paper we review our results on BaF_2 films. This includes a discussion of the preparation parameters which are particularly important for determining the structural and electrical quality of the films. We also review the correlations which we have observed between film structure and superconducting properties and point out the various length scales which are important in determining such correlations. Finally, we discuss the possible origin of the difference between the properties of BaF_2 process films and *in situ* films.

EXPERIMENTAL PROCEDURES

FILM PREPARATION. Y, Cu, and BaF_2 are co-evaporated onto $LaAlO_3$(100) substrates in a vacuum deposition chamber as described previously.[4] Films 100 nm thick are simultaneously deposited onto $LaAlO_3$ (100) and Si substrates which are not intentionally heated. Rutherford backscattering spectrometry (RBS) with 1.8 MeV $^4He^+$ ions is performed on the Si samples to determine the deposited metal composition. Except in the studies specifically studying film stoichiometry, only films with Ba:Y:Cu stoichiometry within 1% of 2:1:3 are used. In the case of the stoichiometry studies, a series of films have been produced in which two of the three deposited cationic species are within ±1% of the stoichiometric composition, while the third species varies over a range of approximately ±5%.

After growth the films are annealed in a tube furnace by a two-stage process described previously[3] to produce superconducting BYCO films. The first stage of the anneal is performed between 750°C and 950°C in 1 atmosphere flowing O_2 bubbled through deionized water ("wet" O_2). The second stage of the anneal is performed at temperatures between 500°C and 600°C in 1 atmosphere flowing dry O_2 ("dry" O_2).

CHARACTERIZATION. The superconducting transition temperature T_c is measured by the resistive transition as described elsewhere[3] and by the magnetic shielding effect, as appropriate.[5] The latter data have been fit to BCS theory to extract the penetration length, $\lambda(T)$.[5] J_c is determined using a vibrating sample magnetometer to measure the magnetization of films in fields up to 0.9T perpendicular to the film surface. J_c values are extracted using the Bean model,[7] with the appropriate correction due to the anisotropy of J_c.[8] In selected cases, the transport J_c has been used to validate the assumptions underlying the J_c determination from magnetization measurements. The agreement has been found to be within 10%. Film morphology is studied by scanning electron microscopy (SEM). Crystalline quality is determined by RBS ion channeling, where χ_{min} is measured as the ratio of the yield behind the Ba surface peak in a channeling spectrum to the same yield in a random spectrum.

THE IMPORTANCE OF ANNEALING CONDITIONS AND FILM STOICHIOMETRY

ANNEALING CONDITIONS. We have found that three aspects of the anneal are especially important in determining film properties. A time-temperature profile which optimizes the structural properties of 100 nm thick films is shown in Figure 1.[3]

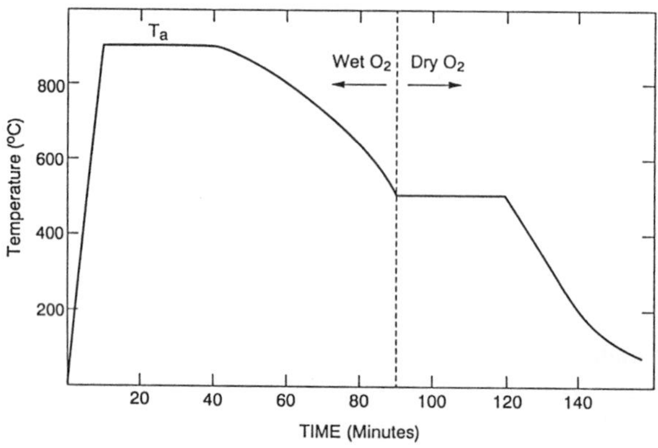

FIGURE 1. Anneal cycle yielding optimized BYCO film properties.

The high temperature annealing stage is particularly important for the development of the 213 crystal structure. This is the stage during which solid phase epitaxy of the film occurs, transforming the layer from its amorphous precursor state into a single crystal epilayer. The morphology of the film is also strongly affected by the temperature (T_a) and duration of this part of the anneal. The superconducting properties are optimized during the low temperature annealing stage. The oxygen content of the film can be adjusted by minor variations in this part of the anneal to maximize the T_c and minimize the width of the superconducting transition. Finally the ambient ("wet" or "dry" oxygen) during various parts of the anneal is extremely important, affecting the structural as well as superconducting properties.

One somewhat surprising finding is that the best quality films are obtained when "wet" O_2 is used from the initiation of the anneal until the second (lower temperature or oxidation) annealing stage, as indicated in Fig. 1. It is straightforward to understand the importance of water vapor during the initial heating to the high temperature annealing stage, since this is when most if not all of the BaF_2 hydrolysis occurs. Initiating the hydrolysis at the lowest possible temperature can be expected to prevent the grain growth of BaF_2 prior to its dissociation. Such grain growth may have an adverse effect on the morphology of the superconducting films and may also lead to other structural inhomogeneities. It is harder to understand the role of the water vapor during cooling from T_a, but a clue may be found in the phase diagram indicating the stability of the 213 vs. 214 phases as a function of oxygen pressure and temperature.[9] Upon cooling in 1 atmosphere of dry oxygen, the sample quickly leaves the region of 213 stability. It may be expected that the presence of water vapor would modify this phase diagram, perhaps enlarging the region of 213 stability and preventing the conversion of some of the 213 material formed during the high temperature part of the anneal into 214 or other phases.

PROPERTIES OF OPTIMIZED FILMS. We have found that the time-temperature profile shown in Fig. 1 results in 100 nm BYCO films with optimized structural properties.[3] They have surface morphology and crystallinity generally associated with high quality *in situ* films. χ_{min} values as low as 2.1% have been obtained (identical to the value which has been calculated for a perfect BYCO crystal). SEM examination shows terraces ~100 nm wide separated by steps and a low density of precipitates. Pinholes ≥50 nm in size are present in films annealed under these conditions with a density (as measured by SEM) $\sim 10^9/cm^2$. Films annealed at lower T_a have rougher morphology and poorer overall crystallinity, but pinholes have not been observed.[5]

The superconducting properties of these films are also excellent,[3] with sharp superconducting resistance transitions at 90-91K. J_c at 77K is $\sim 10^6 A/cm^2$ in zero magnetic field. J_c drops quickly in a magnetic field, however, falling to less than $10^5 A/cm^2$ for $\mathbf{H}_c$=0.9T oriented perpendicular to the ab plane of the film. These results are discussed in more detail below.

STOICHIOMETRY. We have found that films grown by the deposition technique described above are extremely sensitive to the cation stoichiometry of the initially deposited film.[6] In fact, all of the superconducting and structural properties of films deposited with an excess or deficiency of one metal species of greater than 1% are degraded significantly compared with the same properties of "stoichiometric" films prepared under the same conditions. Representative results are shown in Table 1. The precipitate density of non-stoichiometric films is much higher than that of

Film composition		Film properties					
Ba:Y:Cu	Stoichiometry	T_c (K)	ΔT (K)	$J_c(H=0)$ (A/cm^2)	$J_c(H=0.2\,T)$ (A/cm^2)	$H_c(J_c=0)$ (Tesla)	χ_{min} (%)
2.01:1.00:3.00	Correct	90.1	0.2	1×10^6	3×10^5	>0.9	2.1
1.93:1.00:3.00	Ba-poor	86.6	1.9	3×10^5	6×10^4	0.85	10.7
2.07:1.00:3.00	Ba-rich	87.4	2.0	5×10^5	6×10^4	0.5	8.8
2.00:0.96:3.00	Y-poor	85.1	1.2	2×10^5	2×10^4	0.9	7.2
1.98:1.03:3.00	Y-rich	85.3	3.6	1×10^5	4×10^4	0.4	9.5
2.00:1.00:2.89	Cu-poor	88.7	0.7	2×10^5	8×10^4	0.45	8.6
2.00:1.00:3.10	Cu-rich	88.4	0.4	1×10^5	3×10^4	0.7	5.7

TABLE 1. BYCO film properties for various stoichiometric and non-stoichiometric films. J_c values are determined at 77K.

stoichiometric ones. These precipitates have three characteristic shapes. The component phases expected for a given deviation from ideal film composition can be obtained from the equilibrium phase diagram for the high temperature annealing step.[10] Such an analysis has led to the identification of the chemical composition of each of the characteristic precipitate shapes observed. The expected precipitate compositions are indicated in Table 2. These predictions are well correlated with the shapes of precipitates seen in SEM micrographs.

Such extreme importance of the film stoichiometry contrasts sharply with what has been reported by others growing *in situ* films by pulsed laser deposition and molecular beam epitaxy.[11] Such *in situ* growth techniques require growth

Composition	Precipitate Phases
Ba-Poor Y-Rich	BaY_2CuO_5 + CuO
Ba-Rich Cu-Poor	BaY_2CuO_5 + $BaCuO_2$
Y-Poor	$BaCuO_2$ + CuO
Cu-Rich	CuO

TABLE 2. Expected precipitate phases for various BYCO film compositions.

temperatures which are generally 100-200°C lower than the 900°C annealing temperature which we have used in these experiments. This difference is due at least partially to the fact that the *in situ* films require only enough atomic mobility to allow surface diffusion to equilibrium lattice sites, whereas films grown by the BaF_2 process require bulk diffusion to convert the amorphous precursor film into an epitaxial layer by solid phase epitaxy. As long as only surface diffusion is allowed, any precipitates which arise from deviations from ideal stoichiometry may be expected to remain small and very likely embedded within the film (hence not easily observable by SEM), since the atomic mobility is so limited. Once bulk diffusion becomes possible, on the other hand, small precipitates may be expected to coalesce and, depending on the surface and interface energies involved, even to congregate preferentially at the film surface where they are readily apparent by SEM.

CORRELATIONS BETWEEN FILM STRUCTURE AND SUPERCONDUCTING PROPERTIES

χ_{min} AND λ_o. Figure 2 shows the dependence of χ_{min}, which is a measure of the crystalline order in the BYCO films, as a function of T_a, the temperature of the high temperature annealing stage. As previously discussed (Fig. 1), the annealing temperature which gives the best overall crystallinity as measured by RBS is 900°C. Figure 2 shows that above and below this temperature, the crystallinity degrades significantly. Table 3 shows λ_o as a function of T_a.[5] For $T_a \leq 875^{o}$C, λ_o decreases monotonically as both the overall crystallinity and film morphology (measured by both standard and high resolution SEM) improve. For $T_a > 875^{o}$C, however, the temperature dependence of λ deviates from BCS behavior, even though the overall crystallinity continues to improve. This deviation has been modeled by including the effect of defects larger than the coherence length in the ab plane (ξ_{ab}) which reduce

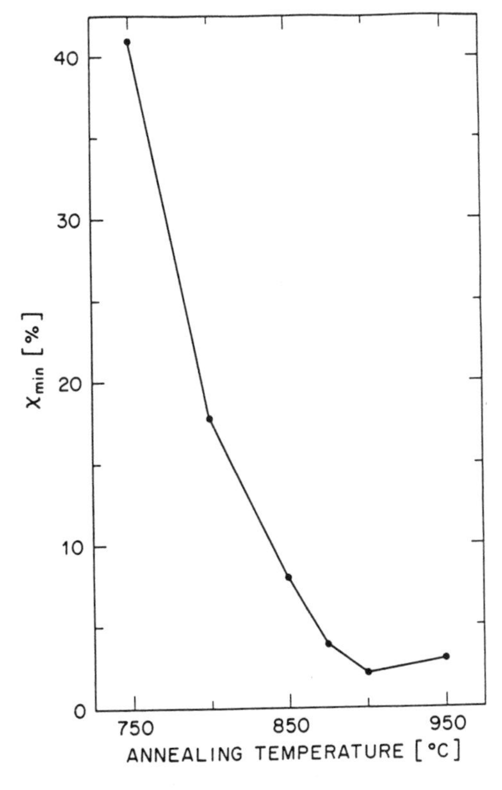

FIGURE 2. χ_{min} as a function of annealing temperature T_a.

TABLE 3. Penetration length (λ_o) at T = 0K for various annealing temperatures (T_a).

T_a(°C)	750	800	850	875
λ_o(Å)	5100	3050	2320	2040

screening.[12] Such defects could take the form of pinholes and microcracks. The density of defects required to model the discrepancy with BCS behavior is $\sim 10^{11}$ cm^{-2}. We attribute the difference between this value and our measured pinhole density of $\sim 10^9 cm^{-2}$ to the presence of pinholes larger than ξ_{ab} but smaller than the HRSEM lateral resolution. As shown in Table 4, the surface resistance measured at 10 GHz using the parallel plate resonator technique is qualitatively consistent with this interpretation, with lower R_s values for films annealed at 850°C than at 900°C.

BEHAVIOR OF J_c(T,H). We have conducted a study of J_c(T,H) of films annealed at a number of values of T_a.[5] The data taken in zero magnetic field give high values of J_c at all temperatures measured and for all T_a values between 800

R_s(10 GHz)		
T_a / T_{meas}	850°C	900°C
4.2K	75μΩ	120μΩ
77K	500μΩ	700μΩ

TABLE 4. Surface resistance measurements (R_s) at 10 GHz for 200 nm BYCO films annealed at different temperatures (T_a) (data taken by R. Taber).

and 900°C, with $J_c(77K,0) \geq 10^6 A/cm^2$ and $J_c(4.2K,0) \geq 10^7 A/cm^2$. Such high values of J_c indicate a low density of weak links in all of these films. $J_c(T,0)$ is optimized at T_a=850°C, which is 50°C lower than the value of T_a which gives the lowest χ_{min} and best morphology.

The J_c behavior of these films in an applied magnetic field in some ways mirrors the zero field results. J_c is once again the highest at T_a=850°C. The magnitude of J_c has fallen substantially from the zero field values, however, especially at higher measurement temperatures, e.g. $J_c(77K,0.9T) \sim 10^5 A/cm^2$. The functional dependence has changed from $J_c^{2/3} \propto T$ in zero field to $J_c^{1/2} \propto T$ in finite magnetic field. What is particularly interesting is that the behavior of $J_c(T,0.9T)$ for T_a=900°C is very similar to that for single crystals in both magnitude and temperature dependence. This suggests that other (primarily *in situ*) films with higher J_c at 77K (but with similar J_c at 4.2K) have additional pinning mechanisms which these films lack. In these T_a=900°C films, we seem to have eliminated some source(s) of pinning common to most thin films so that they behave comparably to single crystals.

DISCUSSION. These results show three regimes of correlation between the electrical and structural properties of our BYCO films.[5] If $T_a \leq 850°C$, all structural and electrical properties improve as the annealing temperature increases, resulting in a direct correlation between the structural quality and electrical properties. If 850°C $< T_a \leq 900°C$, the structure and morphology of the films improve with increasing temperature, while the electrical properties begin to degrade. The onset of degradation is slightly different for λ_o and J_c. If $T_a > 900°C$, the structural properties change little except for the onset and development of pinholes and microcracks. The electrical properties degrade monotonically with increasing T_a.

The point to be made is that λ_o and $J_c(H)$ are sensitive to features of different size in the BYCO films. λ is particularly sensitive to features larger than ξ_{ab}, as noted above. The appearance of pinholes and microcracks in the films coincides with the T_a value (900°C) at which $\lambda(T)$ begins to deviate from BCS behavior. χ_{min} is not sensitive to a low areal coverage of such defects, which is why this structural probe is not an accurate determinator of the behavior of λ once such defects become important. It should be noted, however, that χ_{min} and λ_o are correlated under preparation conditions under which such defects are not observed in the films.

In contrast to λ, J_c is sensitive to features which have a size ~ξ_{ab}. As the crystalline order on this length scale improves, pinning is expected to decrease and to approach the level observed in single crystals, as noted above. This finding is well correlated with the improvement in crystal structure revealed by χ_{min}, with the lowest pinning occurring for films which have the same χ_{min} as single crystals. The difference between these results and the findings of other groups is probably due to differences in film preparation techniques. Other studies of the dependence of J_c on magnetic field strength in thin films have generally been carried out on films grown by *in situ* deposition techniques.[13] As pointed out above, these techniques use considerably lower growth temperatures in order to activate the surface diffusion processes necessary for epitaxial film formation. These lower growth temperatures may be expected not to remove stacking faults and other near-atomic-scale disorder from these films, allowing such defects to act as pinning sites. The higher temperatures used in the current studies allow bulk diffusion to become active which facilitates the removal of such atomic scale disorder.

CONCLUSIONS

By careful identification of the important production parameters, the properties of BYCO thin films grown by co-evaporation of Y, Cu, and BaF_2 followed by a two stage anneal can be investigated and optimized. We find that control of the stoichiometry of the deposited film to ≤1% is critical to the optimization of both structural and electrical properties. The temperature and time of the high temperature annealing stage (T_a) control primarily the crystallinity and morphology of the film. In general, film morphology and crystallinity improve as T_a increases. Such improvements are correlated with decreasing penetration depth and flux pinning. At sufficiently high T_a, however, pinholes and microcracks appear. Since these defects are much larger than the coherence length, they do not affect flux pinning, but they have a profound effect on the penetration depth due to reduced screening. These findings demonstrate the critical importance of achieving control of the structural aspects of high T_c superconducting thin films in order to control their superconducting behavior.

ACKNOWLEDGMENTS

A. F. Hebard and R. B. van Dover were invaluable collaborators on the work correlating the electrical and structural properties of the BYCO films. R. C. Farrow provided the HRSEM micrographs. T. H. Tiefel obtained the J_c data. J. H. Marshall provided technical assistance. The $LaAlO_3$ substrates were provided by C. D. Brandle. R. Taber of Hewlett-Packard performed the surface resistance measurements. B. E. Weir provided technical assistance with the RBS measurements.

REFERENCES

1. See, for example, R. G. Humphreys, J. S. Satchell, N. G. Chew, J. A. Edwards, S. W. Goodyear, S. E. Blenkinsop, O. D. Dosser, and A. G. Cullis, Supercond. Sci. Technol. *3,* 38 (1990).

2. P. M. Mankiewich, J. H. Schofield, W. J. Skocpol, R. E. Howard, A. H. Dayem and E. Good, Appl. Phys. Lett. *51,* 1753 (1987).

3. M. P. Siegal, J. M. Phillips, R. B. van Dover, T. H. Tiefel, and J. H. Marshall, J. Appl. Phys. *68,* 6353 (1990).

4. M. P. Siegal, J. M. Phillips, Y.-F. Hsieh, and J. H. Marshall, Physica C *172,* 282 (1990).

5. M. P. Siegal, J. M. Phillips, A. F. Hebard, R. B. van Dover, R. C. Farrow, T. H. Tiefel, and J. H. Marshall, J. Appl. Phys. (in press).

6. D. J. Carlson, M. P. Siegal, J. M. Phillips, T. H. Tiefel, and J. H. Marshall, J. Mater. Res. *5,* 2797 (1990).

7. C. P. Bean, Rev. Mod. Phys. *36,* 31 (1964).

8. E. M. Gyorgy, R. B. van Dover, K. A. Jackson, L. F. Schneemeyer, and J. V. Waszczak, Appl. Phys. Lett. *55,* 283 (1989).

9. D. E. Morris, J. H. Nickel, A. G. Markelz, R. Gronsky, M. Fendorf, and C. P. Burmester, Mater. Res. Soc. Symp. Proc. *169,* 245 (1990).

10. R. S. Roth, C. J. Rawn, F. Beech, J. D. Whitler, and J. O. Anderson, in *Ceramic Superconductors II,* edited by M. F. Yan (The American Ceramics Society, Inc., Westerville, OH, 1988), p. 303.

11. V. Matijasevic, P. Rosenthal, K. Shinohara, A. F. Marshall, R. H. Hammond, and M. R. Beasley, J. Mater. Res. *6,* 682 (1991); also, D. K. Fork, private communication and J. R. Kwo, private communication.

12. A. F. Hebard, A. T. Fiory, M. P. Siegal, J. M. Phillips, and R. C. Haddon, Phys. Rev. B (in press).

13. A. Inam, X. D. Wu, L. Nazar, M. S. Hegde, C. T. Rogers, T. Venkatesan, R. W. Simon, K. Daly, H. Padamsee, J. Kirchgessner, D. Moffat, D. Rubin, Q. S. Shu, D. Kalokitis, A. Fathy, V. Pendrick, R. Brown, B. Brycki, E. Belohoubek, L. Drabeck, G. Gruner, R. Hammond, F. Gamble, B. M. Lairson, and J. C. Bravman, Appl. Phys. Lett. *56,* 1178 (1990).

GROWTH OF 1212 PbSrYCaCuO AND NdCeCuO FILMS

R. A. Hughes, Y. Lu, T. Timusk and J. S. Preston
Institute for Materials Research
McMaster University, Hamilton, Ont., Canada L8S 4M1

ABSTRACT

We report the preparation of $(Pb_{0.75}Cu_{0.25})Sr_2(Y_{1-y}Ca_y)Cu_2O_7$ thin films by laser ablation. The films are grown in situ on (100) $LaAlO_3$ at the relatively low substrate temperature of 620 °C. The target does not have the same stoichiometry as the thin films in that it is lead enriched. The films are highly oriented with the c-axis perpendicular to the substrate and exhibit a surface morphology which is unique to the oxide superconductors. By varying the Y/Ca ratio it is possible to systematically vary the onset of the superconducting transition temperature from 10 K to a maximum of 86 K. We also report the successful preparation of highly oriented $Nd_{1.85}Ce_{0.15}CuO_{4+\delta}$ thin films grown on (100) $SrTiO_3$ substrates. By carefully controlling the film's oxygen content through vacuum anneals it is possible to obtain superconducting transitions which onset at temperatures as high as 23 K.

INTRODUCTION

Laser ablation has become a powerful technique in the production of highly oriented superconducting thin films. The quality of an oxide superconductor is strongly dependent upon the ability to control the oxygen stoichiometry. This attribute provides an excellent match to the laser ablation technique since its control over the oxygen growth parameter is one of its main strengths. $(Pb_{0.75}Cu_{0.25})Sr_2(Y_{1-y}Ca_y)Cu_2O_7$ and $Nd_{1.85}Ce_{0.15}CuO_{4+\delta}$ are two superconductors which are very difficult to grow in the polycrystalline form. In this report, we will demonstrate the relative ease with which laser ablated films of these materials are grown in comparison to their polycrystalline counterparts.

$(Pb_{0.75}Cu_{0.25})Sr_2(Y_{1-y}Ca_y)Cu_2O_7$

The lead based superconductor $(Pb_{1-x}Cu_x)Sr_2(Y_{1-y}Ca_y)Cu_2O_7$ (1212 PSYCCO) was independently discovered by Cava et al.,[1] Subramanian et al.[2] and Lee et al.[3] All of these groups reported that the 1212 phase was nonsuperconducting. However, it was later demonstrated that superconductivity could be induced in polycrystalline samples by carefully manipulating the oxygen stoichiometry through such procedures as high pressure oxygen anneals[4] or liquid nitrogen quenches.[5] The ability to control the oxygen stoichiometry by laser ablation presented the opportunity to explore this material in greater detail.

The ablation chamber used for the film growth is described elsewhere.[6] Briefly, the beam from a Lumonics excimer laser (λ=308 nm, full width half maximum (FWHM)=30 ns) is masked and then focused onto a rotating target. The spot size is approximately 1 x 2.5 mm^2 which corresponds to a target fluence of 2 J/cm^2. The plume is pink in colour and appears narrower than the plume of $YBa_2Cu_3O_{7-\delta}$. The ablated material lands upon a heated substrate located 4 cm from the target. The substrate temperature used to grow these films was the relatively low value of 620 °C as measured by a Pt/PtRh thermocouple attached to the surface of the furnace. The substrate used for this work was (100) $LaAlO_3$ as it has a close lattice match to the tetragonal 1212 PSYCCO phase. In order to obtain in situ growth it was necessary to leak oxygen into the chamber during the growth process. A 300 mTorr oxygen atmosphere was used during growth followed by an increase to 1/2 atmosphere during the cooldown process.

In order to obtain superconducting films the target had to be lead enriched. This deviates from the usual tendency of laser ablation where the stoichiometry of the target is transferred to the film. This is due to the low sticking coefficients of lead and its oxides to the substrate at high temperature. The degree to which the target must be enriched is intimately connected to other growth parameters such as substrate temperature, repetition rate and the oxygen partial pressure within the chamber. Typically between 5 and 10 % of the lead within the target ends up in the film as measured by ICP. The target was prepared in a two step process. A precursor of $Sr_2Y_{0.5}Ca_{0.5}Cu_2O_x$ was made from stoichiometric amounts of $SrCO_3$, Y_2O_3, $CaCO_3$ and CuO. The precursor was sintered at 900 °C for 12 hours in air, after which the furnace was shut off and allowed to cool to room temperature. The precursor was then reground and PbO was added. The target was then annealed at 700 °C for 12 hours in flowing nitrogen at which time it was allowed to cool to room temperature. Because of the lead enrichment, the targets are not superconducting and are insulating.

The $(Pb_{0.75}Cu_{0.25})Sr_2(Y_{1-y}Ca_y)Cu_2O_7$ superconductor allows for the systematic variation of the Y/Ca ratio. Its effect is very pronounced as indicated by the resistivity curves given in Fig. 1 and the variation in the superconducting transition temperature given in Fig. 2. The onset of the superconducting transition of 86 K with zero resistance at 80 K for the y=0.6 sample are the highest reported values for a 1212 PSYCCO superconductor. It is clearly seen that as Ca is substituted for Y there is a systematic increase in T_c. In addition, there is an obvious progression from semiconducting to metallic behavior. The above results can be explained by an increase in the hole carrier concentration in the CuO_2 planes arising from the substitution of Ca^{2+}on the Y^{3+} site. Tang et al.[4] shows similar results on polycrystalline samples, except that the transition temperatures of our films are consistently higher, and our T_c versus Ca content plot shows a better defined trend. This disagreement may originate from differences in the oxygen content between the films and polycrystalline samples. There is, however, agreement in

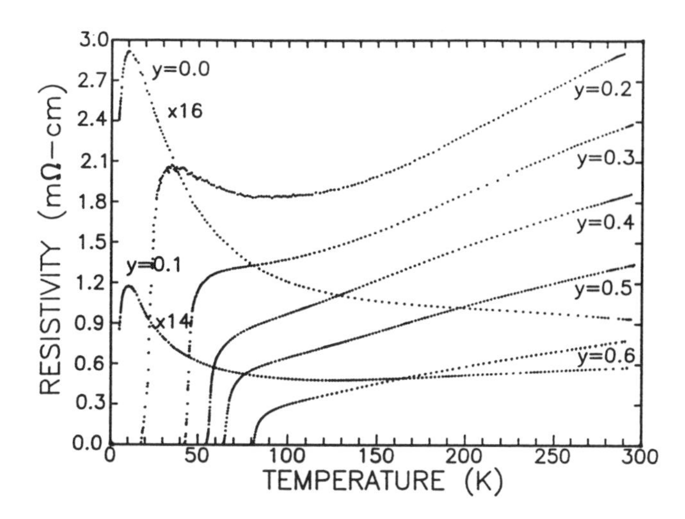

Figure 1 Resistivity versus temperature for $(Pb_{0.75}Cu_{0.25})Sr_2(Y_{1-y}Ca_y)Cu_2O_7$ films for various values of y. There is a progression from semiconducting to metallic behavior as the Ca content in the film is increased.

the fact that the solubility limit of Ca in this structure is reached for $y \geq 0.7$ as indicated by a deterioration in both the resistive transitions and x-ray spectra. It is worthwhile to point out that films which contain no calcium (y=0) still show the onset of a superconducting transition. A short vacuum anneal destroyed the transition implying that oxygen is the source of the remaining carriers.

The $\theta - 2\theta$ x-ray diffraction scans indicate that the films are highly oriented with the c-axis perpendicular to the substrate as illustrated by Fig. 3. These (00l) reflections yield a lattice parameter of 11.90 Å. In addition, we have removed films from the substrate in the form of small flakes and have used these flakes to obtain a powder x-ray pattern using a Guinier camera with Cu $K_{\alpha 1}$ x-rays. This measurement indicated a tetragonal unit cell with a=3.821(5) Å.

Figure 4 shows an SEM micrograph corresponding to a film with a stoichiometry of $(Pb_{0.75}Cu_{0.25})Sr_2(Y_{0.4}Ca_{0.6})Cu_2O_7$. It exhibits an array of well defined squares approximately 700 Å on a side. The squares originate from the tetragonal crystal structure and indicate the presence of columnar grains. The alignment of the columns is a clear indication that the films are growing epitaxially. We believe this surface morphology is distinct among the cuprate superconductors.

In summary, we have prepared highly oriented 1212 PSYCCO thin films by laser ablation at the relatively low substrate temperature of 620 °C. By varying the Y/Ca ratio we have varied T_c from 10 K to 86 K.

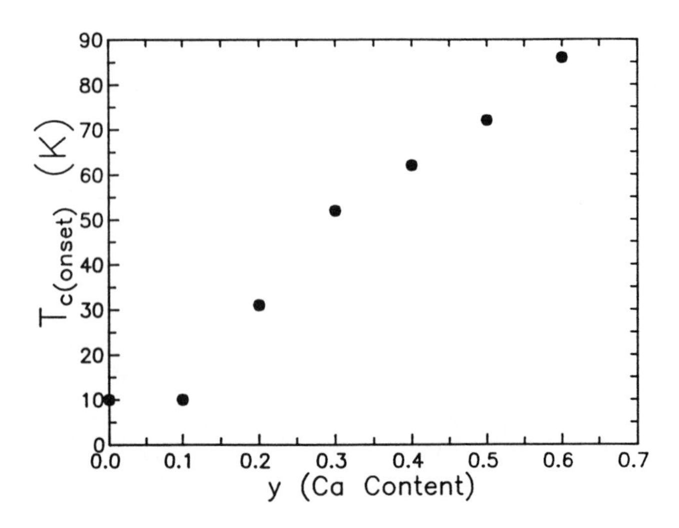

Figure 2 Onset of the superconducting transition temperature as a function of Ca content within the film.

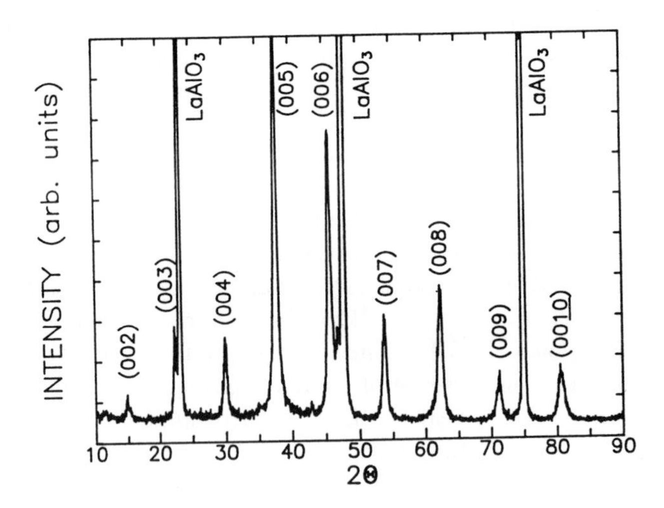

Figure 3 $\theta - 2\theta$ diffraction scan for a 1212 PSYCCO film which indicates that the film is oriented with the c-axis perpendicular to the substrate.

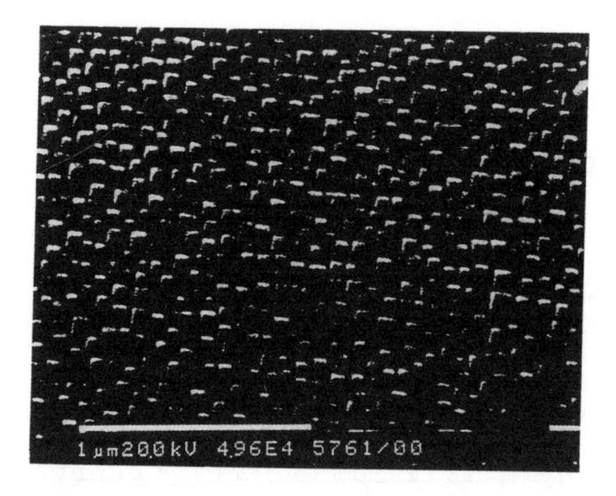

Figure 4 SEM micrograph of a 1212 PSYCCO film grown on a $LaAlO_3$ substrate. The surface is characterized by well alligned squares indicating the presence of epitaxial growth.

$Nd_{1.85}Ce_{0.15}CuO_{4+\delta}$

The oxide superconductor $Nd_{1.85}Ce_{0.15}CuO_{4+\delta}$[7] gained immediate notice due to the fact that superconductivity could be induced by substituting Ce^{4+} on the Nd^{3+} site which made this material the first candidate for n-type superconductivity among the cuprate superconductors. Growth of high quality material can only be achieved by carefully controlling the oxygen stoichiometry through high temperature anneals as a small excess of oxygen can render the sample nonsuperconducting. Homogeneously reducing bulk and single crystal samples at high temperatures is a delicate task and is usually accompanied by a deterioration in surface quality. By comparison, thin films are easier to grow as there large surface area to volume ratio provides optimal conditions for the reduction of samples.

Initially, the growth of $Nd_{1.85}Ce_{0.15}CuO_{4+\delta}$ proceeds in a manner very similar to all other high-T_c superconductors. The laser pulses (fluence = 2 J/cm^2) are directed onto a stoichiometric target. The resulting plume, which is blue in colour, lands upon a (100) $SrTiO_3$ substrate heated to 780 °C. During the deposition oxygen is leaked into and out of the chamber while maintaining a constant pressure of 300 mtorr. The growth of $Nd_{1.85}Ce_{0.15}CuO_{4+\delta}$ is unique in that oxygen must be removed from the film to induce superconductvity. This is accomplished by cooling the sample to 600 °C and evacuating the chamber for a period of approximately 4 hours. At this point the furnace was shut off and the sample was allowed to cool to room temperature.

Many of the above growth parameters were choosen to coincide with those given by Gupta et. al [8] who produced $Nd_{1.85}Ce_{0.15}CuO_{4+\delta}$ films by laser ablation. Their

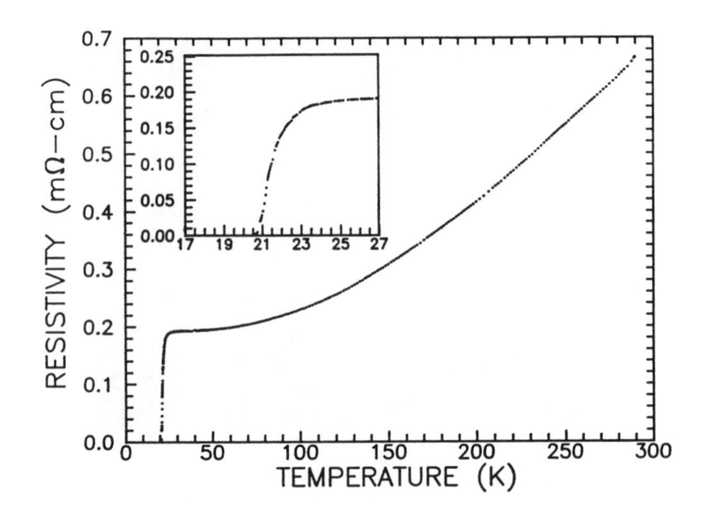

Figure 5 Resistivity versus temperature for a $Nd_{1.85}Ce_{0.15}CuO_{4+\delta}$ film showing the characteristic T^2 temperature dependence. The inset shows an expanded view of the superconducting transition.

films showed excellent transport properties, but they expressed much concern over the surface quality and showed that the anneal rendered the films deficient in copper. Two changes were made to our growth parameters in an attempt to eleviate this problem. Firstly, it has been shown that increasing the oxygen pressure to 300 mTorr during the growth stabilizes the copper within the film.[9] Secondly, we have replaced a short (20 min) high temperature anneal (780 °C) with a long low temperature anneal. This eliminates the possibility of copper loss during the anneal process.

Figure 5 shows a resistivity curve for a film annealed for 4 hours. It exhibits the characteristic T^2 normal state temperature dependence unique to this compound. The superconducting transition onsets at a temperature of 23 K which is consistent with the best results on bulk samples for the Ce concentration choosen. The width of the transitions is 1.5 K.

The $\theta - 2\theta$ x-ray diffraction scans show that the samples are highly oriented with the c-axis perpendicular to the substrate as illustrated by Fig. 6. These (00l) reflections yield a lattice parameter of c=12.07 Å which is consistent with single crystal results for the Ce concentration used.[7] Two lines corresponding to (110) and (220) reflections indicate the presence of non-c-axis growths which were also apparent in other laser ablated films.[8] These growths became more prevalent as the film thickness was increased.

The films are black in color and exhibit optically smooth surfaces. Figure 7 shows a scanning electron microscope image which shows the superconducting grains. An expanded view of the surface shows the presence of particulates and a few micron-size pinholes. These films provide excellent surfaces for optical mea-

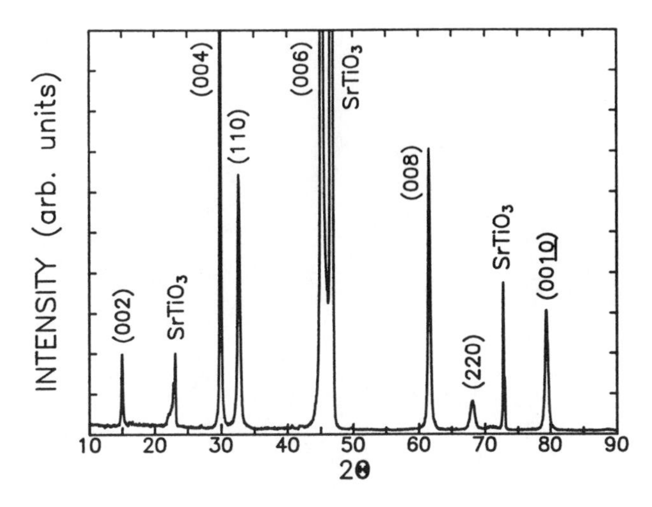

Figure 6 $\theta - 2\theta$ diffraction scan of a $Nd_{1.85}Ce_{0.15}CuO_{4+\delta}$ film showing predominantly (00l) lines.

surements. The far infrared reflectivity (Fig. 8) in both the superconducting and normal states is as high as the best $YBa_2Cu_3O_{7-\delta}$ samples and indicates some very revealing features. The ledge in the reflectivity spectrum at 400 cm^{-1}, which appears above and below T_c, is also seen in YBCO,[10] BSCCO [11] and TBCCO[12] samples. This feature has been attributed to a superconducting energy gap[13] as well as a bound electron phonon interaction.[14] These results argue against the energy gap scenario since the gap would have to be present in the normal state and that its size must be independent of T_c with values of $2\Delta/k_BT_c$ as high as 25.

In summary, we have prepared highly oriented $Nd_{1.85}Ce_{0.15}CuO_{4+\delta}$ thin films by laser ablation. By increasing the oxygen pressure during the deposition and lowering the anneal temperature we have produced samples of superior quality with transition temperatures as high as 23 K.

CONCLUSIONS

In conclusion, we have successfully prepared both $Nd_{1.85}Ce_{0.15}CuO_{4+\delta}$ and $(Pb_{0.75}Cu_{0.25})Sr_2(Y_{1-y}Ca_y)Cu_2O_7$ thin films by laser ablation. The ease with which these two very different materials are grown is a testiment to the versatility of the laser ablation process.

ACKNOWLEDGEMENTS

The authors wish to acknowledge W. H. Gong, K. Teeter and J. Hudak for technical assistance. We thank J. Greedan, J. Xue, A. Dabkowski and H. Dabkowski for helpful discussions. This work is supported by the Ontario Center for Materials

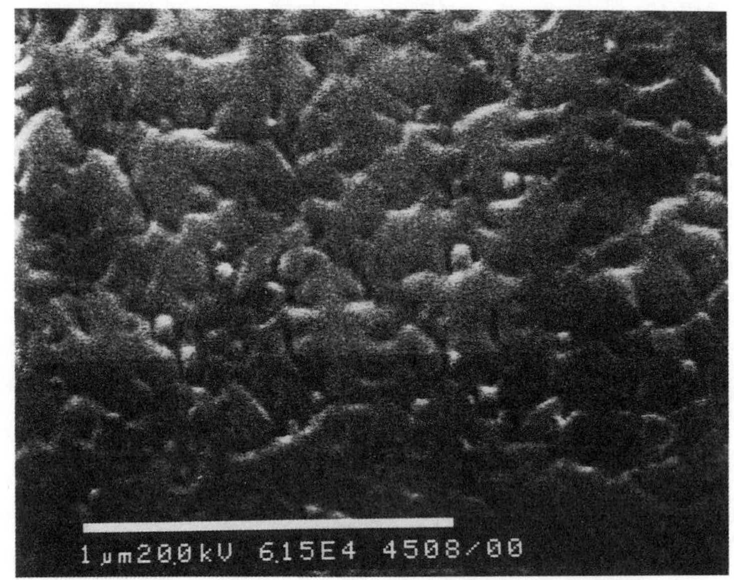

Figure 7 SEM micrograph showing the superconducting grains of a $Nd_{1.85}Ce_{0.15}CuO_{4+\delta}$ film.

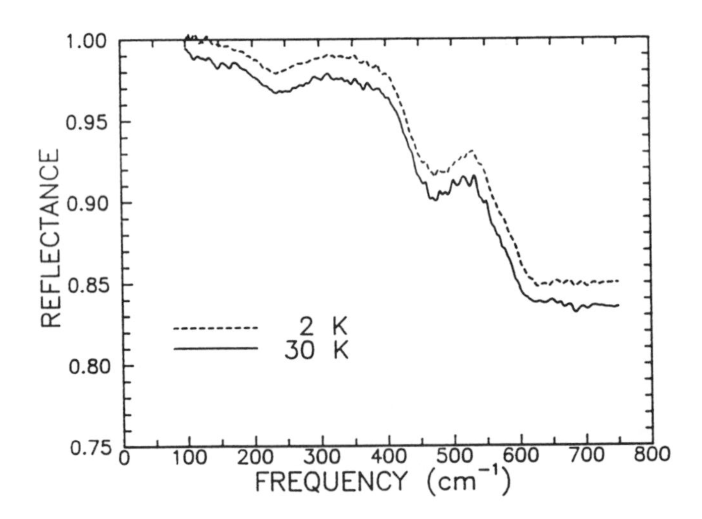

Figure 8 Reflectivity versus frequency for a $Nd_{1.85}Ce_{0.15}CuO_{4+\delta}$ film for temperatures above and below T_c. Note the ledge in the reflectivity at 400 cm^{-1}.

Research and the Natural Science and Engineering Research Council of Canada.

REFERENCES

1. R. J. Cava, B. Batlogg, J. J. Krajewski, L. W. Rupp, L. F. Schneemeyer, T. Siegrist, R. B. vanDover, P. Marsh, W. F. Peck Jr., P. K. Gallagher, S. H. Glarum, J. H. Marshall, R. C. Farrow, J. C. Waszczak, R. Hull, and P. Trevor, Nature 336, 211 (1988).

2. M. A. Subramanian, J. Goparakrishnan, C. C. Trardi, P. L. Gai, E. D. Boyes, T. R. Askew, R. B. Flippen, W. E. Farneth, and A. W. Sleight, Physica C 159, 124 (1989).

3. J. Y. Lee, J. S. Swinnea and H. Steinfink, J. Mater. Res. 4, 763 (1989).

4. X. X. Tang, D. E. Morris and A. P. B. Sinha, Phys. Rev. B 43, 7936 (1991).

5. T. Maeda, K. Sakuyama, S. Koriyama, H. Yamauchi and S. Tanaka, Phys. Rev. B 43, 7866 (1991).

6. R. A. Hughes, Y. Lu, T. Timusk and J. S. Preston, Appl. Phys. Lett. 58, 762 (1991).

7. Y. Tokura, H. Takagi, and S. Uchida, Nature 337, 345 (1989).

8. A. Gupta, G. Koren, C. C. Tsuei, A. Segmüller, and T. R. McGuire, Appl. Phys. Lett. 55, 1795 (1989).

9. J. S. Horwitz, D. B. Chrisey, M. S. Osolofsky, K. S. Grabowski and T. A. Vanderah, J. Appl. Phys. 70, 1045 (1991).

10. Z. Schlesinger, R. T. Collins, D. L. Kaiser and F. Holtzberg, Phys. Rev. Lett. 59, 1958 (1987).

11. M. Reedyk, D. Bonn, J. D. Garrett, J. E. Greedan, C. V. Stager, T. Timusk, K. Kamaras and D. B. Tanner, Phys. Rev. B 38, 11981 (1988).

12. C. M. Foster, K. F. Voss, T. W. Hagler, D. Mihailovic, A. J. Heeger, M. M. Eddy, W. L. Olson, and E. J. Smith, Solid State Comm. 76, 651, (1990).

13. Z. Schlesinger, R. T. Collins, F. Holtzberg, C. Field, G. Koren and A. Gupta, Phys. Rev. B 41, 11237 (1990).

14. T. Timusk, C. D. Porter and D. B. Tanner, Phys. Rev. Lett. 66, 663, (1991).

POST-ANNEALING OF THIN FILMS OF YBCO AT TEMPERATURES AND OXYGEN PARTIAL PRESSURES USED BY IN SITU METHODS

A. Mogro-Campero and L.G. Turner
GE Research and Development Center, Schenectady, NY 12301

ABSTRACT

Post-deposition annealing of thin films of $YBa_2Cu_3O_7$ (YBCO) at temperatures of 850-900°C and in atmospheric pressure oxygen has yielded films with values of critical current density and microwave surface resistance over large areas comparable to those obtained by in situ methods. A recent improvement in the properties of post-annealed films has come about by using low oxygen partial pressures, which allow annealing temperatures to be lowered as has been done by in situ methods. Surface smoothness, critical current density, and ac losses are all improved by using the new post-annealing conditions. These results mean that high quality thin films of YBCO can be made over large areas and with double-sided coating by ambient temperature film deposition and post-annealing in a tube furnace.

INTRODUCTION

Thin films of high temperature superconductors were first formed in two steps: ambient temperature deposition and subsequent furnace annealing (the post-annealing method). Later work, particularly on $YBa_2Cu_3O_7$ (YBCO), combined these steps by depositing the film on heated substrates; this procedure is known as the in situ method. YBCO thin film formation is the subject of a recent review.[1]

Work on the optimization of the post-annealing process for thin films of YBCO using atmospheric pressure oxygen[2] has shown that the optimum annealing temperature is around 900°C. On the other hand, in situ processes have obtained excellent films at considerably lower temperatures of around 750°C. The in situ methods are carried out at typical oxygen partial pressures of 26 Pa.

Both the post-annealing and in situ methods are able to produce YBCO thin films of excellent quality: zero resistance transition temperatures (T_c) around 90 K, critical current densities (J_c) at 77 K in excess of 1 MA cm^{-2}, and values of surface resistance (R_s) at 77 K considerably below those of normal metals at microwave frequencies (say around 10 GHz).

In this paper we point out that for the larger areas (several cm^2) desired for microwave applications, a comparison between YBCO films made by the post-annealing and in situ methods gives similar results for R_s measured by the same technique.

Nevertheless, the fact that post-annealing under atmospheric pressure oxygen is carried out at considerably higher temperatures than those used by in situ methods, is in general disadvantageous. Although high quality films of YBCO can be made at these higher temperatures on selected substrates such as single crystal $SrTiO_3$ and $LaAlO_3$, other desirable substrates (such as sapphire and silicon) benefit from lower processing temperatures.

The major topic of this paper is the very recent advance in the post-annealing method: thin films of YBCO made at temperatures and oxygen partial pressures used by in situ methods. This development has allowed the production of post-annealed films close to the YBCO thermodynamic stability line, yielding generally superior properties. Thus, an advantage of the post-annealing method (ambient temperature film deposition) can be used to make large area and double-sided YBCO films with superior properties.

SURFACE RESISTANCE OVER LARGE AREAS

Some of the lowest values of R_s have been measured on small area samples (typically about 1 cm^2) of YBCO film made by in situ methods.[3] Larger area samples (at least a few cm^2) are desirable for passive microwave devices. By scaling a given deposition process to cover larger areas, film quality may not be the same as when small area samples are produced. In fact, in a study of R_s of YBCO films on $LaAlO_3$, it was noted that R_s increased with increasing film area.[4] Therefore, when comparing values of surface resistance it is best to compare measurements over equal film areas, and if possible the measurements should be performed by the same experimenter and technique. Such a comparison is shown here between YBCO films post-annealed in atmospheric pressure oxygen and YBCO films made in situ by laser ablation.[5]

Figure 1 shows R_s as a function of frequency. The solid symbols correspond to our coevaporated and post-annealed YBCO films,[4,6], and the open circle is for YBCO films made in situ by laser ablation.[5] All circles correspond to data taken by the cavity end-wall replacement method on samples of area 5-6 cm^2. The bar represents a range of values measured on resonators patterned from the coevaporated films.[6] The figure shows that for these large area films, values of R_s at 77 K are similar for post-annealed and in situ samples.

The data in Figure 1 is for YBCO films on $LaAlO_3$. This substrate allows high annealing temperatures (850 to 900°C for the post-annealing method using atmospheric pressure oxygen), and yet produces films of acceptable quality. There are two main reasons to try to reduce the annealing temperature: to be able to use more difficult substrates (such as silicon and sapphire), and to improve thin film properties. Recent developments which allow a significant reduction in the post-annealing temperature are the subject of the remainder of this paper.

POST-ANNEALING OF YBCO FILMS AT LOWER TEMPERATURES

In situ YBCO film fabrication is customarily carried out by vacuum techniques (laser or conventional evaporation and sputtering). Since annealing is done simultaneously with film deposition, these methods operate at low gas pressure during the high temperature portion of the process (the oxygen partial pressure is increased substantially during cool-down, i.e., after the deposition is complete). It was found that good quality YBCO films were made at temperatures in the range of about 600 to 780°C. The higher temperatures were found to be acceptable at the higher oxygen partial pressures compatible with the deposition technique (laser ablation or sputtering), and lower temperatures were found for lower oxygen partial pressures (used e.g., in evaporation methods).

An understanding of the relationship between oxygen partial pressure and annealing temperature evolved from studies of high temperature chemistry and thermodynamics of the YBCO system.[7-11] An explicit connection between these studies and YBCO thin film processing was made by correlating in situ growth conditions with thermodynamic stability criteria.[12-14] In essence, this comparison showed that the oxygen partial pressures and annealing temperatures for the best YBCO films made by in situ methods was close to the thermodynamic critical stability line for YBCO.

It took a relatively long time for these ideas to be taken up by those working on the post-annealing method of forming YBCO films. At first, a low oxygen partial pressure was used during part of the post-annealing cycle, but the maximum annealing temperature was kept at 900°C.[15]

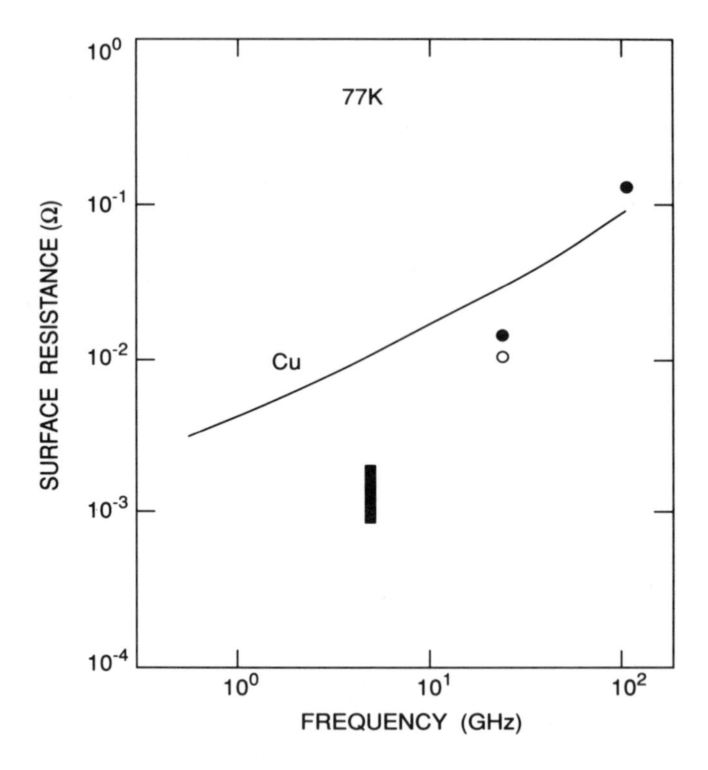

Fig. 1. Surface resistance as a function of frequency; depositions were all YBCO on 5-6 cm^2 areas of $LaAlO_3$. The circles are measurements on unpatterned films by the cavity end-wall replacement method[4,5]; the solid circles are for our films, and the open circle is for a film made in situ by laser ablation. The bar covers a range of values obtained from resonators patterned from our films.[6]

As shown in Figure 2, the standard post-annealing conditions using atmospheric pressure oxygen lie far from the critical stability line for YBCO (which, as discussed above, is also close to the best conditions found experimentally for the production of in situ YBCO thin films). Excellent results were obtained[16] when post-annealing was tried at a temperature (750°C) and oxygen partial pressure (29 Pa) which lie on the YBCO stability line (these temperature and oxygen pressure conditions are shown in Figure 2). These relatively thick 0.6 μm films, which had a critical current density at 77 K around 10^5 A cm^{-2} when post-annealed at 850°C in atmospheric pressure oxygen, were found to improve by about an order of magnitude in J_c when post-annealed at 750°C and 29 Pa of oxygen.[16] In addition, optical micrographs of the surface of these films showed an improved surface smoothness and less grains with the c-axis in the plane of the film for the samples post-annealed at low oxygen partial pressure. This work showed that the lower annealing temperatures found possible for the in situ formation of YBCO films were also applicable to post-annealed films, so the authors concluded that because film deposition can be carried out at ambient temperature, this new development in post-annealing meant that large area and double-sided YBCO films of the highest quality can be easily made.[16]

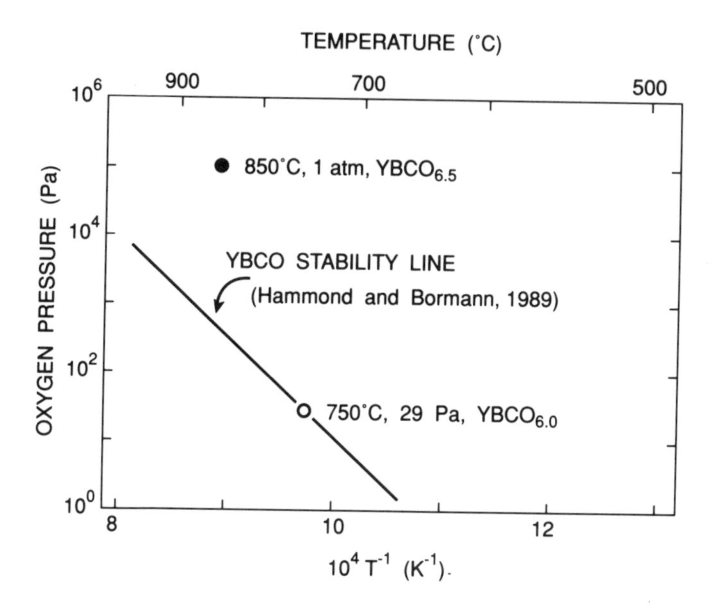

Fig. 2. Oxygen partial pressure versus temperature, showing the YBCO stability line.[14] Points corresponding to the post-annealing conditions we have used are also shown.[18] Oxygen stoichiometries[9] are also indicated.

A large number of post-annealing temperature and oxygen partial pressure conditions were explored, and it was also noted that enhanced c-oriented epitaxy occurred at 740°C and 26 Pa of oxygen.[17] From these observations, annealing lines were derived on the temperature-oxygen partial pressure diagram for either c- or a-axis oriented YBCO film growth on $SrTiO_3$.[17] In a comparison of YBCO films post-annealed at 850°C and atmospheric pressure oxygen and others at 750°C and 29 Pa of oxygen, it was found that for the sample annealed at higher temperature, the room temperature resistance was a factor of 2 higher and J_c at 77 K a factor of 3 lower than for the sample annealed at 750°C (using $LaAlO_3$ substrates).[18] The samples post-annealed at lower temperature and oxygen partial pressure were also found to have a much sharper eddy-current transition (measured at 25 MHz), so that ac losses were also lower.[18]

DISCUSSION

The improvement in YBCO thin film properties by processing (either in situ or by post-annealing) at values of temperature and oxygen partial pressure in proximity to the thermodynamic stability line[12-14] suggests that an important factor is the increased atomic mobility which is present near a phase boundary. Because the improvement is observed for YBCO films made in situ or by post-annealing, an increase in surface mobility on the growing surface seems to be less important. A lower oxygen stoichiometry (6.0 versus 6.5 in Figure 2) near the stability line may also be favorable during YBCO crystal growth.

A summary of the most recent YBCO stability measurements results in a stability line in the temperature-oxygen partial pressure diagram which is parallel to that published earlier[12-14] but lower in oxygen partial pressure by an order of magnitude.[17] High quality films were made under conditions in between these two lines, and it has been suggested that the conditions for best in situ films and best post-annealed films may not be coincident.[17] However, a recent study of the optimization of post-annealing of YBCO films at an oxygen partial pressure of 29 Pa has concluded that 750°C is the optimum temperature (precisely on the original stability line as shown in Figure 2).[19] Further work should clarify this point.

CONCLUSIONS

Thin films of YBCO post-annealed at atmospheric oxygen pressure can be made with values of J_c at 77 K exceeding 1 MA cm^{-2}, i.e., which is comparable to the results obtained by in situ methods. When a comparison is made of the surface resistance values in the microwave regime, comparable values are also found for post-annealed and films made in situ for large area samples (several cm^2).

A recent and very significant development in the processing of YBCO films by the post-annealing route is the finding that lower temperatures and partial pressures of oxygen can be used to produce YBCO films of higher quality.[16-19] This means that substrate/film interdiffusion is minimized, so that a wider variety of substrates can be used for YBCO films using the post-annealing method. In addition, even in the case of robust substrates such as $SrTiO_3$ and $LaAlO_3$, which resist high temperature processing, dropping the temperature and oxygen partial pressure can improve film quality.

The best values of processing temperature and oxygen pressure are close to those found by in situ methods, and close to the thermodynamic YBCO stability line.

Double-sided YBCO film deposition is advantageous for certain applications (e.g., to provide ground planes in microwave device stripline and microstrip configurations). In the post-annealing method double-sided deposition is simple because it can be carried out at ambient temperature. Ambient temperature deposition also makes it relatively easy to cover large areas. Large area and double-sided deposition is very difficult when using in situ methods because of the need of good thermal contact between the wafer and the heating stage and the problem of uniform heating over large areas.

REFERENCES

1. R.G. Humphreys, J.S. Satchell, N.G. Chew, J.A. Edwards, J.W. Goodyear,S.E. Blenkinsop, O.D. Dosser, and A.J. Cullis, Supercond. Sci. Technol. 3, 38 (1990).
2. M.P. Siegal, J.M. Phillips, R.B. van Dover, T.H. Tiefel, and J.H. Marshall J. Appl. Phys. 68, 6353 (1990).
3. A. Inam, X.D. Wu, L. Nazar, M.S. Hegde, C.T. Rogers, T. Venkatesan, R.W. Simon, K. Daly, H. Padamsee, J. Kirchgessner, D. Moffat, D. Rubin, Q.S. Shu, D. Kalokitis, A. Fathy, V. Pendrick, R. Brown, B. Brycki, E. Belohoubek, L. Drabeck, G. Grüner, R. Hammond, F. Gamble, R.M. Lairson, and J.C. Bravman, Appl. Phys. Lett. 56, 1178 (1990).
4. D.W. Cooke, E.R. Gray, P.N. Arendt, N.E. Elliott, A.D. Rollett, T.G. Schofield, A. Mogro-Campero, and L.G. Turner, J. Appl. Phys. 68, 2514 (1990).
5. D.W. Cooke, P.N. Arendt, E.R. Gray, R.E. Muenchausen, B.L. Bennett, S.R. Foltyn, R.C. Estler, X.D. Wu, G.A. Reeves, N.E. Elliott, P.R. Brown, A.M. Portis, R.C. Taber, and A. Mogro-Campero, Proc. Int. Conf. Electronic Mater. - 1990, R.P.H. Chang, T. Sugano, and V.T. Nguyen, ed. (Materials Research Society, Pittsburgh, 1991) p. 93.
6. M.R. Namordi, A. Mogro-Campero, L.G. Turner, and D.W. Hogue, IEEE Trans. Microwave Theory Tech. 39, 1468 (1991).
7. T.B. Lindemer, J.F. Hunley, J.E. Gates, A.L. Sutton, Jr., J. Brynestad, C.R. Hubbard, and P.K. Gallagher, J. Amer. Ceram. Soc. 72, 1775 (1989).
8. K.W. Lay, and G.M. Renlund, J. Amer. Ceram. Soc. 73, 1208 (1990).
9. D.E. Morris, A.G. Markelz, B. Fayn, and J.H. Nickel, Physica C 168, 153 (1990).
10. T. Wada, N. Suzuki, A. Ichinose, Y. Yaegashi, H. Yamauchi, and S. Tanaka, Appl. Phys. Lett. 57, 81 (1990).
11. T.B. Lindemer, F.A. Washburn, C.S. MacDougall, and O.B. Cavin, Physica C 174, 135 (1991).
12. R. Bormann, and J. Nölting, Appl. Phys. Lett. 54, 2148 (1989).
13. R. Bormann, and J. Nölting, Physica C 162-164, 81 (1989).
14. R.H. Hammond, and R. Bormann, Physica C 162-164, 703 (1989).
15. S.W. Filipczuk, R. Driver, and G.B. Smith, Physica C 170, 457 (1990).
16. A. Mogro-Campero, and L.G. Turner, Appl. Phys. Lett. 58, 417 (1991).
17. R. Feenstra, T.B. Lindemer, J.D. Budai, and M.D. Galloway, J. Appl. Phys. 69, 6569 (1991).
18. A. Mogro-Campero, and L.G. Turner, Physica C 176, 429 (1991).
19. A. Mogro-Campero, and L.G. Turner, J. Superconduct., in press (1991).

STRUCTURAL AND ELECTRICAL PROPERTIES OF $YBa_2Cu_3O_{7-x}$ FILMS ON $SrTiO_3(110)$ SUBSTRATES

J.P. Zheng, S.Y. Dong, D. Bhattacharya and H.S. Kwok
Institute on Superconductivity
State University of New York at Buffalo
Amherst, New York 14260

ABSTRACT

The structural and electrical properties of in-situ laser deposited thin films of $YBa_2Cu_3O_{7-x}$ on $SrTiO_3(110)$ substrates were studied as a function of deposition temperature. It was found that at low temperatures, the films with (110) orientation dominated. At higher temperatures, the (103) orientation was obtained. The lattice mismatch model can qualitatively explain the orientation of the films at different deposition temperatures, but the transition point for the orientation of the film was found to shift as much as 60 °C to higher temperature when compared to the predictions of the lattice mismatch model. The J_c was found to be extremely anisotropic. For (103) oriented films at 70 K, the J_c's were found to be 0.85×10^6 A/cm^2 and 5×10^2 A/cm^2 along the substrate <100> and <011> directions respectively.

1. INTRODUCTION

$YBa_2Cu_3O_{7-x}$ (YBCO) is one of the strong candidates for electronic applications of high temperature superconductors because stable, high quality films can readily be made in-situ on many substrates. Much effort and great advances have been achieved in preparing YBCO films with large critical currents by various methods such as laser deposition[1-7], magnetron sputtering[8,9] and evaporation[10]. Recently, films with either c-axis perpendicular (c⊥) or a-axis perpendicular (a⊥) to the substrates have been controllably grown[9,11]. While c⊥ films have extremely large J_c along the film surface, a⊥ films are unique because they have larger coherence lengths perpendicular to the films. Such a⊥ films can be exploited in making large area tunnel junctions and superlattices where the currents are required to flow perpendicular to the films[6,12-14]. All of these films were grown on substrates with a (100) surface.

In this paper, we would like to report a systematic study of films deposited on substrates with (110) surfaces. It was believed that this is the first quantitative study of the grain orientation on $SrTiO_3(110)$ substrates. It is well-known that on such surfaces, the YBCO film can have either the c-axis on the plane or at 45° to the substrate surface[10,15,16]. Such films are different from the a⊥ or c⊥ films on (100) substrates because they are structurally and electrically anisotropic along the surface of the film, specifically, along the <100> and <011> directions of the substrate surface. Such anisotropy can be exploited in making novel single layer planar devices. One such example is the recently studied anisotropic photovoltaic effect in YBCO[17].

In terms of the dependence of the structural and electrical anisotropy of such films, Enomoto et al reported a factor of 180 in J_c along two orthogonal directions on the film surface[15]. Gupta et al found that the film was not even superconducting in one direction while the T_c was 82 K in the other[16]. These electrical anisotropies are larger than those measured with single crystal samples. For example, in zero field, the ratio of J_c between the [001] and [100] directions in single crystal YBCO is found to be only 30 [18]. It is the purpose of this study to examine systematically the substrate temperature dependence of the structural and electrical anisotropy of such YBCO films.

2. SAMPLE PREPARATION

$SrTiO_3$(110) was chosen as the substrate in this study. Pulsed laser deposition was used to grow the in-situ films. An ArF laser at 193 nm and a fluence of 2 J/cm^2 was used. The target-substrate separation was 10 cm with a deposition pressure of O_2 at only 12.5 mtorr. The deposition rate was 0.2 nm/sec and the films were typically 200 nm thick.

3. TEXTURE OF THE FILMS

Fig. 1 shows the θ-2θ scan of the YBCO film grown at 610 °C. The results are similar to Enomoto et al [15] except that we cannot claim that the <110> axis is perpendicular to the substrate. It is because the (103) and (110) diffraction peaks are difficult to distinguish in such θ-scans. To measure the ratio of <110>⊥ and <103>⊥ grains in the film, other diffraction geometries are needed [16,19] as shown in Fig. 2, by tilting the sample to $\alpha = 10.2°$, it is possible to observe the (225) and (108) peaks which are proportional to the amount of <110>⊥ and <103>⊥ grains respectively. The corresponding 2θ angles are 81.4° and 68.7° which are easily distinguishable.

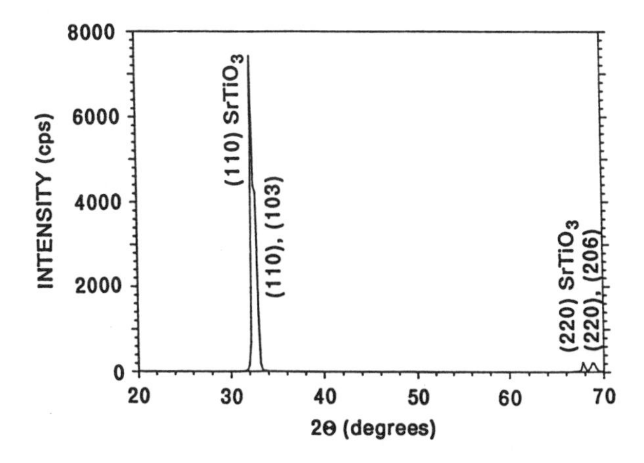

Fig. 1. X-ray diffraction pattern of the 610 °C film.

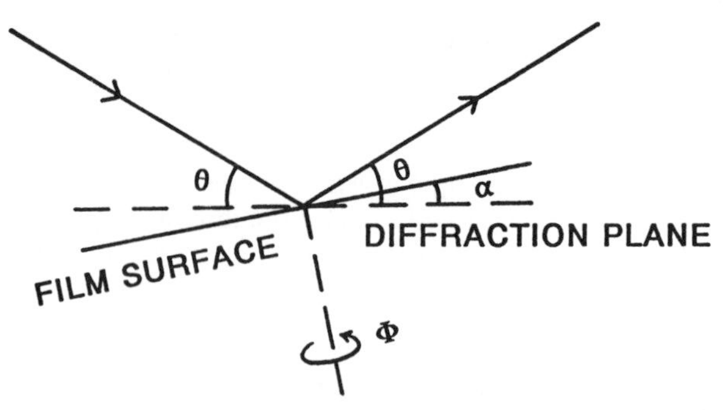

Fig. 2. Field-Merchant geometry for texture analysis. The plane of diffraction and film surface are not the same as in Fig. 1. Φ is varied while θ is fixed.

Fig. 3 shows the percentage of (103) and (110) oriented grains for the 6 substrate temperatures investigated. It can be seen that at high temperatures, the film tends to be <103>⊥ which corresponds to the YBCO c-axis being at 45° to the substrate and a-axis parallel to the <100> direction of the substrate. The c-axis, however, can point in either direction of the <011> axis of the substrate. At lower deposition temperatures, <110>⊥ grains dominate. This alignment corresponds to the YBCO c-axis lying on the film surface and parallel to the substrate <100> direction and the YBCO a-axis at 45° to the film surface. In both cases, there is an ubiquitous a-b twinning. This is because during deposition, the tetragonal phase is formed which transforms to the orthorhombic phase when the film cools down in an O_2 ambient[20]. During this transformation, the stretching and contraction of the tetragonal a-axis should be random in order not to induce too much stress at the film/substrate interface.

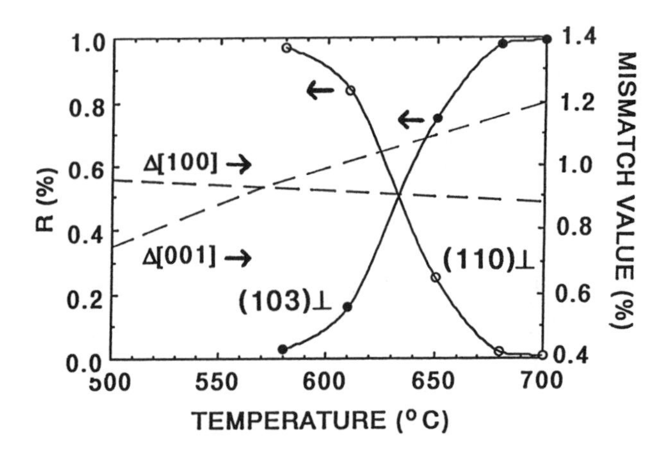

Fig. 3. Grain orientations of the films deposited at different temperatures. The dashed lines represent lattices mismatch reported in Ref. 10.

The texture of these films was also analyzed by x-ray pole figure measurements. It was confirmed that for both types of orientations, the texture was anisotropic along the surface of the film as well. This was predicted by the (110) orientation of the substrate. This property is unlike the (100) substrate case in which the c- or a- axes can lie along either the [001] or [010] directions resulting in isotropic electrical and structural properties.

The results on the structure of the YBCO film and its dependence on temperature is similar to those reported by Terashima et al[10]. They explained the occurrence of the two alignments by the different thermal expansion coefficients of the YBCO a- and c-axes. At T < 580 °C, the lattice mismatch between YBCO[001]/3 and $SrTiO_3$[100] is the smallest, resulting in <110>⊥ alignment. At higher temperatures, the mismatch between YBCO[100] and $SrTiO_3$[100] is smaller, resulting in the <103>⊥ alignment. The dashed lines in Fig. 3 represent the lattice mismatch value reported in Ref. 10. The mismatch values were defined by $\Delta[100]=|a_t-a_s|/a_t$, $\Delta[001]=|(c_t/3)-a_s|/(c_t/3)$, where a_t and c_t are the lattice constants of YBCO in the tetragonal symmetry, and a_s is the lattice constant of $SrTiO_3$.

It can be seen from Fig. 3 that the transition from (110) film to (103) film occurs at a temperature about 60 °C higher than that calculated from the lattice mismatch model. After analysing the results of other groups[9,16,21], it was found that similar shift occurs for c⊥ and a⊥ films. We believe that the lattice mismatch relationship must play an important role in determining the orientation of the films. However some other temperature dependent factors such as the mobility of the atoms and the minimum surface energy for these different oriented grains could also depend on the substrate temperature during the early stages of film growth. These need to be further studied.

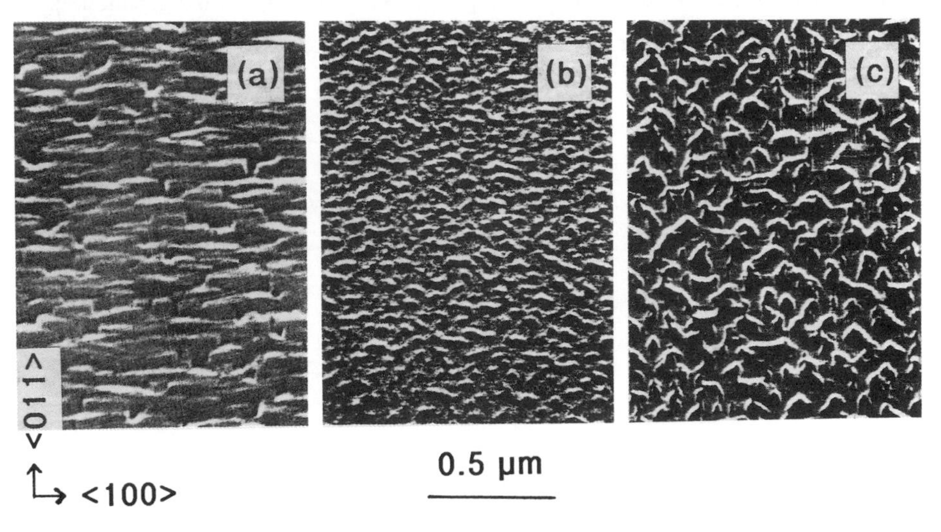

Fig. 4. Scanning electron micrographs of the films deposited at temperature of (a) 700 °C, (b) 620 °C and (c) 580 °C.

The surface morphology of these films has been investigated by Scanning Electron Microscopy (SEM). Fig. 4 shows the surface morphology for the films deposited at three different temperatures. Compared with the films deposited on $SrTiO_3$ (100) substrates, the surfaces of the films deposited on $SrTiO_3$ (110) are much rougher.

4. ELECTRICAL PROPERTIES OF THE FILMS

The electrical anisotropy of these films is closely related to their structural anisotropy. The T_c's and J_c's were measured using a standard 4-probe technique with 20-100 μm wide microbridges. The 1 μV/cm criteria was used with 1 mm long bridges. Fig. 5 shows the superconducting transition for the <103>⊥ film in the two orthogonal directions. It was found that the T_c was independent of the measurement direction. But there is a factor of 5 difference in the normal state resistivity. As a comparison, we also show the case of a $SrTiO_3$(100) film which has the highest quality.

The J_c dependence on temperature is strongly anisotropic. Fig. 6 shows the results. For the best <103>⊥ film, there is a factor of 10^3 in the ratio of J_c along the two orthogonal <011> and <100> directions of the substrate. It is reasonable to expect that the J_c along the basal plane is larger. However, the surprising fact is that this anisotropy is even larger than that of single crystals which only show a factor of 30 between the [100] and [001] directions[18]. For the <110>⊥ films, the J_c anisotropy is reversed, as expected from the c-axis alignment. Table I summarizes the anisotropic properties of the six films grown at different temperatures. It can be seen that the superconducting properties are diminished for the <110>⊥ films. This is believed to be due to the lower growth temperatures and incomplete oxygenation[20].

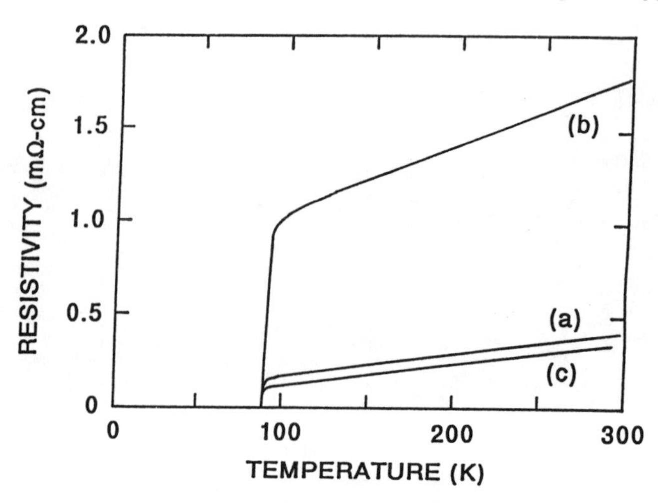

Fig. 5. Superconducting transition of the YBCO film (a) along the substrate <100> direction, (b) along the substrate <011> direction and (c) on $SrTiO_3$(100) substrate.

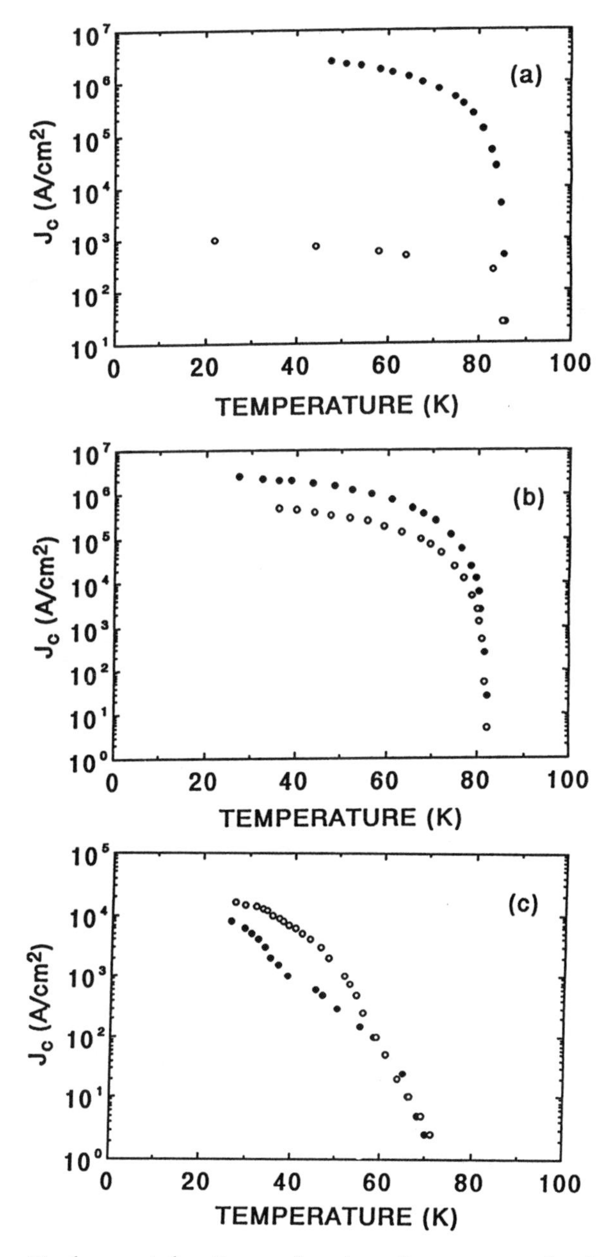

Fig. 6. The critical current density as a function of temperature for the films deposited at temperatures of (a) 700 °C, (b) 650 °C and (c) 610 °C. The closed and opened circle represent the Jc measured along <100> and <011> directions of the substrates, respectively.

TABLE I Summary of the characteristics of YBCO thin films on $SrTiO_3(110)$

Growth temperature(°C)	<103>⊥	<110>⊥	T_c(<100>)/ T_c(<011>)(K)	J_c(<100>) (A/cm^2)(70 K)	J_c(<011>) (A/cm^2)(70 K)
580	3 %	97 %	< 10	—	—
610	16 %	84 %	72/72	$3x10^2$**	$1.5x10^3$**
630	—	—	79/78	$9x10^4$	$9x10^4$
650	75 %	25 %	83/83	$3x10^5$	$8x10^4$
680	98 %	2 %	84/83	$7x10^5$	$1.5x10^4$
700	> 99 %	< 1 %	88/88	$8.5x10^5$	$5x10^2$

** Critical current densities were measured at 50 K.

Surveying the literature, it was found that the anisotropy in J_c is generally larger in thin films than in single crystals. Enomoto et al[15] observed a difference of 180x while Eom et al[9] noted a difference of 70x between a⊥ and c⊥ films. The factor of 10^3 measured here is the largest reported. We believe that the reason is due to the grain structures in all of these "epitaxial" films. Unlike single crystals, there exists a greater amount of grain boundaries even in the highest quality epitaxial YBCO films as demonstrated in Fig. 2(a). It can be seen that there are more grain boundaries in the <011> direction than the <100> direction. Since there are 45° grain boundaries along the <011> direction, it is therefore not too surprising to have a large reduction in J_c in this direction[22].

5. CONCLUSION

In summary, controllable <103>⊥ and <110>⊥ films can be grown on $SrTiO_3(110)$ substrates. These films are well textured and uniform. They are extremely anisotropic in electrical properties and should have applications as novel planar devices. Another important application of these films may be in the growth of superlattices and large area junction devices similar to the a⊥ films[9-14]. Since the c-axis is at 45° to the substrate surface for the <103>⊥ films, the coherence length can be large perpendicular to the substrate, an advantage enjoyed by a⊥ films. Moreover, the J_c can be very different along the <100> and <110> directions of the substrate, a property which is not shared by the a⊥ films. There should be interesting electrical devices that can be fabricated with these unique film properties.

This research was supported by National Science Foundation grant no. ECS 9006847.

6. REFERENCES

1. see for example M. Leskela, J.K. Truman, C.H. Mueller and P.H. Holloway, *J.Vac.Sci. Tech.* A7, 3147 (1989).
2. Q.Y. Ying and H.S. Kwok, *Appl. Phys. Lett.* 56, 1478 (1990).
3. A. Inam, C.T Rogers, R. Ramesh, K. Remschnig, L. Farrow, D. Hart, T.

Venkatesan and B. Wilkes, *Appl. Phys. Lett.* 57, 2487 (1990).
4. R.K. Singh, J. Narajan, A.K. Singh and J. Krishnaswamy, *Appl. Phys. Lett.* 54, 2271 (1989).
5. B. Roas, L. Schultz and G. Endres, *Appl. Phys. Lett.* 53, 1557 (1989).
6. R. Gross, A. Gupta, E. Olsson, A. Segmuller and G. Koren, *Appl. Phys. Lett.* 57, 203 (1990).
7. D.P. Norton, D. Lowndes, J.D. Budai, D.K. Christen, E.C. Jones, K.W. Lay and J.E. Tkaczyk, *Appl. Phys. Lett.* 57, 1164 (1990).
8. Q. Li, O. Meyer, X.X. Xi, J. Geerk and G. Linker, *Appl. Phys. Lett.* 55, 1792 (1989).
9. C.B. Eom, A.F. Marshall, S.S. Laderman, R.D. Jacowitz and T.H. Geballe, *Science*, 249, 1550 (1990).
10. T. Terashima, Y. Bando, K. Iijima, K. Yamamoto and K. Hirata, *Appl. Phys. Lett.* 53, 2232 (1988).
11. T. Venkatesan, A. Inam, B. Dutta, R. Remesh, M.S. Hegde, X.D. Wu, L. Nazar, C.C. Chang, J.B. Barner, D.M. Hwang and C.T. Roger, *Appl. Phys. Lett.* 56, 391 (1990).
12. J.M. Triscone, O. Fischer, O. Brunner, L. Antognazza, A.D. Kent and M.J. Karkus, *Phys. Rev. Lett.* 64, 804 (1990).
13. C.L. Jia, B. Kabius, H. Soltner, U. Poppe, K. Urban, J. Schubert and Ch. Buchal, *Physica* C 167, 463 (1990).
14. H. Obara, S. Kosaka and Y. Kimura, *Appl. Phys. Lett.* 58, 298 (1991).
15. Y. Enomoto, T. Murakami, M. Suzuki and K. Moriwaki, *Jap. J. Appl. Phys.* 26, L1248 (1987).
16. A. Gupta, G. Koren, R.J. Basement, A. Segmuller and W. Holber, *Physica* C162-164, 127 (1989).
17. H.S. Kwok, J.P. Zheng and S.Y. Dong, *Phys. Rev.* B43, 6270 (1991).
18. E.M. Gyorgy, R. B. Van Dover, K.A. Jackson, L.F. Schneemeyer and J.V. Waszczak, *Appl. Phys. Lett.* 55, 283 (1989).
19. J.P. Zheng, S.Y. Dong and H.S. Kwok, *Appl. Phys. Lett.* 58, 540 (1991).
20. H.S. Kwok and Q.Y. Ying, *Physica C*, 177, 122 (1991).
21. C.B. Eom, J.Z. Sun, B.M. Lairson, S.K. Streiffer, A.F. Marshall, K. Yamamoto, S.M. Anlage, J.C. Brarman, and T,H, Geballe, *Physica C*, 171, 354 (1990).
22. A. Dimos, P. Choudhari, J. Mannhart, and F.K. Legoues, *Phys. Rev. Lett.* 61, 219 (1988).

SUPERCONDUCTING Tl-Ba-Ca-Cu-O THIN FILMS

by

M. Young, X-L. Yang, Y.F. Li, Y.G. Tang, Z.Z. Sheng
G.J. Salamo, R.J. Anderson*, and F.T. Chan

Physics Department/HIDEC
University of Arkansas
Fayetteville, Arkansas 72701

ABSTRACT

Tl-Ba-Ca-Cu-O superconducting thin films have been prepared using both laser ablation and "fog" spray techniques. The zero resistance temperature varied between 92 K and 119 K.

INTRODUCTION

The potential application of high T_c superconductors in the field of advanced microelectronics depends on the development of low cost high quality superconducting thin films. In this paper we will describe results obtained using two techniques to produce Tl-Ba-Ca-Cu-O superconducting thin films. The first technique is laser ablation while the second involves the use of a chemical "fog" spray.

LASER ABLATION

Results using the laser ablation technique on an alumina substrate are shown in Figure 1. A very low resistance is maintained until a T_c of up to 92 K. The technique used to fabricate these films employs a simple two-step process[1, 2]. In the first step a target of $Ba_2Ca_2Cu_3O_x$ is placed in a vacuum chamber. The output of an excimer laser is then focused into the chamber and onto the target. The high laser intensity at the target then results in ablation of the target material. The resulting spray of material then deposits onto an alumina substrate. The experimental apparatus for the deposition of the thin film is shown in Figure 2. In the second step the sample is removed from the chamber and placed in an oven along with a boat of $Tl_2Ba_2Ca_2Cu_3O_x$ powder. The sample is then heated at 830° C for approximately five minutes and then cooled and removed for testing.

*National Science Foundation, Washington, D.C.

The excimer laser shown in Figure 2 was operated using an AF gas mixture producing about 100 millijoules in a 10 nanosecond pulse at a wavelength of 193 nanometers. The output of the laser was focused, using a quartz lens of focal length 50 mm, into the vacuum chamber housing the Ca-Ba-Cu target and alumina substrate. The focused laser spot size at the target was about 100 $\mu \times 200\ \mu$. At the laser repeation rate of 20 Hz the time for forming a 2 micron thick film was about 40 minutes giving an average deposition rate of about 0.8 nanometers per second. The vacuum chamber was maintained at a background pressure of 10-5 torr while the alumina substrate was maintained at a temperature of 200° C. The angle θ between the incident laser beam and the normal to the substrate was fixed at 30° and the distance between the target and substrate was 3 cm. These values were chosen to optimize the uniformity of the thin film formed on the substrate. Several alumina substrates were tried of smoothness varying from 1.0 μ to .25 μ. Results were similar to that shown in Figure 1 in each case. However, a higher T_c was generally found for smoother samples. For example, Figure 3(a) and (b) shows plots of the surface smoothness for two different substrates before laser deposition. Substrate (a) yielded the higher T_c of 92 K shown in 3c while substrate (b) gave the results in 3(d). Figure 4, shows both the substrate smoothness and the corresponding film smoothness for our best alumina samples (Fig. 3(a)). Note that the film smoothness is similar to that of the substrate. The improved performance with increased substrate smoothness is also supported by our measured T_c of 110° K when polished sapphire was used as a substrate. Similar high quality thin films have also been obtained using MgO as a substrate. Results from a scanning electron microscope, shown in Figures 5(a), (b), and (c) show in more detail the degree of smoothness for the alumina superconducting samples.

As mentioned earlier, once the $Ca_2Ba_2Cu_3O_x$ thin film is prepared it is placed in an oven and heated along with a boat containing about 10 grams of $Tl_2Ba_2Ca_2Cu_3O_x$ powder. The oven temperature is maintained at 830° C for approximately five minutes and then allowed to cool to room temperature. In this way, the interaction between the thalium and substrate is only possible, at elevated temperatures, for a period of 5 minutes. This relatively brief interaction time is believed to be responsible for the successful growth of the films without the need for buffer layers. In fact, prolonged heating for periods longer than five minutes results in the loss of the T_c behavior of Figure 1.

The microstructure of the thalium thin films on alumina was also characterized using x-ray diffraction. Figure 6 shows a typical x-ray diffraction pattern for a thin film after baking in thalium vapor. The data reveals that the superconducting films are composed of the (2212) low phase and the (2223) high phase. Each phase is equally present. This observation agrees with the fact that the measured T_c is about 92 K which is between the bulk T_c temperatures of 80 K and 120 K for the two different phases. Corresponding data for sapphire and MgO substrates have not yet been taken.

CHEMICAL FOG SPRAY

The second technique used to produce thin films involves the use of a chemical "fog" spray. This technique is one of the simplest but useful methods of fabricating thin film superconductors. A solution of Tl-Ba-Ca-Cu-O (Tl: Ba: Ca: Cu = 2: 2: 2: 3) and .05 M concentration was prepared using high-purity $Tl\ NO_3$, $Ba\ (NO_3)_2$, $Ca\ (NO_3)_2$, and $Cu\ (NO_3)_2$. The solution was sprayed on using a fog sprayer on a Mg O substrate which was heated to 400 to 500° C using a hot plate. The substrate was sprayed many times in order to obtain a 1 to 2 μm uniform film. The Tl-Ba-Ca-Cu-O film was then subjected to an annealing process in flowing oxygen heated to 890 to 900° C. After a short annealing period the furnace was allowed to cool slowly to room temperature. A thalium source was placed near the film to compensate for Tl loss during the annealing process. Figure 7 shows resistance-temperature curves for two films on Mg O substrates. Film (A) is for $Tl_2 Ba_2 Ca_2 Cu_3 O_x$ while film (B) is for Tl_2 Ba $Ca_3 Cu_3 O_x$. Both films give resistance-temperature curves typical to bulk Tl-Ba-Ca-Cu-O samples. The films have remained stable in a closed system, such as, a small covered bottle.

CONCLUSION

In summary, we have demonstrated two techniques capable of producing high T_c superconducting thin films on various substrates. Measurements of the critical current density and the critical field for these thin films are presently underway in our laboratory.

The authors are grateful to Y. Xin and J.W. Lai for their help in the early stages of this work. Professor Z.Z. Sheng acknowledges support by the Department of Energy Ar-DNRG/AEO-UAF-88-005.

REFERENCES

1. B. Johs, D. Thompson, N.J. Ianno, John A. Woollam, S.H. Lion, A.M. Hermann, Z.Z. Sheng, W. Kiehl, Q. Shams, X. Fei, L. Sheng, and Y.H. Liu, Appl. Phys. Lett. **54**, 1810 (1989).

2. Z.Z. Sheng, L. Sheng, H.M. Su, and A.M. Hermann, Appl. Phys.Lett. **53**, 2686 (1988).

FIGURE CAPTIONS

Fig. 1 Resistance versus temperature for a typical laser ablated superconducting thin film. Data shows zero resistance until about 92° K.

Fig. 2 Experiment apparatus for deposition or step 1 of the process reported in this letter. The output of the ArF excimer laser was focused into a vacuumed chamber and onto the precursor material. The ablation then caused a thin film to form on the substrate which was heated to 200° C.

Fig. 3 (a) Smoothness of the sample with the T_c of 92° K shown in (a); (b) Smoothness of the sample with the lower T_c value shown in (c); (c) Resistance versus temperature giving a T_c of 92° K; (d) Resistance versus temperature giving a $T_c < 92°$ K

Fig. 4 On the left-hand side the surface smoothness of the substrate is given while the right-hand side shows the corresponding smoothness of the thin film

superconducting film with a T_c of 92° K. The figure also gives a measure of the film thickness by comparison of the right- and left-hand sides of the figure.

Fig. 5 (a) Shows the surface quality of the superconducting thin film with a T_c of 92° K using a scanning electron microscope with x1000 resolution; (b) Same as (a) but x3000 resolution; (c) same as (a) but x5000 resolution

Fig. 6 Shows results from x-ray diffraction of the superconducting thin film with a T_c of 92° K. Results indicate that the high and low phases are equally present.

Fig. 7 Resistance versus temperature for chemical spray films. Film (A) is for $Tl_2\ Ba_2\ Ca_2\ Cu_3\ O_x$ while film (B) is for $Tl_2\ Ba\ Ca_3\ Cu_3\ O_x$.

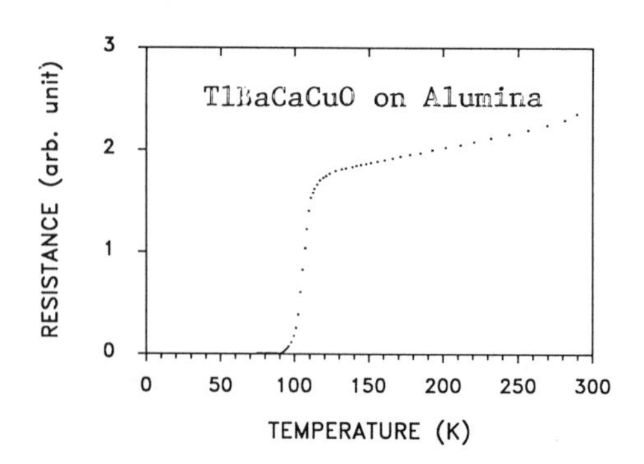

FIGURE 1

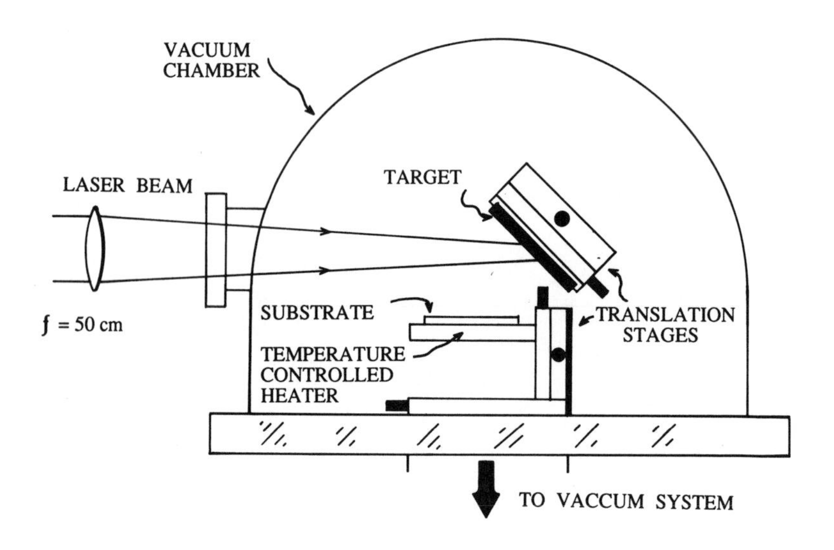

FIGURE 2

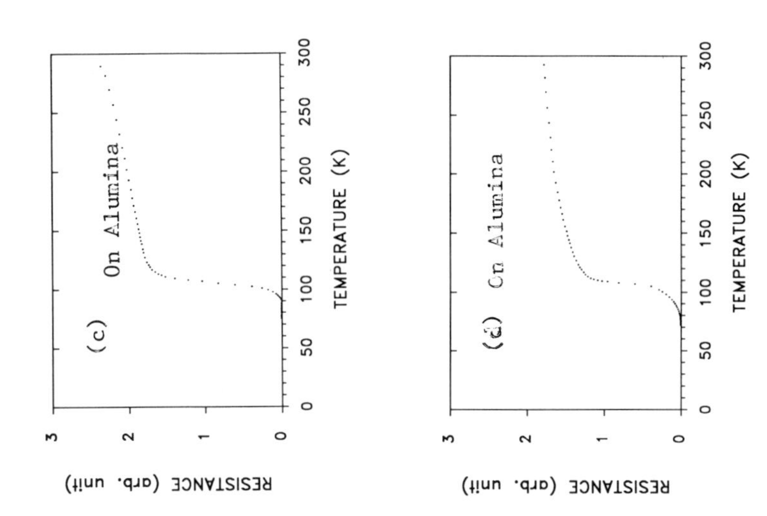
(c)
On Alumina
RESISTANCE (arb. unit)
TEMPERATURE (K)
0
1
2
3
0
50
100
150
200
250
300
(d)
On Alumina
RESISTANCE (arb. unit)
TEMPERATURE (K)

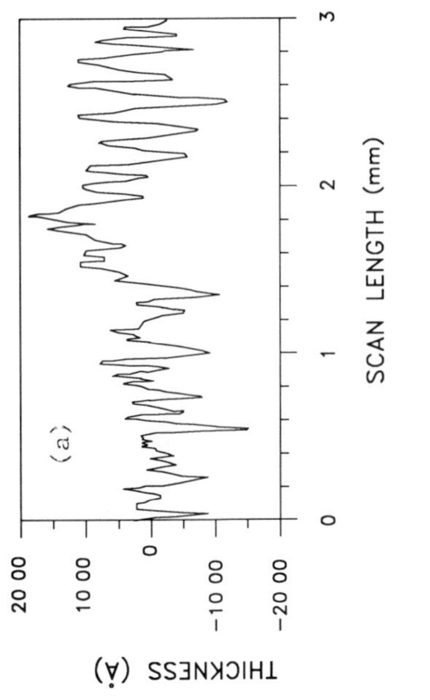
(a)
THICKNESS (Å)
SCAN LENGTH (mm)
20 00
10 00
0
-10 00
-20 00
0
1
2
3

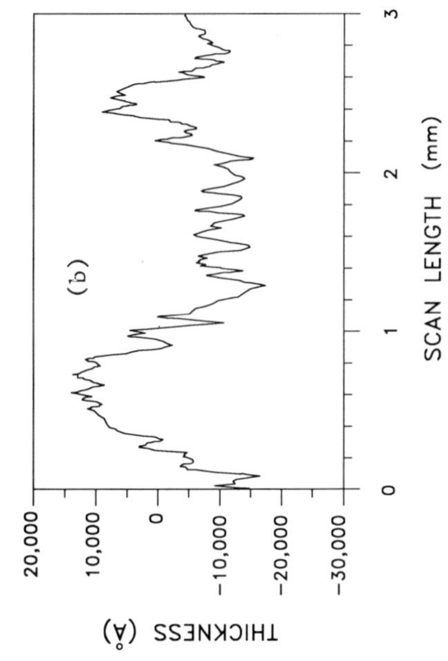
(b)
THICKNESS (Å)
SCAN LENGTH (mm)
20,000
10,000
0
-10,000
-20,000
-30,000
0
1
2
3

FIGURE 3

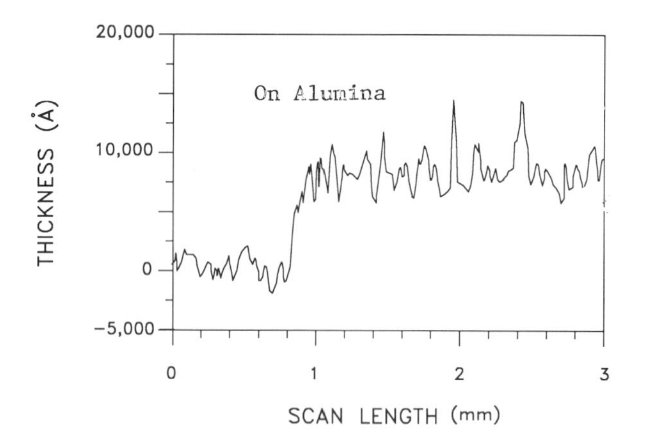

FIGURE 4

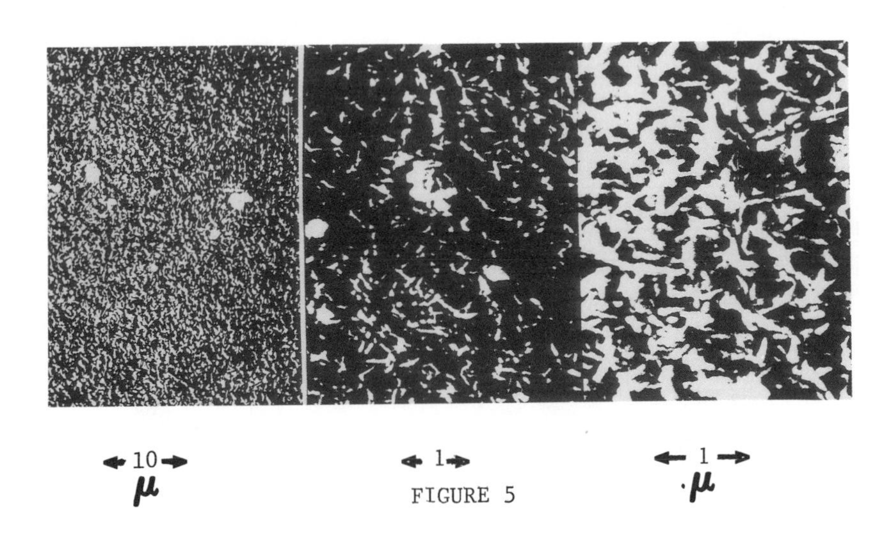

FIGURE 5

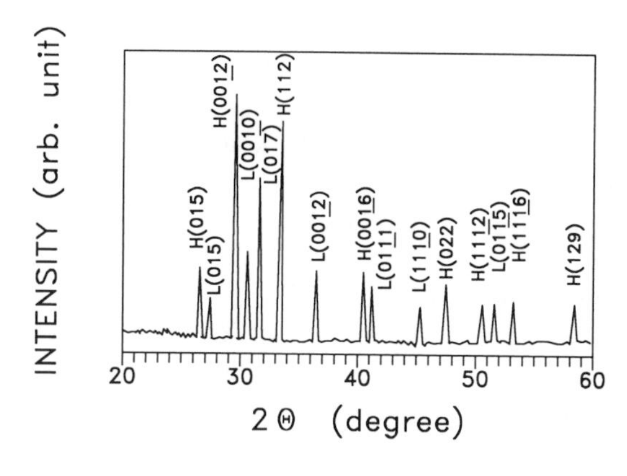

FIGURE 6

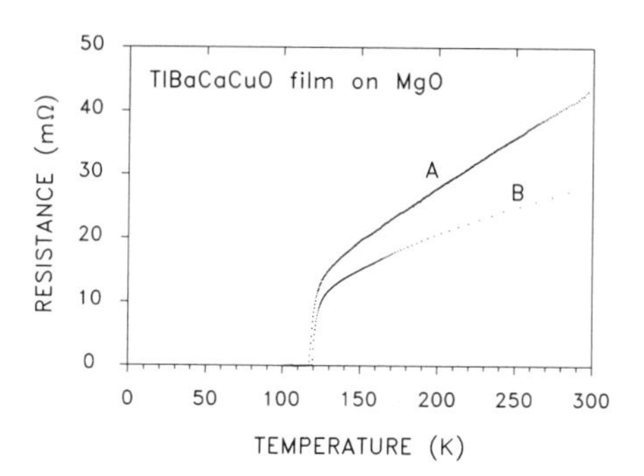

FIGURE 7

SUPERCONDUCTING $YBa_2Cu_3O_{7-x}$ THIN FILMS ON GaAs SUBSTRATES USING A DOUBLE BUFFER LAYER STRUCTURE

Q.X.Jia, S.Y.Lee, W.A.Anderson, and D.T.Shaw
State University of New York at Buffalo
New York State Institute on Superconductivity
Department of Electrical & Computer Engineering
Amherst, NY 14260

ABSTRACT

High temperature superconducting $YBa_2Cu_3O_{7-x}$ (YBCO) thin films on GaAs were realized by using a double layer of YSZ/Si_3N_4 between YBCO and GaAs. The barrier layers are required not only to prevent the interdiffusion between YBCO and GaAs but also to protect the GaAs surface at relatively high temperature and oxygen rich ambient. By optimizing the deposition conditions for different layers, high performance superconducting YBCO thin films were obtained by laser ablation. The advantages of using double buffer layers for growth of high temperature superconducting film on semiconductor substrate are discussed.

INTRODUCTION

Since the discovery of high temperature superconducting materials, much work was done to deposit YBCO film on semiconductor substrates. The significance of depositing the superconductor on a semiconductor substrate lies in the possibility of integration of superconductor with semiconductor for electronic device applications. However, it is well documented that high temperature superconductor materials are very sensitive to foreign doping. In fact, people have purposefully utilized this reaction property to pattern the high temperature superconducting thin films, such as YBCO on Si [1,2]. Apparently, it is very desirable to insert a buffer layer between YBCO and semiconductor substrate to reduce the interdiffusion. The diffusion of other elements from the buffer layer or YBCO into the semiconductor substrate should also be considered in choosing buffer material since the introduction of any trace elements into the semiconductor can also seriously influence the electrical properties of semiconductor substrates [3]. The effectiveness of the buffer layer to protect the semiconductor surface, in particular for III-V compound semiconductors, is another very important issue for practical applications. The integration might be useless if the semiconductor material degrades very much after high temperature superconductor thin film deposition.

To prevent the interdiffusion between YBCO and GaAs and to enhance the nucleation of YBCO on GaAs, the most successful buffer layer previously used was Al_2O_3 [4] although other buffer materials were also utilized [6,8,9]. The best results reported for YBCO film on GaAs showed a zero resistance temperature of 80K and a critical current density of $10^6 A/cm^2$ at 5K [4]. In this paper, we discuss the most important issue in depositing YBCO thin films on GaAs substrates. Our success in depositing YBCO on GaAs substrates using a double buffer layer of YSZ/Si_3N_4 is also illustrated.

EXPERIMENTAL DETAILS

Wafers of n-GaAs were used as the substrates. An amorphous Si_3N_4 layer was deposited on the GaAs surface using plasma enhanced chemical vapor deposition at a substrate temperature less than 300°C. The YSZ and YBCO layers were subsequently deposited using in situ laser ablation. The following table outlines the most important parameters used for YSZ and YBCO layer depositions.

Table 1 YSZ and YBCO deposition conditions *

Paramete	YSZ	YBCO
Laser Frequency (Hz)	9	5
Power (J/cm^2)	2.0	2.0
O_2 pressure (mTorr)	10-50	200
T_s(°C)	350-700	650

* 193nm ArF Excimer Laser Deposition.

The electrical properties of the YBCO thin films were characterized using standard four-probe resistance vs temperature measurement. The critical current density was determined at a voltage drop across the film of $1\mu V/cm$. The structural properties of the films were analyzed using X-ray diffraction (XRD), scanning electron microscopy (SEM), and Auger electron spectroscopy (AES).

RESULTS AND DISCUSSION

The most obvious observations through this study are the enhancement of YBCO thin film growth on GaAs and the preservation of the GaAs substrate at high temperature and oxygen rich ambient. The processing parameters either for YSZ or YBCO layers could be independently optimized due to the introduction of a sub-bottom buffer layer of Si_3N_4 on the GaAs surface. Figure 1 shows the resistance vs temperature curves for the YBCO films deposited on GaAs with YSZ/Si_3N_4 as a buffer layer, where the YSZ was deposited at 350°C for curve (a) but 700°C for curve (b). The increase of the YSZ deposition temperature can greatly improve the YBCO film quality. Nevertheless, it must be pointed out that YSZ cannot adhere to the GaAs surface, due to the serious damage of the GaAs surface, at a deposition temperature of 700°C if Si_3N_4 was not used as a intermediate layer between GaAs and YSZ. Two-step processing of the YSZ layer using low temperature for the initial deposition followed by high temperature deposition could partially solve the problems mentioned above. Nevertheless, it is still quite difficult to obtain high quality YBCO film and to preserve the GaAs surface.

Using our technique, the zero resistance temperature of the films was around 85.5K. A metallic R-T curve was clearly demonstrated above the transition temperature. The role of buffer layers for superconducting YBCO thin films on GaAs substrates has been previously published [5]. The superior properties of the YBCO films due to using a double buffer layer could be clearly seen by comparing the present results with those utilizing a single buffer layer, such as CaF_2 [6], ITO [7], and YSZ [8]. Serious degradation of the GaAs substrate was reported if YBCO was directly deposited on GaAs substrate. The zero resistance temperature of the film was only around 20K in this case [9].

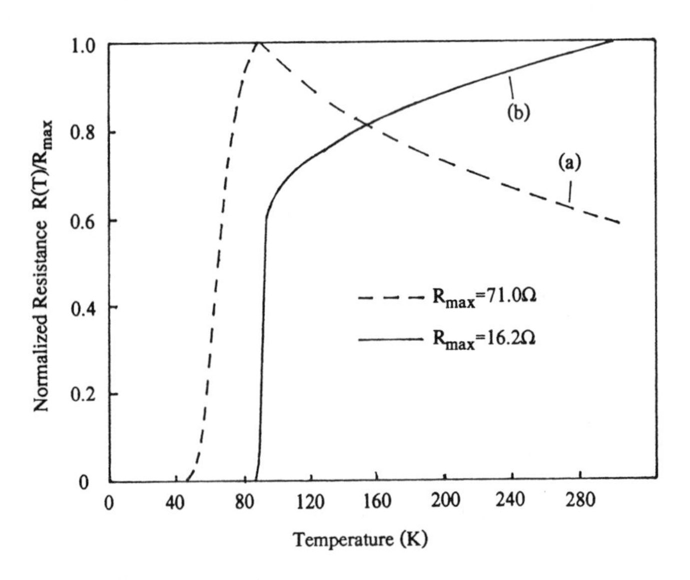

Fig.1 R-T curves for YBCO thin films on GaAs substrates with YSZ/Si_3N_4 as a buffer layers.

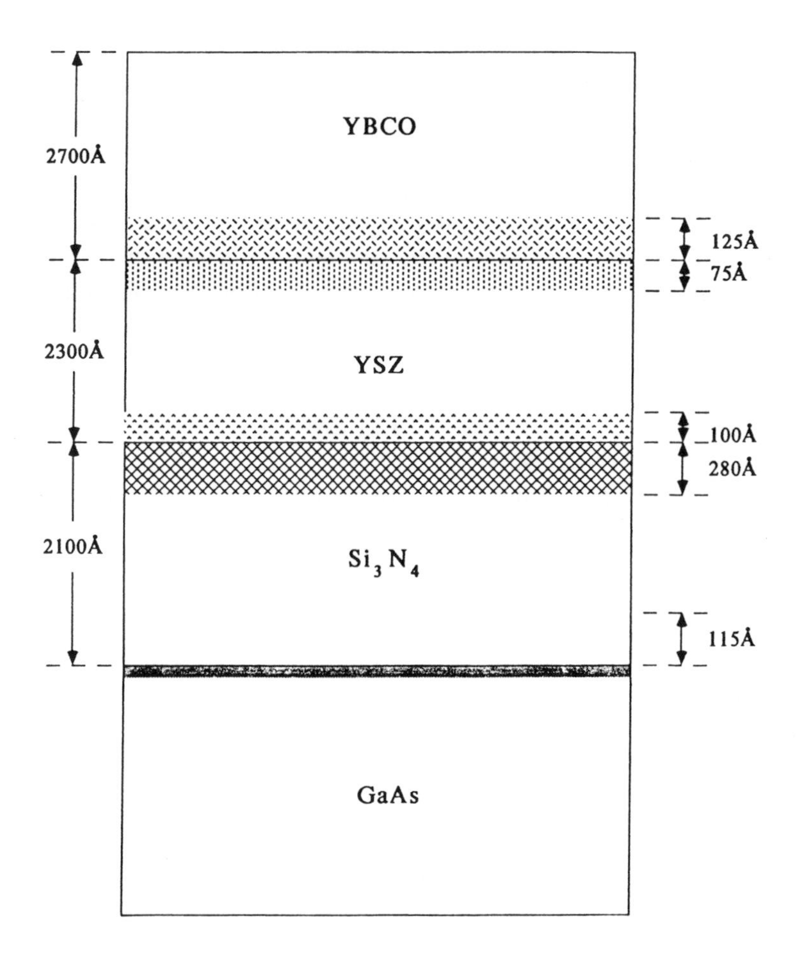

Fig.2 Quantitative characterization of the interdiffusion among different layers using a combination of cross-sectional SEM and AES depth profiling technique.

The structural analysis of the films using XRD showed the YBCO film to be highly c-axis oriented if the YSZ was deposited at a temperature of 700°C. A very sharp interface among different regions could also be clearly seen from cross-sectional SEM analysis of the samples [10]. In fact, a combination of cross- sectional SEM and AES depth profiling analysis on the sample revealed quite weak interdiffusion. Figure 2 gives the quantitative data for the diffusion regions which occurred in different layers, where a maximum thickness, characterized using the deepest element diffused in different regions, was used in all cases.

CONCLUSION

Enhancement of electrical and structural properties of superconducting YBCO thin films on GaAs substrates was realized by using a double buffer layer of YSZ/Si_3N_4 between YBCO and GaAs. The use of Si_3N_4 as a sub-bottom buffer layer can preserve the GaAs surface at a deposition temperature as high as 700°C and in an oxygen rich ambient. Utilizing a sub-top buffer layer of YSZ can serve as a seed layer to enhance the growth of high quality superconducting YBCO thin films.

REFERENCES

[1] Q.Y. Ma, E.S. Yang, G.V. Treyz, and C.A. Chang, Appl. Phys. Lett. 55, 896(1989)
[2] D.K. Fork, A. Barrera, T.H. Geballe, A.M. Viano, and D.B. Fenner, Appl. Phys. Lett. 57, 2504(1990)
[3] Q.X. Jia, K.L. Jiao, and W.A. Anderson, J. Appl. Phys., Sept. 1991
[4] J. Schewchun, Y. Chen, J.S. Holder, and C. Uher, Appl. Phys. Lett. 58, 2704(1991)
[5] Q.X. Jia, S.Y. Lee, W.A. Anderson, and D.T. Shaw, Appl. Phys. Lett. 59, 1120(1991)
[6] K. Mzuno, M. Miyauchi, K. Setsune, and K. Wasa, Appl. Phys. Lett. 54, 383(1989)
[7] B.J. Kellett, A. Gauzzi, J.H. James, B. Dwir, D. Pavuna, and F.K. Reinhart, Appl. Phys. Lett. 57, 2588(1990)
[8] P. Tiwari, S. Sharan, and J. Narayan, Appl. Phys. Lett. 59, 357(1991)
[9] M.R. Rao, E.J. Tarsa, L.A. Samoska, J.H. English, A.C. Gossard, H. Kroemer, P.M. Petroff, and E.L. Hu, Appl. Phys. Lett. 56, 1905(1990)
[10] S.Y. Lee, Q.X. Jia, W.A. Anderson, and D.T. Shaw, J. Appl. Phys., Oct. 1991

CRITICAL CURRENT DENSITIES IN BI - TEXTURED POLYCRYSTALLINE SUPERCONDUCTING THIN FILMS

D.H. Kim, S.Y. Dong, and H.S. Kwok
Institute on Superconductivity, State University of New York at Buffalo, Amherst, New York 14260

ABSTRACT

The Jc of a polycrystalline film was studied experimentally and modeled by a percolation calculation. The superconducting thin film was modeled as a single layer of grains so that current limitation was considered as a linear optimization problem. Since the degree of in plane a - axis texturing of the grains in *in situ* Y-Ba-Cu-O thin films was known, the critical current densities in polycrystalline superconducting thin films could be calculated. Comparison of transport and SQUID critical current densities from the experimental data and the calculated values in terms of intragrain critical current density,gave the estimation of J_C reduction across the 45° grain boundary.

INTRODUCTION

Many applications of high T_C material require high critical current density and/or single crystal like good quality thin films. While all the YBCO thin films are polycrystalline , YBCO on various substrates include SrTiO3, LaAlO3, MgO, and YSZ, all cut with (001) direction, have high degree of texturing with c-axis perpendicular to the substrates [1]. And it is believed that there is a good relation between the high J_C and the degree of texturing. In this study, we investigated the critical current density over the 45° misorientation grain boundary which occurs naturally on many substrates, particularly YSZ.

Dimos et al.[2] reported their studies on critical current density across a grain boundary as a function of the misorientation angle between the two grains. Specially, the critical current over the 45° misoriented grain boundary in the basal plane with c-axis perpendicular to the substrate surface, were suppressed 50 times compared with the critical current density inside the grain. The data were obtained from the artificial grain boundary (GB) produced by film growing over a bi-crysral. The question is "What is the reduction of Jc across natural GB, specially for 45° misoriented GB ?". To answer the question, we have studied the transport Jc in bi-textured polycrystalline superconducting thin film in two ways: i) Calculate the transport Jc based on the characteristic of the individual GB with proper modeling of the superconducting film, ii) measuring the intragrain critical current j_{co} and transport critical current density J_c. In our experiments, we used YSZ(001) as substrates. Because of the well known facts that c-axis perpendicular films are obtained in high deposition temperature and the a-axis of the YBCO grains is aligned along either the YSZ [100] or YSZ [110] due to the lattice matching condition [1].

PERCOLATION CALCULATION

Since the typical size of the grains [3,4] (0.1 - 0.5 μm) is much smaller than that of the microbridge (20 μm x 1mm) which is used in the transport J_c measurements, we can safely assume that the numbers of grains in the microbridge are quite large. So we considered a superconducting thin film as a single layer of grains[1] with m x n hexagonal lattice (the length of each side of a grain be 1). Fig. 1 shows our model schematically. The model contains n grains horizontally and m grains vertically with current flowing upwords. Fig. 1 shows the first few grains. Then, the critical current Ic can be determined by solving the following linear problem [5],

$$\text{i) } I_c = \max \sum_{j=1}^{n} i_{3n-1} + i_{3n-2}$$

ii) $-i_{jc} < i_j < i_{jc}$
iii) The sum of the currents flow into a grain equals the sum of the currents flow out from the grain, i.e. $i_3 + i_4 + i_5 = i_6 + i_{3n+1} + i_{3n+2}$.
iv) The sum of the currents on the bottom boundary and the sum of the currents on the top boundary are the same.

This linear problem has been solved by using a simplex method [6]. The i_c's on left and right boundaries are set to zero which means that no current is permitted to flow over the boundaries. The i_c's of the top and bottom grains are assigned to be equal to the intragrain critical current density j_{co}. Here we use the terms critical current and critical current density equivalently since the each side of the grain has unit length. The misorientation angle between neighboring grains is characterized by the angle θ between the a - axes, and can either be 0° or 45°.

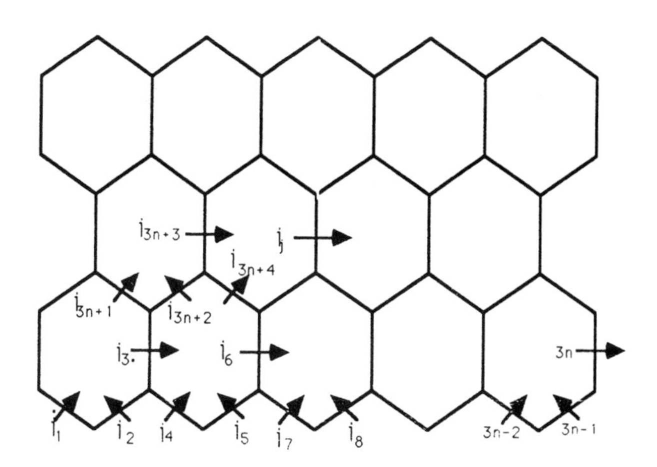

Fig. 1 Modeling the superconducting thin film with single layer of m x n hexagonal lattice. The angle θ (0° or 45°) between a - axes of neighboring grains is assigned randomly with the random number generator. The critical current (density) over the boundary i_{jc} is determined by the angle difference $\phi = \theta - \theta'$ between the adjacent grains. No current is allowed to flow over left and right boundaries.

Since it has been known [1] that a - axis of the grains of $YBa_2Cu_3O_{7-\delta}$ thin film have [100] or [110] direction on the YSZ substrate dominentely with c axis perpendicular to the substrate surface, we have considered only two directions for the grains orientation, i.e. $\theta = 0°, 45°$. Therefore there exist only two kinds of GB. One is 0° misoriented GB and the other is 45° misoriented GB. Thus only two kinds of Jc exist across the grain boundaries. In our calculation we considered no critical current density reduction over 0° misoriented GB. And the reduction ratios of Jc across 45° misoriented GB (γ) is taken as the adjustable parameter.

EXPERIMENT

YBCO films were deposited *in situ* using the ArF laser on YSZ substrates at a temperature of 680 °C. At this temperature all the grains are c - axis perpendicular to the substrate [7]. X - ray pole - figure measurements were used to determine the in - plane texturing of the YBCO grains. The transport Jc's were measured with the criteria of 1μV/cm, with the microbridges of 20 μm wide and 1 mm long. By measuring the magnetization hysteresis loop (M-H loop) in the SQUID, an estimate of the intragrain critical current density j_{co} can be made assuming that the Bean's model applies.

RESULTS

J_c/j_{co} was calculated as a function of texturing, defined as [100]/([100] + [110]). γ was used as a parameter. Fig 2 shows the results of the calculation. θ is assigned to each grain randomly with the given ratio of [100] direction. The J_c is calculated by dividing I_c with the transverse width of the sample. Since j_c depends only on the angle difference between the grains, J_c/j_{co} is expected to be symmetry about the 50% [100] texture point. Each data point is the average value of J_c/j_{co} for different simulation run. The standard

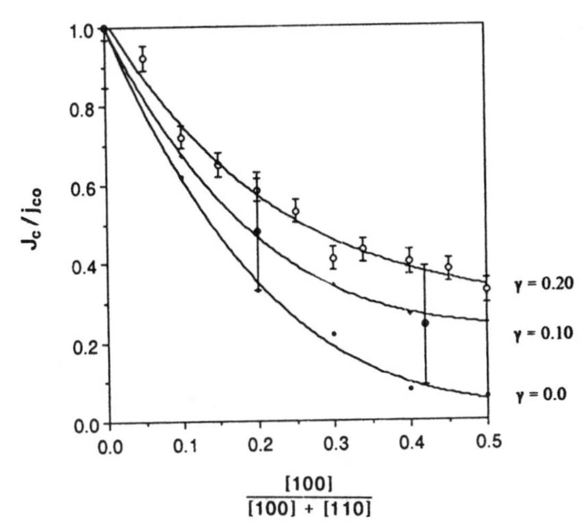

Fig. 2 Results of the Jc calculation for various ratios of a axis of grains parallel to [100] direction on the substrate with different γ. 15 x 15 hexagonal lattice is used for the calculation. Averages are taken over the several different runs and the standard deviations from the average values are also shown on the plot. The deviations are from small size (15 x 15) of the samples. The normalized experimental results (•) are also shown.

deviations of the data are also plotted on the upper curve. The other curves also have comparable amount of deviation. The deviations are from the random assignment of grain orientation and the relatively small size (15 x 15) of the sample. One should note that Jc/jco for a 50% texturing sample, which is the worst case in bi-texturing material, is reduced down only to 1/20. This result is somewhat predictable since not every grain boundary has 45° misorientation. In other word, it is possible there exists supercurrent path even though the sample is not totally textured.

Table 1 shows the experiment data for various texturing at 70K. J_C is the transport J_C and J_{CO} is the SQUID J_C. Unfortunately, the transport J_C are smaller than the intragrain J_{CO} even for the totally

Table 1. Critical current densities ($10^6 A/cm^2$) at 70K with various texturing.

Sample	Texturing	T(K)	J_{co}	J_c	J_c/J_{co}
DSY 136a	0 % [100]	89.0	15.6	2.2	0.14
DSY137a	100 % [100]	89.0	11.97	6.0	0.50
DSY138b	0 % [100]	88.0	24.6	6.1	0.25
DSY144b	20 % [100]	85.6	4.1	0.53	0.13
DSY149	42 % [100]	85.1	2.2	0.03	0.06

textured samples and Jco are also differ from sample to sample even if they have same texturing. We believe that there are other factors contributing to the critical current density reduction other than misoriented GB. So all the J_c/J_{co} are normalized to the average J_c/J_{co} for totally textured samples. The normalized results are plotted on Fig. 2. It shows that the data are in agreement with a value of 1/50 - 1/10. For more quantitative value of γ, more experimental data and more cautious data analysis, i.e. estimation of J_{co}, temperature effect on Jc and Jco, are needed.

SUMMARY

The transport Jc were calculated from the characteristics of the single grain boundary. The calculation showed the Jc would decrease only 20 times even for the worst case in ideal bi-textured polycrystalline thin film. Comparison between experiments and calculations suggested the γ value in the range of 1/50 - 1/10 where more studies are needed for more quantitative analysis of critical current density across the 45° misoriented grain boundary.

REFERENCES

1. J.P. Zheng, S.Y. Dong, and H.S. Kwok, Appl. Phys. Lett. 58,

540 (1991)

2. D. Dimos, P. Chaudhari, and J. Mannhart Phys. Rev.B41, 4038 (1990)
3. D.M. Huang, T. Venkatesan, C. C. Chang, L. Nazar, X.D. Wu, A. Inam, and M.S. Hegde, Appl. Phys. Lett. **54**, 1702 (1989)
4. R. Ramesh, D. Hwang, T.S. Ravi, A. Inam, J.B. Barner, L. Narzar, S.W. Chan, C.Y. Chen,B. Dutta, T. Venkatesan, and X.D. Wu,Appl.Phys.Lett.**56**,2243(1990)
5. J, Rhyner, and G. Blatter, Phys.Rev.B **40**,829 (1989)
6. W.H. Press, B.P. Flanney, S.A. Teukolsky, and W.T. Vetterling, Numerical Recipes, Cambridge Univ. Press, Cambridge, 1986
7. Q. Li, O. Meyer, X.X. Xi, J. Geerk, and G. Linker, Appl.Phys.Lett. **55**, 1792 (1989)

SUPERCONDUCTING $YBa_2Cu_3O_{7-X}$ THIN FILMS ON METALLIC SUBSTRATES PREPARED BY RF MAGNETRON SPUTTERING USING $BaTiO_3$ AS A BUFFER LAYER

Q.X. Jia and W.A. Anderson
State University of New York at Buffalo
New York State Institute on Superconductivity
Department of Electrical and Computer Engineering
Bonner Hall, Amherst, NY 14260

ABSTRACT

Superconducting $YBa_2Cu_3O_{7-x}$ (YBCO) thin films were deposited on metallic substrate of Hastelloy C-276 using in situ RF magnetron sputtering. A buffer of $BaTiO_3$ was proven to be an effective barrier layer between YBCO and Hastelloy. Highly c-axis oriented films with a zero resistance temperature of 81.3K were obtained when the films were in situ deposited at 640^oC. The clear interfaces among different regions were confirmed by cross- sectional scanning electron microscopy analysis and Auger electron spectroscopy depth profiling.

INTRODUCTION

The discovery of an oxide superconductor with zero resistance temperature exceeding the liquid nitrogen temperature promised wide application in future technology including energy or magnets. One of the most desirable configurations for a useful superconductor is as a wire, tape, or cable product. In order to realize the above forms, it is necessary to develop techniques that can make thin or thick films on flexible metallic substrates. The preference of using thin film forms lies in its much higher critical current density. The more controllable and reproducible processing to deposit thin films also makes thin films widely used compared to thick films or bulk forms. Nevertheless, high temperature superconducting thin films, such as $YBa_2Cu_3O_{7-x}$ (YBCO), on metallic substrates still show relatively poor superconducting properties compared to the films on single crystal substrates, such as $SrTiO_3$ [1] and MgO [2].

Most of the previous work on depositing YBCO on metallic substrates was done using laser ablation [3-5]. Chemical vapor deposition was also tried to deposit YBCO on metallic substrates [6]. To prevent the interdiffusion between metallic substrate and YBCO, MgO [3], YSZ [4], Ag [5], and $SrTiO_3$ [6] have been used as a buffer layer. A double layer of YSZ/Pt was also recently tried [7-8]. The best results for the YBCO on metallic substrate using such a buffer structure showed a zero resistance temperature of 90K and a critical current density of $4x10^4$ A/cm^2 at zero field and 77K [8]. In this work, we used RF magnetron sputtering to deposit superconducting YBCO thin films on metallic substrates with $BaTiO_3$ as a buffer layer. The advantage of sputtering is the possibility of uniform large area deposition.

EXPERIMENTS

A metallic substrate of Hastelloy C-276 (Ni-Cr-Mo alloys) was used as the substrate in this study. A buffer layer of $BaTiO_3$ was deposited on Hastelloy using reactive RF magnetron sputtering from a composite $BaTiO_3$ target, where oxygen was used as a reactive gas. The oxygen partial pressure during sputtering was controlled in the range of 0.5-1mTorr. The substrate temperature was changed from 350^oC to 600^oC in order to investigate the buffer layer deposition conditions on the properties of YBCO thin films. The YBCO thin film was in situ deposited using RF magnetron sputtering from a single stoichiometric YBCO target at a substrate temperature of around 640^oC. A perpendicular arrangement of the substrate with respect to the target was used to avoid resputtering effects and to obtain stoichiometric YBCO thin films. Detailed descriptions of the deposition procedure with this configuration have been published [9]. The YBCO thin films were in situ annealed at 640^oC in oxygen ambient ($>$2Torr) for 30min after film deposition.

The structural properties of the films were analyzed by X-ray diffraction (XRD). The microstructure of the films was investigated by scanning electron microscopy (SEM). The interface properties among different layers were characterized by Auger electron spectroscopy (AES). The electrical properties of the films were characterized using standard four-probe resistance vs temperature measurement.

RESULTS AND DISCUSSION

The obvious observation of the YBCO on Hastelloy with $BaTiO_3$ as a buffer layer was the improvement on the film morphology. The YBCO film surface appeared shiny black regardless of the $BaTiO_3$ thin film deposition temperature from 350^oC to 600^oC although high temperature deposition of $BaTiO_3$ film resulted in a slightly rough $BaTiO_3$ surface. No cracks or peeled regions were found from SEM surface survey in a quite large area of the film surface.

Structural analysis using XRD showed very different texture properties for $BaTiO_3$ and YBCO films, respectively. As shown in Figure 1(a), the $BaTiO_3$ film deposited at a temperature of 350^oC with a thickness of around 150nm seems to be featureless with an amorphous structure. However, the YBCO film deposited on such a substrate showed highly textured structures as shown in Figure 1(b). Pure c-axis oriented grain growth perpendicular to the substrate surface can be clearly deduced from Figure 1(b).

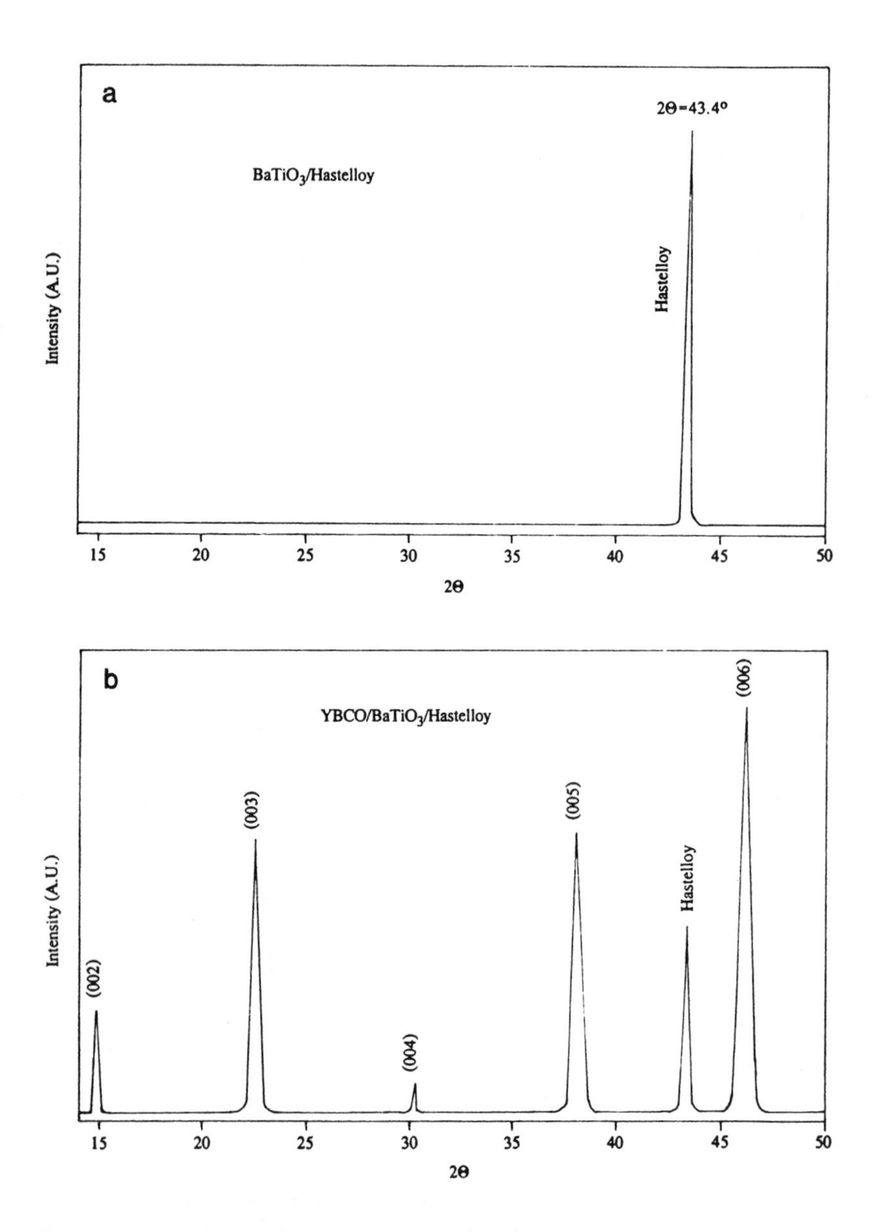

Fig.1 XRD patterns for (a) BaTiO3 on Hastelloy, and (b) YBCO on BaTiO3/Hastelloy

Surface analysis of the YBCO film using AES surface scan revealed only expected elements. This demonstrated $BaTiO_3$ to be an effective barrier layer to prevent the interdiffusion between YBCO and Hastelloy. No serious interdiffusion among different regions was also confirmed from AES depth profiling as shown in Figure 2. The YBCO composition through the film thickness was quite uniform. However, oxidation of the Hastelloy surface might occur during $BaTiO_3$ deposition or if oxygen diffuses through the $BaTiO_3$ layer during YBCO film deposition since a relatively broad transition region appears between $BaTiO_3$ and Hastelloy as demonstrated in Figure 2. The reactive gas of oxygen during either $BaTiO_3$ or YBCO film deposition might react with the Hastelloy substrate, in particular, at a relatively high temperature.

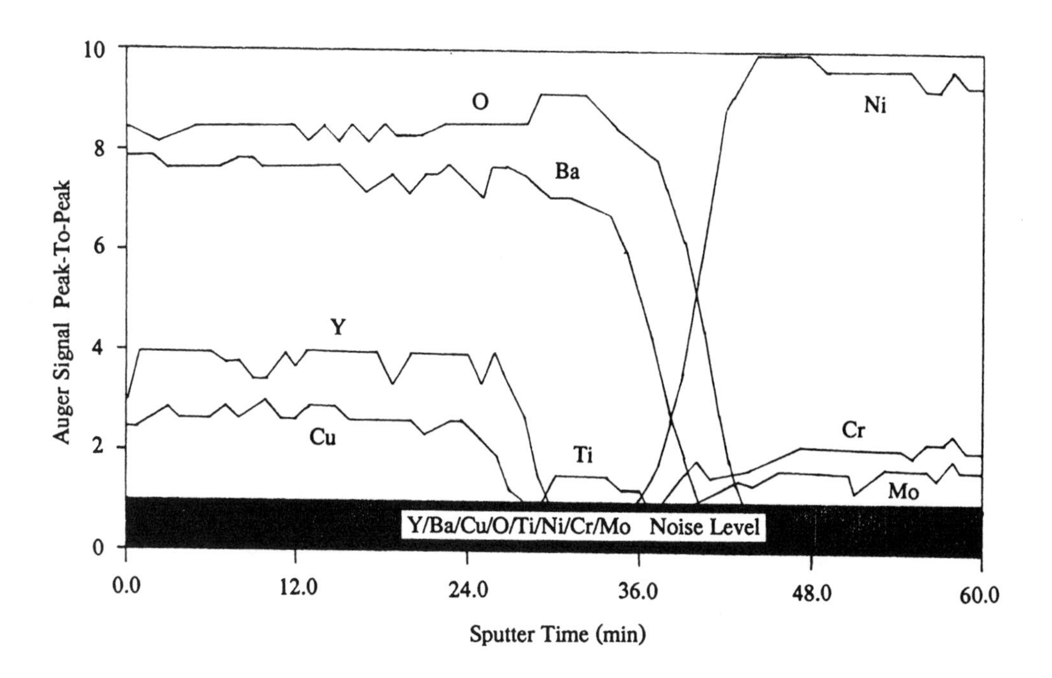

Fig. 2 AES depth profiling of the YBCO film on Hastelloy using $BaTiO_3$ as a buffer layer.

Electrical measurement on the YBCO films showed the metallic nature of R-T curves. Films with a thickness of around 250nm and a buffer layer thickness of 150nm became superconductive above liquid nitrogen temperature. Our best data for zero resistance temperature of the films was 81.3K. Figure 3 shows a typical R-T curve for a YBCO film on Hastelloy using $BaTiO_3$ as a buffer layer. The critical current density of the film at 77K was around $7.4x10^3 A/cm^2$ on a film with a width of 3mm.

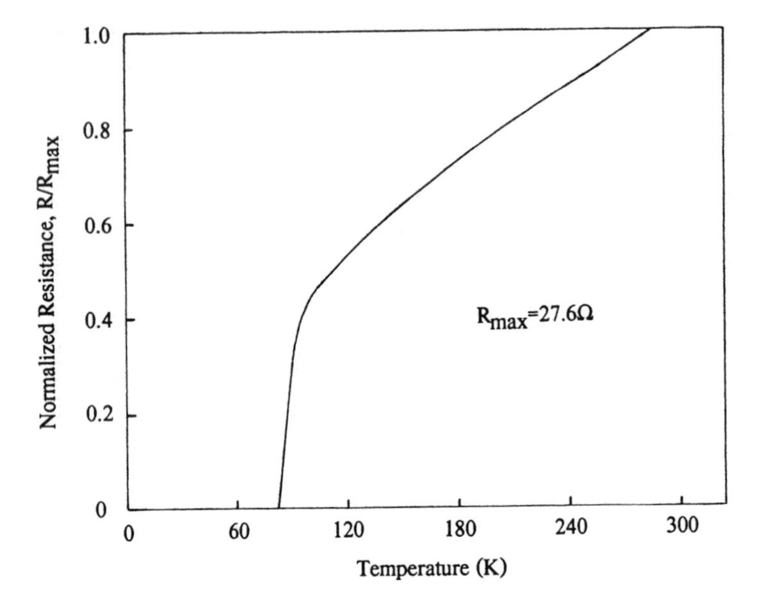

Fig. 3. Resistance vs temperature characteristics for a YBCO film on Hastelloy using $BaTiO_3$ as a buffer layer.

SUMMARY

High temperature superconducting YBCO thin films have been deposited on metallic substrates of Hastelloy C-276 using in situ RF magnetron sputtering. The interdiffusion between substrate and YBCO during processing can be effectively reduced using $BaTiO_3$ as a buffer layer. Highly c-axis oriented superconducting YBCO thin films with zero resistance temperature of 81.3K and a critical current density of $7.4x10^3 A/cm^2$ have been obtained in our experiments. Further increasing critical current density and zero resistance temperature of the YBCO films by optimizing the processing parameters is underway in our lab.

ACKNOWLEDGMENT

We wish to thank P. Bush and R. Barone for their technical assistance in AES and SEM analysis. This work was supported by a Link Foundation Energy Fellowship and the New York State Institute on Superconductivity.

REFERENCES

[1] X.D. Wu, R.E. Muenchausen, S. Foltyn, R.C. Estler, R.C. Dye, A. Garcia, N.S. Nogar, P. England, R. Ramesh, D.M. Hwang, T.S. Ravi, C.C. Chang, T. Venkatesan, X.X. Xi, Q. Li, and A. Inam, Appl. Phys. Lett. 57, 523(1990)

[2] R.H. Moeckly, S.E. Russek, D.K. Lathrop, R.A. Buhrman, J. Li, and J.W. Mayer, Appl. Phys. Lett. 57, 1687(1990)

[3] J. Saitoh, M. Fukutomi, Y. Tanaka, T. Asano, H. Maeda, and H. Takahara, Jpn. J. Appl. Phys. 29, L1117(1990)

[4] E. Narumi, L.W. Song, F. Yang, S. Patel, Y.H. Kao, and D.T. Shaw, Appl. Phys. Lett. 56, 2684(1990)

[5] R.E. Russo, R.P. Reade, J.M. McMillan, and B.L. Olsen, J. Appl. Phys. 68, 1354(1990)

[6] T. Yamaguchi, S. Aoki, N. Sadakata, O. Kohno, and H. Osanai, Appl. Phys. Lett. 55, 1581(1989)

[7] E. Narumi, L.W. Song, F. Yang, S. Patel, Y.H. Kao, and D.T. Shaw, Appl. Phys. Lett. 58, 1202(1991)

[8] J. Saitoh, M. Fukutomi, K. Komori, Y. Tanaka, T. Asano, H. Maeda, and H. Takahara, Jpn. J. Appl. Phys. 30, L898(1991)

[9] Q.X. Jia and W.A. Anderson, Appl. Phys. Lett. 57, 304(1990)

SINGLE CRYSTAL GROWTH AND PROPERTIES OF $NdGaO_3$, $Nd_yPr_{1-y}GaO_3$ AND $LaAl_{1-x}Ga_xO_3$ AS SUBSTRATES FOR HIGH TEMPERATURE SUPERCONDUCTORS.

H. Dabkowska, A. Dabkowski, J.E. Greedan, R. Hughes, Y. Lu, D. Poulin, J. Preston, T. Strach and T. Timusk,
McMaster University, Hamilton, Ontario, L8S 4M1, Canada.

ABSTRACT

Preparation and properties of perovskite-related single crystals of $NdGaO_3$, $Nd_yPr_{1-y}GaO_3$ and $LaAl_{1-x}Ga_xO_3$ as substrates for superconducting films of $YBa_2Cu_3O_{7-\delta}$ and $Pb_2Sr_2(Y,Ca)Cu_3O_{8+\delta}$ have been studied. $LaAl_{1-x}Ga_xO_3$ single crystals up to 4.5x4x2mm^3 have been grown from $PbO-MoO_3-B_2O_3$ flux. Powder X-ray diffraction was used to determine x values which were 0.16, 0.19, and 0.80 and the lattice mismatch between substrate and film is about half that for $LaAlO_3$ and $LaGaO_3$. $NdGaO_3$ and $Nd_yPr_{1-y}GaO_3$ single crystals have been grown from a similar flux in the temperature range 1260-1080°C so as to avoid the cubic to orthorhombic phase transition at 1350°. Twin-free crystals up to 4x4x2 mm^3 have been obtained. The lattice constants of gallates suitable for depositing films of both $YBa_2Cu_3O_{7-\delta}$ and $Pb_2Sr_2(Y/Ca)Cu_3O_{8+\delta}$ on (001) and (110) planes are investigated.

INTRODUCTION

The problem of deposition of epitaxial films of high temperature superconductors is closely related to the problem of obtaining high quality substrates. Without a "proper" substrate it is not possible to obtain an epitaxial film of reasonable thickness and expected physical properties.

An epitaxial film is a single crystalline film being a continuation of a single crystalline substrate. In the best known cases, where extremely high quality epitaxial films have been obtained (i.e. Si, II-V semiconductors) a doped, thin film was grown on a substrate of the same material.

Also, very high quality epitaxial films of magnetic garnets have been grown on non-magnetic garnet substrates, chosen so that the structure and cell dimensions of the substrate and of the film are similar.

Recently extensive application for superconducting oxide films have forced many laboratories to look for potential substrates. Table I summarizes some properties of pseudoperovskite substrates already used, as well as methods for obtaining them.

During the search for promising substrates the following criteria should be considered.

1. in the temperature range between liquid He and thin film growth the mismatch of lattice constants of film and substrate should be small enough to reduce stress.
2. the crystallographic structure of the film and the substrate should be similar (or the same).
3. the film and the substrate should have similar thermal expansion coefficients.
4. phase transitions in the temperature range investigated should be avoided - usually they cause twinning of both the substrate and the film.
5. the substrate should be obtainable in single crystal form, preferably large, easy to cut and polish, with no cleavage and no defects.

TABLE I: PSEUDOPEROVSKITE SUBSTRATES FOR HTS

Space Group[1] & symmetry of substrate	Lattice Mismatch[2] [%] 1 2 3	Lattice Mismatch[2] [%] 2 2 1 3	Phase Transition [°C]	Average Thermal Expansion Coefficient ppm/K	Dieletical Constant ϵ at RT	Melting Point [°C]	Reference concerning Data for HTS films
$Sr\ Ti\ O_3$ Pm3m Cubic	$\lambda_a = -1.8$ $\lambda_b = -0.3$ (100)	$\lambda_a = -2.5$ $\lambda_b = -0.8$ (100)	~ 380? ~ 110K str. 65K	12.3	300	1920	[1],[2]
$LaGaO_3$ Pbnm Orthorhombic	$\lambda_a = -1.03$ $\lambda_b = -0.23$ (110)	$\lambda_a = -1.90$ $\lambda_b = -1.75$ (001)	145^{I} str. 875 1180	8.8	25	1720	[3]
$PrGaO_3$ 1) Orthorhombic	$\lambda_a = -0.88$ $\lambda_b = -0.25$ (110)	$\lambda_a = 1.44$ $\lambda_b = -1.27$ (001)	not found up to 1000 800^{II}	9.5	24		[4]
$NdGaO_3$ Or thorhombic	$\lambda_a = -0.74$ $\lambda_b = 0.73$ (110)	$\lambda_a = -0.88$ $\lambda_b = -1.83$ (001)	1350^{str} 950 320	9.1	20–25	1600	[5]
$LaAlO_3$ R$\bar{3}$C	$\lambda_a = 1.2$[3] $\lambda_b = 2.4$	$\lambda_a = 0.4$[3] $\lambda_b = 1.2$	~ $475^{II str}$ 435±25 527±10	9.1	15	2180	[2],[3],[6]
$Y\ Al\ O_3$ $P6_3/mmc$	$\lambda_a = 4.3$ $\lambda_b = 5.6$ (100)	$\lambda_a = 3.5$ $\lambda_b = 4.3$ (100)	~ 950		16	1870	[7]

1) For low (R.T.) temperature phase

2) $\lambda_a = \frac{a_f - a_s}{a_s}$ where for 1 2 3 $a_f = 3.836$ $b_f = 3.883$[8] and for 2 2 1 3 $a_f = 5.383$ $b_f = 5.423$[9]
a_s, b_s are parameters either of the original unit cell or of a related one.

3) pseudo–cubic.

6. the chemical stability of the substrate near the film growth temperature should prevent reactions and diffusion at the substrate-film interface.
7. the electrical, optical and magnetic properties of the substrate should not affect measurements and/or applications [i.e. low dielectric constant, low dielectric losses are desirable].

In this work we present the flux grown single crystals of the $REGaO_3$ family as substrates for superconducting films of $YBa_2Cu_3O_{7-\delta}$ and $Pb_2Sr_2(Y/Ca)Cu_3O_{8+\delta}$. Two members of this family, $NaGaO_3$ and $PrGaO_3$ are of interest as substrates for both oxide superconductors mentioned above[4,5].

There are two possible substrate orientations (001) and (110) which fit respectively to $Pb_2Sr_2(Y/Ca)Cu_3O_{8+\delta}$($T_c$=80K) and $YBa_2Cu_3O_{7-\delta}$ (T_c=90K) respectively. On both substrates the film is under tension, but because of its relatively low thermal expansion coefficient, $NdGaO_3$ fits better at low temperature, whereas $PrGaO_3$ fits better at the film growth temperature. To obtain a better fit of substrate to film, preliminary experiments with solid solutions of $NdGaO_3$-$PrGaO_3$ have been performed.

EXPERIMENTAL

Single crystals of $NdGaO_3$, $Nd_yPr_{1-y}GaO_3$ and $LaAl_{1-x}Ga_xO_3$ have been grown by the flux method in a Carbolite furnace with Eurotherm controller; the crucibles of 20cm^3 capacity were of pure platinum, with closely fitting lids. The flux contained PbO and was similar to that used by Watts et al[10]. The chemicals were JM 4N La_2O_3, Al_2O_3, Ga_2O_3 and PbO; JM 3N PbO and PbF_2, RC 3N Pr_6O_{11} and Nd_2O_3; JM 99.95 MoO_3 and CERAC 3N B_2O_3.

The furnace was heated at 100Kh^{-1} to the maximum temperature and held there for the soak period of 1 to 4h before the slow cooling program was commenced. The rate of cooling was from 4 to 1°/h. When the cooling program had been completed the melts were hot-poured while the flux was still liquid. Then, crystals were mechanically separated and cleaned in dilute HNO_3 (1:10). In Table II the starting compositions, temperature programs and results obtained are presented. Typical single crystal is shown in Fig.1. The larger crystals obtained feature interior dendritic growth as well as sector growth. This can be improved with a modified temperature program, and these experiments are in progress. However, the surface of the crystal was often good and it was possible to cut a thin substrate of reasonable quality.

CHARACTERIZATION

EDAX analysis shows that traces of Pb are present in all the single crystals, and traces of Al are also found in $NdGaO_3$. TEM investigation did not reveal any twinning, neither did X-ray spectroscopy. Further investigation of the transition temperature of $Nd_yPr_{1-y}GaO_3$ solid solution is in progress. Single crystals of $LaAl_xGa_{1-x}O_3$ have been found to be twinned, when viewed in polarized light.

ORIENTATION, CUTTING AND POLISHING

The single crystals of the gallates investigated have the form of cuboids. Smaller, submillimeter crystals show triangular facets on some corners. With the help of a 4-axis diffractometer (110),

TABLE II: CONDITIONS FOR FLUX GROWTH OF GALLATE SUBSTRATES FOR HTS FILMS

	Compound	Starting Composition (g)	Soak temp (°C)	Soak time (h)	Cooling rate ($K h^{-1}$)	Final temp. hp	Product & crystal size (mm)
1	$NdGaO_3$	Nd_2O_3 8.5 Ga_2O_3 3.5 PbO 11.7PbF_2 4.1$PbO_2$0.5 M_0O_3 0.5 B_2O_3 0.5	1260	1	4	1100	maroon, up to 6x4x1
2	$LaGaO_3$	La_2O_3 5.2 Ga_2O_3 1.0 PbO 2.3 PbF_2 7.8 $PbO_2$0.6 MoO_3 0.5 B_2O_3 0.5	1260	1	1	800	brown, transparent up to 4x3x2
3	$LaAl_{(1-x)}Ga_xO_3$ x=0.16;0.19;0.80	La_2O_3 6.9 Al_2O_3 0.5 Ga_2O_3 0.4 P604 $P6F_2$ 11 MoO_3 1 B_2O_3 0.7 *	1300	1	3-1.5	980	brown, transparent 4.5x4x2
4	$PrGaO_3$	Pr_4O_7 5.0 Ga_2O_3 2.2 PbO 8 PbF_2 3.0 PbO_2 0.3 MoO_3 0.2 B_2O_3 0.2	1260	1	4	1100	green 4x3x3
5	$(Nd_yPr_{1-y})Ga\ O_3$	Nd_2O_3 6.0 Pr_4O_7 2.0 Ga_2O_3 3.5 PbO 11.7 PbF_2 4.1 PbO_2 0.5 MoO_3 0.5 B_2O_3 0.5	1280	1	1	1100	greenish 4x3x2

* x = 0.19
y = 0.82

(001) and (101) facets have been found on some single crystals. To produce (001) and (110) oriented substrates, the <001> axis has been identified on the larger crystals by means of exmination with a polarizing microscope, then the natural habits of crystals have been used.

DISCUSSION

Flux-grown single crystals sometimes lack the high quality of Czochralski grown materials but they usually can be obtained more easily, and they have the great advantage of being grown at lower temperatures. This is very important for Nd gallates, which can be grown from a temperature below the phase transition by the flux method. The crystals are twin free and they are suitable for deposition of both $YBa_2Cu_3O_{7-\delta}$ and $Pb_2Sr_2(Y/Ca)Cu_3O_{8+\delta}$. The flux method also enables solid solutions of different end members to be made, giving new possibilities for continuous change of physical properties of the substrates.

Recently, films of $YBa_2Cu_3O_{7-\delta}$ have been grown on $NdGaO_3$ and $Nd_yPr_{1-y}GaO_3$ substrates, similar in quality to those grown on $SrTiO_3$. Measurements of physical properties are now in progress, but it is clear that the films obtained are epitaxial.

Fig. 1. Single crystal of $NdGaO_3$.

CONCLUSIONS

1. Flux-grown single crystals of rare-earth gallates can be used as substrates for superconducting oxide films of $YBa_2Cu_3O_{7+\delta}$ and $Pb_2Sr_2(Y/Ca)Cu_3O_{8+\delta}$.

2. Single crystals of $NdGaO_3$, $PrGaO_3$ and their solid solutions are grown from below the phase transition temperature and so their quality is better: no twinning has been observed under TEM or polarizing microscope.

3. The preparation of solid solutions of the rare-earth gallates and/or rare earth aluminates offers extensive possibilities for obtaining single crystalline substrates with low mismatch to HTS films and low dielectric constant.

4. The quality of superconducting films obtained on these flux grown substrates is comparable to that of films grown on $LaAlO_3$ and $SrTiO_3$ substrates obtained by other methods.

ACKNOWLEDGEMENTS

The authors are grateful to John Hudak for EDAX work and to Barbara Wanklyn for helpful comments. This work was supported by NSERC and OCMR grants.

REFERENCES

1. J. Narayan, S. Sharan, R.K. Singh, K. Jagannadham, Mats. Sci. & Eng. B2, 333 (1989).
2. A. Koren, A. Gupta, E.A. Giess, A. Segmuller, R.B. Laibowitc, Appl. Phys. Lett, 54 (11) 1054 (1989).
3. C.W. Nieh, L. Anthony, J.Y. Josefowicz, F.g. Krajenbring, Appl. Phys. Lett. 56 (21) 2138 (1990).
4. M. Sasaura, M. Mukaida, S. Miyazawa, Appl. Phys. Lett. 57 (25) 2728 (1990).
5. M. Mukaida, S. Miyazawa, M. Sasaura, M. Kuroda, Jpn. J. Appl. Phys. 29 (6) L936 (1990).
6. R.W. Simon, C.E. Platt, A.E. Lee, G.S. Lee, K.P. Daly, M.S. Wire, J.A. Luine, Appl. Phys. Lett. 53 (26) 2687 (1988).
7. H. Asano, S. Kubo, O. Michikami, M. Satoh, T. Konaka, Jpn. J. Appl. Phys. 29 (8) L1452 (1990).
8. K. Brodt, H. Fuess, E.F. Paulus, W. Assmus, J. Kowalewski, Acta Cryst. C46, pt.3, 354 (1990).
9. J.S. Xue, M. Reedyk, Y-P. Lin, C.V. Stager, J.E. Greedan, Physica C 166 29, (1990).
10. B.E. Watts, H. Dabkowska, B.M. Wanklyn, J. Cryst. Growth 94 125 (1989).

PROPERTIES OF CRITICAL CURRENT DENSITY IN HETEROSTRUCTURES OF $Y_1Ba_2Cu_3O_{7-y}/Y_1Ba_2(Cu_{1-x}Ni_x)_3O_{7-y}$ SUPERCONDUCTING FILMS

Sang Yeol Lee, Eiki Narumi and David T. Shaw
New York State Institute on Superconductivity, Department of Electical and Computer Engineering, State University of New York at Buffalo, Buffalo, NY. 14260

ABSTRACT

Heterostructures of $Y_1Ba_2Cu_3O_{7-y}/Y_1Ba_2(Cu_{1-x}Ni_x)_3O_{7-y}$ superconducting thin films have been grown epitaxially on (100) ZrO_2 substrates by pulsed laser deposition technique. Variations of critical current density were investigated as a function of temperature in magnetic fields up to 5 T. Heterostructures of thin films are shown to result in significant changes in the magnetic field dependence of transport properties without degrading critical temperatures. Magnetic field dependence of normalized critical current density Jc(H)/Jc(0) is improved by growth of Ni doped multilayer structures possibly due to an increase of flux pinning force. The critical current density of heterostructures in 1 T magnetic field has maintained more than 50 % of Jc(B=0) at 77 K.

INTRODUCTION

High critical current densities above 10^6 A/cm^2 at 77 K are easily obtained for c-axis oriented $Y_1Ba_2Cu_3O_{7-y}$ films epitaxially grown on a variety of substrate materials in these days. Considerable work has been achieved on critical current density by a number of groups with the pulsed laser ablation technique for in-situ deposition of high quality epitaxial films of $Y_1Ba_2Cu_3O_{7-y}$ [1] since the discovery of high Tc superconductors [2-3]. Films of $Y_1Ba_2Cu_3O_{7-y}$ show sharp transition of critical temperature and high critical current densities.

Even though the films grow epitaxially without forming of secondary phases, a lot of microscopic defects and stacking defects are shown in these films [4]. Some of crystalline defects probably acts as flux pinning centers, which could result in higher Jc values. In order to explore the effectiveness of different types of defects to act as pinning centers, a systematic study of thin films about controlling of formation of defects will be required. The microstructure of high Tc superconducting thin films is known to be significantly different from that in bulk high temperature superconductors. The grains in thin films are usually smaller, and tend to be oriented with the c-axis perpendicular to the film surface. In low magnetic field, the drastic reduction of the critical currents of these films is mainly due to the weak link behavior between the grains, which is also related to the flux pinning [5]. However, the critical current density in high Tc superconductors is limited by flux creep by the Anderson-Kim creep model [6].

Heterostructures of $Y_1Ba_2Cu_3O_{7-y}$ superconductors and Ni doped $Y_1Ba_2(Cu_{1-x}Ni_x)_3O_{7-y}$ superconductors have been grown by the laser ablation technique to enhance critical current density in magnetic field by introducing superconductor and weak superconductor layers. Artificial layering of different copper oxide superconductors or insulators can result in modification of the superconducting properties of the high Tc films [7]. Interfaces and strain in the layered materials can result in additional defects, which may play an important role with respect to flux pinning and critical current density enhancement [8]. Multilayer films have been used successfully to provide periodic flux pinning centers in low Tc superconductors, which results in increase of the critical current density.

In addition to the artificial layering structure, substitutions of different atoms for some of Cu have been widely used as a technique for controlling the physical properties of superconductors. In all cases of Cu substitution in $Y_1Ba_2Cu_3O_{7-y}$, using both magnetic and non magnetic atoms, Tc is significantly decreased at relatively low concentrations ~5 at. % [9]. However, multilayer structure of Ni doped layers between high quality of $Y_1Ba_2Cu_3O_{7-y}$ films has been shown sharp transition temperature and high critical current density even though Ni doping degrades the superconducting properties of single layer films and hence introduces weak superconducting films [10]. Measurements of transport properties could reveal much useful information on the effect of Ni doping and artificial layering structures through better understanding of flux pinning. Here, we report on the superconducting transport properties of heterostructures of $Y_1Ba_2Cu_3O_{7-y}$ and $Y_1Ba_2(Cu_{1-x}Ni_x)_3O_{7-y}$ thin film on (100) ZrO_2 substrates.

EXPERIMANTAL PROCEDURE

The $Y_1Ba_2Cu_3O_{7-y}/Y_1Ba_2(Cu_{1-x}Ni_x)_3O_{7-y}$ were grown by laser deposition from ceramic targets of $Y_1Ba_2Cu_3O_{7-y}$ (123) and $Y_1Ba_2Cu_3O_{7-y}$ with 4% of Cu replaced by Ni (Ni-123). Laser deposition from these targets was carried out in oxygen partial pressure which was controlled through a precision leak valve. In the present investigations, the target assembly and processing chamber were adjusted to obtain multitarget in situ processing. There are two target holders mounted on a disk which can be rotated by an external feedthrough to bring the desired target in line of the laser beam for laser ablation.

The $Y_1Ba_2Cu_3O_{7-y}/Y_1Ba_2(Cu_{1-x}Ni_x)_3O_{7-y}$ multilayers were prepared by sequential deposition using the laser ablation technique starting with $Y_1Ba_2Cu_3O_{7-y}$ layer and ending with $Y_1Ba_2(Cu_{1-x}Ni_x)_3O_{7-y}$ layer. By depositing alternating layers of thickness A(123) and B(Ni-123) and repeating same process N times, multilayers estimated (A/B)xN are obtained. Finally, top layer was ending with A(123) layer in order to maintain high quality superconducting properties as described before. Typical heterostructures have a total thickness of 2000 Å with fixed thickness of B=60 Å, estimated based on the deposition rate per pulse for the given deposition condition.

Films were deposited on single crystal ZrO_2 substrates by an in-situ laser ablation technique using an ArF excimer laser with λ=193 nm at 650~700 ^oC substrate temperatures. At the end of the deposition cycle, post-deposition annealing at 450 ^oC in flowing oxygen with a dwell time of 30 minutes was following. Particularly, substrate temperature plays important role among other parameters to obtain high transition temperature superconducting multilayer films. We have observed the films which were grown at 700 ^oC have slightly higher Tc than the films deposited at 650 ^oC due to the increase of uniformity in the early stage of deposition process for layer structures.

RESULT AND DISCUSSION

The basic properties of these films were characterized by X-ray diffraction and rocking curves, scanning electron micrograph (SEM), energy dispersive X-ray analysis (EDX), and temperature dependence of resistivity. Comparison of X-ray diffraction spectrum for $Y_1Ba_2Cu_3O_{7-y}$ single layer thin film and the heterostructure of 300Å(123)/60Å(Ni-123) thin film is shown in Fig.1. X-ray studies reveal that the heterostructure thin films are completely single phase like pure 123 single layer films, with the c-axis largely oriented perpendicular to the film surface. The degree of alignment of heterostructures was investigated on scanning across (006) peak by the X-ray rocking curve technique and it indicates slightly larger value of full width at half maximum(FWHM) compared to that of $Y_1Ba_2Cu_3O_{7-y}$ single layer thin films. Within the resolution limit of our instruments, however, no clear trace of sattellite peak for multilayers was found. SEM shows very smooth surface of heterostructures same as that of $Y_1Ba_2Cu_3O_{7-y}$ single layer thin film with very few small pinholes.

Critical temperatures of heterostructures showed similar transitions with Tc of $Y_1Ba_2Cu_3O_{7-y}$ single layer films close to 88 K unless the thicknesses of A(123) are less than120 Å when B(Ni-123) is given as 60 Å. The critical temperature, however, was degraded to the range of 82~85 K when the thickness of A(123) was less than 120 Å, almost twice of B(Ni-123). All measurements of critical temperatures and critical current densities were made using a conventional four terminal method. Critical temperature was determined at the bottom of the resistive transition in the temperature dependence of the resistivity curve. Critical current densities of these films were measured using a 20-μm-wide bridge patterned by a cw Q-switched Nd:YAG laser as described elsewhere [11]. Jc was determined with a threshold of voltage at 1 μV level which is equivalent to 10~20 μV/cm in the I-V plot. A typical plot of reduced critical current density Jc(T)/Jc(0) vs. T of a [A(123)/B(Ni-123)] x 5 film with A=300 Å and B=60 Å is shown in Fig. 2 with Tc=87 K and Jc=2.34×10^6 A/cm^2 at 77 K.

As derived by Tinkham [12] in a flux creep controlled process expression, $Jc(t)=Jc(0)[1-\alpha(B)t-\beta t^2]$ where the reduced temperature t=T/Tc and the coefficient $\alpha(0)=[k_BTc/F_p(0,0)]\ln(E_o/E)$. Here, E is electric field criterion used for the Jc measurement, E_o is a characteristic electric field [5], and Tc is the critical temperature of the film. The coefficient β comes from the temperature dependence of the free energy difference between pinned and unpinned flux quanta as expressed, $F_p(B,t)\sim F_p(B,0)(1-\beta t^2)$. In high Tc superconducting material α is close to 1, as described by Tinkham [12]. Best fit of our data for heterostructures to the flux creep expression of Tinkham gives α=0.822 and β=0.178 as the solid line shown in Fig. 2. Similar curve fitting to $1-\alpha t-\beta t^2$ for high quality single layer $Y_1Ba_2Cu_3O_{7-y}$ films on ZrO_2(100) substrate gives α=0.925 and β=0.093 with Tc=88 K and Jc=1.7×10^6 A/cm^2 at 77 K as shown in Fig. 3. The decrease of α in heterostructures could result from the increase of flux pinning energy F_p due to the presence of new pinning centers in the films after introducing weak superconductors including Ni and interfaces between layers.

The variation of Jc with magnetic field in $Y_1Ba_2Cu_3O_{7-y}$ /$Y_1Ba_2(Cu_{1-x}Ni_x)_3O_{7-y}$ multilayer films supports a basis for understanding the effect of introducing heterostructures of magnetic/non-magnetic superconducting thin films. The field dependence of Jc at 77 K and 70 K for a 300Å(123)/60Å(Ni-123) heterostructure film and pure $Y_1Ba_2Cu_3O_{7-y}$ film is shown in Fig. 4. All critical current densities were measured with magnetic fields parallel to the film surface. Even though Jc suffers an initial drastic reduction in a magnetic field, which can be explained by an average of the field dependence of transport Jc(H) of weak links, it is interesting to note that normalized critical current density Jc(H)/Jc(0) shows a slower reduction in magnetic field for the heterostructure. Critical current density of the heterostructure has maintained more than 50% of Jc(0) in 1 T magnetic field at 77 K and more than 60% at 70 K. This observation is consistent with the suggestion that heterostructures have actually higher flux pinning energy to prevent reduction of Jc(H) from Jc(0) values than $Y_1Ba_2Cu_3O_{7-y}$ single layer films.

In summary, we have shown that the critical current density behavior is changed significantly in magnetic fields after introducing heterostructures of $Y_1Ba_2Cu_3O_{7-y}$/$Y_1Ba_2(Cu_{1-x}Ni_x)_3O_{7-y}$ superconducting thin films. The temperature dependence of Jc suggests one possibility for increasing the flux pinning force in multilayer superconducting thin films including ferromagnetic elements.

The magnetic field dependence of Jc reveals that the critical current densities of heterostructures are less sensitive to magnetic field and maintained more than 50% of initial Jc(0) at 77 K. Normalized critical current density is enhanced by the presence of magnetic/non-magnetic multilayers.

We would like to thank S. Patel, M. Pitsakis, J.P. Zheng and L.W. Song. This research was supported by the New York State Institute on Superconductivity.

REFERENCES

[1] P. Chaudhari, R. H. Koch, R. B. Leibowitz, T. R. McGuire, and R. J. Gambino, Phys. Rev. Lett. 58, 2684(1987); B. Roas, L. Schultz, and G. Endres, Appl. Phys. Lett. 53, 1557(1988).
[2] J. C. Bendorz and K. A. Muller, Z. Physics. B64, 189(1986)
[3] M. K. Wu, J. R. Ashburn, C. J. Thorn, P. H. Hor, R. L. Meng, L. Gao, Z. J. Huang, Q. Wang, and C. W. Chu, Phys. Rev. Lett. 58, 907(1987).
[4] D. M. Huang, T. Venkatesan, C. C. Chang, L. Nazar, X. D. Wu, A. Inam, and M. S. Hedge, Appl. Phys. Lett. 54, 1702(1989).
[5] J. Mannhart, P.Chaudhari, D. Dimos, C.C Tsuei, and T. R. McGuire, Phys. Rev. Lett. 61, 2476(1988).
[6] P.W. Anderson, Phys. Rev. Lett. 9, 309(1962).
[7] J. M. Triscon, M. G. Karkut, L. Antognazza, O. Brunner, and O. Fischer, Phys. Rev. Lett. 63, 1016(1989); T. Venkatesan, A. Inam, B. Dutta, R. Ramesh, M. S. Hedge, X. D. Wu, L. Nazar, C. C. Chang, J. B. Barmer, D. M. Wang, and C. T. Rogers, Appl. Phys. Lett. 56, 391(1990).
[8] R. Gross, A. Gupta, E. Olsson, A. Segmuller, and G. Koren, Appl. Phys. Lett. 57, 203(1990).
[9] J. M. Tarascon, P. Barboux, P.F. Miceli, L.H. Greene, and G.W. Hull, Phys. Rev. B. 37. 7458(1988).
[10] S. Witanachchi, S. Y. Lee, L. W. Song, Y. H. Kao, and D. T. Shaw, Appl. Phys. Lett. 57, 2133(1990).
[11] Z. P. Zheng, H. S. Kim, Q. Y. Ying, R. Barone, P. Bush, D. T. Shaw, and H. S. Kwok, Appl. Phys. Lett. 55, 1044(1989).
[12] M. Tinkham, Helv. Phys. Acta. 61, 443(1989); Introduction to Superconductivity (Kreiger, Melbourne, 1975).

FIGURE CAPTIONS

FIG. 1 X-ray diffraction patterns: (a) [300Å(123)/60Å(Ni-123)] x 5 multilayer film with a 300Å(123) top layer, (b) $Y_1Ba_2Cu_3O_{7-y}$ single layer film.

FIG. 2 Temperature dependence of the critical current density of a 300Å(123)/60Å(Ni-123) multilayer thin film in zero magnetic field. The solid line is a fit to $1-\alpha t-\beta t^2$ with $\alpha=0.822$ and $\beta=0.178$ according to the flux creep model. $Jc(0)=1.7x10^7$ A/cm^2.

FIG. 3 Typical temperature dependence of the critical current density of a $Y_1Ba_2Cu_3O_{7-y}$ single layer thin film in zero magnetic field with $\alpha=0.925$ and $\beta=0.093$. $Jc(0)=1.3x10^7$ A/cm^2.

FIG. 4 Comparison for magnetic field dependence of Jc(H,T) in $Y_1Ba_2Cu_3O_{7-y}$ single layer and heterostructure thin films. Squares indicate 300Å(123)/60Å(Ni-123) multilayer samples and circles the $Y_1Ba_2Cu_3O_{7-y}$ single layer films. White and black are 77 K and 70 K, respectively. Critical current densities in zero magnetic field are Jc(77K)= $1.7x10^6$ A/cm^2, Jc(70K)=$2.8x10^6$ A/cm^2 for a $Y_1Ba_2Cu_3O_{7-y}$ single layer and Jc(77K)=$2.34x10^6$ A/cm^2, Jc(70K)=$4.8x10^6$ A/cm^2 for a heterostructure film.

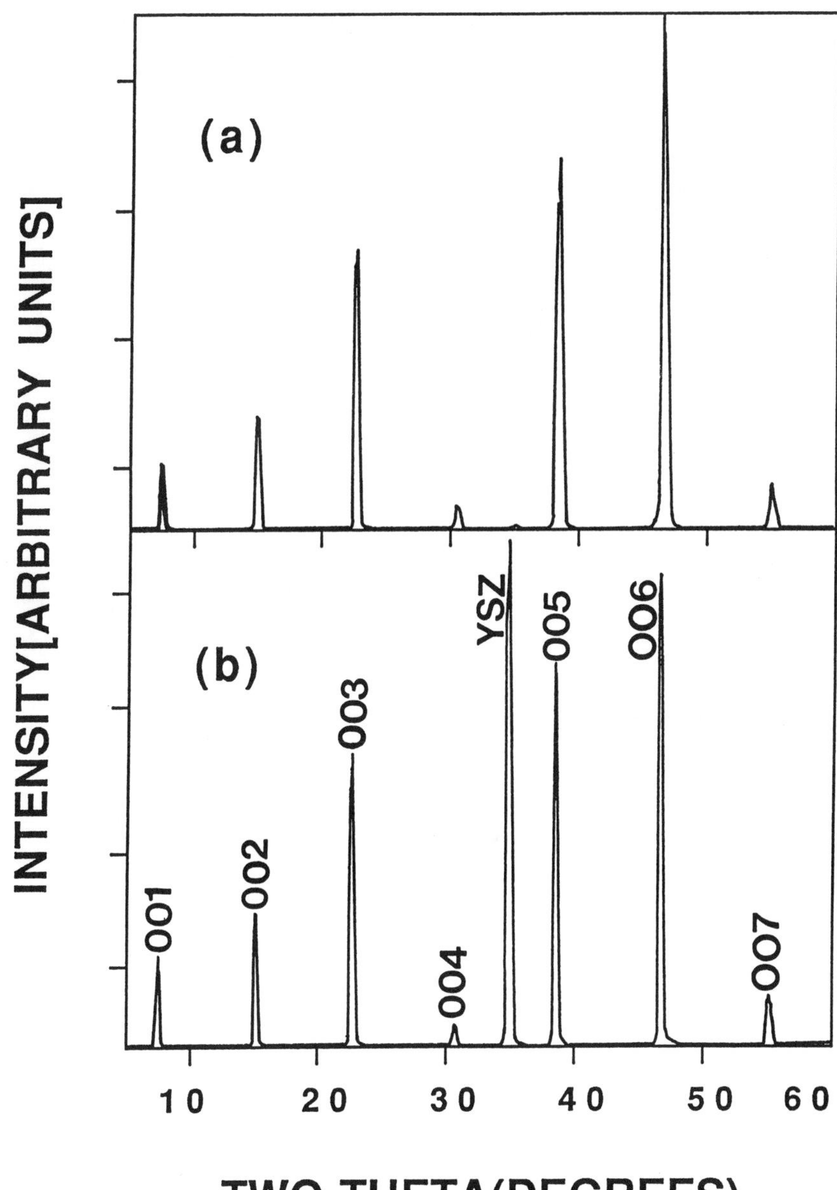
(a)
(b)
INTENSITY[ARBITRARY UNITS]
001
002
003
004
YSZ
005
006
007
10
20
30
40
50
60
TWO-THETA(DEGREES)

FIG. 1.

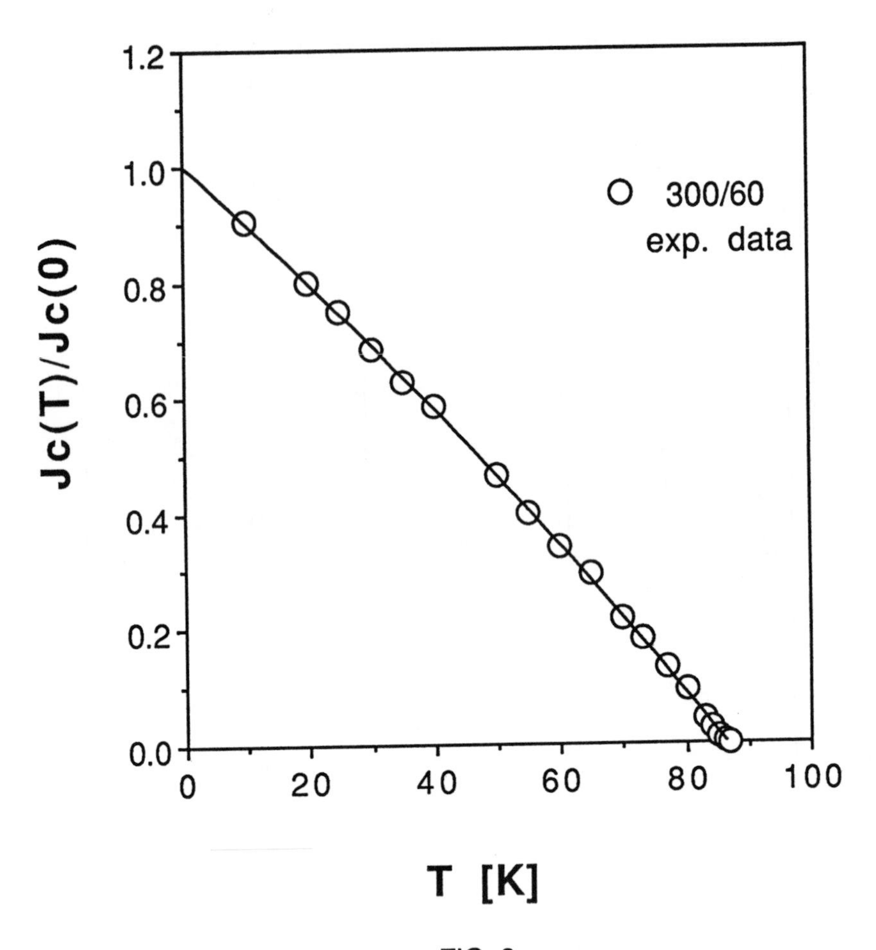

FIG. 2.

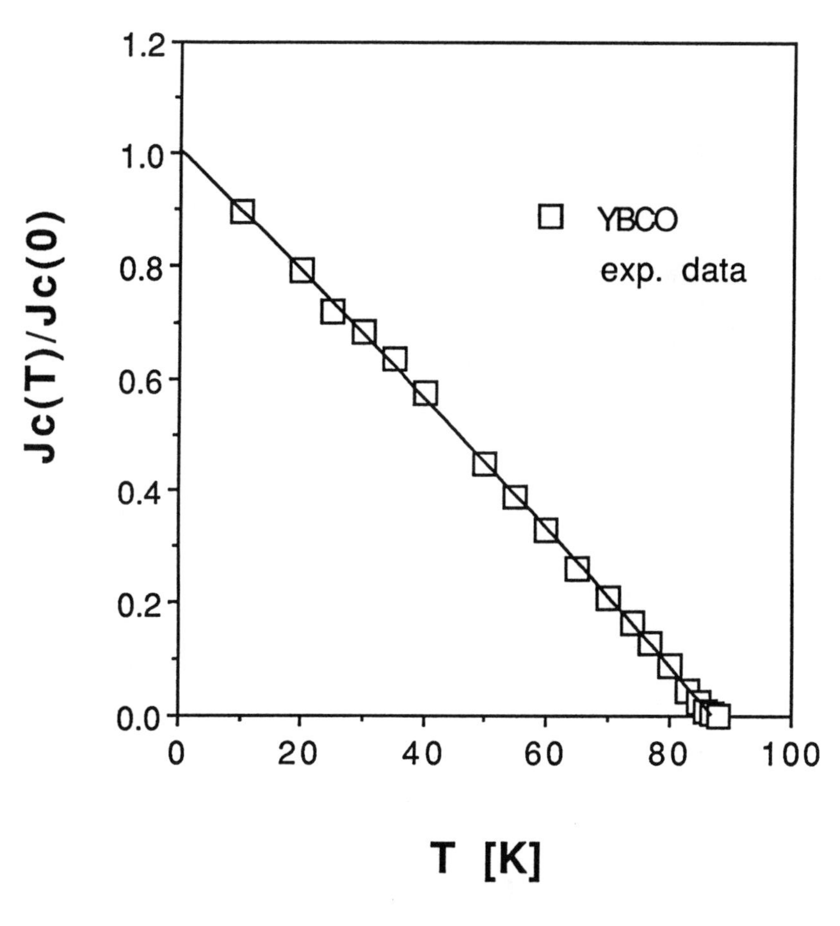

FIG. 3

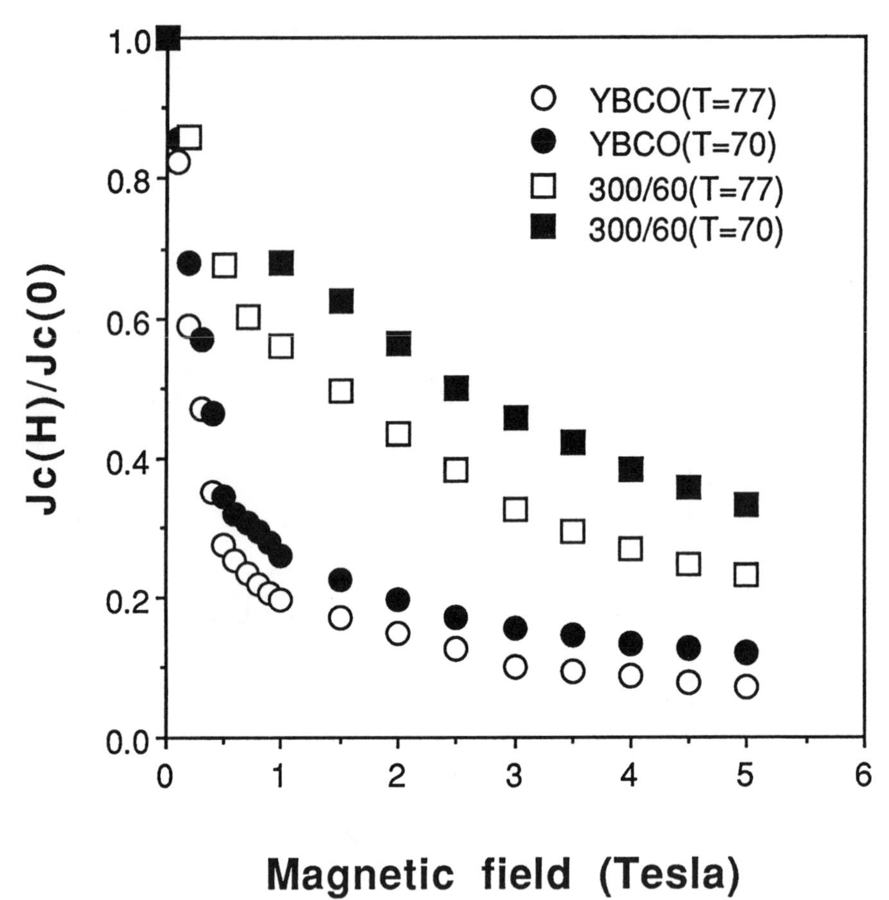

YBCO(T=77)
YBCO(T=70)
300/60(T=77)
300/60(T=70)
Jc(H)/Jc(0)
1.0
0.8
0.6
0.4
0.2
0.0
0
1
2
3
4
5
6
Magnetic field (Tesla)

FIG. 4

COMPUTER SIMULATION OF THIN FILM GROWTH WITH THERMAL HOPPING OF ATOMS

J. Schutkeker, L. Chen, F. Wong, S. Patel, and D. T. Shaw
State University of New York at Buffalo
Amherst, NY

ABSTRACT

A three-dimensional computer simulation that manipulates individual particles has been developed to model the early phases of the deposition of epitaxial films of a material with a specified lattice structure. The model is implemented with the simple cubic structure and the on-lattice positioning scheme, and calculations have been performed for a variety of experimental conditions, including temperature and flux. A comparison has been made between this method, which uses the fixed time stepping (FTS) method of Outlaw and Heinbockel, and a more accurate, more efficient method that uses an event based scheme. Agreement between the methods is good. Three dimensional particle migration and surface evaporation have been included in the FTS model, but have been found unimportant to its final results. The model shows a transition similar to the one predicted by Muller, but at more realistic temperatures, due to an improvement in the description of the surface migration energy barrier. Packing densities and surface roughnesses have also been calculated.

I. INTRODUCTION

Because they perform calculations on the smallest length scales and the shortest time scales, atomistic simulations are inherently constrained to model phenomena that take place at a microscopic scale far below almost all scales of engineering interest. No matter what the application of the modelling, it is always very difficult to bridge the gap to the more macroscopic scales. In the case of thin film growth, the scales of interest are time scales particular to the total film formation and length scales at least as large as the grain size of the material in question. When the scale of a simulation is as restricted as atomistic simulation, it is very difficult to identify any quantity of relevance whose behavior can be modelled.

In our work, as well as proceeding work [1], our results are presented in terms of simple quantities such as atomic packing density (filling fraction) and surface roughness. Although our improved time stepping method allows us to model longer times, our simulations are still constrained to atomistic length scales, so it is not possible to improve greatly over the modelling of departures from perfect crystallinity on an extremely microscopic scale. Xiao [2] has describes a much broader variety of specific morphological features very clearly, but his methods are different from ours in two ways. His approaches cannot be used to show the effects of varying temperature on morphology, since his treatment of temperature does not well reflect its true nature, and he has built a more detailed model, using off-lattice particle positioning, whereas we have only used on-lattice positioning, since we are only interested in testing our novel time stepping method, not in actually generating physical results.

II. TEST MODEL

Muller [1] modelled by two-dimensional atomistic simulation the growth of thin films of FCC metals, using Nickel as his test case. He has chosen a fixed time step (FTS) time stepping scheme with the time step given by the average hopping time for an isolated surface adatom, evaluated by Outlaw and Heinbockel [3] to be

$$\Delta t = \nu_o^{-1} e^{Q/kT}, \tag{1}$$

where ν_o is the surface vibration frequency, given by $\nu_o = 2kT/h$, and Q is the migration energy barrier for an isolated surface adatom, given by

$$Q = \left(5 + \frac{20}{3}\frac{T}{T_m}\right) kT_m, \tag{2}$$

with T_m being the melting temperature of the simulated material. If an atom is not isolated, its energy barrier must also include the energy necessary to break all bonds. At each time step, an exponentially distributed random number is used to generate a Monte Carlo energy, ϵ, for each atom of the simulation, according to

$$\epsilon = -kTln(1 - R^{1/\nu_o \Delta t}), \tag{3}$$

where R is a random number uniformly distributed on the interval [0,1]. For each particle, this energy is compared to the energy barrier from migration and bond breaking, and if it is large enough, the particle migrates to a randomly chosen neighboring site. Depending on the ratio of the migration time to the time between the arrival of new particles, either the migration process will occur many times for each new particle, or the arrival process will occur many times in the time interval between migrations.

This approach is not economical in processing time, because at each time step only a small fraction of atoms will migrate to neighboring sites, so many unused evaluations of the random number and logarithm functions must be performed. Not only that, but the method is slightly inaccurate, because it uses an "average" time step for migration, thereby ignoring that small minority of particles that might migrate more than once in each time step. This loss of accuracy, however, is expected to be so small as to be undetectable. Another problem (and this is true for any such Monte

Carlo modelling of atomic hopping, including the event driven model) is that there is uncertainty over how best to choose the destination site of a migrating atom, based on the physics of the migration process. A good argument can be made for any of the following methods, and without justification, we have used the first of these methods.

1) Choose the direction uniformly from among all possible directions along the principle crystalline axes, regardless of whether the destination is occupied.

2) Choose only from among those directions along the principle axes that will lead only to unoccupied destinations.

3) Choose preferentially from among those directions along the principle axes that will lead to destinations where there will be an increase in bonding with neighboring particles.

We have implemented this approach in three dimensions, using an on-lattice scheme for positioning atoms, in which each site is represented in computer memory by a single bit which tells whether that site is either filled or empty. Since atomic positions cannot take on a continuum of values, our codes cannot model the formation of certain kinds of defects, such as dislocations or grain boundaries, for which the energy barrier calculation would become inaccurate, anyhow. Simple cubic coordination of the atoms is used. This method is clearly a brute force model, and it has all of the bad CPU performance generally associated with brute force calculations. Table 1 shows just how bad its CPU performance is. Below the onset of thermal annealing, where $\Delta t_{migration} = \Delta t_{between-arriving-particles}$, the code's performance is dominated by particle precipitation, but above this value, CPU time increases at an apparently exponential rate. Notice that the CPU time consumed is in *hours* of VAX processor time. The highest T runs took well over a month of wall clock time to execute. Upgrading this code from Muller's 2-D version to the complete 3-D version has degraded the performance so severely that it is not possible even to make a scan of a temperature range nearly as wide as Muller's. This code cannot be used to examine temperatures high enough that the smoothing effects of annealing completely dominate the roughening effects of random precipitation. It is thought that a 3-D simulation might show surface smoothing at lower temperatures than a 2-D simulation because of the increased degrees of freedom for surface mobility, but this has not been tested. Results will be discussed in Sec. IV.

III. UPGRADED MODEL

Our improved time stepping scheme is an event driven (ED) scheme, motivated by techniques to model queueing of jobs in computer operating systems [4]. We store in memory an event queue in which two possible kinds of event can take place – particle migration or particle arrival. Each event has three data values associated with it: the type of event, the location of the particle experiencing that event, and the time that the event takes place. The queue is ordered in ascending values of the time until each event occurs. For ease in sorting, the event queue is a linked list. At the start of the code, the queue is initialized by filling it with all particle arrival events that will occur over the duration of the simulation. Then, at each pass through the outer loop of the code, the first event in the queue is retrieved and performed, and the times of all coming events are decremented. Then the times until the next migration events are generated for all particles affected by the event just processed, and the events are inserted into the queue in the appropriate time sequence. Particle arrivals are spaced evenly in time, and the time until the next migration event for each particle are generated according to

$$\Delta t = \frac{ln(R)}{\nu_o ln\left(1 - e^{-Q/kT}\right)}, \tag{4}$$

with all parameters being the same as in the test simulation.

IV. RESULTS AND COMPARISON

For the simulated material, we have used the same parameters as Muller, who chose Nickel as his material of interest. The melting temperature, T_m is 1450 ^oC, which gives kT_m =148 meV, and the chemical bonding energy is 740 meV. The greatest departure from true Nickel is in the simple cubic coordination, which is not important, since we are not studying Nickel, just using it as a test case to see if the ED model agrees with the FTS model.

Figs. 1 and 2 show surface morphologies for different temperature values, as generated by the FTS and ED models, respectively. Agreement is at least qualitative, and it is clear, since only the ED scheme is able to generate results in the temperature domain where bulk annealing occurs, how easily it can outperform the FTS model. Figs. 3 and 4 show scans of packing density and roughness vs. temperature, with the packing density plots showing exact agreement. These figures again show how well the ED model outperforms the FTS model, since sufficient wall clock time was available to run higher deposition rates with the ED model. Considering how much CPU time the FTS model consumes, running any further data points would be completely impractical.

As new work, the ED model has been used to simulate interrupted growth, whose conditions are roughly equivalent to the experimental laser film deposition. In interrupted growth, periods of material deposition alternate with periods of annealing, in which only thermal migration occurs. Fig. 5 shows the results, which are very interesting. Two step edges are present, which demonstrate that there are two widely disparate characteristic times associated with the annealing process. The shorter characteristic time is where particle migrations that break a single chemical bond become significant, and the longer time is where double bond breaking begins to play a role. Although the ED model could not be run at even higher times, the figure also seems to suggest the existence of a third characteristic time, corresponding to the breaking of three bonds. Figs. 6 are cross-sections of the simulated films, which further show the improved annealing at higher temperature.

Our results depart drastically from Muller's as a result of a slight, but important, oversight on his part. Muller took, as the energy for a particle at a step edge (ie. having at least one bond) to hop up the edge and become a free adatom, $\Delta E = (1 - N_i)\phi$, where ϕ is the bond energy and N_i is the number of bonds before hopping. He should have added Q, the migration energy barrier, to this term, so the temperatures at which his simulation predicts good annealing are much too low. His predictions are for near perfect crystallinity at 175 ^{o}C, which is obviously an unrealistically low temperature. We do not include this barrier hopping, either uncorrected or corrected, and our results predict much higher annealing temperatures, in the vicinity of 700 ^{o}C.

V. CONCLUSIONS

The increase of efficiency of this time stepping scheme has allowed three significant gains to be realized simultaneously in this atomistic simulation. Temperatures high enough to be interesting for thin film growth can be modelled routinely. Three dimensional simulations can be performed without difficulty. And time periods long enough to model interrupted growth or simple annealing can be treated easily. Also, some new physical phenomena have been postulated based on this modelling of interrupted growth. The general approach of Muller's work has also been shown to be valid, to within it's theoretical limitations, and a small oversight in the implementation that led to drastically inaccurate results has been corrected. This time stepping method should prove useful for more rigorous, atomistic Monte Carlo simulations, such as the off-lattice model, and if implemented on a more powerful computer, such as a Cray, it may be possible to trade some of the gains in time scale for improvements in spatial resolution, perhaps allowing more macroscopic, and therefore more useful, simulations. It is also worth noting the sensitivity of the physical results to the barrier energy and chemical bond energy values. When we were making our final comparison of the model, we found that a discrepancy of 40 meV in the bond energy ed to a shifting of the curves of packing density vs. temperature by about 100 ^{o}C.

VI. FUTURE WORK

Further work relevant to superconducting materials can proceed from here in a number of directions. Applying the new time stepping scheme to an off-lattice simulation would allow the modelling of relevant defect structures such as screw dislocations in FCC metals. The most important modification in this case would be to develop an energy barrier equation for the hopping of adatoms to neighbors slightly displaced from their exact lattice sites. We have discussed at great length the possibility of building an EAM model [5], which would allow the modelling of compound materials rather than single elements. The approach we suggest would be to use energy band structure calculations to solve for the electronic charge density throughout the material and then to use this charge density calculation in an extended EAM model. Improvement in charge density calculations should lead to significant improvements in the EAM model, as charge density is one of the more important quantities ion that model. This approach, however, would rely on accurate band structure models for the High T_c materials, which are not yet available. Until such an improvements can be made, applying these models to High T_c materials can only be made within the most gross approximation, that is, that one atom in the simulation corresponds to one entire unit cell of of the actual material. Any such approximation would make results highly questionable, at the very least.

BIBLIOGRAPHY

[1] K. H. Muller, J. Appl. Phys. **58**(7), pp. 2573–2576, 1985.

[2] R. F. Xiao *et al*, Phys. Rev. A **43**(6), pp. 2977–2992, 1991.

[3] R. A. Outlaw and J. H. Heinbockel, Thin Solid Films **108**, pp. 79–86, 1983.

[4] J. Banks and J. S. Carson II, **Discreet-Event System Simulation,** Prentice Hall, NJ, 1984.

[5] M. I. Baskes, J. S. Nelson and A. F. Wright, Phys. Rev. B **40** (9), pp. 6085–6100, 15 Sept. 1989-II.

Table I - Performance of FTS Code

Temperature	VAX CPU time
252 K	15 hours
297 K	145 hours
312 K	180 hours
356 K	330 hours

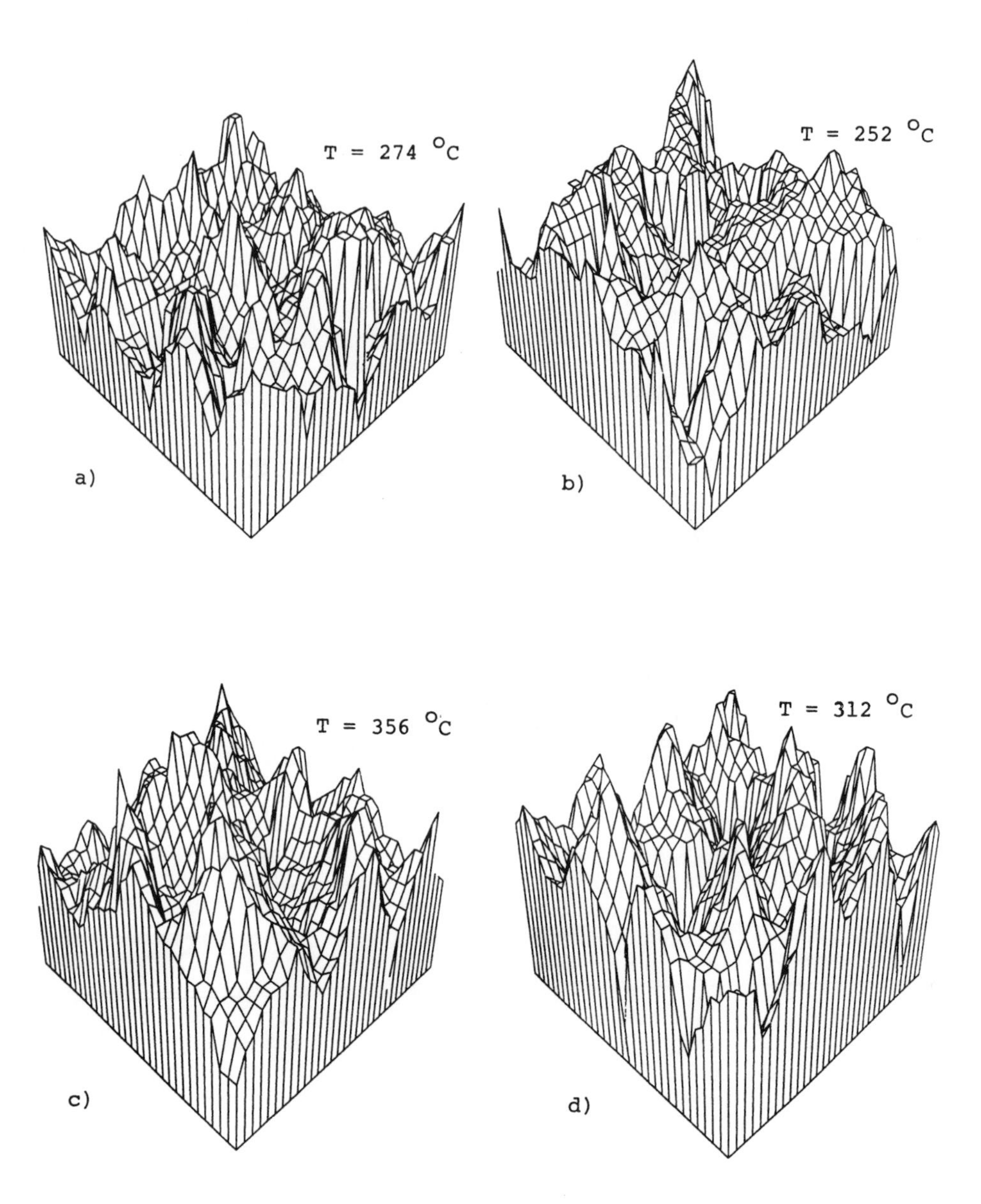

Figs. 1 - Surface textures of FTS simulated films

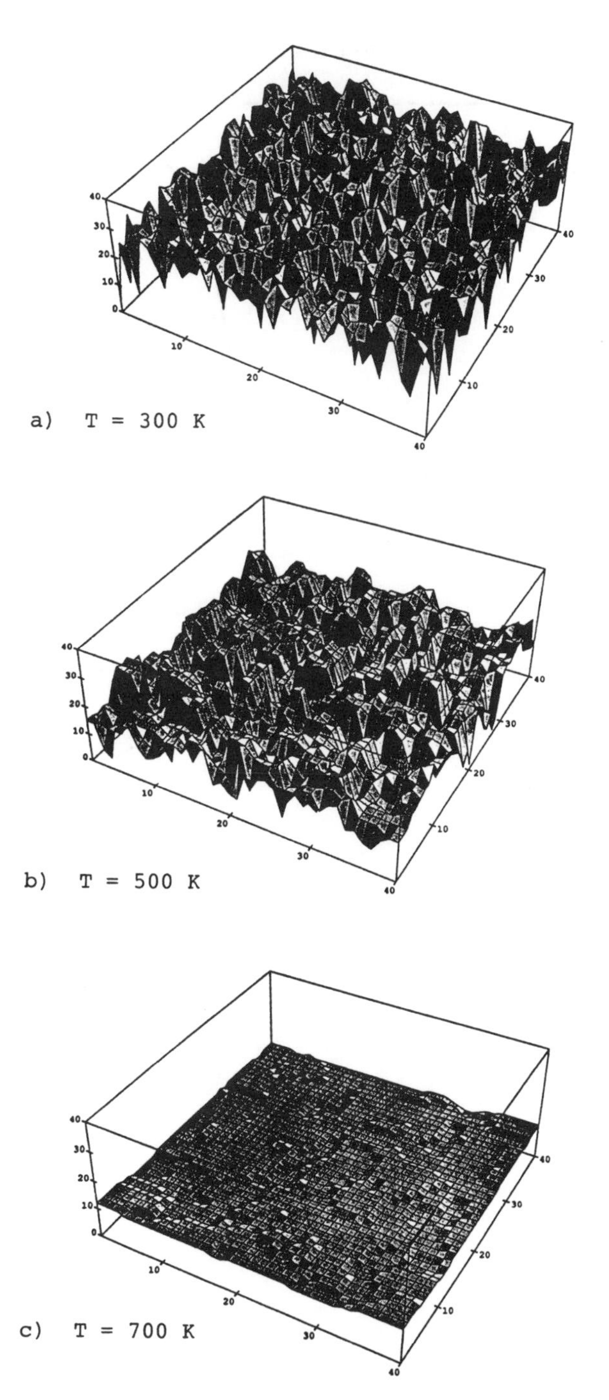

Figs. 2 - Surface textures of films generated by ED code

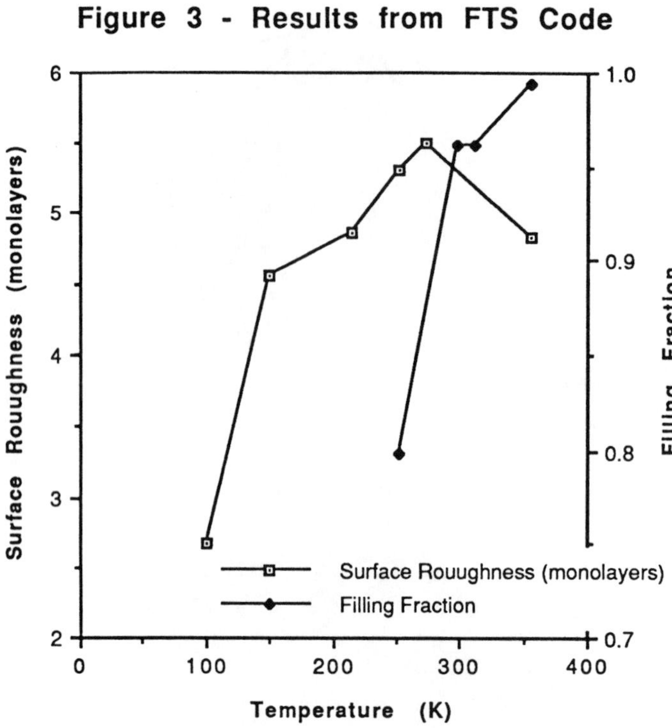
Figure 3 - Results from FTS Code
6
5
4
3
2
1.0
0.9
0.8
0.7
0
100
200
300
400
Surface Rouughness (monolayers)
Filling Fraction
Temperature (K)
Surface Rouughness (monolayers)
Filling Fraction

a)

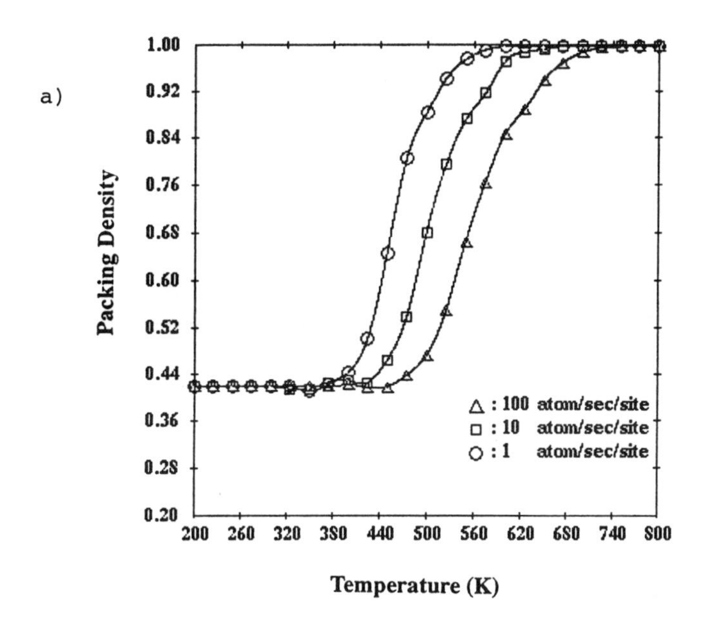

Figs. 4 - Results of parameter scans with ED model

b)

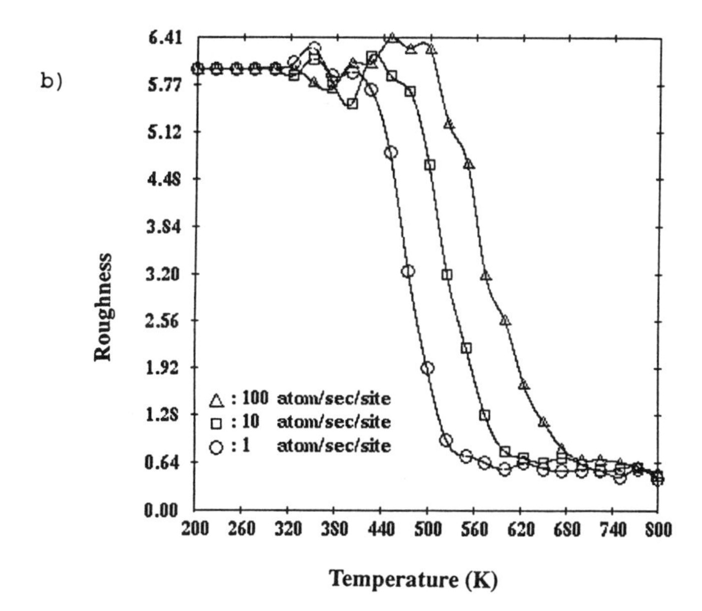

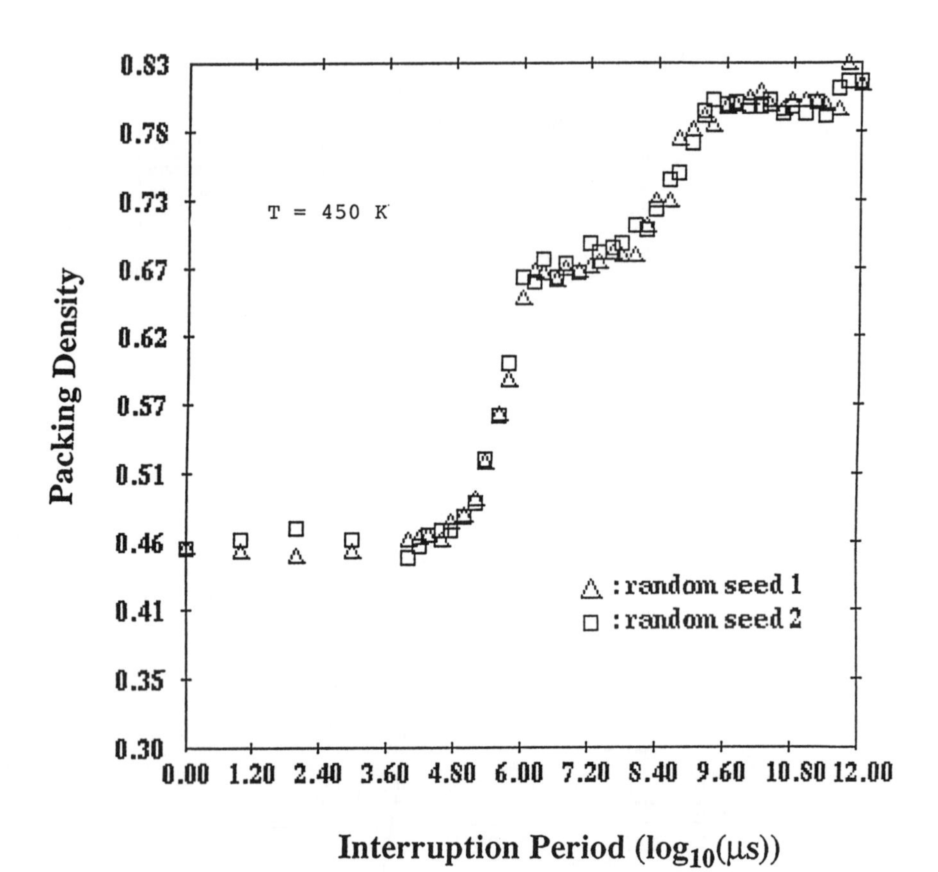

Fig. 5 - Interrupted growth data from ED code

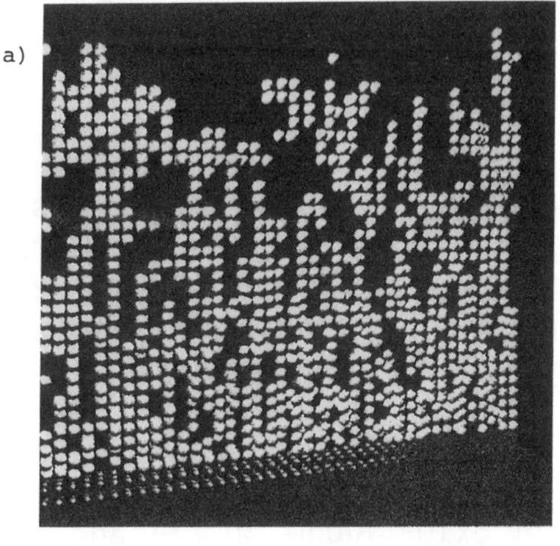

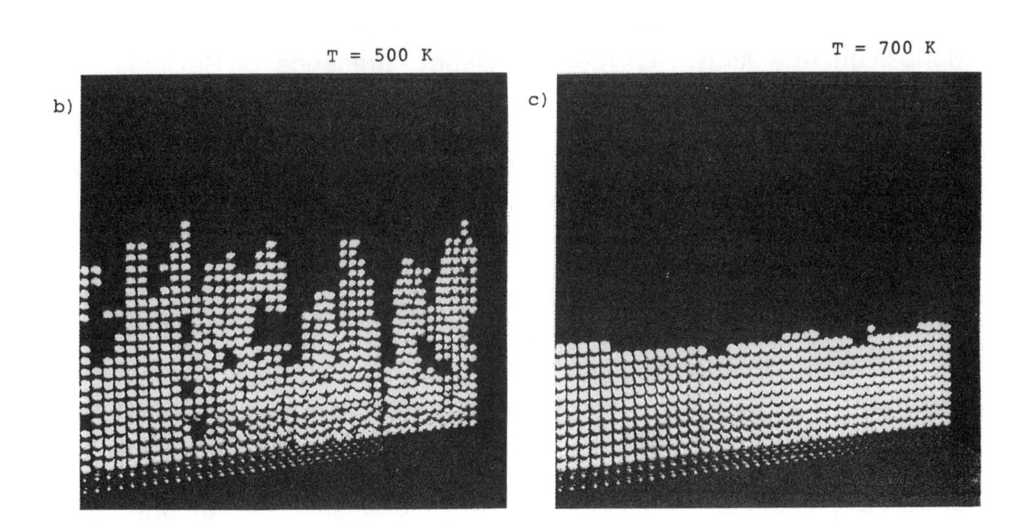

Figs. 6 - Generated film cross sections from ED code

DEGRADATION MECHANISMS OF HIGH DEPOSITION RATE $Y_1Ba_2Cu_3O_{7-x}$ THIN FILM PROPERTIES

A. Shah, R. Barone, S. Patel and D.T. Shaw
State University of New York at Buffalo
Amherst, New York14260

ABSTRACT

Superconducting $Y_1Ba_2Cu_3O_{7-x}$ thin films have been grown at a high deposition rate of 10,000Å/min, in-situ, at atmospheric pressure. The rf plasma deposition technique used has been scaled up by more than a 100 times when compared to a technique such as laser ablation. After a 30 second deposition, a 0.8μm thick, high deposition rate film has the same critical temperature of 87K as the lower rate, 100Å/min, film. On the other hand, the critical current is an order of magnitude lower than the critical current density of $6x10^5$ A/cm^2, at 77K and zero field, of the lower rate film. Degradation of the high rate film properties, such as crystal structure, homogeneity and current carrying capacity have been investigated with respect to the low deposition rate films. Oxygen deficiency, presence of $BaCuO_2$ impurities and incomplete orientation of the film have been specifically identified as some of the reasons responsible for the degradation . Scanning Auger Electron Spectroscopy (AES), Auger depth profile, EDS mapping, x-ray diffraction and electron microscopy have been used to analyze and detect these defects.

INTRODUCTION

In situ superconducting $Y_1Ba_2Cu_3O_{7-x}$ thin films have been grown by a variety of vapor deposition techniques but in most of these the deposition rate is small. High deposition rate films have been grown by a few groups using techniques such as MOCVD[1], plasma deposition[2] and laser deposition[3]. For practical applications such as superconducting tapes, where the requirement is that the film thickness should be 1-10μm, very few deposition techniques have any hope of succeeding. We have developed a rf plasma deposition technique to make high deposition rate films. This technique can easily be scaled up to make practical conductors. The films are grown at atmospheric pressure and hence expensive vacuum equipment is not required. Again, as the deposition rate is 1μm/min the total deposition time is short. These are advantages when the question of scale up arises. As mentioned before the growth process is in situ and no post annealing processing steps are required. When films are made with this process at a low deposition rate of 200-300Å/min the film quality is quite high with a critical temperature and critical current density of 87K and $6x10^5 A/cm^2$ at 77K and zero field. However, these properties could not be maintained in the high deposition process and degradation in the film properties was observed. To improve the properties the degradation and its causes have to be identified so that they can be eliminated. The degradation is observed in the surface morphology, critical current density and film structure. The critical temperature does not seem to be effected. Besides the unavoidable defects found even in the low deposition rate films such as twinning, grain boundary problems, etc., other microscopic defects such as precipitates or growth of unwanted phases have also been identified. In this investigation we have identified conclusively the presence of the insulating $BaCuO_2$ phase, increase in surface roughness and oxygen deficiency in the films.

EXPERIMENT

The deposition system has been described elsewhere[4]. A few changes have been made to the initial low deposition system. The quartz nozzle above the plasma torch has been changed from a cylindrical tube to a converging nozzle. The aerosol generator used at present is an ultrasonic nebulizer from Medisonic. The concentration of the precursor, metal nitrate, solution has been increased to Y: Ba : Cu :: 0.082M : 0.165M : 0.224M. This requires a higher rf input power level of 5.5kW, an increase of 1kW, and also an increase in the torch input gases. The aerosol flow rate has been decreased to 1.5 lpm of oxygen from 4.1 lpm, the flow used in the low deposition process. The rationale behind this decrease is to increase the residence time of the input precursor in the plasma reaction zone. After deposition, a slow cool down in flowing oxygen was carried out at atmospheric pressure.

RESULTS AND DISCUSSION

The films grown at 1μm/min are made in situ and at atmospheric pressure. The best films have a critical temperature of 87K with a critical current density of $2x10^5$ A/cm^2 at 77K and zero field. The critical temperature and current density are measured by the standard four probe measurement method with a transport current estimated by using the 1μV/cm criterion. The usual film thickness is 0.8μm. The films are completely c-axis oriented perpendicular to the substrate as shown in figure 1. The substrate used to grow these films is single crystal Yttria stabilized Zirconia (YSZ).

The surface morphology of the high rate films is significantly different from the low rate films as seen in figure 2a - 2c. The low deposition rate films have a much smoother surface morphology. On the other hand, the surface of the high deposition rate films is more irregular. There are more particulates on the surface. Large 5-10μm mounds and smaller particals are evenly distributed on the surface and embedded in the film. This is sketched in figure 3. Even though the x-ray diffraction pattern, figure 1, shows the high rate films to be predominantly c-axis oriented, a powder peak is present in the diffraction pattern at $2\theta = 32.8°$. This can be explained by the rough surface morphology.

The x-ray diffraction pattern further shows the presence of a $BaCuO_2$ impurity peak. The presence of this phase could be a reason for the decrease in the critical current density in the films. The critical temperature does not seem to be effected by the presence of $BaCuO_2$. This is important because it means that only a small amount of this insulating phase is present, not enough to completely degrade the films. The exact location of this precipitate in the film will be discussed later. A contrasting feature of the high rate films is the precise two theta position of the <00l> peaks in the diffraction pattern with reference to the low deposition rate films. The high deposition

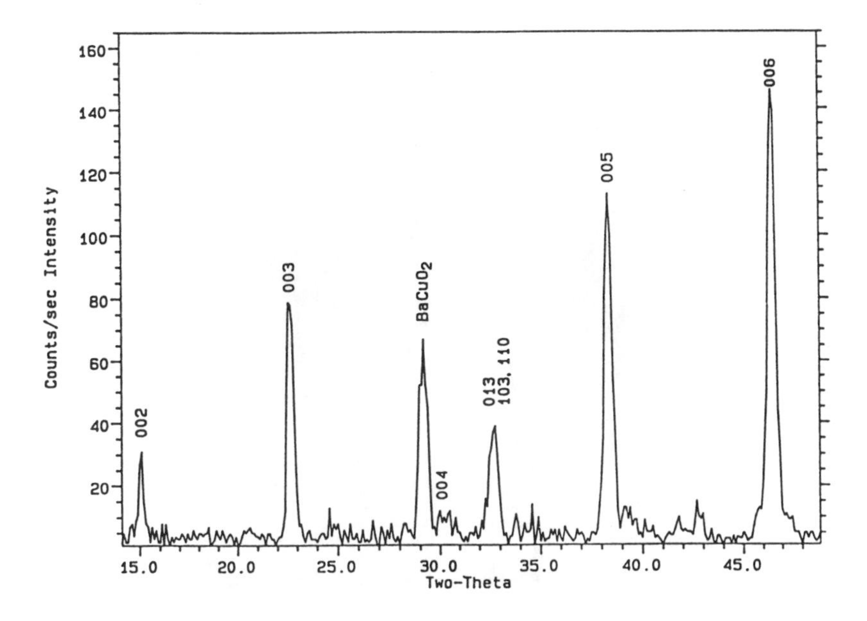

Fig. 1. X-ray diffraction pattern of a 1μm/min deposition rate, c-axis oriented, $Y_1Ba_2Cu_3O_{7-x}$ film.

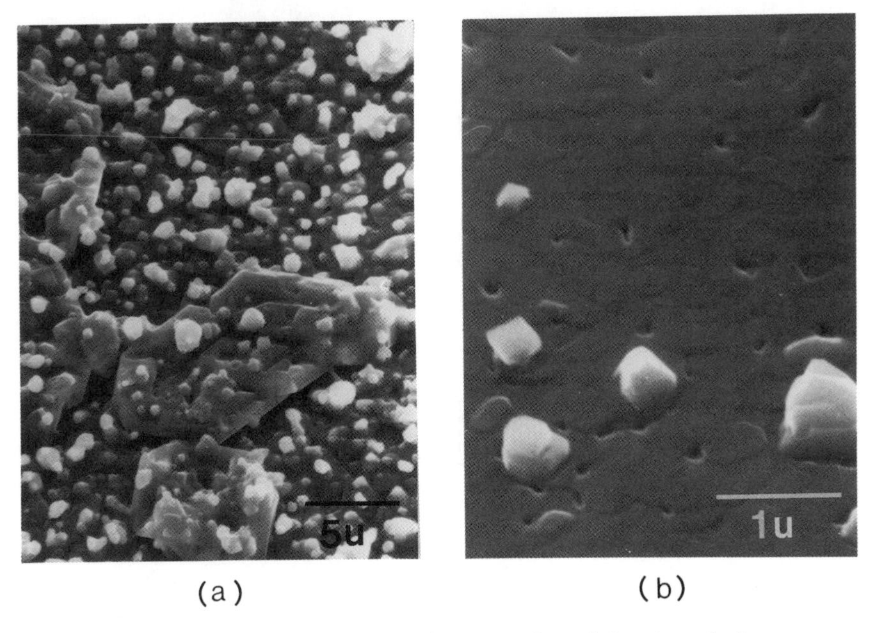

Fig. 2. $Y_1Ba_2Cu_3O_{7-x}$ film micrograph. (a) 1μm/min high deposition rate film. (b) low deposition rate film.

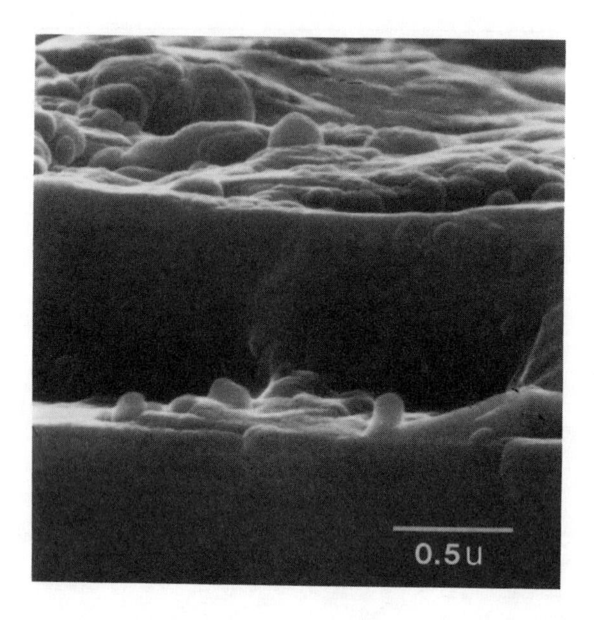

Fig. 2c. Cross-section of a 1μm/min deposition rate $Y_1Ba_2Cu_3O_{7-x}$ film.

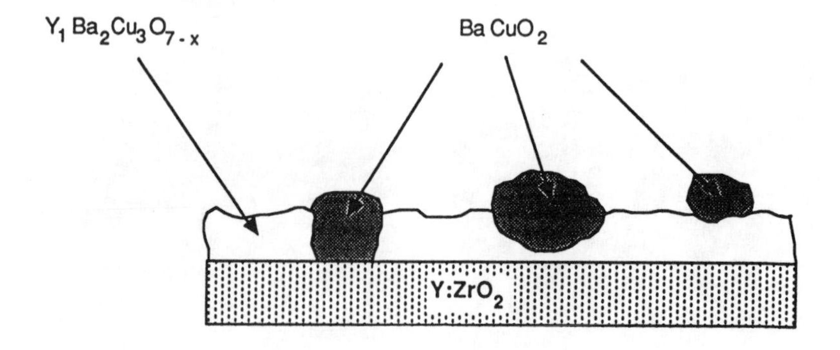

Fig. 3. Possible growth origins of the $BaCuO_2$ phase.

rate film's <006> peak position is usually at $2\theta = 46.5°$, at least 0.1° lower than the low rate film shown in figure 4. This translates to a high rate film c-axis length of 11.71Å, a little longer than the the c-axis length of a high quality $Y_1Ba_2Cu_3O_{7-x}$ film of 11.68Å. According to Ono[5] this is due to oxygen disorder in the film, that is, they may be oxygen deficient. The short deposition time may be responsible for this effect in the 0.8μm $Y_1Ba_2Cu_3O_{7-x}$ film. Again this may be responsible for the low critical current density in the high rate films. On carrying out a rocking curve measurement on the high rate films, the FWHM are found to be usually 1.7 degrees wide compared to an average of 0.55 degrees for the low rate films as seen in figure 5. This indicates that the grains are relatively less aligned.

Observed in the x-ray diffraction pattern is the presence of a $BaCuO_2$ peak. It was suspected that the large mounds seen in figure 2a could be $BaCuO_2$. On conducting energy dispersive spectroscopy (EDS) on the mounds it is found that the barium and copper ratios are nearly 1:1 and barely any signal from Yttrium is seen. But in the smoother sections of the film the ratios are more characteristic of a $Y_1Ba_2Cu_3O_{7-x}$ film with the yttrium, barium and copper ratios being 1:2:3. On carrying out a EDS x-ray map of the films, the mounds turned out to be barium and copper rich, further confirming that the mounds are $BaCuO_2$. To further confirm this, after 1 minute of sputtering of the film (a 5 kV rastered beam of Ar ions yielded a sputter rate of 200Å/min), an Auger spot analysis was taken of the mounds and the smooth area, respectively. Again the mound area did not show the presence of yttrium while the latter survey showed a clear yttrium signal. This is shown in figure 6. Again a high resolution scan of the oxygen peak was carried out by AES and this revealed that the oxygen content of the mound was significantly less than that in the smoother valley area. This would indicate that the mound is $BaCuO_2$ having less oxygen than the smooth or $Y_1Ba_2Cu_3O_{7-x}$ area. The question would then arise as to the origin of the $BaCuO_2$ phase. Three possible modes are depicted in figure 3. Growing from the substrate, embedded in the film or as particulates lying on the surface. To investigate this an Auger depth profile was performed on one of the large mound areas. The profile, figure 6,

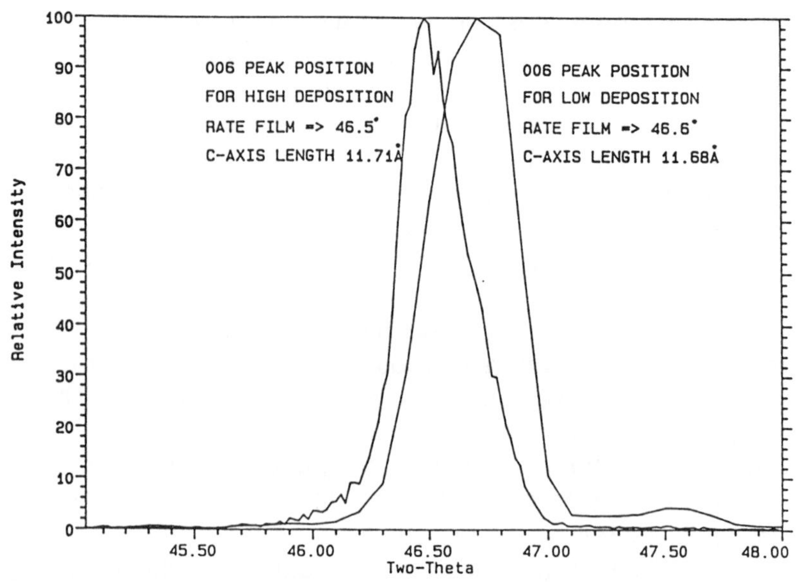

Fig. 4. 006 reflection peak positions for high and low deposition rate $Y_1Ba_2Cu_3O_{7-x}$ films indicating different c-axis lengths.

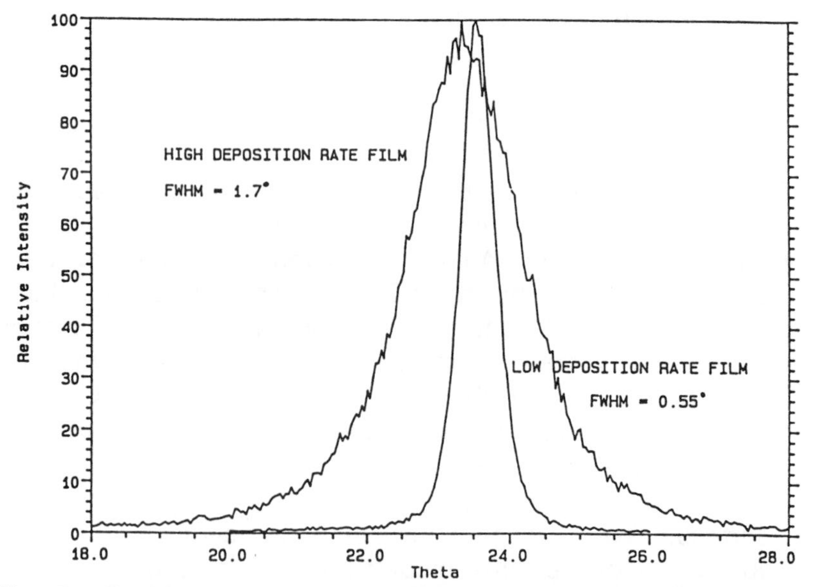

Fig. 5. Rocking curve measurement on the 006 peaks for high and low deposition rate $Y_1Ba_2Cu_3O_{7-x}$ films showing relative difference in alignment.

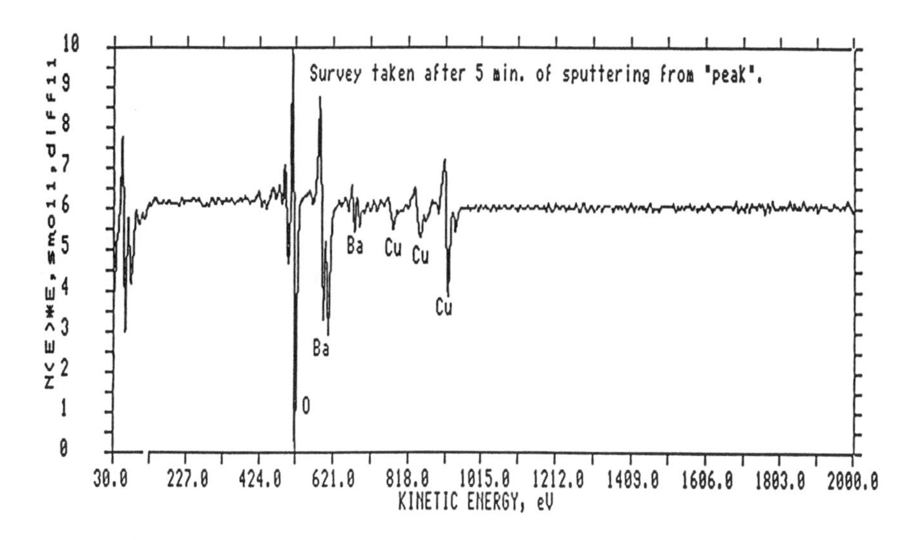

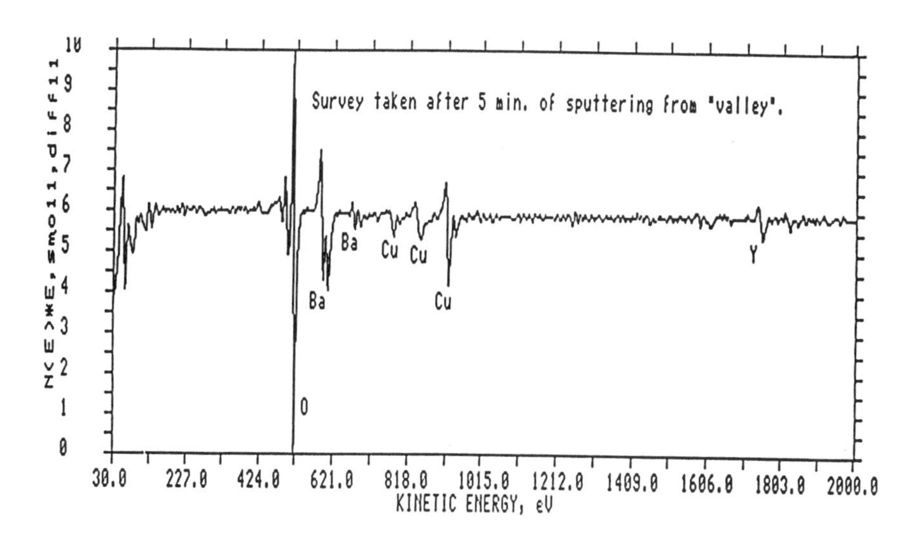

Fig. 6. AES survey of (a) $BaCuO_2$ mound or peak showing the absence of yttrium. (b) AES survey of a valley or flat area showing a typical spectrum of $Y_1Ba_2Cu_3O_{7-x}$.

seemed to indicate that a thin layer of $Y_1Ba_2Cu_3O_{7-x}$ is still present under the mound because a yttrium signal is present after a substantial thickness of the film had been sputtered away. Even the oxygen signal seems to rise along with the yttrium signal indicating the presence of $Y_1Ba_2Cu_3O_{7-x}$. This is not conclusive as a depth profile would have to be obtained for every mound on the film to confirm this. This hypothesis may still be valid because the EDS map and the scan of the mound still showed the presence of a small yttrium signal.

CONCLUSION

High deposition rate films have been grown consistently at 1μm/min, in situ, at atmospheric pressure. The films are c-axis oriented with critical temperatures of 87K and a critical current density of 10^5 A/cm^2 at 77K and zero field. The electrical properties are thus lower than that of the low rate films. This is due to a number of reasons, some of which are film structure degradation due to short deposition time, misalignment of the grains and oxygen deficiency. Microscopic impurities in the films such as $BaCuO_2$ have been identified. These precipitates are embedded in the film and do not seem to nucleate from the substrate. An important note is that only a degradation of the critical current is observed in contrast to the critical temperature of the films, which remains at a high value.

ACKNOWLEDGEMENT

This work was supported in part by the New York State Institute on Superconductivity.

REFERENCES

1. H. Abe, T. Tsuruoka and T. Nakamori, Jpn. J. Appl. Phys. **27** L1473 (1988).
2. K. Terashima, K. Eguchi, T. Yoshida and K Akashi, Appl. Phys. Lett. **52**, 1274 (1988).
3. X. D. Wu, R.E. Muenchausen, S. Foltyn, R.C. Estler, R. C. Dye,

C. Flamme, N. S. Nogar, A. R. Garcia, J. Martin and J. Tesmer, Appl. Phys. Lett. **56**, 1481 (1990).
4. A. Shah, S. Patel, E. Narumi and D.T. Shaw, Appl. Phys. Lett. **57**, 1452 (1990).
5. A. Ono and Y. Ishizawa, Jpn. J. Appl. Phys. **26** L1043 (1987).

$YBa_2Cu_3O_7$ FILMS GROWN BY METAL COSPUTTERING

Richard N. Steinberg, James D. McCambridge, and Daniel E. Prober
Yale University, New Haven, CT 06520

Bruce M. Guenin
Olin Metals Research Laboratories, New Haven, CT 06511

ABSTRACT

Superconducting $YBa_2Cu_3O_7$ films have been grown *in situ* by simultaneously sputtering from Y, BaCu, and Cu targets. One advantage of such metal cosputtering is the higher deposition rate compared to oxide target sputtering. Another advantage is the ability to control the individual element rates to vary composition or to substitute for any of the metals without interrupting film growth and without making additional composite targets. To prevent film damage due to oxygen ion bombardment during film growth which was observed when the sputter guns faced the substrate (on-axis sputtering), an off-axis geometry was used. One disadvantage we found with *in situ* metal cosputtering was that reproducibility of stoichiometries was difficult because of the presence of oxygen at the targets. To minimize the oxygen partial pressure at the targets during sputtering, the chamber was differentially pumped. Films grown in the off-axis geometry with a substrate temperature near 700 °C, a chamber pressure of 7.5 mT, and an O_2:Ar flow ratio of 1:50 had zero resistance at 85 K. Results for on-axis, composite target magnetron sputtering with a high-strength magnet are also presented. These results are not promising.

INTRODUCTION

Several processes have been successfully used to grow thin films of the high temperature superconductor $YBa_2Cu_3O_7$ (YBCO). They include composite target sputtering,[1,2] pulsed laser deposition,[3,4] and several multisource techniques[5-12] discussed below. In this study, we have grown superconducting YBCO films *in situ* by simultaneously sputtering from three metal targets (Y, BaCu, and Cu) with a heated substrate in a partial pressure of oxygen. Films grown *in situ* are preferred over films oxygenated after deposition.[13]

Metal cosputtering in Ar has been a successful technique for depositing multi-element films. High deposition rate with good control of composition is possible. However, for *in situ* YBCO deposition, the presence of high temperature and oxygen adds complications. These complications include oxygen ion bombardment of the growing film[14] and sensitivity of individual gun rates to deposition parameters. The sensitivity of gun rates is due to the fact that if a target starts to oxidize, its deposition rate changes. We have addressed these complications by differentially pumping our deposition chamber and pointing our sputter guns away from the substrate.

Codeposition of YBCO films can be compared to deposition from a single source. One advantage of codeposition is a much higher deposition rate if metal sources are used. Our deposition rate for metal cosputtering was about 100 Å/min, which is more than an order of magnitude higher than we obtained with sputtering a single composite target. A second advantage of codeposition is its flexibility. Codeposition readily allows for controlling film stoichiometry by adjusting the relative rates of each element. Also, any of the elements can be easily substituted. These changes can be made without interrupting film growth; one can either change individual rates or use several shuttered sources. The main disadvantages of codeposition are the difficulty of achieving the high pressure of oxygen at the growing film needed for *in situ* growth, and any process complications that may result from the presence of oxygen. For our technique of metal cosputtering, a higher oxygen pressure leads to undesirable target oxidation. For e-beam coevaporation, the electron beam guns may not operate well at a high pressure of oxygen. Another disadvantage of codeposition, and in particular of metal cosputtering, is the strong interdependence of many variables, as will be discussed later. This makes it difficult to achieve reproducible stoichiometries. Several groups have been able to grow superconducting YBCO films by codeposition. Most demonstrated codeposition techniques involve a post-deposition oxygen anneal.[5-7] Demonstrated *in situ* codeposition techniques include coevaporation in oxygen[8-11] ($T_{R=0}$ = 70 K to 90 K), and cosputtering from oxide targets in both an on-axis and off-axis geometry[12] ($T_{R=0}$ = 84 K for off-axis).

YBCO FILMS GROWN BY METAL COSPUTTERING

YBCO films were deposited by simultaneously sputtering from an Y metal target, a BaCu alloy target, and a Cu target onto a heated substrate in the presence of oxygen. Torus 2C magnetron sputter guns were used.[15] The BaCu alloy was chosen because it does not oxidize as readily as pure Ba.[16] A schematic of our deposition system is shown in Fig. 1.

Films were cosputtered in Ar-O_2 at a high temperature onto a rotating stage. For each run, the chamber was brought to between 6 and 8 mT total pressure. Argon was fed into the sputter gun region at a rate of 40 - 90 sccm and oxygen was introduced at the substrate at a rate of 1 - 3 percent of the Ar flow. We found that if a metal target oxidized while sputtering, its rate, and hence the film stoichiometry, became difficult to control. For this reason, the substrate region was pumped separately from the sputter gun region. Deposition was through a grid with roughly 6 square centimeters of total open area. The substrates ($SrTiO_3$ or MgO), were mounted with Ag paste to a 5-centimeter Haynes alloy[17] pallet, and were heated to 700 °C. This was the estimated temperature of the pallet. The sputter guns were about 15 cm from the substrate.

Each deposition run began with about a 20 minute presputter to stabilize conditions and to sputter through the oxide on each target surface. We judged whether we had sputtered through the oxides by waiting for the gun voltages to stabilize for a fixed power and for the plasmas to turn the appropriate colors. In particular, the BaCu gun had a purple plasma when the target surface was oxidized and a bright green plasma when it was not. BaCu oxidizes the most easily of the three targets. With the above conditions, the color of the BaCu gun plasma remained green during the deposition, indicating that the target was not oxidized. We chose an input power high enough that the BaCu target did not oxidize, yet low enough that the target did not melt; the BaCu gun power was kept between 125 and 150 W (rf). The Y and the Cu were sputtered at powers between 30 and 75 W (dc). These two values were adjusted according to measurements of film stoichiometry.

Film stoichiometries were measured by the inductively coupled plasma emission spectroscopy (ICP) technique at Olin Metals Research Laboratories. Composition measurements were accurate to better than 5 percent. Rutherford Backscattering Spectroscopy (RBS) performed at University of Connecticut at Storrs produced results in reasonable agreement with ICP measurements. The ICP measurements were carefully calibrated and thus are quoted here. To improve on uniformity, the substrate pallet was rotated during deposition. After a run, the substrates were cooled over several hours in ≈10 Torr of oxygen.

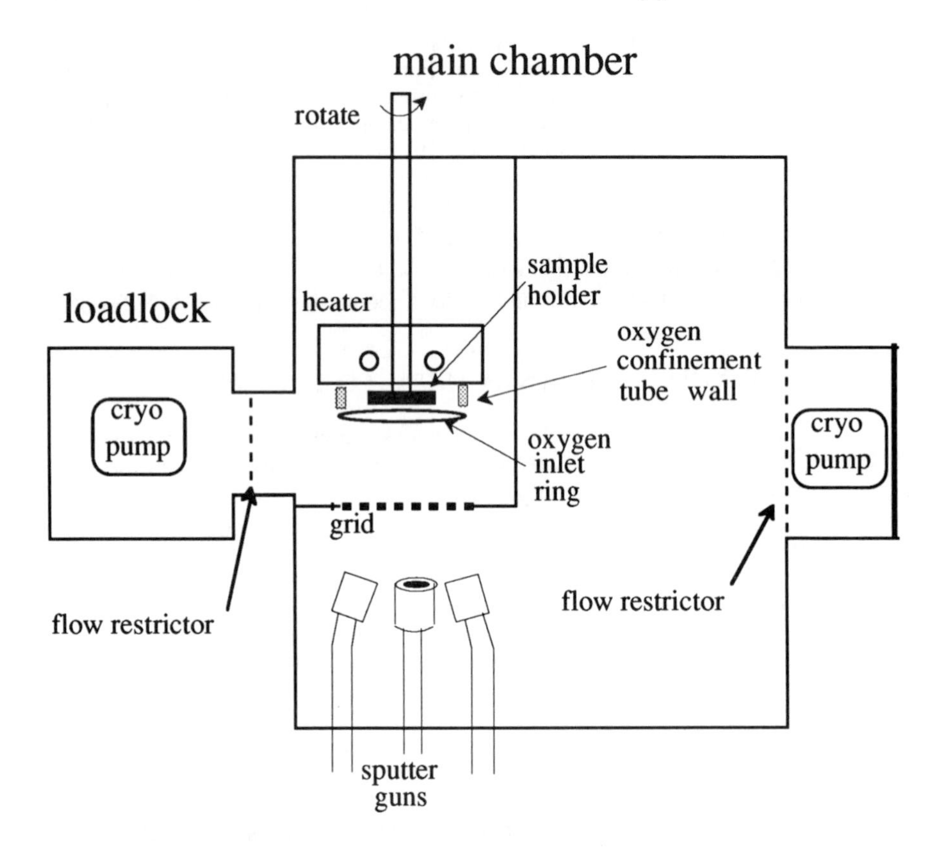

Figure 1. Deposition system schematic. Three targets - Y, Cu, and a BaCu alloy, were used. This figure shows the on-axis geometry, with sources pointing at the substrate. Off- axis sources, pointing away from the substrate, were used to obtain better film quality.

The concentration of O_2 in the substrate region was considerably higher than in the sputter gun region. During a run, the total pressure was measured to be roughly 3 mT in the substrate region when the sputter gun region pressure was set to 7 mT. Based on these pressures and the measured flow rates, the percentage of oxygen in the substrate region was calculated to be roughly 9 percent. This corresponds to a partial pressure of oxygen of 0.3 mT. The actual oxygen pressure at the substrate was higher because the oxygen flow was directed at the substrate, and the region between this inlet ring and the heater box was surrounded by a 10-cm diameter, 2-cm tall confinement tube, as shown in Fig. 1. The percentage of oxygen in the sputter gun region was estimated to be 0.7 percent or 0.05 mT.

In the on-axis geometry, resistive transitions of films near the correct stoichiometry were broad and had low zero-resistance temperatures, near 60 K. Because of these poor results for T_c and the suspicion of film damage due to oxygen ion bombardment in this geometry, we switched to an off-axis geometry. This was achieved by rotating the guns illustrated in Fig. 1 such that they pointed several centimeters away from the substrate.

RESULTS AND DISCUSSION - OFF AXIS GEOMETRY

Results for T_c were significantly better for films grown in the off-axis geometry. Fig. 2 shows the resistance as a function of temperature for a good film grown in the off-axis geometry and for a good film grown in the on-axis geometry. These two films were grown in very similar deposition conditions (roughly 700 °C, 7 mT total pressure, 2.5% oxygen flow rate) and have similar Y:Ba:Cu ratios (1.0:1.9:3.3 for the film grown in the on-axis geometry and 1.0:1.8:3.3 for the film grown in the off-axis geometry). However, in switching to the off-axis geometry, T_c for $R = 0$ improved from 57 K to 85 K, the resistivity ratio improved from 1.0 to 1.8, and the normal state resistivity just above the transition was lower by a factor of 4. Unfortunately, reproducibility of composition was a problem for the off-axis geometry as well as the on-axis geometry, as discussed below.

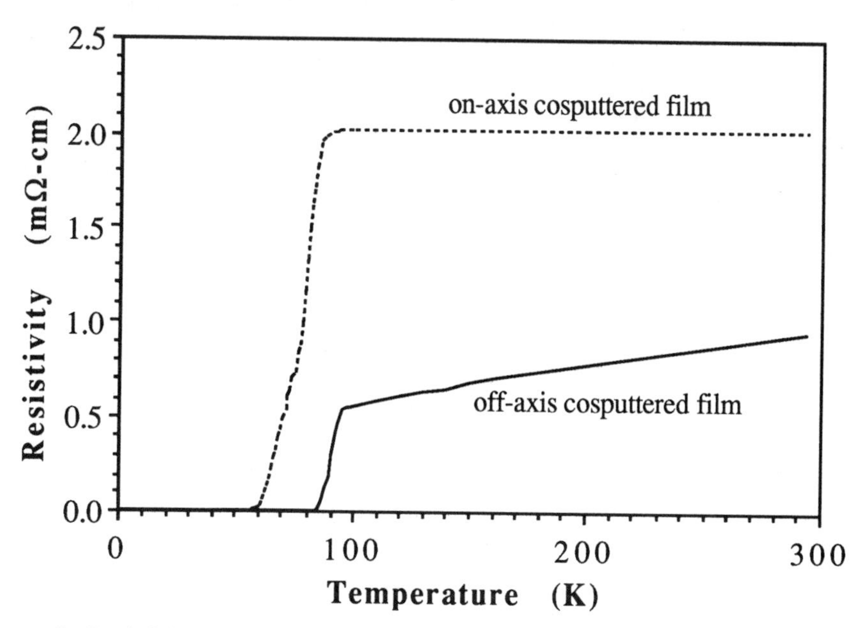

Figure 2. Resistivity vs. temperature for films grown in the 2 different deposition geometries, compositions close to 1:2:3 (see text).

Our proposed explanation of why film properties improved in switching from the on-axis to the off-axis geometry is as follows. The effect of pointing the sputter guns away from the substrate was to lessen the effects of ion bombardment of the growing film. This bombardment preferentially resputtered the Ba, changing the stoichiometry of the film.[14] We were able to compensate for this Ba resputtering by adjusting sputter gun powers, and could grow films of nearly correct stoichiometry in each geometry. The electrical properties of nearly stoichiometric on-axis films were not as good as for off-axis films. We thus conclude that there was damage done to the film in the on-axis geometry. A similar observation of damage to sputtered films has been reported by Enami *et al.*;[18] this is similar to the lattice damage done by ion bombardment.[19,20]

Compared to other techniques,[1-4, 8-12] the oxygen partial pressure, $P(O_2)$, was small for our technique. Evidence of low $P(O_2)$ at the substrate during deposition was seen in the fact that we observed an improvement in T_c for Ba deficient films. An on-axis film with an Y:Ba:Cu ratio of 1.0:1.1:3.2 had T_c = 79 K (compared to T_c = 57 K for the on-axis film in Fig. 2 grown in the same deposition conditions but with cation ratio of 1.0:1.9:3.3). A comprehensive study of coevaporated films grown at low oxygen pressures (0.2-10 mT) was reported by Matijasevic *et al.*[21] In that study they also found that stoichiometric films have reduced T_c values, and that Ba-deficient films have higher values of T_c. They attribute this effect to an oxygen-deficiency-induced cation disordering. We note that our films were deposited near T = 700 °C and in $P(O_2) < 10$ mT, yet are superconducting. This is a thermodynamically unstable region of $P(O_2)$ - T phase space for YBCO.[21,22] Sputter deposition is, of course, not an equilibrium process.

With our cosputtering technique, the film composition depended quite sensitively on deposition parameters, especially when compared to other techniques. To illustrate this, Fig. 3 shows the measured Y:Ba composition ratio as a function of Y gun power for a fixed BaCu sputter gun power. (These results are from off-axis cosputtering but are indicative of both geometries.) Even though the deposition conditions were changed only slightly, the dependence of Y content on Y gun power changed dramatically. We also found that the appearance of the plasma over the BaCu source changed somewhat with changes in substrate temperature and in oxygen flow.

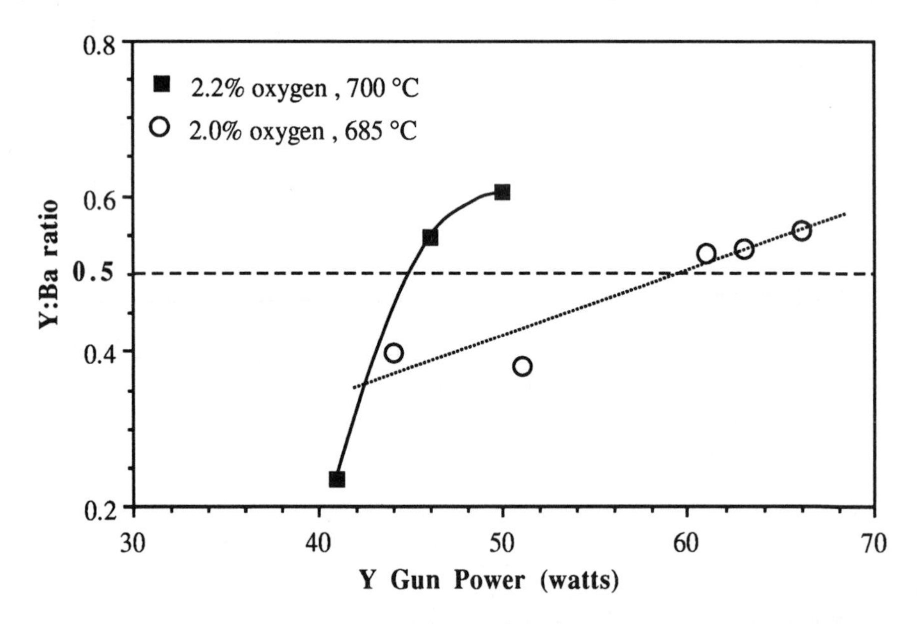

Figure 3. The Y:Ba composition ratio as a function of Y gun power for a fixed BaCu gun power = 140 W.

MAGNETRON SPUTTERING WITH A HIGH STRENGTH MAGNET

An alternate approach was attempted for alleviating the problem of resputtering and ion bombardment of the growing film. Migliuolo *et al.* reported[23] that stoichiometric YBCO films can be deposited by on-axis, composite target sputtering if a high strength magnet is used in the magnetron sputter gun instead of a conventional strength magnet. They measured deposition rate and stoichiometry as a function of substrate position relative to the magnetic field, and reported that using a high strength magnet resulted in little or no resputtering. They attributed this effect to lower operating voltages, and hence lower oxygen ion energy.

Our results differ from those of Migliuolo *et al.* We found little improvement in film composition with use of the stronger magnet. Films were still Ba deficient. For films grown with the high strength magnet in 20 mT Ar at room temperature, the Ba deficiency was larger at the greater target-to-substrate distance (see Table I). We expect the deposition rate to fall off faster with distance than the oxygen ion bombardment rate, which is more directional. Therefore, the smaller Ba content at the greater distance is indicative of the greater importance of resputtering at this distance. In addition, when the deposition was done at an elevated temperature (roughly 600° C), the Cu content was greatly reduced (Table I). These results are not encouraging for the use of on-axis composite target sputtering of high temperature superconductors, particularly onto heated substrates.

Table I. Stoichiometries of films sputtered with high strength magnet.

Target-Substrate Distance	Substrate Temperature	Y : Ba : Cu ratio
9 cm	ambient	1.0 : 1.3 : 3.5
15 cm	ambient	1.0 : 0.7 : 3.4
15 cm	600 °C	1.0 : 0.6 : 0.1

CONCLUSIONS

We have grown superconducting YBCO films by off-axis metal cosputtering. Results for T_c compare well with other codeposition techniques, but are not as good as those obtained with a composite target. In the on-axis geometry, we have observed film degradation caused by oxygen ion bombardment, even for stoichiometric films. This was eliminated by changing to an off-axis geometry. Further improvement in both film properties and reproducibility should be possible with improvements in the ability to differentially pump. This would allow a higher oxygen pressure at the substrate and a lower oxygen pressure at the sputter guns.

ACKNOWLEDGEMENTS

We thank Mr. Ed Cooney of Olin Metals Research Laboratories for measuring film stoichiometry and for useful suggestions. We also thank Dr. Michele Migliuolo of Kurt J. Lesker Co. for use of a high strength magnet and discussions on the high strength magnet work, Professors Joeseph Budnick and Fred Otter at University of Connecticut for RBS measurements, and Dr. Dean Face of DuPont for RBS analyses and many useful suggestions. This research was supported by Olin Corporation, and by the CT Department of Higher Education under grants 89-303, 89-315, and 91-401.

REFERENCES

1. C. B. Eom, J. Z. Sun, K. Yamamoto, A. F. Marshall, K. E. Luther, T. H. Geballe, and S. S. Laderman, Appl. Phys. Lett. **55**, 595 (1989).
2. H. C. Li, G. Linker, F. Ratzel, R. Smithey, and J. Geerk, Appl. Phys. Lett. **52**, 1098 (1988).
3. C. C. Chang, X. D. Wu, A. Inam, D. M. Hwang, T. Venkatesan, P. Barboux, and J. M. Tarascon, Appl. Phys. Lett. **53**, 517 (1988).
4. S. Witanachchi, H. S. Kwok, X. W. Wang, and D. T. Shaw, Appl. Phys. Lett. **53**, 234 (1988).
5. P. Chaudhari, R. H. Koch, R. B. Laibowitz, T. R. McGuire, and R. J. Gambino, Phys. Rev. Lett. **58**, 2684 (1987).
6. B. Oh, M. Naito, S. Arnason, P. Rosenthal, R. Barton, M. R. Beasley, T. H. Geballe, R. H. Hammond, and A. Kapitulnik, Appl. Phys. Lett. **51**, 852 (1987).
7. K. Char, A. D. Kent, A. Kapitulnik, M. R. Beasley, and T. H. Geballe, Appl. Phys. Lett. **51**, 1370 (1987).
8. R. M. Silver, A. B. Berezin, M. Wendman, and A. L. deLozanne, Appl. Phys. Lett. **52**, 2174 (1988).
9. D. K. Lathrop, S. E. Russek, and R. A. Buhrman, Appl. Phys. Lett. **51**, 1554 (1987).
10. T. Terashima, K. Iijima, K. Yamamoto, Y. Bando, and H. Mazaki, Jpn. J. App. Phys. **27**, L91 (1988).
11. N. Missert, R. H. Hammond, J. E. Mooij, V. Matjasevic, P. Rosenthal, T. H. Geballe, A. Kapitulnik, M. R. Beasley, S. S. Laderman, C. Lu, E. Garwin, and R. Barton, IEEE Trans. Magn. **25**, 2418 (1989).
12. N. Akutsu, M. Fukutomi, K. Katoh, H. Takahara, Y. Tanaka, T. Asano, and H. Maeda, Jpn. J. App. Phys. **29**, L604 (1990).
13. M. R. Beasley, Proc. IEEE **77**, 1155 (1989).
14. S. M. Rossnagel and J. J. Cuomo, AIP Conf. Proc. **165**, 106 (1988).
15. Kurt J. Lesker Company, Clairton, PA.
16. M. Scheuermann, C. C. Chi, C. C. Tsuei, D. S. Yee, J. J. Cuomo, R. B. Laibowitz, R. H. Koch, B. Braren, R. Srinivasan, and M. M. Plechaty, Appl. Phys. Lett. **51**, 1951 (1987).
17. Haynes International, Inc., Kokomo, IN.
18. H. Enami, T. Shinohara, N. Kawahara, S. Kawabata, H. Hoshizaki, and T. Imura, Jpn. J. App. Phys. **29**, L782 (1990).
19. G. J. Clark, F. K. LeGoues, A. D. Marwick, R. B. Laibowitz, and R. Koch, Appl. Phys. Lett. **51**, 1462 (1987).
20. J. M. Valles, Jr., A. E. White, K. T. Short, R. C. Dynes, J. P. Garno, A. F. J. Levi, M. Anzlowar, and K. Baldwin, Phys. Rev. B **39**, 11599 (1989).
21. V. Matijasevic, P. Rosenthal, K. Shinohara, A. F. Marshall, R. H. Hammond, and M. R. Beasley, J. Mater. Res. **6**, 682 (1991).
22. R. Bormann and J. Nolting, Appl. Phys. Lett. **54**, 2148 (1989).
23. M. Migliuolo, R. M. Belan, and J. A. Brewer, Appl. Phys. Lett. **56**, 2572 (1990).

SUPERCONDUCTING TlBaCaCuO THIN FILMS FROM BaCaCuO PRECURSORS

H.M. Duan, B. Dlugosch, and A.M. Hermann
Department of Physics
University of Colorado, Boulder, CO. 80309
X.W. Wang, J. Hao, and R.L. Snyder
Institute of Ceramic Superconductivity
New York College of Ceramics at Alfred University
Alfred, NY 14802

Abstract

We report new results of a two-step process for making superconducting TlBaCaCuO thin films. The first step is to make a precursor film with the stoichiometry of $Ba_2Ca_2Cu_3O_7$ in an ambient atmosphere environment, using an RF plasma deposition technique. The second step is to thalliate (i.e. insert thallium) into the precursor by heating it in the presence of a thalliation agent, forming the $Tl_2Ba_2Ca_2Cu_3O_{10+\delta}$ phase. The zero resistance temperature of one such film is 108K.

1 Introduction

Among all known high T_c superconductors, the $Tl_2Ba_2Ca_2Cu_3O_{10+\delta}$ superconductor has the highest transition temperature reported to-date[1-3]. There are only two methods which have been reported for fabricating TlBaCaCuO superconducting films. The first method involves two steps: (1) prepare a TlBaCaCuO film in a vacuum chamber by evaporation or sputtering techniques; (2)

post-anneal the film in a sealed furnace in the presence of oxygen or air[4-10]. The second method also involves two steps: (1) prepare a BaCaCuO precursor film in a vacuum chamber; (2) thalliate the precursor with a thallium vapor releasing material like thallium oxide, TlBaCuO, or TlBaCaCuO[11-13]. When a film is prepared in a vacuum chamber, the film size is limited. Thus, these vacuum technologies are not suitable for film coatings on large surface-areas. In order to develop a non-vacuum technique, a collaborative agreement was reached between University of Colorado and Alfred University during the Annual NYSIS Conference in 1990[14-15]. Since then, we have pursued the merging of two technologies developed at each institution. The team at University of Colorado had developed procedures for introducing thallium into ternary precursor films to produce TlBaCaCuO superconducting phases[11,13-24]. The team at Alfred University had developed an RF plasma technique which can lay down films onto a large substrate in an atmospheric enviornment[15-17]. Here, we report some preliminary results of the collaborative work.

2 Experimental Procedure

2.1 Plasma Evaporation of BaCaCuO Precursor Films

A plasma evaporation setup consists of: an RF plasma reactor[18], an ultrasonic nebulizer[19], and a substrate assembly. Starting material was an aqueous solution of ion in the stoichiometric ratio of Ba:Ca:Cu, 2:2:3. In preparing such solution, appropriate amount of reagent grade $Ba(NO_3)_2$, $Ca(NO_3)_2 \cdot 4H_2O$ and $Cu(NO_3)_2 \cdot 3H_20$ were dissolved in distilled water with the concentration held to 100 total grams per liter or less. The solution was placed into a plastic mist chamber which was inserted into an ultrasionic nebulizer. Under the pressure of a carrier gas, the solution, now in the from of an aerosol mist generated in the mist chamber, is carried into the plasma zone of the RF plasma reactor. Ions resulting from the decomposition of aerosol droplets in the reactor, are then deposited onto a single crystal MgO (100) substrate with a substrate temperature of 600°C . The

Table 1: Plasma conditions.

Gas	Flow Rate
Primary plasma gas: Ar	10 liter/min
Plasma sheath gas: Ar	20 liter/min
O_2	10 liter/min
Mist carrier gas: Ar	0.9 liter/min

distance between the substrate and the bottom of the torch was approximately 12 cm. Gas flow rates are listed in Table 1.

As-deposited films were black and film thicknesses ranged from one to 30 μm. Energy dispersive X-ray analysis (EDAX) revealed that all three metal elements are present in the films. X-ray diffraction measurements indicated the co-existence of several phases including $BaCO_3$ and CuO.

2.2 Thalliation of Precursor Films

A thalliation set up is illustrated in Fig. 2. The system consists of a quartz tube, a furnace, an oxygen supply, and a safety Tl vapor transport system.

After the temperature inside the quartz tube reached thalliation temperature, a sample assembly on an alumina crucible (not shown) was inserted into the tube.

The sample assembly consists of a TlBaCaCuO pellet in the stoichiometric ratio of 2:2:2:3, a copper ring, and the BaCaCuO precursor film (facing down). The pellet was press-formed from a mixture of an appropriate amount of Tl_2O_3 powder and reacted (at 950°C for 48 to 72 hours) BaCaCuO powder in the stoichiometric ratio of 2:2:3. Since the pressed TlBaCaCuO pellet is to be used as a Tl vapor source it is not heat treated before the thalliation process.

The thalliation temperature was 800-900°C , and thalliation time was 2-

5 minutes. After this step, the temperature of the sample assembly was lowered to 600°C at a rate of 3-10 °C /min. The film was then annealed for 30 minutes at this temperature. After the annealing process, the furnace was then cooled to room temperature. During the whole heat treatment process, a continuous oxygen flow was provided.

3 Results

All films treated by this process were superconductive with an onset temperature of approximately 125K. However, the zero resistance temperature, T_o, of each film depends on the thalliation temperature, T_t, when other experimental conditions are held constant. For a film with T_t = 800°C , T_o = 89K as shown in Fig. 3(a). For another film with T_t increased to 890°C , T_o correspondingly increased to 108K as shown in Fig. 3(b).

X-ray diffraction studies show that phase formation also depends on T_t. Figure 4 shows a diffraction pattern of a film with T_t = 800°C . The single superconductive phase is $Tl_2Ba_2Ca_1Cu_2O_{8+\delta}$ with certain 00L orientation. Figure 5 shows the powder pattern of another film with T_t = 890°C . The dominant superconductive phase is $Tl_2Ba_2Ca_2Cu_3O_{10+\delta}$, and the minor phase is $Tl_2Ba_2Ca_1Cu_2O_{8+\delta}$, with 00L orientations. In Figs. 5-6, peaks located at 59-60 degrees are not identified.

4 Discussion

The two-step process discussed here has produced superconductive Tl-BaCaCuO films with T_o of 108K. Currently, we are working on the optimization of heat treatment conditions to increase T_o. From other experiments carried out at the University of Colorado, we observed that the 2212 phase can be formed at thallination temperatures higher than 780°C , and the 2223 phase may be formed at temperatures higher than 850°C , but lower than the melting temperature of $Ba_2Ca_2Cu_3O_7$ (936°C , bulk). This ongoing study may allow us to better understand the 2223 phase formation conditions in the films produced by this technique, and to develop high T_c film coating on large areas.

5 Acknowledgement

The work at Alfred University was partially supported by the Center for Advanced Ceramic Technology, and New York State Science and Technology Foundation. The authors at Alfred would also acknowledge S. Bayya and M. Rodriguez for their help.

6 References

1. For review articles, see, for example, A.W. Sleight, Science 242, 1519 (1988).

2. Z.Z. Sheng, and A.M. Hermann, Nature 332, 138 (1988).

3. Z.Z. Sheng, W. Kiehl, J. Bennett, A.EI Ali, D. March, G.D. Mooney, F. Arammash, J. Smith, D. Viar, and A.M. Hermann, Appl. Phys. Lett. 52, 1738, (1988).

4. S.H. Liou, K.D. Aylesworth, N.J. Lanno, B. Johs, D. Thompson, D. Meyer, J.A. Woolam, and C. Barry, Appl. Phys. Lett. 54, 760 (1989).

5. G. Subramanyam, F. Radpour, and V.J. Kapoor, Appl. Phys. 56, 1799(1990).

6. Y. Ichikawa, H. Adachi, K. Setsune, S. Hatta, K. Hirochi, and K. Wasa, Appl. Phys. Lett. 53, 919 (1988) .

7. D.S. Ginley, J.F. Kwak, R.P. Hellmer, R.J. Baughman, E.L. Venturini, and B. Morosin, Appl. Phys. Lett. 53, 406 (1988).

8. W.Y. Lee, V.Y. Lee, J. Salem, T.C. Huang, R. Savoy, D.C. Bullock and S.S.P. Parkin, Appl. Phys. Lett. 53, 329 (1988).

9. M. Hong, S.H. Liou, D.D. Bacon, G.S. Grader, J. Kwo, A.R. Kortan and B.A. Davidson, Appl. Phys. Lett. 53, 2102 (1988).

10. M. Nakao, R. Yuasa, M. Nemoto, H. Kuwahara, H. Mukaida and A. Mizukami, Jpn. J. Appl. Phys. 27, L849 (1988).

11. B. Johs, D. Thompson, N.J. Ianno, J.A. Wollam, S.H. Liou, A.M. Hermann, Z.Z. Sheng, W. Kiehl, Q. Shams, X. Fei, L. Sheng, and Y.H. Liu, Appl. Phys. Lett. 54, 1810 (1989).

12. S.I. Shah, N. Herron, C.R. Fincher, and W.L. Holstein, Appl. Phys. Lett. 56, 782 (1990).

13. D.G. Naugle, P.S. Wang, X.Y. Shao, and A.M. Hermann, J. Appl. Phys. 68, 1399 (1990).

14. A.M. Hermann, H.M. Duan, W. Kiehl, S.Y. Lin, L. Lu, and D.L. Zhang, in "Superconductivity and Its Applications" eds. Y.H. Kao, P. Coppens, and H.S. Kwok, Am. Inst. of phys., New York, (1991).

15. H.H. Zhong, X.W. Wang, J. Hao, R.L. Snyder, in "Superconductivity and Its Applications" eds. Y.H. Kao, P. Coppens, and H.S. Kwok, Am. Inst. of phys., New York, (1991).

16. X.W. Wang, H.H. Zhong, and R.L. Snyder, Appl. Phys. Lett. 57, 1581 (1990).

17. X.W. Wang, H.R. Zhong, and R.L. Snyder, in "Science and Technology Of Thin Film Superconductors 2", eds. R.B. McConnell, and R. Noufi, Plenum, New York, (1990).

18. Lepel-Tafa Model 56 plasma reactor, Tafa, Inc., Concord, New Hampshire, 4 MHz, 30kw.

19. Ultra-Neb 99 Ultrasonic Nebulizer, BeVilbiss Health Care Inc. Somerset, Pennsylvania. 1.63 MHz, 70w.

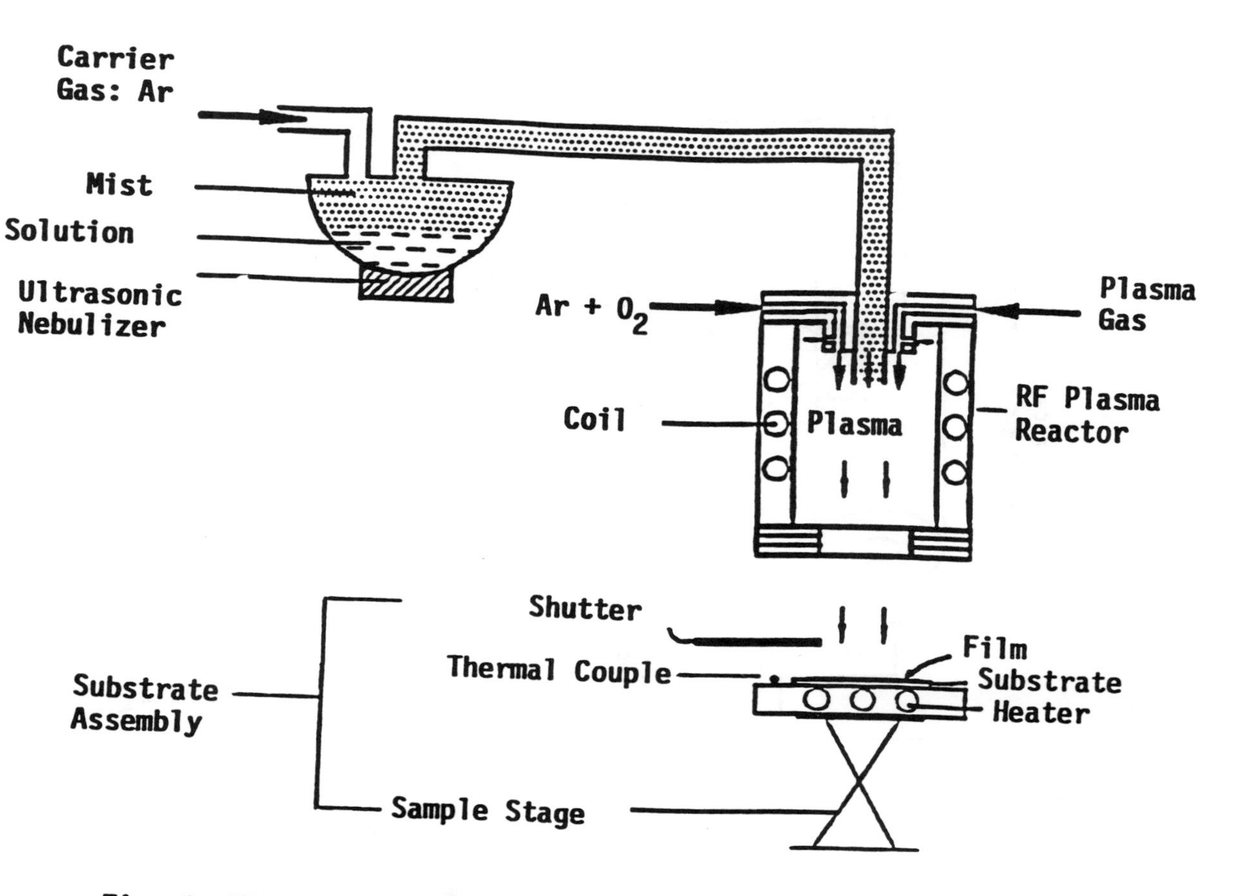

Fig. 1. Plasma Evaporation Set Up

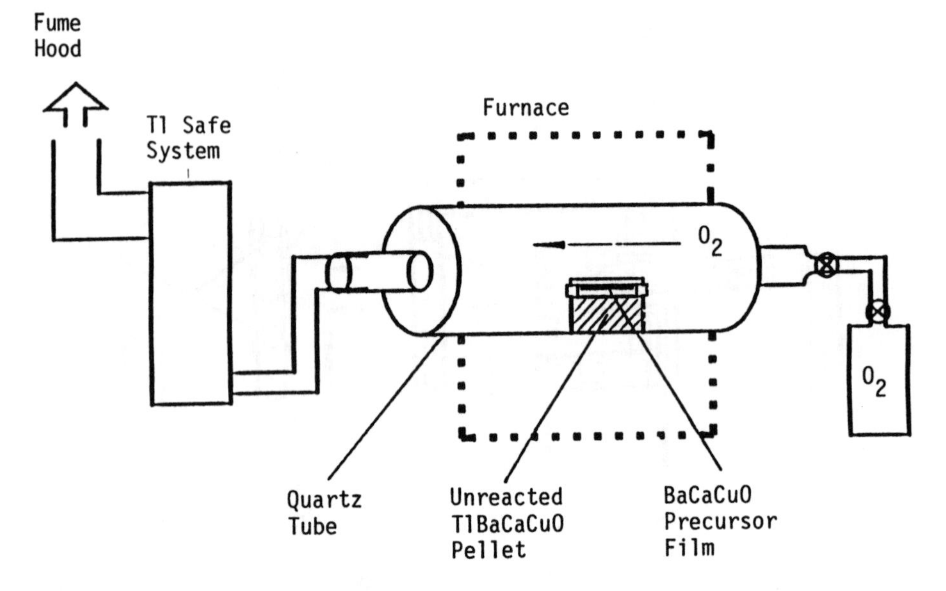

Fig. 2. Thallination Set Up

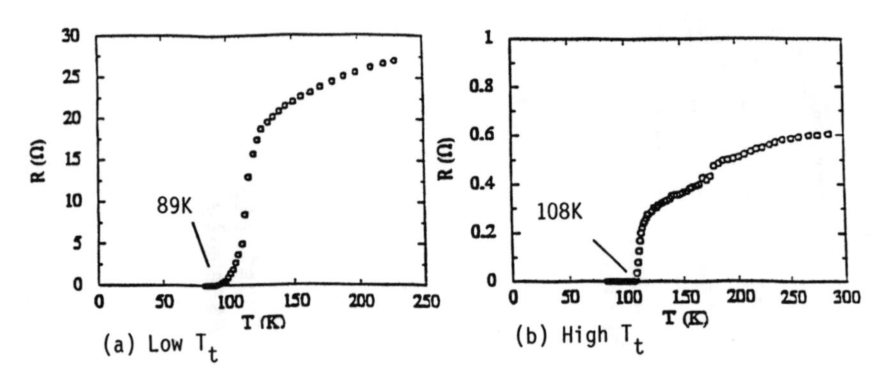

Fig. 3. Resistance vs. Temperature

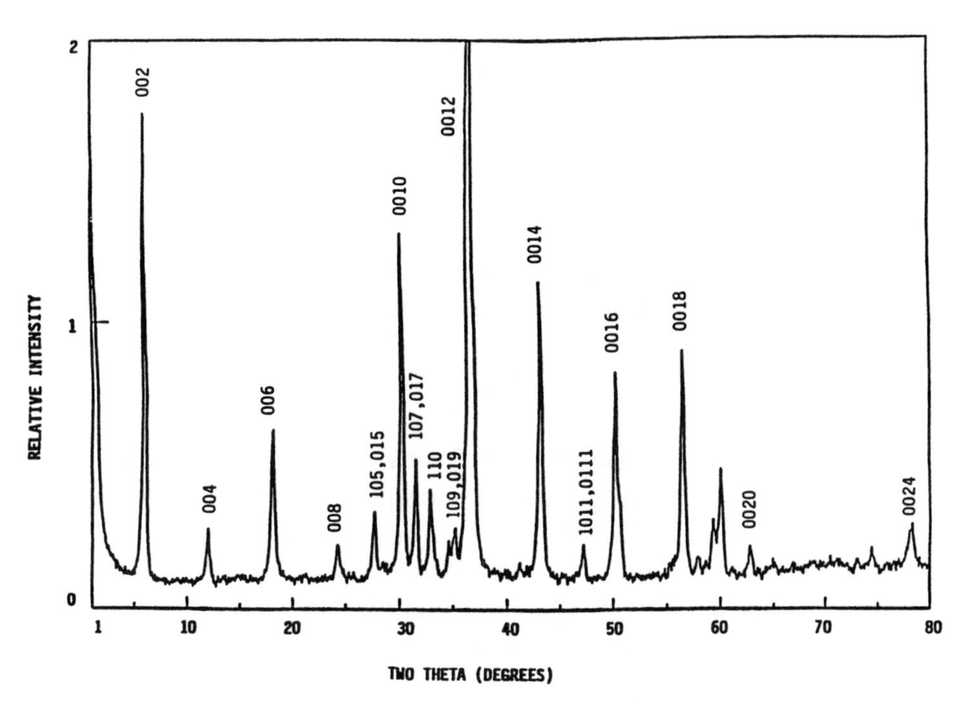

Fig. 4. X-ray Diffraction Pattern (Low T_t)

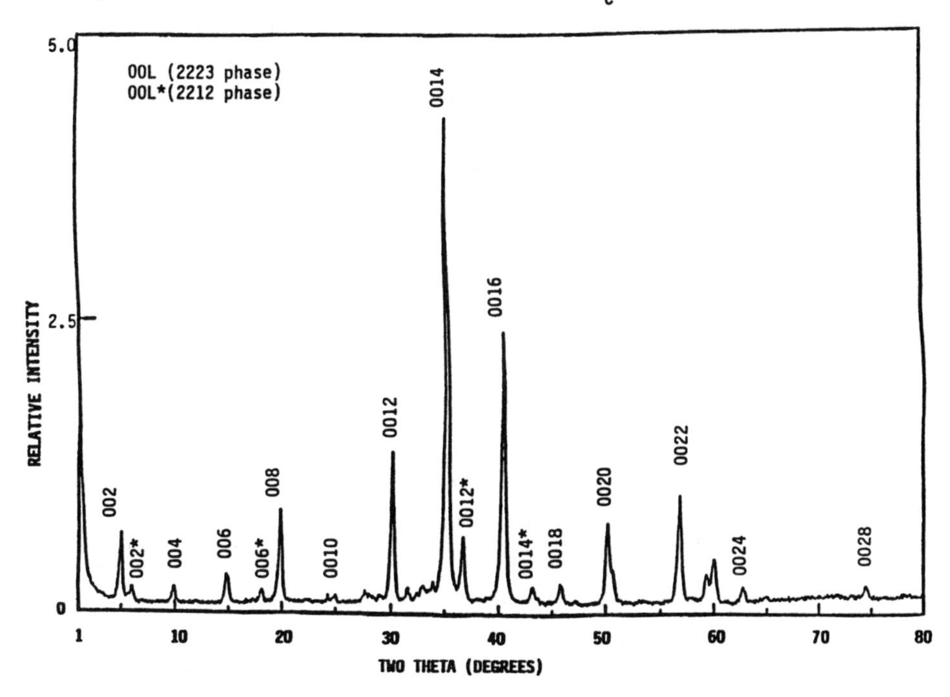

Fig. 5. X-ray Diffraction Pattern (High T_t)

EFFECTS OF PROCESSING PARAMETERS ON LASER ABLATED THIN FILMS OF $Y_1Ba_2Cu_3O_{7-x}$

S. Mahajan, R. L. Cappelletti and D. C. Ingram

Department of Physics and Astronomy
Condensed Matter and Surface Sciences Program
Ohio University, Athens, OH 45701-2979

ABSTRACT

The effects of processing parameters such as substrate temperature, oxygen pressure, laser repetition rate and substrate crystallinity on the properties of resulting thin films of $Y_1Ba_2Cu_3O_{7-x}$ (YBC) are presented. The films were prepared on single crystal $SrTiO_3$ (100) substrates using the technique of laser ablation. Depending upon the deposition conditions, it was found that the resulting films had either their c-axis or a-axis oriented normal to the plane of the substrate. Some films had a mixture of a and c axis grains so oriented. Critical currents of these films were determined from vibrating sample magnetometry and their transition temperatures were determined using the conventional four probe resistance technique. X-ray diffraction analysis was used to study the growth conditions of the films and to determine the oxygen content of the samples.

INTRODUCTION

The technique of laser ablation has been used by several researchers[1-5] to obtain high quality thin films of $Y_1Ba_2Cu_3O_{7-x}$ (YBC). The technique is simple and quite versatile. It provides an efficient way of fabricating *in situ* superconducting films of complex stoichiometric compounds like the new high T_c materials. There are several processing parameters involved which have to be optimized before achieving the desired properties. High quality YBC films need to have a correct stoichiometry, appropriate crystal structure and adequate oxygen content. Each of these parameters can be controlled *in situ* without the need for removal and further treatment of the sample. The deposition is performed at elevated temperatures followed by a subsequent cooling period.

We have performed several depositions to investigate the effects of substrate temperature and crystallinity, oxygen pressure and laser repetition rate on the deposition process. Depositions were performed at substrate temperatures ranging from 625-850° C. Oxygen pressure was varied between 50-200 mTorr while the laser repetition rate was kept between 2-4 Hz. It was found that films fabricated at such oxygen pressures were superconducting. Higher temperatures favored c-axis orientation while low temperature growth of film resulted in a-axis being oriented normal to the plane of the substrate. At higher temperatures the orientation of the film also depended upon the deposition time and the repetition rate. At such temperatures it was found that films with thickness less than 4000 Å were c-axis oriented and for thickness greater than 4000 Å, the films had a mixture of a and c axis grains so oriented.

EXPERIMENT

Thin films of YBC were prepared using a KrF (248nm) excimer laser. A stainless steel chamber was specially designed for the purpose. A superconducting pellet of YBC prepared by the conventional sintering process[6] was used as a target for ablation. The substrates used were single crystals of $SrTiO_3$ (100) and were kept 4 cm away from the target. They were glued to an inconel strip using silver paint. This strip was fastened between two copper rods and was heated to high temperatures by resistive heating. The temperature of the substrate was measured using a chromel-alumel thermocouple. The thermocouple was mounted at the back of the substrate. The temperatures were further confirmed by a Leeds and Northrup optical pyrometer and the temperatures reported here are accurate to ± 20° C.

Before the substrate was heated, the pressure in the chamber was brought down to ~ 2×10^{-5} Torr by means of a liquid nitrogen-trapped oil diffusion pump. The effect of base pressure (i.e. pressure before oxygen injection) on the superconducting properties of YBC films has already been studied in our previous publication[4]. The substrate was heated very slowly (approximately one hour from room temperature to deposition temperature) to avoid any thermal shock. The chamber was flushed with oxygen several times before the deposition temperature was reached. At the deposition temperature the chamber was sealed to an oxygen pressure between 50-200 mTorr. The laser was fired at this point at a prefixed repetition rate for a certain time. Due to out-gassing the pressure in the chamber rose during deposition. When it reached 200 mTorr, the chamber was pumped to 25 mTorr and oxygen was admitted bringing the pressure back to between 50-200 mTorr. This process was repeated approximately every ten minutes. After the deposition, oxygen pressure was immediately raised to 100 Torr and the temperature reduction was simultaneously begun bringing it down to 450° C in about 10 minutes. The pressure was then raised to 1 atm of oxygen and the sample was held at this temperature and pressure for one hour. After that, the sample was slowly cooled to room temperature in about 20 minutes. The resulting films were superconducting and no post heat treatment was required.

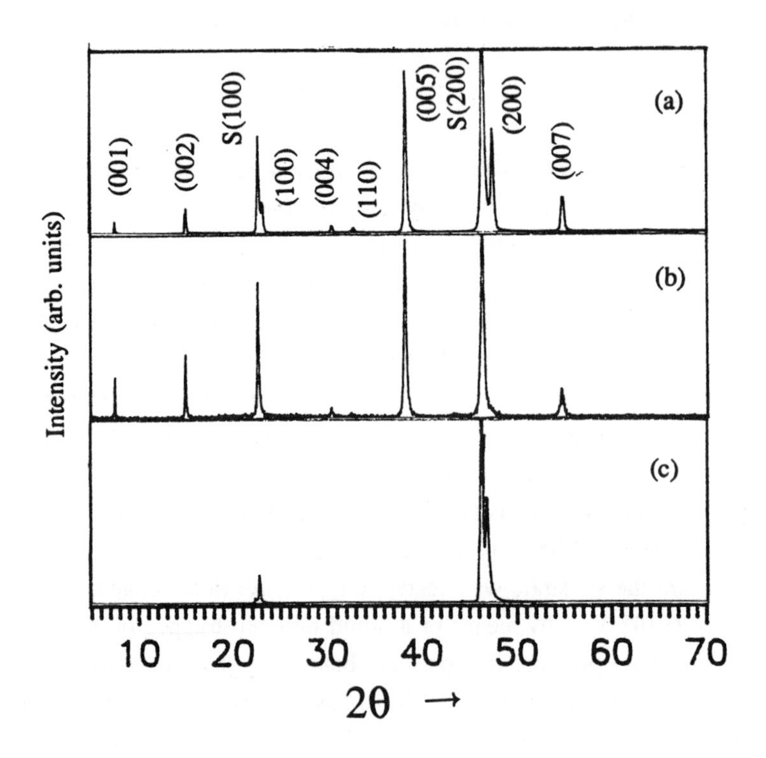

Fig. 1 XRD pattern of a) mixed a+c (Film B1), b) c-axis (Film A2) and c) a-axis (Film E) oriented films. 'S' denotes substrate peaks.

X-ray diffraction (XRD) was employed to determine the orientation and the oxygen content of the as prepared films. Fig. 1 a, b and c are respectively the XRD patterns of a-axis, c-axis and a+c axis oriented films. The composition and thickness was determined to ± 10 % using Rutherford back-scattering (RBS). A 3 Mev rather than a 2 Mev He^{2+} ions beam was found to give better resolution for the oxygen peak and was used to determine the oxygen content of the sample. The transition temperature was measured using the conventional four probe resistance technique. DC magnetization measurements were used to determine the critical currents using a PAR (Model 4500) vibrating sample magnetometer. The surface of the samples was also studied under a scanning electron microscope but the results won't be presented here.

RESULTS

The results of the deposition process are presented in Table 1. From the table it is clearly observed that laser ablation is an effective technique to fabricate thin films as it preserves the stoichiometry of the compound. The composition of all the films was found to be close 1:2:3. Films A1, A2 and B1, B2 were prepared under identical deposition conditions and as expected identical properties are observed for the respective films confirming the reproducibility of the technique. The oxygen pressure during all of the runs was kept between 50-200 mTorr. All films were superconducting after deposition except for films E and F. This will be discussed in a later section. The superconducting properties of the films are tabulated in Table 2.

DISCUSSION

It is clear from Table 1 that the lattice constants of all the films as calculated from the XRD patterns are approximately the same as that of the bulk sample. However, the critical current density of the films in comparison to bulk samples is several orders of magnitude higher. This suggests a different type of defect structure in these films. The critical current in the presence of applied magnetic field is also much higher in these thin film samples than a bulk sample. This indicates that the currents in thin films might not be limited by weak links as in bulk. A low value ($\sim 10^3$ A/cm^2) of critical current density was measured for Film F after post-annealing at 950° C in flowing oxygen. Such temperatures are high enough to cause recrystallization of the film [7,8] and the

Film	T_{sub} ° C	Rep. Rate (Hz)	Thickness (Å)	Compo-sition	a (Å)	b (Å)	c (Å)
A1	720	2.2	2600	1:2:3	-	3.88	11.69
A2	720	2.2	2600	-	-	3.88	11.69
B1	750	4	4200	1:1.8:3	3.83	3.88	11.69
B2	750	4	4200	-	3.83	3.88	11.69
C	720	2.4	-	-	3.83	3.88	11.69
D	675	4	4775	1:1.8:3	3.83	3.88	11.69
E	625	3	5100	1:1.8:2.6	3.87	3.84	11.75
F	850	4	6000	-	Multiphase		
Film F after post annealing at 950° C					3.82	3.90	11.68
Bulk	-	-	-	-	3.82	3.90	11.68

Table 1. Results of the deposition process. All of the films were made at O_2 pressure between 50-200 mTorr. See text for details.

resulting properties could be similar to the bulk samples.

We will now discuss the effects of the processing parameters on the crystal structure and superconducting properties of the prepared films.

Oxygen Content

Following the work of Ref. 9 and 10 we have used the relation $c = 12.771 - 0.1551x$, to determine the oxygen content in our samples. The above relation relates the oxygen content to the c-lattice parameter. Using the calculated value of the c-parameter from Table 1, the oxygen content in samples A through D is ~6.97. RBS measurements were also used to estimate the oxygen content in the samples. A 3.0 Mev He^{2+} ion beam was used for enhanced resonance scattering of oxygen (Fig. 2). This way the oxygen content was determined to be 7 with an accuracy of ± 10% (The accuracy can be increased to 1-2% by comparing the samples to a standard which was not

FIlm	Orientation	T_{co} (K)	T_{ce} (K)	J_c(20K,0Oe) (A/cm^2)	J_c(20K,1T (A/cm^2)
A1	c-axis	89	86.2	8×10^6	2.3×10^6
A2	c-axis	88.3	85.8	4×10^6	1.3×10^6
B1	a+c	91.5	87.5	6×10^6	2.4×10^6
B2	a+c	90.8	88.0	1.5×10^7	5×10^6
C	a-axis	-	-	-	-
D	a+c	88.5	Below 77K	9×10^5	3.4×10^5
E	a-axis	-	-	-	-
F	a+c, after post-annealing	-	-	$\sim10^3$	-
Bulk	Poly-crystalline	92	90	10^3	-

Table 2. Superconducting properties of films prepared under various deposition conditions.

available for the present experiment). For our depositions, 50-200 mTorr of oxygen pressure was found to be adequate to achieve the superconducting properties indicated in Table 2.

Film Thickness

The deposition time for all the films was between 40-50 minutes. The thickness of films prepared by laser ablation can be very conveniently controlled by controlling the repetition rate at which the laser is fired. For films A1 and B1 the deposition time was 41 minutes whereas the rep. rates were respectively 2.2 and 4 Hz. The deposition rate (defined as the total thickness divided by the deposition time) was 1 Å/sec for film A1 and 1.7 Å/sec for film B1 while for both the films ~0.5 Å of the material was ablated with each laser pulse.

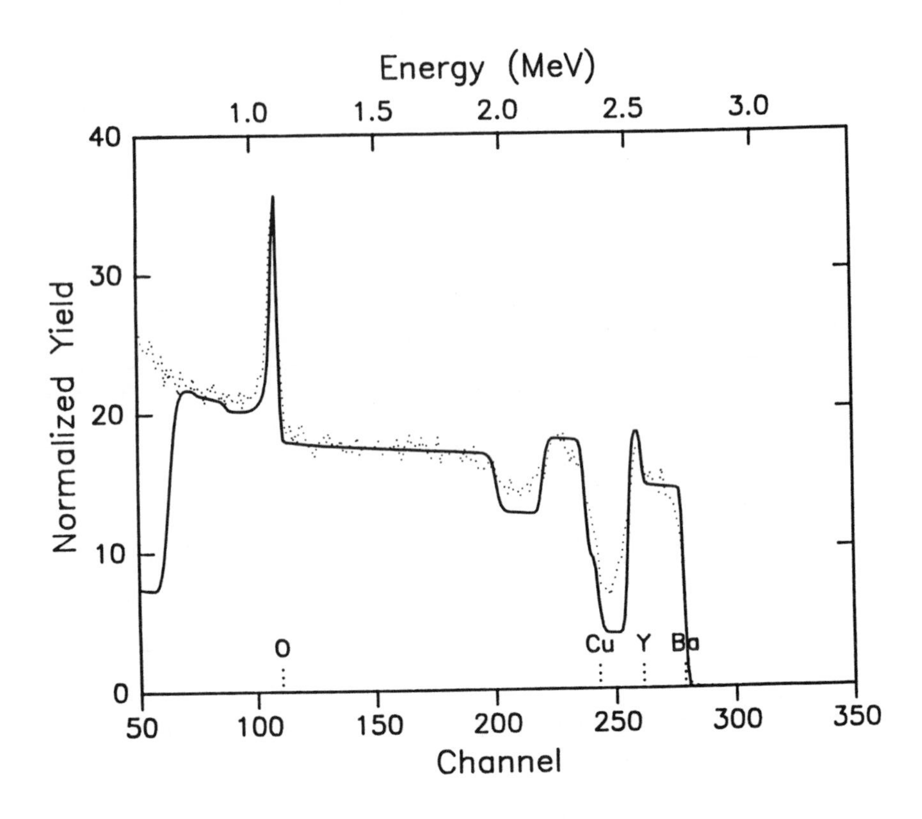

Fig. 2 RBS spectrum of film A1. A 3.02 Mev He^{2+} ion beam was employed to achieve a better resolution for oxygen peak. The solid line is a simulated curve corresponding to an oxygen content of 7.

Real time synchrotron X-ray diffraction studies[11] of growth of YBC films on $SrTiO_3$ by sputtering have revealed that at high temperature c-axis grains grow during the early stages of deposition and after approximately 4000 Å of thickness, a-axis oriented grains begin to nucleate on c-axis films. This may account for the mixed a+c nature of the thicker film B1 compared to c-axis orientation of the thinner film A1.

Comparison of J_c of the two films indicate a higher J_c for film A1 than B1 at zero field. However, J_c drops far more rapidly in film A1 in applied field than for film B1. Perhaps the mixed orientation granularity of film B1 is responsible for providing more effective pinning of flux lines than in film A1.

Substrate Crystallinity

Some YBC films were also prepared on repolished single crystals of $SrTiO_3$. These substrates had already been used for deposition and the deposited material on the crystal was removed off by polishing the crystal on Buehler polishing cloth wetted by a mixture of polishing alumina and glycerine. An XRD pattern of repolished crystal showed only (h00) peaks having FWHM several times of those of a fresh substrate. Films C and D were fabricated on such crystals. T_{ce} for film D was below our measuring capability of 77 K. This film was also granular and the J_c was an order of magnitude lower than film A1 and B1 prepared on new crystals. Film C was a-axis oriented as opposed to c-axis orientation of film A1 which was deposited at the same temperature and laser repetition rate. Magnetization measurements on Film C show no evidence of superconductivity down to 10 K.

Substrate Temperature

Having optimized the oxygen pressure, substrate crystallinity and the repetition rate, the crystal structure and superconducting properties of the films were found to be dependent upon substrate temperature. From our results we find that depositions performed at temperatures ranging from 675-750° C result in superconducting films. The dependence of crystal structure on temperature is observed by comparing Films B1 and E. Both the films are thick enough to have mixed orientation as mentioned before. However, low temperature (625° C) growth of Film E results in an a-axis orientation rather than a+c orientation of film B1 prepared at 750° C. The lattice constants of this film as determined from the XRD pattern are also analomous and no evidence of superconductivity was observed in this film to temperatures as low as 10 K. At temperatures 675-750° C superconducting films were achieved. At much higher temperatures as for sample F, several metastable phases were found. This film was found to be Cu and Ba rich and the XRD pattern confirmed the presence of $BaCuO_2$ phase in addition to 123 phase (Fig. 3a). A post heat treatment at 950° C in an oxygen furnace made the film superconducting however, the critical current density (J_c)

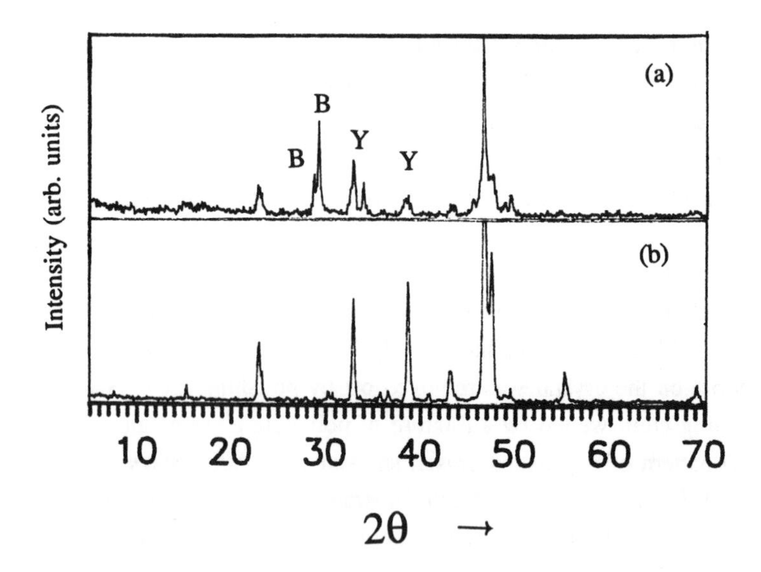

Fig. 3 (a) XRD pattern of Film F prepared at T_s = 850° C showing the presence of several phases (B-$BaCuO_2$, Y-YBC). (b) XRD pattern of the same film after post annealing at 950° C in flowing oxygen.

was quite low (~ 10^3 A/cm^2 at 20 K). Crystallinity of the film after post annealing improved greatly as the XRD pattern showed prominent c-axis orientation (Fig. 3b).

SUMMARY

The effects of several processing parameters on the crystal structure and superconducting properties of laser ablated thin films of YBC have been discussed. It was found that films with a high degree of orientation, critical temperature and current density could be fabricated using this technique by the proper optimization of the processing parameters.

ACKNOWLEDGEMENTS

We would like to thank Dr. J. A. Butcher Jr. of the Department of Chemistry for use of the laser.

REFERENCES

1) D. Dijkkamp, T. Venkatesan, X. D. Wu, S. A. Shaheen, N. Jisrawi, Y. H. Min-Lee, W. L. Mclean and M. Croft, Appl. Phys. Lett. 51, 619 (1987).

2) G. Koren, E. Polturak, B. Fisher, D. Cohen and G. Kimel, Appl. Phys. Lett. 53, 2330 (1988).

3) D. H. A. Blank, D. J. Adelerhof, J. Flokstra and H. Rogalla, Physica C 167, 423 (1990).

4) S. Mahajan, R. L. Cappelletti, R. W. Rollins, D. C. Ingram and J. A. Butcher jr., AIP Conference Proceedings 219 569 (1990).

5) R. K. Singh, J. Narayan, A. K. Singh and J. Krishnaswamy, Appl. Phys. Lett. 54, 2271 (1989).

6) M. K. Wu, J. R. Ashburn, C. J. Torng, P. H. Hor, R. L. Meng, L. Gao, Z. L. Huang, Y. Q. Wang and C. W. Chu, Phys Rev. Lett. 58 908 (1987).

7) K. Char, A. D. Kent, A. Kapiltunik, M. R. Beasley and T. H. Geballe, Appl. Phys. Lett. 51, 1370 (1987).

8) C. B. Eom, J. Z. Sun, K. Yamamoto, A. F. Marshall, K. E. Luther, T. H. Geballe and S. S. Laderman, Appl. Phys. Lett. 56, 595 (1989).

9) R. J. Cava, B. Batlogg, C. H. Chen, E. A. Rietman, S. M. Zahurak and D. Werder, Phys. Rev. B 36, 5719 (1987).

10) M. Ohkubo, T. Kachi, T. Hioki and J. Kawamoto, Appl. Phys Lett. 55, 899 (1989).

11) J. O. Zheng, X. K. Wang, M. Shih, S. Williams, J. So, S. J. Lee, P. Dutta, R. P. H. Chang and J. B. Ketterson, IEEE Transactions on Magnetics 27, 1025 (1991).

CHEMICAL VAPOR DEPOSITION

MOCVD AND PE-MOCVD OF HTSC THIN FILMS

Peter S. Kirlin
ATM, Inc., 7 Commerce Drive, Danbury, CT 06810

ABSTRACT

High quality YBaCuO and TlBaCaCuO thin films were deposited on MgO, $LaAlO_3$ and Ag substrates by standard thermal and plasma enhanced MOCVD. The growth was done in inverted vertical reactors designed to achieve stagnation point flow and extremely uniform deposition rates were achieved (± 0.5 %) over large areas (5 cm^2). The films were characterized by SEM-EDX, X-ray diffraction, four point probe, critical current density, dynamic impedance, and surface resistance measurements. C-axis oriented films with resistive transitions (R ≤ 0.1 μV/cm) exceeding 110 K and 85 K were routinely obtained for the Tl- and Y-based superconductors grown on single crystal substrates. The best films had inductive transition widths less than 1 K and critical current densities (ambient field) as high as 10^6 A/cm^2 at 77 K. The surface resistance of the films was measured using a cavity end wall replacement method and values as low as 10 mΩ were observed at 78 K and 35 GHz on both $LaAlO_3$ (100) and Ag substrates.

INTRODUCTION

Recent progress in the *in-situ* growth of high temperature superconductor (HTSC) thin films by plasma enhanced and standard thermal metalorganic chemical vapor deposition (PE-MOCVD and MOCVD) has been dramatic.[1,2] YBaCuO films grown by CVD on single crystal substrates exhibit dc critical current densities[3] and microwave surface resistivities[4] comparable to the best films deposited by physical vapor deposition methods.[5,6,7] In addition, c-axis oriented HTSC films have been grown by PE- and thermal MOCVD on polycrystalline ceramic[8] and metallic[9] substrates which is the first step towards utilizing thin films in bulk applications because these materials can be produced in complex shapes such as accelerator cavities or wires.

Despite these advances, several challenges remain to be overcome in the CVD of HTSC thin films. Present process reproducibility is poor which is caused by the decomposition of the organometallic source reagents, particularly the barium source.[10,11,12] The *in-situ* deposition of uniform, high quality HTSC thin films on large area substrates (> 2.5 cm diameter) and the *in-situ* deposition of TlBaCaCuO films have not been reported. The latter two challenges are germane to all HTSC thin film deposition techniques. CVD is a likely candidate to overcome these problems based on its proven performance in the growth of large area

(20 cm diameter) electronic thin films and its operational pressure regime which encompasses the equilibrium pressure of Tl_xO_y over TlBaCaCuO.

EXPERIMENTAL

Synthesis and handling of all organometallic complexes was performed under nitrogen atmosphere using standard Schlenk techniques or in a Vacuum Atmospheres inert atmosphere glove box. Tetraglyme was distilled over Na/K alloy and stored over calcium hydride. Barium hydride, 1,1,1,5,5,5-hexafluoro-2,4-pentanedione, 1,1,1-trifluoro-2,4-pentanedione (Strem Chemicals), 1,1,1,2,2,3,3-heptafluoro-7,7-dimethyl-octane-4,6-dione and 2,2,6,6-tetramethyl-3,5-heptanedione (Aldrich Chemical Co.) were used as received. The compounds were routinely identified by their ^1H nmr and ^{13}C nmr spectra which were measured with an IBM WP200SY spectrometer. A detailed description of the synthesis, purification and x-ray crystal structure determination of the compounds described below is reported elsewhere.[13]

$YBa_2Cu_3O_7$ thin films were deposited on MgO(100) single crystal substrates by PE-MOCVD. The source reagents used for the *in-situ* deposition of the YBaCuO films were yttrium and copper 2,2,6,6-tetramethyl-3,5-heptanedionate ($Y(thd)_3$ and $Cu(thd)_2$) and barium (2,2,6,6-tetramethyl-3,5-heptanedionate)·tetraglyme ($Ba(thd)_2 \cdot tg$). The PE-MOCVD experiments were carried out in a standard vertical reactor with a downstream N_2O plasma. A range of growth rates, 0.5 to 1 μm/hr, were obtained at reactor pressures between 0.5 and 5 torr and a substrate temperature of 700°C. Typical run conditions are listed in Table I; a more detailed description of the reactor system is reported elsewhere.[10]

The TlBaCaCuO thin films were prepared via a thermal process and the BaCaCuOF precursor films were deposited on Ag and $LaAlO_3$(100) substrates by MOCVD. The source reagents used for the deposition of the precursor films were barium and calcium 1,1,1,2,2,3,3-heptafluoro-7,7-dimethyloctane-4,6-dionate ($Ba(fod)_2$ and $Ca(fod)_2$) and copper hexafluoroacetylacetonate ($Cu(hfacac)_2$). The MOCVD experiments were carried out in an inverted vertical reactor designed to achieve stagnation point flow (Figure 1) and growth rates ranging from 0.1 to 0.4 μm/hr were achieved at a susceptor temperature of 500°C in a pressure regime of 0.5 to 4 torr. Typical run conditions are listed in Table II.

TABLE I. Deposition conditions for YBaCuO films.

N_2O flow rate (sccm)	Ar flow rate (sccm)	Source reagent transport rate 10^{-6}mol/m		
		$Y(thd)_3$	$Ba(thd)_2 \cdot tg$	$Cu(thd)_2$
100 ~ 200	20	1 ~ 2	3 ~ 6	4 ~ 8

TABLE II. Deposition conditions for BaCaCuOF films.

O_2 flow rate (sccm)	Ar flow rate (sccm)	Source reagent transport rate 10^{-6}mol/m		
		$Ba(fod)_2$	$Ca(fod)_2$	$Cu(hfac)_2$
170	190 ~210	12 ~ 18	6 ~ 15	0.35 ~ 0.56

The fluorine was stripped from the BaCaCuOF films in the second step by annealing in wet O_2 at 785°C. The fluorine content of the post-annealed films was determined by energy dispersive X-ray analysis (EDX) and the rate of fluorine removal was monitored by measuring the pH of the contents of an effluent gas scrubber placed at the exit of the annealing tube. C-axis oriented $Tl_2Ba_2CaCu_2O_x$ films were formed in the final step by annealing the BaCaCuO films in dry O_2 at temperatures between 820 and 870°C in the presence of $Tl_2O_3/Ba_2Ca_2Cu_3O_x$ powder. The physical properties of the films were characterized by SEM, XRD, and EDX and the electrical properties were evaluated with four point probe, rf eddy current, and microwave cavity measurements.

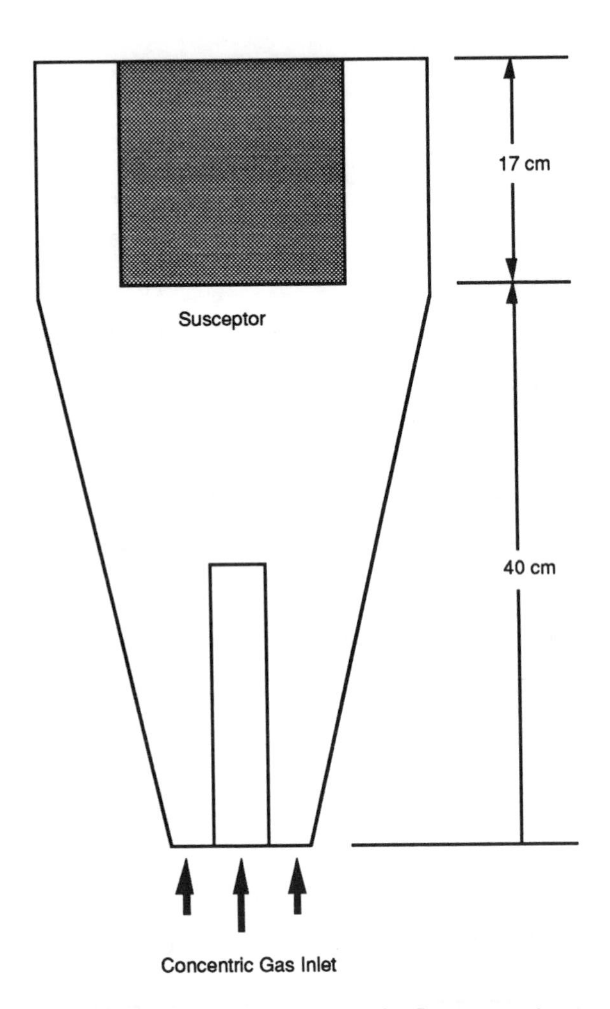

Figure 1. Schematic of inverted vertical reactor (not to scale).

The CVD reactors were modeled to improve the film uniformity and expedite process development. In general, MOCVD reactor models consist of equations representing the conservation of momentum, energy, total mass and individual species. The general forms of these partial differential equations are given in standard transport phenomena and reaction engineering texts.[14,15] Because of the non-linearity and interdependence of the equations, numerical procedures are used to for their solution. The simulations were solved with Fluent, a computation fluid dynamics program marketed by Creare. Fluent uses a finite difference algorithm in which the governing equations are integrated over a control volume surrounding each grid point using an interpolation function for the dependent variable between adjacent grid points. The interpolation function is a power law[16] which is used to relate the value of a dependent variable at the control volume faces to the adjacent grid points. This scheme allows the finite difference counterparts to the governing partial differential equations to be obtained as simple algebraic equations which relate the value of a dependent variable at a given node to its value at the surrounding nodes. This set of algebraic equations is then solved iteratively using a line-by-line under-relaxation method.[17] The detailed modeling equations and boundary conditions used will be presented elsewhere.

Measurements of the surface resistance (Rs) were made at a frequency of 35 GHz and from a temperature of 10 K to above the transition temperature. The measurement apparatus consisted of a cylindrical TE_{011} mode cavity in which the sample served as one end wall. By comparing the temperature dependent Q and resonant frequency data for a superconducting sample with similar data obtained for a copper standard, the Rs for the superconductor as a function of temperature was calculated.[18] The experimental uncertainty that results from using the difference between two Q values of similar magnitude limits the resolution to approximately 8 mΩ. In order to fit the cavity, 1 cm x 1 cm squares were cut from the 2.5 cm diameter discs using photoresist to protect the films during sawing. A 820 MHz Nb TEM cavity was used to study the rf field dependence of the Rs; the sample was placed inside a Teflon cup inside the cavity.[19] Both sides of the substrate are exposed to the microwave fields in this configuration; consequently, the sensitivity of the measurement was limited to the Rs of Consil which was ~ 2 mΩ at 4 K.

RESULTS AND DISCUSSION

The primary obstacle to the widespread application of MOCVD to the growth of high T_c superconducting oxide thin films (YBaCuO, BiSrCaCuO, TlBaCaCuO) is the absence of an effective alkali metal, in particular barium, source reagent. Barium prefers high coordination numbers, up to 12,[20] and the lack of steric saturation of many barium organometallic complexes leads them to be strongly associated in the solid state; barium organometallics often undergo thermal decomposition before vaporization.[21] The current MOCVD precursors for barium $Ba(fod)_2$ and $Ba(thd)_2$ undergo partial decomposition at the temperatures (>200°C) required to sustain sufficient usable vapor pressure.[10,11,12]

Improved stability of barium transport has been accomplished by saturating the carrier gas stream going to the $Ba(thd)_2$ source with the 2,2,6,6-tetramethyl-3,5-heptanedione vapor.[22] Similarly, saturating the carrier gas with oxygen or nitrogen Lewis bases such as tetrahydrofuran can lower the required bubbler temperature.[23] Excess hydrocarbons can lead to the co-deposition of carbon and compete for surface sites thus interfering with the *in-situ* deposition of the HTSCs. Consequently, a mononuclear barium source reagent with a vapor pressure of ~1 Torr at 150°C which is thermally stable at its vaporization temperature would simplify MOCVD process development and increase process reproducibility and control.

We have synthesized several volatile barium ß-diketonate:polyether adducts and evaluated their potential as CVD sources. The most volatile complexes of this general class of compounds contain partially fluorinated ß-diketonate ligands. Both thermal and plasma assisted dissociation of these compounds results in large amounts of fluorine contamination, as BaF_2, in the as-deposited films.[13] $Ba(thd)_2 \cdot tg$ (Figure 2) and analogous polyether adducts were synthesized to circumvent fluorine contamination. Of the non-fluorinated compounds synthesized to date, $Ba(thd)_2 \cdot tg$ exhibits the highest volatility and was used to deposit highly c-axis oriented $YBa_2Cu_3O_7$ films at 700°C on MgO(100) as described below.

The inverted vertical reactor configuration was selected to optimize the uniformity of the as-deposited films. Standard 1-D analytical analysis predicted that stagnation point flow[24] could be achieved under standard reduced pressure MOCVD conditions. A reactor chamber was built and tested with a joint empirical/modeling approach A 2 D numerical grid of one half of the reactor(s) was developed with geometric decreasing spacing before and increasing spacing after the susceptor and the gas injector. This approach to grid construction maximizes resolution and minimizes wasted CPU time; a typical simulation required 5 days of dedicated time on the IBM RISC 6000 to reach convergence. A solution was considered to be converged when the values of the pressure and flow residuals[25] reached 10^{-5} and the enthalpy and species residuals fell below 10^{-6}. A solution was accepted as legitimate when overall mass and energy balances closed to within 1%. The simulation was compared to experimental results to establish the validity of the model and then used to help guide the experimental approach.[26] $Ba_2CaCu_2O_xF_y$ with thickness uniformities less than (±5%) over 2 in diameter wafers were grown using this methodology (Figure 3).

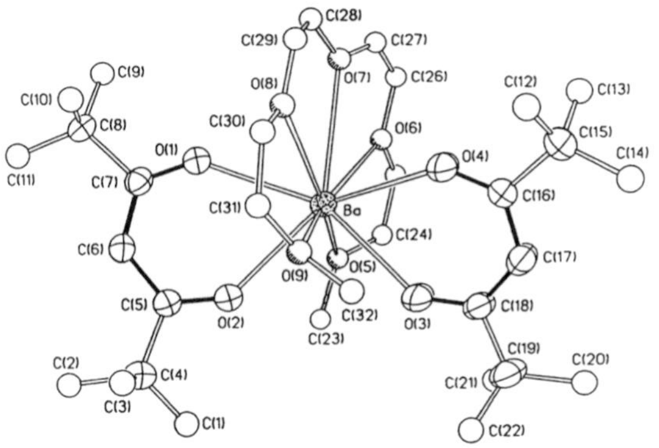

Figure 2. ORTEP plot of $BaC_{20}H_{60}O_9$. H atoms are omitted for clarity.

YBaCuO thin films were grown by PE-MOCVD in a remote plasma configuration using $Ba(thd)_2$·tetraglyme as the Ba source. The glow discharge of pure N_2O was sustained at 100 watts and highly c-axis oriented $YBa_2Cu_3O_7$ films were grown on MgO(100) at substrate temperatures between 700 and 750°C (Figure 4). The films were pin-hole free and smooth with an average rms roughness of 500 Å.

Superconducting $Tl_2Ba_2CaCu_2O_x$ films were fabricated using the three step process described above. Highly c-axis oriented $Tl_2Ba_2CaCu_2O_x$ films were formed on both $LaAlO_3$ (100) (Figure 5) and Ag substrates. In the latter case thicker films and lower annealing temperatures were required to prevent Ag/Tl interdiffusion and to retain the structural integrity of the films. Optimized protocols gave highly uniform and phase pure films as demonstrated by the extremely narrow, <1 K, inductive transitions (Figure 6). The low field microwave properties of the films were measured using a cavity end wall replacement method and the Rs of 4 μm films grown on Ag substrates starts to drop at ~ 105 K and reaches that of Cu at ~ 100 K (Figure 7). The 88 K Rs of 19 mΩ at 35 GHz scales to 10 μΩ at 1 GHz which is the lowest value reported to date for any HTSC film on a metallic substrate. The rf field dependence of the Rs of these films was evaluated by placing the coated Ag alloy discs inside a 820 MHz Nb TEM cavity. The Rs of the samples did not exceed that of the Ag alloy background (2 mΩ) from 0.1 to 200 Gauss at 4 K.

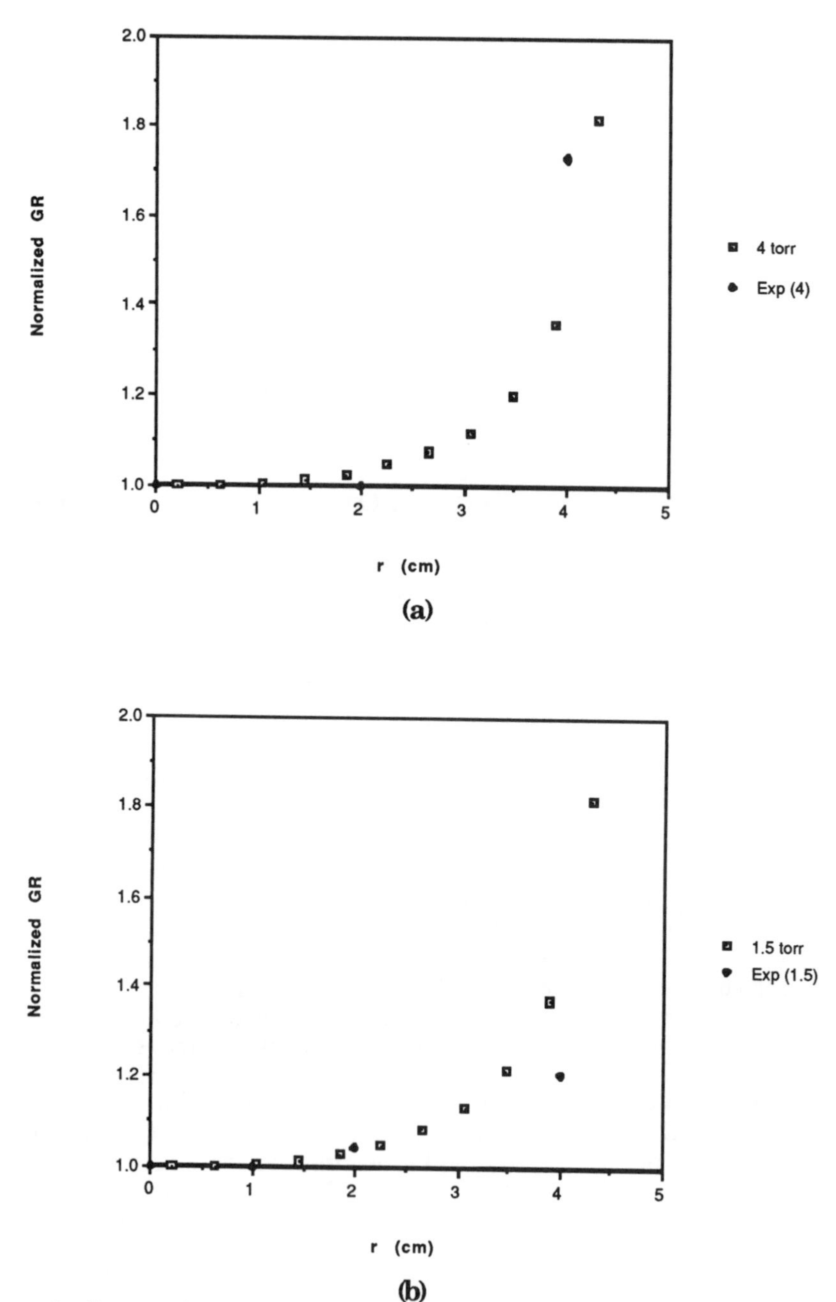

Figure 3. Comparison of predicted (squares) and experimental (diamonds) growth rates at (a) 4 torr and (b) 1.5 torr and 500°C.

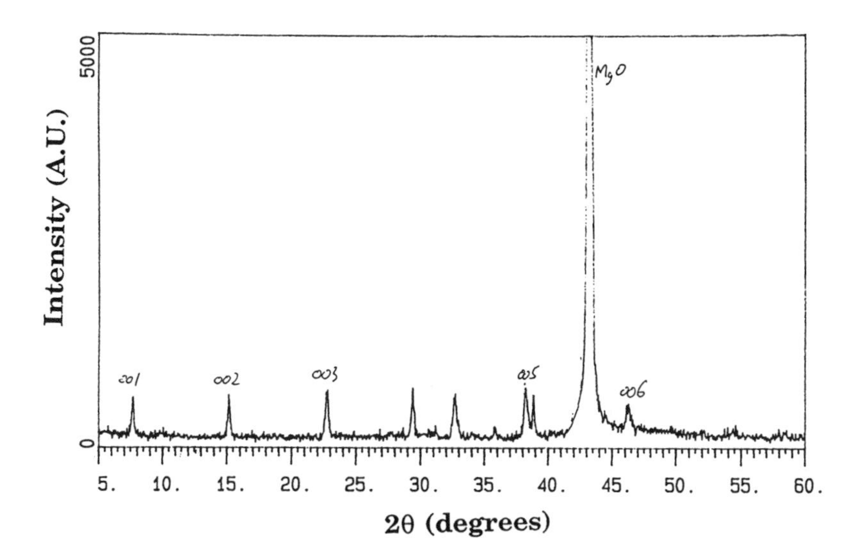

Figure 4. XRD of $YBa_2Cu_3O_7$ film grown on MgO (100) by PE-MOCVD at 700°C.

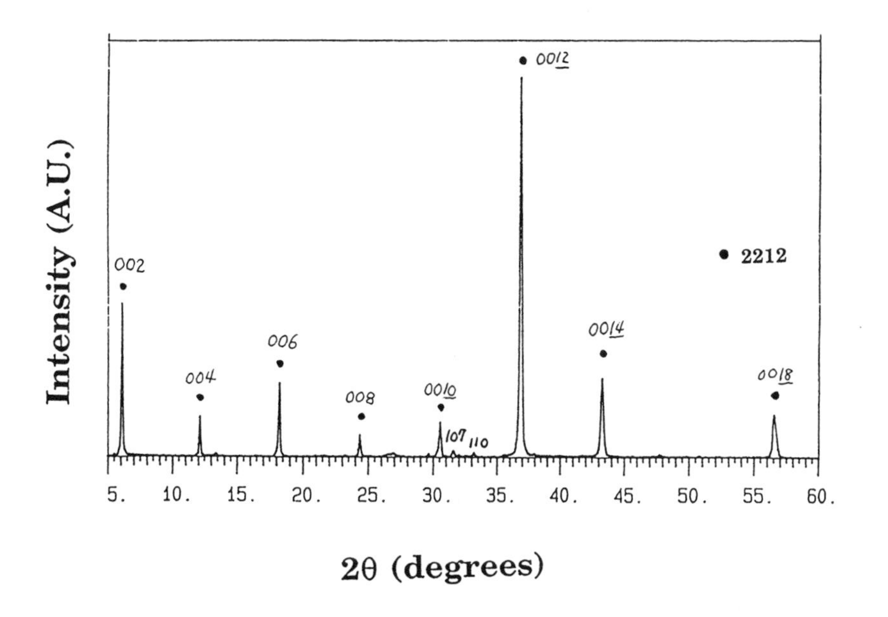

Figure 5. XRD of $Tl_2Ba_2CaCu_2O_x$ film grown on $LaAlO_3$ (100).

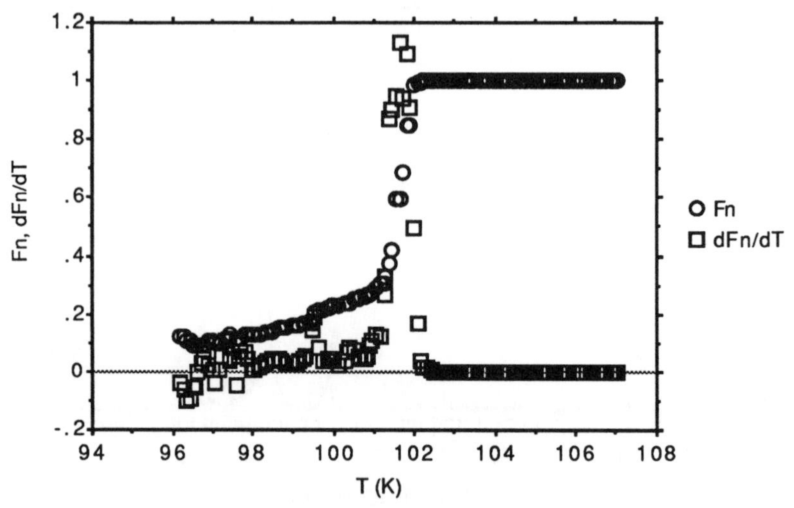

Figure 6. Inductive transition of $Tl_2Ba_2CaCu_2O_x$ film grown on $LaAlO_3$

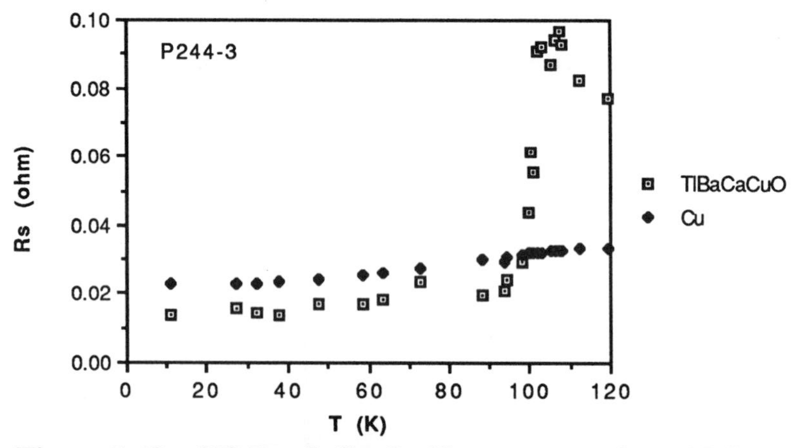

Figure 7. R_s of $Tl_2Ba_2CaCu_2O_x$ film grown on Ag at 35 GHz.

SUMMARY

Novel barium complexes have been synthesized and tested as MOCVD source reagents for the *in-situ* growth of high quality HTSC thin films. CAD tools have been used in conjunction with standard empirical CVD process development to achieve state-of-the-art HTSC film uniformities on 2 in diameter wafers. TlBaCaCuO films have been grown on Ag with the lowest R_S and rf field dependence of R_S reported to date for HTSC films prepared by any deposition method on metallic substrates. Efforts are ongoing to synthesize new barium complexes with enhanced stability and future film growth work is focused on extending the uniformities achieved on 2 in to 3 in diameter wafers and leveraging the PE-MOCVD process developed for the *in-situ* growth of YBaCuO to the TlBaCaCuO film system. A broad range of active and passive microwave components such as bandpass filters and phase shifters will be fabricated to test the efficacy of the deposition and fabrication processes.

ACKNOWLEDGEMENTS

We gratefully acknowledge Dr. Bill Wilber (Fort Monmouth) and Dr. William Kennedy (Argonne) for the low and high field Rs measurements, respectively. This was supported in part by DARPA under N00014-90-C-0201 and NSF under contracts ISI 9002399 and ISI 8960636.

REFERENCES

[1]J. H. Takemoto, C. M. Jackson, H. M. Manasevit, D. C. St. John, J. F. Burch, K. P. Daly, R. W. Simon, Appl. Phys. Lett., **58**, 1109 (1991).

[2]B. Schulte, M. Maul, W. Becker, E. G. Schlosser, P. Hausler, H. Adrian, Appl. Phys. Lett., **59**, 869 (1991)

[3]S. Matsuno, F. Uchikawa, K. Yoshizaki, Jpn. J. Appl. Phys., **29**, L947 (1990).

[4]R. Hiskes, S. A. DiCarolis, J. L. Young, S. S. Laderman, R. D. Jacowitz, R. C. Taber, Appl. Phys. Lett., **59**, 606 (1991).

[5]J. D. Klein, A. Yen, S. L. Clauson, J. Appl. Phys., **67**, 6389 (1990).

[6]T. Venkatesan, X. D. Wu, B. Dutta, A. Inam, M. S. Hedge, D. M. Hwang, C. C. Chang, L. Nazar, B. Wilkens, Appl. Phys. Lett., **54**, 581 (1989).

[7]J. Saitoh, M. Fukutomi, K, Komori, Y. Tanaka, T. Asano, H. Maeda, H. Takahara, Jpn. J. Appl. Phys., **30**, L898 (1991).

[8]A. Feng, L. Chen, T. W. Piazza, H. Li, A. E. Kaloyeros, D. W. Hazelton, L. Luo, R. C. Dye, Appl. Phys. Lett., **59**, 1248 (1991).

[9]J. M. Zhang, B. W. Wessels, L. M. Tonge, T. J. Marks, Appl. Phys. Lett., **56**, 976 (1990).

[10]P. S. Kirlin, R. Binder, R. Gardiner, D. W. Brown, Proc. SPIE: Processing of Films for High Tc Superconducting Electronics, **1187**, 115 (1989).

[11]G. Malandrino, D. S. Richeson, T. J. Marks, D. C. DeGroot, J. Schindler, C. R. Kannerwurf, Appl. Phys. Lett., **66** , 444 (1989).

[12] T. Nakamori, H. Abe, T. Kanamori, S. Shibata, S. Jpn. J. Appl. Phys., **27**, L1265 (1988).

[13]R. Gardiner, D. W. Brown, P. S. Kirlin, A. L. Rheingold, Chem. Mater., in press.

[14]R. B. Bird, W. E. Stewart, E. N. Lightfoot, Transport Phenomena (Wiley, New York, 1960) 71.

[15] G. F. Froment, K. E. Bischoff, Chemical Reactor Analysis and Design (Wiley, New York, 1979) 347.

[16]S. V. Patankar, Numerical Heat Transfer and Fluid Flow (McGraw Hill, New York, 1980).

[17]T. Jasinski, Proc. Nat. Heat Trans. Conf., 1988, July 24, Houston, TX.

[18]J. S. Martens, J. B. Beyer, and D. S. Ginley, Appl. Phys. Lett., **52**, 1822 (1988).

[19]J. R. Delayen, C. L. Bohn, C. T. Roche, Rev. Sci. Instru., **61**, 2207 (1990).

[20] A. F.Wells, Structural Inorganic Chemistry, 5th ed. (Clarendon: Oxford, 1984)

[21] W. J. Evans, Adv. Organomet. Chem., **24**, 131 (1985).

[22] P. H. Dickinson, T. H. Geballe, A. Sanjuro, D. Hildenbrand, G. Craig, M. Zis Collman, J. Appl. Phys., **66**, 444 (1989).

[23] A. R. Barron, J. M. Buriak, L. Cheatham, R. Gordon, J. Electrochemical Soc., 1 225C (1990).

[24] Velocity vectors perfectly parallel to the surface of the film with equal magnitud across the breadth of the susceptor.

[25] Residuals measure the imbalance in the conservation equations and the criteria refer to the sum over all 3762 nodes in the computation domain.

[26] An additional test of the validity of the solution is to establish that it is independe the grid.

TRANSPORT PROPERTIES OF $YBa_2Cu_3O_{7-x}$ THIN FILMS FORMED BY PLASMA-ENHANCED METALORGANIC CHEMICAL VAPOR DEPOSITION AT REDUCED TEMPERATURES

J. Zhao, P. Norris

EMCORE Corporation, 35 Elizabeth Ave., Somerset, N.J. 08873

T.L. Peterson, I. Maartense

Wright Laboratory, Wright Patterson AFB, OH 45433

C.S. Chern , P.Lu, B. Kear

Rutgers, The State University, New Brunswick, N.J. 08854

Y.Q. Li , B. Gallois

Stevens Institute of Technology, Hoboken, N.J. 07030

ABSTRACT

$YBa_2Cu_3O_{7-x}$ superconducting thin films with a high transition temperature of 90.7K and a high critical current density of 4.1 x 10^6 A/cm^2 at 77.7 K and 0 T were prepared by a plasma-enhanced metalorganic chemical vapor deposition process. The films with sharp transition and critical current density > 2 x 10^6 A/cm^2 were also formed in-situ on $LaAlO_3$ at a substrate temperature as low as 640°C. Magnetic susceptibility measurements of the as deposited films show sharp superconducting transition temperatures with narrow widths < 0.3K.

INTRODUCTION

Metalorganic chemical vapor deposition (MOCVD) is promising to become the deposition choice for large-scale fabrication of high T_c superconducting (HTSC) thin films due to its scale up capability and the ability to deposit at a high rate. MOCVD has become the manufacturing standard for the fabrication of device-quality thin films of compound semiconductors. To date, superconducting transition T_c >89K and critical current densities exceeding $1x10^6$ A/cm^2, measured at liquid nitrogen temperature in zero field, have been reproducibly achieved in high quality epitaxial $YBa_2Cu_3O_{7-x}$ (YBCO) thin films prepared by MOCVD methods.[1-4] Recent advances[4] in reduction of the required deposition temperature in conventional, thermal, MOCVD via plasma-enhancement have further increased its competitiveness in the fabrication of oxide films of HTSC. The ability to form high quality HTSC films at low temperatures will permit the deposition of HTSC films on many technologically important materials. In this paper we report a detailed study of the transport properties of high quality YBCO thin films formed in-situ at reduced temperatures of 640°C and 710°C by plasma-enhanced MOCVD (PE-MOCVD).

EXPERIMENT

The films were prepared in a commercial-scale EMCORE system described in earlier publications.[5] Films were formed in situ on (100)$LaAlO_3$ at substrate temperatures of 640 to 710°C. Two samples were studied in this paper. Sample1 and Sample 2 were prepared at 640 and 710°C, respectively. N_2O was used as the oxidizer and 120 Watts microwave plasma was used during the growth. Substrates of 1x1 cm^2 and 2.5 cm diameter were placed on a high-speed rotating disk susceptor heated resistively and mounted inside a vertical stainless steel reaction chamber. The organometallic precursors of $Y(dpm)_3$, $Ba(dpm)_2$, and $Cu(dpm)_2$ (dpm = dipivaloymethanate) were used as yttrium, barium, and copper sources, respectively. The films were about 150nm thick as measured by a profilometer. After deposition, the films were slow cooled to room temperature under one atmosphere of oxygen.

The superconducting properties of the films were measured by both four-probe dc resistivity and ac magnetic susceptibility measurements with magnetic fields from 0.04 to

3.6 Oe. The critical current density (J_c) was measured by the four-probe dc transport method on a 120μm x 40μm bridge patterned by a conventional wet photolithographic technique. A positive photoresist (AZ1350J) and a 5% phosphoric acid solution in water were used. Low-resistance contacts were prepared by annealing evaporated silver films at 500°C for 1 hour in a pure oxygen ambient. The values of the critical current density were determined from the variations of current with voltage with a criterion of 1μV/cm.

RESULTS AND DISCUSSIONS

Several characterization techniques were used to examine the structural properties of the films. As described in detailed by another paper [4], the films were epitaxially grown with c axes perpendicular to the $LaAlO_3$ surface. Typical rocking curves (ω scan) obtained from the YBCO (005) reflection displayed a narrow full-width at half maximum of about 0.3°, and typical Rutherford backscattering spectroscopy of the films shown channeling yield of about 10% (determined from Ba spectrum). The films also have a high degree of in-plane crystallinity with twinned structures, as shown in Fig.1 of a typical planar high-resolution transmission electron microscopy micrograph. The structure and the effect of the nano-precipitates which appear in Fig.1 has also been described elsewhere. [6] Fig.2 and Fig.3 are the normalized resistances of the films grown at 640 (sample1) and 710°C (sample2), revealing that the films have a sharp zero resistance transition temperature (T_c) of 88 and 90.7K, respectively. The insert in Fig.2 is unnormalized resistance of sample 2, showing the detail of the sharp transition. The resistances were measured on patterned films with Line a and Line b normal and parallel to the direction of the twins of the substrates, respectively. As indicated in Fig.2 and 3, no twinning effect on the temperature dependent resistance was observed. The films exhibit a good metallic behavior in the normal state with ratios of R(300K)/R(100K) = 3.1 and 3.0 for the films grown at 640°C and 710°C, respectively. We found that the ratio does not provide qualitative information about the transport properties for the thin films with sharp superconducting transitions. Fig.4 and Fig.5 are the imaginary and real parts of field dependent magnetic susceptibilities. The near-zero field susceptibilities, as shown in Fig.4 and Fig.5, show sharp superconducting transitions with a narrow transition width of about 0.3K, indicating the films have an excellent homogeneity. Sample 1, however, has an added component as revealed by the field-dependent behavior below the main transition in Fig.4. Such a component, when visible, usually originates in a narrow region of the film adjacent to the edge of the substrate. The critical superconducting current densities measured by the transport method at 77K in 0 field were 2.5 and 4.1 $\times 10^6$ A/cm^2 for sample 1 and 2, respectively. Fig. 6 shows the temperature dependence of J_c of sample 1 and sample 2.

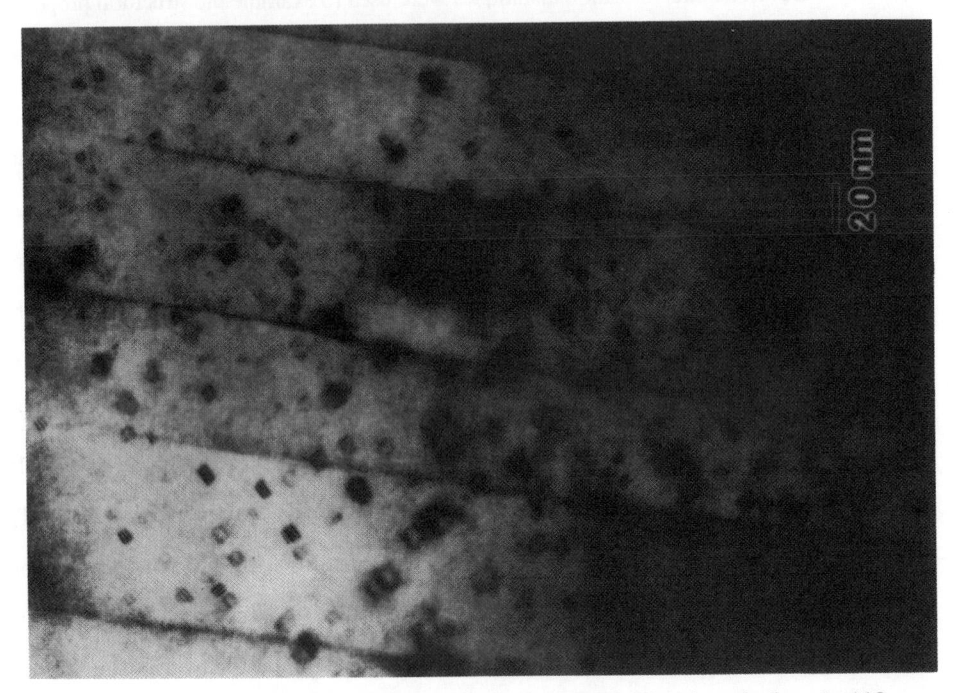

Fig. 1 HREM planar-section micrograph of the YBCO film deposited on $LaAlO_3$ at 670°C.

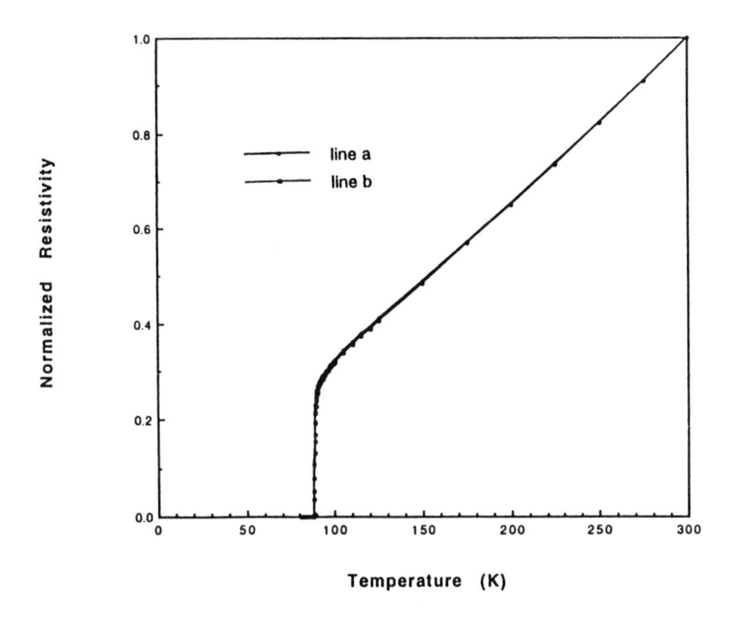

Fig. 2 Normalized resistance as a function of temperature of as-deposited YBCO films grown on $LaAlO_3$ at 640°C by PE-MOCVD.

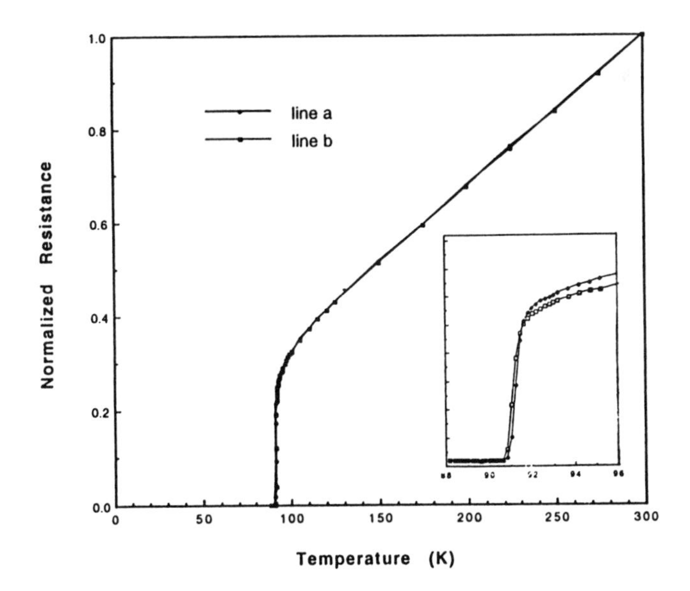

Fig. 3 Normalized resistance as a function of temperature of as-deposited YBCO films grown on $LaAlO_3$ at 710°C by PE-MOCVD.

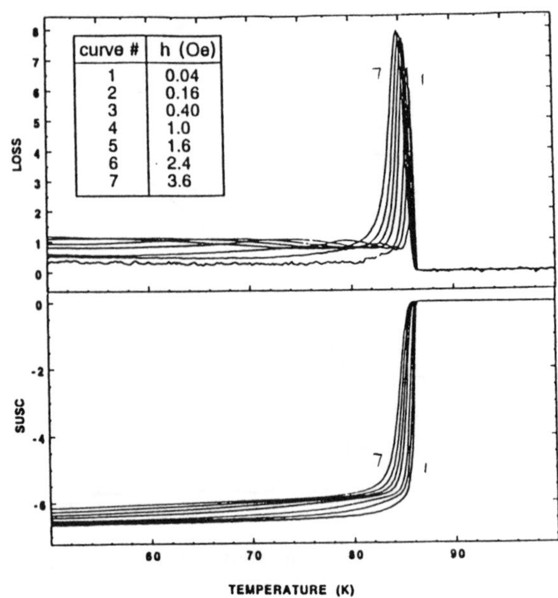

Fig. 4 Field dependent ac magnetic susceptibility measurement of YBCO films grown on $LaAlO_3$ at 640°C by PE-MOCVD. The upper curve represents the imaginary part, and the lower curve represents the real part.

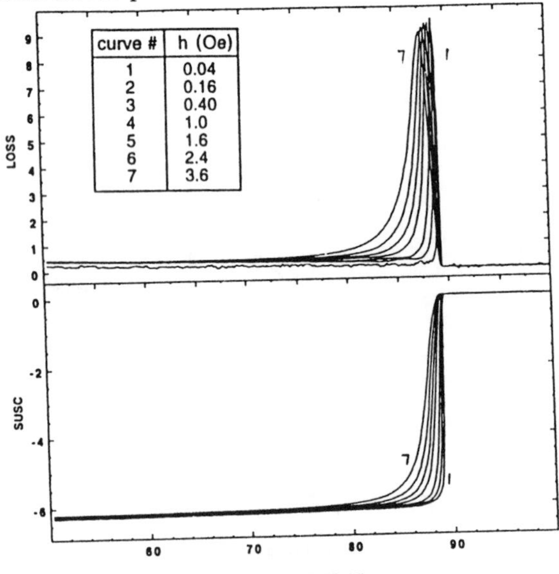

Fig. 5 Field dependent ac magnetic susceptibility measurement of YBCO films grown on $LaAlO_3$ at 710°C by PE-MOCVD. The upper curve represents the imaginary part, and the lower curve represents the real part.

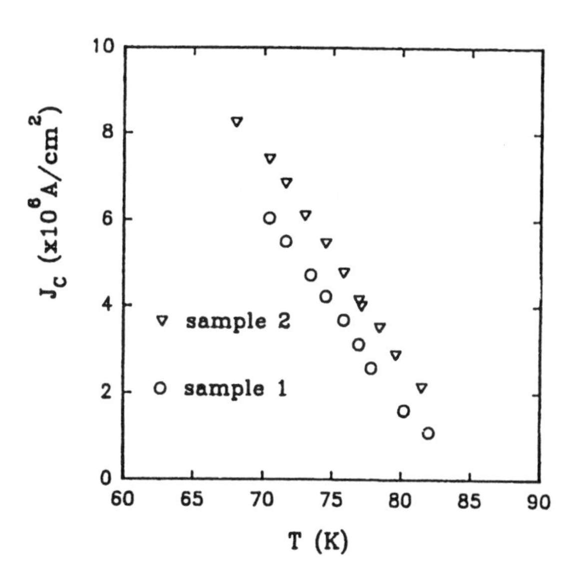

Fig.6 Critical current densities of YBCO films grown on $LaAlO_3$ at 640 and 710°C by PE-MOCVD.

SUMMARY

High-quality YBCO thin films with sharp transition width < 0.3 K and high J_c (77K) > $2x10^6 A/cm^2$ were epitaxially grown on $LaAlO_3$ substrates by PE-MOCVD at reduced substrate temperatures between 640 to 710°C. Our results indicate that PE-MOCVD is a very promising film deposition process for high T_c superconductors. Additional exploration of growth conditions and the many other possible reactants will likely result in further improvement of the film quality and reduction of the deposition temperature.

Th work at EMCORE was partially supported by the U.S. Air Force under contract F 33615-91-C-5662

REFERENCES

1. H.Yamane, H.Kurosawa, T.Hirai, H.Iwasaki, K. Watanabe, H. Iwasaki, N. Kobayashi, and Y.Muto, Supercond. Sci. Technol 2, 115 (1988).
2. S. Matsuno, F. Uchikawa and K. Yoshizaki, Jpn. J. Appl. Phys. L947 (1990).
3. Y.Q. Li, J. Zhao, C. Chern, W. Huang, G.A. Kulesha, P. Lu, P. Norris, B. Gallois, and B. Kear, Applied Physics Letters, 58, 648 (1991).
4. J. Zhao, Y.Q. Li, C. Chern, W. Huang, P.Lu, P. Norris, B. Gallois, B. Kear, F. Cosandey, X.D. Wu, R.E. Muenchausen, and S. Garison. Applied Physics Letters, Vol.59,1254, (1991)
5. J. Zhao, D.W. Noh, C. S. Chern, Y.Q. Li, P. Norris, B. Gallois, and B. Kear, Appl. Phys. Lett. 56, 2342 (1990).
6. P. Lu, Y.Q. Li, J. Zhao, C. Chern, P. Norris, B. Gallois, and B. Kear, unpublished.

CHEMICAL VAPOR DEPOSITION (CVD) OF HIGH-T_C $YBa_2Cu_3O_{7-\delta}$ FILMS

M.Sommer, L.Csajagi-Bertok, H.Oetzmann, F.Schmaderer
Asea Brown Boveri AG, Corporate Research, 6900 Heidelberg, FRG

W.Becker, H.Klee
Hoechst AG, Central Research II, 6230 Frankfurt, FRG

B.Schulte
Institut f. Festkörperphysik, TH Darmstadt, 6100 Darmstadt, FRG

ABSTRACT

High quality $YBa_2Cu_3O_{7-\delta}$ (YBCO) films were prepared from β-diketonate complexes of yttrium, barium, and copper on single crystalline $SrTiO_3$ and MgO substrates at substrate temperatures between 800 and 850 °C. The deposition experiments were carried out with two different experimental arrangements operating with and without carrier gas, respectively. The deposited $YBa_2Cu_3O_{7-\delta}$ films exhibit excellent crystallographic orientation and superconducting properties ($T_c \geq 90$ K, and j_c(0 T, 77 K) $\geq 10^6$ A/cm^2).

INTRODUCTION

The use of CVD for preparing thin superconducting films offers many advantages, such as high deposition rates, excellent film uniformity, and oriented or even epitaxial grain growth. In addition, CVD is a chemical process which is not limited to line-of-sight deposition, thus allowing the uniform coating of large areas, or complex three-dimensional structures. However, the main problem in the preparation of High-T_c superconductors via CVD seems to be the development of suitable volatile precursors for each particular materials system.

PRECURSORS FOR CVD

Although we have investigated the evaporation behavior of a variety of CVD precursors[1] (e.g. $Y(thd)_3$, $Cu(hfac)_2$, $Cu(acac)_2$, $Cu(thd)_2$, $Ba(fod)_2$, $Ba(thd)_2$, and $Ba(thd)_2$xL) in this report we focus on the use of 2,2,6,6-tetramethylheptanedionates (thd) of yttrium, barium, and copper as source materials for the deposition of YBCO films. These compounds are synthesized either in aqueous solutions from the metal nitrates and tetramethyl-heptanedione, or from alkoxides in nonaqueous solvents[2-6] according to

$$Me(NO_3)_2 + 2\ H(thd) + 2\ NaOH \rightarrow Me(thd)_2 + 2\ NaNO_3 + 2\ H_2O \quad (1)$$

$$Me(OR)_2 + 2\ H(thd) \rightarrow Me(thd)_2 + 2\ HOR \quad (2)$$

In Fig. 1 the molar evaporation rates $\dot{n}_{ev}$ of $Cu(thd)_2$, $Y(thd)_3$, and $Ba(thd)_2$ are plotted as functions of inverse absolute evaporation temperature T_v. It is apparent from this figure that both $Y(thd)_3$ and $Cu(thd)_2$ are sufficiently volatile (i.e. $\dot{n}_{ev} > 10^{-6}$ mole/min) at relatively low temperatures (100-150 °C), while $Ba(thd)_2$ requires temperatures exceeding 200 °C to achieve similar evaporation rates. In addition, $Y(thd)_3$ and $Cu(thd)_2$ evaporate without considerable thermal decomposition, and, hence, their evaporation rates remain constant for several hours. Generally, there exists a problem with $Ba(thd)_2$ as has been reported previously[6,7]. Since $Ba(thd)_2$ partially decomposes during the evaporation process its volatility decreases significantly with time. Furthermore, the evaporation behavior of $Ba(thd)_2$ varies from batch to batch. These observations lead us to examine the coordination chemistry of $Ba(thd)_2$ in more detail.

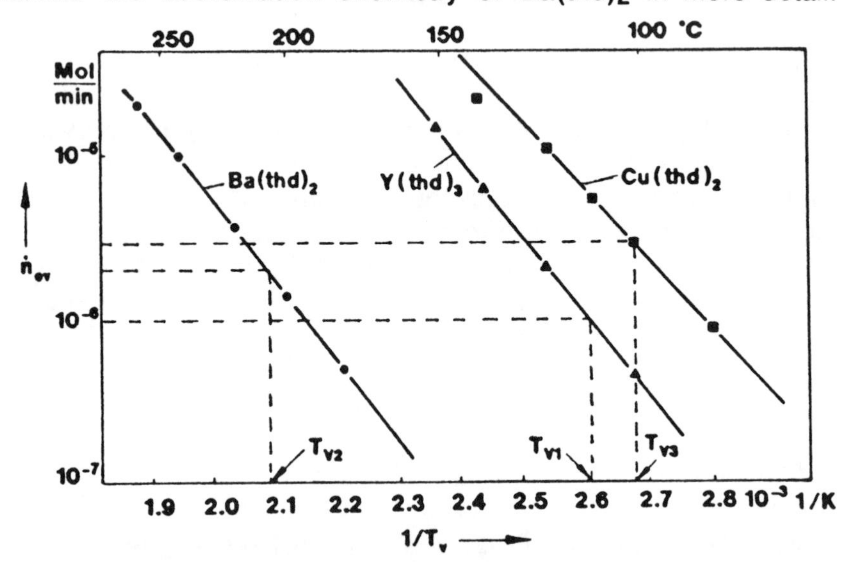

Fig. 1 Molar evaporation rates $\dot{n}_{ev}$ of $Y(thd)_3$, $Cu(thd)_2$, and $Ba(thd)_2$ vs. inverse absolute temperature T_v.

Nuclear magnetic resonance studies show that the barium compound principally crystallizes with solvents in order to saturate the coordination sphere of the barium cation. Depending on which synthesis conditions are used (Eqs. 1 or 2) the solvent content can vary dramatically. In addition, small amounts of free β-diketone H(thd) can be detected. Through recrystallization from diethylenglycoldimethylether (DGL) or 1,2-dimethoxyethane (DME) it is possible to partially, or even completely, replace the coordinated solvent.

Figure 2 presents the combined results of Thermogravimetric (TG) and Differential Thermal Analysis (DTA) for two different barium compounds: (a) $Ba(thd)_2$ * 0.7 H_2O, and (b) $Ba(thd)_2$ * 1.45 DGL * 0.89 H(thd). The two features at 95 and 116 °C in the DTA signal of the ether adduct (Fig. 2(b)) can be related to the evaporation of the ether and the β-diketone, respectively. The shoulder on the low temperature side of the DTA curves is caused by ethanol impurities in the sample remaining from the synthesis process. The initial decrease in the TG signals near 100 °C reveals that heating of the barium compound leads first to the evaporation of the solvent. Assuming that the decrease of 5% in the TG curve at 90 °C in Fig. 2(a) is due to the evaporation of pure water one can calculate a water content of 1.5 formula units per formula unit $Ba(thd)_2$. NMR analysis, however, yields a water content of 0.7 formula units. This observation leads to the conclusion that the gases evaporating at 90 °C actually consist of a mixture of solvent and barium. This explanation is also consistent with the results obtained from CVD experiments where barium was deposited during the initial heating period.

Upon examination of Fig. 2(b) one observes a decrease of 21% in the TG curve at approximately 100 °C. Assuming that the solvent, whose content is known quantitatively from NMR analysis, evaporates completely at this temperature one would expect the TG signal to decrease by 42%. This means that 50% of the solvent remains in the substance. Consequently, the ether adduct and the hydrate show quite different behavior at higher temperatures. For example, the melting points of the hydrate and the ether adduct are 192 °C and 219 °C, respectively. Although their heats of vaporization are quite similar (119 kJ/mole and 127 kJ/mole for the hydrate and the ether adduct, respectively) the rate of evaporation at 250 °C of the ether adduct is twice as high as that of the hydrate (0.2 mg/min compared to 0.1 mg/min).

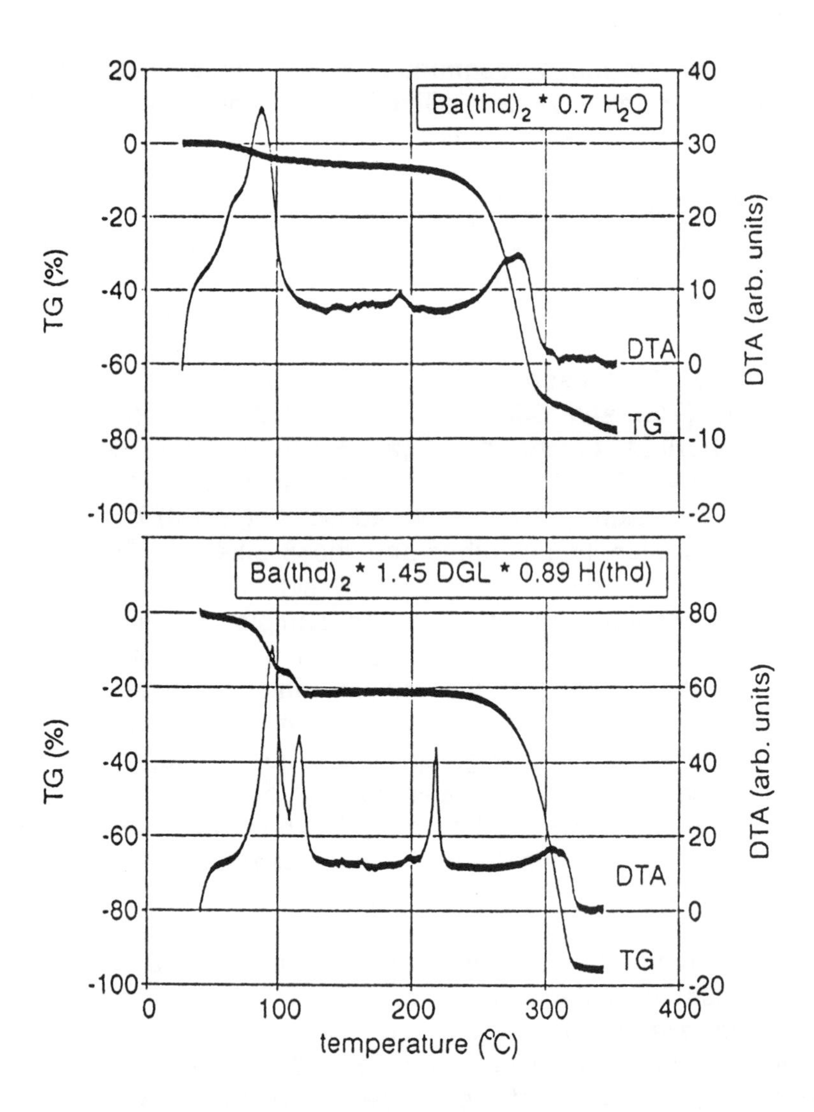

Fig. 2 Combined Thermogravimetric /Differential Thermal Analysis of (a) $Ba(thd)_2$ * 0.7 H_2O, and (b) $Ba(thd)_2$ * 1.45 DGL * 0.89 H(thd). The mass change of the sample is given in percentage of the starting mass (TG). The temperature difference between sample and reference (Al_2O_3) is given in arbitrary units (DTA). Positive values correspond to endothermal processes. The heating rate was 2 °C/min, the pressure was 3 mbar, and the flow rate of argon carrier gas was 10 ml/min (STP).

At present, the reasons for the differences in the evaporation behavior of the various solvates are unknown. It is clear, however, that in order to perform reproducible CVD experiments identical starting conditions are necessary, i.e. we prepare and use $Ba(thd)_2$ with constant solvent content.

EXPERIMENTAL SETUP AND PROCEDURES

Figures 3 and 4 show schematic drawings of the two different experimental arrangements used in the present study for the deposition of thin $YBa_2Cu_3O_{7-\delta}$ films. More detailed descriptions of the individual experimental setups can be found elsewhere[7,8].

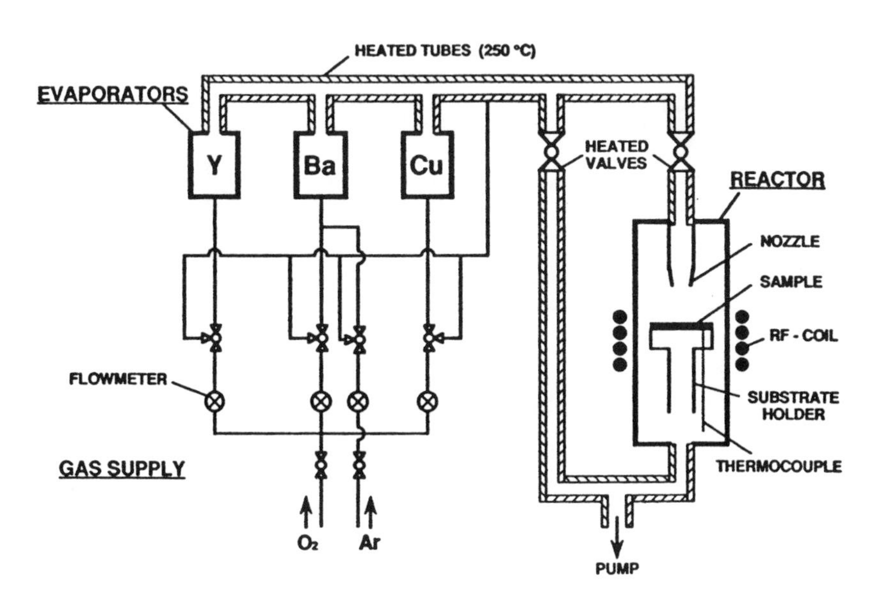

Fig. 3 Schematic diagram of the CVD arrangement (with carrier gas) used for the deposition of $YBa_2Cu_3O_{7-\delta}$ films.

Briefly, the setup in Fig. 3 is that of a cold-wall stagnation flow reactor which, under suitable flow conditions, allows homogeneous mass transport to the substrate surface, and, which minimizes reactions between reactor walls and reactive species. The substrates are heated inductively with the aid of a rf-generator (200 kHz, 2.5 kW). The temperature T_S of the substrate holder is measured with a NiCr/Ni thermocouple inserted into the holder. It should be noted that the temperature of the substrate surface is lower than the measured value by approximately 100 °C at T_s = 950 °C.

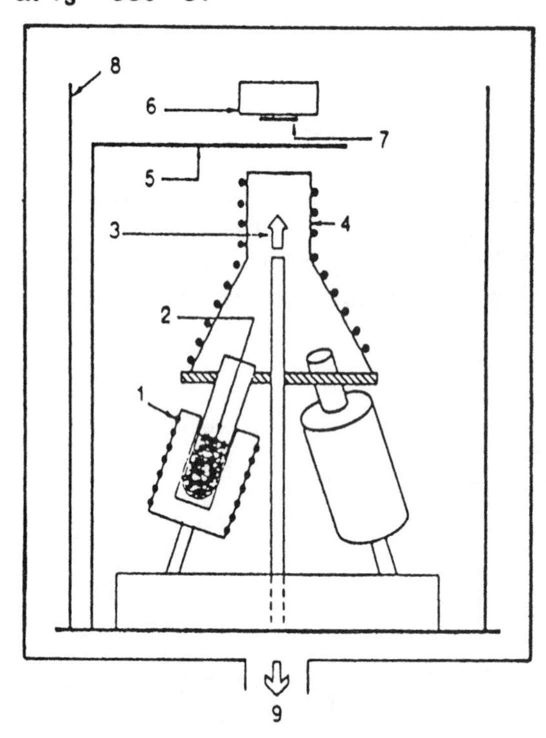

Fig. 4
Schematic diagram of the CVD apparatus operating without carrier gas.
1 = vapor sources
2 = precursor
3 = oxygen inlet
4 = heated chimney
5 = shutter
6 = sample holder
7 = substrate
8 = quartz cylinder to direct mass flow to substrate
9 = to pump.

The source materials, i.e. $Y(thd)_3$, $Ba(thd)_2$, and $Cu(thd)_2$, are contained in metal crucibles, placed inside three stainless steel evaporators. The temperature of each evaporator can be controlled independently. If it is assumed that a ratio of $Y(thd)_3$: $Ba(thd)_2$: $Cu(thd)_2$ = 1:2:3 in the vapor phase leads to stoichiometric $YBa_2Cu_3O_{7-\delta}$ films on the substrate surface the temperatures at which the source materials need to be evaporated can be extracted from Fig. 1 where three such temperatures labeled T_{v1}, T_{v2}, and T_{v3} corresponding to $Y(thd)_3$, $Ba(thd)_2$, and $Cu(thd)_2$, respectively, have been included.

Deposition with this particular arrangement was carried out at a system pressure of 10 mbar with a carrier gas transporting each precursor to the substrate. The flow rate through each evaporator is 20 l/h (STP); while oxygen is used for the transport of $Cu(thd)_2$ and $Y(thd)_3$, argon is used for $Ba(thd)_2$. Since the evaporation behavior of the source materials is not stable during the initial heating period (≈30 min) the gas is passed by the reactor during this initial stage thus avoiding uncontrolled deposition conditions. To prevent premature condensation of the reactive species en route to the substrate surface all transport lines and valves are heated to approximately 250 °C. Growth rates varied from 1 to 1.5 μm/h.

The apparatus presented in Fig. 4 offers an alternative approach to the deposition of high quality $YBa_2Cu_3O_{7-\delta}$ films. Here, the three temperature-controlled crucibles containing the precursors are inside the same vacuum chamber. The vapors are mixed together and guided to the substrate surface without the use of a carrier gas. Oxygen which is necessary for the growth of the High-T_c superconducting oxide is coaxially supplied at a rate of 12 l/h (STP) close to the substrate. During the preheating stage a shutter prevents uncontrolled deposition on the substrate surface. The main advantage of not using a carrier gas is the high concentration of active species in the vapor phase, thus leading to high growth rates (> 2 μm/h).

Typical deposition times ranged from 30 to 60 minutes. (100) $SrTiO_3$ as well as (100) MgO single crystals were used as substrates. Following deposition two different annealing protocols were employed for the different experimental setups. For the reactor shown in Fig. 3 the samples were cooled to room temperature at a rate of 15 °C/min in 1 atm of pure oxygen, for the reactor presented in Fig. 4 the samples were cooled in 1 atm of oxygen from 800 to 500 °C within 5 min, then held at 500 °C for 40 min, and then cooled to room temperature within 10 min.

RESULTS AND DISCUSSION

Figure 5 shows the dependence of the logarithmic YBCO deposition rate $\dot{s}$ on the reciprocal substrate holder temperature T_S. Note that these results were obtained with the stagnation flow reactor presented in Fig. 3. It is obvious from Fig. 5 that there are two different regimes corresponding to two main categories of a typical CVD process: At low temperatures (T_S < 600 °C) the deposition of $YBa_2Cu_3O_{7-\delta}$ is temperature dependent with an activation energy of E_A = 82 kJ/mole. In this region deposition is controlled by the kinetics of the thermal decomposition of the source materials on the hot substrate surface. This temperature regime should be favored if complex three-dimensional shapes, such as fibers, have to be coated homogeneously[9]. However, at these low temperatures ($T_s < 600$ °C) the surface mobility of adsorbed molecules is low and the supersaturation of the gas phase is high. This leads to a high nucleation rate but low crystal growth rate, and therefore results in a fine-grained deposit with poor superconducting properties.

At temperatures higher than $T_s = 600$ °C the YBCO deposition rate becomes independent of the substrate temperature. In this regime the deposition rate is limited by the mass transport of reactive species towards the substrate surface. Maximum deposition rates with the given set of experimental parameters ($T_{v1} = 112$ °C, $T_{v2} = 208$ °C, $T_{v3} = 117$ °C, $p_{tot} = 10$ mbar) were 20 nm/min. The deposition rate, however, can be increased by increasing the evaporation rates of the source materials. Deposition in the mass transport controlled regime has the advantage that the deposition rate is high and, moreover, that the surface mobility increases with increasing temperature T_s. This, together with a low supersaturation of the gas phase, provides suitable conditions for highly textured, or even epitaxial films.

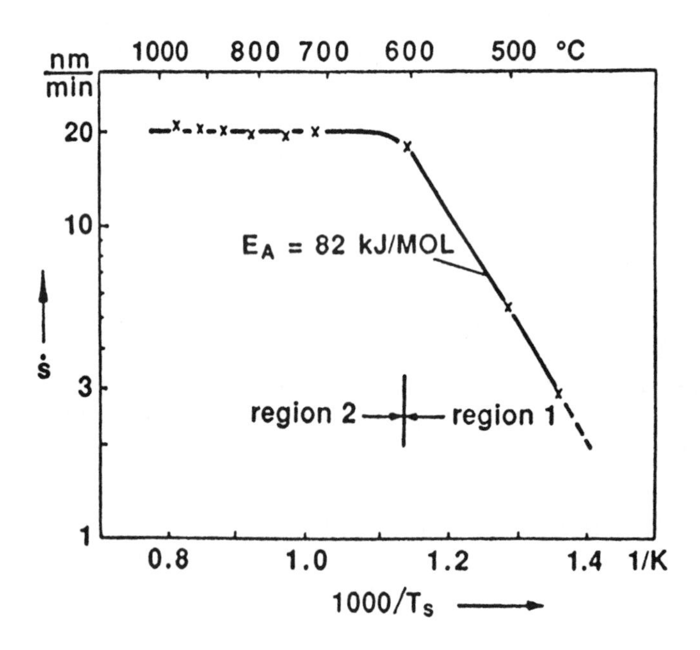

Fig.5 $YBa_2Cu_3O_{7-\delta}$ growth rate $\dot{s}$ as function of inverse absolute deposition temperature T_s.

The composition of the deposited films was determined by energy dispersive x-ray analysis (EDX) within a rectangular area of 45 x 100 μm, by comparison (linear extrapolation) with a stoichiometric $YBa_2Cu_3O_{7-\delta}$ standard. It has been found that for deposition conditions which fall within the mass controlled regime the composition of the films almost coincides with the composition of the vapor phase. This is not true, however, when deposition takes place in the kinetically controlled regime. Deviations between the composition of the deposited film and the composition of the gas phase increase with decreasing temperature T_S. This behavior can be explained by the different decomposition kinetics of the source materials with temperature[7].

Figure 6 shows a scanning electron micrograph of a typical $YBa_2Cu_3O_{7-\delta}$ film deposited at T_s = 950 °C onto a (100) $SrTiO_3$ single crystalline substrate. EDX analysis demonstrates that this film consists of a dense, smooth $YBa_2Cu_3O_{7-\delta}$ matrix with inserts of Ba-free precipitates. Examination of a cross section of the

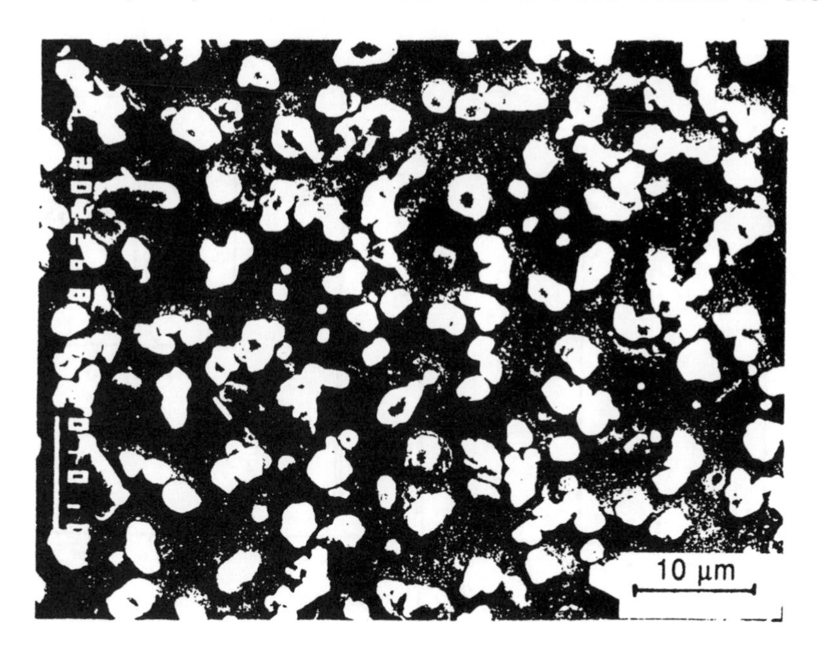

Fig. 6 SEM image of a typical film deposited at T_s = 950 °C onto (100) $SrTiO_3$. The smooth matrix consists of $YBa_2Cu_3O_{7-\delta}$, the precipitates are composed of Ba-free material.

film reveals that these precipitates reach from the substrate surface through the complete film, indicating that they were formed at a very early stage of the growth process. These precipitates are likely to be caused by variations in the vapor phase due to the strong temperature dependence of the evaporation process of the source materials (according to Fig. 2 a fluctuation of 1 °C in evaporation temperature results in a 10 % change of evaporation rate).

Figure 7 shows the x-ray diffraction pattern corresponding to the sample described above. The intense ($\approx 10^6$ cps) and narrow (00ℓ) reflections indicate significant preferential orientation of crystallite c-axes perpendicular to the surface of the substrate. A Θ-scan of the (007) peak gives a FWHM (full width at half maximum) value of 0.35° which is consistent with epitaxial growth. Calculations of the c-axis lattice parameter yield an oxygen content[10] of x = 6.9-7.

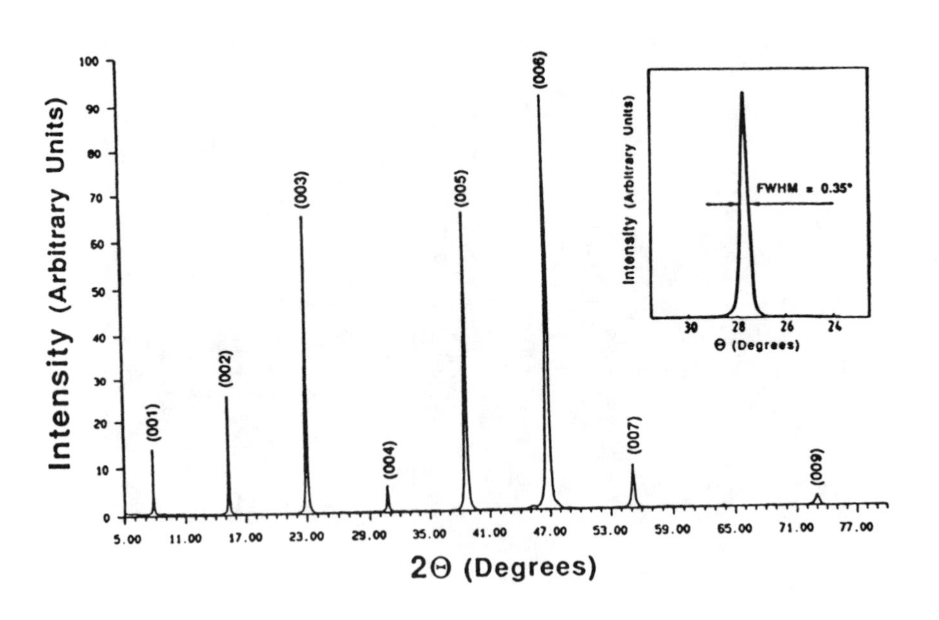

Fig. 7 X-ray diffraction pattern of $YBa_2Cu_3O_{7-\delta}$ deposited onto (100) $SrTiO_3$. The insert shows the rocking curve of the (007) peak.

$YBa_2Cu_3O_{7-\delta}$ films with a high degree of c-axis orientation exhibit a resistance vs. temperature behavior as presented in Fig. 8. Transition temperatures T_c exceeding 90 K are characteristic of such high quality films. Above T_c the (normalized) resistance shows metallic behavior, i.e. $dR/dT > 0$, and the linear extrapolation of the normal state is seen to pass through the origin. It should be added, however, that for $YBa_2Cu_3O_{7-\delta}$ films with inclusions of a, b-oriented grains the intercept with the vertical axis is not zero.

For measurements of the critical current densitiy, j_c, the films were patterned by wet chemical etching. At liquid nitrogen temperature, and in zero magnetic field, j_c values above 10^6 A/cm^2 were routinely obtained for high quality films, having thicknesses between 1 and 2 μm.

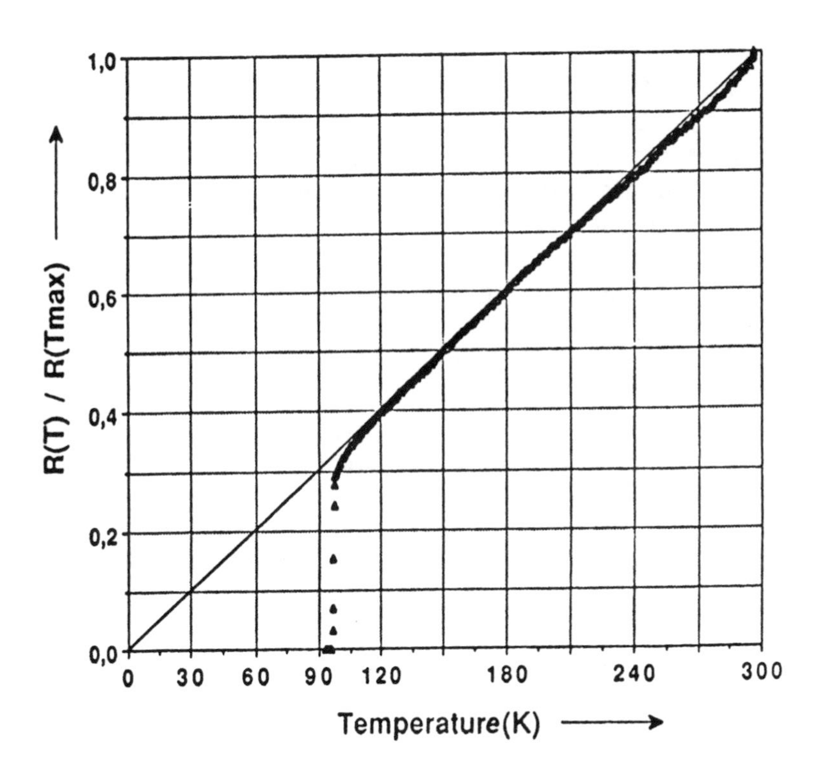

Fig. 8 Temperature-dependence of the (normalized) resistance of $YBa_2Cu_3O_{7-\delta}$ on (100) $SrTiO_3$.

CONCLUSION

It has been demonstrated that chemical vapor deposition (CVD) is a suitable technique to deposit High-T_c superconductors with good uniformity and electrical properties (T_c > 90 K, j_c(0T, 77 K) > 10^6 A/cm^2). Two different experimental reactors, operating with and without carrier gas, respectively, have been successfully employed for the growth of high quality $YBa_2Cu_3O_{7-\delta}$ films. At present, a drawback is the unstable evaporation behavior of the precursors which leads to the deposition of YBCO films with varying amounts of non-superconducting phases. Hence, the design of suitable precursors with stable, reproducible evaporation rates remains one of the most challenging areas of research.

REFERENCES

1. F. Schmaderer, R. Huber, H. Oetzmann, and G.Wahl, Appl. Surf. Sci. **46**, 53 (1990).
2. R.C. Mehrotra, R. Bohra, and D.P. Gaur, "Metal ß-Diketonates and Allied Derivatives", Academic Press, London 1978.
3. K.J. Eisentraut and R.E. Sievers, J. Amer. Chem. Soc. **87**, 5254 (1965).
4. K.J. Eisentraut and R.E. Sievers, J. Inorg. Nucl. Chem. **29**, 1931 (1967).
5. G.S. Hammond, D.C. Nonhebel, and C.-H.S. Wu, Inorg. Chem. **2**, 1973 (1963).
6. H. Zama and S. Oda, Jpn. J. Appl. Phys. **29**, L1072 (1990).
7. F. Schmaderer, R. Huber, H. Oetzmann, and G.Wahl, Metall **44**, 638 (1999).
8. B. Schulte, M. Maul, W. Becker, E.G. Schlosser, S. Elschner, P. Häussler, and H. Adrian, Appl. Phys. Lett. **59**, 869 (1991).
9. G. Wahl and F. Schmaderer, J. Mater. Sci. **24**, 1141 (1989).
10. R.J. Cava, B. Battlog, C.H. Chen, E.A. Riethman, S.M. Zahurak, and D. Werder, Nature **329**, 423 (1989), and Phys. Rev. **B36**, 5719 (1987).

SUPERCONDUCTING FILMS BY MOCVD ON IRREGULAR SURFACES

Anton C. Greenwald
Spire Corporation, One Patriots Park, Bedford, MA 01730

ABSTRACT

Spire is using an isothermal reactor to deposit YBaCuO and a diffusion barrier layer of zirconia at atmospheric pressure. The objective is to coat the inside of cylindrical microwave cavities.

Zirconia was found to be a good diffusion barrier between silica and YBaCuO, but coating efficiency of ZrO_2 on an unpolished ceramic was not complete. We tested atomic layer epitaxy (ALE) of YBaCuO to obtain a "seed" layer for later, faster epitaxial growth for an aligned crystal layer. ALE is defined as the deposition of a material one atomic layer at a time. Growth rates of Ba, Cu, and Y oxides were reduced to less than 100 Å per hour, implying deposition of a monolayer per minute. Growth of each layer was separated by two minutes, purging with an inert gas. Analysis showed excess yttrium in the final film, implying that the sticking coefficients of the source compounds or partially decomposed intermediates are significantly different.

In other tests, Spire showed that CVD of YBaCuO (with all sources running simultaneously) at atmospheric pressure in a uniformly heated furnace could be achieved without gas phase nucleation, which had been expected to be a potential problem. We also tested the use of ammonia as a carrier gas for transport of the barium TMHD compound, but found it to be sensitive to water content of the source material.

INTRODUCTION

The objective of this work is to produce a superconductive coating on the inside of cylindrical or irregularly shaped cavities for use in particle accelerators, radar oscillators, etc.[1] As the cavity wall cannot have a correctly aligned single crystal structure over the entire surface, some method must be found to align the structure of the superconducting material.

The concept of atomic layer deposition, where each atomic layer of the $YBa_2Cu_3O_7$ (YBCO) crystal is deposited separately, was tested for creating a seed layer on the inside of a cylinder. Atomic layer growth of the structure:

Cu - Ba - Cu - Y - Cu - Ba - Cu - Ba - Cu - Y - Cu - Ba - Cu ...

where each layer is a metal oxide has been shown to be feasible on a planar substrate using MBE.[2,3] Extremely smooth, well aligned superconducting films were achieved. The technique was also used to produce engineered compositions for the thallium and bismuth compounds. However, these techniques cannot be used to coat the inside of a cylinder.

Atomic layer growth of GaAs and related compounds has been demonstrated by metalorganic chemical vapor deposition (MOCVD). MOCVD can coat the inside of a cylinder (Figure 1) and it was the purpose of this research to test this technique as a suitable means of producing a seed layer for later epitaxial growth at higher film deposition rates.

Figure 1 *Coated cylinder split in half after processing in MOCVD reactor (left) and uncoated cylinder (right).*

The MOCVD process for GaAs deposited one layer at a time is effective because certain of the relevant source alkyls for gallium will be adsorbed on arsenic coated surfaces, but not adsorbed on gallium coated surfaces. The reverse is true for arsine gas. This chemical driving force to produce monolayer growth is not available for YBCO synthesis. All available source compounds are adsorbed on the oxide of the same metal. It is easy to deposit BaO_x, CuO_x, and Y_2O_3. To grow the desired seed layer extremely good uniformity and low deposition rates are required.

EXPERIMENT DESCRIPTION

MOCVD of YBCO, continuously without atomic layer growth has been demon-strated by many researchers.[3,4] Most of the published work used a cold wall reaction chamber where there was a thermal gradient over the heated substrate. For the cylindrical geometry, Spire had to use an isothermal reaction chamber, a tube furnace. The conditions used for this research are shown in Table I. When all sources were run simultaneously, the film was 500Å thick. Deposition time was eight hours at atmospheric pressure. RBS data from this film, Figure 2, shows correct stoichiometry.

Table I *Flow parameters.*

Source Chemical	Source Temperature (°C)	Carrier Flow (sccm)
Y $(TMHD)_3$	130	190
Ba $(TMHD)_2$	260	840
Cu $(TMHD)_3$	130	90

Oxygen flow 200 sccm
Total flow (argon added) 5,000 sccm
Temperature 550 °C

Spire worked with Harvard University to achieve transport of the barium compound at lower temperatures. While the reported results were initially encouraging,[1] transport of Ba TMHD with ammonia as a carrier gas was extremely sensitive to the hydration of the initial compound. Reproducibility was improved through the use of pure argon as a carrier gas, though it required higher temperatures.

Note the extremely low deposition rate for the film in Figure 1. Each metal, using these flow rates as a basis, could be deposited at a rate of one monolayer per minute. This was tested with the sequence of: 1 minute deposition; two minute purge with no metalorganics in the gas flow (but inert gas and oxidant and total flow remain constant); one minute deposition; etc. The sequence was repeated so that seven complete crystal layers of YBCO were deposited. The resulting film, approximately 50Å thick, was analyzed by Auger Spectroscopy, Figure 3.

The films deposited by MOCVD with attempts at atomic layering were yttrium rich. The presence of silicon in the surface signal implies the presence of pin holes in the film. However, the film appears to be perfectly flat and featureless at high magnification. The composition may be easy to adjust empirically; to improve uniformity of coating on a very fine scale, the experiment should be performed at low pressure.

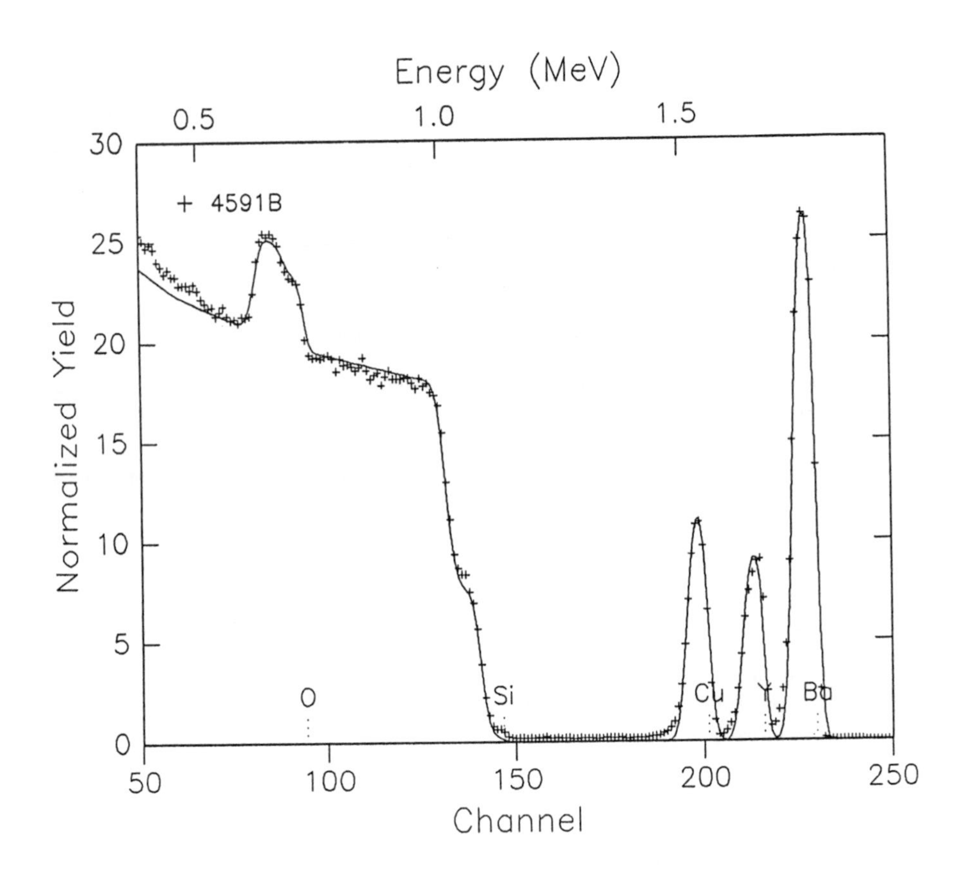

Figure 2 *Rutherford backscattering spectroscopy of thin film with all three metals.*

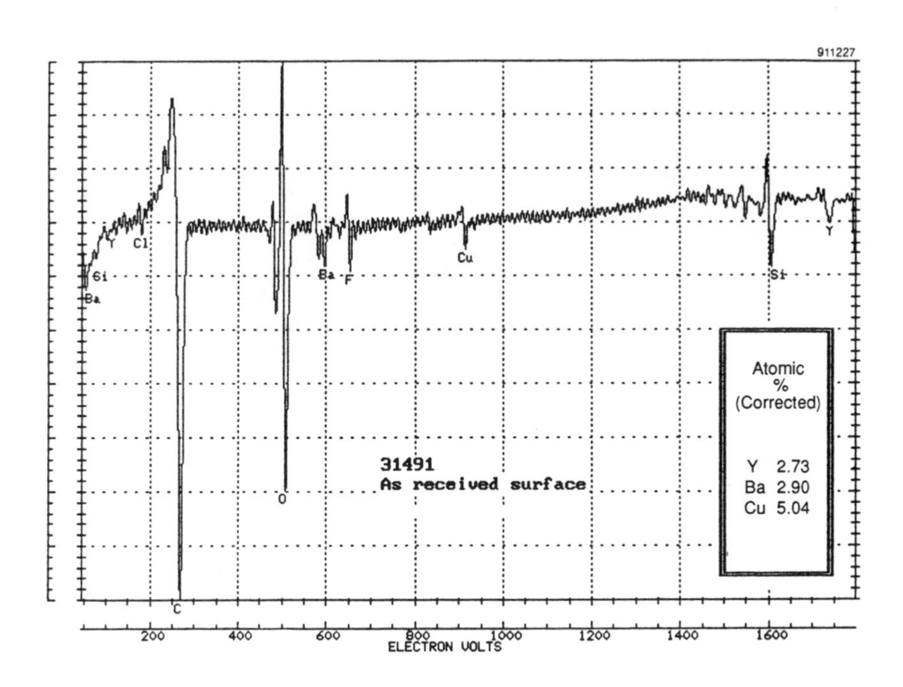

Figure 3 *Auger spectroscopy analysis of atomic layer MOCVD film.*

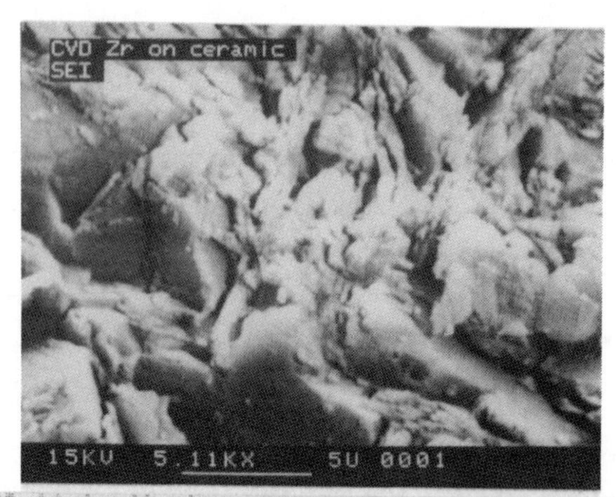

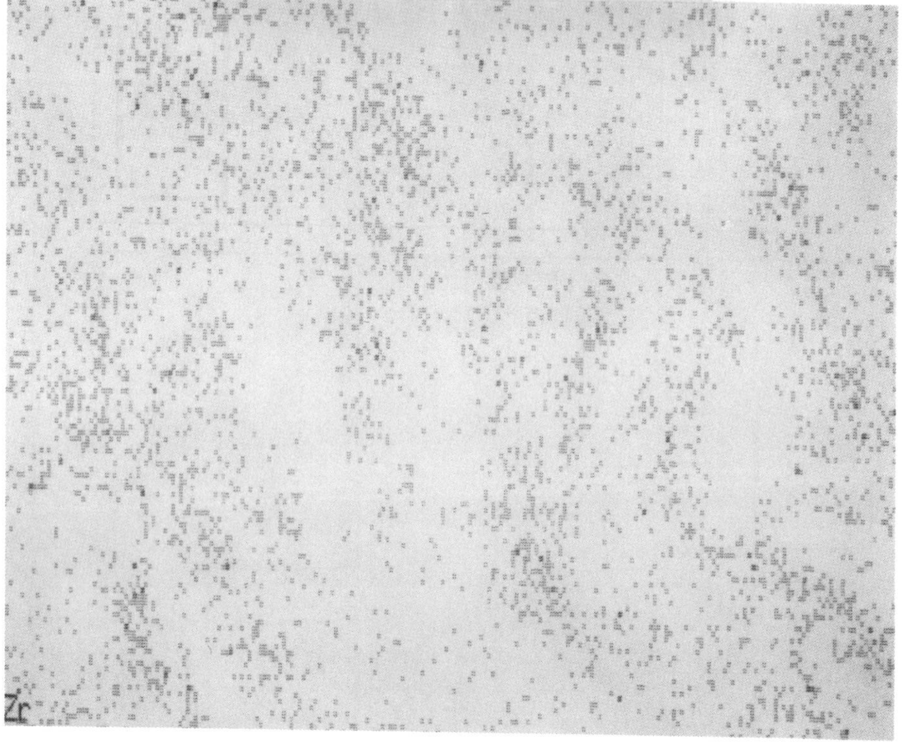

Figure 4 *Microphotograph of polished ceramic fabricated from pressed powders and Auger surface scan of Zr concentration, 0.5 μm CVD film.*

The planar substrates used for this work were all silicon or sapphire, highly polished with a ZrO_2 buffer layer deposited by ion beam assisted deposition.[2] To coat cylindrical substrates, zirconia, if used as a diffusion barrier, would also have to be deposited by MOCVD. Adequate coating rates could be obtained using zirconia trifluoro-acetylacetonate with argon carrier gas and added oxygen at atmospheric pressure at 550°C. The coating was not capable of sealing ceramics fabricated from pressed powders and polished. The uniformity of such a coating is shown in Figure 4.

CONCLUSIONS

MOCVD could coat the inside surface of cylindrical surfaces. Depositing one monolayer at a time of this stoichiometry was not possible at the given conditions. This may be due to the use of high pressure, surface adsorption kinetics, or surface reactions. Highly polished substrates formed from melt grown ceramics are necessary for continuous coatings less than one micron thick.

This work was performed with partial support from DOE.

REFERENCES

1. U.S. Patent 4,918,049
2. J.N. Eckstein *et al*, Appl. Phys. Lett. 57, 931 (1990).
3. A. Schuhl *et al*, Appl. Phys. Lett. 57, 819 (1990).
4. A. Greenwald, in Microelectronics Manufacturing Technology (May 1991).
5. Y.Q. Li *et al*, Appl. Phys. Lett. 58, 648 (1991).
6. J.M. Buriak *et al*, Mat. Res. Soc. Symp. Proc. v204, 545 (1991).
7. A. Greenwald *et al*, Mat. Res. Soc. Symp. Proc. v.169, 1153 (1990).

HIGH T_c CONDUCTORS BY MOCVD ON SILVER SUBSTRATES

D.W. Hazelton[*+], L. Chen[+], T.W. Piazza[+], A. Sweeney[+], A.E. Kaloyeros[+],

*Intermagnetics General Corporation, Guilderland, NY 12084
+State University of New York at Albany, Albany, NY 12222

ABSTRACT

Temperature-controlled chemical vapor deposition (TC-CVD) processing was employed to grow highly c-axis oriented $YBa_2Cu_3O_{7-x}$ (YBCO) superconducting thin films directly on silver substrates. Films were grown in-situ utilizing either an isothermal or a three stage fabrication scheme. While comparable J_c values were measured for the two different films, the three-stage TC-CVD processed film was much denser. It is postulated that in the first step, a seed layer of YBCO is nucleated on the Ag substrate. In the second step, the deposition is discontinued and the substrate temperature is quickly raised with a short hold to promote densification and oriented growth of the YBCO film. In the third step, the substrate temperature is lowered and the deposition continued with the additional film growth driven by the underlying oriented (c-axis perpendicular) YBCO seed layer. The samples were analyzed using dynamic impedance (DI), four point resistivity probe, x-ray diffraction (XRD) and scanning electron microscopy (SEM). These studies indicated that the films had a high degree of c-axis normal orientation, superconducting zero resistance transition temperatures exceeding 85 K and critical current densities of 1.4 - 1.6 x 10^4 A/cm^2 (77 K, self field). Thicknesses of the films were determined to be between 6000 Å and 6500 Å. These properties represent an appreciable improvement over those reported for YBCO films on silver produced by other chemical and physical deposition processes.

INTRODUCTION

This paper reports on the initial progress to date of a NYSIS sponsored program to develop an MOCVD-based high temperature superconducting (HTS) tape technology for use in power applications, particularly power transmission. Since the discovery of high temperature superconductors, numerous research and development efforts pertaining to the growth of HTS films have been conducted. Most of this research, however, has focused on HTS film growth on insulating single crystal substrates for electronics applications[1-4]. Fewer studies have been conducted on the deposition of HTS films on technologically useful metallic substrates for use as conductor in power applications[5-9]. Of the candidate metallic substrate materials, silver (Ag) is one of the most promising due to its chemical compatibility[10], its useful mechanical properties and its availability at moderately low cost. Based on these factors, work by the present investigators has focused on the direct growth by MOCVD processes of YBCO thin films on silver substrates. Direct growth of YBCO films on Ag substrates is appealing since it eliminates the need for a buffer material, thereby making the process more economical and easier to incorporate into a manufacturing process amenable to long length fabrication.

There are two requirements to be considered when producing conductor for power applications by thin film processes. These are the need to produce "thick" films with uniform properties so that useful currents can be carried and the need to produce dense, mechanically rugged films able to withstand the fabrication and operational stresses present in the power application environment. Using the TC-CVD technique, we have been able to produce comparatively dense films of increasing thickness when compared to isothermal techniques.

SAMPLE PREPARATION

The samples were prepared using temperature controlled CVD (TC-CVD)[11,12] in a custom, vertical, cold wall CVD system using the β-diketonate precursors $Y(tmhd)_3$, $Ba(tmhd)_2$ and $Cu(tmhd)_2$ (where tmhd = 2,2,6,6-tetramethyl-3,5-heptanedionato) as the elemental sources for yttrium, barium and copper, respectively. Pure (99.9%) Ag substrates were prepared using Ag sheet and were approximately 20 mm x 5 mm x 0.127 mm in size. The surface of the Ag substrates was polished with 1 micron diamond polish before each TC-CVD run.

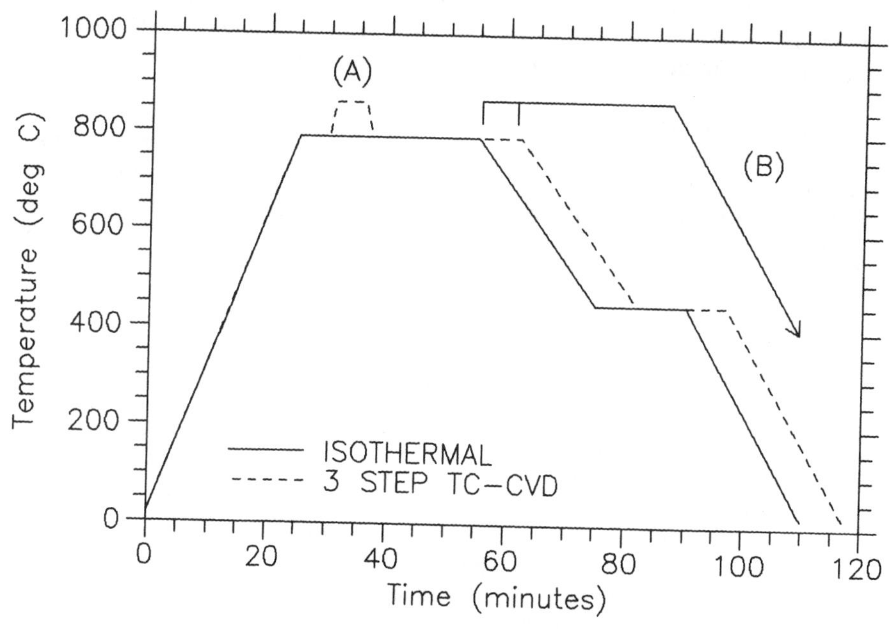

Figure 1. Temperature-time profile of the two sample preparation techniques. Reactor working pressure is 5 to 10 torr (principally N_2O) except as noted. Region (A) working pressure is 50 torr dry O_2. Region (B) working pressure is 760 torr dry O_2.

Samples were prepared using either an isothermal or a three stage TC-CVD fabrication scheme. The two processes are compared in Figure 1. During the deposition steps, the N_2O carrier gas flow rates were set at 400 sccm for the Ba precursor and 100 sccm for the Y and Cu precursors. The precursors were sublimed at 240°C, 150°C and 120°C respectively for the Ba, Y and Cu. The reactor working pressure was held between 5 to 10 torr except as noted. The substrate was heated from room temperature to 790°C in 20 to 30 minutes in N_2O before deposition was initiated. In the case of the isothermal sample, the deposition was run for 30 minutes at 790°C followed by a slow cool in 760 torr of dry oxygen to room temperature with an intermediate hold at 450°C for 15 minutes. For the three stage TC-CVD sample, the first stage consisted of a 5 minute deposition at 790°C. At this point, the deposition was discontinued and the substrate temperature quickly raised to 860°C in 50 torr of dry oxygen with a 5 minute hold to promote densification and oriented growth of the YBCO film. In the third step, the substrate temperature was quickly lowered back to 790°C and the deposition continued for an additional 25 minutes. At this point, the sample was slow cooled in 760 torr of dry oxygen to room temperature with an intermediate hold at 450° for 15 minutes.

RESULTS

Measurements of critical temperature were carried out using dynamic impedance (DI) and four point resistivity probe measurements. The dynamic impedance technique yields similar results to AC susceptibility except that the DI technique uses a single inductive coil and measures the out of phase (reactive) component at a set drive frequency. This technique gives a direct measure of the impedance change in the coil caused by the coupling between the coil and the eddy currents induced in the film. The onset of the drop in impedance of the HTS film generally coincides with the point of zero resistance in a standard four point probe measurement. However, the DI technique samples a much larger volume of material and is not limited to just a percolating pathway[13]. The dynamic impedance curve (Figure 2) for the isothermal film indicates that the film has a zero transition temperature T_c of 87 K with a 3 K transition width to a totally superconducting film volume. The dynamic impedance curve for the 3 stage TC-CVD film indicates that the film has a zero transition temperature T_c of 87 K with a 6 K transition width to a totally superconducting film volume. The very smooth and sharp transition of the curves indicates that the films consist primarily of single phase superconducting material.

X-ray diffraction measurements (Figure 3) of the films showed only ($00l$) peaks (where $l = 1,2,...$) implying that the YBCO films are highly oriented with the c-axis normal to the substrate surface. Examination by SEM (Figure 4) corroborated the XRD results showing a high degree of a-b platelet formation with c-axis normal orientation. The SEM examination also showed that the isothermal film (Figure 4a) exhibited some porosity while the 3 stage TC-CVD film (Figure 4b) exhibited a much denser structure. Energy Dispersive

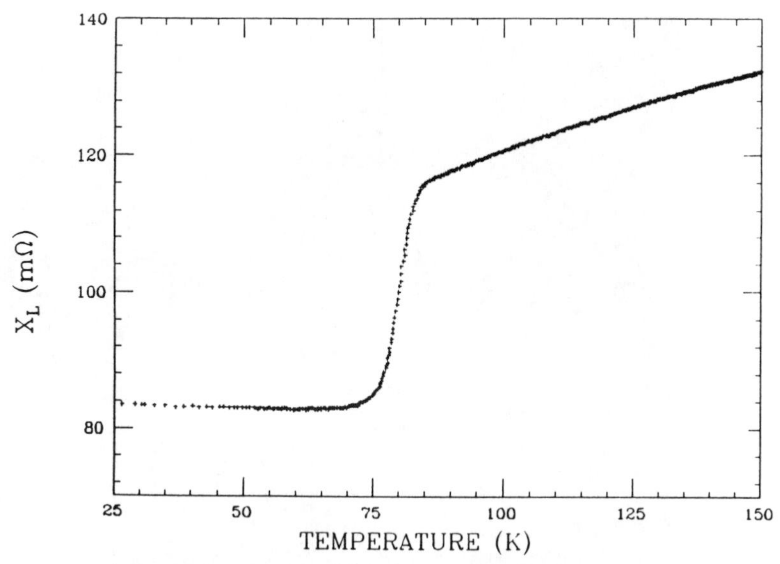

Figure 2. T_c determination using dynamic impedance versus temperature. This curve for the isothermal sample is typical of the samples tested.

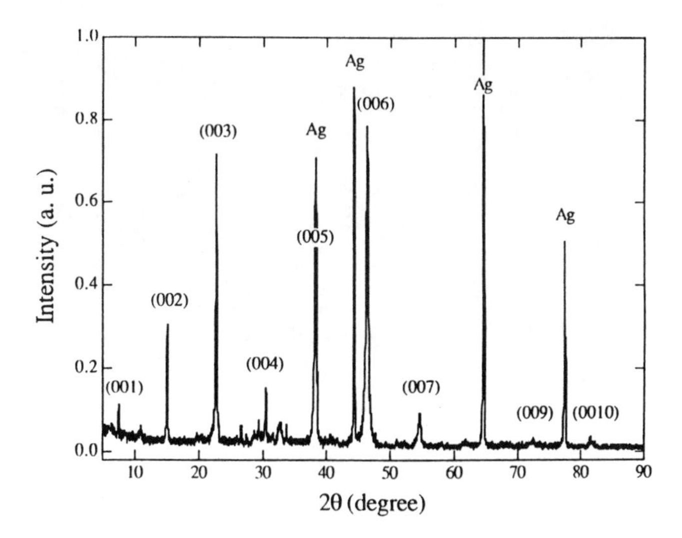

Figure 3. XRD trace of the three-stage TC-CVD sample exhibiting a high degree of c-axis normal orientation.

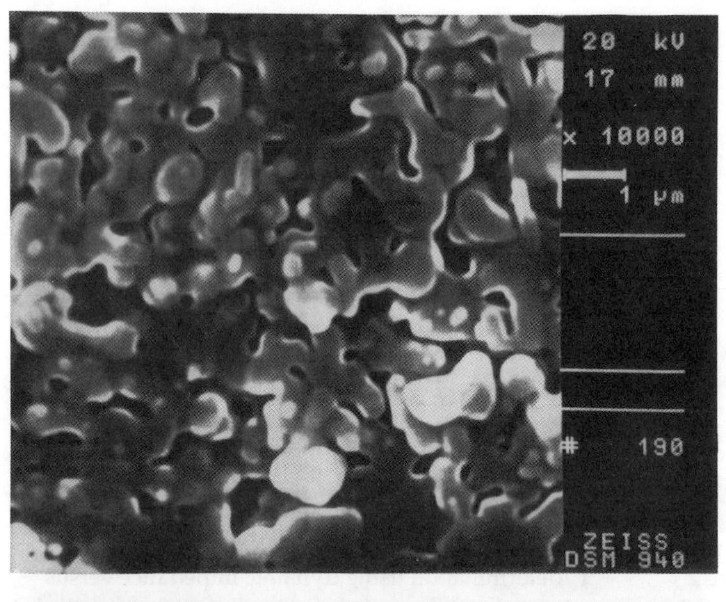

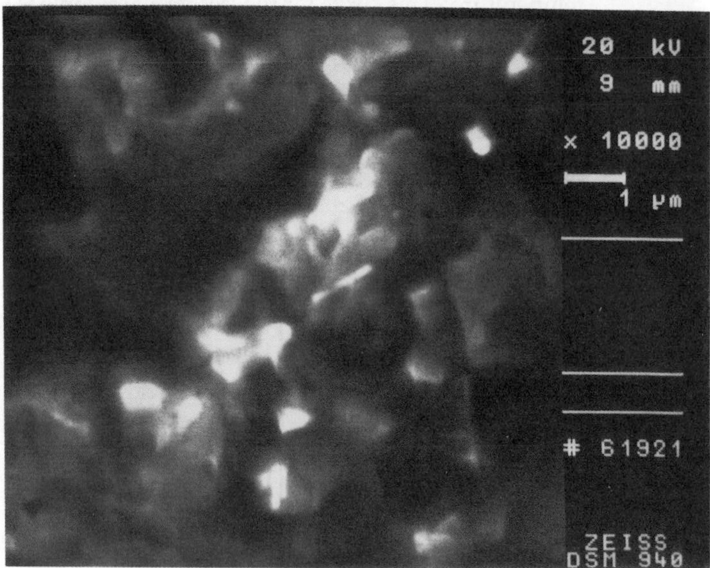

Figure 4. (a) Surface of isothermal film exhibiting porosity. (b) Surface of three-stage TC-CVD film exhibiting a much denser structure. Note high degree of a-b platelet formation with c-axis normal.

Spectroscopy (EDS) measurements yielded film cation ratios comparable to ones obtained with a reference $YBa_2Cu_3O_{7-x}$ sample. Rutherford Backscattering Spectroscopy (RBS) analysis yielded film thicknesses of about 6000 A for the isothermal film and 6500 A for the 3 stage TC-CVD film. This gives an average film growth rate of about 200 Å/min. for the isothermal film and 175 Å/min. for the 3 stage TC-CVD film.

The critical current densities (J_c) of the films were measured using an adaptation of the dynamic impedance technique used to measure the critical temperatures. DC transport currents are passed through the samples immersed in liquid nitrogen. When the current level reaches J_c, the sample becomes normal with a distinct change in the impedance signal. The correlation between these measurements and four point J_c measurements on numerous samples has been very good[14].

The 77 K, self field critical current densities for the two films are 1.6×10^4 A/cm^2 and 1.4×10^4 A/cm^2 for the isothermal and 3 stage TC-CVD films respectively. J_c measurements in field are planned in order to determine if the grains are weak linked coupled or exhibit strong coupling with good J_c vs. H behavior.

DISCUSSION

YBCO films produced by both the isothermal and TC-CVD techniques exhibit high superconducting transition temperatures and higher critical current densities than those reported in the literature for YBCO films grown on silver by Plasma Enhanced CVD (PE-CVD) and Physical Vapor Deposition (PVD) processes at constant substrate temperatures[5-9]. For example, Zhao et al[5] report a T_c of 85 K with a transition width of 8 K on a sample prepared using PE-CVD on Ag, although no J_c results are given. Similarly, Russo et al[7] and Witanachchi et al[6] report, respectively, a T_c of 84 K and a J_c of around 10^3 A/cm^2 (at 67 K, B=0) for laser-deposited YBCO on a Ag buffered Pt substrate and a T_c of 83 K and J_c around 4×10^3 A/cm^2 (at 40 K, B=0) for laser-evaporated YBCO on Ag buffered stainless steel.

The improved density of the YBCO films produced using TC-CVD is believed to enhance the mechanical stability of the films by eliminating regions of high stress concentration within the film which could lead to crack formation. There are indications in the literature[15,16] that improving the density of HTS materials enhances the J_c vs. strain tolerance of the superconducting phase.

A potential model for the nucleation and growth by TC-CVD of improved quality YBCO films is presently under investigation. Additional studies are presently underway to optimize the TC-CVD process and to further understand the underlying physical mechanisms that control the growth of high quality HTS films on Ag.

The following mechanism is postulated to explain the differences seen in the isothermal and three-stage TC-CVD process films. First, consider the silver substrate surface structure. The starting Ag foil substrate has a definite texture due to the rolling used to form the foil. However, as the Ag substrate is heated before deposition, the Ag surface is cleaned, annealed and recrystallized forming equiaxed grains on the order of 20-100 μm in size. It is postulated that the recrystallization of the Ag results in the formation of steps on the exposed surface of the Ag grains. As deposition is initiated at 790°C, these surface steps, in addition to the Ag grain boundaries, provide preferential nucleation sites for the formation of the initial YBCO nuclei. Evidence for such a nucleation mechanism has been demonstrated by Norton et al[17] in the growth of YBCO films on MgO. The preferential c-axis orientation of YBCO at higher temperatures has been demonstrated in a number of studies[18-20]. The rough (stepped) surface of the Ag substrate limits the sideways growth of the c-axis oriented nuclei. This results in the YBCO crystallization being dominated by nucleation rather than by growth, leading to a high c/a axis length ratio in the widely spaced nuclei. Basu et al[21] have presented a similar argument for YBCO films deposited on $SrTiO_3$ with different surface conditions. In the isothermal case, as deposition is continued and the film thickness increases, the influence of the substrate is diminished and a-b platelet growth of the YBCO begins to dominate. However, since this growth is initiated from many different YBCO sites, the result is a porous film as the growing YBCO platelets begin to impinge on each other.

In the case of the three-stage TC-CVD film, it is speculated that as the temperature is raised to 860°C (along with the pO_2 to maintain YBCO phase stability[22]), the increased surface mobility at the higher temperature minimizes the limiting effect of the Ag surface steps and the YBCO crystallization is dominated by growth. The YBCO nuclei with high c/a axis length ratios deposited during the first stage are driven by surface energy considerations to transform into a dense film of low c/a axis length ratio platelets. When the temperature is lowered and the deposition continued, the influence of the substrate is diminished and the YBCO film formation is dominated by growth from the low c/a axis length ratio platelets resulting in improved film density.

CONCLUSION

Initial results of MOCVD deposition of YBCO on Ag are encouraging with the fabrication of dense films exhibiting high T_c's and useful critical current densities. The use of polycrystalline Ag as a viable substrate material has been demonstrated provided proper processing is utilized. Additional characterization, process optimization and understanding of the nucleation and growth mechanism is required to move the process forward.

ACKNOWLEDGEMENTS

The authors would like to thank L. Luo, R.C. Dye and C.J. Maggiore of the Los Alamos National Laboratory for discussions and XRD, T_c and J_c measurements. This work was supported in part by the New York State Institute on Superconductivity (NYSIS) under grant #88F032 and grant #90F028 and in part by the National Science Foundation (NSF) Presidential Young Investigator Award #DMR-9157011.

REFERENCES

1. X.D. Wu, R.E. Muenchausen, S. Foltyn, R.C. Dye, A.R. Garcia, N.S. Nogar, P. Enland, R. Ramesh, D.M. Hwang, T.S. Ravi, C.C. Chang, T. Venkatesan, X.X. Xi, Q. Li and A. Inam, Apply. Phys. Lett. 57, 523 (1990).
2. D.M. Hwang, T.S. Ravi, R. Ramesh, S.W. Chan, C.Y. Chen, L. Nazar, X.D. Wu, A. Inam and T. Venkatesan, Appl. Phys. Lett. 57, 1690 (1990).
3. Y.Q. Li, J. Zhao, C.S. Chern, W. Huang, G.A. Kulesha, P. Lu, B. Gallois, P. Norris, B. Kear and F. Cosandey, Appl. Phys. Lett. 58, 648 (1991).
4. M.P. Siegal, J.M. Phillips, R.B. vanDover, T.H. Tiefel and J.H. Marshall, J. Appl. Phys. 68, 6353 (1990).
5. J. Zhao, Y.Q. Li, C.S. Chern, P. Norris, B. Gallois, B. Kear and B.W. Wessels, Appl. Phys. Lett. 58, 89 (1991).
6. S. Witanachchi, S. Patel, Y.Z. Zhu, H.S. Kwok and D.T. Shaw, J. Mater. Res. 5, 717 (1990).
7. R.E. Russo, R.P. Reade, J.M. McMillan and B.L. Oslen, J. Appl. Phys. 68, 1354 (1990).
8. J. Saitoh, M. Fukutomi, Y. Tanaka, T. Asano, H. Maeda and H. Takahara, Jpn. J. Appl. Phys. 29, L1117 (1990).
9. A. Kumar, L. Ganapathi, S.M. Kanetkar and J. Narayan, Appl. Phys. Lett. 57, 2594 (1990).
10. C.T. Cheung and E. Ruckenstein, J. Mater. Res. 4, 1 (1989).
11. A. Feng, L. Chen, T.W. Piazza, A.E. Kaloyeros, D.W. Hazelton, L. Luo and R.C. Dye, Appl. Phys. Lett. 59(9) (1991).
12. A.E. Kaloyeros, L. Chen and A. Feng, "A Temperature-Controlled Chemical Vapor Deposition Process," Patent Pending (1991).
13. H.L. Libby, "Introduction to Electromagnetic Non-Destructive Test Methods," Robert E. Krieger Publ. Co., Florida (1971).
14. L. Luo, R.C. Dye, Los Alamos National Laboratory, private communication.

15. K. Sato, T. Hikata, H. Mukai, M. Ueyama, N. Shibuta, T. Kao, T. Masuda, M. Nagata, K. Iwata and T. Mitsui, IEEE Trans. Mag. 27(2), 1231 (1991).

16. T. Kuroda, presented at the Intl. Cryogenic Mat'l. Conf., Huntsville, AL, June 1991.

17. M.G. Norton, L.A. Tietz, S.R. Summerfelt and C.B. Carter, "Materials Research Society Symposium Proceedings, Vol. 169," Materials Research Society, Pittsburgh, PA, 509 (1990).

18. C.C. Schuler, F. Greuter and P. Kluge-Weiss, "Materials Research Society Symposium Proceedings, Vol. 169," Materials Research Society, Pittsburgh, PA, 1255 (1990).

19. H. Asano, H. Yonezawa, M. Asahi and O. Michikami, IEEE Trans. Mag. 27(2), 844 (1991).

20. S.E. Russek, B. Jeanneret, D.A. Rudman and J.W. Ekin, IEEE Trans. Mag. 27(2), 931 (1991).

21. S.N. Basu, N. Bordes, A. Rollett, M. Cohen and M. Nastasi, "Materials Research Society Symposium Proceedings, Vol. 169," Materials Research Society, Pittsburgh, PA, 1157 (1990).

22. R. Bormann and J. Noelting, Appl. Phys. Lett. 54, 21 (1989).

VAPOR PHASE PROCESSING OF HTS FILMS FOR TECHNOLOGICAL APPLICATION

Ling Chen, Timothy W.Piazza, Jean E.Kelsey, Hongwen Li, and Alain E.Kaloyeros
Department of Physics, The University at Albany-SUNY, New York 12222

ABSTRACT

Work by the present investigators has focused on the growth of HTS films on technologically useful substrates for energy-related and cryoelectronic applications. In this paper, results from two such studies are presented: (1)Temperature-controlled chemical vapor deposition (TC-CVD) processing was used to grow $YBa_2Cu_3O_{7-\delta}$ (YBCO) thin films directly on sapphire substrates. Films were grown by employing a three-step substrate temperature ramping scheme: first, film was deposited at 730°C for 5 minutes; then the substrates were heated up to 825°C rapidly and held at that temperature for 5 minutes with no deposition; and finally the film was deposited for 10 more minutes using a the substrate temperature ramping from 825°C to 730°C. The films were characterized utilizing dc transport four point measurements, and scanning electron microscopy (SEM). These measurements showed the film of highly c-axis oriented with Tc around 90K. It is postulated the intermediate high deposition temperature helps to form a high c-axis seed layer, and subsequent lower deposition temperature diminished the chemical interactions between the film and the substrate. (2)A multi-step CVD process was employed for the growth of YBCO films a sapphire layer on silicon. A sapphire thin film layer of 2000Å (as determined by Rutherford backscattering measurement (RBS)) was first grown on Si by CVD. This was followed by the CVD growth of YBCO on the sapphire layer. Subsequent film characterization showed that these initial studies produced an YBCO film with an onset transition temperature of 90K but with a broad transition region.

INTRODUCTION

Microcircuit design and manufacturing innovations are undergoing rapid evolution towards the realization of higher integration density with smaller design rules in high density interconnect electronic substrates. High density interconnection and packaging schemes exist because electrical conductor resistance and capacitance stand as major limiting factors to the signal processing throughout the new high speed integrated circuits. The current research is based on the expectation that high temperature superconductor (HTS) technology can provide the critical improvements needed in electronic packaging scheme, especially in the 100 MHz to 10 GHz range, due to the exceptional low surface impedance or sheet resistance of superconductors compared to normal conductors.[1-3] For example, Withers and co-workers compared the frequency dependence of surface resistance for YBCO film with copper at 77K. The results show that the surface resistance of the normal conductor varies as $f^{1/2}$, as a result of the skin depth, while that of the superconductor varies as f^2, as a consequence of two-fluid models. As a result, in the frequency range from 100 MHz to 10 GHz the surface resistance of the OFHC copper is one to two orders of magnitude higher than that of YBCO at 77K[1]. The development of cryoelectronic packaging, however, requires solving key material and processing issues which include the identification of a suitable buffer layer/diffusion barrier which is compatible with both superconductor and silicon technologies; the growth of high quality single layered HTS film on such barrier on silicon; and the development of a multi-layered structure of superconductor/insulating layer thin films.

Sapphire is one good candidate for such applications because of its relatively low dielectric constant and its compatibility with silicon technology. Unfortunately, although chemical vapor deposition (CVD) and physical vapor deposition (PVD) techniques have been successfully applied to the fabrication of epitaxial HTS thin films with high critical current densities, in excess of 10^6 A/cm^2 at 77 K in zero magnetic field, on single crystal $SrTiO_3$, YSZ, $LaAlO_3$ and MgO substrates,[4-7] little progress has been obtained on the growth of HTS films on sapphire.[8-10] A major obstacle is the undesirable thermally- and chemically-induced HTS-sapphire interactions which lead to the degradation of superconductor characteristics and performance.

Work by the present investigators has focused on the development of CVD processing scheme for the fabrication of HTS thin films on sapphire with the quality and properties required for incorporation in cryoelectronic packaging scheme. This goal requires a research program which focuses initially on: (1) the development and optimization of a CVD process for the growth of high-quality YBCO films on single crystal sapphire substrates, and (2) the development of a CVD process for the in-situ growth of the multi-layered YBCO/sapphire structure, possibly with up to eight alternating layers, needed for cryoelectronic applications.

In this paper, initial results are reported from both research tasks. A temperature-controlled CVD (TC-CVD) process was employed to grow YBCO films directly on single-crystal sapphire substrates. This process was already successfully employed to fabricate high-quality YBCO thin films on polycrystalline MgO,[11] and on silver [12] without an alien buffer layer. The process has shown some unique advantages in reducing film substrate chemical interactions and improving film quality. Simultaneouslya two-step CVD was used to grow YBCO films on a 2000Å sapphire layer on silicon. Although the initial results from both tasks are promising, much further work is needed to optimize and fully develop both processes.

GROWTH OF YBCO ON SINGLE-CRYSTAL SAPPHIRE BY TC-CVD

The experiment was carried out in a horizontal hot-wall resistively-heated quartz reactor. The substrate temperature was electronically controlled with a programmable temperature control system. Three programmable electronic temperature controllers were also employed for sublimator heating and were interfaced to the substrate temperature controller to permit close coupling of the three sublimator temperatures to that of substrate. The β-diketonate complexes $Y(tmhd)_3$, $Ba(tmhd)_2$, and $Cu(tmhd)_2$ (where tmhd =2,2,6,6-tetramethyl-3,5-heptanedionato) were used as the source compounds for yttrium, barium, and copper, respectively. The sublimed source compounds were carried on to the substrates by N_2O gas with the gas flows for the Y, Ba,and Cu sources set at 490 sccm, 420 sccm, and 560 sccm, respectively.

One unique feature of the TC-CVD process is the controlled ramping of substrate temperature during film growth. The temperature profiles of the substrate and sublimators are as follows. Prior to the actual deposition, the substrates were heated to an initial temperature of 730°C. Meanwhile the sublimators were heated to their precursor sublimation temperatures, namely, 120°C, 240°C, and 115°C respectively for the Y, Ba, and Cu sources, and kept at these temperatures during deposition. A three-step substrate temperature ramping scheme was employed: first, film was deposited at 730°C for 5 minutes; then the substrates were heated up to 825°C rapidly and held at that temperature for 5 minutes with no deposition; and finally the film was deposited for 10 more minutes using a the substrate temperature ramping from 825°C to 730°C. It is believed that during the intermediate high temperature growth process, because of the high surface mobility the initial film shall rearrange, and subsequently a seed layer with highly c-axis oriented grains is formed.[13] After the deposition, the reactor was then filled up to one atmosphere of dry oxygen while the substrates were slowly cooled down to room temperature.

The superconducting transition temperature of the YBCO films on sapphire was measured by conventional dc four point probe. As can be seen from figure 1, the measurements yielded the superconducting transition onset of $T_{c,onset}$ = 90 K and the zero resistance transition temperature of T_{c0} = 81 K. The films also exhibited good metallic behavior in the normal state with R(280K)/ R(90K) = 2. The films thickness were around 1800Å as determined by Rutherford backscattering analysis (RBS).

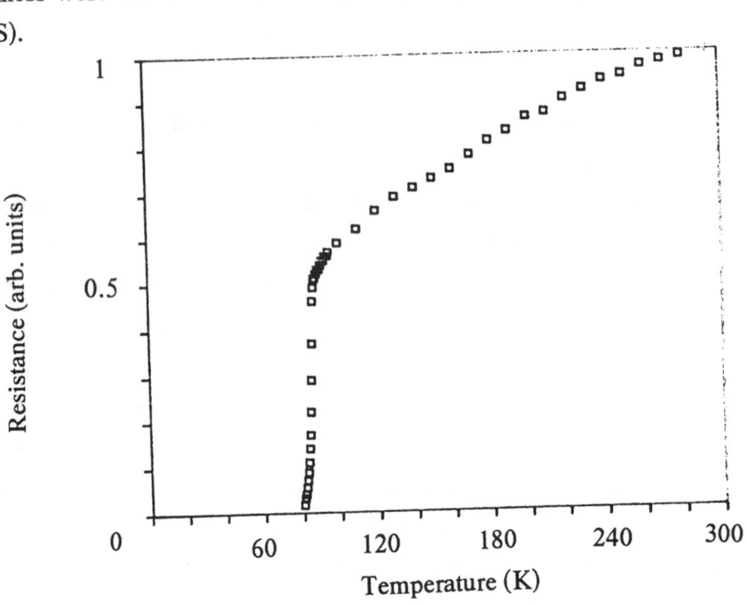

Figure 1. Temperature transition of the YBCO film on sapphire substrate was measured by dc four point probe.

Scanning electron microscope (SEM) was also performed on the films to study the surface morphology. The corresponding micrograph, shown in the figure 2, indicates a texture of superconducting YBCO grains approximately a micron in size.

The improved quality of the YBCO thin films on sapphire can be attributed to the nature of the TC-CVD temperature profile. As pointed out earlier, it is believed that the initial high substrate temperature is preferential to the formation of a dense and highly c-axis oriented YBCO seed layer because of the high atomic surface mobility and the characteristics of the YBCO system.[11-13] According to the RBS experiments, an excellent fit of simulated and experimental spectra was achieved when a two layer structure model of YBCO films was used, namely, a intermediate layer of Y_2BaCuO_5, Y_2O_3 and CuO and a layer of YBCO grown on top of it. The intermediate layer of a mixture of yttrium, barium and copper compounds is believed to play a dual role as non alien seed layer for the growth of the YBCO film and as a buffer layer that minimizes the diffusion of aluminum into the superconductor films. Once the seed layer is formed, film growth can proceed at a relatively lower substrate temperature with the seed layer promoting the growth of the YBCO film with the desirable grain orientation. The reason for this could be that the thermal energy required for the growth of the film is lower than that needed for the initial nucleation. The lower substrate temperature also helps minimize the chemical and thermal interaction between sapphire and YBCO film. However, much further work is needed to improve the quality of YBCO film on sapphire in terms of the temperature transition sharpness and the current density. More extensive systematic investigations of the temperature ramping profile are being conducted to accomplish this goal.

GROWTH OF YBCO ON SAPPHIRE FILM ON SILICON BY CVD

A two-step growth process was carried out in a hot-wall quartz CVD reactor to fabricate YBCO thin films on sapphire layer on silicon. In the first step, the β-diketonate compound $Al(acac)_3$ (where acac = acetylacetonate) was used as the precursor for the deposition of sapphire films. O_2 was used as the carrier gas, and the pressure in the reactor chamber was kept at 5 torr. The sublimator was heated to 120°C to sublime the compound which was introduced with the O_2 carrier gas into the reactor chamber, where it was exposed to the silicon substrates heated to 400°C. The compound subsequently decomposed at this temperature and yielded a sapphire film on the silicon substrates. The deposition time was about 30 minutes, and the deposition rate was around 100Å/min. The film of uniform blue color was then analyzed by RBS shown in figure 4. The RBS results yielded signals for Al and oxygen and indicated some carbon contamination, which might explain the film color. The problem of carbon contamination can be eliminated by using an appropriate carrier gas, such as H_2, or by using different precursors, which is presently under investigation.[14-17] The RBS measurements yielded a film thickness of around 2000Å.

Subsequently, an YBCO film was deposited on the sapphire layer on silicon substrate by using the TC-CVD process described above. A conventional dc four resistivity probe was then used to examine the superconducting transition temperature of the YBCO film thus produced. It showed an onset for the superconducting transition temperature of 90K with a broad transition region ending with a T_{c0} about 60K (see figure.4). Although this preliminary data is promising, it is clear that much further work is needed to optimize and fully develop this process.

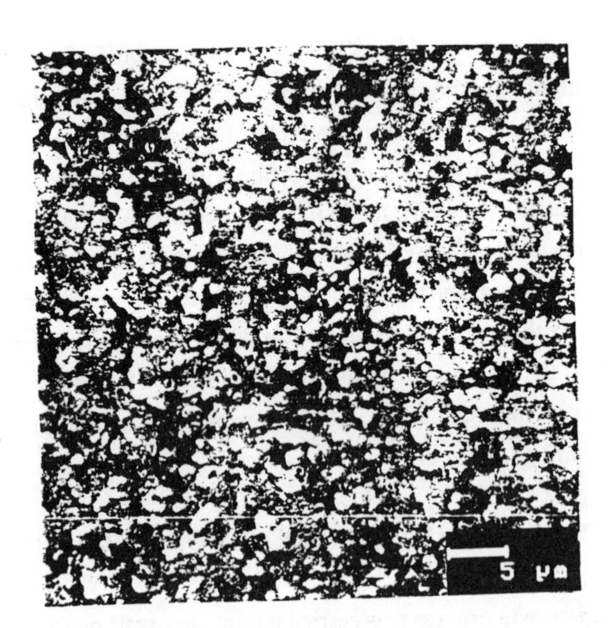

Figure 2. SEM graph of the YBCO on sapphire substrate showed the uniform surface morphology

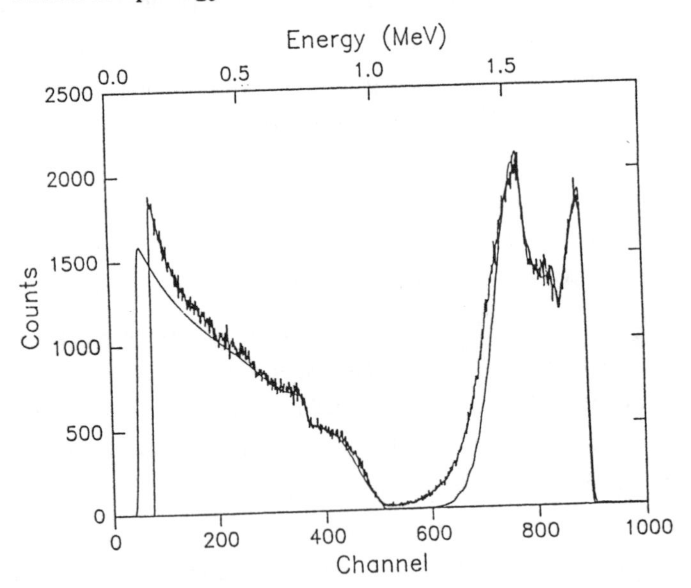

Figure 3. Rutherford backscattering spectra of YBCO film on sapphire substrate. The solid line is the simulated spectrum. Only a two layer structure model ensured good fit with the experimental data.

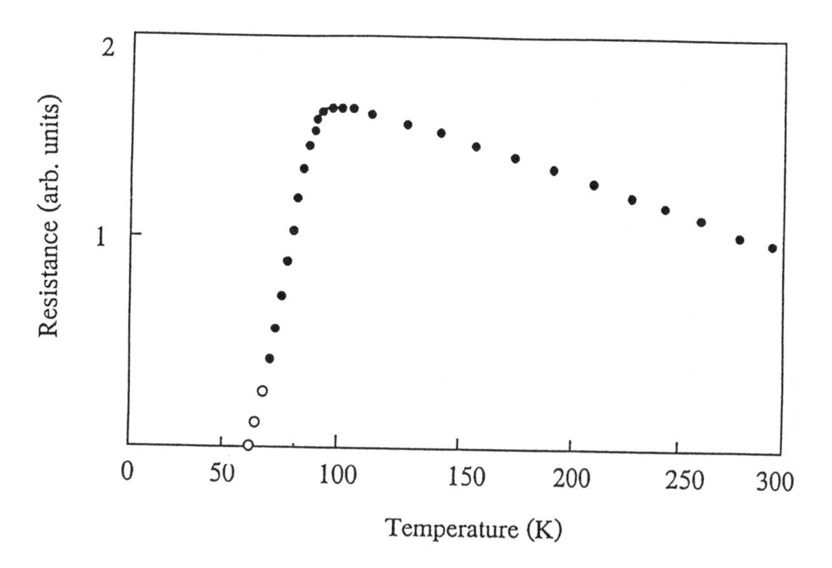

Figure 4. Temperature transition of the YBCO film on sapphire film on silicon was measured by dc four point probe.

CONCLUSIONS

YBCO thin films were grown without alien buffer on single-crystal sapphire substrate and a CVD-grown sapphire layer on silicon by a temperature controlled CVD process. The YBCO film on sapphire exhibited high c-axis orientation and a transition temperature of 90 K. The YBCO film on sapphire film on silicon substrate, on the other hand, were of lower quality, with an onset transition temperature of 90K but an extremely broad transition region ending at T_{co} about 60K.

These results, although very preliminary, seem to indicate that CVD processing is a promising tool for the growth of YBCO thin films on sapphire for cryoelectronic applications. However, additional and much more extensive studies are presently underway to optimize and fully develop the process to the level of incorporation on cryoelectronic applications.

ACKNOWLEDGEMENT

This work was supported by the National Science Foundation (NSF) Presidential Young Investigator under grant #DMR, which is gratefully acknowledged.

REFERENCES

1 R.S.Withers, A.C.Anderson, and D.E.Oates, Solid State Technol. August (1990)p.83

2 J.-Y.Jeongand K.Rose, in Superconductivity and Applications, ed H.Kwok Plenum Press, Now York 1988

3 K.Rose, Cryogenic Packaging for Low-Temperature Electronics, Oct., 1990

4 X.D.Wu, R.E.Muenchausen, S.Foltyn, R.C.Dye, A.R.Garcia, N.S.Nogar, P.Enland, R.Ramesh, D.M.Hwang, T.S.Ravi, C.C.Chang, T.Venkatesan, X.X.Xi, Q.Li, and A.Inam, Appl.Phys.Lett. 57, 523(1990)

5 D.M.Hwang, T.S.Ravi, R.Ramesh, S-W Chan, C.Y.Chen, L.Nazar, X.D.Wu, A.Inam, and T.Venkatesan, Appl. Phys. Lett. 57, 1690(1990)

6 Y.Q.Li, J.Zhao, C.S.Chern, W.Huang, G.A.Kulesha, P.Lu, B.Gallois, P.Norris, B.Kear, and F.Cosandey, Appl. Phys. Lett. 58, 648(1991)

7 M.P.Siegal, J.M.Phillips, R.B.van Dover, T.H.Tiefel, and J.H.Marshall, J Appl. Phys. 68, 6353(1990)

8 A.B.Berezin C.W.Yuan, and A.L.De Lozanne, Appl. Phys. Lett. 57,90(1990)

9 C.S.Chern, J.Zhao, Y.Q.Li, P.Norris, B.Kear, and B.Gallois, Appl. Phys. Lett. 57, 721(1990)

10 M.Schieber, M.Schwartz, G.Koren, and E.Alaroni, Appl. Phys. Lett. 58, 301(1991)

11 A.Feng, L,Chen, T.W.Piazza, A.E.Kaloyeros, D.W.Hazelton, L.Luo, and R.C.Dye, Appl. Phys. Lett. 59(9), (1991)

12 L.Chen, T.W.Piazza, J.E.Kelsey, A.E.Kaloyeros, D.W.Hazelton, L.Luo, R.C.Dye, and C.J.Maggiore, Submitted to Appl. Phys. Lett. (1991)

13 H.Asano, H.Yonezawa, M.Asahi, and O.Michikami, IEEE trans. Mag., 27(2) 931(1991)

14 L.A.Ryabova, and Y.S.Savitskaya, Thin Solid Films, 2, 141(1968)

15 J.A.Aboaf, J. Electrochem. Soc., 114, 948(1967)

16 M.T.Duffy. and A.G.Revesz, J. Electrochem. Soc., 117, 372(1970)

17 L.H.Hall, and W.C.Robinette, J. Electrochem. Soc., 118, 1624(1971)

VOLATILE BISMUTH COMPLEXES - PRECURSORS FOR HIGH-T_c SUPERCONDUCTING FILMS MANUFACTURING VIA CHEMICAL VAPOUR DEPOSITION

A.P.Pisarevsky
Moscow State University, 117234 Moscow, USSR

ABSTRACT

Bismuth carboxylates $Bi(O_2CR)_3$ ($R=C_nH_{2n+1}$) and beta-diketonates $Bi(R^1COCHCOR^2)_3$ ($R=CMe_3$ and/or C_nF_{2n+1}) were synthesized and characterized by means of chemical analysis, DTA, DSC, X-ray diffraction, IR and NMR spectroscopy, mass-spectrometry. Pivalate and all beta-diketonates can be vapourized quantitatively at temperatures between 100 and 200°C (0,05 Torr), while formate is not volatile and carboxylates from acetate to caprylate vapourize with slight decomposition. Bi carboxylates and beta-diketonates essentially extend the variety of Bi compounds applied in CVD technique. They are more stable to hydrolysis than Bi alkoxides and being complexes of oxygen-donor ligands more easily form oxide phase than organobismuth compounds. Fine regulation of volatility can be achieved by substitution of ligands.

INTRODUCTION

Analysis of literature data concerning organometallic chemical vapour deposition of superconducting films reveals the scarcity of information describing preparation of Bi-containing films as compared with that for $YBa_2Cu_3O_x$ films. The principal reason must be associated with volatile compounds applied as precursors in each case. Whereas Y, Ba and Cu precursors are beta-diketonates, Bi is introduced in vapour phase in the form of $BiPh_3$[1-8], rarely as $Bi(OEt)_3$[9]. However Sr, Ca and Cu are applied only in the form of beta-diketonates and reactions of ligand exchange can embarrass the deposition of films with required cation stoichiometry [10]. Moreover the apparent inconvenience of $Bi(OEt)_3$ and other proposed alkoxides[11,12] is their extreme instability to hydrolysis and oxidation in air. On the contrary, $BiPh_3$ is so stable to hydrolysis and thermolysis that one should employ water introduction in oxidizer[2] or plasma assisted CVD[5].

Thus the goal of present investigation is the search for new volatile Bi complexes with oxygen-donor ligands which application in CVD will overcome the aforesaid difficulties. Derivatives of carboxylic acids and beta-diketones were synthesized and examined as possible precursors. It is to be mentioned that one can find very poor literature data concerning Bi carboxylates (their volatility wasn't discovered so far) and practically no information about Bi beta-diketonates (with the exception of $Bi(Me_3CCOCHCOCMe_3)_3$ [12]). Meanwhile the volatility of antimony(III) acetate and some beta-diketonates has already been revealed, so it was attractive to study their bismuth analogues.

SYNTHESIS

The elaboration of technique for synthesis of Bi complexes aimed at the isolation of final products with maximum yields and minimum contaminations.

Most of the authors suggest reaction of Bi_2O_3 with carboxylic acid for synthesis of corresponding carboxylate. But it is far more facile to obtain Bi acetate on the first step and then use it in reaction with stoichiometric amount of carboxylic acid.

$$Bi_2O_3 \xrightarrow{CH_3COOH\ +\ (CH_3CO)_2O} Bi(O_2CCH_3)_3 \quad (1)$$

$$Bi(NO_3)_3 * 5\,H_2O \xrightarrow{(CH_3CO)_2O} Bi(O_2CCH_3)_3 \quad (2)$$

$$Bi(O_2CCH_3)_3 + RCO_2H \longrightarrow Bi(O_2CR)_3 + CH_3CO_2H \quad (3)$$

Reaction (3) gives quantitative yields independently on R since the equilibrium shift is provided with the removal of acetic acid during the synthesis, Only in the case of formic acid the latter should be taken with excess. The advantage of this method is obvious. It is necessary to point out that metathesis reaction of acetates with carboxylic acids was used formerly for various metals but never for Bi and Sb. It led to the idea to synthesize mixed carboxylates by transformation of oxide mixture into the mixture of acetates and then into the mixture of required carboxylates - one more way of chemical homogenization.

$$\tfrac{1}{2}a\ Bi_2O_3 + \tfrac{1}{2}b\ Sb_2O_3 + c\ PbO + d\ CaO + e\ SrO + f\ CuO$$

$$\downarrow CH_3CO_2H + (CH_3CO)_2O$$

$$Bi_aSb_bPb_cCa_dSr_eCu_f(O_2CCH_3)_x$$

$$-CH_3CO_2H \quad \downarrow RCO_2H$$

$$Bi_aSb_bPb_cCa_dSr_eCu_f(O_2CR)_x \qquad (4)$$

As to beta-diketonates two synthetic routes were employed in the case of Bi dipivaloylmethanate as well as Sb beta-diketonates.

$$M(OR)_3 + R^1COCH_2COR^2 \longrightarrow M(R^1COCHCOR^2)_3 + ROH \qquad (5)$$

$$MCl_3 + Na(R^1COCHCOR^2) \longrightarrow M(R^1COCHCOR^2)_3 + NaCl \qquad (6)$$

Drawback of both these methods consists in the instability of $BiCl_3$ and $Bi(OR)_3$ in air. It is simple to use $BiPh_3$ as precursor.

$$BiPh_3 + R^1COCH_2COR^2 \longrightarrow Bi(R^1COCHCOR^2)_3 + C_6H_6 \qquad (7)$$

After the removal of benzene the product is vapourized in vacuo. The yields exceed 95 per cent. It should be noted that this method can result in success if the acidity of beta-diketone is sufficiently great and/or its boiling point is high enough. Besides thermal stability of beta-diketone should allow its prolonged heating. Acetylacetone, for instance, doesn't meet all above-mentioned requirements, hence the attempt of Bi acetylacetonate synthesis failed. Of course $BiPh_3$ can be used as precursor for synthesis of Bi carboxylates but Bi acetate is more accessible.

PHYSICAL PROPERTIES

Whereas all Bi carboxylates and some beta-diketonates are colourless, derivatives of hexafluoroacetylacetone and benzoyltrifluoroacetone are lemon yellow, pivaloylacetonate is yellow and $Bi(C_3F_7COCHCOCMe_3)_3$ has light orange colour. Carboxylates from formate to n-butyrate as well as isobutyrate, isovalerate , pivalate and stearate are crystalline solids, so are all beta-diketonates. Derivatives of $CH_3(CH_2)_3CO_2H$ and $CH_3(CH_2)_4CO_2H$ are liquids at room temperature while derivatives of $CH_3(CH_2)_5CO_2H$ and $CH_3(CH_2)_6CO_2H$ are amorphous solids. Bi carboxylates(with the exception of formate and acetate) are soluble in corresponding acids and non-polar organic solvents, so are all beta-diketonates.

CHEMICAL ANALYSIS

Determination of bismuth, carbon and hydrogen contents confirmed that all considered complexes correspond to formula BiL_3.

THERMAL ANALYSIS

Heating in vacuo (0,05 Torr) leads to vapourization of Bi carboxylates (with the exception of formate and stearate) at temperatures between 160 and 200°C. But only pivalate can be vapourized quantitatively while in other cases the degree of vapourization fluctuates from 70 to 90 per cent owing to partial decomposition. Derivatives of carboxylic acids with branched radicals surpass those with linear radicals in volatility and thermal stability. Heating in nitrogen (1 atm) with the rate of 9°C/min leads to decomposition of all Bi carboxylates to Bi_2O_3, in the case of pivalate decomposition is accompanied by partial vapourization. Mechanism of thermolysis must be sufficiently complicated. Only in the case of acetate the intermediate formation of $OBiO_2CCH_3$ was observed.

$$Bi(O_2CCH_3)_3 \xrightarrow[-\ (CH_3CO)_2O]{290-310^\circ C} OBiO_2CCH_3 \xrightarrow{330-350^\circ C} Bi_2O_3 \quad (8)$$

Table I. Thermal behaviour of Bi carboxylates

Substance $Bi(O_2CR)_3$ R	m.p., °C	b.p., °C (0,05 Torr)	thermolysis in N_2(1 atm) °C	loss of mass under decomp. to Bi_2O_3, % found	calcul.
H	decomp.	decomp.	200–280	32	32
CH_3	decomp.	200–210	290–350	40	40
C_2H_5	165	200–210	290–360	45	46
$CH_3(CH_2)_2$	89	180–190	260–360	50	50
$(CH_3)_2CH$	124	170–180	300–360	51	50
$CH_3(CH_2)_3$	liquid	190–200	260–360	55	54
$(CH_3)_2CHCH_2$	80–82	180–190	270–350	56	54
$(CH_3)_3C$	164	160–170	400–450	58	54
$CH_3(CH_2)_4$	liquid	190–200	260–360	59	58
$CH_3(CH_2)_5$	–	200–210	270–350	61	61
$CH_3(CH_2)_6$	–	210–220	270–370	63	63
$CH_3(CH_2)_{16}$	90	decomp.	300–380	79	79

Table II. Thermal behaviour of Bi beta-diketonates

Substance $Bi(R^1COCHCOR^2)_3$ R^1	R^2	m.p., °C	b.p., °C (0,05 Torr)	vapourization in N_2 (1 atm), °C
CF_3	CF_3	95–97	110–120	200–240
$CF_3CF_2CF_2$	CMe_3	130–132	150–160	270–295
CF_3	CMe_3	96	135–145	260–290
CMe_3	CMe_3	112–115	120–130	260–290
Me	CMe_3	105–107	110–120	220–260
CF_3	Ph	120–123	140–150	260–300

All Bi beta-diketonates vapourize quantitatively in vacuo at temperatures between 100 and 150°C. Heating in nitrogen atmosphere leads to vapourization at temperatures between 220 and 300°C, so no thermolysis products could be identified. Although saturated steam pressure was not precisely determined, temperature of vapourization can be used for approximate estimate of volatility dependence upon the nature of R^1 and R^2 radicals. Thus perfluoroalkyl groups increase the volatility of Bi beta-diketonates as compared with those having hydrocarbon groups. Meanwhile complexes with greater molecular weight are less volatile.

STRUCTURAL INVESTIGATIONS

Undoubtedly all physical properties including volatility can be accounted for the structure of a given substance. Only two crystal structures of Bi carboxylates has been determined so far: $Bi(O_2CH)_3$ [13] and $Bi(O_2CCH_3)_3$[14]. The latter is presented in fig.1. In both cases the coordination number of Bi atom is 9, but Bi formate molecules form a three-dimensional network while Bi acetate molecules are united into layers by means of additional bridging through oxygen atoms. Such decrease of the intermolecular interaction causes the difference in volatility.

Unfortunately X-ray single crystal investigations of other Bi complexes haven't yet been accomplished, so there is nothing but use the indirect spectroscopic data. IR spectroscopic investigations of Bi carboxylates indicate that carboxylic groups have asymmetric bidentate coordination to their "own" Bi atoms and form bridging bonds with neighbouring Bi atoms. Thus $\nu^{as}_{C=O}$ and $\nu^{s}_{C=O}$ frequencies are observed at 1525-1565 and 1400-1410 cm^{-1}, respectively. ^{13}C NMR spectroscopic data confirm the ionic nature of Bi-O bonds in carboxylates: carbonyl signals are shifted downfield by 3-4 ppm as compared with free acids.

Prediction of Bi beta-diketonates structures is an ungrateful problem since coordination number of Bi atom must be more than 6, its lone electron pair is stereochemically active and ligands contain bulky radicals. Shift of $\nu_{C=O}$ bands to lower frequencies as compared with free beta-diketones is the evidence for coordination of ligands through O atoms. Different values of H and C chemical shifts of CH group are determined by R^1 and R^2 effects.

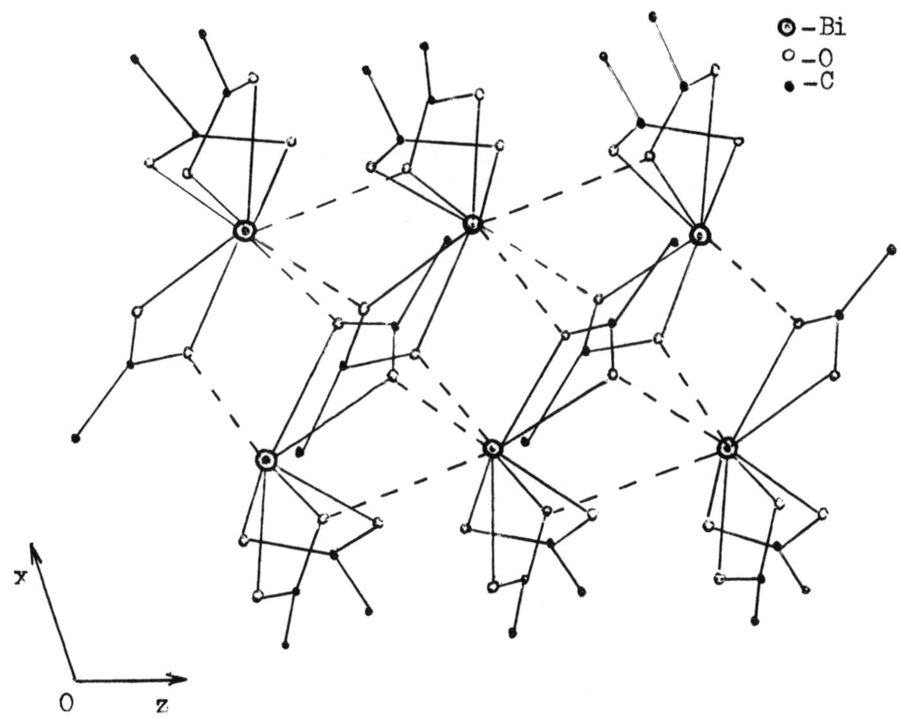

Fig. 1. xOz projection of $Bi(O_2CCH_3)_3$ crystal structure.

Mass-spectroscopy presents valuable information concerning polymerization in vapour phase. Bi carboxylates apparently exsist in oligomeric form while molecules of Bi beta-diketonates aren't associated. This must be the reason for the freater volatility of the latter.

APPLICATIONS

It is obvious that among Bi carboxylates only pivalate and maybe isobutyrate can be used in CVD of superconducting films since Ca and Sr derivatives of other carboxylic acids aren't volatile. Bi beta-diketonates are undoubtedly more promising. Both carboxylates and beta-diketonates are free of above-mentioned drawbacks of Bi alkoxides and organobismuth compounds. Application of Bi, Pb, Ca, Sr, Cu complexes with the same ligand will completely exclude ligand-exchange reactions in vapour phase and therefore simplify CVD process. One shouldn't also forget that apart from CVD all given Bi complexes can serve as precursors for other techniques of superconductor production - spray pyrolysis or sol-gel, for instance.

REFERENCES

1. A.O.Berry, R.T.Holm, E.J.Cukauskas et al, J. Cryst. Growth 92, 344(1988).
2. J. Zhang, J.Zhao, H.O.Marcy et al. Appl. Phys. Lett. 54, 1166 (1989).
3. J.M.Zhang, B.W.Wessels, L.H.Tonge et al. Appl. Phys. Lett. 56, 976 (1990).
4. K.Natory, S.Yoshizawa, Y.Yoshino et al. Japan. J. Appl. Phys. Part 2. 29, L930 (1990).
5. K,Kobayashi, S.Ichikawa, K.Ohmore et al. J. Mater. Sci. Lett. 9, 945 (1990).
6. K.Endo, S.Hayashida, J.Ishiai et al. Japan. J. Appl. Phys. Part 2. 29, L294 (1990).
7. N.Hamaguchi, J.Vigil, R.Gardiner et al. Japan. J. Appl. Phys. Part 2. 29, L596 (1990).
8. N.Masanori, Y.Mitsugu, J. Mater. Res. 5, 1 (1990).
9. H. Yamane, H.Kurosawa, H.Iwasaki et al. Japan. J. Appl. Phys. Part 2. 28, L827 (1989).
10. C.Sant, O.Poncelet, M.Laugt et al. Vide, Couches mines 45, 205 (1990).
11. M.A.Matchett, M.Y.Chiang, W.E.Buhro, Inorg. Chem. 29, 358 (1990).
12. L.G.Hubert-Pfalzgraf, M.C.Massiani, R.Papiernic, Colloq. phys. 50, C5-981 (1989).
13. C.-I.Stalhandske, Acta chem. scand. 23, 1525 (1969).
14. S.I.Troyanov, A.P.Pisarevsky (in press).

BULK PROCESSING AND PROPERTIES

PROGRESS IN BULK HIGH-TEMPERATURE SUPERCONDUCTORS

S. Jin

AT&T Bell Laboratories, Murray Hill, New Jersey 07974

ABSTRACT

Some recent advances in materials processing, properties, and fabrication of the bulk high-temperature superconductors are reviewed with respect to the weak link removal and flux pinning enhancement. Remarkable progress has been made during the past few years; however, the present status is far from ideal, especially in obtaining high critical currents in strong magnetic fields. Further innovations are needed to improve the superconductor properties and to accelerate progress toward major applications.

I. INTRODUCTION

For major bulk applications of the new high temperature superconductors (HTSC), desirable superconducting properties and physical shapes have to be obtained.

While the fabrication of the ceramic superconductors into wires has been successfully demonstrated using a number of different processing methods, the progress toward the much-expected applications of bulk HTSC has been hindered by their low critical currents, especially in strong magnetic fields. By contrast, thin films epitaxially grown on single crystal substrates typically exhibit high J_c values in excess of 10^6 A/cm^2 at 77K.

Two major causes for the low critical currents at 77K in bulk, polycrystalline HTSC, are; i) weak links at the grain boundaries, and ii) flux line movement within the grains due to insufficient flux pinning in Y-Ba-Cu-O and severe flux creep in Bi-Sr-Ca-Cu-O and Tl-Ba-Ca-Cu-O. In this paper, some recent progress made in these two problem areas will be discussed.

II. TEXTURE FORMATION

The weak link problem in HTSC can be greatly reduced by elimination or minimization of high angle grain boundaries in the paths of the transport current. This can be achieved by a preferential crystallographic alignment of grains parallel to the a-b conduction planes (CuO_2 planes). Several approaches with differing degrees of success have been reported, as described below.

(1) Melt-Texture Processing

Melt-texture processing[5,6] is basically a directional solidification of the $YBa_2Cu_3O_{7-\delta}$ phase from the melt or partial melt state. While there are many possible variations, a typical melt-texture technique consists of heating the

$YBa_2Cu_3O_{7-\delta}$ (Y-123) compound to a temperature of ~ 1030 – 1100 °C (above the peritectic temperature of ~ 1010 °C in an oxygen atmosphere) for decomposition into Cu-rich liquid and Y_2BaCuO_5, and slow cooling (e.g. at 2-10°C/h) from the peritectic temperature to ~ 900 °C for nucleation and growth of the Y-123 phase, followed by additional slow cooling (at 2-10°C/h) to below ~ 400 °C for oxygenation.

Because of the natural preference for crystal growth in the a-b direction (as compared to the c-direction), the melt-textured Y-123, even in the absence of a temperature gradient, tends to crystallize as large parallel plates if sufficiently slow cooling rates are used. The presence of a temperature gradient during the crystallization period plays the important role of eliminating undesirable, multiple nucleation of these parallel plates at various locations along the sample length. This effect of temperature gradient on the crystal alignment in melt-textured Y-Ba-Cu-O is schematically illustrated in Fig. 1(a) and (b). The local microstructure and J_c behavior within each of the aligned regions (often 1-10 mm size) in Fig. 1(b) is essentially the same as those in Fig. 1(a); the parallel Y-123 grains (thin plate shape) exhibit near-perfect alignment, separated by low angle grain boundaries, and no high angle grain boundaries (weak links) are encountered along the alignment direction.

The structure in Fig. 1(b) would exhibit low transport $J_c(H)$ if measured across the high angle grain boundaries but high $J_c(H)$ if the voltage leads are placed within the locally melt-textured region. Although the long-range melt-textured sample of Fig. 1(a) is always more desirable, the locally melt-textured structure in Fig. 1(b) may still find some useful applications (such as in magnetic bearings or microwave cavity applications) by virtue of much larger magnetic hysteresis and the much smaller number of weak links than in the sintered structure.

The melt-textured Y-Ba-Cu-O exhibits significantly improved J_c and greatly reduced field dependency of J_c as shown in Fig. 2. The figure includes recent data from grain oriented samples obtained by a number of researchers using the melt-texture type processing, or related, improved melt processing, e.g., by Salama, et al,[3] Murakami, et al.,[4] McGinn, et al,[5] and Meng, et al.[6] $J_c(H)$ values for melt-textured Y-Ba-Cu-O, in the range of $10^4 - 10^5$ A/cm^2, are improved by orders of magnitude over the values for the random-grained Y-Ba-Cu-O, especially in magnetic fields.

For the $YBa_2Cu_3O_{7-\delta}$ superconductor system, the melt-texture type processing is the only proven technique for eliminating the severe weak-link behavior. However, progress in the fabrication of long wires by melt-texture processing has been slow because of the complicated phase relationships, difficulty of material handling in the high temperature processing, and slow processing speed. Considerable progress has recently been reported in the zone-melting type continuous processing for longer fiber samples [6,7].

(2) Magnetic Field Alignment

A strong (00ℓ) texture has been obtained in the various RE-123 superconductors prepared by magnetic field alignment.[8,9] However, the field alignment (e.g. c-axis alignment) does not differentiate the a- vs b-orientation. As a result, many of the in-plane aligned, plate-like grains tend to have a twist relationship relative to each other (rotation of grains around the c-axis), producing a considerable number of high angle grain boundaries. Unless the a,b misorientation (in-plane twist) is reduced by some means, it may be difficult to completely eliminate high-angle (weak-link) grain boundaries and achieve high $J_c(H)$ by this technique.

(3) Deformation Texture

This technique is based on the metal-clad (metal-sheathed) wire fabrication method, and has been most successfully utilized for the Bi-Sr-Ca-Cu-O[10,11] and the Tl-Ba-Ca-Cu-O[12] superconductors. Typical processing procedures are as follows. The superconductor powder (e.g., 105K phase $Bi_{1.6}Pb_{0.4}Sr_2Ca_2Cu_3O_x$) is poured into a silver tube (e.g., 1.2-2.5 cm in dia.), which is then wire drawn to 1-2 mm dia., cold rolled into a ribbon shape, sintered at 820°-860°C for ~ 100 h, cold rolled (or pressed) into ~ 0.1 mm thick × ~ 3 mm wide ribbons and sintered again for ~ 100 h.

The thin superconductor ribbon thus prepared exhibits a highly textured microstructure with the a-b conduction plane parallel to the ribbon surface. X-ray diffraction analysis on the superconductor core reveals predominantly (00ℓ) peaks indicating a high degree of crystallographic alignment. The formation of such a strong texture is of course beneficial for obtaining high J_c. The exact mechanism for the texture formation in the Ag-sheathed Bi-Sr-Ca-Cu-O wire is not clearly described in the literature. The following three possibilities can be considered: i) deformation-induced alignment of basal-plane-fractured grains which are simply coarsened into long, parallel, plate-shaped grains during the subsequent sintering of the wire, ii) annealing texture which is created as the internal stresses accumulated during the wire drawing and rolling processes are relieved during the sintering stage, and iii) silver-sheath-induced texture (nucleation and growth of aligned grains from the silver-superconductor interface in the ribbon). Additional research is required to elucidate the mechanism and enhance the texture formation.

The textured Bi-Pb-Sr-Ca-Cu-O wires[10] exhibit high J_c values of ~ 30,000 A/cm^2 at 77K, H = 0. The J_c values are maintained in low magnetic fields. In high magnetic fields, however, significant decrease in $J_c(H)$ is observed. The most recent J_c values obtained by Sumitomo researchers are ~ 53,000 A/cm^2 at H = 0 and ~ 12,000 A/cm^2 at H = 1T. For an unfavorable field direction (perpendicular to the ribbon surface) the loss of J_c is so drastic that there is practically no critical current at $H \geq 1T$.

This severe field dependent deterioration of J_c in high magnetic fields in the textured Bi-Pb-Sr-Ca-Cu-O superconductor is caused mainly by thermally activated flux creep rather than by the presence of weak-links, as such a severe deterioration is not seen at a lower temperature of 4.2K even at H > 20T. The most recent data by Furukawa Electric Co. include remarkable J_c values at 4.2K of ~ 2×10^5 A/cm^2 at 30T field. The J_c values at H > 20T in these HTSC are significantly higher than are possible with the low T_c superconductor wires such as Nb-Ti or Nb_3Sn, thus offering hope of achieving higher-field magnets.

Thermally activated flux creep in HTSC drastically increases the electrical resistivity sometimes even to the level of copper resistivity in high magnetic fields

(Fig. 5). The flux creep is a recognized problem in all HTSC materials, especially for the more anisotropic Bi-Sr-Ca-Cu-O and Tl-Ba-Ca-Cu-O systems. This difference among the various HTSC compounds is generally attributed to the reduced tilt modulus (rigidity) of the flux lines caused by stronger electronic anisotropy,[13] as well as the lack of sufficiently strong flux pinning centers. Unless the flux creep is reduced by enhanced flux pinning, the operating temperature and field will have to be reduced, e.g., to $T \leq 30K$ for the Bi-Sr-Ca-Cu-O system, or to $H \leq 5T$ for the Y-Ba-Cu-O system.

(4) Substrate-Induced Texture

The beneficial effect of silver-substrate-induced texture on J_c is much more noticeable in the Bi-Sr-Ca-Cu-O superconductor ribbons. These ribbons are prepared by doctor blade technique[14] or by spray coating and cold rolling technique,[15] followed by partial melt processing. The superconductor material near the silver/superconductor interface apparently experience more extensive melting and layer-like growth (c-axis texture), which is aided by the two-dimensionality of the interface.

An exemplary microstructure of the c-axis textured Bi-Pb-Sr-Ca-Cu-O ribbon[15] (~ 2 μm thick superconductor on ~ 8 μm thick Ag) is shown in Fig. 4. The in-plane misorientation of the (00ℓ) textured grains produces a large number of a-b grain boundaries in the ribbon. Unlike the Y-Ba-Cu-O system, these high angle grain boundaries in the Bi-Sr-Ca-Cu-O system are not weakly coupled, and hence excellent transport J_c can be achieved (Fig. 5). J_c values as high as 2.3×10^5 A/cm^2 at 4.2K, $H_{\perp ab}$ = 8T have been obtained.[15]

As demonstrated by recent studies of the critical current behavior in bulk, polycrystalline Bi-Sr-Ca-Cu-O, the high angle grain boundaries in partial-melt-processed, random-grained, Bi-Sr-Ca-Cu-O are not severe weak links. The mild but still noticeable field-dependence of J_c in these samples (Fig. 5), however, implies, on the average, a "mild" weak link behavior of the high angle boundaries,[16] indicating the necessity for grain texturing if the full J_c capability of the Bi-Sr-Ca-Cu-O type superconductors is to be exploited.

III. FLUX-PINNING ENHANCEMENT

Even though the weak-link problem in bulk materials can be avoided through grain alignment, the high-field J_c still seems to be limited to ~ 10^4 A/cm^2 at 77K. Many of the potential bulk applications of the high T_c superconductors depend on the capability of the materials to carry sufficiently large currents, i.e., J_c > ~ 10^5 A/cm^2 at 77K in H > 1 Tesla. Considering that weak-link-free, epitaxial thin films of $YBa_2Cu_3O_{7-\delta}$ are known to exhibit $J_c(H)$ of more than 10^6 A/cm^2, Fig. 3, it is most likely to be the low density of strong flux-pinning sites in the bulk Y-Ba-Cu-O which limits $J_c(H)$ to ~ 10^4 A/cm^2.

Effective flux-pinning requires the presence of extremely fine defects (regions with reduced order parameter) with the size scale comparable to the superconducting coherence length (about 4-30 Å in Y-Ba-Cu-O). Several processing techniques have been shown to induce fine defects and improve the flux pinning as discussed below.

(1) Irradiation

It is well known that irradiation with fast neutrons improves the flux pinning in HTSC materials. The most striking effect has been reported by van Dover, et al[17] in a $YBa_2Cu_3O_{7-\delta}$ single crystal. A hundred fold improved critical current density (magnetization J_c estimated from the M-H curve using the Bean model) to 0.6×10^6 A/cm^2 has been obtained at 77K, H = 0.9 tesla, which is the highest in bulk HTSC and approaches those in epitaxial thin films. Proton irradiation has also proven effective in flux pinning enhancement. Because of the charge on the proton, the penetration into HTSC is somewhat limited.

Civale, et al[18] utilized high-energy, heavy-ion (Sn) irradiation to introduce a high density of pinning defects in the form of discontinuous columns of damaged tracks (~ 50 Å dia × ~ 150 μm long) in Y-Ba-Cu-O crystals. A large increase in J_c as well as improvement in irreversibility line (a shift toward higher temperature and field) was obtained.

(2) Phase Decomposition

This is based on a decomposition of $YBa_2Cu_4O_8$ ("124" phase) precursor into $YBa_2Cu_3O_7$ ("123" phase) containing a high density of phase-transformation-induced defects.[19] Since Cu atoms have to come out from every unit cell, extremely fine-scale defects are anticipated. A tenfold improvement in flux pinning was obtained, which is attributed to the presence of extremely fine scale stacking defects (stacking faults) on the order of 10-30 Å thick, and other associated microstructural features, such as fine twin structure (100-500 Å spacing), and coarse CuO particles (~ 2500 Å in average dia., too large for efficient flux pinning).

(3) Other Methods

Nonsuperconducting second phase inclusions have been reported to be useful for pinning enhancement; for example, the Y_2BaCuO_5 ("211" phase) particles in the melt-processed $YBa_2Cu_3O_{7-\delta}$ superconductor,[20] although the flux pinning effect of the '211' inclusions have been disputed in more recent investigations.[21,22] The beneficial effect of the Ca_2CuO_3, $CaSrPbO_4$, and $Sr_2Ca_2Cu_3O_x$ precipitates in the Bi-Sr-Ca-Cu-O superconductors[23] have been reported. These particles are typically on the order of submicron to micron size.

Chemical doping may also be employed to locally disturb the crystal structure of the superconductor at the unit-cell or subunit-cell level and create coherence-length-scale defects. Shock wave loading[24] has also been used to enhance flux pinning in Y-Ba-Cu-O.

IV. SUMMARY

Some recent advances in bulk high T_c superconductor materials have been reviewed. Significant progress has been made in materials science and all aspects of processing techniques for wire fabrication, weak-link-removal, and flux-pinning enhancement. However, further processing innovations are needed in order to accelerate progress toward major bulk applications, thus offering many challenges and research opportunities to the materials scientists.

REFERENCES

[1] S. Jin, T. H. Tiefel, R. C. Sherwood, M. E. Davis, R. B. van Dover, G. W. Kammlott, R. A. Fastnacht, and H. D. Keith, Appl. Phys. Lett. **52**, 2074 (1988).

[2] S. Jin, R. C. Sherwood, E. M. Gyorgy, T. H. Tiefel, R. B. van Dover, S. Nakahara, L. F. Schneemeyer, R. A. Fastnacht, and M. E. Davis, Appl. Phys. Lett. **54**, 584 (1989).

[3] K. Salama, V. Selvamanickam, L. Gao, K. Sun, Appl. Phys. Lett. **54**, 2352 (1989).

[4] M. Murakami, M. Morita, K. Doi, K. Miyamoto, Jap. J. Appl. Phys. **28**, 1189 (1989).

[5] P. J. McGinn, W. Chen, N. Zhu, U. Balachandran, and M. T. Lanagam, Physica C. **165**, 480 (1989).

[6] R. L. Meng, C. Kinalidis, Y. Y. Sun, L. Gao, Y. K. Tao, P. H. Hor, and C. W. Chu, Nature **345**, 326 (1990).

[7] J. D. Hodge, L. J. Klemptner, and M. V. Parish, unpublished results (1990).

[8] D. E. Farrel, B. S. Chandrasekhar, M. R. DeGuire, M. M. Fang, V. G. Kogan, J. R. Clem, and D. K. Finnemore, Phys. Rev. **B36**, 4025 (1987).

[9] J. D. Livingston, H. R. Hart, Jr., and W. P. Wolf, J. Appl. Phys. **64**, 5806 (1988).

[10] T. Hikata, M. Ueyama, H. Mukai, and K. Sato, Cryogenics **30**, 924 (1990).

[11] N. Enomoto, H. Kikuchi, N. Uno, H. Kumakura, K. Togano, and K. Watanabe, Japn. J. Appl. Phys. **29**, L447 (1990).

[12] M. Okada, R. Nishiwaki, T. Kamo, T. Matsumoto, K. Aihrara, S. Matsuda, and M .Seido, Jpn. J. Appl. Phys. **27**, L2345 (1988).

[13] T. T. M. Palstra, B. Batlogg, R. B. van Dover, L. F. Schneemeyer, and J. V. Waszczak, Phys. Rev. **B41**, 6621 (1990).

[14] J. Kase, K. Togano, H. Kumakura, D. R. Dietdrich, N. Irisawa, T. Morimoto, and H. Maeda, Japn. J. Appl. Phys. **29**, L1096 (1990).

[15] S. Jin, R. B. van Dover, T. H. Tiefel, J. E. Graebner, and N. D. Spencer, Appl. Phys. Lett. **58**, 868 (1991).

[16] S. Jin, T. H. Tiefel, R. B. van Dover, J. E. Graebner, T. Siegrist, G. W. Kammlott, and R. A. Fastnacht, Appl. Phys. Lett. **59**, 366 (1991).

[17] R. B. van Dover, E. M. Gyorgy, L. F. Schneemeyer, J. W. Mitchell, K. V. Rao, R. Puzniak, and J. V. Waszczak, Nature **342**, 55 (1989).

[18] L. Civale, A. D. Marwick, T. K. Worthington, M. A. Kirk, J. R. Thompson, L. Krusin-Elbaum, Y. Sun, and J. R. Clem, Phys. Rev. Lett. **67**, 648 (1991).

[19] S. Jin, T. H. Tiefel, S. Nakahara, J. E. Graebner, H. M. O'Bryan, R. A. Fastnacht, and G. W. Kammlott, Appl. Phys. Lett. **56**, 1287 (1990).

[20] M. Murakami, S. Gotoh, N. Koshizuka, T. Matsushita, S. Kambe and K. Kitazawa, Cryogenics, **30**, 390 (1990).

[21] P. J. McGinn, W. Chen, N. Zhu, S. Sengupta, and T. Li, Physica C **176**, 203 (1991).

[22] S. Jin, T. H. Tiefel, and G. W. Kammlott, Appl. Phys. Lett. **59**, 540 (1991).

[23] D. Shi, M. S. Boley, U. Welp, J. G. Chen, and Y. Liao, Phys. Rev. **B40**, 5255 (1989).

[24] S. T. Weir, W. J. Nellis, M. J. Kramer, C. L. Seaman, E. A. Early, and M. B. Maple, Appl. Phys. Lett. **56**, 2042 (1990).

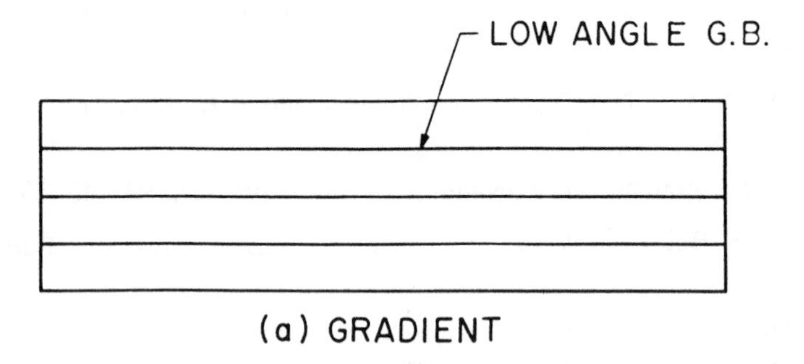

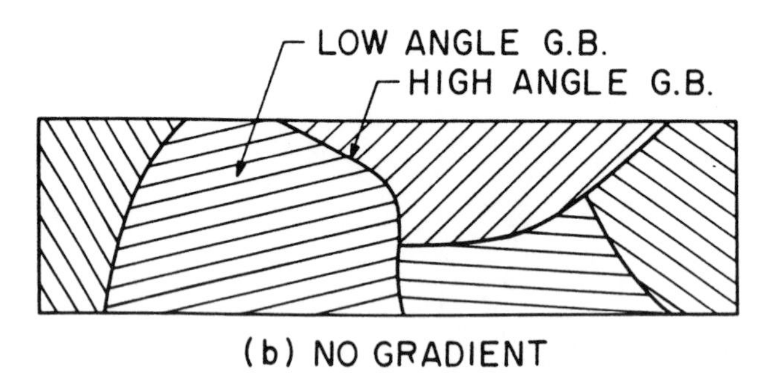

Fig. 1 Effect of temperature gradient on the melt-textured microstructure. (a) Long range grain alignment obtained with temperature gradient. (b) Locally melt textured structure obtained without temperature gradient.

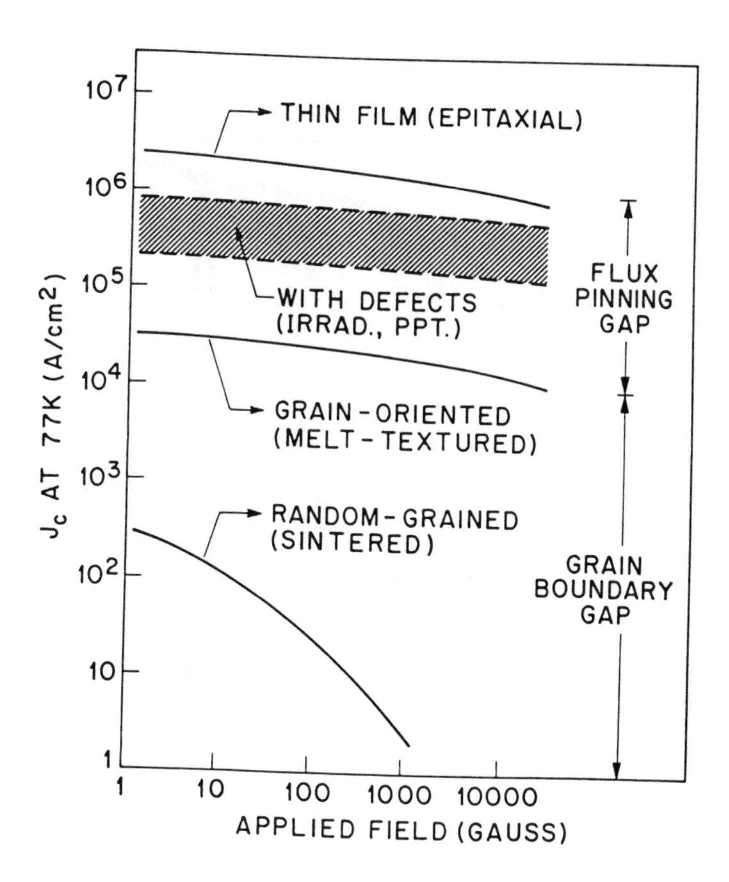

Fig. 2 Transport J_c vs H in Y-Ba-Cu-O. The shaded area represents the improved intragrain J_c values in bulk Y-Ba-Cu-O with added flux pinning defects.

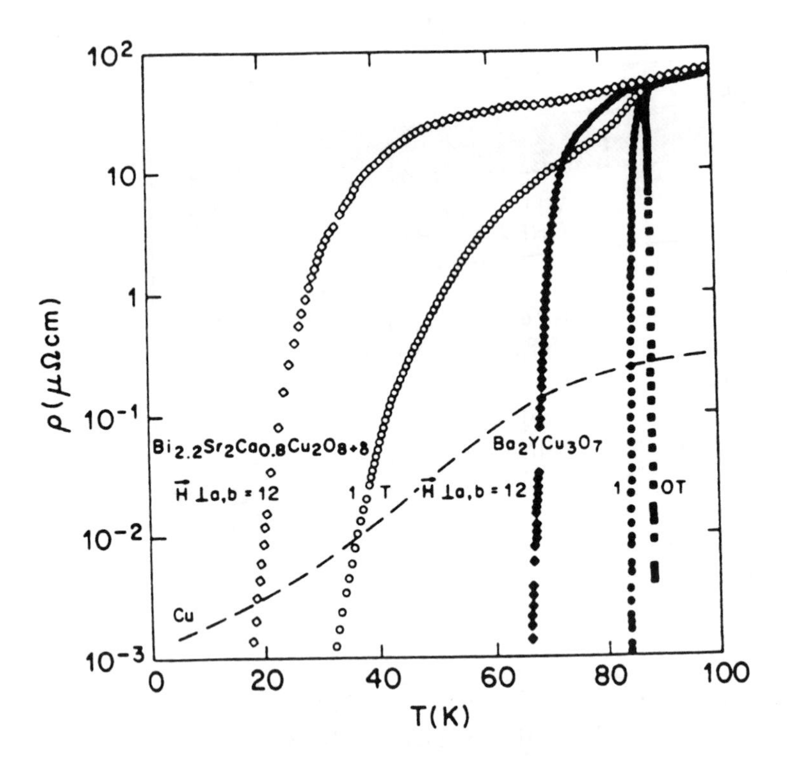

Fig. 3 Temperature and field dependence of the electrical resistivity for the $Bi_2Sr_2Ca_1Cu_2O_8$ and $YBa_2Cu_3O_{7-\delta}$ single crystals. The curve for copper is included for comparison.[13]

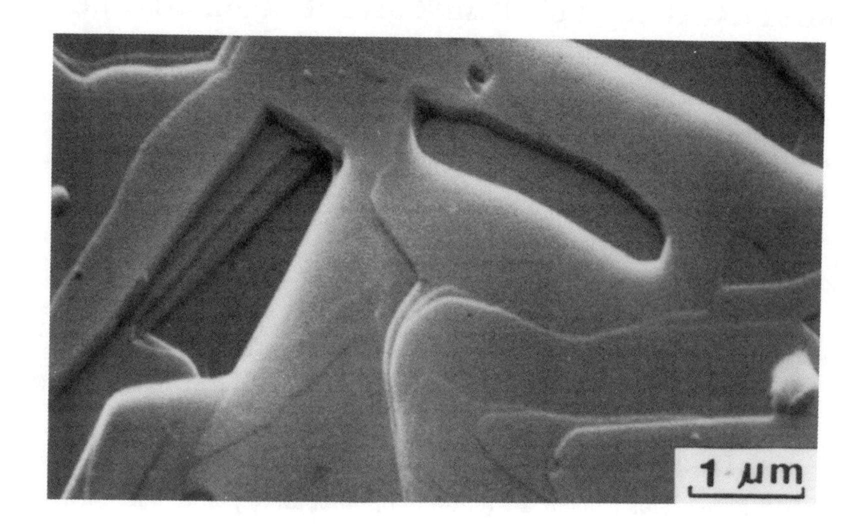

Fig. 4 Microstructure of the melt-processed Bi-Pb-Sr-Ca-Cu-O ribbon[15] (top view).

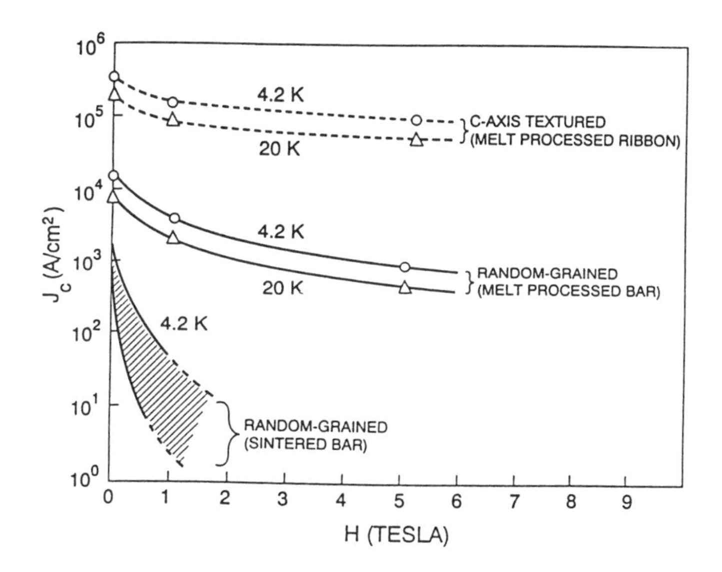

Fig. 5 Transport J_c vs H in textured and random-grained Bi-Sr-Ca-Cu-O samples.

TOWARDS PRACTICAL HIGH-TEMPERATURE SUPERCONDUCTORS

Robert L. Snyder
Institute of Ceramic Superconductivity
New York State College of Ceramics at Alfred University
Alfred, NY 14802

INTRODUCTION

It has now been four years since the discovery of high temperature superconductivity in ceramic oxides. The initial flush of enthusiasm has long since drained away leaving researchers attempting to solve the problems associated with fabrication of high current carrying materials. Most of the work in this area has been directed toward thin film applications because of the early successes in fabricating high J_c films. This has left a window of opportunity open to a few academic laboratories and a larger number of Japanese companies and consortia to investigate the problems associated with bulk ceramic superconductors in the absence of much competition or, U.S., funding.

The Institute for Ceramic Superconductivity at the New York State College of Ceramics was started in the spring of 1987. The major thrust of the Institute is to coordinate research on this new class of ceramic materials with the goal of developing the fundamentals of the ceramic science required to make practical bulk devices. In the last four years we have focused over a million, nonfederal, dollars in sponsored research, published over 80 research papers,[1] presented over 250 talks and have eleven patents awarded or pending, most representing significant technical advances in this field. Our research has been primarily directed toward understanding the fundamental chemical and physical principles needed to fabricate bulk ceramic superconductors. In this paper I will summarize the main approaches taken within the institute toward fabricating practical conductors.

MELT PROCESSING

The currently most promising method for improving the current density in ceramic superconductors is to hold the material above the peritectic melting point for a period of time and then at a lower temperature, to allow the development of a highly oriented microstructure. This procedure was discovered at Alfred by

Prof. Jenifer Taylor[2] along with two graduate students, who have recently been awarded a patent for this process.[3] It has been extensively studied by Jin[4-6] and a number of other groups[7] all over the world. At the moment this remains the only technique which has been proven to produce large current densities in bulk materials. However, it is not clear that any of these pieces conduct much below their surfaces.

The various trial and error studies over the years have produced only modest gains in the current densities obtainable via this technology. We have therefore begun an extensive program of dynamic characterization in order to determine the kinetics and dynamics of microstructure formation with the goal of microstructure design. One such study[8] is reported in this volume. By dynamic characterization we mean the use of special hot-stage techniques in XRD, thermal, spectroscopy, optical and electron microscopy which allow the direct observation of the formation of a microstructure. We have developed procedures primarily for XRD and optical microscopy which allow direct observation of the peritectic reactions and microstructure development of high temperature superconducting phases.[9,10]

Currently these techniques are being extended to measurement of the kinetics and reaction dynamics in the $YBa_2Cu_3O_{7-\delta}$, $YBa_2Cu_4O_{8-\delta}$, $Bi_2Sr_2CaCu_2O_8$, $Bi_2Sr_2Ca_2Cu_3O_{10}$, $Tl_2Ba_2CaCu_2O_8$ and $Tl_2Ba_2Ca_2Cu_3O_{10}$ systems.[11] The hot-stage optical microscopy is proving the most powerful tool in understanding the mechanism of texture formation. It is clear from the studies to-date that the texturing is *not* an epitaxial phenomenon. Figure 1 shows the peritectic melting of $Bi_2Sr_2CaCu_2O_8$. Note that in the XRD scan just prior to the complete peritectic melting of $Bi_2Sr_2CaCu_2O_8$, the (00ℓ) reflections grow in intensity accompanied by a strong decrease in all other diffraction intensities. This indicates a "free surface epitaxy" where the formation of the peritectic liquid phase allows the remaining crystallites to move and turn their lowest free energy face outward.

GLASS-CERAMIC SUPERCONDUCTORS

One of the most promising technologies for producing practical bulk ceramic superconductors is the glass-ceramic route. The first exploration of this approach was done at Alfred and resulted in the first patent awarded for this technology.[12] The patent covers a series of compositions of, and procedures for, preparing a stable glass in the Y_2O_3-BaO-CuO-B_2O_3 system, that when heat treated, precipitates superconducting crystals. The glass-ceramic material has near zero porosity and is protected from atmospheric degradation. The potential

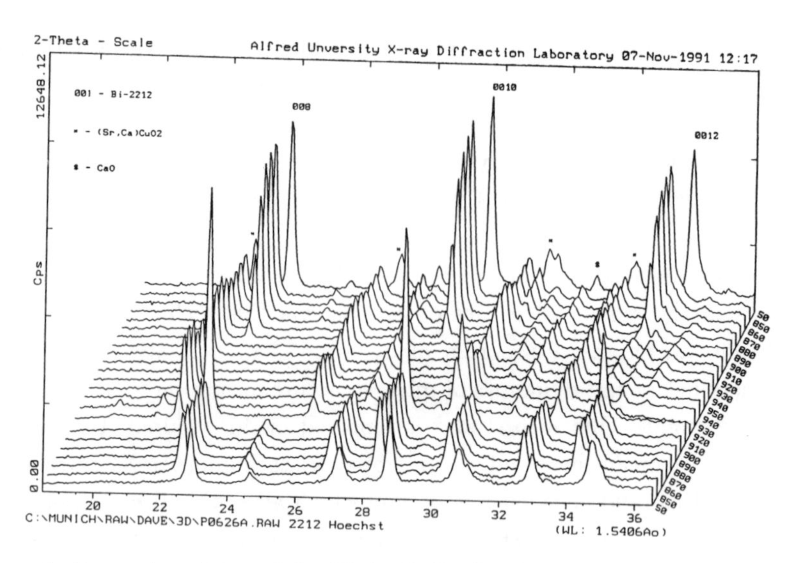

Figure 1: Dynamic melting of the $Bi_2Sr_2CaCu_2O_8$ phase as show by X-ray diffraction.

applications for these materials is in the field of electronics and small size devices with an emphasis on applications as magnetic shielding and levitation devices.

The principal drawback of the 123-borate system is its strong reducing nature which leaves most of the 123 in the non-superconducting tetragonal form. Presently, work is concentrated on increasing the critical current by increasing the concentration of $YBa_2Cu_3O_{7-\delta}$ and increasing the oxygen activity in the glass.[13,14] A recent study of the interface reaction between 123 and the glass precursor showed that it is possible to crystallize the 123 phase from the glasses by a surface induced crystallization technique.[15]

Glass-ceramic studies in the BSCCO system have been concentrated on the $Bi_2Sr_2CaCu_2O_8$ phase,[16,17] although some Bi-2223 studies have been reported.[18–20] The bismuth phases permit some glass-ceramic fabrication in the absence of any added glass former due to the limited glass forming ability of bismuth. However, bismuth by itself permits too small a region of glass formability (T_x -T_g). Glass-ceramic processing of Bi-system superconductors has been improved through the addition of intermediate glass formers such as SiO_2, B_2O_3 and most recently Ga_2O_3 and In_2O_3.[21] The Ga and Ga/In glasses will crystallize $Bi_2Sr_2CuO_6$ and $Bi_2Sr_2CaCu_2O_8$. The T_x -T_g difference for the Ga and Ga/In glasses are 73K and 89K respectively for molar ratios of 1:1 (2223:Ga) and 1.0:0.9:0.1 (2223:Ga:In).[22]

Current Bi system glass-ceramic studies range from the science side where we are examining the effect of dopants on the oxidation state of the Cu ions to EXAFS studies of the structure of the glass before crystallization,[23] to technological preparations of oriented coatings of glass ceramics on metal wires.[24-26]

TAPE CASTING, SCREEN PRINTING, COATINGS AND EXTRUSION

In order apply the traditional ceramic fabrication techniques of tape casting, screen printing and extrusion to high temperature superconductors, large crystallites must be grown in order to promote orientation. Schulze and his colleagues have developed a molten-salt synthesis technique[27-29] which produces large platy crystallites on the order of 50 to 100 μm in size. These crystallites, when prepared as a viscous slurry, can be used in tape casting, inks, extrusion and coating wires and foils, to produce highly oriented layers.[30] Recently these procedures have been applied, with similar success, to the thallium system.[31]

AEROSOL PLASMA THIN FILMS

The only techniques which have been developed to-date for producing practical devices, are those which produce highly oriented thin films. Many research teams around the world have developed methods for preparing such films. While these films will not carry currents large enough for traditional applications, like motors which require ampere turns, most researchers agree that these thin films will result in the first practical products. The difficulty is that all methods for making these films involve a very high vacuum and are limited to sizes on the order of 1 cm^2. The only procedure for making thin (or thick) films that is suitable for a mass production system and capable of laying down films in ambient atmosphere at rates of meters per minute uses plasma vapor deposition.[34-37]

Superconducting $YBa_2Cu_3O_{7-\delta}$ films have been produced by a radio frequency (RF) plasma aerosol evaporation technique at atmospheric pressure without post-annealing. Aqueous solutions containing Y, Ba, and Cu are generated as an aerosol which is then injected into the plasma region. The ionized species are deposited onto substrates outside of both the plasma and flame regions forming nanophases. The deposition rates may be as high as 0.01-100 μm/(min cm^2), and film thickness of 1-200 μm may be achieved. For an "as deposited" film on a single crystalline MgO substrate (100) with substrate temperature of 600°C, an onset temperature of the superconducting transition is 100K, with a transition width (10% - 90%) of 3K, and zero resistance at 93K. The critical current density of the film is 0.8×10^4 A/cm^2 at 77K. The as-deposited films compare well with other post-annealed films (850°C , 1 hour). Two patents have been allowed for this process.[32,33] The process has been extended to the production of thallium phase films by depositing the Ba-Sr-Cu-O precursor material with the plasma, as a nanophase and then post-annealing in a thallium atmosphere.[39,38]

POWDER IN TUBE

A number of workers worldwide have reported considerable success in the preparation of the Bi superconductors in silver tubes. We are currently investigating the catalytic effect that silver seems to have on the formation of the $Bi_2Sr_2Ca_2Cu_3O_{10}$ phase. However the most exciting possibility for this technology is to prepare the precursor powders for the thallium 2223 phase, in a tube already bent to the desired conductor shape, and then run an SHS (self propagating high temperature synthesis) reaction through the tube, to produce a bulk superconductor. We have shown that $Tl_2Ba_2Ca_2Cu_3O_{10}$ can be prepared by an SHS reaction.[31]

CHARACTERIZATION

Lastly, when bulk high J_c superconductors are developed, there will be a serious problem in characterizing their current carrying capacity due to ohmic heating of the contacts. A new contactless procedure has been developed[40,41] which permits the rapid (approximately 10 min.) determination of J_c *vs.* magnetic field. This inductive procedure gives a true estimate of the transport J_c .

ACKNOWLEDGMENTS

We would like to thank the New York State Science and Technology Foundation and their Center for Advanced Ceramic Technology along with the New York State College of Ceramics for their sponsorship for this work.

References

1. R. L. Snyder, "Toward a Practical High-Temperature Superconductor: Research at the Institute for Ceramic Superconductivity at Alfred University", in Superconductivity and Ceramic Superconductors II, K. M. Nair et al. editors, American Ceramic Society Inc., Westerville, OH, 607-619 (1991).

2. J.A.T. Taylor, P. Sainamthip, and D.F. Dockery, "Sintering Time and Temperature For $Ba_2YCu_3O_{7-\delta}$ Superconductors," *Mater. Res. Soc. Symp. Proc.* **99**, 663-66 (1988).

3. D. F. Dockery, P. Sainamthip and J. A. T. Taylor, "Melt Texturing Method for Fabricating $YBa_2Cu_3O_{7-\delta}$", U.S. Patent awarded (1990)
4. S. Jin, T.H. Tiefel, R.C. Sherwood, M.E. Davis, R.B. van Dover, G.W. Kammlott, R.A. Fastnacht, H.D. Keith, *Appl. Phys. Lett.* **52**[24], 2074, (1988).
5. S. Jin, T.H. Tiefel, R.C. Sherwood, R.B. van Dover, M.E. Davis, G.W. Kammlott, *Phys. Rev. B.* **37**, 7850 (1988).
6. S. Jin, R.C. Sherwood, E.M. Gyorgy, T.H. Tiefei, R.B. van Dover, S. Nakahara, L.F. Schneemeyer, R.A. Fastnacht and M.E. Davis, *Appl. Phys. Lett.* **54** [6], 584 (1989).
7. K. Salama, V. Selvamanickam, L. Gao, and K. Sun, *Appl. Phys. Lett.* **54** [23], 2352 (1989).
8. B. J. Chen, M. A. Rodriguez and R. L. Snyder, "Dynamic Studies of Texturing in $YBa_2Cu_3O_{7-\delta}$ Superconductors", Y. H. Kao, P. Coppens and H. S. Kwok editors, Superconductivity and Its Applications, Vol 4, Am. Phys. Soc. in press (1992).
9. R. L. Snyder, M. A. Rodriguez, and B. J. Chen, H. E. Göbel, G. Zorn, and F. B. Seebacher, "Analysis of $YBa_2Cu_3O_{7-\delta}$ Peritectic Reactions and Orientation by High Temperature XRD and Optical Microscopy", Adv. X-ray Anal. in press (1991).
10. M. A. Rodriguez, R. L. Snyder, B. J. Chen, D. P. Matheis, S. T. Misture, and V. D. Frechette, "The Sequence of Peritectic Reactions for $YBa_2Cu_3O_{7-\delta}$", *J. Amer. Ceram. Soc* in press (1992).
11. D. P. Matheis, R. L. Snyder and C. R. Hubbard, "High Temperature Reactions in the Processing of Ceramic Superconductors via the Glass ceramic Method", Superconductivity and Its Applications, Vol 3, Y. H. Kao, P. Coppens and H. S. Kwok editors, Am. Phys. Soc. 582-588 (1991).
12. A. Bhargava, A. K. Varshneya and R. L. Snyder, "Superconducting glass-ceramics made by controlled crystallization from glass", U.S. Patent Number 4,970,195 awarded Nov. 13, (1990).
13. B. J. Chen, M. A. Rodriguez, and R. L. Snyder, "Glass Formation and Textured Crystallization in the Y_2O_3-BaO-CuO-B_2O_3 and Y_2O_3-BaO-CuO-P_2O_5 Systems", Superconductivity and Its Applications, Vol 3, Y. H. Kao, P. Coppens and H. S. Kwok editors, Am. Phys. Soc. 589-598 (1991).

14. A. Bhargava, A. K. Varshneya and R. L. Snyder, "Crystallization of Glasses in the System BaO-Y_2O_3-CuO-B_2O_3", pages 124-129, The Proceedings of the Second Annual Conference on Superconductivity, edited by David T. Shaw and Hoi Sing Kwok, Elsevier Scientific Publishing Co., July (1988).

15. A. Bhargava, A. K. Varshneya and R. L. Snyder, "On the Stability of Superconducting $YBa_2Cu_3O_{7-\delta}$ in a Borate Glass-ceramic Matrix", *Materials Letters*, **8** [1-2], 41-45 (1989).

16. D. Matheis and R. L. Snyder, "The Crystal structures and Powder Diffraction Patterns of the High T_c Ceramic Superconductors", *Powder Diffraction*, **(5)** [1], 8-25 (1990).

17. A. Bhargava, R. L. Snyder and A. K. Varshneya, "Preliminary Investigations of Superconducting Glass-ceramics in the Bi-Sr-Ca-Cu-O System", *Matl. Ltrs.* **8** [10], 425-431 (1989).

18. D. Shi, M. Blank, M. Patel, D. G. Hinks, A. W. Mitchell, K. Van der Voort, H. Claus, "Superconductive Phases in the Bi- Sr-Ca-Cu-O System," Physica C 156, 822 (1988).

19. T. Komatsu, R. Sato, C. Hirose, K. Matusita, and T. Yamashita, "Preparation of High T_c Superconducting Bi-Pb-Sr- Ca-Cu-O Ceramics by the Melt Quenching Method," Japanese Journal of Applied Physics Letters, Vol. 27, L2293 (1988).

20. H. Zheng, J.D. Mackenzie, "$Bi_{1.84}Pb_{0.34}Ca_2Sr_2Cu_4O_y$ Superconducting Tapes with Zero Resistance at 100K Prepared by the Glass-to-Ceramic Route," Journal of Non-Crystalline Solids, Vol. 113 (1989).

21. H. Aβmann, A. Günthier, B. Steinmann, "Preparation and Properties of Superconducting Crystallized Glasses and Enamels Based on the Bi-Sr-Ca-Cu-Oxide System," Icmc' 90 Conference "High-Temperature superconductors, Materials Aspects" May 9-11, 1990 Garmisch-Partenkirchen, FRG.

22. B. J. Reardon and R. L. Snyder, "The Use of Ga_2O_3 and In_2O_3 in Forming Superconducting Glass Ceramics", Y. H. Kao, P. Coppens and H. S. Kwok editors, Superconductivity and Its Applications, Vol 3, Am. Phys. Soc. 599-609 (1991).

23. S. S. Bayya, R. Kudesia and R. L. Snyder, "EXAFS Studies of $Bi_2Sr_2CaCu_2O_8$ Glass and Glass-Ceramic", Y. H. Kao, P. Coppens and

H. S. Kwok editors, Superconductivity and Its Applications, Vol 3, Am. Phys. Soc. 306-314 (1991).

24. A. Bhargava, A. K. Varshneya and R. L. Snyder, "Synthesis of Superconducting Ceramic Coated Metal Wires by a Glass-Ceramic Technique", *Matl. Ltrs.* **11** 313-316 (1991).

25. A. Bhargava, M. A. Rodriguez and R. L. Snyder, "Metal-Ceramic Composite Superconducting Wires" *J. Mat. Res. Soc.* submitted (1991).

26. A. Bhargava, "Synthesis of Superconducting Ceramic Coated Metal Wires by a Glass-Ceramic Technique", U.S. Patent pending (1992).

27. C.T. Decker, V.K. Seth and W. A. Schulze, "Feasibility of Synthesis of Anisotropic Morphology of $Ba_2YCu_3O_{7-x}$ Powder by Molten Salt Technique," p. 169-176 Advanced Superconductors II, Man F. Yan editor, The Am. Ceram. Soc., Westerville, OH, (1988).

28. S. Gopalakrishnan and W.A. Schulze, "Grain Orientation in High T_c Superconductors by Molten Salt Synthesis", NASA Conference Publication 3100, AMSAHTS '90, ed., L.H. Bennett, Y. Flom, and K. Moorjani, p. 173-180 (1990)

29. S. Gopalakrishnan and W. A. Schulze, "Grain Orientation in High T_c Ceramic Superconductors", Superconductivity and Its Applications, Vol 2, pp. 411-418, H. S. Kwok editor, Plenum Press (1990).

30. S. Gopalakrishnan and W. A. Schulze, "Grain Orientation in High T_c Superconductors by Molten Salt Synthesis" U.S. Patent pending (1992).

31. S. S. Bayya, G. C. Stangle and R. L. Snyder, "Synthesis of Superconducting Phases in Tl-Ba-Ca-Cu-O System" Superconductivity and Its Applications, Vol 4, Y. H. Kao, P. Coppens and H. S. Kwok editors, Am. Phys. Soc. in press (1992).

32. R. L. Snyder, X. W. Wang and H. Hsong, "Process for Preparing Superconducting Films by Radio-Frequency Generated Aerosol-Plasma Deposition in Atmosphere" U.S. Patent allowed 07/528,147 (1991).

33. R. L. Snyder, X. W. Wang and H. Hsong, "Atmospheric Plasma Vapor Deposition of Superconducting Thin Films", U.S. Patent pending 07/510,011 (1990).

34. X. W. Wang, H. H. Zhong, and R. L. Snyder, "RF Plasma Aerosol Deposition of Superconductive $YBa_2Cu_3O_{7-\delta}$ Films at Atmospheric Pressure", *Appl. Phy. Lett.* **57** [15], 1581-1583, (1990).

35. J. Hao, X. W. Wang and R. L. Snyder, "RF Plasma Aerosol Deposition of Superconductive $YBa_2Cu_3O_{7-\delta}$ Films at Atmospheric Pressure", Science and Technology of Thin Film Superconductors 3, Plenum Press, New York in press (1991).

36. H. H. Zhong, X. W. Wang, and R. L. Snyder, "Deposition of Superconductive $YBa_2Cu_3O_{7-\delta}$ Films at Atmospheric Pressure by RF Plasma Aerosol Technique", Superconductivity and Its Applications, Vol 3, Y. H. Kao, P. Coppens and H. S. Kwok editors, Am. Phys. Soc. 531-542 (1991).

37. X. W. Wang, H. H. Zhong, and R. L. Snyder, "Superconducting $YBa_2Cu_3O_{7-\delta}$Films by RF Plasma Aerosol Evaporation at Atmospheric Pressure", Science and Technology of Thin Film Superconductors 2, Plenum Press, New York 311-317 (1990).

38. R. L. Snyder, X. W. Wang, H. M. Duan and A. A. M. Hermann, "Preparation of Thallium Superconducting Films Using an Atmospheric Plasma Vapor Deposition Process", U.S. Patent allowed 07/640,221 (1991).

39. H. M. Duan, A. M. Hermann, X. W. Wang, J. Hao and R. L. Snyder, "Superconducting TlBaCaCuO Thin Films with BaCaCuO Precursors", Superconductivity and Its Applications, Vol 4, Y. H. Kao, P. Coppens and H. S. Kwok editors, Am. Phys. Soc. in press (1992).

40. L. M. Fisher, V. S Gorbashev, N. V. Il'in, N. M. Makarov, I. F. Voloshin, V. A. Yampol'skii, R. L. Snyder, J. A. T. taylor, V. W. R. Amarakoon, A. M. M. Barus, J. G. Fagan, M. A. Rodriguez, D. Matheis, and S. Misture, "The Universal Magnetic Field Dependance of the Local Critical Current Density in High-T_c Ceramics", *Phys. Rev B* submitted (1991).

41. L. M. Fisher, V. S Gorbashev, N. V. Il'in, N. M. Makarov, I. F. Voloshin, V. A. Yampol'skii, R. L. Snyder, J. A. T. Taylor, V. W. R. Amarakoon, M. A. Rodriguez, S. T. Misture, D. P. Matheis, A. M. M. Barus and J. G. Fagan "The Universal Magnetic Field Dependance of the Local Critical Current Density in High-T_c Ceramics", Superconductivity and Its Applications, Vol 4, Y. H. Kao, P. Coppens and H. S. Kwok editors, Am. Phys. Soc. in press (1992).

SYNTHESIS OF SUPERCONDUCTING PHASES IN TL-BA-CA-CU-O SYSTEM

S.S. Bayya, G.C. Stangle and R.L. Snyder
Institute of Ceramic Superconductivity
New York State College of Ceramics at Alfred University
Alfred, NY 14802

Abstract

This paper describes various novel processing techniques for the synthesis of superconducting phases in the Tl-Ba-Ca-Cu-O system. A Self propagating high temperature synthesis technique has been used to synthesize phase pure 2212 and 2223. Various engineering parameters are identified for this process. A glass-ceramic (melt quench) technique with subsequent post heat-treatment produced pure 2201 and 2212 phases. Tl_2O_3 itself is not a very good glass former and the addition of other glass formers is necessary to form stable glasses. Only the gallate glass system has been found to stabilize the 2201 and 2212 superconducting phases. Molten salt synthesis studies showed that the superconducting phases in the thallium system are stable in the NaCl-KCl eutectic salt system. Highly textured 2201 grains (about 60 μm x 60 μm platelets) were grown by this technique. Various potential applications of these techniques are also discussed.

Introduction

Thallium based superconductors hold the current record high T_c with zero-resistance at 125K. Since the discovery of superconductivity in a thallium containing phase,[1-3] a number of other Tl-based Ruddelsden-Popper superconducting compounds have been identified.[4-10] However, due to the high toxicity of thallium, many of the non-chemist researches have declined to work in this system. Hence, not many processing techniques have been explored for the synthesis of bulk superconductors in this system. Tl_2O_3 melts at 717°C and is very volatile. Researchers have tried to overcome this problem of high volatility of Tl_2O_3 by carrying out reactions in sealed gold tubes,[8,14] reacting pellets by wrapping them in gold foils[11-13] or by carrying out reactions in a very short time[6,15] (approximately 5 minutes). These reactions often result into off-stoichiometric products. Little work has been done to tailor the processing of these compounds, to overcome the problems of thallium loss and to develop controlled microstructure and textured grain growth.

This paper discusses some novel ceramic processing techniques used to solve these problems with thallium synthesis.

Self-propagating High-Temperature Synthesis (SHS)

Self-propagating high-temperature synthesis (SHS) of ceramic materials was first developed by A. G. Merzhanov in the Soviet Union in 1967. Since then it has been extensively used to synthesize intermetallics, oxide and non-oxide ceramics.[16] SHS involves highly exothermic reactions where the reaction, once initiated by a thermal pulse, self propagates through the reactants, as a combustion wave driven by the ΔH of the reaction, forming products. These reactions are very fast and occur in reactions times about 2 orders of magnitude less than those for conventional oxide synthesis techniques. The velocity of the combustion front is typically 0.1-1 mm/sec. SHS reactions employ a metal-fuel and oxygen sources, where one of the reactants is a metal powder whose oxidation provides the heat required for SHS. Typical SHS reaction in the Tl-Ba-Ca-Cu-O system can be written as:

$$2Cu + 2BaO_2 + CaO + Tl_2O_3 \rightarrow Tl_2Ba_2CaCu_2O_8 + Q \tag{1}$$

where Q = heat of reaction. Similar reaction can also be written for $Tl_2Ba_2Ca_2Cu_3O_{10}$.

The major problem in carrying out these reactions by SHS is the small metal-fuel to oxide ratio. The combustion front in the SHS reaction has two zones, namely a heating zone which is followed by a reaction zone. Due to the small metal-fuel to oxide ratio, the heat generated from the oxidation of copper is lost in heating up the reactants at the tip of the combustion front. This problem may be avoided, either by increasing the metal-fuel to oxide ratio or by providing an extra amount of energy to heat the reactants.

Metal-fuel to oxide ratio can be increased either by adding another component as metal powder (probably thallium) or by carrying the SHS reaction in two steps: reacting BaO_2, CaO and Cu by SHS first and thalliating it in a second step. Addition of thallium as a fine powder could be hazardous, so the reaction was carried out in two steps. BaO_2, CaO and Cu were mixed, pressed in the form of a pellet and reacted by the SHS process at room temperature. This reaction

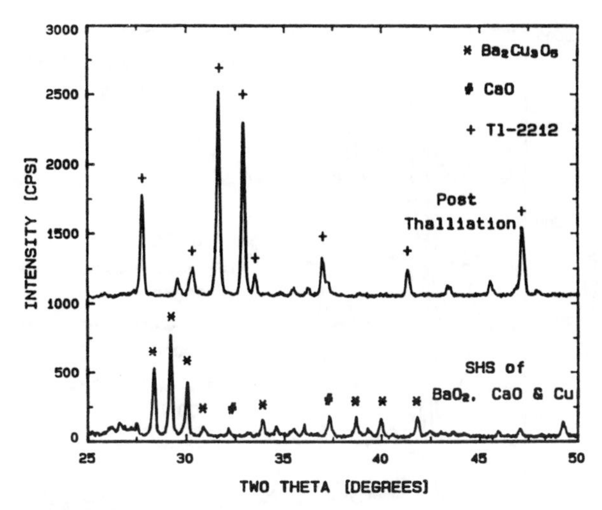

Figure 1: SHS of the BaO_2, Cao & Cu and post thalliation to form Tl-2212

resulted in the formation of $Ba_2Cu_3O_5$ and CaO. The product was mixed with a stoichiometric quantity of Tl_2O_3, pressed into pellet and fired at 850°C for 20 min. to form phase pure 2212 (figure 1).

In another set of experiments, stoichiometric quantities of Tl_2O_3, BaO_2, CaO and Cu were mixed, pressed in the form of pellets, preheated at different temperatures and ignited to carry out the SHS reaction. Figures 2 and 3 summarize the results of the SHS reaction in the 2212 and 2223 systems. In these two figures one can see the disappearance of the reactants and the formation of the superconducting phases systematically with increasing pre-heating temperatures. Pre-heating the reactants to higher temperatures did help in forming the superconducting phases but the temperatures achieved during the SHS reaction were not high enough to complete the reaction in that short a time. The SHS products were heat-treated at 850°C and 900°C for 20 min. to form 2212 and 2223 phases respectively (figure 4).

The following engineering parameters were found to be critical in the SHS reaction:

- particle size of the reactants
- mixing
- compaction density
- combustion atmosphere

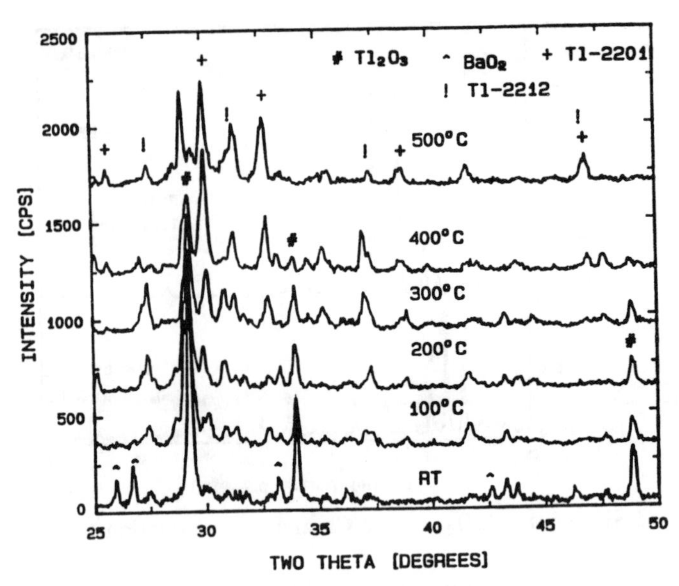

Figure 2: Effect of pre-heating the reactants in SHS of Tl-2212 superconductor.

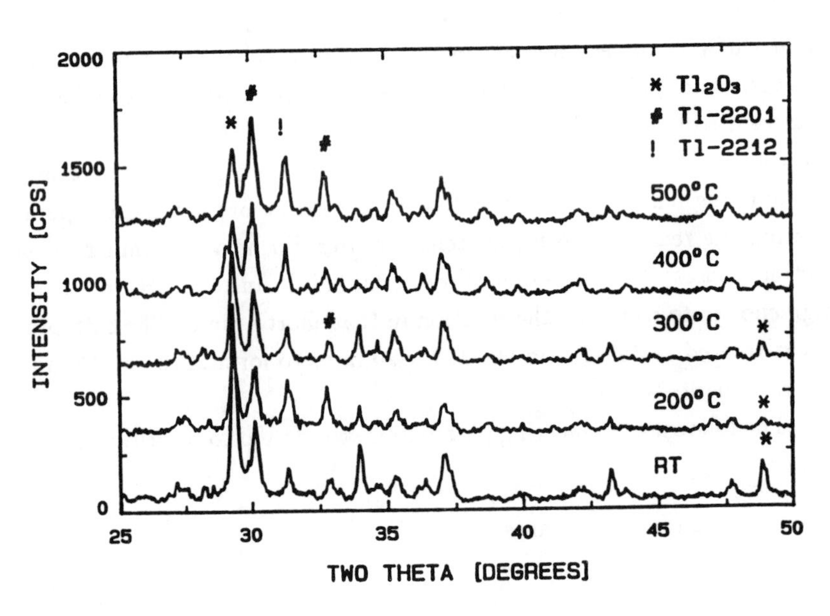

Figure 3: Effect of pre-heating the reactants in SHS of Tl-2223 superconductor.

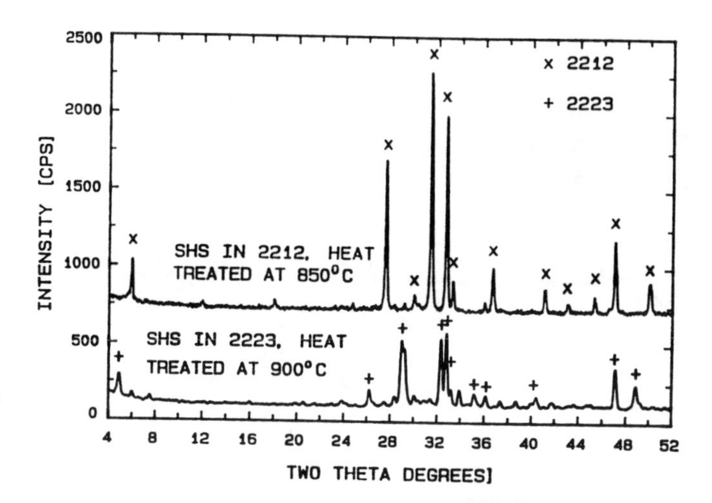

Figure 4: SHS of Tl-2212 and Tl-2223 superconductors with post process heat treatment.

- oxygen pressure
- pre-heating temperature

A better control over these parameters should provide suitable conditions for the formation of superconducting phases in a single step. This would reduce the loss of thallium by volatilization due to the decrease in the reaction time to less than a minute. This synthesis technique also has potential application in powder-in-tube technology.

Molten Salt Synthesis

Molten salt synthesis of powders has been extensively used in electroceramics.[17-1] It has also been used in Y-Ba-Cu-O[20,21] and Bi-Sr-Ca-Cu-O[22] systems. This synthesis technique often results in highly textured grain growth. Since diffusivities of the reactants are much higher in the molten salt system compared to solid state reactions these often require shorter reaction time and temperatures.

The first step in this process is to find a suitable salt system compatible with the oxide system, which would facilitate the formation of the desired phase, enhance textured growth of the grains, not react with the phase or any other component and separate out completely after the salt synthesis. In the present study a eutectic mixture of NaCl-KCl was used (1:1 mole ratio). In all the cases a 1:1 weight ratio of the precursor batch (stoichiometric mixture of Tl_2O_3, BaO_2, CaO & CuO) to the total salt was used.

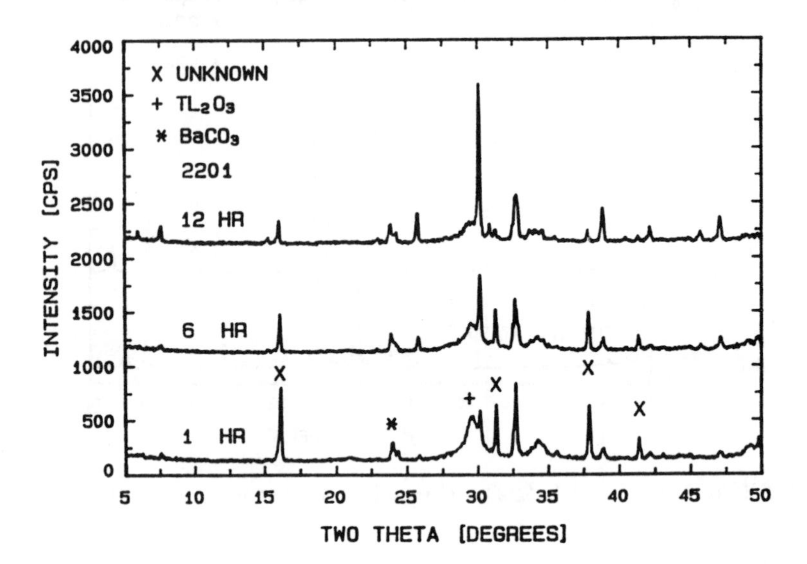

Figure 5: Effect of reaction time at 800°C on the molten salt synthesis in the 2201-salt system.

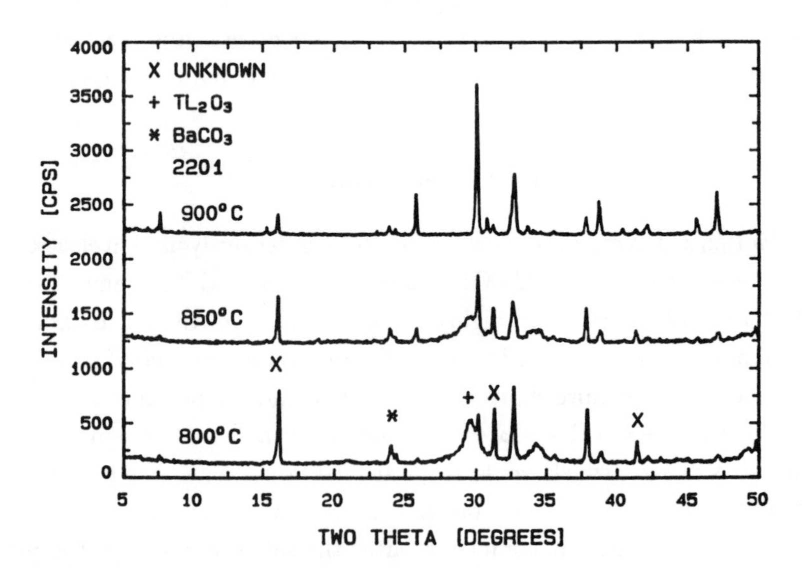

Figure 6: Effect of reaction temperature for 1 hour reaction time on the molten salt synthesis in the 2201-salt system.

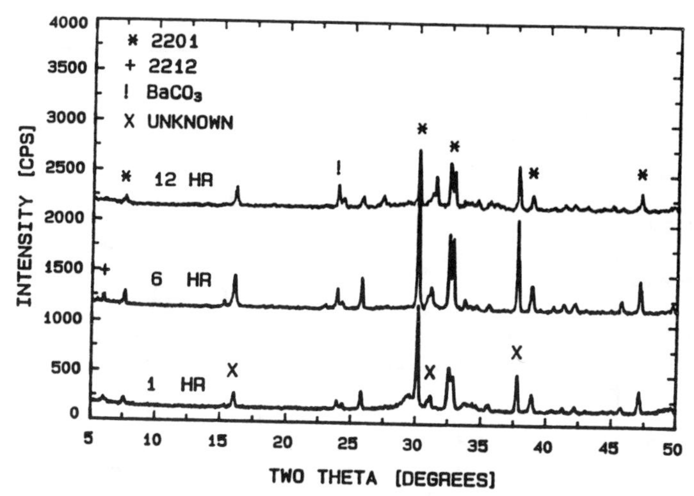

Figure 7: Effect of reaction time at 800°C on the molten salt synthesis in the 2212 precursor-salt system.

Different temperature-time studies were carried out in (2201) precursor-salt and (2212) precursor-salt systems. In the case of the Tl-2201 system, it was observed that even a 12 hour reaction time at 800°C was not sufficient to completely react Tl_2O_3 (fig. 5). The 2201 particles did show anisotropy which increased with reaction temperature and also with reaction time at 800°C (fig. 6). It was also observed that shorter reaction times and higher temperatures drive the reaction faster towards completion (figure 6). However, results were very interesting in the case of 2212 precursor-salt system. In this case the major phase formed was always 2201 (fig. 7 and 8). With increasing reaction time at 800°C the anisotropy of the grains increases up to 6 hrs and degraded by 12 hrs. The best condition for the enhanced grain growth of 2201 was found to be at 850°C for 6 hrs. Under these conditions, platey grains of size about 60 μm × 60 μm were formed (fig. 9).

This powder with textured grains can be used to make an ink for screen printing or to spin coat substrates or to dip coat wires and can be used for tape casting followed by lamination to produce a textured superconductor ceramic with high critical current densities.

Melt Quench and Glass-Ceramic Synthesis

Since the discovery of high temperature superconductivity in ceramic systems, we have extensively explored, with some success, the possibility of forming

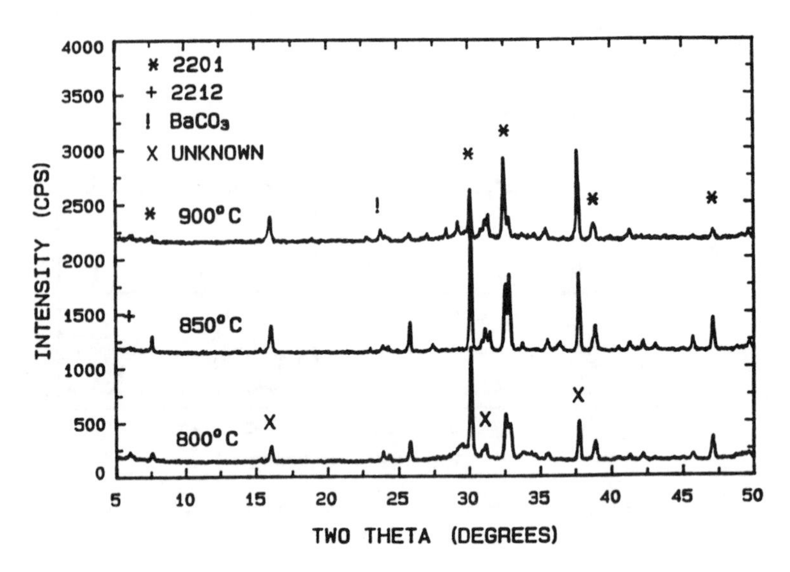

Figure 8: Effect of reaction temperature for 1 hour reaction time on the molten salt synthesis in the 2212 precursor-salt system.

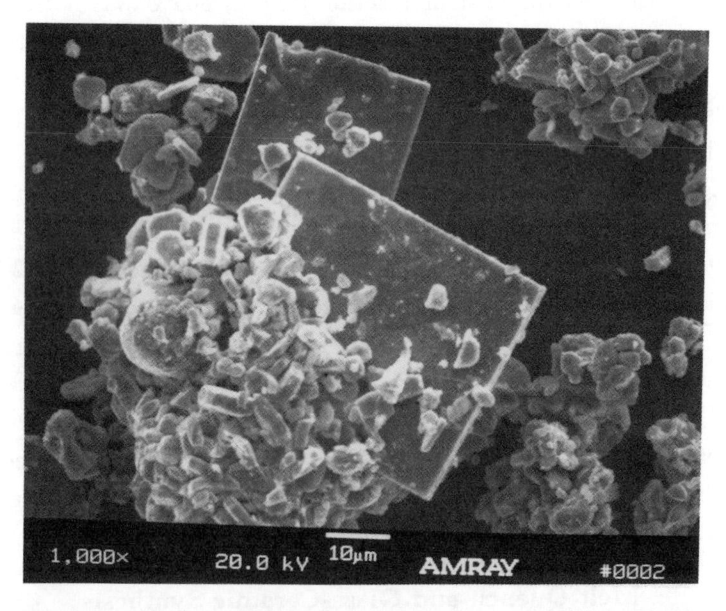

Figure 9: SEM picture showing large 2201 grains developed by salt synthesis in the 2212 precursor-salt system reacted at 850°C for 6 hrs.

a glass-ceramic superconductor. Advantages of using the glass-ceramic route for the synthesis of superconductors has been discussed by Bhargava *et al.*[23] The 123 phase in the Y-Ba-Cu-O system was successfully crystallized out of borate based glass.[24] Bi_2O_3 itself is a good glass former and easily crystallizes the superconducting 2201 and 2212 phases in the Bi-Sr-Ca-Cu-O system, from the melt.[25] These phases also have been crystallized out of gallia based glass systems.[26,27] The glass-ceramic technique in the bismuth system has also been used to draw glass fibers followed by crystallization of the superconducting phases and also to coat wires.[28,29]

BORATE SYSTEM

Composition Tl,Ba,Ca,Cu-B	Melting	XRD results	Heat treatment	XRD results
2201-0	1150°C 3.5min	mostly amorphous	600°C 10hr	2201 (major phase) $Ba_2Cu_3O_5$
2201-1	1150°C 4min	mostly amorphous	600°C 10hr	Tl_2O_3 (major Phase) $BaCuO_2$, CuO
2201-2	1150°C 4min	$Tl_6B_2O_6$	600°C 10hr	Tl_2O_3, BaB_2O_4, CuO
2201-3	1150°C 4min	BaB_2O_4	600°C 10hr	X
2201-4	1150°C 4min	amorphous	600°C 10hr	X
2201-5	1150°C 4min	amorphous	600°C 10hr	X
2201-10	1150°C 4min	amorphous	600°C 10hr	X
2212-0	1025°C 10min	partly crystalline	700°C 8hr	2212 (major phase) CuO, CaO
2212-1	did not	melt at 1200°C	in 1hr	
2212-2	did not	melt at 1200°C	in 1hr	
2212-3	1100°C 5min	CuO_2, mostly amorphous	700°C 16hr	CuO, X
2212-4	1100°C 4min	CuO_2, $Tl_6B_2O_6$	700°C 16hr	$Ba_2Cu_3O_5$ CuO, X
2212-5	1100°C 4min	CuO_2, $Tl_6B_2O_6$ mostly amorphous	700°C 4hr	$Tl_6B_2O_6$, CuO, X
			700°C 16hr	$Ba_2Cu_3O_5$, CuO, X
2212-10	1100°C 4min	amorphous	700°C 4hr	CuO, X
			700°C 16hr	CuO, X

GALLATE SYSTEM

Composition Tl,Ba.Ca,Cu-Ga	Melting	XRD results	Heat treatment	XRD results
2201-0	1150°C 3.5min	mostly amorphous	600°C 10hr	2201 (major phase) $Ba_2Cu_3O_5$
2201-1	1165°C 5min	$Ba_3Ga_2O_6$	600°C 16hr	$Ba_2Cu_3O_5$, X
			700°C 16hr	2201 (major phase) $Ba_2Cu_3O_5$, X
2201-2	1165°C 3min	crystalline (X)	700°C 12hr	2201, X
2212-0	1025°C 10min	partly crystalline	700°C 8hr	2212 (major phase) CuO, CaO
2212-1	1165°C 5min	mostly amorphous	700°C 16hr	2212 (major phase) CuO, X
2223-0	1025°C 10min	CaO, CuO_2, X	700°C 8hr	2212 (major phase) CuO, CaO
2223-1	1170°C 4min	mostly amorphous	700°C 16hr	2212 (major phase) CuO, X
2223-2	1170°C 4min	mostly amorphous	600°C 12hr	Tl_2O_3, CuO CaO, X
			700°C 12hr	$Ba_3Ga_2O_6$, CuO, X
2223-3	1170°C 3min	Partly crystalline	600°C 12hr	Tl_2O_3 (major phase) CuO, X
2223-4	1170°C 3min	$BaGa_2O_4$, X	600°C 12hr	Tl_2O_3 (major phase) $BaGa_2O_4$, CuO

Tl_2O_3 is not a very good glass former by itself and does not form a very good glass on melt quenching the batch compositions in 2201, 2212 or 2223 system. It often crystallizes Cu_2O and CaO phases. On heat-treating these melt quench samples 2201 and 2212 phases crystallize out as single phases. However, for using glass-ceramic technique as a ceramic processing technique, one needs a stable glass with a large difference between T_g (glass transition temperature) and T_x (first crystallization temperature after T_g). This can be attained by the addition of external glass formers. Crystallization of different phases was studied in borate and gallia base glass systems. Table 1 & 2 summarize the results of melt quench and glass ceramic synthesis in 2201 and 2212 systems. It can be seen that small additions of B_2O_3 puts the composition in an immissibility dome where it becomes difficult even to melt the batch. At higher additions of B_2O_3 stable glasses were observed but resulted into the crystallization of stable binary borates on devitrification. 2201 and 2212 were devitrified from gallate

seen that small additions of B_2O_3 puts the composition in an immiscibility dome where it becomes difficult even to melt the batch. With higher additions of B_2O_3 stable glasses were observed but they crystallized stable binary borates phases rather than the desired superconductors. 2201 and 2212 were devitrified from gallate based glasses on smaller additions of Ga_2O_3 which somewhat improved the glass formability. Higher additions of gallia resulted in the crystallization of binary gallates.

References

1. Z.Z. Sheng, A.M. Hermann, A.El Ali, C. Almason, J. Estrada, T. Datta and R.J. Matson, *Phys. Rev. Lett.* **60** 937 (1988).
2. Z.Z. Sheng and A.M. Hermann *Nature* **332** 55 (1988).
3. Z.Z. Sheng and A.E. Hermann *Nature* **332** 138 (1988).
4. C.C. Torardi, M.A. Subramanian, J.C. Calabrese, J. Gopalakrishnan, K.J. Morrissey, T.R. Askew, R.B. Flippen, U. Chowdhry and A.M. Sleight, *Science* **240** 631 (1988).
5. S.S.P. Parkin, V.Y. Lee, E.M. Engler, A.I. Nazzal, T.C. Huang, G. Gorman, R. Sovoy and R. Beyers, *Phys. Rev. Lett.* **60** 2539 (1988).
6. R.M. Hazen, L.W. Finger, R.J. Angle, C.T.Prewitt, N.L. Ross, C.G. Hadidiacos, P.J. Heaney, D.R. Veblen, Z.Z. Sheng, A.El Ali and A.M. Hermann, *Phys. Rev. Lett.* **60** 1657 (1988).
7. L. Gao, Z.J. Huang, R.L. Meng, P.H. Hor, J. Bechtold, Y.Y. Sun, C.W. Chu, Z.Z. Sheng and A.M. Hermann *Nature* **332** 623 (1988).
8. C.C. Torardi, M.A. Subramanian, J.C. Calabrese, J. Gopalakrishnan, E.M. McCarron K.J. Morrossey, T.R. Askew, R.B. Flippen, U. Chowdhry and A.M. Sleight, *Phys. Rev. B* **38** 225 (1988).
9. M.A. Subramanian, J.C. Calabrese, C.C. Torardi, J. Gopalakrishnan, T.R. Askew, R.B. Flippen, K.J. Morrossey, U. Chowdhry and A.M. Sleight, *Nature* **332** 420 (1988).
10. R. Beyers, S.S.P. Parkin, V.Y. Lee, A.I. Nazzal, R. Savoy, G. Gormn, T.C. Huang and S. La Placa, *Phys. Rev. Lett.* **61** 750 (1988).
11. R. Beyeres, S.S.P. Parkin, V.Y. Lee, A.I. Nazzal, R. Savoy, G. Gormn and T.C. Huang, *Appl. Phys. Lett.* **53** (5) 432 (1 August 1988).

12. M. Eibschütz, L.G. Van Uitert, G.S. Grader, E.M. Gyorgy, S.M. Glarum, W.H. Grodkiewick, T.R. Kyle, A.E. White, K.T. Short and G.J. Zydzik, *Appl. Phys. Lett.* **53** (10) 911 (5 September 1988).

13. S.S.P. Parkin, V.Y. Lee, A.I. Nazzal, R. Sovoy, R. Beyers and S.J. La Placa *Phys. Rev. Lett.* **61** (6) 750 (1988).

14. A.W. Hewat, P. Bordet, J.J. Capponi, C. Chaillout, J. Chenavas, M. Godinho, E.A. Hewat, J.L. Hodeau and M. Marezio, *Physica C* **156** 369 (1988).

15. Z.Z. Sheng, L. Sheng, X. Fei and A.M. Hermann, *Physical Review B* **39** (4) 2918 (1989).

16. A.G. Merzhanov, Self-propogating High-temperature Synthesis: twenty years of search and findings preprint, ISMAN, Chernogolovka, (1989).

17. M. Granahan, M. Holmes, W. Schulze and R.E. Newnham, *J. Am.Ceram. Soc.* **64** [4] C68 (1981).

18. R.H. Arendt, J.H. Rosolowski and J.W. Szymaszek, *Mat. Res. Bull.* **14** 703 (1979).

19. T. Kimura and T. Yamaguchi, *Advances in Ceramics* Vol. 21 *Ceramic Powder Science, eds. G.L. Messing, K.S. Mazdiyasni, J.W. McCauley and R.A. Haber, Am. Ceram. Soc. Westerville, OH* 169 (1987).

20. S. Gopalakrishnan and W.A. Schulze, *Proceedings of the conference on Advances in Materials Science and Applications of High Temperature Superconductors, NASA, MD, eds. L.H. Bennett, Y. Flom and K. Moorjani,* 173, (1990).

21. D.B. Knorr and C.H. Raeder, *Proceedings of the conference on superconductivity and aplications, Buffalo, NY, eds. H.S. Kwok, Plenum* (1989).

22. S. Gopalakrishnan and W.A. Schulze, *Presentation at fourth annual conference on Superconductivity and applications, Buffalo, NY* Sept. 24-26 1991.

23. A. Bhargava, A.K. Varshneya and R.L. Snyder, *Proceedings of the conference on Superconductivity and applications, Buffalo, NY, eds. H.S. Kwok and D.T. Shaw, Elsevier, Amsterdam* 124 (1988).

24. A. Bhargava, A.K. Varshneya and R.L. Snyder, *Mat. Lett.* **8** No. 41, (1989).

25. A. Bhargava, A.K. Varshneya and R.L. Snyder, *Mat. Lett.* **8** No. 425, (1989).

26. H. Assmann, A. Günthier, B. Steinmann, Icmc' 90 Topical Conference "High-Temperature superconductors, Materials Aspects" May 9-11, 1990 Garmisch-Partenkirchen, FRG.

27. B.J. Reardon and R.L. Snyder, *Proceedings of the conference on Superconductivity and it's Applications, Buffalo, NY, eds. Y. Kao, P. Coppens, H. Kwok, AIP, NY* 599 (1990).

28. A. Bhargava, A.K. Varshneya and R.L. Snyder, *Mat. Lett.* **11** No. 10,11,12, 313 (1991).

29. A. Bhargava, M. A. Rodriguez and R. L. Snyder, *J. Mat. Res. Soc.* submitted (1991).

NEW PREPARATIVE METHODS TO ENHANCE PHASE PURITY AND PHYSICAL PROPERTIES OF CUPRATE SUPERCONDUCTORS

R.E.Salomon, R.Schaeffer, J.Macho, Allan Thomas and G.H. Myer
Temple University and the Ben Franklin Superconductivity Center,
Philadelphia, Pa. 19122

N.V.Coppa
Los Alamos National Laboratory, Los Alamos, NM. 87545

ABSTRACT

Several methods which avoid the problems inherent in solid state reactions have been developed. These methods include freeze drying, liquid ammonia based processes and a novel xerogel process. They are applicable to both 123, 124 and BISCCO based superconductors and have as their goal the atomic mixing of precursors in order to reduce inhomogeneity in the final product.

The three methods are described and compared to each other and to conventional methods of synthesis. The products prepared by these methods are fully characterized by elemental analysis, XRD, TGA, DSC, resistivity and magnetic susceptiblity versus temperature and by SEM.

INTRODUCTION

Several methods which avoid the problems inherent in solid state reactions for the preparation of bulk powders of high T_c superconductors have been developed by us[1,2,3,4] and others.[5-10] Solid state reactions start with mechanical mixtures which must react at grain boundaries and these grain boundaries quickly become laden with products thereby slowing the rate of reaction. Accordingly, attempts are made to grind samples to finer particle sizes and these long grinding processes lead to contamination of the final product with material abraded from the grinding apparatus. If the grinding times are reduced, then the particles are not as well mixed and longer times and higher temperatures are required to produce the desired final product. The disadvantages that accrue to these longer processing times and higher processing temperatures are obvious. One well known method of enhancing mixing in a non-mechanical manner is to co-precipitate the precursor metal ions as a salt which can be decomposed to form the oxide. Organic acids and carbonates are the favored precipitating agents although there is concern about the contamination of product with carbon at the grain boundaries. Furthermore, co-precipitation, doesn't really lead to atomic mixing although if done carefully, it does lead to an intimate mixture of small particles. Finally, it should be mentioned that control of stoichiometry is difficult in coprecipitation because of finite solubilities of the precipitates.

For these reasons, a general study of alternate methods of mixing which would lead to near atomic mixtures was carried out over the past few years. With atomically mixed precursors, one has the opportunity to have diffusionless reactions. With these diffusionless reactions, high processing temperatures are in principal only needed to decompose the anion to the oxide. In general, both the processing time and temperature can be reduced and ideally the phase purity can be enhanced.

In the following, a description of three different methods which have been used to achieve a near atomic mixture of precursor ions will be described. These methods are 1)a freeze drying technique 2)a set of liquid ammonia based methods and 3) a xerogel method.

EXPERIMENTAL

For many of these methods, starting materials in the form of water soluble nitrates and acetates are needed. These materials often contain varying amounts of water of hydration. Determination of this water of hydration by TGA is not always an acceptable method of determining metal to weight ratios since overlap between the loss of water and the decomposition of the anion can occur. Accordingly, analytical methods to determine the exact metal ion content of starting materials are needed. Copper ion can be determined most readily and accurately by electrogravimetry wherein a solution of the copper salt is electrolyzed between platinum electrodes. The copper deposits quantitatively at the cathode in a period of 10 or more hours. The weight gain of the platinum electrode gives the mass of copper in the solution. Yttrium nitrate can be decomposed to the oxide which loses water of hydration at relatively low temperatures. Barium and yttrium can also be determined by atomic absorption spectroscopy. Bismuth, copper and lead solutions can also be prepared by dissolution of the pure metal in nitric acid. Calcium and strontium solutions can be readily analyzed by EDTA titrations. In the following, it will be assumed that 123 material ($YBa_2Cu_3O_{7-x}$) is being produced unless otherwise stated.

The first method of preparation to be described is the freeze drying method. The goal of this method of preparation is a precursor material that is a random solid solution of Y, Ba and Cu nitrates in the appropriate stoichiometry. To achieve this end, an aqueous nitrate solution is ultrasonically aerosolyzed and allowed to impinge on the surface of liquid nitrogen at near zero velocity head. The micron sized droplets freeze throughout, thereby trapping the metal ions. The "snow" containing precursor ions is rapidly transported to a freeze dryer with a shelf temperature maintained at ca. -25 C and a condenser temperature at -55 C. The salt containing ice sublimes in a period of days. During this period of sublimation, adjustment of the shelf temperature is carried out to facilitate this rate process without allowing melting to occur. The samples are eventually transferred to a vacuum drying oven where the last traces of water are removed without allowing melting in water of hydration with subsequent phase separation.

Materials prepared in this manner have some unusual properties the most interesting of which is the large heat of solution which conforms to the interpretation of a random solid solution which has a higher enthalpy than the corresponding enthalpy of the phase separated system. X-ray diffraction studies also do not reveal the presence of any crystalline phases. Of course, both results could also be explained by the presence of amorphous pure solid phases.

Even though we have succeeded in our initial goal of preparing atomically mixed precursors, it turns out that during the subsequent thermal processing some unmixing occurs. Samples have been taken from partially thermally processed materials and examined by XRD and the results show that some phase separation occurs. Nevertheless, this unmixing occurs at the sub-micron level and after a final high temperature anneal, phase and chemically pure product is obtained. More detailed reports on the kinetics and mechanism of the thermal decomposition of this random solid solution will be provided in a forthcoming publication[11]. Recently, this technique has been applied to the preparation of pure and doped 124 high T_c superconductors.[12] It appears that this freeze dry method of preparing precursors is particular useful for synthesizing materials that are thermally sensitive or which do not readily lead to the desired product for thermodynamic or kinetic reasons. The one negative aspect of this technology, especially when applied to the preparation of BISSCO materials, is that some freeze drying equipment may not be compatible with the type of strongly acidic solutions needed to dissolve bismuth. The use of acid traps in the freeze dryer can overcome some of these problems. Recently Song et al.[13] have prepared some bismuth based superconductors by a freeze drying methodology.

Liquid ammonia is a medium dielectric solvent which can dissolve a variety of inorganic salts. The heat of vaporization of liquid ammonia is about half of that of water. It also seems that the energy required to remove ammonia from inorganic solids is smaller than that required to remove water of hydration. This smaller energy facilitates a more rapid removal of the solvent which aids in avoiding phase separation. Whether or not this rationale is correct, we have found that the spray pyrollysis of these liquid ammonia solutions readily yields intimate mixtures of the precursor salts which can be formed into final product with lower temperatures and processing times. We are aware of the fact that spray pyrollysis of aqueous solutions has recently made its way into the commercial market but it is not clear whether this will be competitive. In the method described here, the ammonia is used only as a solvent and transport agent and it can be recycled. Attempts to utilize the flammability of ammonia to develop a flame spray method to produce thin films of superconductor are underway.

In the liquid ammonia method described here, gaseous ammonia is condensed on a cold finger cooled with dry ice-acetone in a suitable three-necked flask which contains a total of approximately 6 g of the starting reagents. The flask contains a stirring bar and is itself surrounded by a dry-ice acetone cold bath. Typically, 400 ml of liquid ammonia are collected in this fashion and the resulting solution is stirred for an hour to ensure dissolution. In one case, the solution is sprayed onto a heated plate (ca. 600 C) where ammonia is lost and some decomposition/reaction occurs. In the second case, gaseous carbon dioxide is bubbled into the liquid ammonia solution resulting in the solidification of the solvent and precipitation of the metal ions. Solidification is a result of the formation of carbamates (mostly ammonium carbamate). Any water present could lead to the formation of the less desirable carbonates. Further details are provided elsewhere.[4]

The carbamates are especially easy to process since ammonium carbamate can be readily sublimed. Both nitrate and acetate anions can be used in this ammonia based methodology. One special advantage which accrues to the use of liquid ammonia is the ease of dissolving bismuth salts, which are so difficult to dissolve in aqueous solutions. The liquid ammonia method also lends itself to the preparation of films of superconductors. If the ammonia solution is sprayed through an artists air brush (maintained at dry ice temperatures) onto the surface of a MgO substrate heated to 600 C, reasonably thick films of superconductor are obtained.

Another way to achieve the goals of atomic mixing, without the need for much elaborate equipment, is to immobilize the ions of the solution in a gel and to remove the water from this gel to produce a dry gel or "xerogel." This gel can be dehydrated at room temperature or at slightly elevated temperatures to speed up the process.

A variety of gelling agents, in pure form and as mixtures, have been tested and it was found that for the preparation of 123 and 124, ordinary gelatin, 2-3 % by weight of the solution, serves admirably. In practice, a stoichiometric acetate solution (caution: nitrates form explosive gels) is prepared and then 2-3 % by weight of the gelling agent is added and the solution is allowed to gel. Once the gel is formed, it is dried under vacuum or in a circulating dry air chamber. The xerogel so formed, is then thermally decomposed to yield the desired product. The only disadvantage of this method is the need to introduce the organic gelling agent which ultimately could lead to contamination of the sample with carbon.

This xerogel method is quite distinct from the now classic "sol-gel" method. In the sol-gel, method one must first prepare a colloidal precursor solution which is followed by dispersion of the colloidal particles to form the sol. This is followed by gelation of the colloidal dispersion which is achieved by hydrolysis or polycondensation of the colloidal precursors. This last step involves aging and drying under ambient conditions. One of the major advantages of the xerogel method over the sol-gel method is that colloidal particles are not needed and soluble precursors (acetates,formates,nitrates, etc.) can be used. Additionally, the rigid control of conditions (concentration, temperature and pH) required in the sol-gel process are not required in the xerogel method.

RESULTS

The XRD results on 123 and 124 materials indicates that very high phase purity material can be obtained. The elemental composition of both precursors and final products demonstrates the close control of stoichiometry that is possible with these methods. For example, average metal ratios for 123 from liquid ammonia methods (carbamate process and spray pyrollysis), as determined by atomic absorption, are within the error range of the instrument. Phase purity was established with XRD (Rigaku) using standard conditions. Samples of 123 were consistently produced with phase purities in excess of 98 %. One such pattern, prepared using the ammonia process (carbamate) is given in Fig. 1. Figure 2 shows results of resistivity and magnetization on the same sample of 123. The transition temperature at midpoint is about 93 K.

Both the carbamate and xerogel methods produce precursors which are very stable even under high atmospheric humidity. This could prove quite useful in the preparation of bulk superconducting devices and wires. These precursors are much less frangible than ceramic superconductors and accordingly, the firing step could be saved as the final step. In the case of 123 produced by the xerogel method, an additional indication of the high phase purity is the extrapolated value of the the pre-transition resistivity versus temperature to absolute zero.

The 124 powders made by the xerogel process show the same high phase purity as the 123 according to XRD results. This is confirmed by TGA

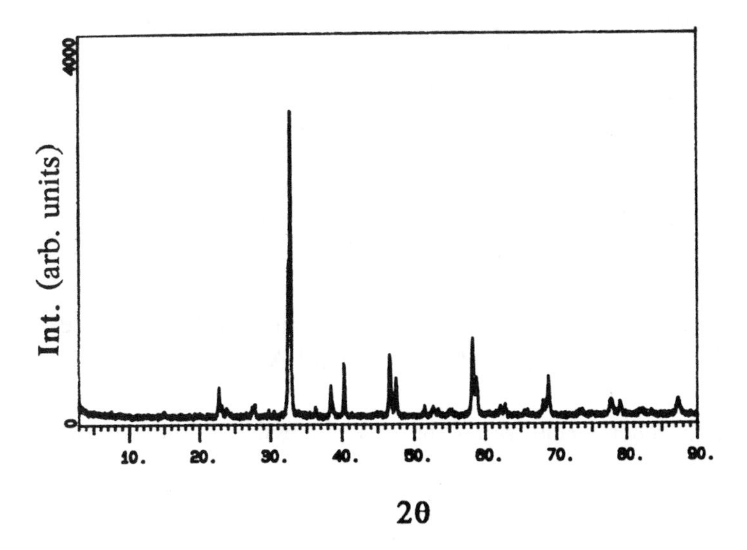

Fig.1 XRD Pattern of 123 Produced By Carbamate Method in Liquid Ammonia

studies in which no weight loss occurs from 0 to 800 C during increasing and decreasing temperature regimes. One of the main advantages of the xerogel process to the preparation of 124 material is that all the processing can be carried out under 1 atm of oxygen rather than the 200 atm traditionally needed.

All three methods lead to very small particle size (as revealed by SEM) and the freeze drying method leads to an exceptionally narrow distribution of particle sizes.

As previously mentioned, it seems to be the case with all three methods, that some unmixing occurs as the temperature is raised during the thermal processing step. Of course, once the oxide is formed, no additional unmixing

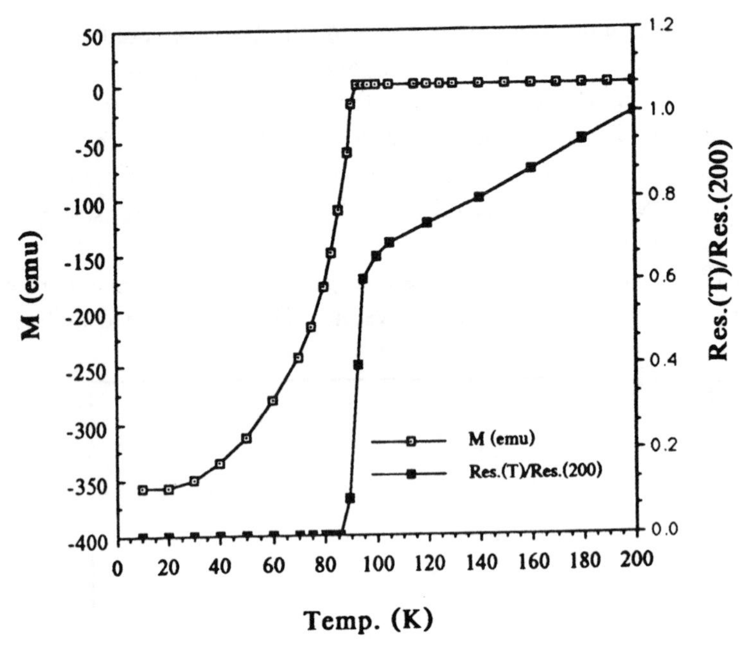

Fig.2 Resistivity and Magnetization as a Function of Temperature for 123 Produced by the Carbamate Method in Liquid Ammonia

occurs. As a general rule, rapid thermal processing minimizes subsequent phase separation.

CONCLUSION

The goal of achieving diffusionless reactions in the preparation of high T_c superconductors has been only partially reallized. The tendency for phase separation is so strong that materials which are initially randomly mixed tend to unmix as the processing temperature is raised. It is for this reason that attempts are being made to choose alternate anions (e.g. formates) which decompose to the oxide at lower temperatures. Alternatively, it is found that a rapid thermal ramp to the decomposition temperature results in the least amount of phase separation.

Of the three methods described in the above, the freeze drying method should lead to the least amount of contamination. It also is especially useful for preparing thermally sensitive compounds such as the 124 material. It does require the most elaborate equipment and careful handling of precursors. The liquid ammonia based methods, offer a great deal of promise for scale-up and for the preparation of thin films. It also is particularly useful for the preparation of bismuth based superconductors. The carbamate route, may be especially useful for the preparation of wires if a method of allowing for the escape of ammonia and carbon dioxide can be found. Needless to say, great care is needed in the handling of liquid ammonia. The xerogel method is in principle the easiest to utilize. As mentioned previously, the effective phase purity of 123 made by this technique is outstanding (based on resistivity vs temperature studies). One possible drawback is the presence of undesirable impurities such as sulfur or phosphorous in some of the gelling agents. The formation of stable sulfates and phosphates must be avoided. Both the xerogel method and the freeze drying method require a time consuming dehydration step. In this regard the xerogel method has a major advantage in that the rate of dehydration is proportional to the vapor pressure of the solid and since the xerogel can be dehydrated at or slightly above room temperature, the vapor pressure of the xerogel will be at least 1000 times greater than that of ice at ca. -20 C.

It seems likely that commercialization of high temperature superconductors will rely on several different synthetic paths dependent upon the nature of the desired product and other factors such as cost and safety considerations. It has been shown that the three methods described in the above can be used to produce good quality high T_c superconducting powder and in the case of the liquid ammonia method, to produce superconducting thick films. It is too early to tell if the latter will be competitive with other methods of producing thick films.

REFERENCES

1. N.Coppa, D.Nichols, J.Schwegler, R.E.Salomon, G.H.Myer and J.E.Crow, J. Mat. Res.,**4**, 1307 (1989).
2. R.W.Schaeffer, J.Macho, R.E.Salomon, G.H.Myer and J.E.Crow, Cer. Trans. Vol. XVIII, Am. Cer. Soc., Westerville, OH (1991).
3. J.Macho, R.W.Schaeffer, R.E.Salomon, G.H.Myer and J.E.Crow, Cer. Trans. Vol. XVIII, Am. Cer. Soc., Westerville, OH (1991).
4. R.W.Schaeffer, J.Macho, R.E.Salomon, G.H.Myer and J.E.Crow, J.Superconductivity (in press).
5. A.M.Kini, V.Geiser, H.C.I.Kao, K.D.Carlson, H.H.Wang, M.R.Monaghan and J.M.Williams, Inorg. Chem.,**26**,1836 (1987).
6. P.Pramanik, S.Biswas, D.Bhattachangh, D.Sen, S.Ghataic, T.K.Dey and K.L. Chopra, Rev. Sol. State Sci., **2**,**2 & 3**,157 (1988).

7. D.Bahadur, A.Banerjee, A.Das, K.P.Gupta, T.Mathews, A.Mitra, M.Tewari and A.K.Majumdar, Rev. Sol. State Sci.,**2,2&3**,177(1988)
8. R.Ford (Ed.), Mat. and Processing Report, MIT Press, Cambridge,3-12 (1989).
9. D.Christen, J. Narayan, L. Schneermeyer, Ed., High Temperature Superconductors: Fundamentental Properties and Novel Materials Processing, Mat. Res. Soc. Proc. Vol. 169, Part III, MRS, Pittsburgh (1990).
10. B.J.Zelinski, C.J.Brinker, D.E.Clark and D.R.Ulrich, Eds., Better ceramics Through Chemistry, Mat. Res. Soc. Symp. Proc. Vol.180, Parts IX-XI, MRS, Pittsburgh (1990).
11. N.Coppa, G.H.Myer, R.E.Salomon, A.Bura, J.W.O'Reilly and J.E.Crow J. Mat. Res. (Submitted September 1991).
12. K.Song, H.Liu, A.Dou and C.C. Sorrel, J. Am. Ceram. Soc., **73**, 1771-1773 (1990).
13. E.L. Brosha, E. Sanchez, P.K. Davies, N.V. Coppa, A. Thomas and R.Salomon, Physica (Submitted August 1991).

ACKNOWLEDGMENT

Support from the Ben Franklin Technology Center and from DARPA/ONR contract #00014-44-0587 is gratefully acknowledged.

ANNEALING STUDY OF $YBa_2Cu_3O_x$ AND $YBa_2Cu_4O_x$ SCREEN PRINTED THICK FILMS

M. Fardmanesh, K. Scoles, A. Rothwarf
Ben Franklin Superconductivity Center, and
Drexel University
Electrical and Computer Engineering Department
Philadelphia, PA 19104

E. Sanchez, E. Brosha, P. Davies
Ben Franklin Superconductivity Center, and
University of Pennsylvania
Department of Materials Science and Engineering
Philadelphia, PA 19104

ABSTRACT

The effects of firing screen printed pastes of $YBa_2Cu_3O_x$ and $YBa_2Cu_4O_x$ on polycrystalline MgO substrates is studied as a function of annealing temperatures and annealing gas ambient. Firing in argon rather than oxygen during the high temperature portion of the anneal is shown to improve the room temperature resistance, lower the optimum peak annealing temperature, produce metallic resistivity behavior above the onset temperature, and produce improved critical transition temperatures.

INTRODUCTION

This work represents the results of an investigation of processing methods to produce high quality superconducting thick films. The materials investigated include $YBa_2Cu_3O_x$ and $YBa_2Cu_4O_x$ on polycrystalline MgO substrates. In this paper we will refer to these materials as 123 and 124 respectively. $YBa_2Cu_4O_x$ is an interesting material for use in superconducting applications due to the stability of its oxygen stoichiometry under environmental exposure.[1] Application of 124 material to thick film processing will require the exposure of the 124 films to temperatures sufficiently high to change the stoichiometry of the film. These temperatures are necessary to promote sintering of the powders and adhesion of the film to the substrate. Conversion of 124 material to 123 occurs at approximately 850 °C (figure 1), with the precipitation of a uniform distribution of CuO in the resulting film. The CuO sites may act as flux pinning centers that will increase the critical current in these converted 124 films relative to those printed from 123 powders.

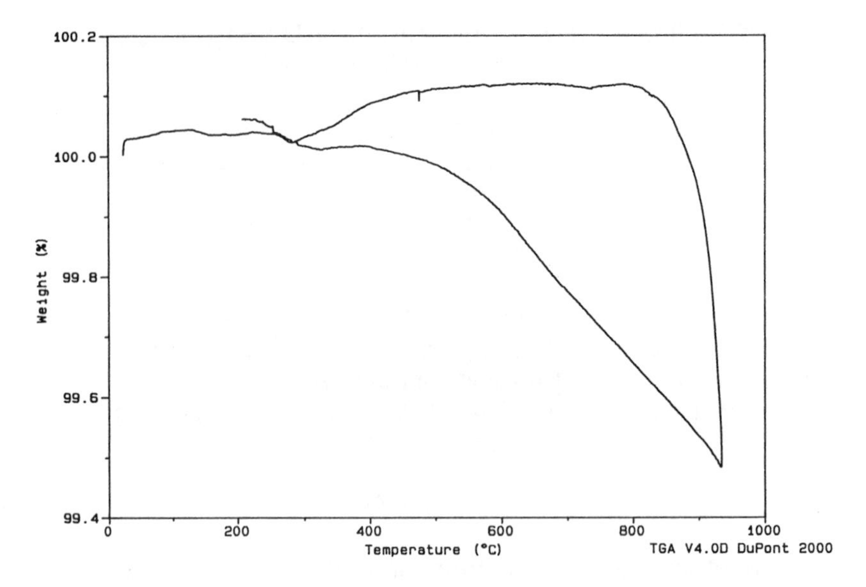

Figure 1. Thermogravimetric analysis (TGA) of 124 powder showing a weight loss at 850 °C. X-ray analysis of the post-TGA film indicates that the film has transformed to 123.

FILM PREPARATION

Thick film pastes were formed by mixing powders of the superconducting materials with a printing vehicle composed of terpineol and ethyl cellulose. A three roll mill (Netzsch 272.00) mixed the compositions for at least one hour. We obtained the 123 powder from a commercial source (High T_C Superconco SP-5, 0.3 - 2.0 µm diameter).

The 124 material was prepared in a freeze-drying process.[2] An aqueous solution of Y, Ba, and Cu nitrates, with a stoichiometric ratio of 1:2:4 respectively, was ultrasonically aerosolyzed and allowed to impinge on liquid nitrogen. The frozen material was rapidly moved to a commercial freeze-dryer with a shelf temperature of -25 °C.

To produce the 124 oxide, the freeze-dried material was subjected to a five step thermal treatment. First, they were heated in a flowing oxygen atmosphere to 180 °C for one hour. Then the temperature was increased to 300 °C for another hour. The next temperature was 450 °C for 3 hours. Then the sample was quickly introduced to 800 °C and soaked for 12 hours. At this point it was removed from the oven for regrinding. Finally, it was reintroduced into the oven at 800 °C for 48 additional hours. This thermal treatment is based on the thermal gravimetric analysis of the precursor. The resulting oxide was single phase 124, as determined by x-ray analysis.

We obtained the pressed-powder MgO substrates used from Trans-

Tech. MgO was chosen for its low reactivity with the 123 material. Magnesia has a dielectric constant of approximately 9.5, a number similar to alumina and significantly lower than materials such as strontium titinate. The loss tangent is approximately 10^{-4}. These properties makes MgO a candidate for microwave applications.

Films were screen printed through 280 or 325 mesh stainless steel screens. After a short leveling period, a box oven was used to dry the films at 120 °C for one hour. A horizontal 75 mm (3 in) diameter quartz tube furnace equipped with a programmable temperature controller annealed the superconducting films. Films were heated to peak temperature at rates of 10 or 100 °C/min. Peak temperatures of 850 to 980 °C were held for 10 - 30 minutes. Cooling occurred at a rate of 2 °C/min. We investigated annealing atmospheres of oxygen and argon. In both cases the flow rate was 2 l/min.

TGA analysis of evolution of organic materials from the paste determined the protocol for the initial heating of the films. To prevent problems such as rapid gas release that might reduce film density, we avoided exposing the films to high temperatures until most of the organics were removed.

Exposing films to oxygen while cooling was expected to restore the oxygen stoichiometry to nearly $YBa_2Cu_3O_7$. Those films annealed in argon were also cooled in argon until their temperature dropped below 700 °C. We believe Ar was effective in reducing the melting point of the powder, producing more effective sintering.

DIAGNOSTICS

We printed films in geometries suitable for measurement of resistance vs. temperature, critical current, and crystal structure. The resistance vs. temperature pattern was a serpentine line of 250 μm (0.010 in) width and 82.6 mm (3.215 in) length, 0.89 mm (0.035 in) pitch and 30-35 μm in thickness. Four pads were available for current injection and voltage measurement. Sputter coating with 800 to 850 Å of gold prepared the pads for attachment of wires. Contacts were not annealed. Conductive epoxy attached the fine copper wires to the gold pads.

Low temperature characterizations were performed on a cryopump modified to provide access to the cold finger. The characterization system uses a computer for closed loop control of temperature, display of results, and recording of data. A computer-controlled reversal of dc bias current direction through the sample removed the effects of temperature-induced voltages.

TGA analysis for evolution of organics and oxygen loss or uptake was performed on a DuPont 2000 TGA. X-ray diffraction for structure, grain size, CuO observation was preformed on a Rigaku diffractometer using copper

Kα-1 radiation.

RESULTS - OXYGEN ANNEALING

123 Pastes

Thick films of 123 material on MgO substrates were heated in oxygen using the following protocol: heat at 100 °C/min to the peak temperature, hold for 10 min, cool at 2 °C/min to 450 °C, hold for 2 hours, slow cool to room temperature. As the annealing temperature rises, films heated in oxygen show an improved $T_{c(zero)}$ and a more abrupt transition (figure 2). Yet, the thermal processing has not restored the oxygen stoichiometry to $YBa_2Cu_3O_7$, as evidenced by the low $T_{c(zero)}$ temperatures.

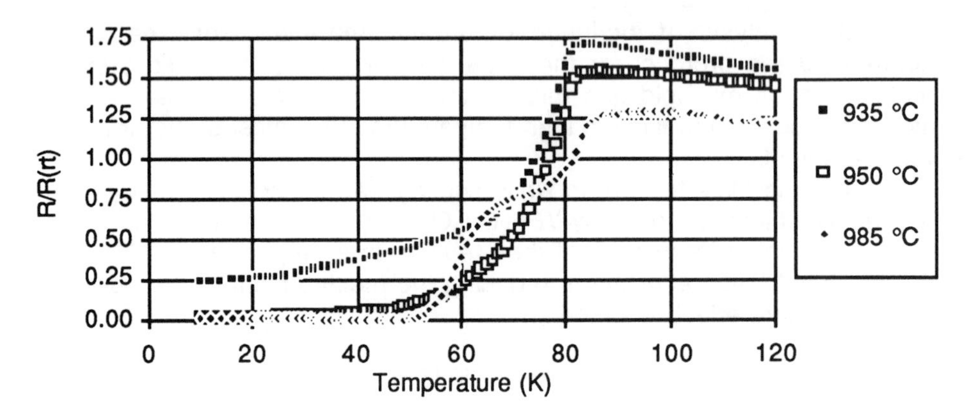

Figure 2. Resistance vs temperature for 123 films on MgO substrates annealed in oxygen for 10 minutes at various peak temperatures. Resistance is normalized to its room temperature value.

Samples heated to 985 °C in oxygen have the highest onset temperature, but show a knee near the middle of their transition indicating the presence of a second phase with a lower onset temperature.

X-ray diffraction measurements on the oxygen annealed films show a peak at approximately 43° 2θ, characteristic of the MgO substrate. Since we would not expect to see this peak in dense films with this thickness (25 to 35 μm), we interpret the presence of the 43° peak as being an indication of porous or poorly sintered powders. This conclusion has been confirmed by SEM.

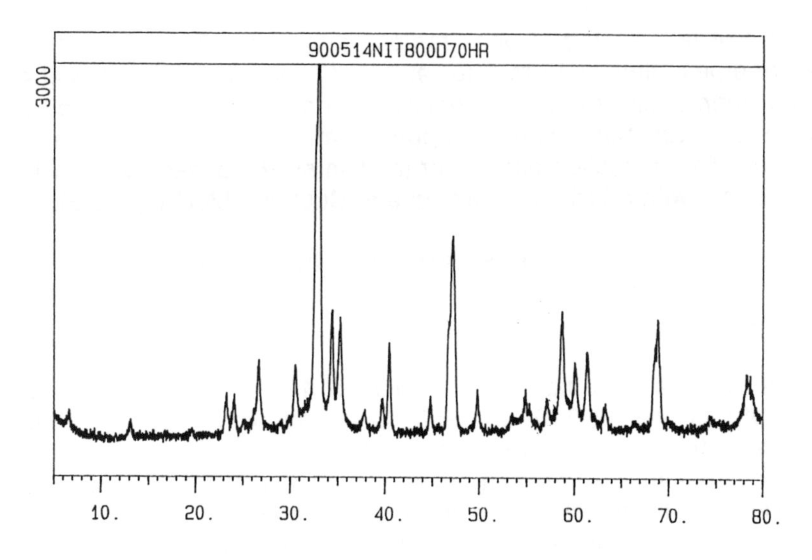

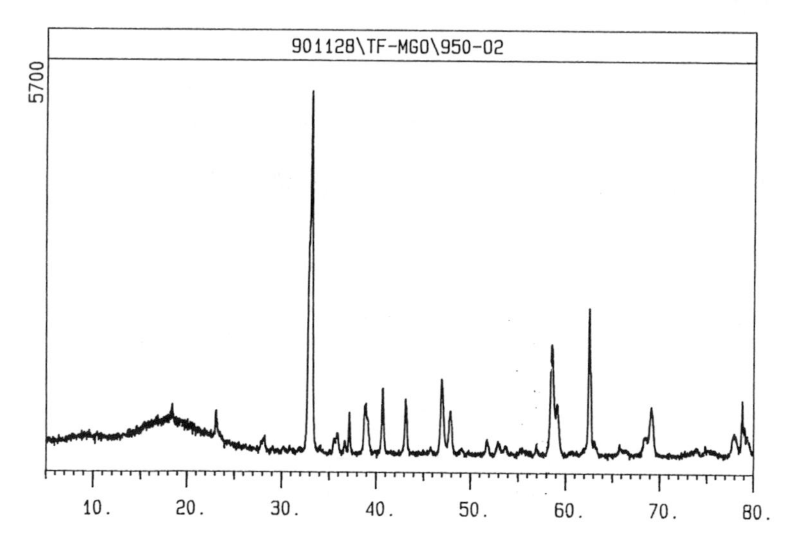

Figure 3. X-ray diffraction results for 124 starting powder (top), and 124 powder screen-printed, dried, and heated to a peak temperature of 950 °C (bottom). Note the disappearance of the low angle peaks after heating, and the presence of 123, CuO, and MgO lines in the heated sample.

124 Pastes

A 124 film heated to peak temperature at approximately 10 °C/min, fired for a peak time of 15 minutes at 950 °C and held for 4 hours at 450 °C shows nonmetallic resistance behavior above onset, a long tail on the resistance transition, with T_C approximately 25 K, and a knee in the transition. The diffraction pattern for the film shows peaks due to a 123 film composition, with additional peaks due to CuO and MgO (figure 3).

RESULTS - ARGON ANNEALING

123 Pastes

Resistance vs temperature results for 123 pastes annealed at peak temperatures of 910, 940, and 950 °C are shown in figure 4. Diffraction peaks from the MgO substrate were again visible in the studies of the 123 Ar-annealed films. Films heated at an initial rate of 100 °C/min have improved metallic behavior above onset and a higher $T_{c(zero)}$. Diffraction patterns of the rapidly heated films do not show any peaks which can be attributed to MgO.

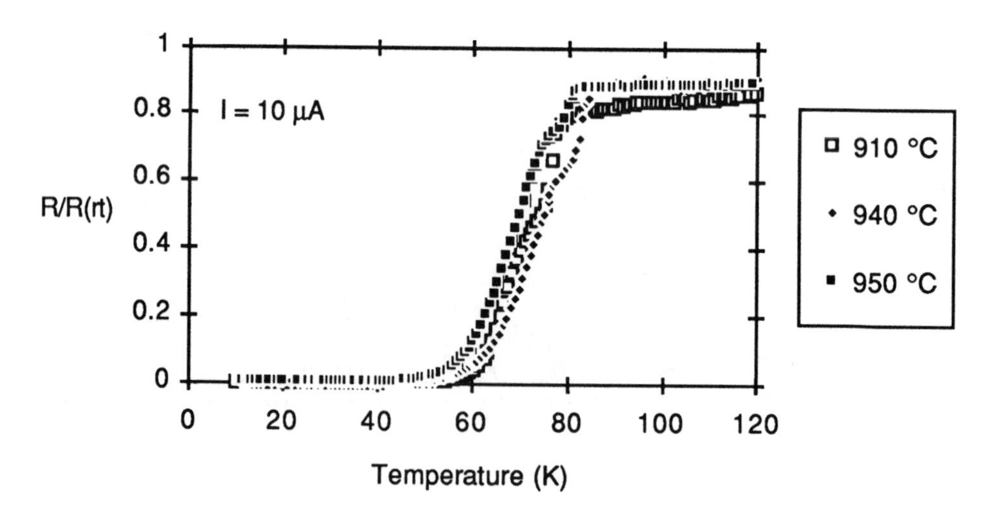

Figure 4. Normalized resistance vs temperature for 123 films on MgO substrates, annealed in argon. These films were heated to peak temperature at approximately 10 °C/min.

124 Pastes

The stoichiometry of 124 films changes to 123 plus dispersed sites of CuO at temperatures of approximately 850 °C in air. This conversion takes place at lower temperatures in argon, as confirmed by TGA (figure 5). Thick films of the 124 paste on MgO substrates were annealed at peak temperatures of 930 and 950 °C. Films were heated at a rate of 100 °C, held at peak for 10 minutes, cooled to 450 °C in 4 hours, and held at this temperature for 4 hours. The resulting film had an onset of about 80 K, a smooth transition, and a $T_{c(zero)}$ of about 73 K (figure 6). Additional time at 450 °C further improved $T_{c(zero)}$ and reduced the normal state resistivity.

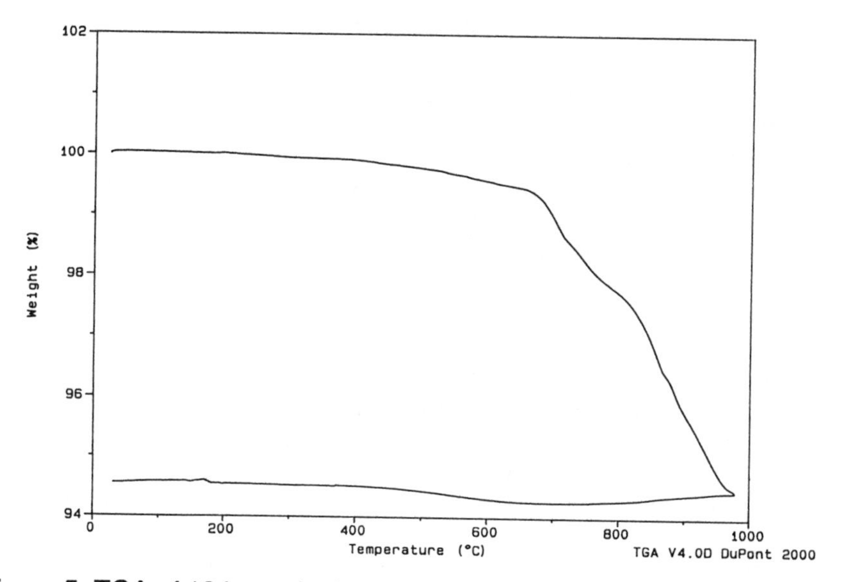

Figure 5. TGA of 124 powder heated to 950 °C at 10 °C/min in argon

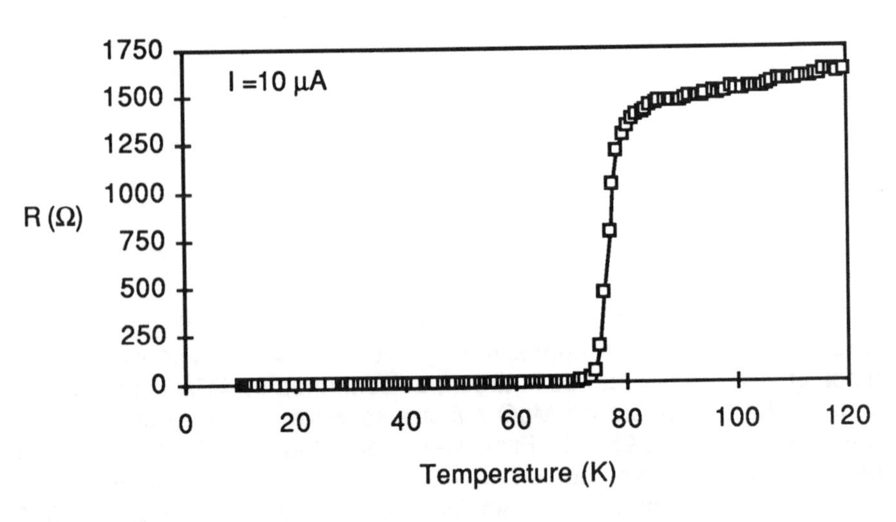

Figure 6. Resistance vs temperature for a 124 paste annealed in argon at 950 °C for 10 min.

DISCUSSION

Firing films in an argon ambient was motivated by the need to improve the conductivity through grain boundaries, both by improving the sintering and reducing the production of undesirable, low $T_{c(zero)}$ phases. Flowing Ar through the open-ended annealing tube reduced the oxygen partial pressure. This changes the ambient from oxidizing to reducing, consequently decreasing the melting temperature of the material. The original protocol for argon annealing was the following: warm at 10 °C/min to the peak temperature, hold for 30 min, cool at 2 °C/min to 450 °C, hold for 4 hours, slow cool to room temperature. The resistance vs temperature results for a 123 paste heated with this protocol are shown in Figure 4.

The effect of reduced oxygen partial pressure on the post-annealing of thin $YBa_2Cu_3O_x$ films has been studied.[3] In summary of reference 3, the findings on annealing thin $YBa_2Cu_3O_x$ films in reduced p_{O_2} include the possibility of lower annealing temperatures, improvements in film surface morphology, and the ability to modify the c-axis orientation.

We cannot quantify the p_{O_2} used in our annealing system, yet we do observe major differences in the temperature dependence of the normal state resistance of both 123 and 124 films as a function of the high temperature sintering atmosphere. Those films annealed in argon show metallic-like resistance above T_c and in the sharper resistive transitions and higher $T_{c(zero)}$, evidence of improved sintering.

Matsuoka, et al[4] studied the effect of annealing 123 films screen printed on yttrium stabilized zirconia under flowing helium. In agreement with Matsuoka, et al, we find a reduced optimum annealing temperature and reduced room temperature resistance in the inert gas ambient.

CONCLUSIONS

We have annealed 123 and 124 pastes in ambients of flowing oxygen and argon. Our best results were found in argon annealing of 124 materials at temperatures which convert the material to 123. All 123 and 124 films heated in oxygen show non-metallic resistance vs temperature above their onset temperature. Additional work will be required to confirm that 124 pastes converted to 123 plus CuO have critical currents that exceed those of 123 pastes. No evidence for preferential alignment of the c-axis was seen under any of our firing conditions.

ACKNOWLEDGEMENTS

The authors would like to thank Dr. Robert Solomon and co-workers at Temple University for preparation of precursors used in the fabrication of the 124 powders, High Tc Superconco for 123 powders and sputtering targets, and ITT Defense Inc., GE Astrospace, and the Ben Franklin Partnership of the Commonwealth of Pennsylvania for financial support.

REFERENCES

1. U. Balachandran, M.E. Biznek, G.W. Tomlins, B.W. Veal, and R.B. Poeppel, "Synthesis of 80 K superconducting $YBa_2Cu_4O_8$ via a novel route," Physica C **165**, 335-339 (1990).
2. N. Coppa, J.W. Schwegler, I. Perez, G.H. Myer, R.E. Solomon, and J.E. Crow, Mat. Res. Soc. Symp. Proc., "Preparation, thermal processing behavior of powdered and bulk $YBa_2Cu_3O_{7-x}$ from freeze dried nitrate solutions," **180**, 935-940 (1990).
3. R. Feenstra, T.B. Lindemer, J.D. Budai, and M.D. Galloway, "Effect of oxygen pressure on the synthesis of $YBa_2Cu_3O_{7-x}$ thin films by post deposition annealing," J. Appl. Physics **69**(9), 6569-6585 (1991).
4. Y. Matsuoka, E. Ban, and H. Ogawa, "Properties of high-T_c screen-printed Y-Ba-Cu-O films prepared under flowing helium," Supercond. Sci. Technol. **2**, 300-303 (1989).

SILVER-CLAD Bi-2223 PROCESSING

K. W. Lay, R. H. Arendt, J. E. Tkaczyk, and M. F. Garbauskas
GE Corporate Research & Development
Schenectady, NY 12301

ABSTRACT

One of the most promising techniques for the production of long lengths of high Tc superconductors utilizes the mechanical deformation and subsequent heat treatment of silver-clad bismuth cuprates. Some recent experiments on $Bi_{1.7}Pb_{0.3}Sr_2Ca_{2+x}Cu_{3+x}O_y$, with x = 0.2, 0.4, and 0.75, processed by packing powder in a silver tube followed by swaging, drawing, and rolling are described. The effect of the initial powder composition, both cation ratios and phase composition, on the critical current of the composite tapes are explored. Different final deformation methods and heat treatment schedules also are discussed. A set of guidelines for the production of high Jc 2223 tapes is given.

INTRODUCTION

A potential method for making long conductors for use in magnets, motors, or generators is the Powder In Tube (PIT) technique. This involves loading superconductor powder into a silver tube and fabricating a long tape by standard metallurgical techniques. Among the advantages of the PIT process are the formation of a relatively dense, aligned polycrystalline superconductor, protection of the superconductor by the silver cladding, and the ability to make long lengths. This approach has been especially applicable to the bismuth 2223 phase, which is usually doped with lead, and whose composition is approximately given by $Bi_{2-x}Pb_xSr_2Ca_2Cu_3O_{10+y}$. Several papers have discussed to formation of 2223 tapes by PIT techniques[1-8]. The 2223 phase is of interest for long conductors because it has been shown that the bismuth and thallium based cuprate superconductors show less intergranular weak link behavior than the $YBa_2Cu_3O_{7-x}$ phase. Unfortunately they exhibit more flux creep at high temperatures[9] so that the use of the bismuth and thallium based materials may be limited to temperatures below 77K, especially if appreciable magnet fields are present.

Many details of the processing of PIT tapes remain to be clarified even though there has been considerable effort worldwide on their fabrication. In this paper some experiments exploring the effect of both cation ratios and phase composition of the initial powder on the critical current of the composite tapes will be described. During our early attempts to make high Jc 2223 PIT tapes we went to considerable effort to make phase pure 2223 powders. The Jc values of tapes made from these powders were low. In a parallel study we found that high Jc tapes could be made from special powder formulations which were a mixture of the 2212 phase as well as other, separately reacted, powders, such that the overall composition would react to form the 2223 phase during final tape processing[10]. It will be shown here that, in a similar manner, it is important to have the 2212 phase present in partly reacted powders which can react to form the 2223 phase during the final processing of tapes. Tapes made from partly reacted powders result in high Jc values in contrast to tapes made from fully reacted powders. The effect of different final deformation methods and heat treatment schedules on Jc also will be discussed. Finally a set of guidelines for the optimization of Jc in silver-clad 2223 tapes will be given.

EXPERIMENTAL

Powders were made by calcining mixtures of lead, bismuth, and copper oxides and calcium and strontium carbonates. The oxides and carbonates were wet milled together in water for 2 hours. The dried mixtures were calcined in alumina trays in air at 860°C. Ball milling in heptane for 2 hours was done after the heat treatments. Four different powders will be discussed in this paper. The overall compositions can be expressed by $Bi_{1.7}Pb_{0.3}Sr_2Ca_{2+x}Cu_{3+x}O_y$ with x equal to 0.2, 0.4, or 0.75. The calcining time for the three partially reacted powders was 25 hours. The fourth powder was fully reacted x = 0.75 material calcined 3 times for 25 + 50 + 50 hours.

Powders were loaded into silver tubes and then swaged, drawn, and rolled into tapes with a final thickness of 0.254 mm and a width of about 3 mm. The initial silver tubes had an OD of 6.25 mm and a wall thickness of 1.0 mm. The superconductor to silver weight ratio of the tapes ranged from 0.27 to 0.34.

The final processing of the tapes involved sequential heat treatments and mechanical deformations. All heat treatments were done in air at 830°C for 48 hours. Pressings were done at a pressure of about 140,000 psi on the total tape area. Heat treating, pressing, or rolling operations are designated as H, P, or R respectively in this paper. For example RHPH would designate an as-rolled tape which was subsequently heated twice with an intermediate pressing operation.

Four point transport measurements of the current-voltage curves at 77K were recorded with the sample immersed in liquid nitrogen. Current contacts to the samples were made by soldering to the Ag sheath. An electric field criterion of 0.5 microvolts/cm was used to define the critical current.

X-ray diffraction was used to determine the amount of aligned 2223 material in the tapes. One silver surface of a tape was peeled back from the superconductor core after gluing the tape to a glass slide and cutting along both edges. Copper K alpha radiation was used to obtain a diffraction pattern from the superconductor surface. Integrated peak areas were analysed for 2 theta values in the range 19 to 37°. Lower angle lines were not used due to occasional shadowing resulting from the imperfect surface of the superconductor core revealed after the silver was removed. Higher angle lines were omitted to avoid lines from the silver. The region about 29° was also omitted since there is a serious overlap of the 2212 (00$\underline{10}$) and the 2223 (00$\underline{12}$) lines. The sum of the 2223 (00$\underline{10}$) and (00$\underline{14}$) lines at about 24.15^{o} and 34° were used as a measure of the amount of aligned 2223 phase. The sum of the 2212 (008) and (00$\underline{12}$) lines at about 23.3^{o} and 35.2^{o} were used as a measure of the amount of aligned 2212 phase. The remaining lines were called "other" and included non-superconducting phases, as well as misaligned 2212 and 2223 phases. The values reported are ratios of the particular integrated peak intensities to the total integrated intensity in this region.

RESULTS

One set of experiments was performed to ascertain the importance of the 2223 phase content in the initial powder. The composition used was $Bi_{1.7}Pb_{0.3}Sr_2Ca_{2+x}Cu_{3+x}O_y$ with x equal to 0.75. The excess calcium and copper were known to accelerate the formation of the 2223 phase[11]. Tapes were processed by the PIT route with final thicknesses of 0.254 mm. One set of tapes was made with a powder which had been reacted in air at 860°C for 125 hours until the 2223 to 2212 (002) x-ray intensity ratio was about 95:5. The second tape was made from a powder which had been reacted at 860°C for 25 hours and consisted principally of the 2212 phase. Both tapes were then heated, pressed, and heated again. The critical currents, Jc, at 77K of the tapes were 1850 and 4700 A/cm^2 for the fully reacted and partially reacted powders respectively. X-ray diffraction patterns of the two materials after final processing are shown in Figure 1. Even though the tape made from the fully reacted material contains a higher fraction of aligned 2223 phase, it has a lower Jc than the tape made from a partially reacted powder. Similar results have been found for other compositions. Fully reacted 2223 powders do not result in high Jc values in PIT processed tapes. The explanation will be discussed later.

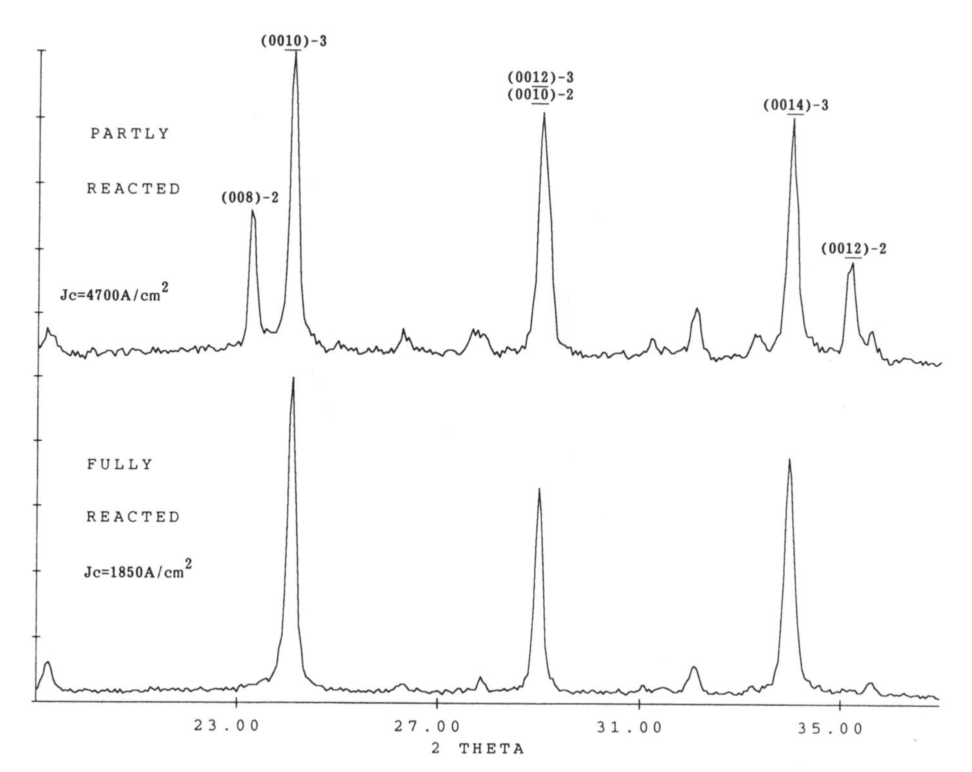

Figure 1 X-ray diffraction patterns for tapes after heating, pressing, and heating. Patterns were taken from the surface of the superconductor core after removing one silver outer layer. The sample labeled partly reacted was made from powder initially calcined for a short time and consisted principally of the 2212 phase before the final two heat treatment steps. The fully reacted sample was made from powder calcined to react totally to the 2223 phase.

A second set of experiments was done to determine the effect of using starting powders which have different reaction rates for the formation of the 2223 phase. Three different $Bi_{1.7}Pb_{0.3}Sr_2Ca_{2+x}Cu_{3+x}O_y$ powders were used with x values of 0.2, 0.4, and 0.75. All powders were reacted only for a short time so that the carbonates were decomposed and the oxides had started to react. The principal superconducting phase in all cases was the 2212 phase. PIT tapes were heated, pressed, heated, pressed, and heated. The resulting phase compositions and Jc values are shown in Figure 2 and the x-ray patterns in Figure 3. In this case there is a direct correlation between the amount of aligned 2223 phase and the critical current.

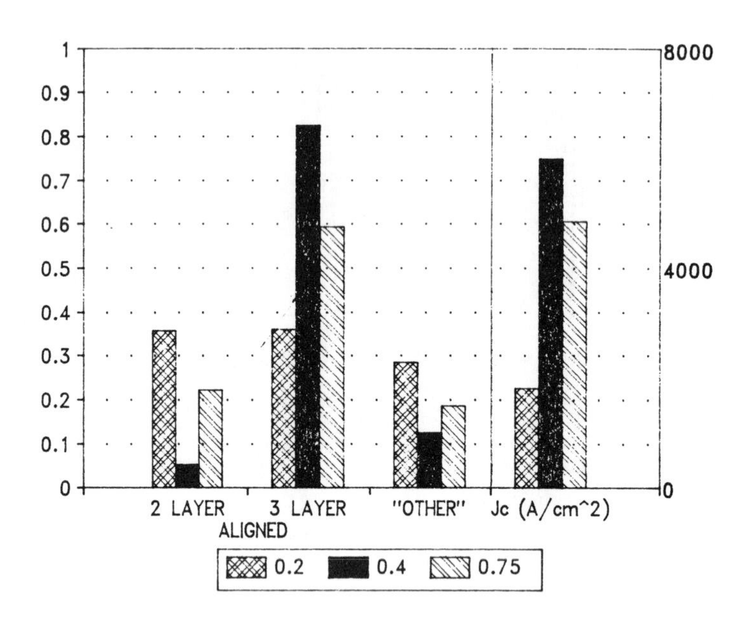

Figure 2 Relative amounts of aligned 2212 and 2223 phases and 77K Jc values for three different powders. The overall compositions of the powders were $Bi_{1.7}Pb_{0.3}Sr_2Ca_{2+x}Cu_{3+x}O_y$ *where x = 0.2, 0.4, or 0.75. All powders were calcined such that the principal phase initially was the 2212 phase. "Other" designates non-superconducting phases or misaligned 2212 or 2223.*

Finally a set of experiments were conducted to demonstrate the effect of different final processing steps on Jc. In this case all tape samples were made from the same partially reacted PIT tape which had excess calcium and copper of 0.75. The as-rolled 0.254 mm thick tape was heated three times with intermediate pressings or rollings performed for some of the samples. The aim was to elucidate the reasons for the final rolling or pressing steps used in making 2223 tape conductors. The x-ray results for the amount of aligned 2223 and 2212 phase and the 77K Jc values

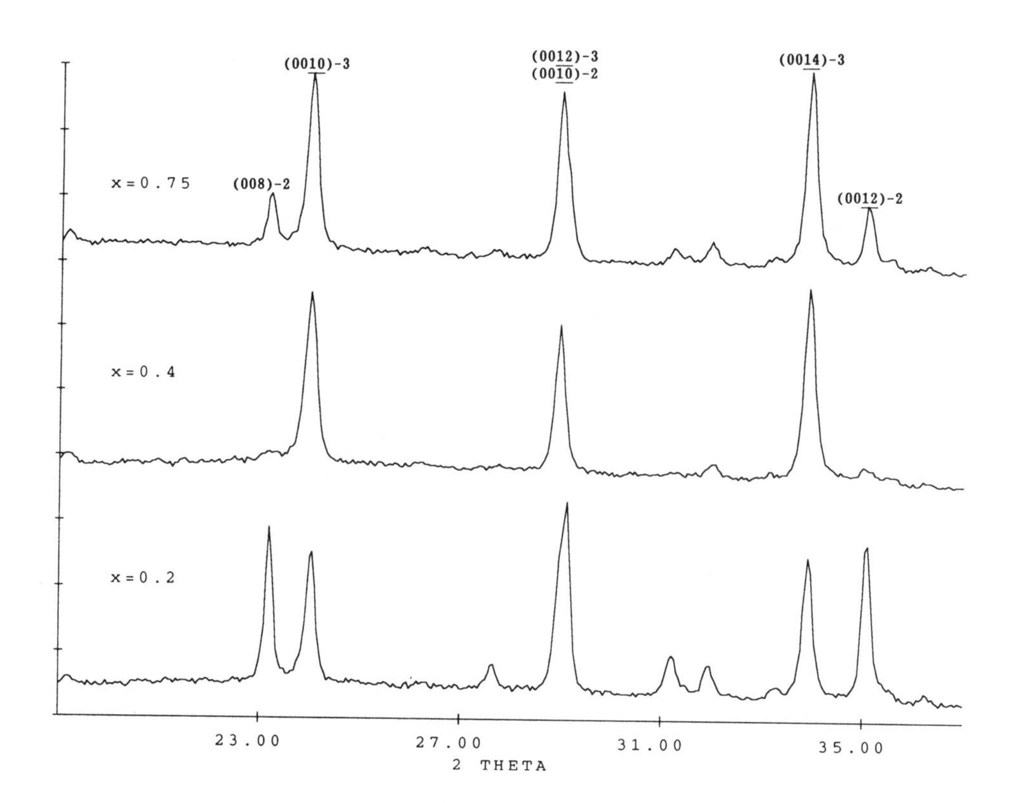

Figure 3 X-ray diffraction patterns for tapes after heating, pressing, heating, pressing, and heating. Patterns were taken from the surface of the superconductor core after removing one silver outer layer. The overall compositions of the powders were $Bi_{1.7}Pb_{0.3}Sr_2Ca_{2+x}Cu_{3+x}O_y$ where x = 0.2, 0.4, or 0.75. All powders were calcined such that initially the principal phase was the 2212 phase.

are shown in Figure 4. It can be seen that the phase compositions for all the sample sets are essentially the same. The Jc values, on the other hand, are quite different. The best Jc is obtained with two intermediate pressings. Rolling as an alternative deformation method to pressing gives lower Jc values. The lowest Jc is obtained with no intermediate pressings.

Further understanding of the results shown in Figure 4 is obtained if the change in the tape thickness during the final processing is considered. The per cent change in tape thickness relative to the initial 0.254 mm value during the final processing is shown in Figure 5. It is seen that the

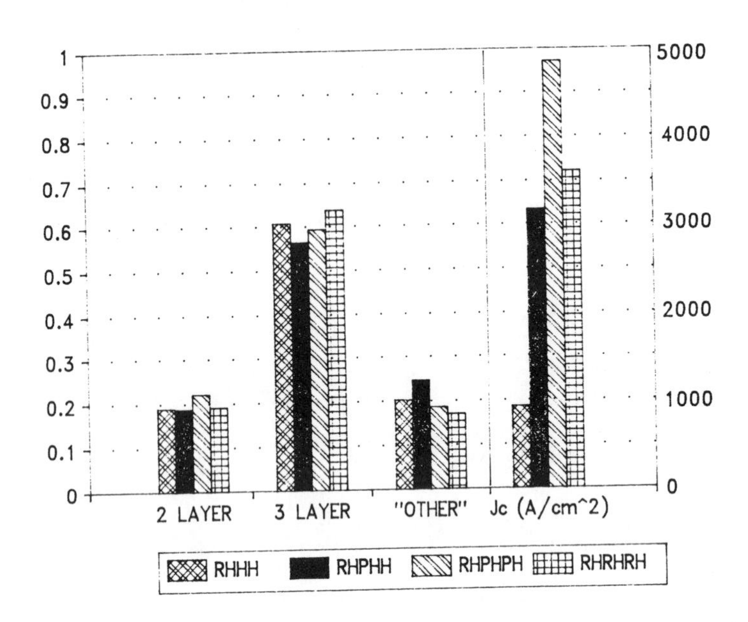

Figure 4 Relative amounts of aligned 2212 and 2223 phases and 77K Jc values for tapes after different final processing. The overall composition of the initial powder was $Bi_{1.7}Pb_{0.3}Sr_2Ca_{2.75}Cu_{3.75}O_y$. The powder was calcined such that the principal phase initially was the 2212 phase. "Other" designates non-superconducting phases or misaligned 2212 or 2223. R, H, and P designate rolling, heating, and pressing respectively.

overall tape thickness, including both the superconductor core and the two outer silver layers, decreased during the pressing operations and the first and third heating operations. The thickness expanded during the second heating operation. The tape with no intermediate pressing operations showed an overall expansion after the three 48 hour heat treatments. The Jc values of the different tapes are related to the density of the superconducting core regions, as will be discussed more fully.

DISCUSSION

A specification for a high Jc 2223-based superconductor tape or wire would probably include: "A dense, well-aligned superconductor region with a high fraction of 2223 grains which are connected by a large fraction of clean grain boundaries". Each of the underlined separate elements of this specification are necessary, but

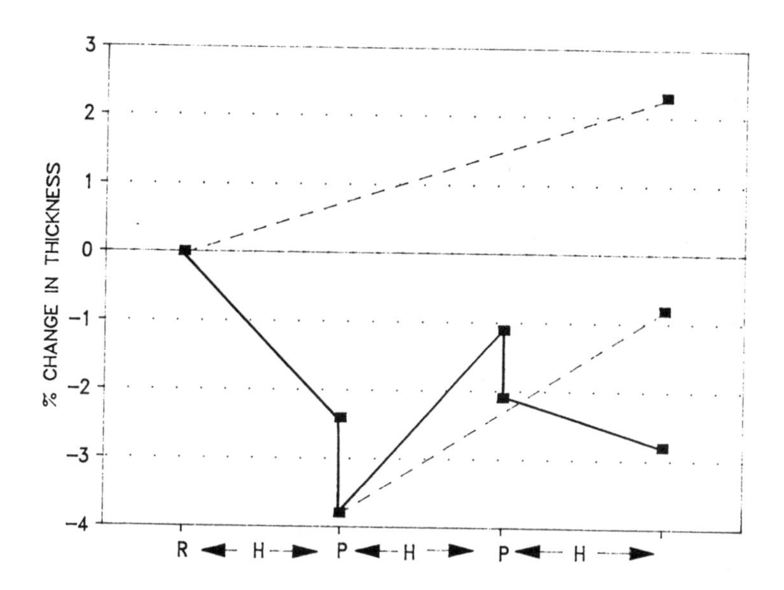

Figure 5 A schematic representation of the relative change in thickness of the same tapes as for Figure 4. The relative thickness changes are with reference to the initial tape thickness of 0.254 mm. R, H and P designate rolling, heating, and pressing respectively. The solid line shows a tape with processing history RHPHPH. The two dashed lines are for RHHH and RHPHH tapes.

not sufficient, conditions to obtain high Jc values. Three separate PIT tape processing aspects, powder composition, tape forming, and final tape processing will be discussed in-turn.

The importance of starting with a powder containing mainly the 2212 phase was confirmed here. This agrees with our companion study using 2212 powders mixed with PbO and prereacted alkaline earth and copper oxides[10]. The importance of not using fully reacted 2223 powder can also be inferred from looking at the calcining times (when this information is given) in papers where high tape Jc values were obtained. Generally, the times are much too short to obtain a high fraction of the 2223 phase in the starting powder. The 2212 phase forms rapidly[12] and will tend to be the majority phase in powders calcined for a short time. The 2212 phase is important since this phase is aligned during the final rolling operation of the tape forming process. The 2223 which is formed subsequently in the final heat treatment(s) of the tape will also be aligned. The overall composition of the initial powder is important since, among other things, this will determine the rate of the 2212 to 2223 conversion during the final heat treatment(s)[11,13,14].

The details of the deformation process by which a cylindrical silver tube filled with superconductor powder is deformed into a long tape were not considered in this paper. It is important to maintain a uniform cross-section of the composite along the conductor. A convenient measure is the superconductor to silver weight ratio. This ratio should remain constant throughout the metallurgical deformation processing and should be constant along the conductor. As stated earlier, the alignment of the 2212 phase occurs during the final tape rolling operation. This can be seen by noting the well aligned grains in the RHHH sample of Figure 4. This sample did not experience any further deformation steps such as pressing or rolling, yet it exhibits alignments as good as samples which did receive these treatments.

The final heat treatment and deformation cycles are very important. It is here that the 2223 phase is formed. This is also where the final microstructure is developed. There are two processes occurring simultaneously during the high temperature treatments, sintering densification which reduces the porosity of the sample and conversion of the 2212 phase to the 2223 phase. The first of these processes will result in reduction in the volume of the superconductor region of the tape. The 2212 and 2223 phases, however, do not sinter readily to high density[14]. This is one reason the intermediate pressing or rolling operations are needed to increase the superconductor core density. In addition, the reaction to form the 2223 phase results in expansion in the volume which must be compensated for by a mechanical deformation step[14, 15]. Another essential aspect of the final heat treatment is healing mechanical damage which has occurred during the tape forming steps. The presence of such damage probably helps to explain the higher Jc values of tapes with final pressing operations as compared with those which are rolled. During pressing the width of the tape is increased and cracks formed will tend to be along the long dimension of the tape. Conversely, rolling tends to elongate the tape and cracks will tend to form across the tape. Obviously the most detrimental configuration for limiting current along the tape occurs from rolling.

ACKNOWLEDGEMENTS

Brady Jones prepared the powders and did the final processing of the tapes. John Hughes and his colleagues in the Metals Processing Operation did the PIT processing. Peter Bednarczyk did the Jc measurements. Funding was provided by DARPA/ONR contract N00014-88-C-0681.

REFERENCES

1 "Microstructure and Jc-B Characteristics of Ag-Sheathed Bi-Based Superconducting Wires" ,M. Ueyama, T. Hikata, T. Kato and K. Sato, J. Jour. Appl. Phys. 30, L1384 (1991)

2 "Critical Currents and Magnetization in C-Axis Textured Bi-Pb-Sr-Ca-Cu-O Superconductors", S. Jin, R. B. van Dover, T. H. Tiefel, J. E. Graebner and N. D Spencer, Appl. Phys. Lett. 58, 868 (1991)

3 "Critical Currents and Processing of Wound Coils of Ag-Sheathed Bi-2223 High Tc Tape: Microstructural and Pinning Effects", L. R. Motowidlo, E. Gregory, P. Haldar, J. A. Rice and R. D. Blaugher, Appl. Phys. Lett. 59[6], 736 (1991)

4 "Critical Current Density and Magnetization of Ag-Sheathed Tapes of Bi(Pb)-Sr-Ca-Cu-O", A. Oota, A. Yata, K. Hayashi and K. Ohba, Supercond. Sci. Technol. 4, S244 (1991)

5 "Effect of Cold rolling on Microstructure, Grain Alignment and Transport Critical Current Anisotropy of Silver Clad $(Bi,Pb)_2Sr_2Ca_2Cu_3O_x$ Superconducting Tapes", W. Lo and B. A, Glowacki, Supercond. Sci. Technol. 4, S361 (1991)

6 "Critical Current Density in Superconducting Bi-Pb-Sr--Ca-Cu-O Wires and Coils", S. X. Dou, H. K. Liu, M. H. Apperley, K. H. Song and C. C. Sorrell, Supercond. Sci. Technol. 3, 138 (1990)

7 "Properties of Ag-Sheathed Bi-Pb-Sr-Ca-Cu-O Superconducting Tapes Prepared by the Intermediate Pressing Process", Y. Yamada, K. Jikihara, T. Hasebe, T. Yanagiya, S. Yasuhara, M. Ishihara, T. Asano and Y. Tanaka, J. Jour. Appl. Phys. 29, L456 (1990)

8 "Temperature-Dependent Critical Current Density of Bi(Pb)-Sr-Ca-Cu-O Tapes in Fields up to 20 Tesla",

Supercond. D. P. Hampshire, S S. Oh, K. Osamura and D. C. Larbalestier, Supercond. Sci. Technol. 3, 138 (1990)

9 "Critical Currents in Aligned YBCO and BSCCO Superconductors," K. W. Lay, AIP Conference Proceedings 219, Eds. Y-H Kao, P. Coppens & H-S Kwok (AIP, New York, 1991)

10 R. H. Arendt, M. F. Garbauskas, K. W. Lay and J. E. Tkaczyk, To be published.

11 "The Chemistry and Superconducting Properties of Species in the System Bi-Ca-Sr-Cu-O," R. H. Arendt, M. F. Garbauskas and L. L. Schilling, J. Mater. Res. 5, 33 (1990)

12 "High Temperature Neutron Diffraction Study of the Formation Reaction of Bismuth Superconductors," M. F. Garbauskas, R. H. Arendt, J. D. Jorgensen, and R. L. Hitterman, Mat. Res. Soc. Symp. Proc. Vol 169, p. 129 (1990)

13 "An Alternate Preparation for $(Bi,Pb)_2Ca_2Sr_2Cu_3O_z$," R. H. Arendt, M. F. Garbauskas and P. J. Bednarczyk, Physica C176, 126 (1991)

14 "An Alternate Preparation for Grain Oriented Structures of $(Bi,Pb)_2Ca_2Sr_2Cu_3O_z$," R. H. Arendt, M. F. Garbauskas, K. W. Lay, and J. E. Tkaczyk, Physica C176, 131 (1991)

15. "Growth of the 2223 Phase in Leaded Bi-Sr-Ca-Cu-O System," T. Hatano, K. Aota, S. Ikeda, K. Nakamura and K. Ogawa, J. Jour. Appl. Phys. 27, L2055 (1988)

PREPARATION OF $YBa_2Cu_3O_{7-x}$/SILVER COMPOSITE RIBBONS BY SPRAY PYROLYSIS

N. L. Wu, Y. K. Wang, and S. H. Lin
Department of Chemical Engineering,
National Taiwan University, Taipei, Taiwan, R.O.C.

ABSTRACT

$YBa_2Cu_3O_7$-on-Ag composite ribbons were prepared by using spray pyrolysis technique and the effects of process variables, including (1) the chemical stoichiometry of the sprayed solution, (2) the spraying, calcination and sintering temperatures, and (3) the film thickness, on the superconducting properties of the ribbon were investigated. Spray at substrate temperatures above 180 ^oC with 3% excess Cu in the sprayed solution followed by calcination and sintering at 920 ^oC produced highly textured $YBa_2Cu_3O_{7-x}$ films with T_c at 90 K and the highest J_c. The transport J_c (78 K, zero field) was 5000~800 A/cm^2 for 1 um film but decreased rapidly to 1500-2000 A/cm^2 as thickness exceeded 2 um. The decrease in J_c with film thickness was due to an increase in the amount of cracks within the oxide films and to the loss of the preferred orientation of c-axis normal to the metal surface. Formation of multi-layered wire from a single-layered ribbon was demonstrated.

INTRODUCTION

Since the discovery[1] of the high-T_c superconducting oxides, a great deal amount of research has been conducted to produce these materials in a more flexible form. One way to achieve this goal is to combine the oxides with metals to form composites, in which the metal part provides necessary mechanical support and acts as stabilizer, while the oxide part is to carry superconducting current. Typical examples of such composites are metal-sheathed oxide wires, mostly prepared by the powder-in-tube method, and oxide-on-metal (i.e. oxide film on metal substrate) ribbons[2-5]. While the powder-in-tube method was most studied, improvement in critical current (J_c) under high magnetic fields has recently advanced rapidly for the oxide-on-metal ribbon system. Due to its open structure, the oxide-on-metal approach has the advantages of easy oxygenation and processing versatility. Furthermore, it is not restricted to producing wires but suitable for making ribbons and/or sheets of various geometries.

For preparing the oxide-on-metal ribbons, the oxide has previously been applied to the surface of metal substrate by doctor-blade, screen-print and, most recently, spray coating methods. These methods were in common in the sense that the oxide films were not 'grown' on the metal surfaces, Rather, in these methods, already-formed superconducting oxides, mixed with organic binders, were applied to the metal surface and additional heat treatment was employed after the application to burn off the binder and to sinter the oxide powders. The oxide powders applied were randomly oriented and, in order to achieve texture in the oxide film, which has been thought to be one of the pre-requisites for high-J_c, melt-and-crystallization heat treatment was often employed. This approach was demonstrated for the Bi-based superconducting oxides [4,5] but not for $YBa_2Cu_3O_{7-x}$, which is also known as the 123 compound, due to its high melting point relative to the metal substrate, mostly silver. In view of a much higher J_c exhibited by 123 in film form than in the bulk and of an enhanced texture of 123 films on metal substrates in the earlier studies, we prepared 123-on-Ag composite ribbons where the oxide film was simultaneously grown and sintered on the metal substrate.

The oxide film was prepared by the spray pyrolysis method, in which nitrate solution was sprayed onto the silver substrate and the sample was subsequently heated at high temperatures to synthesize and sinter the oxide film. The 123 films with thickness ranging from 1 to 10 um have been prepared in a total processing time of a few hours, which is much shorter than the powder methods. The J_c of the oxide film depended on the spraying and sintering temperatures and the film thickness. For oxide films calcined and sintered at the highest temperature, 920 ^oC, which is limited by melting of Ag in oxygen, the transport J_c (78 K, zero field) was 5000~800 A/cm^2 for 1 um film but decreased rapidly to 1500-2000 A/cm^2 as thickness exceeded 2 um. The decrease in J_c with film thickness was due to the presence of cracks within the oxide films and to the loss of the preferred orientation of c-axis normal to the metal surface.

EXPERIMENTAL

In the spray pyrolysis process, nitrate solution containing Y, Ba and Cu was sprayed directly onto a silver foil by using a spraying gun. The solution was prepared by dissolving Y_2O_3, $Ba(NO_3)_2$ and CuO in solution of nitric acid. After certain amount of the reactant materials was sprayed, the foil was placed inside a tubular reactor in flowing oxygen for calcination. Calcination of the foil was conducted at 900-920 ^oC for 5-10 minutes. The spray-and-calcination cycle was repeated until the film reached the desired thickness. Finally, the ribbons were further sintered above 900^oC for different periods of time.

The transport J_c was measured by the four-point method at liquid nitrogen temperature under zero field, and the 1 uv/cm standard was adopted. On the other hand, the intra-grain J_c was measured by SQUID and calculated according to Bean[6] model. The crystal structure and the extent of texture were characterized by XRD and the morphology and composition of the film were examined by using SEM equipped with EDX.

RESULTS AND DISCUSSION

A typical 123-Ag composite ribbon was prepared at the rate of approximately 0.5 um per spray-and-calcination cycle. After each calcination, the film had a uniform black color, and XRD studies showed only the 123 compound without any impurity phase. The films prepared by using a solution of stoichiometric proportion: Y:Ba:Cu = 1:2:3 showed a T_c at 80 K, which was approximately 10 K lower than that reported for the bulk sample. Comparison between the EDX spectrum acquired from a 123 bulk and that from the film indicated an lower Cu content in the film sample. SEM studies on the ribbon which was only partially covered by the oxide film showed that, after several heating and cooling cycles, there were large CuO crystals at the silver grain boundary regions. These result suggested that Cu was depleted from the oxide film by dissolution into the silver substrate during heating (and subsequently precipitated upon cooling). It was found that adding excess Cu in the sprayed solutions by 3% produced the highest T_c (~90 K). By adopting the 1 uV/cm standard, the transport J_c's of the as-calcined film was less than 100 A/cm^2. However, as shown below, the J_c of the films deposited at high substrate temperatures increased dramatically after further sintering above 900 ^oC.

For calcination at the same temperature in the subsequent stage, the films sprayed at substrate temperatures below 150 °C showed net-like structures with large amount of pores within the films (Fig. 1a). Increase in spraying temperature to above 180 °C not only produced films with plate-like grains and a higher density (Fig. 1b) but also enhanced the extent of preferred orientation of c-axis normal to the metal surface (Fig. 2a).

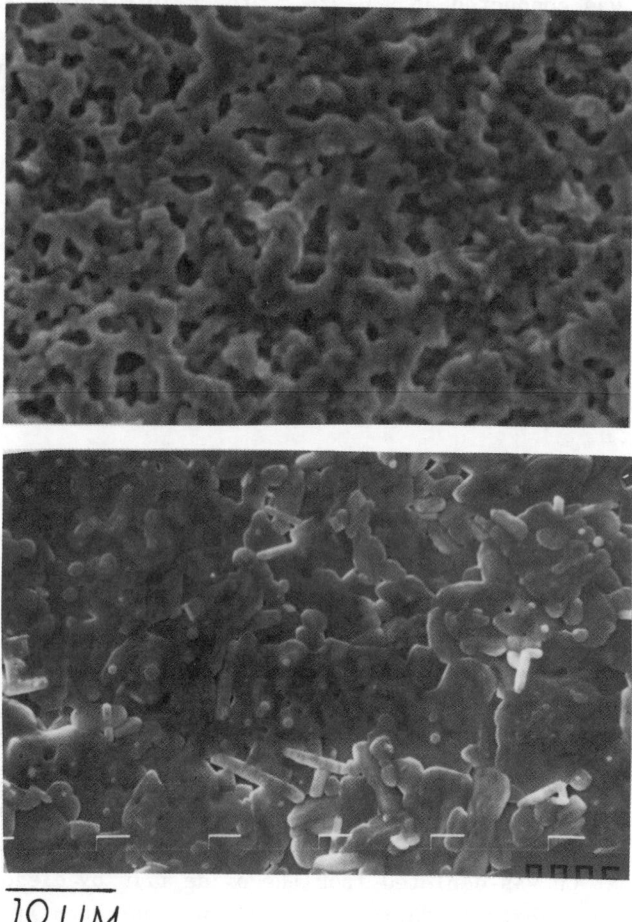

Fig. 1a and b: Morphologies of the 123 oxide films prepared by spraying at (a) below 150 °C and (b) above 180 °C.

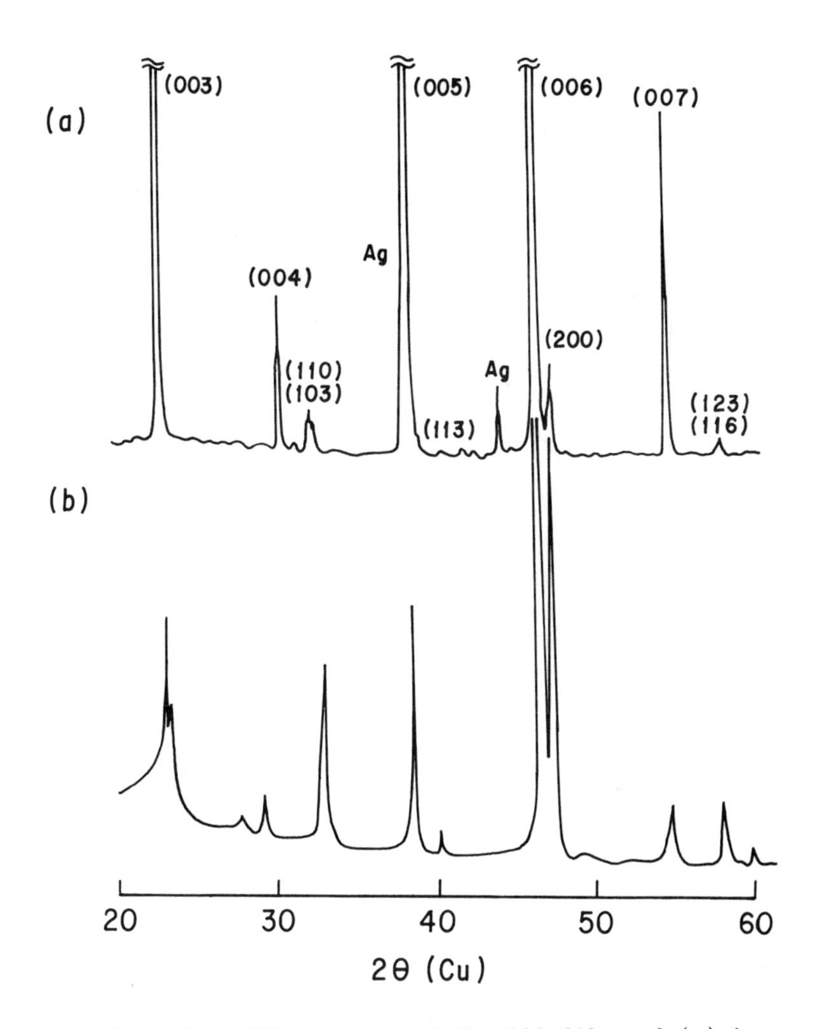

Fig. 2a and b: XRD patterns of the 123 films of (a) 1 um and (b) 5 um thick, showing a change in texture from c-axis to a-axis preferential orientation normal to the metal surface as the film thickness increased.

For films sprayed at the substrate temperature above 180 ^{o}C, the transport J_c depended heavily on the sintering temperature. For example, as shown in Fig. 3, a 20 ^{o}C difference between 900 and 920 ^{o}C could cause an 3~4 fold increase in J_c. The highest calcination and sintering temperatures were limited by the melting of the silver substrate at 920 ^{o}C.

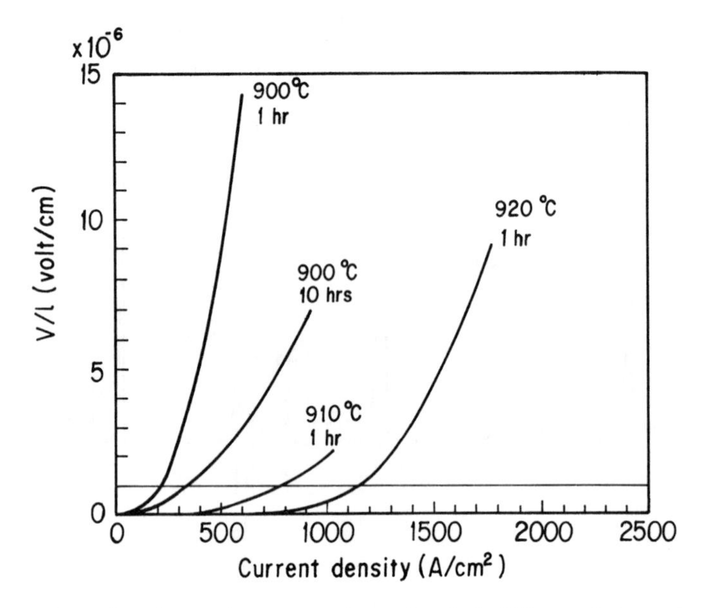

Fig. 3. The V-I curves for the 123 films sintered between 900 and 920 $^{\circ}$C.

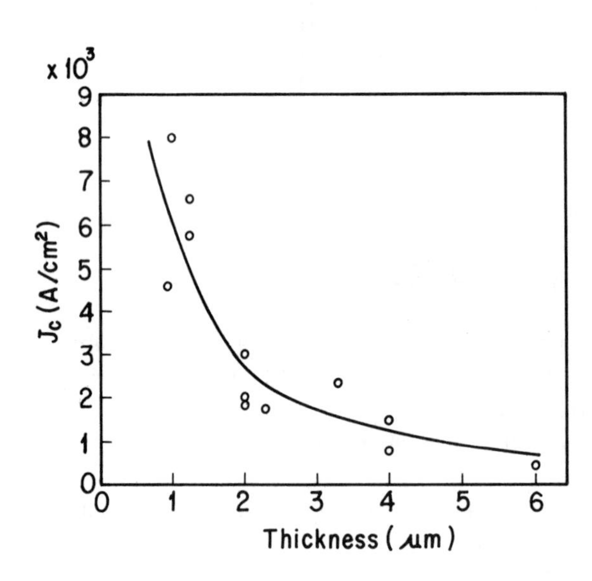

Fig. 4. The transport J_c vs. film thickness for the 123 films sintered 920 $^{\circ}$C.

For sintering at 920 $^{\circ}$C, the transport J_c saturated rapidly after 2~3 hrs of the heat treatment and the saturation value was found to decrease as the film thickness increased (Fig. 4). Transport J_c above 5000 A/cm^2 at 78 K, zero-field was routinely obtained for film thickness of ~1um, while it dropped rapidly to ~1000 A/cm^2 as at film thickness

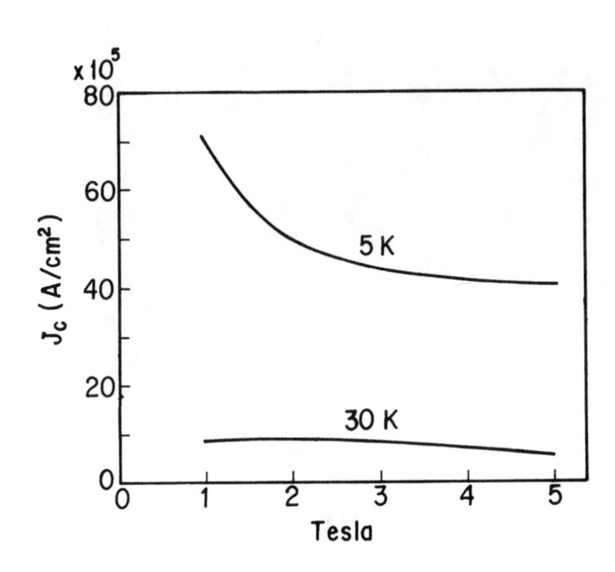

Fig. 5. The intra-grain J_c vs. magnetic field for the 1-um 123 film.

increased to 6 um. For films with thickness less than 2 um, calculation according to Bean model showed intra-grain J_c in the range of 10^5-10^6 A/cm^2 below 78 K (Fig. 5). The fact that the intra-grain J_c is much higher than the transport J_c suggests that the transport J_c of the thin (1-2 um) films is limited by the weak-link effect.

For the thicker (>2 um) films, XRD studies indicated that, as the film thickness increased, the texture changed gradually from c-axis to a-axis preferential orientation normal to the metal substrate (Fig. 2b). It was noted that similar change in texture orientation with film thickness has previously been reported for the 123 films deposited via vacuum-based techniques[7]. In addition, cracks were frequently observed to traverse thick films (Fig. 6). The presence of cracks was believed to be due to the mis-match in thermal expansion coefficients between 123 and silver, which caused stress during cooling.

To form multi-layer composite (Fig. 7), the calcined 123-Ag sheet could be folded and compressed (or hammered) prior to the final sintering stage. The folding pattern can be designed to facilitate oxygenation during cooling. After the same period of sintering time, the transport J_c of the multi-layer composite is either the same as or, in some cases, slight higher than that of the single-layer film. SEM observations indicated less cracks in the oxide films within multi-layer composite, which was compressed or hammered prior to sintering, than in the single-layer film.

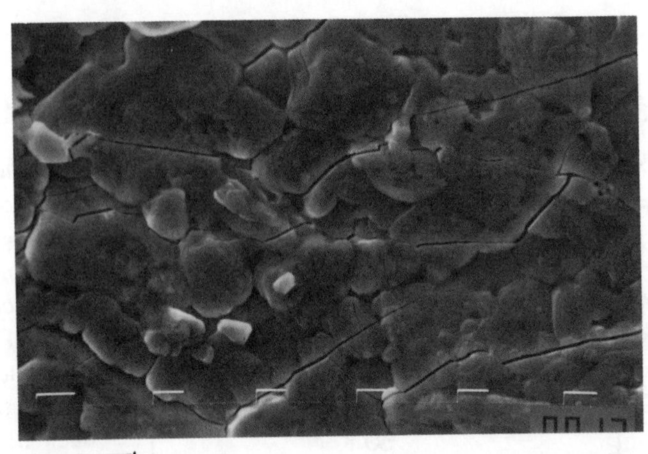

Fig. 6. The presence of cracks in the thick films.

Fig. 7. SEM cross-section micrograph of the two-layered wire prepared from the single-layered ribbon.

ACKNOWLEDGEMENT

This work was supported by the National Science Council, Republic of China, under Contract No. NSC80-0405-E-002-21.

REFERENCES

1. M. K. Wu, J. R. Ashburn, C. J. Torng, P. H. Hor, R. L. Meng, L. Gao, Z. J. Huang, Y. Z. Wang, and C. W. Chu, Phys. Rev. Lett. 58, 908 (1987).
2. K. Togano, H. Kumakura, H. Maeda, E. Yanagisawa, N. Irisawa, J. Shimoyama and T. Morimoto, Jpn. J. Appl. Phys. 28, L95 (1989).
3. K. Hoshino and H. Takahara, Jpn. J. Appl. Phys. 28, L1214 (1989).
4. S. Jin, R. B. van Dover, T. H. Tiefel, J. E. Graebner and N. D. Spencer, Appl. Phys. Lett. 58, 868 (1991).
5. T. H. Tiefel, S. Jin, G. W. Kammlott, J. E. Graebner and R. B. van Dover and N. D. Spencer, Appl. Phys. Lett. 58, 1917 (1991).
6. C. P. Bean, Phys. Rev. Lett. 8, 250 (1962).
7. M. Naito, R. H. Hammond, B. Oh, M. R. Hahn, J. W. P. Hsu, P. Rosenthal, A. F. Marshall, M. R. Beasley, T. H. Geballe, and A. Kapitulnik, J. Mater. Res. 2, 713 (1987).

HIGH-TEMPERATURE REACTIONS AND ORIENTATION OF $YBa_2Cu_3O_{7-\delta}$

B. J. Chen, M. A. Rodriguez, S. T. Misture, and R. L. Snyder
Institute of Ceramic Superconductivity
New York State College of Ceramics at Alfred University
Alfred, NY 14802

Abstract

Real time observations of the peritectic reactions of $YBa_2Cu_3O_{7-\delta}$ (123) using high temperature XRD reveals a reaction sequence which does not correspond well with the literature. Dynamic *00ℓ* orientation studies show that the optimum temperature for orientation and formation of 123 is 940°C. The rate of formation and orientation of the 123 from the melt is rapid. This orientational growth is believed to be a liquid assisted sintering reaction.

INTRODUCTION

The feasibility of practical devices employing bulk ceramic superconductors depends largely on improvements in critical current density (J_c). Techniques such as melt texturing[1–5] and gradient firing[6–10] show promise for improving the critical current densities in high temperature superconductors. Because high-T_c superconductors exhibit a high degree of anisotropy, the main purpose of the melt texturing, gradient firing and various film fabrication techniques, is to align the high J_c (a-b) plane in the direction of current travel. The degree and rate of textured growth and the mechanism of phase formation are the focus of considerable research.

The aim of this study is to investigate the high temperature reactions and orientation behavior of 123 using in-situ high temperature XRD. Specifically, this paper reports the rate of reaction and degree of orientation of 123 at various temperatures, to determine the optimum temperature for orientational growth. Real time high temperature analysis allows a window into the thermodynamics and kinetics of reactions at high temperature that is quite different from the quenched room temperature view. These observations lead to a better understanding of the mechanism of orientation for this compound.

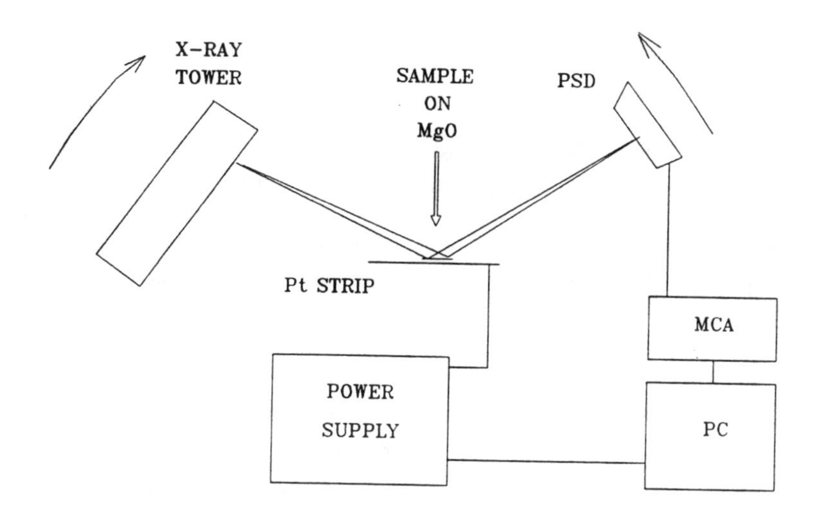

Figure 1: Schematic of dynamic XRD apparatus.

EXPERIMENTAL PROCEDURE

High purity orthorhombic 123 powder was synthesized by the nitrate method[11] using reagent grade Y, Ba and Cu nitrate salts (Alfa Products, 99.95% pure). The nitrates were brought slowly to 800°C and held for 8 hours. The material was reground and sintered at 940°C for 10 hours in air followed by an anneal at 500°C in oxygen to obtain the orthorhombic phase.

A schematic diagram of the hot stage XRD unit is shown in Figure 1. X-ray data were collected on a highly modified, locally automated, Norelco diffractometer using Cu radiation and a Ni filter. The diffractometer was equipped with a curved metal wire position sensitive detector (PSD) and a multichannel analyzer (MCA) set up to be able to scan,[12] allowing rapid data collection. Diffraction patterns for the peritectic melting sequence were collected from 900 to 1170°C with a temperature step of 10°C per scan. The scan rate was typically $5°2\theta$/min. The heating rate between each scan was 20°C/sec. To analyse the orientation effects in 123, samples were heated directly to 1100°C held for 20 minutes then cooled rapidly and held at 930, 940, 950, 960, and 980°C in separate experiments. Continuous X-ray scans began immediately after reaching the holding temperatures. Data for these scans were collected using a $5°2\theta$/min scan rate from 20 to $50°2\theta$, hence, each scan represents 6 minutes of data collection. All the studies were done in air.

The hot-stage and sample setup are as follows.

1. The stage is made of a Pt/Rd resistive heating strip. A Pt-Pt/Rd thermo-

couple is welded to the underside of the heating strip for monitoring the stage temperature.

2. A single or poly-crystalline MgO substrate (≈0.25mm thick) is attached to the Pt/Rd strip using powdered microscopic slide glass as an adhesive.

3. The surface temperature of the MgO substrate is calibrated to the thermocouple by visual observation of well known melting reactions. Ag and Au are used as standards and subsequently removed after calibration.

4. A few milligrams of the specimen powder are mixed with a few drops of cyclohexane and the slurry is ground in a mortar. A pipette is used to draw the slurry from the mortar and deposit it dropwise onto the MgO substrate at room temperature. After the slurry is dried, a 100 to 500μm 123 film is left on the surface of the MgO subtrate.

RESULTS AND DISCUSSION

Figure 2 summarizes the high temperature reaction sequences of 123 in this study. Detailed discussions of the high temperature reactions have been reported elsewhere.[13] Briefly, in the temperature range of 900 to 950°C, there is a $Y_2Cu_2O_5$ phase present, which comes from reaction with atmospheric CO_2, along with the tetragonal 123 compound. At 970°C, $Y_2Cu_2O_5$ vanishes quickly and the $BaCuO_2$ phase appears. In oxygen the $Y_2Cu_2O_5$ phase doesn't appear and peritectic melting proceeds directly to $BaCuO_2$. This reaction to form $BaCuO_2$ is inconsistant with most of the previous reports on the system. The disappearance of both 123 and $BaCuO_2$ phases to form Y_2BaCuO_5 (211) and liquid occurs at 1050°C; this is higher than the values quoted by other authors. The appearance of Y_2O_3 from 211 and liquid is observed at 1150°C. This temperature is much lower than those quoted in the literature.

Figures 3, 4, and 5 display the 3D hot stage XRD plots showing the development of orientation in 123. The samples were heated to 1100°C then rapidly cooled and held at 940, 930, and 960°C respectively to watch for orientation effects. Figure 3 displays the results for 940°C. This figure shows the very rapid reaction of 211 + liquid to 123. Within 6 minutes (ie. the first pattern) the reaction is almost complete. The 211 phase has completely reacted in less than an hour. This plot also illustrates that the orientational growth of the 123 phase from the melt is quite rapid. The appearance of strong *00ℓ* peaks coincides with the appearance of the 100% peak of 123. Within 12 minutes (ie. the second pattern), the *00ℓ* peaks have nearly reached their maximum intensity indicating

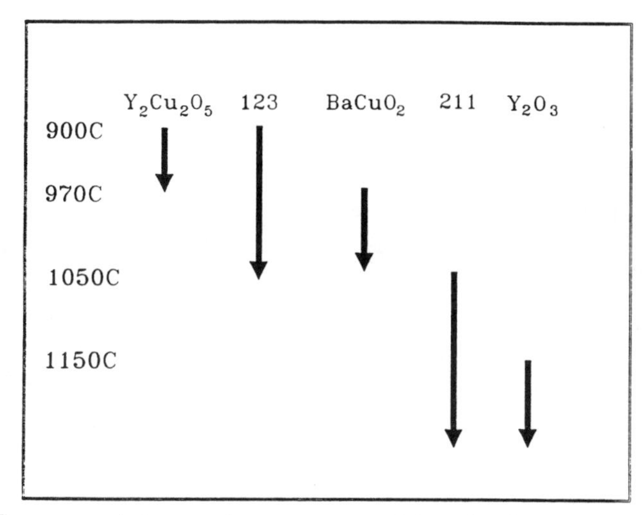

Figure 2: Summary of the reaction sequences of $YBa_2Cu_3O_{7-\delta}$ in this study.

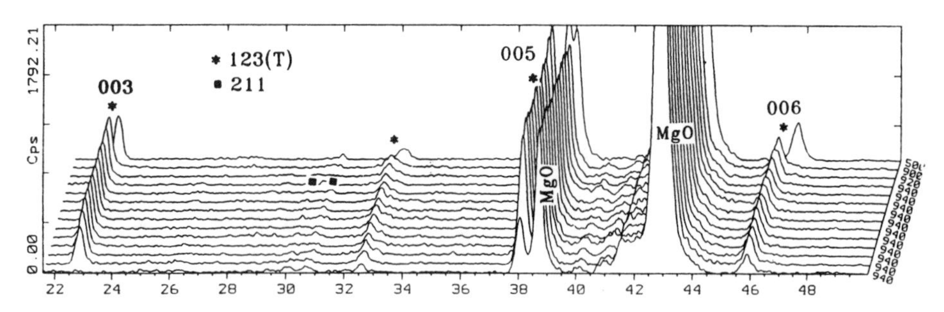

Figure 3: 3-D hot stage XRD plot of 123 at 940°C showing the kinetics of orientation.

a high degree of orientation. The degree of orientation at 940°C, calculated using the Lotgering factor,[14] is near 90%. This suggests that as soon as the 123 phase forms from the melt, the 123 particles will grow in such a way that the a-b plane is oriented parallel to the surface. It therefore seems likely that oriented growth is preferred as long as the liquid phase is present. This supports a "free surface epitaxy" model where the crystallites can turn their lowest free energy face toward the interface due to the liquid lubrication.

Figure 4 is a 3-D plot of 123 held at 930°C. At 930°C, 123 also forms quite fast. Within the first 6 minutes 123 appears and in 12 minutes it reaches its highest intensity. However, the degree of orientation is much lower than at 940°C. Also, there is a large amount of 211 phase present along with the 123 and after one hour, there is no significant decrease of the 211 peaks. This suggests

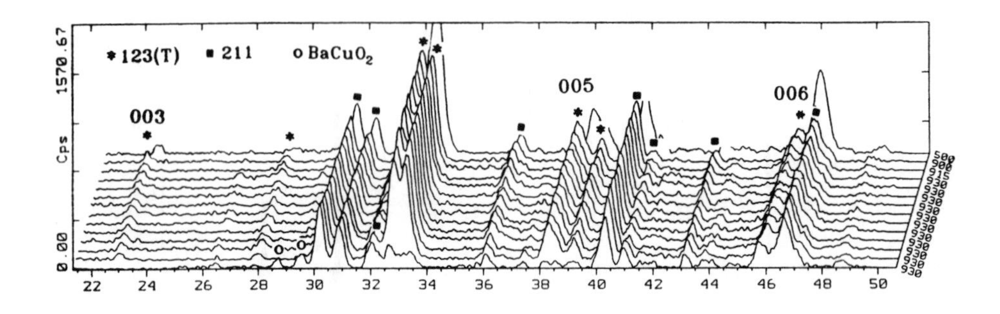

Figure 4: 3-D hot stage XRD plot of 123 at 930°C showing the kinetics of orientation.

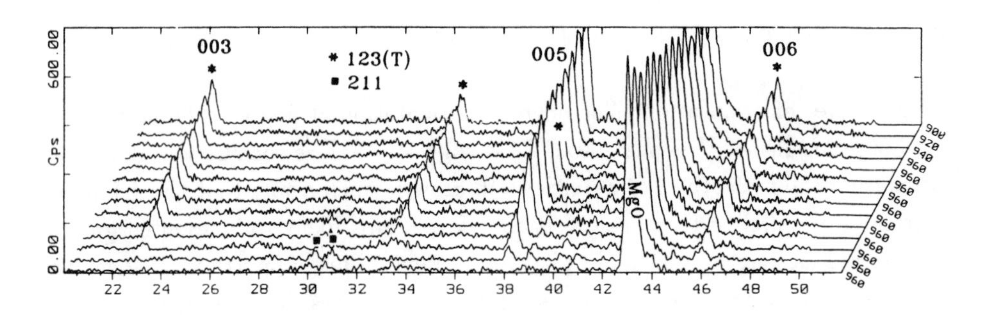

Figure 5: 3-D hot stage XRD plot of 123 at 960°C showing the kinetics of orientation.

that at lower temperature the liquid phase disappears quickly and the reaction occurs purely from the solid state where the reaction rate is much slower than when liquid is present. The very low degree of orientation (Lotgering factors of $\approx$ 20%) provides evidence that the orientational growth depends greatly on the presence of the liquid phase.

At higher temperatures, from 950 to 980°C in this study, the results are all identical. Figure 5 displays a 3-D hot stage XRD plot of 123 at 960°C as a representative sample for the three temperatures. Figure 5 shows that at higher temperature the 211 phase disappears faster than at 940°C, but the growth of 123 peaks are much slower and the degree of orientation is lower than that at 940°C. The Lotgering factors are about 75%. Therefore the orientation, based on Lotgering factors, is maximized at 940°C.

CONCLUSIONS

- The reaction sequences of the 123 composition at high temperature observed in this study are not consistent with the literature.
- The optimum temperature for 123 growth and orientation is 940°C.
- The rate of formation and orientation of the 123 from the melt is rapid. The fast reaction and growth of oriented 123 are due to liquid assisted sintering. The orientational growth depends greatly on the presence of liquid phase.
- This work illustrates the importance of high-temperature analysis in determining the actual sequence of reactions in complicated systems such as peritectic melts.

References

1. J.A.T. Taylor, P. Sainamthip, and D.F. Dockery, "Sintering Time and Temperature For $Ba_2YCu_3O_{7-\delta}$ Superconductors," *Mater. Res. Soc. Symp. Proc.* **99**, 663-66 (1988).
2. S. Jin, T.H. Tiefel, R.C. Sherwood, M.E. Davis, R.B. van Dover, G.W. Kammlott, R.A. Fastnacht, H.D. Keith, *Appl. Phys. Lett.* **52**[24], 2074, (1988).
3. S. Jin, T.H. Tiefel, R.C. Sherwood, R.B. van Dover, M.E. Davis, G.W. Kammlott, *Phys. Rev. B.* **37**, 7850 (1988).

4. S. Jin, R.C. Sherwood, E.M. Gyorgy, T.H. Tiefei, R.B. van Dover, S. Nakahara, L.F. Schneemeyer, R.A. Fastnacht and M.E. Davis, *Appl. Phys. Lett.* **54** [6], 584 (1989).

5. K. Salama, V. Selvamanickam, L. Gao, and K. Sun, *Appl. Phys. Lett.* **54** [23], 2352 (1989).

6. P. McGinn, M. Black, and A. Valenzuela, *Physica C* **156**, 57 (1988).

7. P. McGinn, W. Chen, and M. Black, *Physica C* **161**, 198 (1989).

8. P. McGinn, W. Chen, N. Zhu, U. Balachandran, and M. Lanagan, *Physica C* **165**, 480 (1990).

9. P. McGinn, N. Zhu, W. Chen, M. Lanagan, and U. Balachandran, *Physica C* **167**, 343 (1990).

10. J. Kase, J. Shimoyama, E. Yanagisawa, S. Kondoh, T. Matsubara, T. Morimoto, and M. Suzuki, *Jpn. J. Appl. Phys.* **29**[2], 277-279 (1990).

11. E. C. Berhman, V. W. R. Amarakoon, and S. R. Axelson *Adv. Cer. Mat.* **2** [3B], 539-555 (1987).

12. H. E. Göbel, *Adv. in X-Ray Anal.* **24**, 123-138 (1982).

13. R. L. Snyder, M. A. Rodriguez, B. J. Chen, H. E. Göbel, G. Zorn,and F. B. Seebacher, to be published in *Advances in X-ray Analysis*, Vol. 35.

14. F. K. Lotgering, *J. Inorg. Nucl. Chem.*, **9**, 113-123 (1959).

SYNTHESIS AND PROPERTIES OF NEW FAMILY OF SUPERCONDUCTING COPPER OXIDES BASED ON GaO LAYERS

B. Dabrowski and V. Zhang-McCoy,
Department of Physics, Northern Illinois University,
DeKalb, IL 60115

P. Radaelli, A.W. Mitchell and D.G. Hinks,
Materials Science Division, Argonne National Laboratory,
Argonne, IL 60439

J.T. Vaughey, D.A. Groenke and K.R. Poeppelmeier,
Department of Chemistry, Northwestern University,
Evanston, IL 60208

ABSTRACT

We have discovered the first layered superconducting copper oxide with small, fixed oxidation state cations separating the conducting CuO_2 planes. This material, $GaSr_2Y_{1-x}Ca_xCu_2O_7$, is similar to $YBa_2Cu_3O_7$ with the square planar copper chains replaced by chains of edge-shared GaO_4 tetrahedra. Thus, oxidation can occur only for the copper ion located in square pyramidal coordination in the CuO_2 plane. The undoped parent compound, x = 0, does not show magnetic order above 4K, probably due to the presence of the thick, ionic region separating the CuO_2 planes. However, this ionic region does not suppresses high T_c superconductivity (~70K) for the doped compositions.

INTRODUCTION

All known copper oxide superconductors have an anisotropic structure containing two-dimensional CuO_2 planes with the copper ion in square planar, square pyramidal or octahedral coordination by oxygen. The CuO_2 planes are bounded in the third dimension by metal-oxygen layers, AO, containing large and strongly electropositive (A = Ba, Sr and La-Ga) metal ions, thus forming AO-CuO_2-AO structural blocks. For many superconducting compounds there are frequently multiple CuO_2 planes, separated by metal layers, A′ (A′ = Ca, lanthanides and Y), within the block. All superconducting compounds can be divided into two general structural classes depending on the block stacking sequence. For the first class, blocks are stacked directly together (e.g. La_2CuO_4, Nd_2CuO_4 and $La_2CaCu_2O_6$) [1]. The second class, contains an additional intermediate region, separating the blocks, consisting of mixed oxidation state cations covalently bonded to oxygen (e.g. structures based on Cu, Tl, Pb and Bi) [1].

The best known superconducting copper oxide $LnBa_2Cu_3O_7$ (Ln = lanthanides and Y) contains square planar coordinated copper ions in the intermediate region (frequently referred to as the chain region) between the AO-CuO_2-A'-CuO_2-AO blocks[2]. It was shown for Ln = La that it is possible to substitute this square planar copper site with tantalum and niobium resulting in the distinct, but very similar $LaBa_2TaCu_2O_8$ structure, with octahedraly coordinated Nb and Ta[3]. This latter structure shows the importance of the coordination preference of small ions leading either to the formation of ordered CuO_2 planar compounds or mixing with copper on both sites. Recent discussion of the ionic size and coordination factors that led to the formation of CuO_2 planes in $AA'BCuO_6$ (B = transition or post transition metal) compounds is given by Anderson et al.[4]. Several new layered copper oxides with ionic, fixed oxidation state, cations in the intermediate region between AO-CuO_2-AO structural blocks were found (e.g. single CuO_2 layer $AlSrLaCuO_5$, $GaSrLaCuO_5$ and $SnLa_2CuO_6$, and double CuO_2 layer $AlSr_2LnCu_2O_7$ and $GaSr_2LnCu_2O_7$). However, none of these compounds was superconducting when prepared in air.

The structural similarity of these compounds to the known hole-doped copper oxide superconductors and the possibility of controlling the carrier concentration by varying the oxygen content led us to synthesize the Ca-substituted compositions under high oxygen pressure. The combined calcium and oxygen doping could introduce the necessary charge to the CuO_2 layers. Recently, we have prepared the $GaSr_2Y_{1-x}Ca_xCu_2O_7$ compound that become superconducting after high pressure oxygen annealing[5]. While the undoped parent compound, x = 0, did not show three-dimensional magnetic order above 4K, probably due to the presence of the thick ionic region separating the CuO_2 layers, superconductivity with a high T_c (~70K) was observed for the doped compositions.

EXPERIMENTAL

Polycrystalline samples of $GaSr_2Y_{1-x}Ca_xCu_2O_7$ ($0 \le x \le 0.4$) were synthesized from stoichiometric mixture of oxides and carbonates in air at 980°C followed by fast cooling to room temperature. Samples were fired for 3 weeks with frequent intermediate grindings. High pressure annealing was done for 24 hours in pure oxygen using 200 atm. at 910°C for powdered samples and 300 atm. at 925°C for pressed dense pellets. Lattice parameters were determined using Rietveld refinement of powder x-ray diffraction data. Susceptibility measurements were performed using a Squid (Quantum Design Corp. MPMS) and an a.c. (Lake Shore Cryotronics) susceptometer. Resistivity was measured using a standard four-lead d.c. technique. Oxygen content was determined using thermogravimetric analysis measurements (Cahn TG 171).

RESULTS

The orthorhombic structure of $GaSr_2LnCu_2O_7$ (noncentrosymmetric space group Ima2, No.46) is similar to $LnBa_2Cu_3O_7$ (see Fig.1)[5,6]. The square planar copper chains in $LnBa_2Cu_3O_7$ are replaced by chains of edge-shared GaO_4 tetrahedra. The large lanthanides and Sr are distributed over the A and A′-cation sites and the small lanthanides occupy only the A′ site between the copper planes within the double CuO_2 layer.

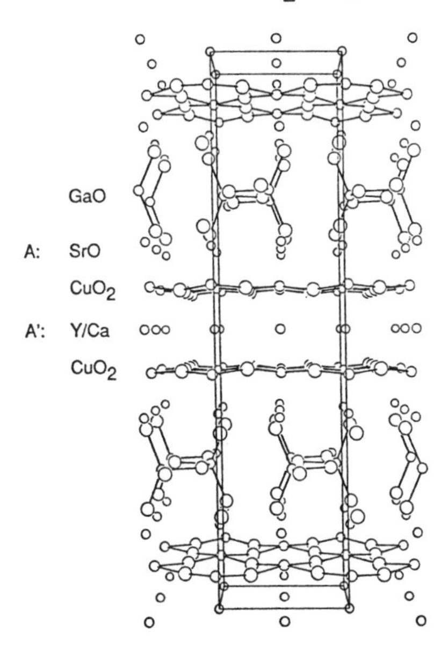

Fig.1 Layered structure of $GaSr_2LnCu_2O_7$ viewed along the b axis.

Air cooled samples with compositions 0≤x≤0.25 were single phase. Larger doping levels led to the presence of small amounts of unidentified impurity phases. In general, a very small contraction of the in-plane and an expansion of the out-of-plane lattice parameters were observed with increasing doping. However, these lattice parameters changes were not continuous which indicates that the oxygen content and the cation distribution of the fast cooled samples may vary. The high pressure annealed samples showed a decreased amount of impurity phase for x≥0.25 and noticeable contraction of the in-plane lattice parameters, indicating an increased hole-doping of the CuO_2 planes.

High sensitivity, zero field cooled Squid susceptibility measurements using 100 Gauss were done for both powder and pellet high pressure annealed samples. For the dense pellets, a gradual development of a superconducting phase with an almost fixed transition temperature, T_c~20-25K, was observed with increased doping as shown on Fig. 2. For powder annealed samples different T_c's were observed with the highest T_c = 73K for x = 0.3 (see insert to Fig.2). The superconducting phase fractions, as determined from these measurements, were only a few percent. The Squid measurements performed for the fast cooled undoped material over an extended temperature range, 4-300K, showed that the parent compound, $GaSr_2YCu_2O_7$, is weakly paramagnetic with no magnetic order above 4K.

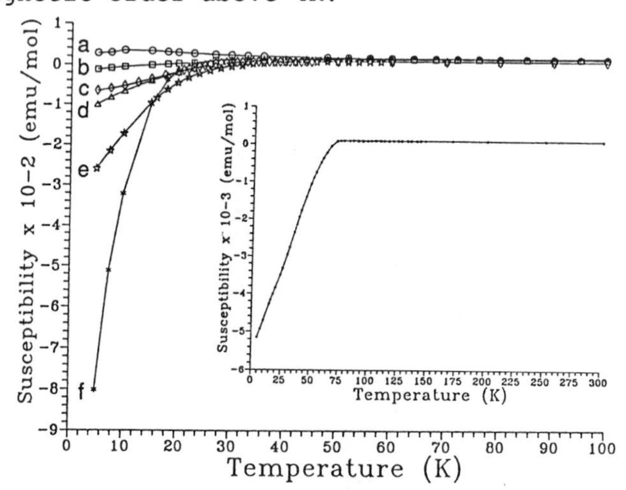

Fig.2 Squid susceptibility for dense x = 0 (a), 0.1 (b), 0.15 (c), 0.2 (d), 0.25 (e) and 0.35 (f) samples annealed under 300 atm. at 925°C. Insert: susceptibility for x = 0.35 powder sample.

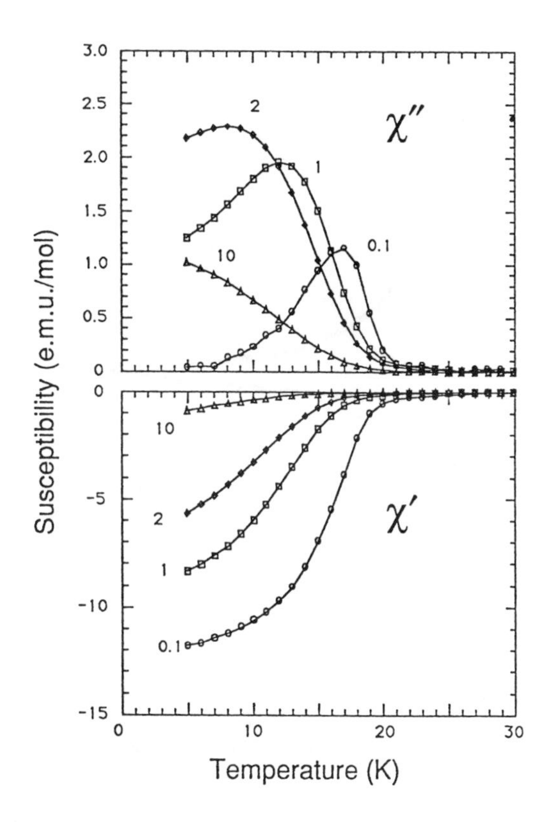

Fig.3 Real (χ') and imaginary (χ'') part of complex a.c. susceptibility for a x = 0.35 dense sample. The magnitude of an a.c. field is denoted on the graph in Gauss.

Additional low field measurements were performed using a.c. susceptibility. Figure 3 shows real and imaginary susceptibility for a x = 0.35 dense sample for various values of the a.c. field. Clearly, for low fields, ≤1 Gauss, the sample shows full diamagnetic behavior, proving that for this composition a large fraction of the sample becomes superconducting. Similar a.c. field dependence of the measured superconducting phase fraction was observed for the other compositions for either dense pellets or loose powders. The a.c. data also showed very good agreement for the onset T_c measured with the Squid.

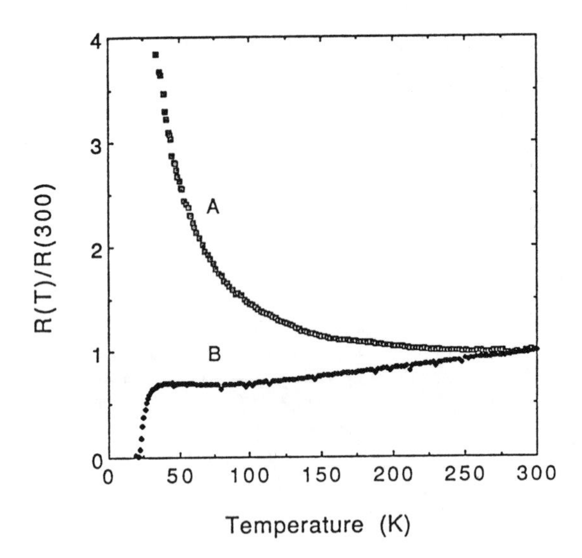

Fig.4 Normalized resistance for the fast cooled (A) and high pressure annealed (B) dense samples.

Resistivity measurements confirmed superconductivity for the high pressure annealed material. Typical R vs. T data for the fast cooled (A) and high pressure oxygen annealed (B) x = 0.35 dense samples is shown on Fig.4. The resistance changes from semiconductor-like to metallic when the sample is annealed under increasingly oxidizing conditions. The almost linear dependence of resistance on temperature for the superconducting sample is the same as observed for all other high temperature superconductors.

DISCUSSION

To verify that the presence of gallium is a necessary condition for superconductivity, we have prepared several samples without Ga and processed them under the same conditions as the Ga-containing material. None of these samples showed any traces of either superconductivity or metallic behavior. Also, in the x-ray diffraction patterns of $GaSr_2Y_{1-x}Ca_xCu_2O_7$ there was no indication for the formation of $YSr_2Cu_{3-x}Ga_xO_7$ impurity that could have superconducting properties. However, the reason the best superconducting properties are observed for $0.25 \leq x \leq 0.40$, i.e. beyond the apparent solubility limit at atmospheric pressure is still uncertain. It is possible, as frequently observed for oxide materials, that the solubility limit depends on the synthesis conditions[7]. Therefore, the solubility limit could extend beyond $x = 0.25$ under high oxygen pressure annealing. The reduced impurity phase content after high pressure annealing indicates that some additional Ca might have been incorporated to the compound, however, the samples did not achieve full equilibrium during the 24 hour anneal. From thermogravimetric analysis there is a clear indication that the oxygen content increases slightly during cooling in oxygen over an extended temperature range for the doped compositions. Thus, the cooling rates are also very important and should be slow, $<1^{o}C/min.$, for maximum oxygen uptake or diffusion of cations, especially for dense samples.

The strong dependence of the superconducting phase fraction on the magnitude of the a.c. field may be related to the expected large anisotropy of the critical magnetic fields. At present, this behavior is not fully understood, but a similar field dependence has been observed for other highly anisotropic copper oxide superconductors (e.g. $Bi_2Sr_2CuO_6$)[8].

X-ray, neutron diffraction and thermogravimetric analysis, indicate that the undoped, x=0, material is stoichiometric in oxygen content and, thus, the copper ion should be in d^9 configuration. The absence of antiferromagnetic order above 4K in this material is significantly different from the parent compounds of other copper oxide superconductors in which the copper ion is unambiguously in a d^9 configuration, (e.g. La_2CuO_4 and Nd_2CuO_4). For these materials, the antiferromagnetic order develops at elevated temperatures (~300K). In fact, the presence of antiferromagnetism in the parent material and superconductivity for doped compositions has led to several theories of high temperature superconductivity based on magnetic coupling[9]. Two-dimensional antiferromagnetic fluctuations are possible in $GaSr_2YCu_2O_7$ but apparently the thick insulating, SrO-GaO-SrO, layer separating the double CuO_2 planes inhibits magnetic coupling between the planes suppressing the development of three-dimensional antiferromagnetic order.

CONCLUSION

We have synthesized the first layered superconducting copper oxide with small, fixed oxidation state cations separating the conducting $AO-CuO_2-A'-CuO_2-$

AO blocks, $GaSr_2Y_{1-x}Ca_xCu_2O_7$. This material may offer advantages for experimental and theoretical study because copper is the only mixed oxidation state ion and occurs only in one coordination. Several similar materials with ionic layers of AlO, NbO and TaO, may become superconducting once properly doped and annealed. These materials can provide important information concerning the nature of the superconducting state, in particular the relation between magnetic coupling and high temperature superconductivity.

This work is supported by the U.S. Department of Energy, Basic Energy Sciences / Materials Sciences under contract No. W-31-109-ENG-38 (BD, AWM, DGH) and the National Science Foundation, Division of Materials Research - Low Temperature Physics Program under grant #DMR 87-19738 (VZ) and the Science and Technology Center for Superconductivity #DMR-8821571 (PR, JTV, DAG, KRP).

REFERENCES

1. I.K. Schuller and J.D. Jorgensen, MRS Bulletin **XIV**, v.1, 27 (1989)
2. M.A. Beno, et al. Appl. Phys. Lett. **51**, 57 (1987)
3. N. Murayama, et al. Jpn J. Appl. Phys. **2**, 27 (1988)
4. M.T. Anderson and K.R. Poeppelmeier Chem. Mater. **3**, 476 (1991)
5. J.T. Vaughey, et al. Chem. Mater. **3**, 935 (1991)
6. G. Roth, et al. J. Phys. I **1**, 721 (1991)
7. J.F. Bringley, et al. J. Solid State Chem. **88**, 590 (1990)
8. R.M. Fleming, et al. Physica C **173**, 37 (1990)
9. P.W. Anderson, Science **235**, 1196 (1987)

STUDY OF SINTERING CHARACTERISTICS USING AN IMPROVED METHOD FOR THE DETERMINATION OF THE OXYGEN CONTENT IN $YBa_2Cu_3O_{7-\partial}$

H. Vlaeminck, F. Persyn and S. Hoste
Laboratory of General and Inorganic Chemistry
Rijksuniversiteit Gent, Krijgslaan 281, 9000 Ghent, Belgium

ABSTRACT

Based on a thorough analysis of the sources of error involved in the iodometric determination of the oxygen deficiency ∂ in $YBa_2Cu_3O_{7-\partial}$, an improved procedure is given below. The crucial amendments to the methods originally published involve better control of the pH and a strict control of the sample dissolution steps. These improvements lead to an accuracy of better than ±0.01 in ∂ and show excellent correlation with data obtained from thermal analysis in a reducing atmosphere on a series of superconducting and non superconducting ceramics exhibiting a large variation in oxygen stoichiometry. The thermal reduction process is discussed and the results are applied to an optimization of sintering conditions, using the oxygen content and relative densification of the material as a guideline.

INTRODUCTION

In view of the crucial importance of the oxygen content on the superconducting properties of a large fraction of the ceramic superconducting oxides[1] many methods have been devised to establish the value of ∂ in $YBa_2Cu_3O_{7-\partial}$ with the highest accuracy and reliability and ease. Furthermore, a rapid method consuming a minimal amount of product is preferable when screening vast amounts of differently treated material. The projected method should consume between 0,01g to 0,1g of material if it is to be of any use in screening tests. This quantity is a rough guideline derived from the requirements of the volumetric analysis, the fact that other complementary and potentially destructive tests are often needed (thermal analysis, application of electrical contacts for resisitivity measurements) and that the preparation of single phase material in quantities exceeding a few grams requires specialized synthesis techniques.

Of all the direct and indirect methods available for the determination of the oxygen deficiency, both iodometry and thermal analysis have probably been most widely used. X Ray diffraction data[2] and Rietveld refinement on Neutron diffraction data[3] can yield similar information albeit with a somewhat lower precision or higher product consumption. Next, the measurement of volume[4] or pressure build up of the gas evolved in various chemical and physical[5] treatments of superconducting ceramics have also been applied.

Thermal analysis has been advocated to allow the determination of ∂ and offers three possible alternatives: the measurement of weight loss when starting from a known mixture of stoichiometrically pure Y_2O_3, $BaCO_3$ and CuO[6,7]; secondly, the measurement of the weight gain by oxygen take-up upon cooling from 950 °C (with a presumed $YBa_2Cu_3O_{6.5}$ composition related to the ideal tetragonal composition) and thirdly, the weight loss accompanying the total reduction of superconducting oxides to a mixture of Y_2O_3, BaO and metallic Cu in a reducing atmosphere. Although of considerable value in exploratory synthesis, the first thermogravimetric approach has the drawback of the inability to determine oxygen contents in existing superconductors and thus lacks in transferability of its results. The second method must be approached with caution as we have found that samples quenched from 900°C show oxygen contents ranging from 6,65 after one hour to 6,30 after 100 hours. depending on the sintering time and atmosphere. Although a value of 6,7 can be inferred from a published stability diagram[8], many thermogravimetric experiments suggest that lower values prevail [9,10]. Grain growth and porosity induced diffusion effects may thus render this value inappropriate for calibration purposes. The third method (total weight difference under reducing atmosphere) mentioned above is to be preferred after some control experiments as will be shown later.

Finally, different titrimetric methods have been proposed based on iodometric[11,12], bromometric[13,14] or ferrous-chromate[15] chemistry. These titrations have the advantage to require only straightforward chemical manipulation but reproducibility and comparison between these and the other[16] techniques mentioned above is rather poor for the moment.

We have therefore extensively studied and refined an existing [12] iodometric procedure by optimizing the sample dissolution step and by restricting the pH range for the titration to the theoretically derived and experimentally verified, correct values. Furthermore, this method has been tested on ultrapure reference CuO material and its results are compared to thermal analysis. It appears that, given the correct procedure outlined below, both methods are in perfect agreement on the value of ∂. Furthermore, this value is routinely obtainable with a maximum probable error of ±0.01 on three samples. Many iodometric determinations on replicate samples showed reproducibility well below this value.

Finally, this rapid and reliable method has allowed us to study the sintering process in deeper detail. By analyzing oxygen content changes as a function of sintering time, sintering temperature and densification processes, we come to the conclusion that a rapid synthesis method employing sintering stages of the order of tens of minutes is usable for the production of fairly large quantities of high quality superconducting ceramic powder samples.

EXPERIMENTAL

Thermal Analysis was performed using a Stanton Redcroft TG-DTA 1500 thermobalance modified to accept samples of 50-150 mg rather than the conventional 10 mg. Thermal runs were recorded in alumina crucibles from room temperature up to 1000°C at a constant speed of 5° min^{-1} in 58 $cm^3.min^{-1}$, 5 volume percent H_2/N_2. The $YBa_2Cu_3O_{7-\partial}$ powders were carefully ground in an agate mortar to maximize homogeneity and minimize grain size and runs were repeated two to three times giving an average deviation of 0.05 % on the weight loss and of 0.015 on the oxygen content, between runs of similar samples. No buoyancy effect was observed in trial runs using repeatedly fired Al_2O_3.

Iodometry was optimized in two steps. First, the volumetric procedure involving the back titration of iodine using thiosulphate in the presence of Cu^{2+} ions was tested at different pH values using 99.999% CuO (Janssen Chimica) CuO. This clearly indicated that, at variance with a previously published recipe[12], the pH must imperatively be kept between 3.0 and 4.5 by a controlled addition of H_3PO_4. This ensures total suppression of side reactions involving the Cu^{2+} catalyzed decomposition of the titrant, of the partial decomposition of the $Cu(NH_3)^{2+}$

complexes initially formed, and of the reoxidation of the I_2, which acts as the fingerprint molecule in the whole procedure.

The second enhancement involves the use of a slender, N_2 capped test tube in the dissolution steps for the finely ground superconducting powders in the KI/HCl mixture. In order to minimize interference with atmospheric oxygen (which reoxidizes I_2, and HCl (which reduces $(Cu\text{-}O)^{p+}$ instead of the intended reducing agent KI) the ceramic powder (40 mg) is covered by a finely ground layer of solid KI(1,5 g) which is carefully moistened with one droplet of HCl solution (6 mole dm^{-3}). This produces a glassy coating covering the mass of the superconducting powder and acts as a diffusion barrier for subsequently added 2 cm^3 HCl. The capped testtube is left standing for 3 hours to ensure total dissolution of the product, followed by the correct titration procedure as outlined above. Further experimental details have been published elsewhere[17].

Relative changes in density were determined from the weigth versus volume ratio measured on disks 2mm thick and 1,3 mm in diameter. These data tend to give sleigthly lower values than methods based on gas or liquid immersion depending on the porosity. The tubular three zone furnaces were calibrated using melting points of In (156,6°C), Sn (232°C), Pb (327,5°C), Zn (419,6°C) and Ag (resistance drop in 0,1 mm diameter wire at 961,9°C).

RESULTS

Iodometric oxygen content: The iodometric oxygen determination applied to a series of 11 samples prepared prepared from spray dried nitrate precursor materials[18] (Y:Ba:Cu in 1:2:3 composition) which were then fired for 10 hours at 1053 K in air. This ensures total decomposition of nitrates to oxides with very low carbonate residues. Changes in time, temperature and atmosphere of the subsequent anneal treatment yielded a large variation in oxygen stoichiometries. To ensure a higher reliability of this oxygen determination method using iodometry we compared each result with the value obtained by thermogravimetry in a reducing atmosphere on a separate portion of the same material. After establishing that Y_2O_3 is not noticeably affected up to 1000°C by the reducing atmosphere used here, we determined the oxygen content from the weight loss between room temperature and 975°C. The chance of reducing either Y^{3+} or Ba^{2+} is much lower than for Cu^{2+} ions since the respective reduction E° potentials are -2.91V, -2,37V and +0.34V. We have chosen this final temperature because partial melting occurs at slightly higher values and the reduction is only completed above 900°C. Our results for 11 samples of $YBa_2Cu_3O_{7-\partial}$ with ∂ ranging from 0 to 1 show an remarkable and gratifying correlation between both iodometry and thermal analysis (correlation coefficient squared = 0.997 for 9 points as seen in figure 1).

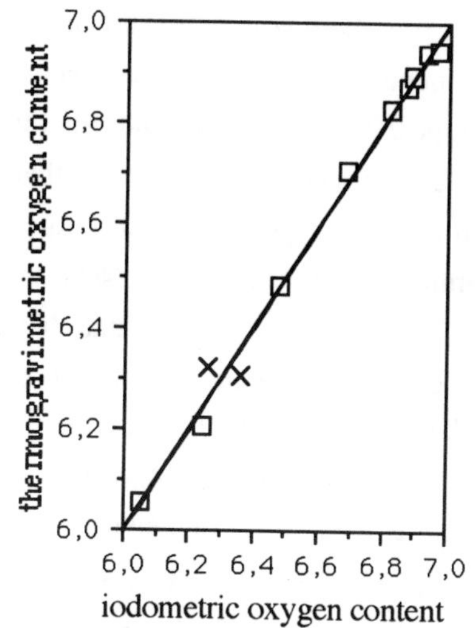

figure 1: correlation observed between oxygen content independently measured using thermal and iodometric methods (x= very inhomogeneous samples)

Thermogravimetry: The reduction curve typically obtained for a superconducting sample with δ=0,10, is shown in figure 2 where the DTA trace is superposed on the TG curve.

Both curves clearly indicate that the reduction proceeds in multiple steps related to the structural differences of the different oxygen species present in these perovskites. Although an earlier study[19] suggested the formation of intermediate hydroxides we have clear indication that the first reduction step occuring between room temperature and 500°C involves the release of oxygen from the (Cu-O) chains until the remaining sceleton is composed of neutral (Cu-O) entities. This intuitive assignment is substantiated by the fact that the weight changes associated with this step account for most the variation observed in the oxygen content the 11 samples as shown in figure 3.

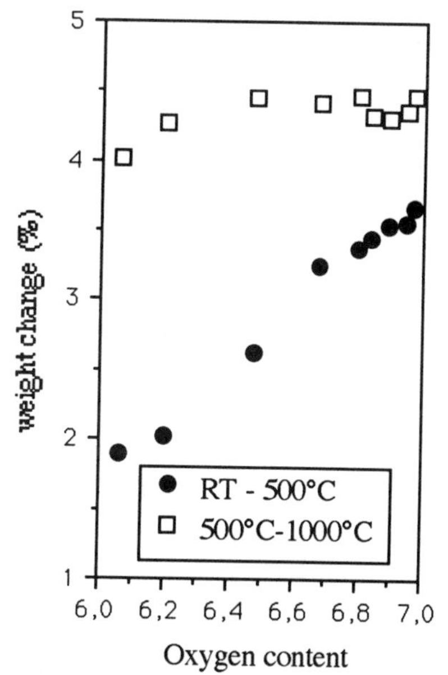

figure 3: thermogravimetric weight change (%) as function of oxygen content for two separate parts of the reduction curve

From the thermogram shown it can furthermore be deduced that all annealing experiments aiming at the enhancement of the oxygen content can only be effective in the range 260°C to 450° as indicated by the onset of the first and third exotherms in figure 2. At higher temperatures further steps can be discerned, their relevance to the superconducting perovskite structure probably diminishes as the Cu-O network structure is slowly degraded. Nevertheless, these steps must be associated with a stepwise reduction in the formal charge of the Cu^{2+} ions yielding several mixed valence structures on the way to full reduction to the metallic state. In particular, the last plateau (between 680°C and 860°C) is associated with a fairly stable $YBa_2Cu_3O_4$ composition. This remarkable stability of, presumably, Cu^{1+} ionic species in the remaining Y_2O_3-BaO framework is in stark contrast to the reduction of pure CuO under identical circumstances, where complete and strongly exothermic reduction to metallic Cu occurs in a single step between 300°C and 370°C. An ESCA study of partially reduced 123 compounds is under way in the hope to clarify this.

Finally, the existence of up to 5 structurally different and separately removable oxygen atom types shown in our DTA trace (figure 2) was confirmed by evolved oxygen pressure measurements[20,21] on bulk samples and thin films in evacuated reaction tube.

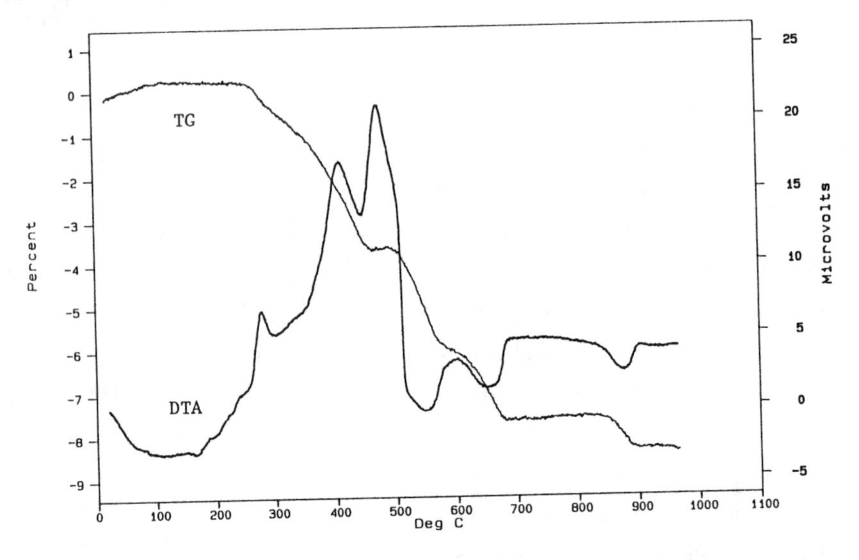

.Figure 3: Thermogram of $YBa_2Cu_3O_{7-\partial}$ with δ=0,10 in 5 volume percent H_2/N_2. Weight change (TG) and differential analysis curve (DTA) both show exhibit clear features indicating that oxygen removal from the CuO_2 chains occurs between 250°C and 450°C

Sintering: The above mentionned oxygen determination methods have allowed us to refine sintering conditions of superconducting ceramics as follows. Due to the large number of parameters influencing the different stages of preparation, a statistical approach (Plackett Burman design[22]) was choosen to assess to their relative importance. A 16 experiment setup was created in order to estimate the influence of 15 parameters: grain size, die pressure, calcination step (temperature, heating rate, time), sintering step (heating rate, temperature, time), cooling rate, cooling atmosphere, annealing step (temperature, time, atmosphere, cooling rate and cooling atmosphere). Using one extreme value and one aceptable or nominal value for each parameter, 16 samples were obtained. These products were then rated using resistance, magnetic susceptibility, density, levitating characteristics, oxygen content, XRD composition, SEM and optical polarizing microscopic examination. Most parameters showed some degree of influence, with sintering temperature as highest and cooling rate as lowest . Details will be published shortly.

This experiment led us to examine the sintering step more closely. We measured the relative densification as a function of sintering temperature (920°C, 940°C, 960°C, 980°C) for reaction times varying from 10 minutes to 200 hrs. The results are given in figure 4. Each curve exhibits final densities approaching 30% above the initial values. The time to obtain these final values however varies tremendously with temperature.

Three parts of the graph are worth discussing in vue of the oxygen contents of the product noted after sintering and its subsequent oxygen uptake after a 'classical' annealing cycle in O_2 (5hr, 450°C) as indicated in the figure. First, the combination of high density and good oxygen uptake is only obtained in the upper right corner, i.e. after very long reaction times at low temperature (920°C). Secondly, for applications where the highest density (and concommitant critical current density) is not required brief reaction times (10-100 minutes) can suffice if relatively higher temperatures are used (lower-middle). Finally, for the shortest reaction times, temperatures of up to 980°C may be employed but these will probably require dedicated annealing procedures (high pressure atmospheres, or plasma techniques) to obtain bulk quantities of material of satisfactory quality. A decrease of critical current density with increasing sintering temperatures and density has been noted before[23] and is associated with the appearance of liquidus phases, obstructing porosities and impeding O_2 transport.

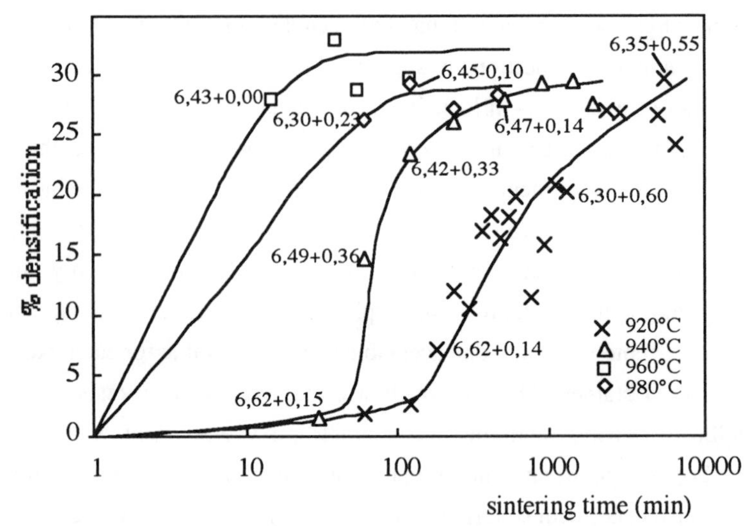

Figure 4: Percentage density increase noted in ceramic superconductors as function of sintering time and temperature. Oxygen contents for selected samples are given as the nominal value obtained directly after the sintering process and its increment observed upon annealing.

CONCLUSIONS

The oxygen content in samples of 40 mg $YBa_2Cu_3O_{7-\partial}$ can be routinely obtained with a precision better than 0.01 using iodometric titrations. These however require strict control of pH (between 3,5 and 4) and dissolution procedure.

Identical information can be obtained from the total weight loss of at least 100 mg ceramic between room temperature and 975°C, in a reducing atmosphere.

Thermal reduction occurs in discrete steps, the first of which corresponds with the removal of oxygen from the CuO_2 chains.

Sintering times of the order of minutes at 940°C-960°C can suffice for the preparation of these high temperature superconducting ceramic oxides in mass production environments.

ACKNOWLEDGEMENTS

The authors would like to thank Prof. D.L. Massart for valuable discussions on the statistical design of some experiments carried out in this research. This work was supported by the Belgian Government through the National Impulse Programme for

High Temperature Superconductors and by the Research Funding Committee of the University of Ghent.

REFERENCES

[1] R.J. Cava , A.W. Hewat , E.A. Hewat , B. Batlogg , M. Marezio , K. Rabe , J.J. Krajewski , W.F. Peck Jr. , L.W. Rupp Jr. , Physica C., 1990, 165, 419

[2] E. Takayama , Y. Uchida , M. Ishii , T. Tanaka and K., Kato Jap. J. Appl. Phys., 1987, 26:7, L1156

[3] H.M. Rietveld , J. Appl. Cryst., 1969, 2 , 65

[4] K. Conder , S. Rusiecki and E. Kaldis , Mat. Res. Bull., 1989, 24 ,581

[5] G.S. Grader and P.K. Gallagher , Adv. Ceramic. Materials, Special Issue , 1987, 2 ,649

[6] P. Kishan , L.K. Nagpaul and S.N. Chatterjee, Solid State Communications, 1988, 65 ,1019

[7] N. Brnicevic, M. Paljevic, Z. Ruzic-Toros, M. Tonkovic, A. Kashta, M. Prester, E. Babic, Physica C, 153-155, 820 (1988)

[8] P.K. Gallagher, Advanced Ceramic Materials, Vol 2, NO 3B, 632(1987)

[9] C. Namgung, J.T.S. Irvine, J.H. Binks, A.R. West, Supercond. Sci. Technol. 1, 169, (1988)169

[10] M.P.A. Viegers, D.M. De Leeuw, C.A.H.A. Mutsaers, H.A.M. van Hal, H.C.A. Smoorenburg, J.H.T. Hengst, J.W.C. de Vries, and P.C. Zalm, J. Mater Res. 2(6), 743 (1987)

[11] e.g.: A. Barkatt , H. Hojaji and K.A. Michael , Adv. Ceramic. Materials, Special Issue , 1987, 2, 701; D.C. Harris , M.E. Hills and T.A. Hewston , J. Chem. Ed., 1987, 64:10 ,847

[12] A.I. Nazzal , V.Y. Lee. , E.M. Engler , R.D. Jacowitz , Y. Tokura and J.B. Torrance , Physica C., 1988, 153-155 ,1367

[13] R.O. Crubellati , D.A. Smichowski , P. Battistoni , G. Polla and E. Manghi , Solid State Communications, 1990, 75:2 ,101

[14] E.H. Appelman , Inorg. Chem., 1987, 26 ,3237

[15] M. Oku , J. Kimura ,M. Hosoya , K. Takada and K. Hirokawa , Fresenius Anal. Chem., 1988, 332, 237; W. Schäfer , J. Maier-Rozenkranz , R. Kiemel and S. Semmler-Sack , J.Less Common Metals, 1988, I38, L25

[16] M.W. Shafer , R.A. de Groot , M.M. Plechaty , G.J. Scilla , B.L. Olson and E.I. Cooper , Mat. Res. Bull., 1989, 24, 687

[17] H. Vlaeminck, H.H. Goossens, R. Mouton, S. Hoste and G.P. Van der Kelen, J. Mater. Chem (1991), accepted for publication.

[18] S. Hoste , H. Vlaeminck , P.H. De Ryck ,F. Persyn , R. Mouton and G. Van der Kelen , Supercond. Sci. Technol. 1989, 1, 238

[19] P.K. Gallagher, H.M. O'Bryan, S.A. Sunshine, D.W. Murphy, Mat. Res. Bull., 22 (1987) 995

[20] H. Strauven, J.P. Locquet, O.B. Verbeke, Y. Bruynsereade, Solid State Communications, 65,4 (1988) 293

[21] B. Wuyts, J. Vanacken, J.P. Locquet, C. Van Haesendonck, I. K. Schuller and Y. Bruynseraede, Physics and Materials Science of High Temperature Superconductors (Kluwer Academic Publishers, Dordrecht, 1990), p307

[22] D.L. Massart, B.G.M. Vandeginste, S.N.Deming, Y.Michotte and L. Kaufman, Chemometrics, Data Handling in Science and Technology, Vol 2 (Elsevier 1988), p 105

[23] M. Reissner, W. Steiner, R. Stroh, S. Hörhager, W. Schmid and W. Wruss, Physica C 167, 495 (1990)

TEXTURE ANALYSIS OF BSCCO TAPES MADE BY
THE POWDER-IN-TUBE METHOD

D.B. Knorr, B. Chan, and D.J. Wilkins
Materials Engineering Department, Rensselaer Polytechnic Institute
Troy, New York 12180-3590

P. Haldar and J.G. Hoehn, Jr.
Intermagnetics General Corporation, Guilderland, New York 12084

L.R. Motowidlo
IGC Advanced Superconductors,Inc., Waterbury, Connecticut 06204

ABSTRACT

Tapes of (Bi, Pb)$_2$ $Sr_2Ca_2Cu_3O_{10}$ are made by the powder-in-tube method where silver tubes are loaded with pre-reacted powder, drawn, and rolled to 0.1 to 0.2mm thickness. High temperature heat treatment after each rolling step forms the superconducting phase. The reduction and heat treatment produces a texture in the ceramic core which is measured by X-ray diffusion. Bragg scans give semi-quantitative information on the strength of the texture while pole figures provide quantitative information on the degree and perfection of the alignment. Textures are found to be nearly fiber in character and are characterized by the spread of the (00ℓ) component. A direct correlation between the strength of the (00ℓ) texture and the high critical current density is found.

INTRODUCTION

The powder-in-tube technique[1-5] has been successfully used to demonstrate high transport critical current densities (J_c's) in silver sheathed $Bi_2Sr_2CaCu_2O_8$ (Bi2212) and (Bi, Pb)$_2$ $Sr_2Ca_2Cu_3O_{10}$ (Bi2223) ceramic superconductors. J_c values as high as $2x10^5$ A/cm^2 at 4.2 K that are sustained at high applied magnetic fields as well as $4.7x10^4$ A/cm^2 at 77K in zero field suggest that the bismuth based high temperature superconductors can be used for practical devices in the near future if long lengths can be manufactured with similar properties. Although other processing techniques such as melt texturing[6] have demonstrated high J_c values, the powder-in-tube method shows the greatest promise for producing long lengths of conductor. Furthermore, powder-in-tube uses conventional technology already developed for fabrication of low T_c superconductors such as NbTi and Nb_3Sn.

The mechanism for achieving high critical current density in silver sheathed Bi2223 is not well understood yet. It has been suggested that the origin of enhanced J_c is related to texture development, densification, grain growth, reduction in size and volume fraction of non-superconducting second phases, and elimination or reduction of the weak links at the grain boundaries. The powder-in-tube process provides most of these qualities. A previous report[7] described the alignment induced by repeated cold-work/annealing treatments for Bi2212 composites, although no attempt was made to quantify the texture. Other work[8] shows that improved alignment of Bi2212 depends on the details of the annealing treatment.

This paper reports texture results on a series of tapes where the superconductor core is examined. Processing histories and microstructures are detailed in a companion paper[9]. Sample preparation and data collection are summarized. Finally, the relation between texture and superconducting properties is discussed.

EXPERIMENTAL

The silver sheathed Bi2223 superconducting composites were prepared as described previously[3]. Precursor powder was prepared by a solid state reaction method with appropriate amounts of Bi_2O_3, PbO, $SrCO_3$, CaO, and CuO by reacting the mixture between 800° and 850°C for 24 to 168 hours. The precursor mix was packed into > 99.9% pure silver tubes and subjected to wire drawing, then rolling to reduce the composite thickness to a tape approximately 0.1mm in thickness. Intermediate heat treatments were used to soften the silver cladding. A final heat treatment at 840°C for 48 to 144 hours was done at the final thickness to fully develop the Bi2223 phase in the core of the conductor. Samples in this study had slightly varying nominal compositions and different precursor powder heat treatments. Transport critical current density was measured at 77K in zero applied field.

Specimens for X-ray diffraction analysis were prepared by gently lapping one side of the tape on a metallographic polishing wheel until some superconductor core appeared. The remaining silver was peeled away and three tape sections aligned next to each other to provide sufficient area for analysis. This procedure was found to minimize disturbance of the ceramic superconducting core.

X-ray diffraction was used to confirm phase content and for texture analysis. $\theta/2\theta$ scans between 15° and 50° in 2θ identified the Bi2223 phase and small spurious peaks associated with either silver or impurity phases. Comparison of relative peak heights also gives some qualitative information on the texture. Texture was quantified by the Schultz reflection technique[10]. The sample is fixed in $\theta/2\theta$ on the $(00\overline{10})$ peak, then tilted and rotated about its normal to record the $(00\overline{10})$ intensity as a function of direction. Tilt angles to 80° from the sample normal direction are covered. The $\theta/2\theta$ position is moved to $2\theta \sim 21°$, which is away from any Bragg peak, to record the background intensity. The data are reduced by subtracting the background, correcting for defocusing, normalizing, and plotting either as pole figures or as $(00\overline{10})$ intensity versus tilt angle measured from the sample normal direction.

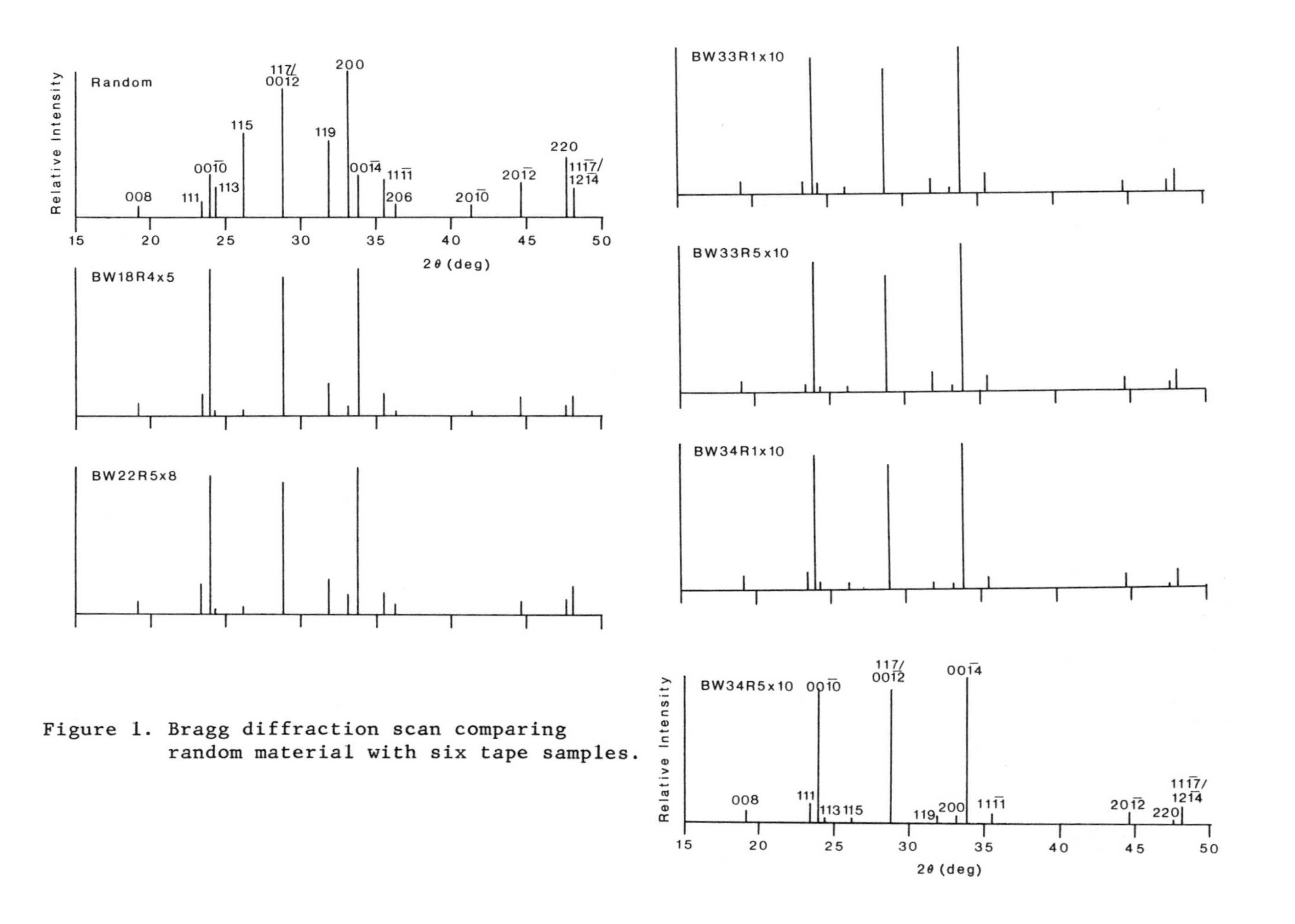

Figure 1. Bragg diffraction scan comparing random material with six tape samples.

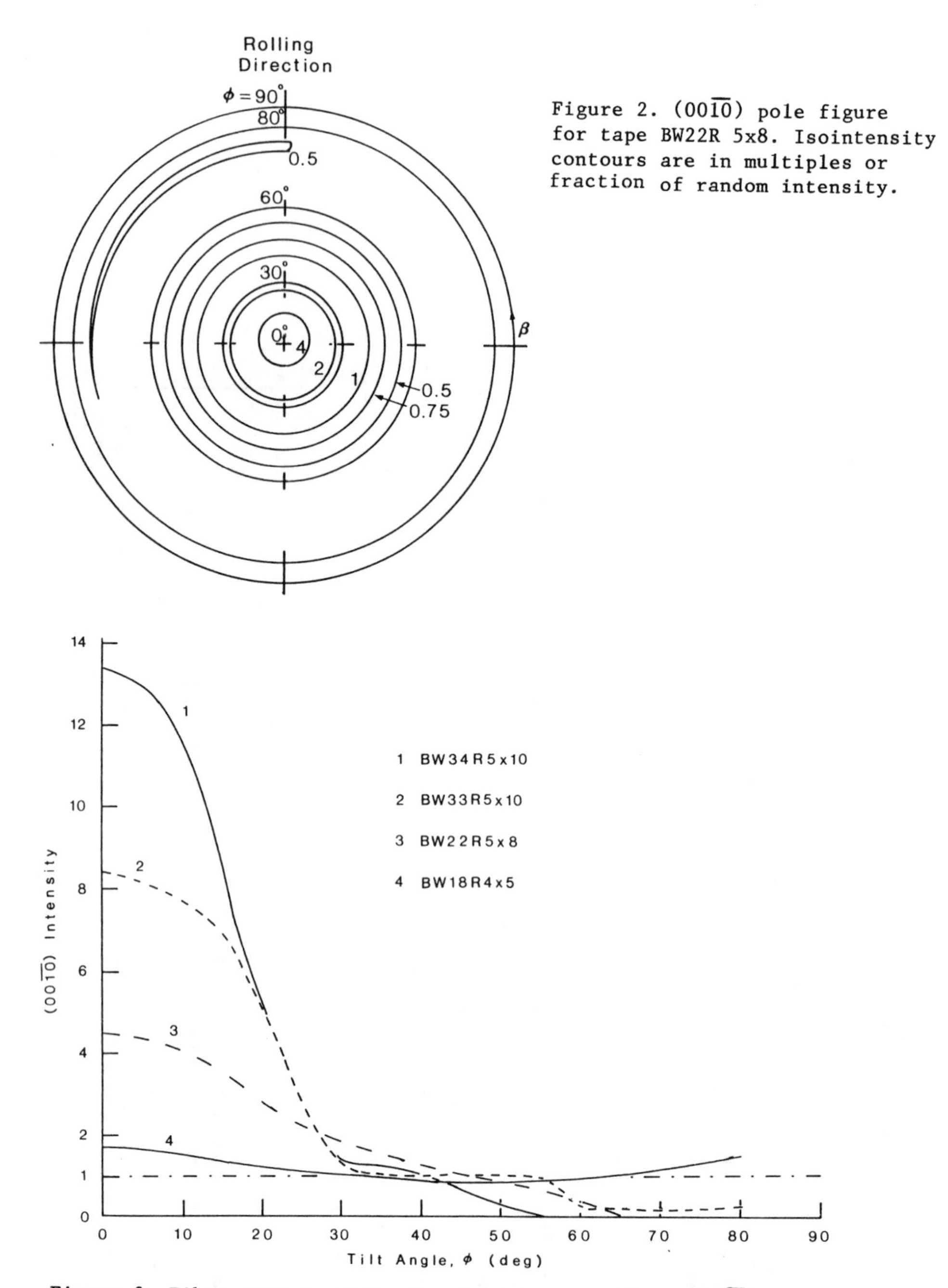

Figure 2. $(00\overline{10})$ pole figure for tape BW22R 5x8. Isointensity contours are in multiples or fraction of random intensity.

Figure 3. Fiber texture plots for four tape samples. $(00\overline{10})$ intensity is normalized to random = 1.

RESULTS

Final heat treatment conditions for six specimens are given in Table I. Sample BW18R4x5 was processed with three cold-roll and anneal cycles. The remaining five samples received two cold-roll and anneal cycles. Critical current density values for the six specimens are also listed in Table I. The BW34 series was able to carry currents from 6 amps to 13 amps corresponding to critical densities of 6000 A/cm^2 to 13000 A/cm^2.

Table I. Details of Tape Samples

Sample Identification	Heat Treatment Time at 840°C (hours)	J_c at 77K (A/cm^2)
BW18R4x5	48	2000
BW22R5x8	96	4600
BW33R1x10	144	5000
BW33R5x10	144	7000
BW34R1x10	144	6200
BW34R5x10	144	8400

The Bragg scans for all six samples are shown in Figure 1 as relative intensity versus 2θ. The calculated relative intensities for a random sample[11] are shown in the top plot. All intensities are normalized to the most prominent peak in the diffraction pattern which is (200) for the random and (00$\overline{14}$) for the tape samples.

Texture data are presented in Figures 2 and 3. Figure 2 gives the (00$\overline{10}$) pole figure for sample BW22R5x8 plotted on a polar stereographic projection. The nearly circular iso-intensity contours indicate that a fiber texture is present. This behavior is typical of all samples. To simplify presentation of the texture information, a plot of (00$\overline{10}$) intensity versus tilt angle for four samples is given in Figure 3. This fiber texture plot is possible because intensity is nearly independent of the rotation angle β shown in Figure 2.

DISCUSSION

The consistency between the θ/2θ scans and the pole figures is quite good. A comparison of the random values with any tape in Figure 1 shows that the (00ℓ) texture is enhanced compared to the other (hkℓ) peaks. The (hkℓ) peaks decrease or disappear as the (00ℓ) becomes stronger as seen by comparing BW18R4x5 with BW34R5x10. The trend is confirmed by the sharpening (00$\overline{10}$) texture evidence in the fiber texture plots of Figure 3.

The occurrence of a near fiber texture in the superconductor phase is somewhat surprising in a rolled piece. Symmetry considerations[12] imply that orthorhombic sample symmetry is enforced by a cold rolling operation. Fiber textures are usually

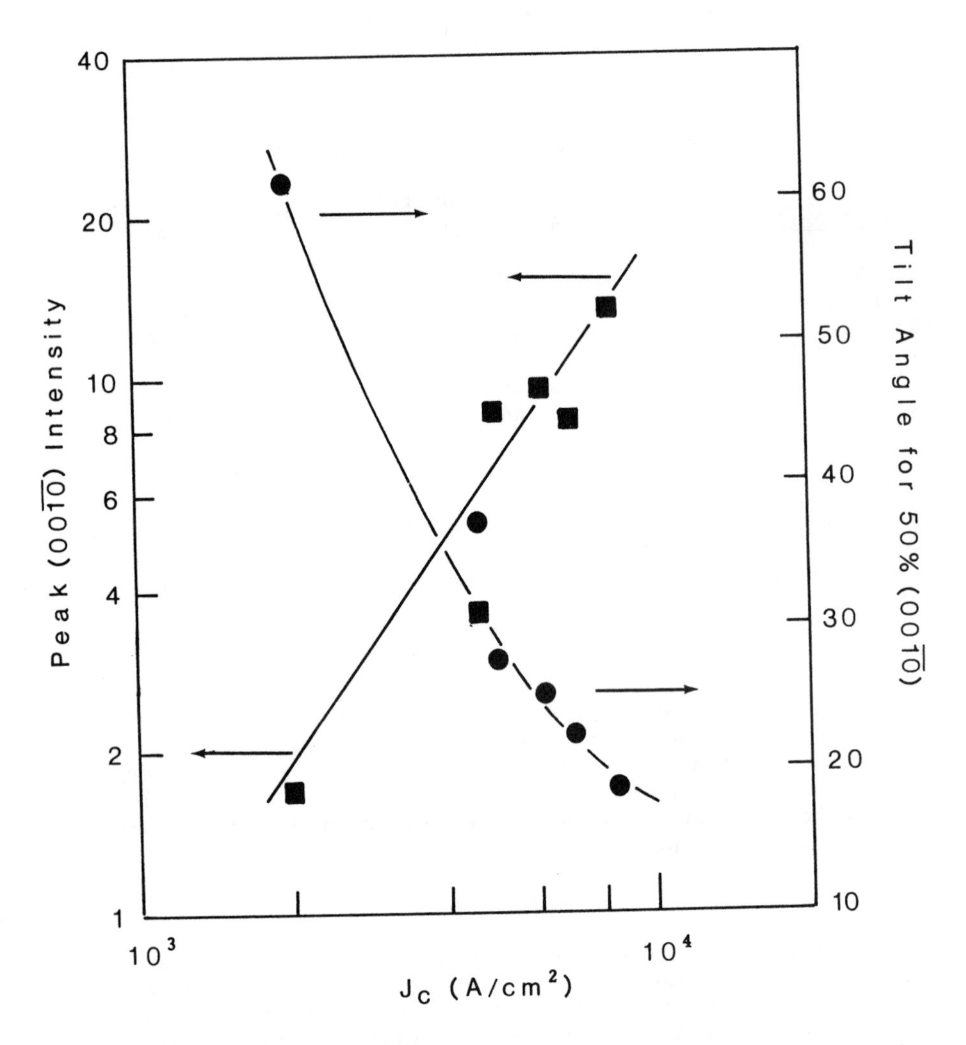

Figure 4. Correlation of J_c with two measures of texture strength. Peak (00$\overline{10}$) intensities are taken from Figure 3 at ϕ = 0°. J_c is measured at 77K in zero applied field.

reserved for drawn wires or thin films formed by vapor deposition normal to the substrate, such as reported for vapor deposited polycrystalline superconducting films[13]. The mechanism of texture development is speculative but probably involves physical alignment of powder particles during thickness reduction and preferential grain growth of the c-axis oriented grains during heat treatment. No plastic deformation of the ceramic core occurs during processing so no in-plane texture is allowed to develop. Additional pole figures from non (00ℓ) planes must be acquired to verify that a fiber texture forms.

Two parameters are used to correlate texture with J_c. The peak $(00\overline{10})$ intensity which occurs in the direction normal to the sample surface is one representation of the texture strength. A second parameter that provides information on the spread of the (00ℓ) plane distribution is the tilt angle for 50% of the $(00\overline{10})$ component. The texture profiles in Figure 3 are replotted on an equal area basis, then integrated starting at the normal direction (ϕ=0°) to give a cumulative volume fraction as a function of ϕ. The ϕ angle for 50% grains is arbitrarily chosen as a metric to compare the six samples. The critical current density is plotted versus both peak $(00\overline{10})$ intensity as a log-log plot and angle for 50% of the $(00\overline{10})$ component as a semi-log plot in Figure 4. A very clear trend emerges where critical current density increases with sharpness of the texture.

A variety of other factors could influence the current density: second phases, chemistry variation in the superconducting grains, impurities in the grain or at grain boundaries, porosity, or microcracks at boundaries. The processing schedules in this study are similar to minimize the variations in superconductor quality which gives microstructures that are quite similar[9]. Diffraction peaks from non Bi2223 phases were always small, in the same 2Θ locations, and approximately equal in magnitude for all samples. No phase identification was done, but the indication is that volume fraction is small. The other important variable in this set of samples is cation chemistry where the relative amounts of bismuth and of copper as an excess are varied by modest amounts. These chemistry variations appear to play a secondary role in current density correlations. For instance, BW34 has an excess of copper compared to BW33. The effect of texture appears to be dominant. The chemistry might play a role by contributing to the texture development during processing but additional studies are required to test this hypothesis.

CONCLUSIONS

Tapes of Bi2223 in silver were made by the powder-in-tube process. Texture was correlated with critical current density leading to the following conclusions:

1. A strong, tight (00ℓ) pole distribution enhances J_c.
2. Texture has a more important effect than chemistry over the

range of cation ratios investigated as long as chemistry variation minimizes grain boundary segregation and second phase formation.
3. Further work is required to determine the mechanisms for texture enhancement during processing.

REFERENCES

1. K. Heine, J. Tenbrink, and M. Thoner, Appl. Phys. Lett., 55, 2441 (1989).
2. K. Sato, T. Mikata, H. Mukai, M. Veyama, T. Kato, T. Masuda, M. Nagata, K. Iwata, and T. Misui, IEEE Trans. Magn., 27, 1231 (1991).
3. L.R. Motowidlo, E. Gregory, P. Haldar, J.A. Rice, and R.D. Blaugher, Appl. Phys. Lett., 59, 736 (1991).
4. K.H. Sandhage, G.N. Riley, Jr., and W.L. Carter, JOM, 43[3], 21 (1991).
5. P. Haldar, J.G. Hoehn, Jr., J.A. Rice, and L.R. Motowidlo, submitted to App. Phys. Lett.
6. S. Jin, R.B. van Dover, T.H. Tiefel, J.E. Graebner, and N.D. Spence, Appl. Phys. Lett., 58, 868 (1991).
7. F. Chen, R. Hidalgo, S.Q. Wang, X.Y. Zhang, R.S. Markiewicz, B.C. Giessen, L.R. Motowindlo, J.A. Rice, and P. Haldar, in Proc. of the Fourth Annual Conference on Superconductivity and Application, ed. by Y.H. Kao, H.S. Kwok, and P. Coppens (Plenum, NY, 1991).
8. R.D. Ray, II and E.E. Hellstrom, Appl. Phys. Lett., 57, 2948 (1990).
9. P. Haldar, J.G. Hoehn, Jr., J.A. Rice, and L.R. Motowidlo, "Thermo-Mechanical Process Influences on the Transport Properties of $(Bi,Pb)_2Sr_2Ca_2Cu_3O_{10}$ Composite Tape Conductors", this conference.
10. L.G. Schultz, J. Appl. Phys., 20, 1030 (1949).
11. H.L. Luo, S.M. Green, Y. Mei, and A.E. Manzi, in Superconductivity and Applications, ed. by H.S. Knok, Y.-H. Kao, and D.T. Shaw (Plenum, NY, 1990), p.389.
12. H.J. Bunge, in Quantitative Texture Analysis, ed. by H.J. Bunge and C. Esling (DGM, Oberusel, Germany, 1982), p.1.
13. A.D. Rollett, H.R. Wenk, F. Heidelbach, T.G. Shofield, R.E. Muenchausen, I.A. Raistrick, P.N. Arendt, D.A. Kozekwa, K. Bennett, and J.S. Kallend, to be published Proceedings of the 9th International Conference on Textures of Materials, September 1990, Avignon, France.

EVOLUTION OF CALCIUM-RICH PRECIPITATES IN MELT-QUENCHED $Bi_2Sr_2Ca_{1.2}Cu_2O_x$

J. Danusantoso and T. K. Chaki
State University of New York, Department of Mechanical Engineering and Institute on Superconductivity, Buffalo, NY 14260

ABSTRACT

The kinetics of evolution of calcium-rich precipitates (such as Ca_2CuO_3) in melt-quenched $Bi_2Sr_2Ca_{1.2}Cu_2O_x$ has been studied with the help of scanning electron microscope. As the time of anneal at $840°C$ was increased from 1 minute to 150 hours, the volume fraction and the average size of Ca-rich precipitates initially increased, reaching maximum values of 27.2% and 1.08 μm respectively at 30 minutes of annealing, and then started to decrease. However, as the annealing time was increased from 30 minutes to 150 hours, $T_{c,zero}$ increased from (35.0 ± 0.5) to (85.0 ± 0.8) K and the transport critical current density (J_{c0}) extrapolated at absolute zero temperature increased from (2.1 ± 0.4) to (107.0 ± 3.3) $A\ cm^{-2}$. An excess of 20 at. % Ca raised the transition temperature at zero resisitance ($T_{c,zero}$) and the transport critical current density (J_c) of the melt-quenched material to (85.6 ± 0.8) K and (18.9 ± 0.5) $A\ cm^{-2}$ at 77 K respectively, compared to (80.2 ± 0.6) K and (10.2 ± 0.7) $A\ cm^{-2}$ at 70 K for melt-quenched material of stoichiometric composition, with both materials annealed at $840°C$ for 100 hours.

INTRODUCTION

Recently, there has been great excitement about bismuth-base oxide superconductors [1], because wires and tapes [2] made of these superconductors can carry large critical current density (J_c). Moreover, they have extremely high critical magnetic fields [3] (H_{c2}). In the sintered materials with random grain orientations the transport J_c is low [4] due to factors such as weak linking at interfaces and no strong pinning. Recently, transport J_c in excess of 10^4 $A\ cm^{-2}$ has been reported [5-7] in melt-textured $Bi_2Sr_2CaCu_2O_x$ wires and tapes at 4.2 K in magnetic fields greater than 20 T. Kase et al. [8] prepared 2212 textured tape by doctor-blade casting followed by partial melting. Transport J_c in this tape at 4.2 K was measured to be as high as 1.4 x 10^5 $A\ cm^{-2}$ in a magnetic field of 25 T. The field was applied perpendicular to the current and parallel to the tape surface. Such high critical current density at such high magnetic fields has brought Bi-base oxide superconductors (2212 phase in particular) to a level of performance comparable to that of metallic superconductors, such as [9] $(Nb,Ti)_3Sn$.

Melt-quenching [10-16] is another possible route for synthesizing Bi-Sr-Ca-Cu-O compounds with a possibility of getting more homogeneous composition than the sintered materials. Formation of glasses of Bi-Sr-Ca-Cu-O is possible by pouring the melt into liquid nitrogen [11] or by twin roller casting [13,14]. Upon annealing, superconducting phases appear in the glass. By melt-processing of $Bi_2Sr_2CaCu_2O_x$, nearly single phase of 2212 can be obtained [17]. In contrast, melt-processing of $Bi_2Sr_2Ca_2Cu_3O_x$ always produces a mixture of 2212 and 2223 phases [18].

Flux lattice depinning [19] and melting [20] are serious obstacles to using Bi-Sr-Ca-Cu-O superconductors for current-carrying applications at temperatures higher than 30 K. Recently, Shi et al. [21] have shown by magnetization measurement that Ca-rich precipitates (such as Ca_2CuO_3) in melt-quenched Bi-Sr-Ca-Cu-O can provide sufficient flux pinning and increase the intragrain critical current density from order of $10^5 A\ cm^{-2}$ to $10^7 A\ cm^{-2}$ at 5 T and at 10 K. Here we study the kinetics of evolution of Ca-rich precipitates in melt-quenched $Bi_2Sr_2Ca_{1.2}Cu_2O_x$ compound having 20 at.% excess Ca over the stoichiometric compound. We also investigate how the Ca-rich precipitates affect T_c and transport J_c in the melt-quenched material.

EXPERIMENTAL PROCEDURE

First, pellets of $Bi_2Sr_2Ca_{1.2}Cu_2O_x$ superconductors with 20 at.% excess Ca were prepared by solid state sintering method. The approximate dimensions of the pellets were 19 mm x 6 mm x 3 mm. For comparisons, pellets with stoichiometric composition (Bi:Sr:Ca:Cu = 2:2:1:2) were also prepared. A few pellets were put in an alumina crucible and then melted by placing the crucible for only 3 minutes in a muffle furnace preheated at $900°C$. The molten material was rapidly solidified by pouring onto a stainless steel plate of dimension 10 cm x 7.5 cm x 1.2 cm. The temperature of the plate was held at $20°C$, by keeping it in contact with water in a bath. It is to be noted that in the present study, sintered superconducting pellets were melted, compared to several other studies [11-14,17,18] in which non-superconducting mixtures of powders of oxides and carbonates were melted. Since the pellets were already well-mixed, the time of melting was kept to only 3 minutes and this minimized the loss of volatile components from the melt and reduced contamination from the alumina crucible. The melt-quenched samples were annealed in air at $840°C$ for periods of time ranging from 1 minute to 150 hours. After annealing, all the specimens were air-quenched.

Various characterizations were performed on melt-quenched samples upon annealing. X-ray powder diffraction, using Cu-Kα radiation, was performed to characterize the phases in the melt-quenched materials. To observe microstructure, some of the samples were ground in SiC abrasive paper and polished to 0.05 μm with a slurry of alumina and water. The polished specimens were examined by a scanning electron microscope (SEM). The compositions of various phases were determined by energy dispersive x-ray (EDX) analysis. To reveal Ca-rich precipitates clearly, a few specimens were etched with 0.1 M perchloric acid for about 30 seconds.

The volume fraction of Ca-rich precipitates after various heat treatments were determined from SEM micrographs by systematic manual count method [22]. A square array of grid points was placed over the micrograph. To calculate the volume fraction, the number of grid points falling over the Ca-rich precipitates was counted and divided by the total number of grid points. 81 grid points over an area of 4 cm x 4 cm of the micrograph were used. For each specimen, counting was made on three SEM micrographs representing different areas. The size of the Ca-rich precipitates was measured also from SEM micrographs with the help of a millimeter scale.

The superconducting properties, such as T_c and transport J_c, were measured on bridge-type spceimens by four probe technique at zero magnetic field. The melt-quenched samples of approximate dimensions of 15 mm x 3 mm x 3 mm were ground by a file to make bridge-type specimens. The cross-section of the neck was approximately 1.5 mm x 1.5 mm and the length was 2 mm. Alternating current with the peak value of 30 mA was used in T_c measurement. The onset of superconductivity took place at the temperature at which the tangent of the resistance *vs.* temperature curve at the normal state during cooling intersected the tangent of the part where the resistance dropped precipitiously. J_c was measured with direct current going up to 1 A and with a voltage criterion of 1 μV/cm. Each value of T_c and J_c was averaged over 3 specimens.

RESULTS

A. X-ray Diffraction: Figs. 1a, b, c, d, e and f show powder diffraction patterns at room temperature of melt-processed $Bi_2Sr_2Ca_{1.2}Cu_2O_x$ in as-quenched condition and after annealing at 840°C for 0.5, 1, 4, 25, and 100 hours respectively. The patterns were compared with works of other investigators [16,23–25] as well as with the diffraction patterns of materials of known composition. The as-quenched samples were not fully amorphous, but contained a small amount of microcrystalline phases (Fig. 1a). This is due to the fact that solidification was not fast enough [11,13,14] in the present study. In the diffraction pattern of the as-quenched sample, there is a broad peak at angle 2θ of about 30° due to presence of microcrystalline Ca_2CuO_3 phase. The peaks due to microcrystalline $Bi_2Sr_2CuO_x$ (2201) phase were also observed. Upon annealing at

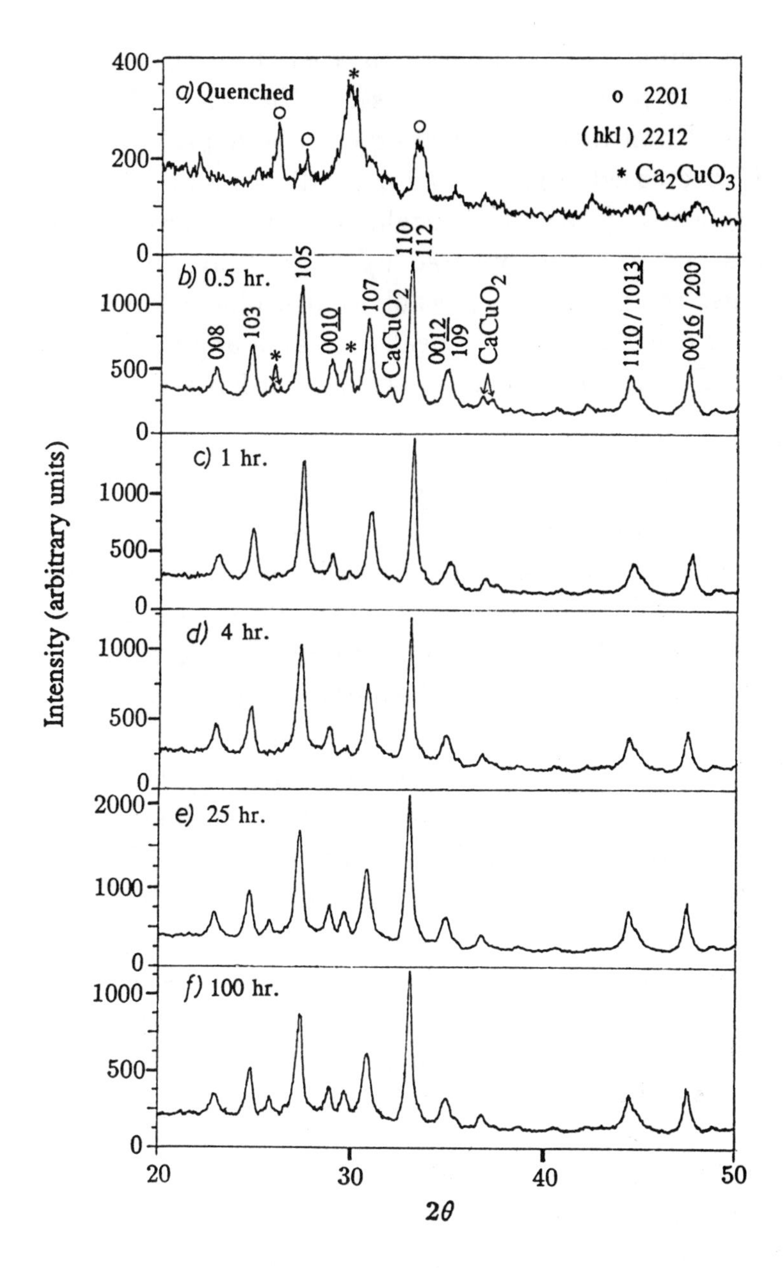

FIG. 1. X-ray powder diffraction patterns of melt-quenched $Bi_2Sr_2Ca_{1.2}Cu_2O_x$ samples upon annealing at $840°C$ for various periods of time. (a) As-quenched. (b) 0.5 hour. (c) 1 hour. (d) 4 hours. (e) 25 hours. (f) 100 hours.

840° for 30 minutes, 2212 phase appeared (Fig. 1b) and the peaks due to Ca-rich precipitates, such as Ca_2CuO_3 and $CaCuO_2$, were sharpened. Upon further annealing Ca_2CuO_3 peaks at $2\theta = 25.8$ and 29.6° decreased (Fig. 1c and d), indicating dissolution of Ca-rich precipitates. Strangely enough, the x-ray diffraction peaks at $2\theta = 25.8$ and 29.6° reappeared upon annealing for 25 and 100 hours (Figs. 1e and f). It is possible that prolonged annealing produced some other phase with peaks at same angles. For example, $Bi_4Sr_3Cu_3O_x$ has diffraction peaks [25] for Cu-Kα x-ray at $2\theta = 25.7$ and 29.7°.

B. SEM Results: Fig. 2a shows SEM micrograph of an unetched sample of melt-quenched 2-2-1.2-2 material without subsequent heat treatment. The needle-like phase is crystals of $Bi_2Sr_2CuO_x$ (2201), the identification of which was confirmed by compositional analysis of EDX. The featureless white matrix is amorphous phase with the composition of 2201. The darker spots are $(Sr, Ca)_2CuO_3$ precipitates. Upon annealing the melt-quenched material at 840°C, the plate-like structures of superconducting 2212 phase grew. Fig. 2b shows such a microstructure of an etched specimen of the melt-quenched material annealed for 30 minutes at 840°C. Ca_2CuO_3 preciitates (appearing brighter in secondary electron) were interspersed among superconducting plates.

During annealing at 840°C, the precipitate size initially increased, but, after 30 minutes, started to decrease, reaching a constant value after 100 hours of annealing (Fig. 3). The average precipitate size reached a maximum of (1.08 ± 0.12) μm at 30 minutes of annealing and decreased to $(0.54 \pm 0{:}06)$ μm at 150 hours of annealing. The volume fraction of the precipitates showed a similar trend (Fig. 4), reaching the maximum value of $(27.2 \pm 2.8)\%$ at 30 minutes and decreasing to $(4.8 \pm 0.5)\%$ at 150 hours annealing at 840°C.

C. T_c and J_c: The sintered pellets with 20 at.% excess Ca had $T_{c,zero}$ of (61.0 ± 0.7) K, $T_{c,onset}$ of (87.6 ± 0.4) K and transport

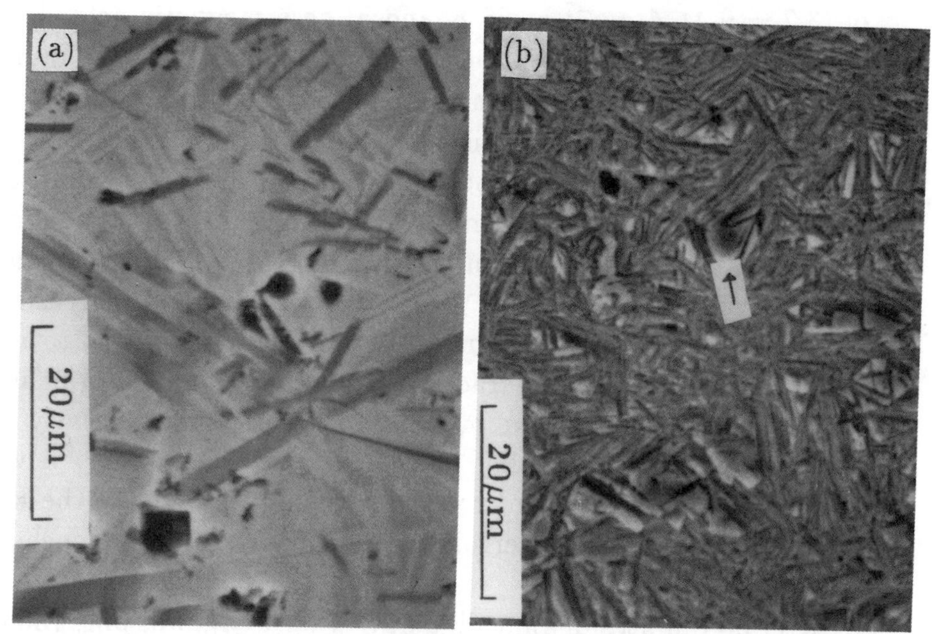

FIG. 2. Scanning electron micrographs of melt-quenched Bi_2Sr_2-$Ca_{1.2}Cu_2O_x$. (a) As-quenched (b) Annealed at 840°C for 30 min.

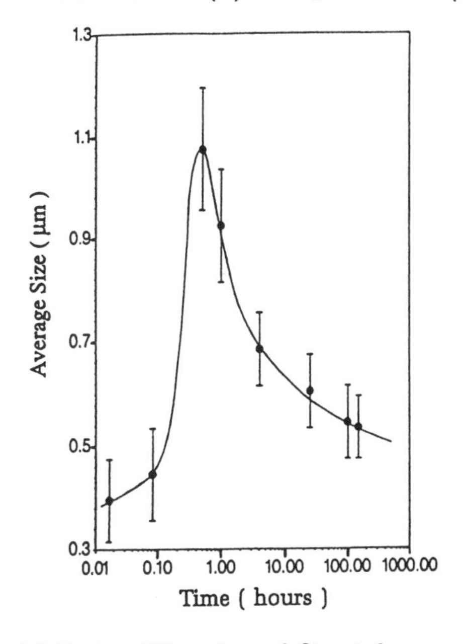

FIG. 3. The size of Ca-rich precipitates against time.

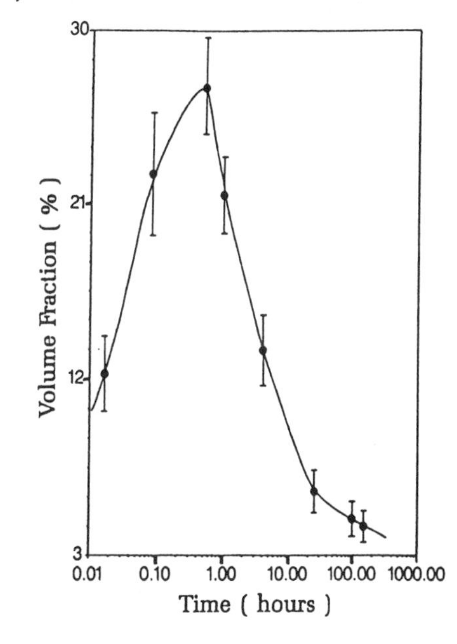

FIG. 4. The volume fraction of Ca-rich precipitates against time.

J_c of (3.7 ± 0.4) $A\ cm^{-2}$ at 55 K and at zero magnetic field. The melt-quenched $Bi_2Sr_2Ca_{1.2}Cu_2O_x$ was non-superconducting in as-quenched condition, but became superconducting upon annealing at 840°C. Fig. 5 shows resistivity versus temperature curves for melt-quenched $Bi_2Sr_2Ca_{1.2}Cu_2O_x$ samples after annealing at 840°C for various periods of time. With the increasing annealing time, not only the transition temperature increased, but the normal state resistivity also decreased, indicating formation of increasing amounts of superconducting phase. However, it is difficult to relate quantitavely the volume fraction of the superconducting phase with the resistivity due to the problem of electrical connectivity of regions of the superconducting phase. In an earlier paper [26], we reported a colored optical microscopic technique to study the kinetics of growth of the superconducting phase in melt-quenched $Bi_2Sr_2CaCu_2O_x$.

After 30 minutes of annealing at 840°C, $T_{c,zero}$ was only (35.0 ± 0.5) K and $T_{c,onset}$ was (81.1 ± 0.7) K. Upon 100 hours of annealing at 840°C, $T_{c,zero}$ of Ca-rich material reached (85.6 ± 0.8) K and $T_{c,onset}$ was (10.0 ± 0.5) K. Fig. 6 shows a plot of transition

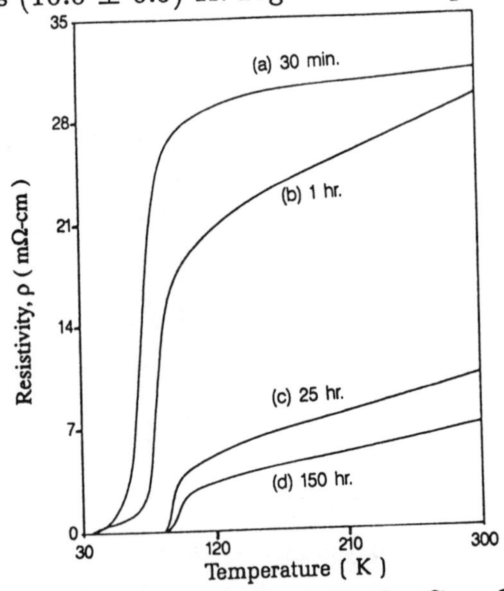

FIG. 5. Resistivity of melt-quenched $Bi_2Sr_2Ca_{1.2}Cu_2O_x$ against temperature for various durations of annealing at 840°C.

temperatures ($T_{c,zero}$ and $T_{c,onset}$) as a function of time of annealing at 840°C. For comparison, it is to be noted that, upon annealing melt-quenched material of stoichiometric composition ($Bi_2Sr_2Ca-Cu_2O_x$) at 840°C for 100 hours, $T_{c,zero}$ was (80.2 ± 0.6) K and $T_{c,onset}$ was (100.0 ± 0.5) K. Thus, by suitable annealing the melt-quenched material with 20 at.% excess Ca can have the transition temperature at zero resistance higher than that of melt-quenched material of stoichiometric composition.

TABLE. Superconducting Properties of Melt-Quenched $Bi_2Sr_2Ca_{1.2}Cu_2O_x$ upon Annealing at 840°C.

Time	$T_{c,zero}$(K)	$T_{c,onset}$(K)	T (K), measurement	J_c(A/cm²)	J_{c0}(A/cm²)
0.5 hr.	35.0±0.5	81.1±0.7	30.0	0.6±0.1	2.1±0.4
1 hr.	36.7±0.5	87.0±0.6	30.0	1.7±0.2	5.0±0.6
4 hr.	77.6±0.6	96.5±0.6	65.0	6.2±0.4	20.7±1.3
25 hr.	82.5±0.6	96.5±0.5	77.0	10.2±0.4	79.4±3.1
100 hr.	85.6±0.8	100.0±0.5	77.0	18.9±0.5	99.0±2.6
150 hr.	85.0±0.8	100.0±0.5	77.0	19.2±0.6	107.0±3.3

The transport J_c of melt-quenched $Bi_2Sr_2Ca_{1.2}Cu_2O_x$ also increased with the annealing time. After 30 minutes of annealing at 840°C, J_c at 30 K and at zero magnetic field was only (0.6 ± 0.1) $A\ cm^{-2}$. But after 150 hours of annealing J_c reached a value of (19.2 ± 0.6) $A\ cm^{-2}$ at 77 K. In order to compare the critical current densities for different samples, an extrapolated critical current density (J_{c0}) at absolute zero temperature was calculated using the following temperature dependence of J_c, in analogy with the magnetic field dependence [27] of J_c:

$$J_c = J_{c0}\ [1 - (T/T_c)^2], \qquad \cdots(1)$$

where T is the temperature at which J_c is measured. The values of J_c and J_{c0} of melt-quenched $Bi_2Sr_2Ca_{1.2}Cu_2O_x$ annealed at 840°C for various durations of time were shown in the Table. J_{c0} of melt-quenched materials increased with the time of annealing, reaching almost a steady value after 150 hours of annealing (Fig. 7). For comparison, it is to be noted that J_c of the melt-quenched material of stoichiometric composition ($Bi_2Sr_2CaCu_2O_x$) annealed at $840°C$ for 100 hours was $(10.2 \pm 0.7)\ A\ cm^{-2}$ at 70 K and extrapolated J_{c0} was $(45.0 \pm 2.9)\ A\ cm^{-2}$. Thus, considering experimental uncertainties, an excess of 20 at.% Ca improved transport J_c in the melt-quenched material only marginally.

CONCLUSION

1. With the increasing time of annealing at $840°C$, the volume fraction and the average size of Ca-rich precipitates in melt-quenched material ($Bi_2Sr_2Ca_{1.2}Cu_2O_x$) with 20 at.% excess Ca increased till 30 minutes and then started to decrease, but T_c and transport J_c increased, eventually reaching steady values.
2. $T_{c,zero}$ and transport J_c of the melt-quenched material with 20 at.% excess Ca were higher than those of the melt-quenched material with stoichiometric composition ($Bi_2Sr_2CaCu_2O_x$), with both materials given identical heat treatment.

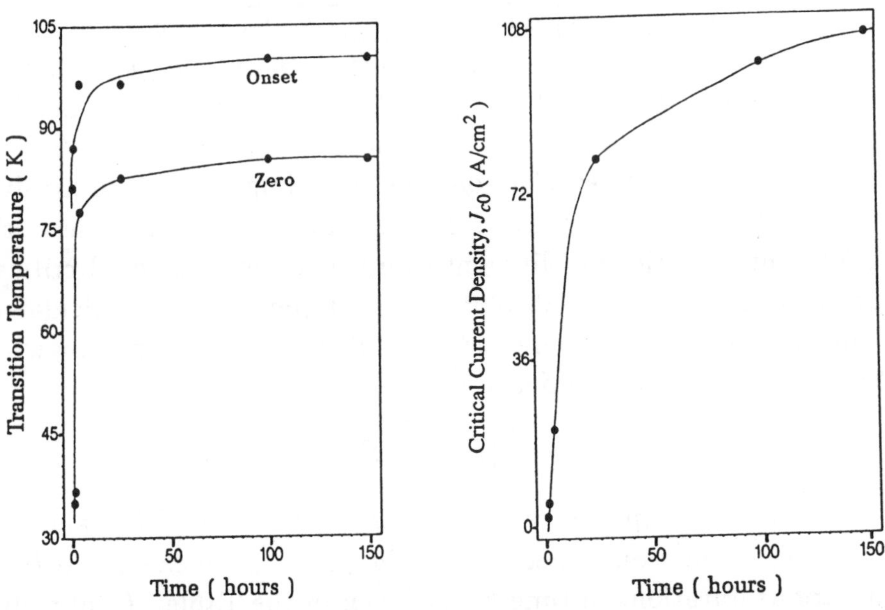

FIG. 6. Transition temperatures against annealing time.

FIG. 7. J_{c0} against annealing time.

REFERENCES

1. H. Maeda, Y. Tanaka, M. Fukotomi, and T Asano, Japan J. Appl. Phys. **27**, L209 (1988).

2. L. R. Motowildo, E. Gregory, P. Haldar, J. A. Rice, and R. D. Blaugher, Appl. Phys. Lett. **59**, 736 (1991).

3. Y. Koike, T. Nakanomyo, and T. Fukase, Japan J. Appl. Phys. **27**, L841 (1988).

4. H. Kupfer, S. M. Green, C. Jiang, YuMei, H. L. Luo, R. Meier-Hirmer, and C. Politis, Z. Phys. **71**, 63 (1988).

5. K. Hein, J. Tenbrink, and M. Thőner, Appl. Phys. Lett. **55**, 2441 (1989).

6. N. Enomoto, H. Kikuchi, N. Uno, H. Kumakura, K. Togano, and K. Watanabe, Japan J. Appl. Phys. **29**, L447 (1990).

7. J. Kase, N. Irisawa, T. Morimoto, K. Togano, D. R. Dietderich, and H. Maeda, Appl. Phys. Lett. **56**, 970 (1990).

8. J. Kase, K. Togano, H. Kumakura, D. R. Dietderich, N. Irisawa, T. Morimoto, and H. Maeda, Japan J. Appl. Phys. **29**, L1096 (1990).

9. K. Tachikawa, T. Asano, and T. Takeuchi, Appl. Phys. Lett. **39**, 766 (1981).

10. D. G. Hinks, L. Soderlom, D. W. Capone II, B. Dabrowski, A. W. Mitchell, and D. Shi, Appl. Phys. Lett. **53**, 423 (1988).

11. T. Komatsu, K. Imai, R. Sato, K. Matusita, and T. Yamashita, Japan J. Appl. Phys. **27**, L533 (1988).

12. F. H. Garzon, J. G. Beery, and I. D. Raistrich, Appl. Phys. Lett. **53**, 805 (1988).

13. M. Tatsumisago, C. A. Angell, S. Tsuboi, Y. Akamatsu, N. Toghe, and T. Minami, Appl. Phys. Lett. **54**, 2268 (1989).

14. M. Tatsumisago, C. A. Angell, Y. Akamatsu, S. Tsuboi, N. Toghe, and T. Minami, Appl. Phys. Lett. **55**, 600 (1989).

15. H. Takei, M. Koike, H. Takeya, K. Suzuki, and M. Ichihara, Japan J. Appl. Phys. **28**, L1193 (1989).

16. M. R. De Guire, N. P. Bansal, and C. J. Kim, J. Am. Ceram. Soc. **73**, 1165 (1990).

17. J. Bock and E. Preisler, Solid State Commun. **72**, 453 (1989).

18. T. Kanai, T. Kamo, and S.-P. Matsuda, Japan J. Appl. Phys. **28**, L2188 (1989).

19. J. Kober, A. Gupta, P. Esquinazi, and H. F. Braun, Phys. Rev. Lett. **66**, 2507 (1991).

20. D. R. Nelson, Phys. Rev. Lett. **60**, 1973 (1988).

21. D. Shi, M. S. Boley, U. Welp, J. G. Chen, and Y. Liao, Phys. Rev. B **40**, 5255 (1989).

22. American Society for Testing and Materials, Annual Book of ASTM Standards 01.02.E562-83 (philadelphia, Pennsylvania, 1984).

23. J. M. Tarascon, W. R. McKinnon, P. Barboux, D. M. Hwang, B. G. Bagley, L. H. Greene, G. W. Hull, Y. LePage, N. Stoffel, and M. Giroud, Phys. Rev. B **38**, 8885 (1988).

24. H. Yamanaka, M. Matsuda, M. Takata, M. Ishii, T. Yamashita, and H. Koinuma, Japan J. Appl. Phys. **28**, L2185 (1989).

25. R. Horyń, I. Filatow, J. Ziaja, M. Wolcyrz, Supercond. Sci. Technol. **3**, 347 (1990).

26. J. Danusantoso and T. K. Chaki, Supercond. Sci. Technol., in press (1991).

27. P. G. De Gennes, *Superconductivity of Metals and Alloys*, (W. A. Benjamin, Inc., New York, 1966)

INCREASING THE CRITICAL CURRENT DENSITY IN THALLIUM-CONTAINING SUPERCONDUCTORS BY TIN OXIDE ADDITION AND COMPRESSIVE DEFORMATION

Jui H. Wang, Min Qi and Zhifeng Ren

Department of Chemistry and New York State Institute on Superconductivity, State University of New York, Buffalo, NY 14214

ABSTRACT

Ceramic samples of $Tl_{1.6}Sn_{0.2}Ba_2Ca_2Cu_3O_{9.8}$ fabricated by heating at 905-907° C and subsequent rapid chilling in liquid nitrogen were found to consist mainly of the 2223-phase and no 2212-phase. Although the transport critical current density (J_c) of these unprocessed samples was only 2 x 10^2A/cm^2, it was raised to 1.1 x 10^4A/cm^2 at 77 K and zero field by compressive deformation of the sample between silver foils. The forced plastic flow of the material was also found to increase the magnetic anisotropy of the sample by a factor of 6 without significant change in its density. Analysis of the data shows that the observed increase in J_c is much larger than that anticipated from simple granular reorientation relative to the macroscopic sample, and is probably due mainly to improved intergranular contact.

INTRODUCTION

The preparation of Tl-Ba-Ca-Cu-O type of superconductors by heating an intimate 2:2:2:3 mixture of the component oxides often led to a product containing both $Tl_2Ba_2Ca_2Cu_3O_{10}$ (2223) and $Tl_2Ba_2CaCu_2O_8$ (2212). Some of the 2223- compound formed at 900-910°C underwent phase transition and decomposed to $Tl_2Ba_2CaCu_2O_8$, CaO and CuO at lower temperature when the sample was slowly cooled down to room temperature. Recently, we found that this undesirable transition could be blocked with the addition of a small amount of tin oxide and by promptly chilling the hot sample in liquid nitrogen [1]. The addition of SnO_2 or $BaSnO_3$ to Y-Ba-Cu-O and Bi-Sr-Ca-Cu-O type of ceramic superconductors has been reported to improve their properties [2-7]. Recently we also observed that the critical current density (J_c) in thallium-containing ceramic superconductors can be greatly enhanced by compressive reorientation [8]. Similar improvements of J_c by compaction [9], by vibrational alignment [10] and by the "powder in tube" method [11-15] in other powdered material have also been reported. The combined effect of SnO_2 addition and compressive deformation of thallium-containing ceramic superconductors is examined and discussed in this report.

FABRICATION AND MEASUREMENT

Prepowder of composition $Ba_2Ca_2Cu_3O_7$ was prepared by heating an intimate mixture of BaO_2, CaO and CuO (molar ratio 2:2:3) at 860-880°C for 24 h twice, with regrinding after each heating period. The resulting prepowder was mixed with equivalent amount of Tl_2O_3 plus a small amount of SnO_2. The mixture was compressed into a pellet, weighed and heated in a capped gold box in a stream of O_2 at 905-907°C for 30-40 min. Then the sample was promptly chilled in liquid nitrogen, dried in O_2 and weighed again. The final thallium content of the pellet was calculated from the fractional weight loss and the initial composition. Without further processing the pellet exhibited a J_c of only $\sim 2 \times 10^2$ A/Cm^2 at 77 K.

About 30-35 mg of the ground pellet were compressed between a folded piece of silver foil at $\sim$ 40 kbar for 5 min, and subsequently trimmed to remove the uneven edges. This process was repeated three times. After the fourth compression, the specimen was weighed and had its dimensions taken. It was then annealed in flowing O_2 at 800°C for 10 h. The cooled specimen was recompressed once more and again annealed in O_2 at 800°C for 6 h. The final silver-clad specimen, with dimensions of 7-10 mm length, 0.3-1.0 mm width and 0.08-0.10 mm total thickness or 0.058-0.078 net thickness, can exhibit a J_c above 1.1×10^4 A/cm^2 at 77 K in the absence of external field.

RESULTS AND DISCUSSION

The powder XRD patterns of Tl-Ba-Ca-Cu-O type of superconducting pellets prepared by different procedures are shown in Fig.1. The pellet of Fig. 1(a) with starting composition $Tl_2Ba_2Ca_2Cu_3O_{10}$ was obtained after slow cooling in the furnace from 905°C to room temperature in 2 h. It contained even more 2212-phase than 2223-phase. The pellet of Fig. 1(b) has approximately equal amount of 2212- and 2223-phases. This pellet had the same bulk composition, but was obtained after rapid quenching of the heated sample in liquid N_2. Fig. 1(c) is the powder XRD pattern of a pellet with starting composition $Tl_2Sn_{0.2}Ba_2Ca_2Cu_3O_{10.4}$ and the hot pellet was also rapidly chilled in liquid N_2. This sample consisted mainly of the 2223-phase, since its reflection peaks due to 2212-phase practically disappeared.

SEM photograph of the pellet of Fig. 1(c), with final composition $Tl_{1.62}Sn_{0.2}Ba_2Ca_2Cu_3O_{9.8}$, shows the presence of three phases: (A) the main phase which is the superconducting 2223-phase according to XRD; (B) a minor phase which is non-superconducting and contains excess Ba according to EDAX analysis; (C) another minor phase which is non-superconducting and contains excess Ca according to EDAX analysis.

The whole area EDAX spectrum of this pellet contains a small peak due to Sn. However, none of the EDAX spectra of phases (A), (B) and (C) in the same pellet has this small Sn peak. Therefore we have to conclude that the added SnO_2 stayed outside of all these observed phases.

The observed transport critical current density (J_c) of silver-clad ceramic superconductor $Tl_{1.64}Sn_xBa_2Ca_2Cu_3O_y$ varies with Sn content as shown in Fig. 2. The addition of increasing amounts of SnO_2 to the prepowder $Ba_2Ca_2Cu_3O_7$ initially increases J_c of the subsequently fabricated sample to about 50% higher than the value without Sn, then decreases it to much lower values. The magnetic J_c measured with SQUID exhibited a similar dependence on Sn content (data not shown),which rules out the possibility that the observed enhancement was due to improved contact between the superconducting material and silver foil.

The enhancement of J_c by compressive deformation of $Ag_8Tl_2Ba_2Ca_2Cu_3O_{10}$ samples was reported previously. After the superconducting specimen was compressed in the Z-direction between two flat steel surfaces while allowing it to expand in the X-and Y-directions and subsequently annealed, its J_c was increased 5- to 11- fold without significant change in sample density [8]. The forced plastic flow of superconducting material was also found to increase the magnetic anisotropy of the specimen by a factor of 6.

Further support of the conclusion that the observed enhancement of J_c is not due to static pressure, but due to plastic flow of the mateial under pressure is given by Table I. The four samples of different net thickness in Table I were recompressed at the same pressure (~4 kbar) for different lengths of time so that they ended up with approximately the same final thickness. But as the net thickness reduction ratio changed from 1 to 0.59, the J_c increased from 6160 to 9030 A/cm^2.

CONCLUSION

The addition of 0.1 - 0.2 SnO_2 per $Tl_2Ba_2Ca_2Cu_3O_{10}$ stabilized the 2223-phase without entering it, and increases the transport J_c by about 50%.

Compressive deformation of silver-clad $Tl_{1.64}Sn_{0.1}Ba_2Ca_2Cu_3O_{9.7}$ raises its transport J_c (77 K) about 50-fold to 1.1 x 10^4 A/cm^2 due mainly to improved intergranular contact.

ACKNOWLEDGEMENT

We thank Mr. Peter J. Bush for taking the SEM and EDAX measurements, and Mr. Gary R. Sagerman for his technical assistance.

REFERENCES

1. Z. Ren, M. Qi and J.H. Wang, submitted to Physica C.

2. K. Osamura, N. Matsukura, Y. Kusumoto, S. Ochiai, B. Ni and T. Matsushita, Jpn. J. Appl. Phys. 29, L1621 (1990).

3. J. Shimoyama, J. Kase, S. Kondoh, E. Yanagisawa, T. Matsubara, M. Suzuki and T. Morimoto, Jpn. J. Appl. Phys. 29, L1999 (1990).

4. T. Lu, L. Shen, X. Shao, X. Jin and N. Yang, Mat. Res. Bull. 25, 315 (1990).

5. P. McGinn, W. Chen, N. Zhu, L. Tan, C. Varanasi and S. Sengupta, Appl. Phys. Lett. 59, 120 (1991).

6. A.S. Nash, K.C. Goretta and R.B. Poeppel, Advances in Powder Metallurgy, Vol. 2, 517 (1989).

7. H.M. Seyoum, J.M. Habib, L.H. Bennett, W. Wong-Ng, A.J. Shapiro and L.J. Swartzendruber, Supercond. Sci. Technol. 3, 616 (1990).

8. M. Qi and J.H. Wang, Physica C 176, 38 (1991).

9. C.L. Seaman, S.T. Weir, E.A. Early, M.B. Maple, W.J. Nellis, P.C. McCandless and W.F. Brocious, Appl. Phys. Lett. 57, 93 (1990).

10. X.G. Zheng, H. Kurlyaki and K. Hirakawa, Supercond. Sci. Technol. 3, 339 (1990).

11. K. Heine, J. Tenbrink and M. Thöner, Appl. Phys. Lett. 55, 2441 (1989).

12. T. Asano, Y. Tanaka, M. Fukutomi and H. Maeda, Jpn. J. Appl. Phys. 29, L1066 (1990).

13. K. Sata, T. Hikada and Y. Iwasa, Appl. Phys. Lett. 57, 1928 (1990).

14. N. Shimizu, K. Michishata, Y. Higashida, H. Yokoyama, Y. Hayami, Y. Kubo, E. Inukai, A. Saju, N. Kuroda and H. Yoshida, Jpn. J. Appl. Phys. 28, L1955 (1989).

15. Y. Matsumoto, J. Hombo, Y. Yamaguchi and T. Mitsunaga, Matter. Res. Bull. 24, 1469 (1989).

Table I. Effect of Recompression of the Net Thickness (t) and Critical Current Density (J_c) in $Tl_{1.64}Sn_{0.1}Ba_2Ca_2Cu_3O_{9.7}$[a]

Specimen	A	B	C	D
Net t before recompression (mm)	0.069	0.082	0.097	0.112
Net t after recompression (mm)	0.069	0.067	0.070	0.067
Thickness reduction factor	1	0.81	0.72	0.59
J_c at 77 K (A/cm^2)	6150	7790	8910	9030

[a] All the compressed and annealed specimens were recompressed at 40 kbar but for different lengths of time so that after recompression they were similar in thickness. The specimens were reannealed in O_2 at 800°C for 6 h before the J_c measurements by the four-point method.

(Captions for Figures)

Fig. 1. Powder x-ray diffraction patterns of superconducting pellets prepared by different procedures. (a) Starting pellet composition $Tl_2Ba_2Ca_2Cu_3O_{10}$ cooled in a furnace under O_2 from 905°C to room temperature in 2 h; (b) Prepared in the same way as (a) but the pellet at 905°C was promptly removed from the tube-furnace and immersed in liquid N_2; (c) The starting pellet composition was $Tl_2Sn_{0.2}Ba_2Ca_2Cu_3O_{10.4}$, and the pellet at 905°C was promptly removed and immersed in liquid N_2. O, 2223-phase; Δ, 2212-phase. The indices of different reflections are indicated by the numbers in parenthesis in (a).[1]

Fig. 2. Dependence of transport critical current density (J_c) on Sn content of ceramic superconductors of bulk composition $Tl_{1.64}Sn_xBa_2Ca_2Cu_3O_y$ at 77 K and zero external magnetic field. Each point represents a separate sample[1]. Each J_c value was determined by the four-point method, with the specimen and silver lead wires all immersed in liquid nitrogen.

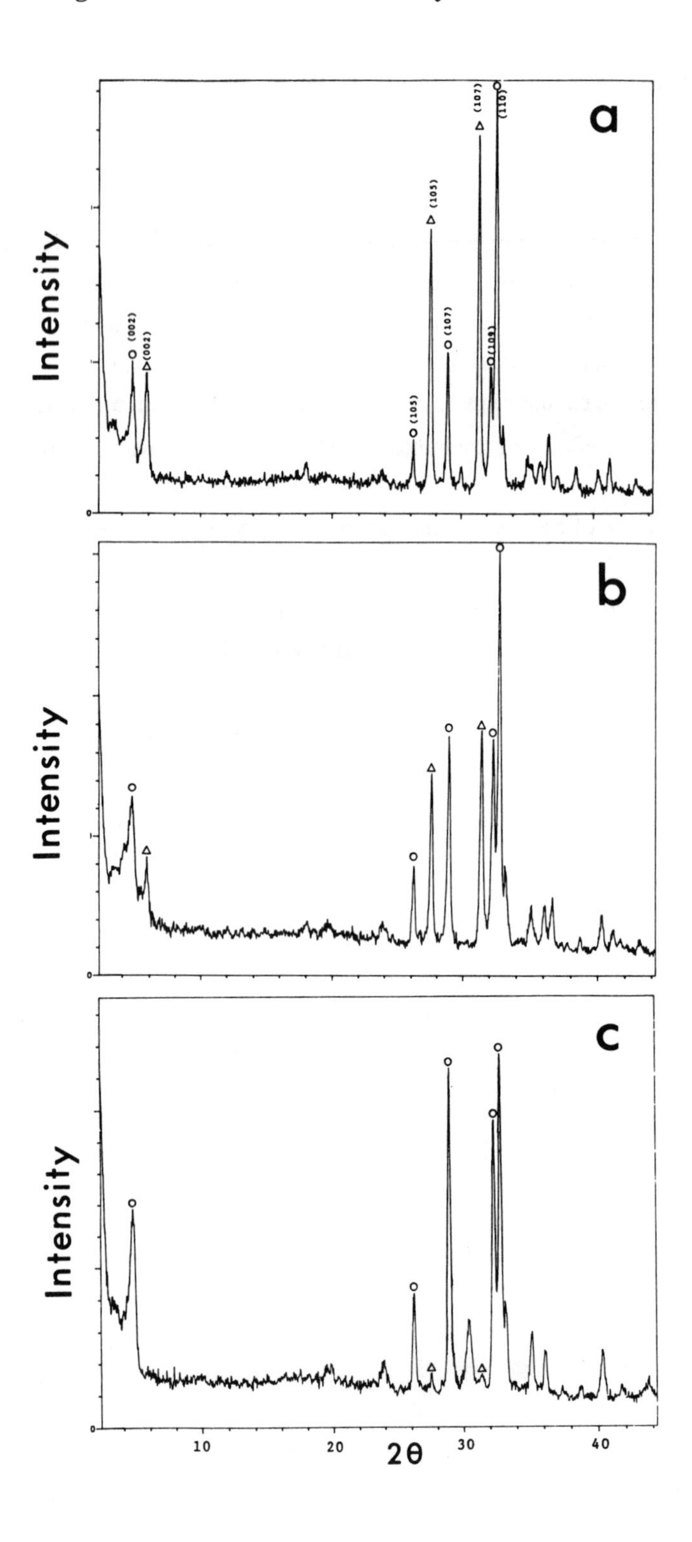
a
b
c
Intensity
Intensity
Intensity
(002)
(002)
(105)
(105)
(107)
(107)
(108)
(110)
10
20
2θ
30
40

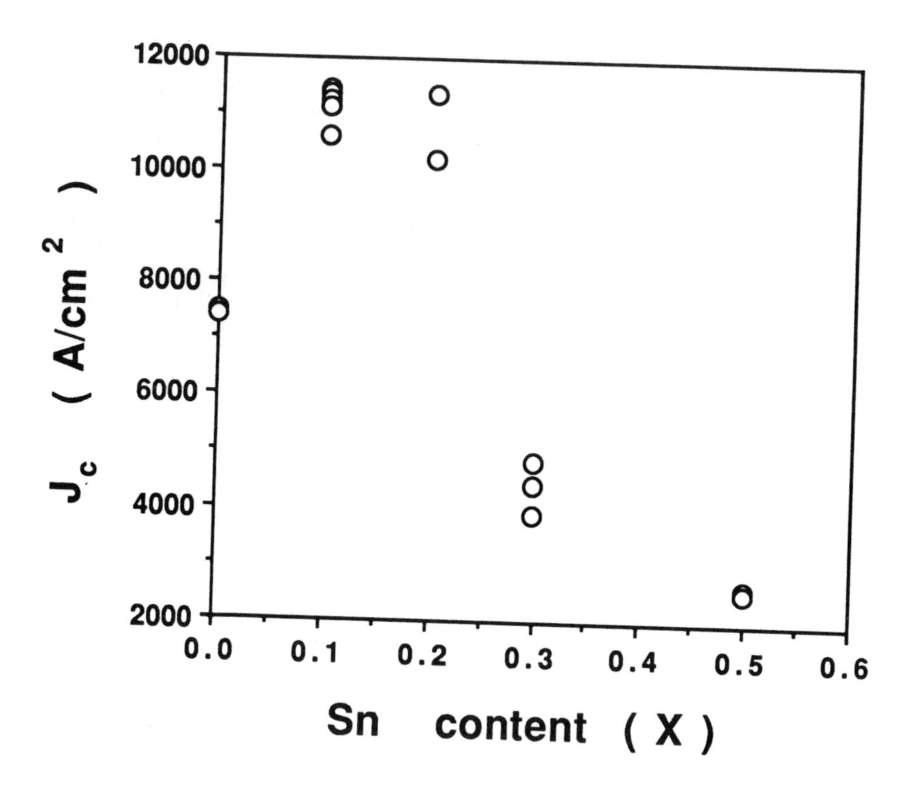

12000
10000
8000
6000
4000
2000
J c (A/cm 2)
0.0
0.1
0.2
0.3
0.4
0.5
0.6
Sn content (X)

STRUCTURAL CHANGES UPON ELECTROCHEMICAL INSERTION OF LITHIUM INTO THE $Bi_2Sr_2CaCuO_{8+y}$ HIGH Tc SUPERCONDUCTOR

Niles A. Fleischer and Joost Manassen, Department of Materials Research
The Weizmann Institute of Science, P. O. Box 26, Rehovot 76100, ISRAEL

Philip Coppens, Peter Lee, and Yan Gao
Chemistry Department, State University of New York at Buffalo
Buffalo, N.Y. 14214

ABSTRACT

The insertion compound $Li_xBi_2Sr_2CaCu_2O_{8+y}$ ($x \leq 2$) was prepared by electrochemical insertion of lithium into $Bi_2Sr_2CaCu_2O_{8+y}$ pellets at room temperature in galvanic cells. The reaction proceeds with retention of the host crystal structure and without the appearance of any new phases. Bulk superconductivity persists, but the superconducting fraction appears to be lower in the reacted samples. The lithiated product was analyzed by crystallographic and chemical measurements. The results show that insertion of Li ions is accompanied by an expansion of the interlayer spacing between Bi-O planes and a small increase in the **a**- and **b**-cell parameters. This change is consistent with some, but not all, of the donated electrons entering the Cu-O planes, with a corresponding reduction in hole concentration. At higher levels of Li content, a substantial breakdown of the lattice is observed.

INTRODUCTION

The $Bi_2Sr_2CaCu_2O_{8+y}$ high-Tc superconductor (2212) has a lamellar structure.[1,2,3] It is therefore important to study how its superconducting properties depend on structural and electronic interactions between layers. The insertion of guest species not only changes the spacing between layers, but in the case of electrochemical reactions where the introduction of guest ions is accompanied by the transfer of electrons, the Fermi level of the host is shifted due to changes in the occupation of the band structure. Thus, the electrochemical technique can be used to accurately modify both the interlayer spacing and electronic structure in order to study how superconducting properties depend on these factors. We report on the electrochemical insertion of lithium into polycrystalline 2212, its effect on superconductivity and on the location of the lithium in the host structure.

EXPERIMENTAL

Electrochemistry

The samples were prepared as 5, 8 or 13 mm diameter polycrystalline pellets (density 4.5 to 4.9 g/cm^3 or 69 - 75% of the crystallographic value) by reacting appropriate amounts of Bi_2O_3, $CaCO_3$, $SrCO_3$ and CuO in air at 790°C for 16 hours, then 825°C for 72 hours, and 845°C for 48 hours and finally 830-835°C for 74 hours, each followed by air quenching and intermediate grindings.

Electrochemical reaction between Li and 2212 was performed in an argon-filled glove box with parallel-plate galvanic cells of the type Li/1.4M $LiAsF_6$ in 2-methyltetrahydrofuran/Li-2212. Cell capacity was cathode limited. Electrical contact to the cathode was via a Ni current collector.

The cell was discharged at a rate of about 35 μA/cm^2. The reaction was reversed by charging with a constant voltage of 4.86 V dropped across 179 kilohms. The extent of reaction was determined by coulometry (accurate to within several %) based on a theoretical 0.030 Ah/g 2212 formula unit per electron transferred.

Discharge of the Li/2212 galvanic cell corresponds to the introduction of lithium ions and electrons into the 2212 cathode. On discharge the lithium metal of the anode is oxidized. Lithium ions move through the electrolyte and enter the 2212 cathode. Electrons travel through the external circuit and also enter the 2212 cathode (as shown schematically in Fig. 1). The reverse reactions occur upon charging the cell. Lithium is removed from the host cathode and is plated at the anode.

Chemical analysis for Li in intercalated materials was performed by standard atomic absorption (AA) spectroscopy techniques. The accuracy of the Li determination is + 0.03 equivalents.

X-ray analysis

Samples for synchrotron x-ray powder diffraction studies were prepared by thoroughly grinding fresh or reacted pellets. Data were collected at the SUNY-X3 beamline at the National Synchrotron Light Source at Brookhaven National Laboratory with λ = 1.2 Å photons. Ge (111) was used as the analyzer crystal. The data were refined with the Rietveld method, using a new computer program GJANA,[4] which treats both the main and satellite reflections according to the scattering formalism for modulated structures.[5] Unit cell dimensions, peak-shape parameters, the **q**-vector of the modulation, as well as the positional, thermal and modulation displacement parameters of the atoms were varied.

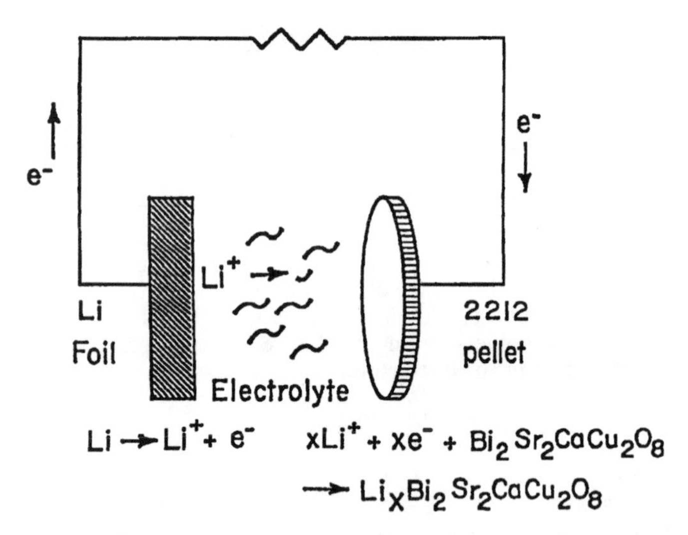

Fig. 1. Schematic representation of the Li/2212 galvanic cell on discharge. In the charging mode, a constant voltage power source is inserted into the circuit between the Li anode and the load resistor. The value of the resistor is also changed so that the charging current will be approximately the same as the discharge current.

Susceptibility measurements

Superconductivity was evaluated by ac magnetic susceptibility measurements at 75 Hz using a diode thermometer. The accuracy of the temperature measurement is about +3 K.

RESULTS

Electrochemical Measurements

It is found that:

a) inserted lithium can be removed from the 2212 cathode (as corroborated by chemical analysis for lithium).
b) after removal of lithium in the first charge cycle, the subsequent discharge profiles are no longer flat but become sloping. After two or three cycles the voltage curves are identical. Cycling curves comparing the first, fifth and sixth cycles to $x = 0.4$ are shown in Fig. 2. This behavior demonstrates that lithium can be reversibly inserted and removed up to a lithium content of $x = 2$ without loss of capacity for a number of cycles.

Some irreversible modifications may occur during the first cycle. But reaction of lithium with a small impurity phase not detected by XRD which is exhausted during the first cycle can not be ruled out.

The voltage profile during discharge to the $x = 2$ level indicates that the reaction proceeds as a unique phase in agreement with the XRD data.

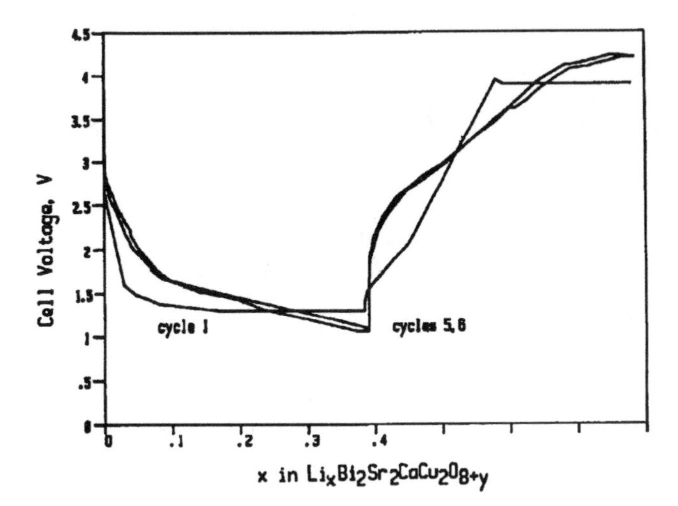

Fig. 2. Comparison of first, fifth and sixth cycling curves for the Li/2212 cell. Discharge was performed on a resistive load corresponding to about 35 μA/cm^2. Charging was performed at a constant voltage of 4.86 volts dropped across 179 kilohms.

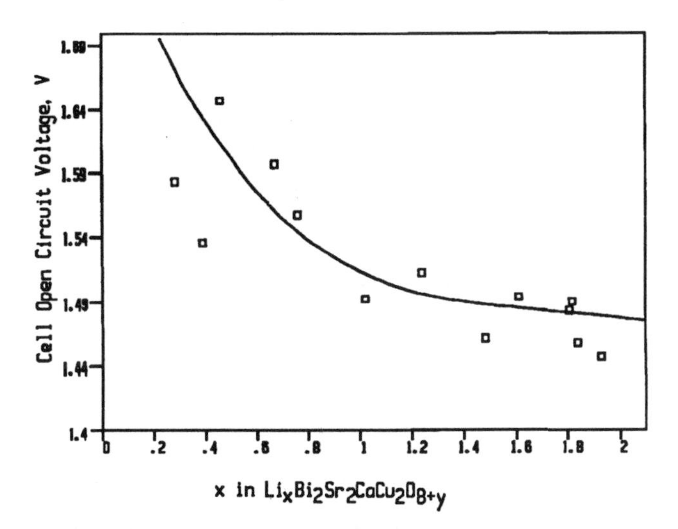

Fig. 3. Coulometric titration curve of $Bi_2Sr_2CaCu_2O_{8+y}$. The Li/2212 cell open circuit voltage is plotted as a function of Li content, x, in $Li_xBi_2Sr_2CaCu_2O_{8+y}$ at various depths of discharge in the first cycle. Each point represents a different individual cell. The average initial open circuit voltage before discharge for the tested cells was 3.339 Volts.

Table I. Lithium content in electrochemically intercalated $Li_xBi_2Sr_2CaCu_2O_{8+y}$ as determined by coulometry compared to chemical analysis values obtained by atomic absorption spectroscopy. Accuracy of coulometric values is several per cent. Accuracy of chemical analysis is +0.03 equivalents.

Value for x in $Li_xBi_2Sr_2CaCu_2O_{8+y}$	
coulometric	chemical analysis
0.06	0.06
0.07	0.09
0.15	0.17
0.30	0.34
1.02	1.03
1.48	1.44
1.84	1.85
2.07	1.96

Coulometric values of the lithium content of samples discharged in the first cycle to a given x value, are identical to within experimental accuracy with chemical values (Table 1) showing that all the measured current leads to the introduction of lithium into the 2212 host.

Variation in the open circuit voltage, E, of the cell as a function of lithium content x corresponding to various depths of discharge in the first cycle is shown in Fig. 3. The sloping decrease in open circuit voltage with increasing lithium content is a characteristic of insertion electrodes. The voltage is related to the continuous change of the chemical potential of the inserted lithium in the 2212 host lattice as compared to the lithium in the anode.

Based on an average value for the observed voltage, the value of the free energy for the introduction of two equivalents of lithium into 2212 is about -70 Kcal/mole 2212. This high free energy is consistent with the formation of an insertion compound.

Superconductivity

Fig. 4 shows ac magnetic susceptibility curves for pure 2212 and reacted pellets. The pellets exhibit a relatively sharp superconducting transition with an onset at about 85-90 K. Bulk superconductivity persists in reacted samples up to the studied lithium composition of x = 2. The onset temperature in lithiated samples is slightly higher, and the transition is broadened as compared to pure 2212. The lower height of the ac magnetic susceptibility transition in reacted samples can be interpreted to mean that the fraction of superconducting material is less in these samples. By removing a portion of the lithium from the sample by charging the cell, some of the transition height can be regained (curve e in Fig. 4). It is also worth noting that the transition height after seven cycles to x = 0.4 is comparable to that for a sample cycled only once to x = 0.4. This shows that in this range of lithium composition, the reduction in transition height is not cumulative with cycling.

X-ray analysis

The powder data show that the reaction proceeds with the retention of crystal structure and without the formation of additional phases. Relevant information is given in Tables II and III. The most pronounced effect of the Li-incorporation is the

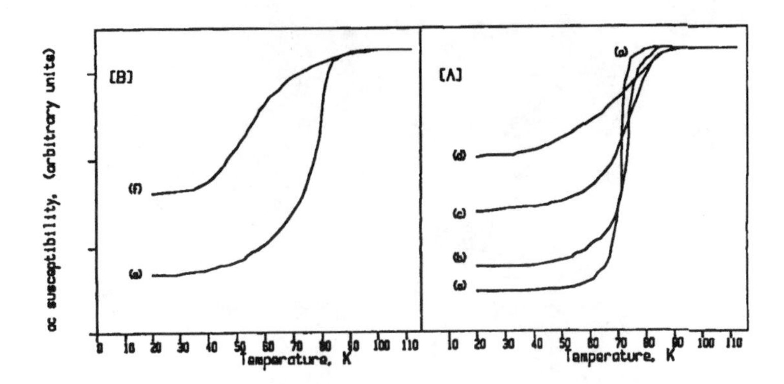

Fig. 4. ac magnetic susceptibility as a function of temperature. [A]: Effect of lithium content in $Li_xBi_2Sr_2CaCu_2O_{8+y}$ for $x = 0$ (a), and for various lithium contents obtained in the first discharge cycle, $x = 0.34$ (b), $x = 1.00$ (c) and $x = 1.93$ (d). [B]: Effect of cycling. Curve (e), sample underwent seven discharge/charge cycles to $x = 0.4$. Curve (f), sample was discharged to $x = 1.8$ and charged back to a lithium content of $x = 1.0$.

Table II. Cell dimensions and agreement factors.

Sample x[1]	0	0.34	1.85
a	5.3996(9)	5.402(1)	5.420(2)
b	5.4014(8)	5.405(1)	5.425(2)
c	30.836(3)	30.887(4)	31.009(6)
q[2]	0.2077(3)	0.2087(6)	0.2057(6)
R_F[3]	0.117	0.097	0.065
R_{powder}[4]	0.154	0.084	0.055

1 in the formula $Bi_2Li_xSr_2Ca_2Cu_2O_{8+y}$

2 in fractions of a*

3 $R_F = \frac{\sum F_o - k|F_c|}{\sum F_o}$, where the sum is over all reflections.

4 $R_{powder} = \frac{\sum I_{i,obs} - I_{i,calc}}{\sum I_{i,obs}}$, where the sum is over all data points.

Fig. 5. Packing diagram of the Bi-O layer viewed along the **c**-axis showing the proposed Li positions. The **a**-axis is vertical and the **b**-axis is horizontal in the drawing. Small circles: Li; medium-sized circles: Bi; large circles: O.

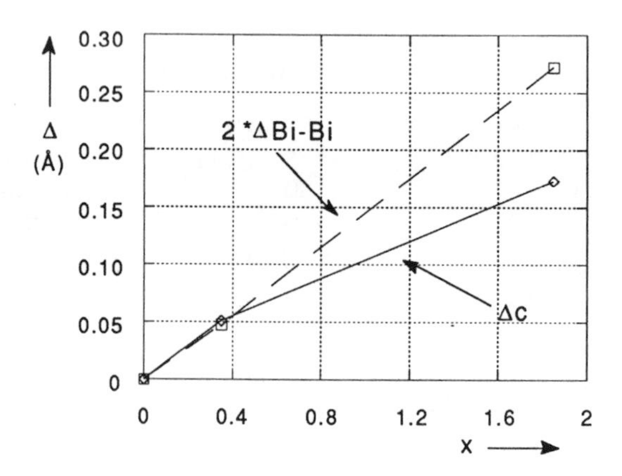

Fig. 6. Variation of **c** and the Bi-O bilayer spacing with Li-content.

expansion of the **c**-axis, but there is also a small but significant increase in the **a**- and **b**-axis dimensions. The **q**-vector is not affected, which indicates that the oxygen concentration in the Bi-O plane is invariant to Li-content.[2, 10]

The Bi-atom z-coordinates show an increase of 0.12(3) Å in the spacing of the Bi-O bilayers. Since there are two such layers in the cell, the **c**-axis expansion of 0.173(7) Å is fully accounted for by this increase in spacing, with a suggestion of contraction of the cell. This is evidence that the Li is inserted between the Bi-O layers (Fig. 6).

There are two distinct vacant interstitial sites, tetrahedrally coordinated by oxygen atoms, which can accomodate the Li atoms, as illustrated in Fig. 5. The lithium atoms at these positions would be 2.23 Å and 2.82 Å from the coordinating O atoms and 2.47 Å and 2.58 Å from the adjacent Bi atoms.

In addition to the expansion of the **c**-axis, an expansion of the **a**- and **b**- lattice parameters is observed. This indicates a lengthening of the Cu-O bonds in the CuO_2 layers in accordance with an increased filling of electron holes in the anti-bonding orbitals upon Li insertion.

DISCUSSION

There are several reports on the increase in the **c**-lattice parameter for decreased oxygen content.[6,7] But the oxygen content of the Bi-O planes (where expansion occurs) on lithiation likely remains the same, as the **q**-vector is unaffected by lithium insertion. Consequently the **c**-axis expansion caused by the lithium insertion is not due to oxygen removal.

Not all of the electrons for lithium contents up to x = 2 can be accommodated in the Cu-O planes. This is seen from the following expression which would relate hole density per Cu atom, p, to lithium content, x

$$p = p_o - x/2 \qquad (1)$$

(where p_o is the initial hole concentration) provided all electrons are accomodated in the CuO_2 planes. (There are two copper atoms per 2212 formula unit; essentially one electron is introduced into the 2212 host per lithium atom for the range of lithium composition studied here). The initial hole concentration, as determined from various measurements[3,8,9] is about 0.35 holes/Cu atom. Based on this value, the maximum number of electrons which the Cu-O planes can accommodate is equivalent to a lithium content of x = 0.7. If all the donated electrons corresponding to x = 0.7 are introduced into the Cu-O planes no holes should remain and superconductivity should be destroyed. However, not only is bulk superconductivity still present at lithium contents greater than x = 0.7 (although the superconducting fraction appears to be lowered), but substantial donation of electrons to the 2212 host continues to occur for x greater than 0.7. Thus the obvious conclusion is that the Cu-O planes do not accomodate all the inserted electrons.

This conclusion is confirmed by the magnitude of the change in the **a**- and **b**-axes lengths, which corresponds, for the x = 1.85 sample, to an increase in Cu-O bond length of less than 0.1 Å. From published dependences of Tc on hole concentration and $r_{Cu\text{-}O}$,[10,11] an order of magnitude estimate of a change of about 0.1 hole per CuO_2 unit is obtained, i.e., a reduction in Cu-valency from 2.35 to 2.25. This is in agreement with the preservation of superconductivity of the samples.

Table III. Fractional coordinates of the Bi, Sr, Ca and Cu atoms.

		x = 0	x = 0.34	x = 1.85
Bi	x	0.5	0.5	0.5
	y	0.233(2)	0.232(2)	0.228(2)
	z	0.0525(2)	0.0528(2)	0.0544(2)
Sr	x	0	0	0
	y	0.248(4)	0.252(4)	0.249(4)
	z	0.1382(3)	0.1386(3)	0.1381(5)
Cu	x	0.5	0.5	0.5
	y	0.25	0.25	0.25
	z	0.1925(5)	0.1925(5)	0.1914(7)
Ca	x	0	0	0
	y	0.25	0.25	0.25
	z	0.25	0.25	0.25

SUMMARY

Electrochemical reduction of 2212 pellets by lithium in galvanic cells occurs with retention of the host crystal structure, via an insertion mechanism for lithium contents up to at least the x = 2 level studied here. Bulk superconductivity persists in the reacted samples. The lithium atoms are essentially completely ionized (from NMR, reported elsewhere[12]), and their insertion occurs with substantial donation of electrons to the 2212 host. The lithium ions occupy interstitial tetrahedral sites between the weakly bonded Bi-O layers causing some expansion in the interlayer spacing. The small increase in the **a** and **b** lattice parameters is consistent with the donation of part of the electrons to the anti-bonding orbitals of the in-plane Cu-O bonds. How the remainder of the electrons are accommodated in the insertion compound is still an open question.

ACKNOWLEDGMENT

Partial support by the Israeli Ministry of Science for this research is gratefully acknowledged. Part of the work was supported by the New York State Institute on Superconductivity (NYSIS90F030) and the National Science Foundation (CHE9021069). The SUNY X3 beamline is supported by the Division of Basic Energy Sciences of the U.S. Department of Energy (DEFG0291ER45231). Research carried out in part at the National Synchrotron Light Source, Brookhaven National Laboratory, which is supported by the U.S. Department of Energy, Division of Materials Sciences and Division of Chemical Sciences.

REFERENCES

1. R. M. Hazen, C. T. Prewitt, R. J. Angel, N. L. Ross, L.W. Finger, C. G. Hadidiacos, D. R. Veblen, P. J. Heaney, P. H. Hor, R. L. Meng, Y. Y. Sun, Y. Q. Wang, Y. Y. Xue, Z. J. Huang, L. Gao, J. Bechtold and C. W. Chu, Phys. Rev. Lett. 60 1174 (1988).
2. D. R. Veblen, P. J. Heaney, R. J. Angel, L. W. Finger, R. M. Hazen, C. T. Prewitt, N. L. Ross, C. W. Chu, P. H. Hor and R. L. Meng, Nature 332 334 (1988).
3. Y. Gao, P. Lee, P. Coppens, M. A. Subramanian and A. W. Sleight, Science 241 954 (1988); V. Petricek, Y. Gao, P. Lee and P. Coppens. Phys. Rev. B 42 387 (1990).
4. GJANA-SUNY Buffalo Crystallographic Computer Program for Refinement of Modulated Structures from Combined Single Crystal and Powder Diffraction Data, Y. Gao, P. Coppens and D. Cox, to be published.
5. V. Petricek, P. Coppens, and P.J. Becker, Acta Cryst. A41 478 (1985).
6. W. A. Groen and D. M. de Leeuw, Physica C 159 417 (1989).
7. S. Miura, T. Yoshitake, T. Manako, Y. Miyasaka, N. Shohata and T. Satoh, Appl. Phys. Lett. 55 1360 (1989).
8. H. Takagi, H. Eisaki, S. Uchida, A. Maeda, S. Tajima, K. Uchinokura and S. Tanaka, Nature 332 236 (1988).
9. R. Ramesh, M. S. Hegde, C. C. Chang, J. M. Tarascon, S. M. Green and H. L. Luo, J. Appl. Phys. 66 4878 (1989).
10. M.-H. Whangbo and C. C. Torardi, Accts. Chem. Res. 24 127 (1991); M.-H. Whangbo, D. B. Kang and C. C. Torardi, Physica C 158 371 (1989).
11. J. Ren, D. Jung, M.-H. Whangbo, J.-M. Tarascon, Y. Le Page, W. R. McKinnon and C. C. Torardi, Physica C 158 501 (1989).
12. N. A. Fleischer, J. Manassen, S. G. Greenbaum, P. Coppens, P. Lee and Y. Gao, Physica C, In Press.

SUPERCONDUCTIVITY AND PHASES IN THE MTlSrCaCuO AND MTlSrCuO SYSTEMS WITH M = Cr, Mo, AND W

Y.F.Li, Z.Z.Sheng, and D.O.Pederson

Department of Physics, University of Arkansas, Fayetteville, AR 72701

ABSTRACT

Cr-, Mo-, or W-substituted TlSrCaCuO and TlSrCuO samples were prepared and investigated by resistance and ac susceptibility measurements and by x-ray diffraction analyses. The results showed that CrTlSrCaCuO, MoTlSrCaCuO, and WTlSrCaCuO systems form a 1212-type phase, but with different Tc: 100 K, 70 K, and 50 K, respectively. Some MoTlSrCaCuO samples exhibit weak superconductivity at 100 K. The Ca-free CrTlSrCuO system is superconducting at 40 K and at 77 K. The 40 K superconducting phase is a 1201-type phase $(Tl,Cr)Sr_2CuO_5$, whereas the 77 K phase may be a new phase and remains to be identified.

INTRODUCTION

Soon after the TlBaCaCuO system was discovered, the TlSrCaCuO system was found to be also superconducting at 20 K and 70 K[1]. However, the superconducting phase $TlSr_2CaCu_2O_7$ is difficult to be synthesized as a pure form, and its Tc ranges from non-superconducting to 80 K[2-5]. When Tl in $TlSr_2CaCu_2O_7$ is partially substituted by Pb or Bi, or/and Sr by La, or/and Ca by a rare earth (R), the 1212 phase can be stabilized, and its Tc is 80-105 K[5-13]. The role of Pb, Bi, La, or R is to adjust the average Cu valence to the optimal value of 2.20 ± 0.05, stabilizing the 1212 phase and improving its superconducting behavior[14]. Recently, we reported a Cr-substituted 1212 phase $TlSr_2(Ca,Cr)Cu_2O_7$ with Tc up to 110 K[15]. This is the first time that a single elemental substitution increases Tc of the 1212 phase to above 100 K. We further carried out substitution experiments on the 1212 phase $TlSr_2CaCu_2O_7$ with Mo and W. We found that the Mo-substitution can also stabilize the 1212 phase $TlSr_2CaCu_2O_7$ with Tc around 77 K. In particular, a superconducting phase with Tc of 100 K also exists in the Mo-substituted samples. The W-substituted 1212 phase has much lower Tc of about 50 K. In this paper, we present systematic results from the Cr-, Mo-, or W-substitution for TlSrCaCuO. In addition, we found that the Ca-free CrTlSrCuO system is superconducting at 40 K and at 77 K. In this paper, we also report the results from resistance and ac susceptibility measurements and powder x-ray diffraction analyses for the Ca-free CrTlSrCuO.

EXPERIMENTAL

The MTlSrCaCuO and MTlSrCuO samples with M = Cr, Mo, or W were prepared using high-purity Tl_2O_3, SrO, CaO, CuO, and Cr_2O_3 or MoO_3 or WO_3. Some Ca-free CrTlSrCuO samples were prepared using Tl_2CO_3, SrO, CuO (or Cu powder), and Cr_2O_3. In a typical procedure, appropriate amounts of component oxides were mixed, ground, and then pelletized. The pellets were put in an alumina crucible with a cover, and were heated in a preheated furnace at about 1000 °C in flowing oxygen for 15 minutes. After heating, the pellets were furnace-cooled to below 700 °C. Resistivity of the samples was measured by the four-probe technique with an ac frequency of 27 Hz. Ac susceptibility was measured using a mutual inductance method (500 Hz). All measurements were carried out in a commercial APD closed cycle refrigerator with computer control and processing. Powder x-ray diffraction measurement was performed by Cu-Kα radiation using a DIANO DTM 1057 diffractometer.

RESULTS AND DISCUSSION

CrTlSrCaCuO

Figures 1 shows resistance-temperature curves for nominal samples $TlSr_2(Ca_{1-x}Cr_x)Cu_2O_y$ with x = 0.0, 0.2, 0.4, 0.6, 0.8, and 1.0. The results are similar to those we reported previously[15]. The samples

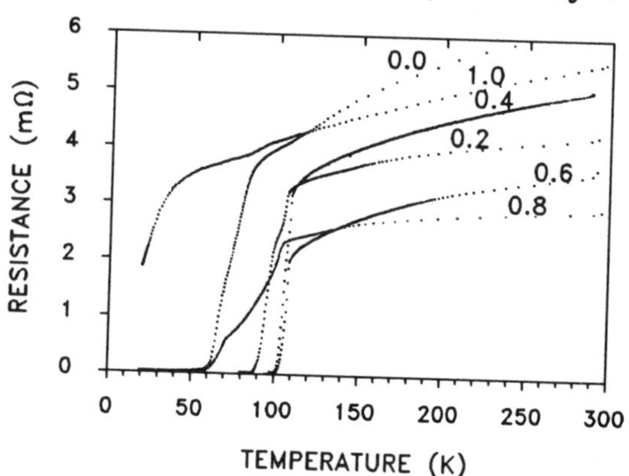

Figure 1 Resistance-temperature curves for nominal samples $TlSr_2(Ca_{1-x}Cr_x)Cu_2O_y$ with x=0.0-1.0.

with x = 0.4 and 0.6 exhibit the best superconducting behavior with a zero-resistance temperature of about 100 K. Powder x-ray diffraction analyses showed that the Cr-samples $TlSr_2(Ca_{1-x}Cr_x)Cu_2O_y$ with x = 0.0, 0.2, 0.4, 0.6, and 0.8 contain a 1212 phase. The Ca-free sample of x = 1.0 contains a 1201-type phase. Note that the Ca-free sample (x = 1.0) is also superconducting at 40 K (see Fig.1). The 1201 phase is responsible for the 40 K superconductivity. The detailed results of the Ca-free samples will be presented later.

MoTlSrCaCuO

Figure 2 shows resistance-temperature curves for nominal samples $TlSr_2(Ca_{1-x}Mo_x)Cu_2O_y$ with x = 0.0, 0.2, 0.4, 0.6, 0.8, and 1.0. The samples with x = 0.2-0.8 are superconducting around 70 K. The samples

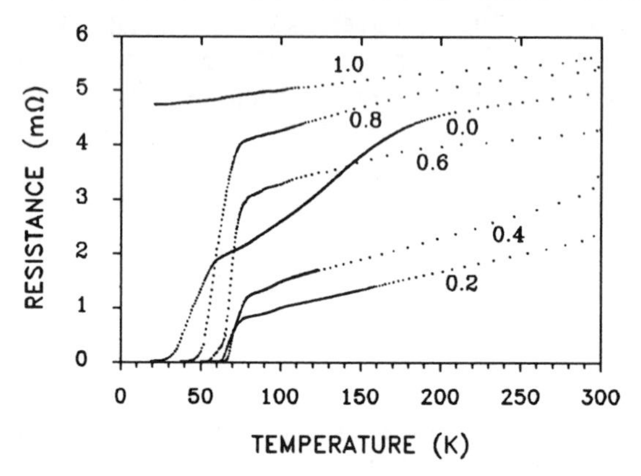

Figure 2 Resistance-temperature curves for nominal samples $TlSr_2(Ca_{1-x}Mo_x)Cu_2O_y$ with x=0.0-1.0.

of x = 0.2 and 0.4 also show a small drop of resistance at 100 K, suggesting a superconducting transition at this temperature. The Ca-free sample (x = 1.0) is metallic but not superconducting down to 20 K. The resistance-temperature curve of the sample $TlSr_2CaCu_2O_y$ (x = 0.0) shows a change of slope at temperature as high as 180 K. This kind of resistance variation was observed in our lab for several times, and is probably a hint of superconducting transition. Further work is needed.

Figure 3 shows resistance-temperature curves for nominal samples $(Tl_{1-x}Mo_x)Sr_2CaCu_2O_y$ with x = 0.0, 0.2, 0.4, 0.6, 0.8, and 1.0. Allsamples are superconducting. The samples of x = 0.2 and 0.4 show two-step superconducting transition at 70 K and 100 K. It is interesting that the Tl-free sample $MoSr_2CaCu_2O_y$ (x = 1.0) is superconducting. But this is an artifact: volatile Tl oxides contaminated this sample and made it superconducting. Separately made $MoSr_2CaCu_2O_y$ samples were semiconducting.

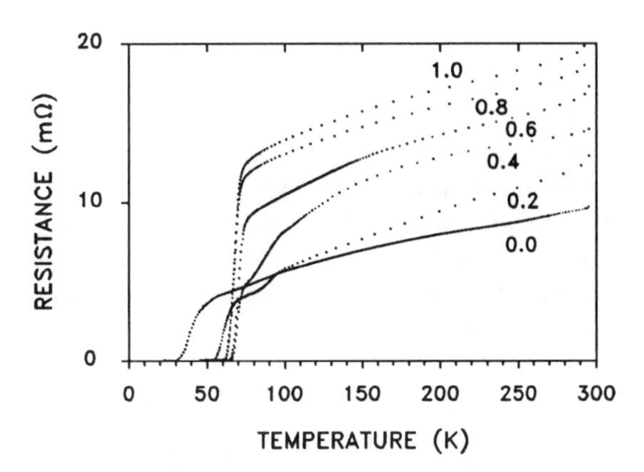

Figure 3 Resistance-temperature curves for nominal samples $(Tl_{1-x}Mo_x)Sr_2CaCu_2O_y$ with x=0.0-1.0.

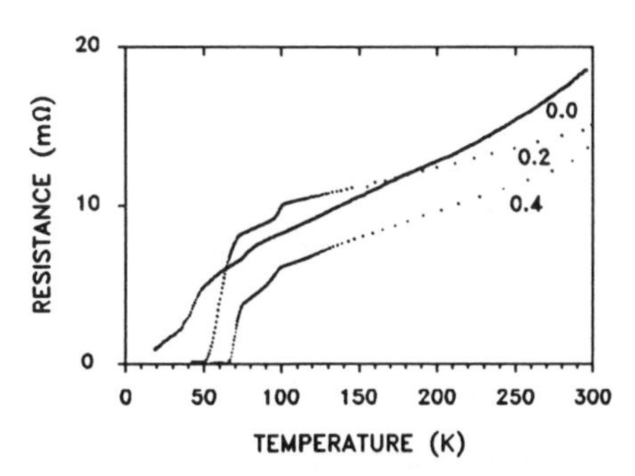

Figure 4 R vs. T for nominal samples $TlSr_2Ca(Cu_{1-x}Mo_x)_2O_y$ with x=0.0, 0.2, and 0.4.

Figure 4 shows resistance-temperature curves for nominal samples $TlSr_2Ca(Cu_{1-x}Mo_x)_2O_y$ with x = 0.0, 0.2, and 0.4. The samples of x = 0.2 and 0.4 exhibit a two-step superconducting transition at 70 K and 100 K. The samples of x = 0.6, 0.8, and 1.0 were also prepared. They are semiconducting, and their resistance-temperature curves are not shown in the figure.

Superconductivity at 70 K and 100 K in the Mo-samples is confirmed by ac susceptibility measurements. As an example, Figure 5 shows the temperature dependencies of ac susceptibility and resistance for a nominal sample $TlSr_2(Ca_{0.8}Mo_{0.2})Cu_2O_y$.

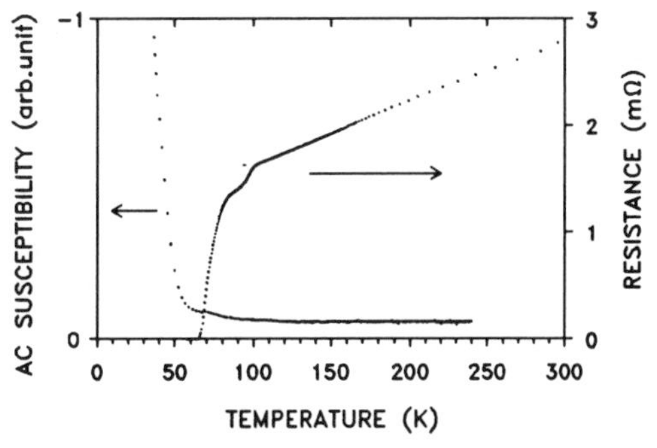

Figure 5 χ and R vs. T for a nominal sample $TlSr_2(Ca_{0.8}Mo_{0.2})Cu_2O_y$.

This sample exhibits a two-step transition: a main transition around 70 K and a minor transition at about 100 K. The results from the ac susceptibility measurement and the resistance measurement are fairly consistent with each other.

Powder x-ray diffraction analyses for the MoTlSrCaCuO samples showed that all samples showing 70 K superconductivity contain a 1212-type phase, while the samples not showing 70 K superconductivity do not contain this phase. Therefore, we concluded that the 1212-type phase is responsible for the 70 K superconductivity. The sample of x = 0.2 in each series is the most pure 1212 phase sample, and the sample $TlSr_2(Ca_{0.8}Mo_{0.2})Cu_2O_y$ is the purest one. Figure 6 shows powder x-ray diffraction pattern for the sample $TlSr_2(Ca_{0.8}Mo_{0.2})Cu_2O_y$. All strong diffraction peaks of this sample can be indexed based on a tetragonal unit cell (space group P4/mmm) with a = 3.801 Å and c = 12.08 Å.

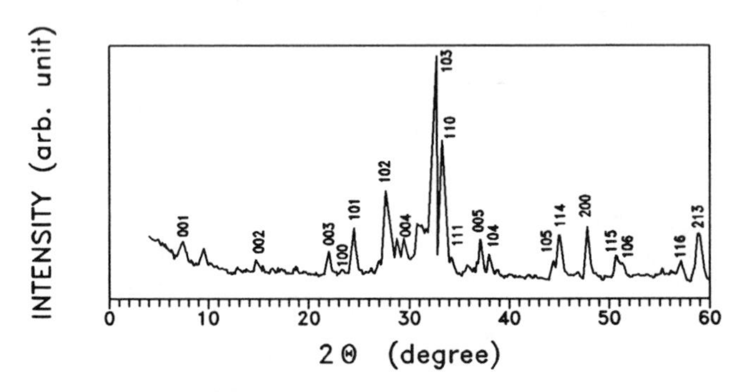

Figure 6 Powder x-ray diffraction pattern for the sample $TlSr_2(Ca_{0.8}Mo_{0.2})Cu_2O_y$ shown in Fig.2.

Although the lattice parameters are almost the same as those of $(Tl_{0.5}V_{0.5})Sr_2CaCu_2O_7$ (a = 3.801 Å and c = 12.02 Å)[16], we would like to tentatively assign to the 70 K Mo-1212 phase a formula of $TlSr_2(Ca,Mo)Cu_2O_7$, similar to the Cr-1212 phase $TlSr_2(Ca,Cr)Cu_2O_7$[15], since its nominal composition is $TlSr_2(Ca_{0.8}Mo_{0.2})Cu_2O_y$. 100 K superconductivity is usually observed in the Mo-samples of x = 0.2 and 0.4, as seen in Figs. 2, 3, and 4. However, the 100 K phase can not be identified alone from powder x-ray diffraction data because of its small amount. Further work is needed.

<u>WTlSrCaCuO</u>

Figure 7 shows resistance-temperature curves for nominal samples $TlSr_2(Ca_{1-x}W_x)Cu_2O_y$ with x = 0.0, 0.2, 0.4, 0.6, and 0.8. With the increase of x, these samples are subjected to conversions from superconducting for the samples of x = 0.0 and 0.2, to metallic for the samples of x = 0.4 and 0.6, and to semiconducting for the samples of x = 0.8 and 1.0 (the resistance-temperature curve of the x = 1.0 sample is not shown in the figure). The sample of x = 0.2 shows the best superconducting behavior with Tc of about 50 K. Powder x-ray diffraction analyses showed that all WTlSrCaCuO samples are not pure phase. The samples exhibiting superconductivity contain a 1212 phase. We believe that this phase is responsible for the observed 50 K superconductivity. Therefore, it seems that the Tc of the Cr-, Mo-, W-substituted 1212 phases decreases in the order of Cr-Mo-W.

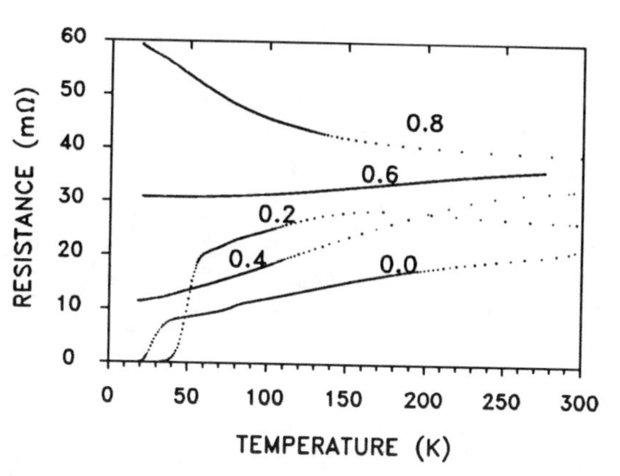

Figure 7 Resistance-temperature curves for nominal samples $TlSr_2(Ca_{1-x}W_x)Cu_2O_7$ with x=0.0-0.8.

Ca-free CrTlSrCuO

As described above, the Cr substituted Ca-free TlSrCuO sample is superconducting, whereas the Mo and W substituted TlSrCuO samples are not superconducting. According to the composition of 1201, we prepared a number of CrTlSrCuO samples. Figure 8 shows resistance-temperature curves for nominal samples $(Tl_{0.5}Cr_{0.5})Sr_2CuO_y$ made using Tl_2O_3 by heating at 800, 850, 900, 950, and 1000 °C for 15 minutes in flowing O_2 followed by quenching. All samples are superconducting, and the sample prepared at 950 °C exhibits the best superconducting

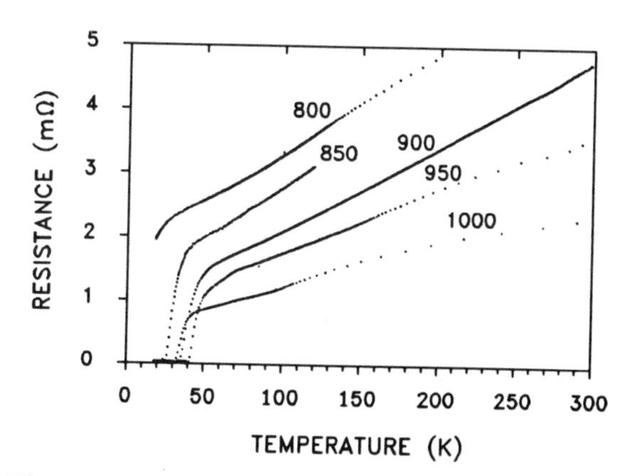

Figure 8 Resistance-temperature curves for nominal samples $(Tl_{0.5}Cr_{0.5})Sr_2CuO_y$ prepared at 800-1000°C.

behavior with an onset temperature of 50 K and a zero-resistance temperature of 40 K. Figure 9 shows temperature-dependencies of ac susceptibility and resistance for the sample $(Tl_{0.5}Cr_{0.5})Sr_2CuO_y$ prepared at 900 °C.

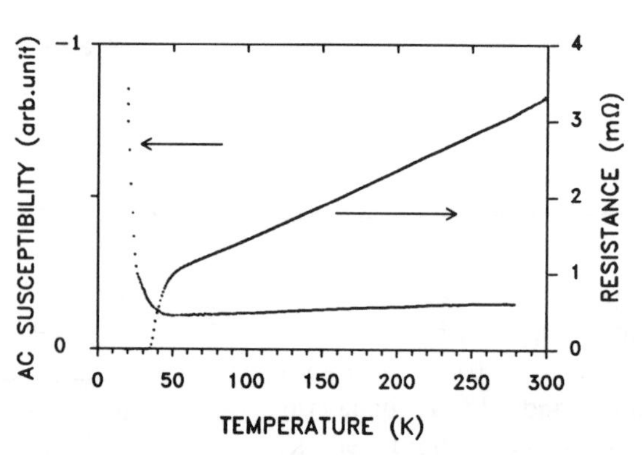

Figure 9 Temperature-dependencies of ac susceptibility and resistance for a sample $(Tl_{0.5}Cr_{0.5})Sr_2CuO_y$ prepared at 900°C.

Powder x-ray diffraction pattern analyses showed that a 1201 phase is responsible for the 40 K superconductivity. Figure 10 shows diffraction pattern for the sample prepared at 900 °C. It can be seen that this sample contains a 1201 phase as its main phase. Based on a tetragonal unit cell (space group P4/mmm) we obtain lattice parameters $\underline{a}$ = 3.795 Å and $\underline{c}$ = 8.880 Å for the 1201 phase. In this figure, the miller indexes are marked for each diffraction peak of the 1201 phase. The unmarked peaks are from impurity.

The Tl-based 1201 phase has three functional sites: Tl, Sr, and Cu. Cr would not substitute for Sr since Cr has ionic radius much smaller than Sr. However, Cr may substitute for Cu since both Cr and Cu are 3d-elements and have similar ionic radii. For examining this possibility, a series of nominal samples $TlSr_2(Cu_{1-x}Cr_x)O_y$ with x = 0.2, 0.4, 0.6, and 0.8 were prepared by heating at 975 °C for 15 minutes. Figure 11 shows resistance-temperature curves for this series of samples. The sample of y = 0.2 shows the best superconducting behavior with an onset temperature of 23 K and a zero-resistance temperature of 18 K. Other samples show only a very small drop of resistance below 20 K. Powder x-ray diffraction analyses showed that only the sample of y = 0.2 contains a

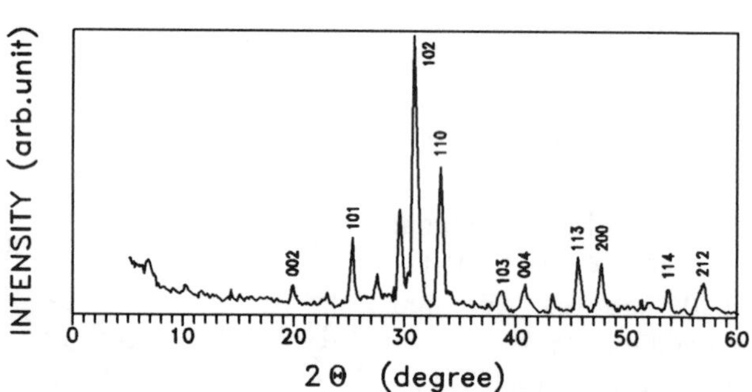

Figure 10 Powder x-ray diffraction pattern for the nominal sample $(Tl_{0.5}Cr_{0.5})Sr_2CuO_y$ prepared at 900°C.

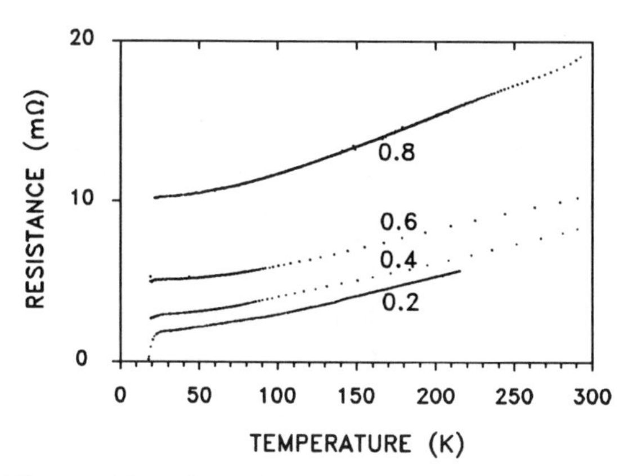

Figure 11 Resistance-temperature curves for nominal sample $TlSr_2(Cu_{1-x}Cr_x)O_y$ with x=0.2-0.8.

small amount of 1201 phase, and other samples do not contain 1201 phase. Therefore, it can be concluded that the 40 K Cr-substituted 1201 phase is $(Tl,Cr)Sr_2CuO_5$. The un-substituted Tl-based 1201 phase $TlSr_2CuO_5$ is orthorhombic and not superconducting (see, for example, ref. 17). Cr-substitution for Tl converts the 1201 phase from orthorhombic to tetragonal, and from metallic to superconducting.

With proper preparation procedures, the CrTlSrCuO samples exhibit superconductivity around liquid nitrogen temperature (77 K). Figure 12 shows resistance-temperature curves for nominal samples $(Tl_{1-x}Cr_x)Sr_2CuO_y$ with x = 0.0-1.0. These samples were prepared using Tl_2O_3 by heating at 1000 °C in O_2 for 30 minutes, followed by decreasing the temperature to 850 °C and holding for about 12 hours. The sample of x = 0.0 is metallic but is not superconducting down to 17 K. The samples of x = 0.2-1.0 all are superconducting with onset temperatures of 71-75 K and zero-resistance temperatures of 68-70 K. It must be noted that the Tl-free sample $CrSr_2CuO_y$ (x = 1.0) is superconducting, but this is an artifact. Since this series of samples were prepared in the same batch, Tl_2O_3 had contaminated the Tl-free sample and made it superconducting. Separately prepared nominal $CrSr_2CuO_y$ samples are not superconducting.

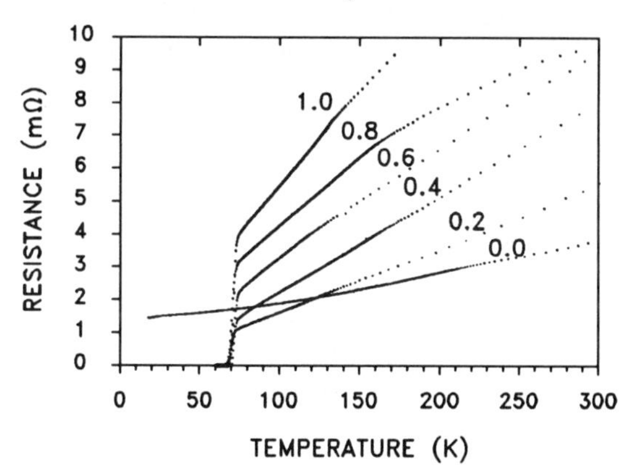

Figure 12 Resistance-temperature curves for nominal samples $(Tl_{1-x}Cr_x)Sr_2CuO_y$ with x=0.0-1.0.

The chemistry of the CrTlSrCuO system is rather complicated.

During the course of experiments, we were aware that when Cr in the 77 K CrTlSrCuO samples reaches a certain amount, the samples were swollen and became porous. This suggests that a strong reaction took place, and a gas released during the heating. We surmise that the gas released is most probably oxygen. Therefore, the 77 K phase may contain, at least partially, Tl^{1+}, Cu^{1+}, or Cr^{2+}. For this reason, we used Tl_2CO_3 (Tl^{1+}) in stead of Tl_2O_3 (Tl^{3+}), and Cu-powder (Cu^{0+}) instead of CuO (Cu^{2+}), as the starting materials. As a result, 77-K CrTlSrCuO samples with even better quality could be made.

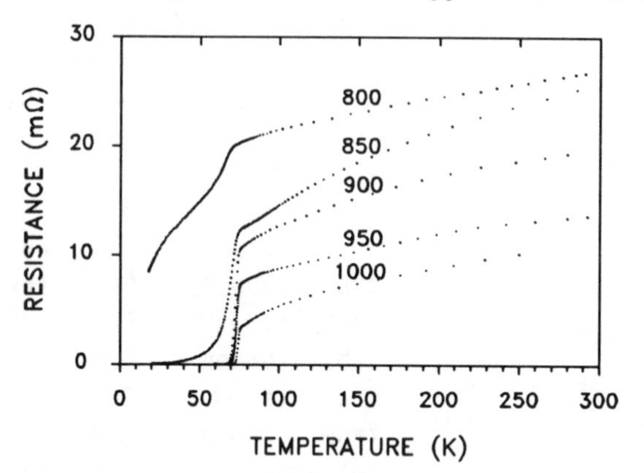

Figure 13 Resistance-temperature curves for nominal samples $(Tl_{0.5}Cr_{0.5})Sr_2CuO_y$ prepared using Tl_2CO_3 at 800-1000°C.

Figure 13 shows resistance-temperature curves for nominal $(Tl_{0.5}Cr_{0.5})Sr_2CuO_y$ samples prepared using Tl_2CO_3, Cr_2O_3, SrO, and Cu-powder by heating at 800-1000 °C in O_2 for 7 min. followed by quenching. The samples prepared at 900-1000 °C show the best superconducting behavior with zero-resistance temperatures up to 72 K. Figure 14 shows resistance-temperature curves for nominal samples $(Tl_{1-x}Cr_x)Sr_2CuO_y$ with x = 0.0, 0.2, 0.4, 0.6, and 0.8 prepared by heating at 1000 °C in oxygen for 7 min. followed by quenching. The sample of x = 0.2 shows the best superconducting behavior with a zero-resistance temperature of 72 K. Note that the Cr-free sample $TlSr_2CuO_y$ (x = 0) shows fairly good superconducting behavior, but this is due to contamination of volatile Cr oxides. The nominal $TlSr_2CuO_y$ samples separately prepared is metallic but not superconducting.

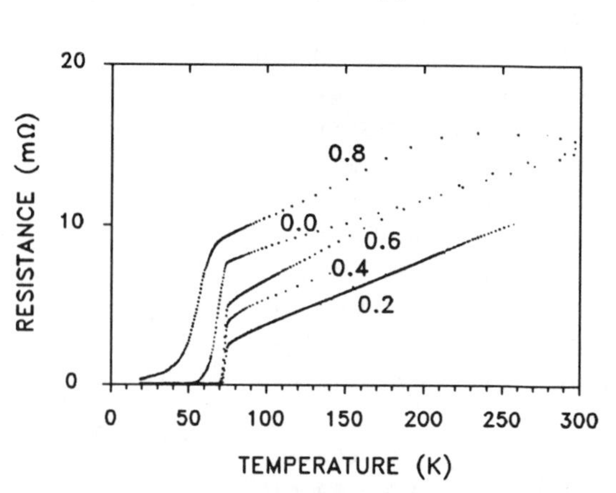

Figure 14 Resistance-temperature curves for nominal samples $(Tl_{1-x}Cr_x)Sr_2CuO_y$ with x=0.0-0.8.

AC susceptibility

measurements for 77 K samples showed superconducting transition temperatures similar to those by resistance measurements. As an example, Figure 15 shows temperature-dependencies of ac susceptibility and resistance for the sample $(Tl_{0.5}Cr_{0.5})Sr_2CuO_y$ prepared at 1000 °C shown in Figure 13.

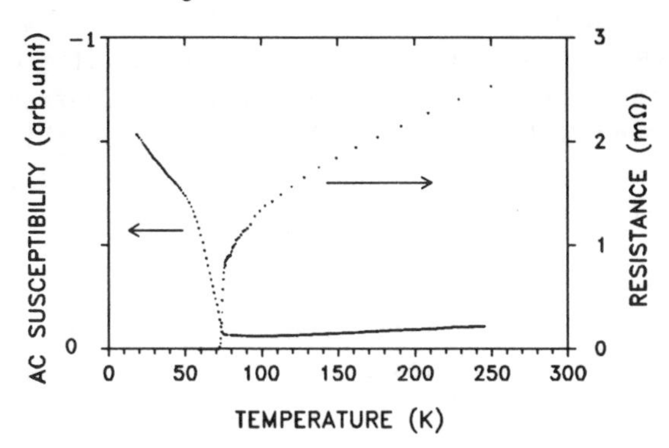

Figure 15 Temperature-dependencies of ac susceptibility and resistance for the sample $(Tl_{0.5}Cr_{0.5})Sr_2CuO_y$.

In view of that the RTlSrCuO system forms two superconducting phases: a 40 K 1201 phase and a 80-90 K 1212 phase[19-23], we surmise that the 77 K phase in the CrTlSrCuO may be a 1212 phase. However, x-ray diffraction showed that the 77 K CrTlSrCuO samples do not contain 1212 phase. The diffraction patterns of the 77 K samples have a common peak at 2θ of about 6°. It is well known that the peak at 2θ of about 6° is the fingerprint peak of the 2212-type phase $Tl_2Ba_2CaCu_2O_8$[24-26]. Therefore, the 77 K phase may be a 2212-type phase. We prepared samples with 2212 starting composition. Figure 16 shows the powder x-ray diffraction pattern for a 77 K sample with nominal composition of $(Tl_{0.8}Cr_{0.2})_2Sr_2SrCu_2O_y$. We failed to index the diffraction pattern within reasonable error either based on a tetragonal 2212 unit cell (space group I4/mmm) or based on a tetragonal 1201 unit cell (space group P4/mmm). Therefore, the phase responsible for the 77 K superconductivity remains to be identified and may be a new phase. According to the preparation condition of the 77 K samples, this new phase would contain some Tl^{1+}, Cu^{1+}, or/and Cr^{2+}. Detailed determinations on structure, composition, and valence for the 77 K phase are needed, and interesting results are expected.

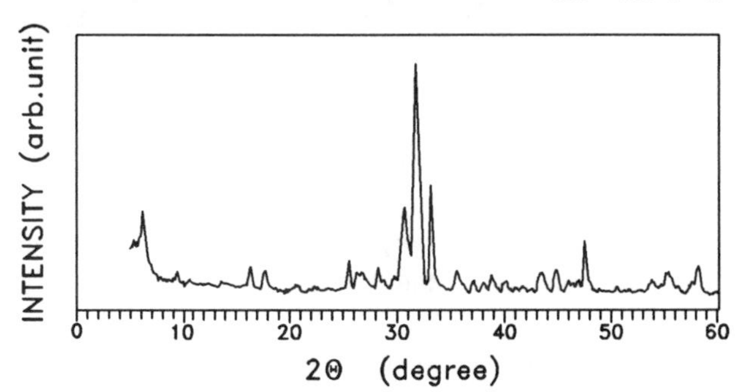

Figure 16 Powder x-ray diffraction pattern for a nominal sample $(Tl_{0.8}Cr_{0.2})_2Sr_2SrCu_2O_y$.

SUMMARY

Cr, Mo, or W substitution for $TlSr_2CaCu_2O_7$ can stabilize the 1212 phase. The Cr-substituted 1212 phase shows the best superconducting behavior with zero-resistance temperature above 100 K. The Mo-substituted 1212 phase shows superconductivity around 77 K. The W-substituted 1212 phase has much lower Tc of 50 K. Some Mo-substituted TlSrCaCuO samples also exhibit a weak 100 K superconductivity. Ca-free CrTlSrCuO samples show Tc around 40 K and 77 K. The phase responsible for the 40 K superconductivity is a 1201-type phase, whereas the phase responsible for the 77 K superconductivity remains to be identified and may be a new phase.

Acknowledgements: We would like to thank Micheal Sorenson, David Ford, and Millicent Megginson for their assistance. This work was supported by a grant of Arkansas Energy Office.

REFERENCES

1. Z.Z.Sheng, A.M.Hermann, D.C.Vier, S.Schultz, S.B.Oseroff, D.J.George, and R.M.Hazen, Phys.Rev.B 38 (1988) 7074.
2. S.Maysuda, S.Takeuchi, A.Soeta, T.Suzuki, K.Alihara, and T.Kamo, Jpn.J.Apll.Phys. 27 (1988) 2062.
3. T.Doi, K.Usami, and T.Kamo, Jpn.J.Apll.Phys. 29 (1990) L57.
4. C.Martin, J.Provost, D.Bourgault, B.Domenges, C.Michel, M.Hervieu, and B.Raveau, Physica C 157 (1989) 460.
5. M.A.Subramanian, C.C.Torardi, J.Gopalakrishnan, P.L.Gai, J.C.Calabrese, T.R.Askew, R.B.Flippen, and A.M.Sleight Science 242 (1988) 249.
6. J.C.Barry, Z.Iqubal, B.L.Ramakrishna, R.Sharma, H.Eckhardt, and F.Raidinger, Appl.Phys.Lett. 65 (1988) 5207.
7. P.Haldar, S.Sridhar, A.Roig-Janicki, W.Kennedy, D.H.Wu, C.Zahopoulos, and B.C.Giessen, J.Superconductivity. 1 (1988) 211.
8. S.Li and M.Dreenblatt, Physica C 157 (1989) 365.
9. Z.Z.Sheng L.Sheng, X.Fei, and A.M.Hermann, Phys.Rev.B 38 (1988) 2918.
10. K.K.Liang, Y.L.Zhang, G.H.Rao, X.R.Cheng, S.S.Xie, and Z.X.Zhao, Solid State Comm. 70 (1989) 661.
11. R.S.Liu, J.M.Liang, Y.T.Huang, W.N.Wang, S.F.Wu, H.S.Koo, P.T.Wu, and L.J.Chen, Physica C 162-164 (1989) 869.
12. Y.T.Hunag, R.S.Liu, W.N.Wang, and P.T.Wu, Physica C 162-164 (1989) 39.
13. M.R.Presland and J.L.Tallon, Physica C 177 (1991) 1.

14. Z.Z.Sheng, Y.F.Li, and D.O.Pederson (see this proceedings).
15. Z.Z.Sheng, D.X.Gu, Y.Xin, D.O.Pederson, L.W.Finger, C.G.Hadidiacos, and R.M.Hazen, Mod.Phys.Let.B 5 (1991) 635.
16. R.S.Liu, W.Zhou, R.Janes, and P.P.Edwards, Solide State Commun. 76 (1990) 1261.
17. T.Kaneko, T.Wada, A.Ichinose, H.Yamauchi, and S.Tanaka, Physica C 177 (1991) 153.
18. Z.Z.Sheng, Y.Xin, J.M.Meason, D.X.Gu, and D.O.Pederson, Supercond. Sci. Technol. 4 (1991) 212.
19. Z.Z.Sheng, C.Dong, Y.H.Lui, X.Fei, L.Sheng, J.H.Wang, and A.M.Hermann, Solid State Comm.71 (1989) 739.
20. T.Itoch and H.Uchikawa, Phys.Rev. B39 (1989) 4690.
21. A.K.Ganguli, V.Manivannan, A.K.Sood, and C.N.R.Rao, Appl.Phys.Lett. 55 (1989) 2664.
22. Z.Z.Sheng, L.A.Burchfield, and J.M.Meason, Mod.Phys.lett. B 4 (1990) 967.
23. Y.Xin, Z.Z.Sheng, D.X.Gu, and D.O.Pederson, Physica C 177 (1991) 183.
24. R.M.Hazen, L.W.Finger, R.J.Angel, C.T.Prewitt, N.L.Ross, C.G.Hadidiacos, P.J.Heaney, D.R.Veblen, Z.Z.Sheng, A.El Ali, and A.M.Hermann, Phys.Rev.Let. 60 (1988) 1657.
25. S.S.P.Parkin, V.Y.Lee, A.I.Nazzal, R.Savoy, and R.Beyers, Phys.Rev.Let. 61 (1988) 750.
26. C.C.Torardi, A.M.Subramanian, J.C.Calabrese, J.Gopalakrishnan, K.J.Morrissey, T.R.Askew, R.B.Flippen, U.Chowdhry, and A.M.Sleight, Science 240 (1988) 631.

PLATELET CONNECTION ACROSS DOMAIN BOUNDARIES AND TRANSPORT CURRENT IN MELT PROCESSED POLYCRYSTALLINE $Y_1Ba_2Cu_3O_{7-\delta}$

R.D. Blaugher*, D.W. Hazelton, P. Haldar, J.A. Rice, J.G. Hoehn, Jr. and M.S. Walker
Intermagnetics General Corporation, Guilderland, NY 12084

This paper reports progress in the achievement of high transport J_c in 3-5 cm long melt processed polycrystalline $Y_1Ba_2Cu_3O_{7-\delta}$ samples. Optical and scanning electron micrographs show apparently clean platelet connection across high angle boundaries between grains (domains) composed of parallel ab platelets. Critical current densities determined by magnetization and transport methods are in agreement at over 10^3 A/cm^3 at 77K in fields above 1 T. Work is underway to more closely examine the condition of these grain boundaries and provide clearer evidence that transport current across them is weak-link-free. Large $Y_2Ba_1Cu_1O_5$ precipitates distributed throughout the matrix are found to produce a factor-of-two increase in critical current density.

INTRODUCTION

The primary technical challenge for applying oxide superconductors to large scale applications depends upon the realization of high transport current in long lengths of practical conductor. High critical current densities, J_c, greater than 10^6 A/cm^2 at 77 K in magnetic fields of 1.0 T have been observed in epitaxial "single grain" $Y_1Ba_2Cu_3O_{7-\delta}$ [123] thin films.[1] Bulk **single crystal** samples of 123, which are flux grown or melt processed, show approximately two orders of magnitude lower current density with typical J_c values of $\approx 10^4$ A/cm^2 in fields of 1.0 T. Transport measurements on melt processed material reported by Jin et al.[2] and Salama et al.,[3] were performed within a single long domain of these aligned a-b platelets. Work by Murakami et al.[4] indicates that by introducing fine [< 1 μm] second phase $Y_2Ba_1Cu_1O_5$ [211] particles, significant enhancements in critical current densities [by an order of magnitude] can be obtained in quench-melt grown samples of $Y_1Ba_2Cu_3O_{7-\delta}$. It was argued that these precipitates acted as efficient flux pinners. More recent work by Alexander et al. suggests that the platelets within a domain comprise a single grain that results from rapid a-b plane growth, with c direction growth enhanced by the presence of 211 precipitates.[5]

The current densities of bulk materials are consistently lower than those of thin films, even for the idealized single grain or single crystal comparison. It is generally observed that for both thin film and bulk materials, the ability to transport current is dramatically reduced for polycrystalline development by high angle grain boundaries. The boundaries produce "weak link" dominated

* National Renewable Energy Laboratory, Golden, CO 80401

current transport characterized by a pronounced decrease in current carrying ability in applied magnetic fields above a few hundred gauss. This polycrystalline result typically occurs in sintered material when one attempts to process lengths longer than a few millimeters. The thin film bi-crystal work of Dimos et al.[6] on $YBa_2Cu_3O_{7-\delta}$ has also clearly shown that mis-oriented high angle boundaries [greater than 10°] produce a Josephson junction type weak link dominated transport mechanism. Thus, the presence of high angle boundaries with possible intergranular precipitates combined with a strong sensitivity of the current flow to crystalline anisotropy would be expected to be a major impediment in realizing high transport properties for bulk materials.

Babcock et al.[7] have reported, however, transport current measured on flux grown bi-crystals which present evidence for **non-weak link** behavior across high angle boundaries. Their results indicate that high current flow may be permitted for certain high angle orientations greater than the 10° limit observed by Dimos et al.[6]

We report here on progress toward increased transport J_c in polycrystalline melt-grown YBCO. This work suggests that clean platelet to platelet connection may be resulting and that high weak-link-free transport J_c's may be present across high angle grain boundaries. Further work is underway to confirm these observations. We also show that by controlling the reaction conditions during melt processing, it is possible to achieve or suppress a distribution of second phases throughout the 123 matrix and that a factor of two increase in critical current density is observed in samples having 211 precipitates.

SAMPLE PREPARATION

Pre-reacted precursor powders of $Y_1Ba_2Cu_3O_{7-\delta}$ [123] obtained from Rhone-Poulenc with a 0.5 - 3.0 µm particle size distribution were compacted into rectangular "bus bars" with lengths of 2.0-5.0 cm, widths of 0.5-1.0 cm and thicknesses of 0.1-0.2 cm. The compacted bars were initially sintered at 900-925°C under pressure to densify and provide stress relief. This sintering treatment was essential for maintaining sample straightness and continuity. Attempts to melt process without the sintering were unsuccessful, resulting in badly distorted bars which often separated into many pieces. The 123 pre-sintered rods were then heat treated using a melt processing method similar to that of Jin et al.[2] and Salama et al.[3] The samples were rapidly raised in temperature to 1050-1100°C and slowly cooled through the peritectic to effect a highly textured 123 development. Various cooling rates of 0.5 to 20°C/hour were used. The slower cooling rates produced material approaching that expected for equilibrium conditions, with essentially complete 123 reaction and almost no residual $Y_2Ba_1Cu_1O_5$ [211] phase. The highly dense melt processed samples required long oxygen equilibration times due to the slow oxygen diffusion.

Fairly large superconducting volume development, inferred from the observed magnetic transitions following flowing O_2 anneals at 480°C x 50 hours, is indicative of well- oxygenated samples. The domains of oriented a-b platelets, typically 1-2 mm, with dispersed 211 phase and the low angle boundaries between the a-b platelets probably provide sufficient grain boundary surface area for enhanced oxygen diffusion. Small sections were taken from the bus bars, typically several millimeters on a side, for magnetization measurements. Sample P0117 is an approximately 15mm X 3mm X 1.5mm slice from a larger bus bar.

MICROSTRUCTURES

The microstructure of sample P0117 was studied in particular detail. Phase, texture and microstructural development were evaluated using an x-ray diffractometer, scanning electron microscopy [SEM] and optical metallography. Energy dispersive x-ray analysis [EDX] was used for detailed analysis of the microstructural composition. Large domains of oriented 123 platelets [1 to 2 mm in length] contain a dispersion of 211 phase. X-ray diffraction analysis by comparing peak heights of the 110 and 005 peak reflections indicates a 50% alignment of the a-b planes along the length of the sample. The high angle boundaries separating the domains, for the most part, appear to be free of any precipitate, as observed by high magnification SEM. Some of the domain boundaries, however, did show the presence of CuO precipitates.

The morphology of a "clean" domain boundary that shows extremely interesting features is presented in Figure 1. The curved interface [top center] between two domains of aligned a-b platelets almost disappears into a region where the a-b plates appear to overlap. This region, detailed in the higher magnification micrograph, shows the barest shadow of a boundary down the center between the platelets of the domain on the left which appear slightly darker, and those of the domain on the right. Polarized optical photomicrographs show many low angle grain boundaries in the region of Figure 1. Additional work is underway toward a higher magnification examination of these boundaries.

MAGNETIC AND TRANSPORT J_c MEASUREMENTS

Superconducting transitions were characterized using a standard four probe DC resistance and AC susceptibility measurements. Susceptibility measurements were made over the range of 25K to 125K using an IGC Superconductor Characterization Cryostat operating at a frequency of 15 Hz with an ac field amplitude of 32 G. Magnetization measurements on small slices from the samples or parallel to the long axis of smaller samples were made using a SHE [SQUID] magnetometer. These measurements were also used to provide one measure of possible critical current densities. The M-H [magnetization versus applied field] curve at different temperatures was measured using a **point by point** method with each value for M obtained after a typical hold time of 10-15 minutes at fixed field. This method assured measurement of magnetization due only to supercurrents after any resistive components of current died away. A conservative minimum value for J_c was estimated using $J_c = 20\Delta M/d$ from the Bean critical state model[8], assuming full coupling and flow of the induced currents across the entire sample. ΔM is the difference in magnetizations for increasing and decreasing magnetic field and the sample width normal to the field is d. Transport current measurements were conducted using a standard four probe method at 1-10 μV/cm sensitivity. A few checks showed essentially no critical current anisotropy for various field orientations normal to the sample long axis. For results presented here samples were mounted with the magnetic field normal to the sample long axis and parallel to the sample's broadest face. Current contacts were applied by a combination of [a] sputter etching followed by deposition of ≈0.5 μm of sputtered Ag, [b] application of silver epoxy and [c] a diffusion reaction at elevated temperature with a final re-oxygenation at 450°C in flowing oxygen.

RESULTS AND DISCUSSION

Susceptibility

The temperature dependent AC susceptibility provides a χ' real component which correlates with the resistive transition and a χ'' [imaginary] out of phase loss component which can provide valuable insights into the intergrain weak link behavior.[9] The superconducting transition for a typical melt processed sample measured by AC susceptibility is shown in Figure 2[b] and is compared with that of the 123 precursor powder [Figure 2(a)], which was sintered at 925°C x 8h in flowing oxygen followed by an oxygen equilibration at 450°C x 12h. The melt processed transition is seen to be considerably sharper than that of the sintered material, with a transition width [1-2°K] comparable to that measured by a resistivity change. The χ' susceptibility result indicates that a large volume fraction of stoichiometric 123 has been produced by melt processing, consistent with x-ray and EDX analysis. Moreover, the melt processed sample shows a sharp peak for the χ'' component at essentially the same temperature as the χ' real component with no evidence at lower temperature for weak links. It is possible, however, that no significant weak link boundaries were included in the induced current path of the measurement, because of the large grain size.

We show in contrast results for the sintered sample. An initial peak for χ'' is followed by a broad lower temperature peak which is consistent with the appearance of intergrain weak link behavior.[9] Note that transport current measurements at 77K for typical sintered samples produced J_c's of ≈200 A/cm^2 current density in zero field and ≈1.0 A/cm^2 at 0.32 T.

Magnetization

Magnetization measurements taken at various orientations on single typical samples revealed essentially no orientation dependence of the resulting calculated J_c. A typical magnetization plot, in this case for a section of sample sample P0120 is shown in Figure 3. Critical current density, J_c, calculated from a similar curve of magnetization at 77 K is shown in Figure 4 for sample P0127.

DC Transport Current

Transport data at 77 K, shown in Figure 4, was obtained on a bus bar [sample P0127] 2.5 cm long, 0.2 cm thick and 1.0 cm wide. The larger bus bars, such as this one, whose contacts were made by applying Ag epoxy followed by a diffusion reaction, showed relatively poor contact resistance with values of 10^{-3} Ω cm^2 as typical. While we at first suspected that this resistance could cause some sample heating at the high currents of these tests, this would not seem to be the case since the J_c's calculated from magnetization at 77 K on a piece of the same sample shows fairly good agreement as shown.

As a way of showing that the low resistances seen in the hard-to-contact bus bars may be due to current transfer and not weak links, two sets of transport current J_c measurements were conducted on P0117, a smaller sample. The first measurement, with contacts made in similar fashion to that on the bus bars, shows a resistive onset with an initial resistive slope (inset of Figure 4). The resistance shown is four orders of magnitude less than the mΩ. The [$\approx$0.01 - 0.1 $\mu\Omega$ cm] corresponding resistivity for this 0.042 cm^2 cross section is believed to be a current transfer resistance associated with the short sample length and the proximity of the voltage contacts to the current leads. Such current transfer resistance has been observed in low temperature superconductors as described in detail by Ekin.[10] The observed zero field resistive slope below the transition did not show any strong change as a function of applied magnetic field up to 8.0 T which tends to argue against the observed resistance as being due to flux flow and for its being due to some kind of contact resistance.

For the second test the contact resistance for this sample was just 5 x 10^{-5} Ω cm^2, obtained using the sputter etched and sputtered Ag method. The same critical current density was measured, but without the long resistive onset shown in the inset of Figure 4. While initially we had thought that this would indicate lack of weak-link boundary resistance, closer examination revealed that the voltage taps on this sample may have spanned single grains. The significance of these measurements regarding boundary resistance thus remains unclear. Further work is aimed at measurement of transport current with improved current contacts and voltage contacts that are clearly across a span of many grains, so as to establish unequivocally the presence or lack of any boundary resistance. The critical current for both measurements on P0117 is in agreement with the magnetization measurements on P0127 at 0.32T, as shown in Figure 4.

INFLUENCE OF 211 PHASE

Figure 5 shows the melt processing profiles for samples P0127 and P0120. Both were cooled through the peritectic at a slow rate [$\approx$1°/hour]. However, P0127 was allowed to undergo solid state reaction between 900°C and 980°C for twice as long as sample P0120. The resulting microstructures are displayed in Figure 6. It can be seen that the longer reaction time enabled all the $Y_2Ba_1Cu_1O_5$ [211] to react and form 123 for sample P0127. However, some precipitates remain unreacted for P0120 and are distributed throughout the matrix. The sizes of these second phases range from 5 to 30 μm.

Sample P0120, which contained 211 precipitates, had critical current density values calculated from magnetization measurements and shown in Figure 7 that are a factor of two times those of P0127, which did not contain 211 precipitates, for the entire range of applied fields used in the measurements. This is in contrast to results reported by Murakami et al.[4] where enhancements of an order of magnitude in J_c were shown by obtaining fine dispersions of 211 by a quench-melt-growth technique. It should be noted that the 211 precipitates reported by Murakami were smaller in size [< 1 μm] and were argued to act as strong flux pinners. It isn't clear from our work whether the 211 acts as a flux pinner (these precipitates seem to be too large to pin effectively) or whether the 211 influences the interplatelet and intergrain connectivity by "soaking up" Ba-Cu-O liquid.

CONCLUSIONS

Critical current densities determined by magnetization and transport methods are in agreement at over 10^3 A/cm^3 at 77K in fields above 1 T. The microstructural features we have observed suggest that the transport current in these large polycrystalline samples is across high angle [>10°] boundaries between domains comprised of aligned a-b platelets. The current would thus be along or across extended a-b platelets which are tilted and misaligned with respect to the platelets that they contact in adjacent domains. Critical currents are high and arguably of sufficiently low resistance to indicate transport free of boundary resistance. However, because of the presence of high contact resistance in transport current measurements in some cases and the possibility of having few or single grains between voltage contacts in others, we cannot say with certainty that this is so. Ongoing work is intended to resolve this question. Finally, the microstructural features of our samples indicate that 211 precipitates in a 123 matrix may be responsible for a factor of two increase in critical current, but the mechanism by which this is accomplished isn't clear.

ACKNOWLEDGEMENTS

The authors thank Dr. Jack Ekin of NIST for advice and the low resistance measurement of J_c on sample P0117 and Drs. Amit Goyal and Don Kroeger of the Oak Ridge National Laboratory for their help in providing polarized optical microscopy on one of the samples. This work was supported in part by the Strategic Defense Initiative Organization under contract SDIO 84-88-C-0049 and in part by the New York State Institute For Superconductivity under contract 90F043.

REFERENCES

1. R.K. Singh and J. Narayan, J. of Metals, **43**, 13 [1991].

2. D. Dimos, P. Chaudhari and J. Mannhart, Phys. Rev. B, **41**, 4038 [1990].

3. S.E. Babcock, X.Y. Cai, D.L. Kaiser and D.C. Larbalestier, Nature, **347**, 167 [1990].

4. S. Jin, T.H. Tiefel, R.C. Sherwood, M.E. Davis, R.B. VanDover, G.W. Kammlott, R.A. Fastnacht and H.D. Keith, Appl. Phys. Lett., **52**, 2074 [1988].

5. K. Salama, V. Selvamanickam, L. Gao and K. Sun, Appl. Phys. Lett., **54**, 2352 [1989].

6. M. Murakami, T. Oyama, H. Fujimoto, S. Gotoh, K. Yamaguchi, Y. Shiohara, N. Koshizuaka and S. Tanaka, IEEE Trans. on Magnetics, **27**, 1479 [1991].

7. H. Kupfer, I. Apfelstedt, R. Flukiger, C. Keller, R. Meier-Hirmer, B. Runtsch, A. Turkowski, U. Wiech and T. Wolf, Cryogenics, **28**, 650 [1988].

8. C.P. Bean, Rev. Mod. Phys., **36**, 31 [1964].

9. J. Ekin, J. Appl. Phys., **49**, 3406 [1978].

10. T.W. Jing, Z.Z. Wang and N.P. Ong, Appl. Phys. Lett., **55**, 1912 [1989].

11. J. Mannhart and C.C. Tsuei, Z. Phys., **B77**, 53 [1989].

12. P. McGinn, W. Chen, N. Zhu, M. Lanagan and U. Balachandran, Appl. Phys. Lett., **57**, 1455 [1990].

13. Z. Yi, S. Ashworth, C. Beduz and R.G. Scurlock, IEEE Trans. on Magnetics, **27**, 1506 [1991].

14. J. Ekin [private communication].

15. K. B. Alexander, A. Goyal, D. M. Kroeger, V. Selvamanickam and K. Salama, "The Microstructure Within Domains of Melt-Processed $YBa_2Cu_3O_{7-}$Superconductors", Submitted for Publication to Physical Review Rapid Communications.

Figure 1: SEM micrographs from sample P0117 showing high angle boundary between domains of oriented 123 a-b platelets higher magneification of the interface between domains of a-b platelets.

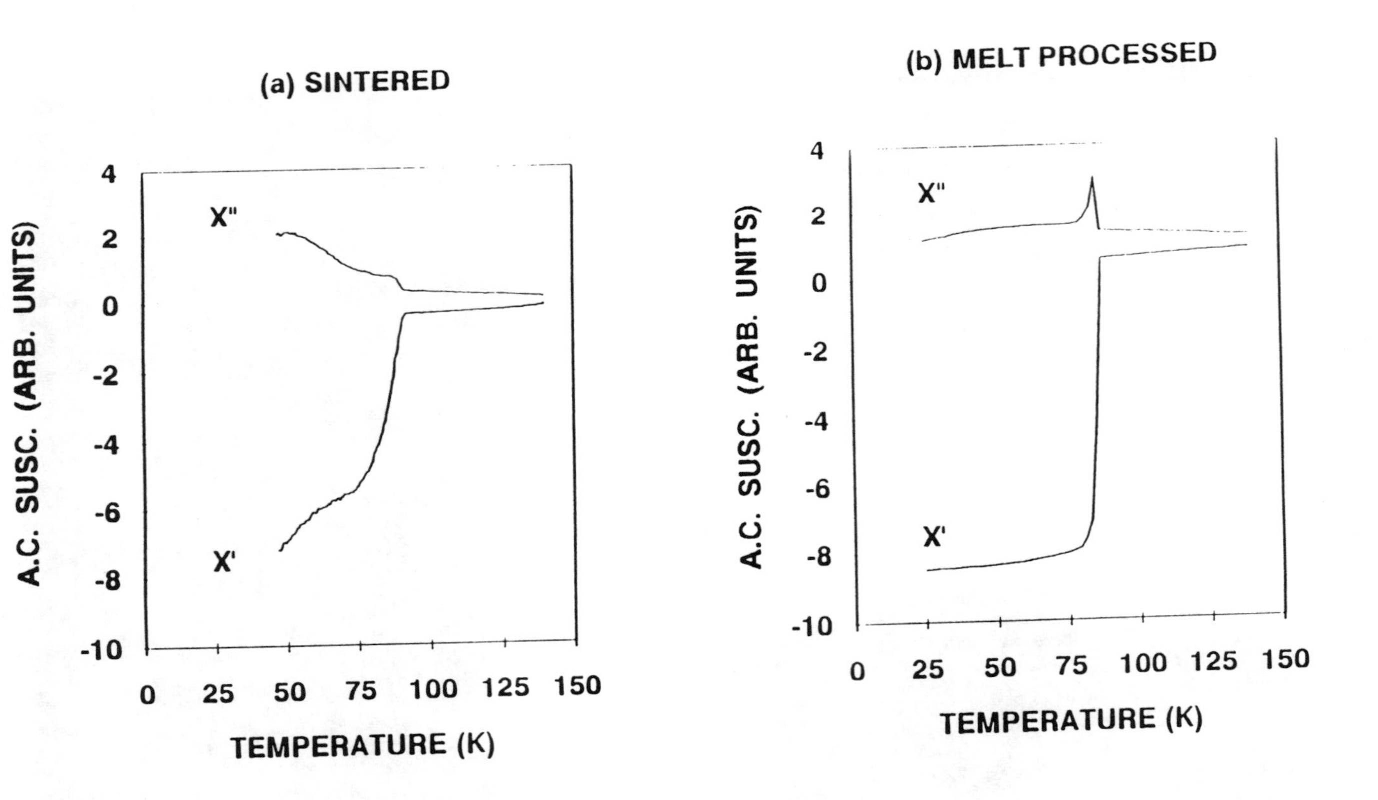

Figure 2: Real χ and imaginary χ'' components of AC susceptibility for (a) sintered sample (b) melt processed sample.

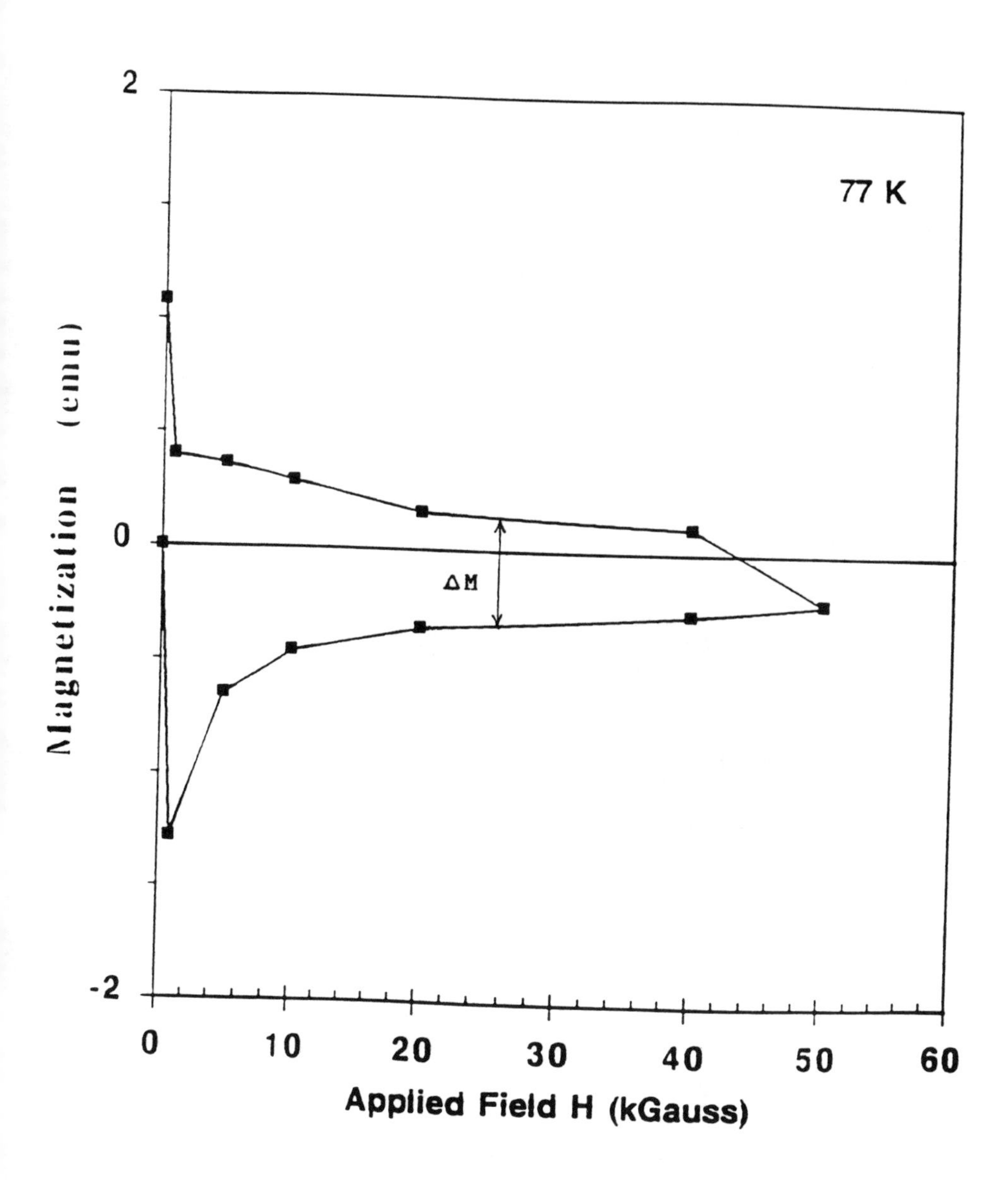

Figure 3: Magnetization versus applied field at 77 K for sample P0120.

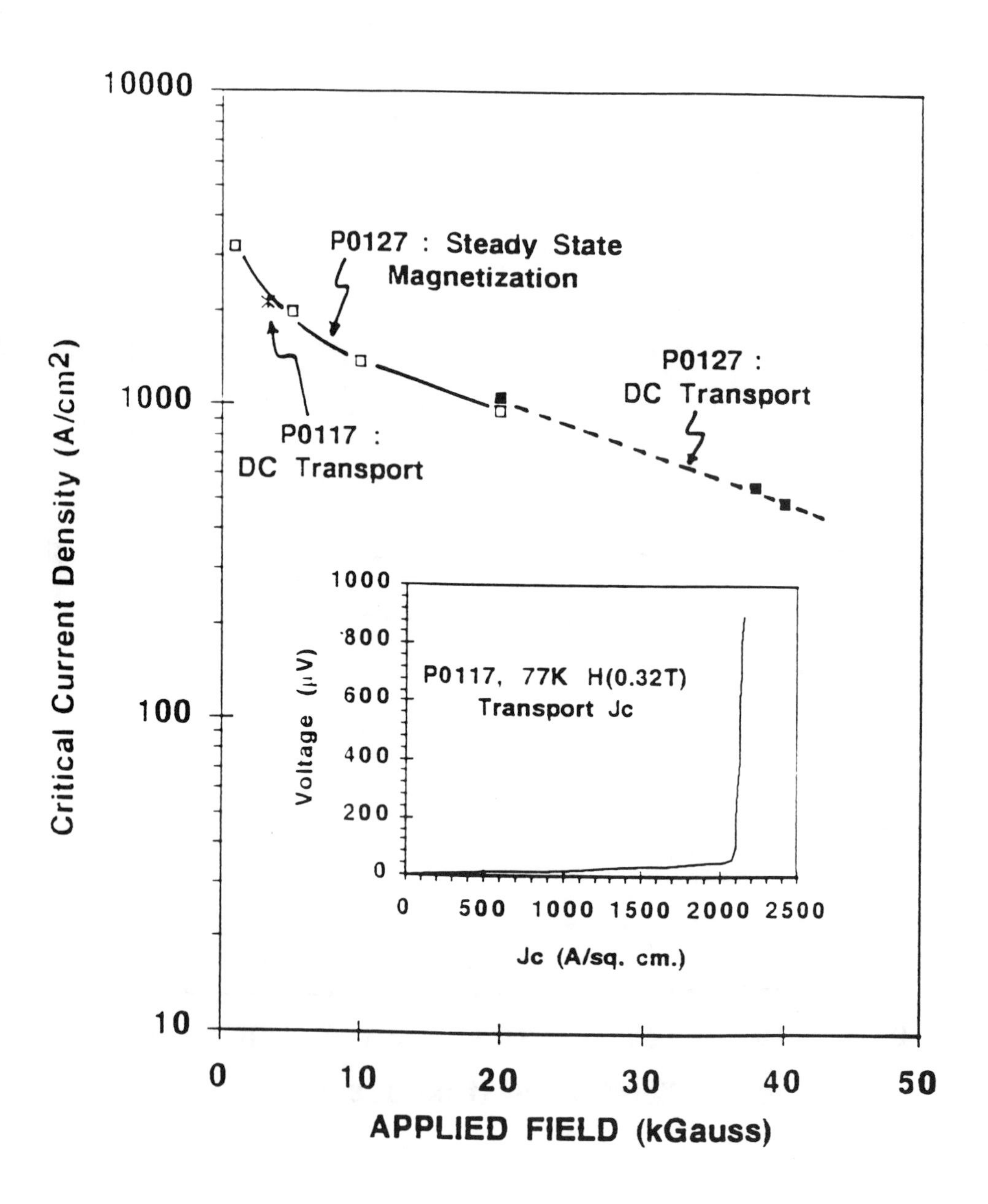

Figure 4: Critical current density as a function of applied magnetic field at 77 K for samples P0117 and P0127 as measured by DC transport and magnetization. Inset shows I-V characteristic of P0117.

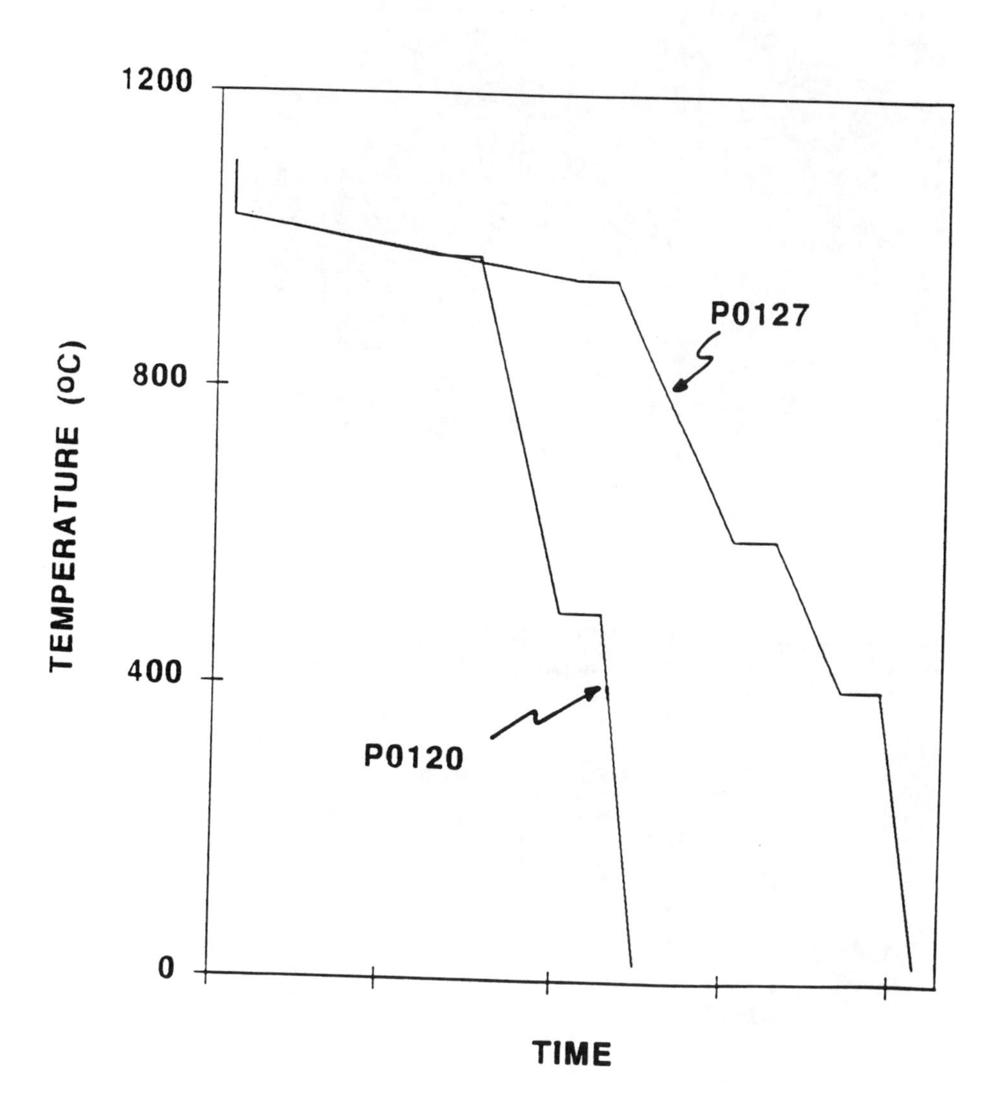

Figure 5: Melt processing profiles for samples P0120 and P0127.

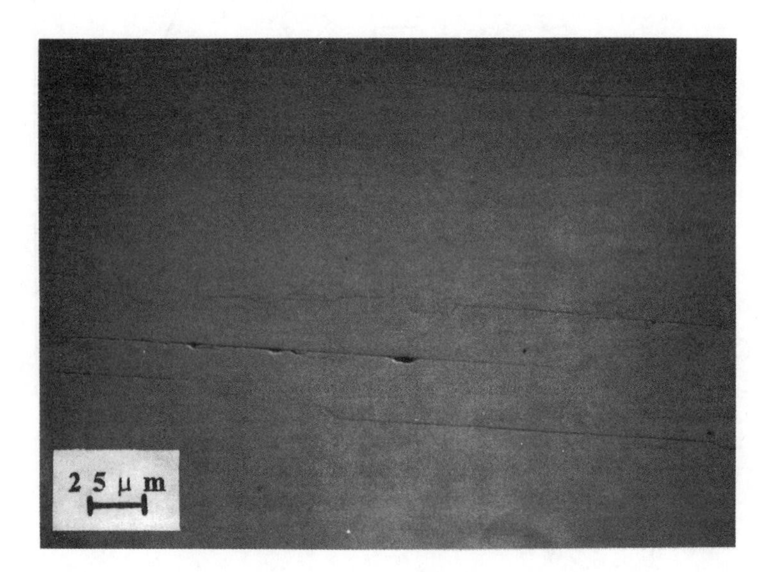

Figure 6: SEM micrographs for (a) P0120 showing the presence of 211 precipitates and (b) P0127 without any 211 phases.

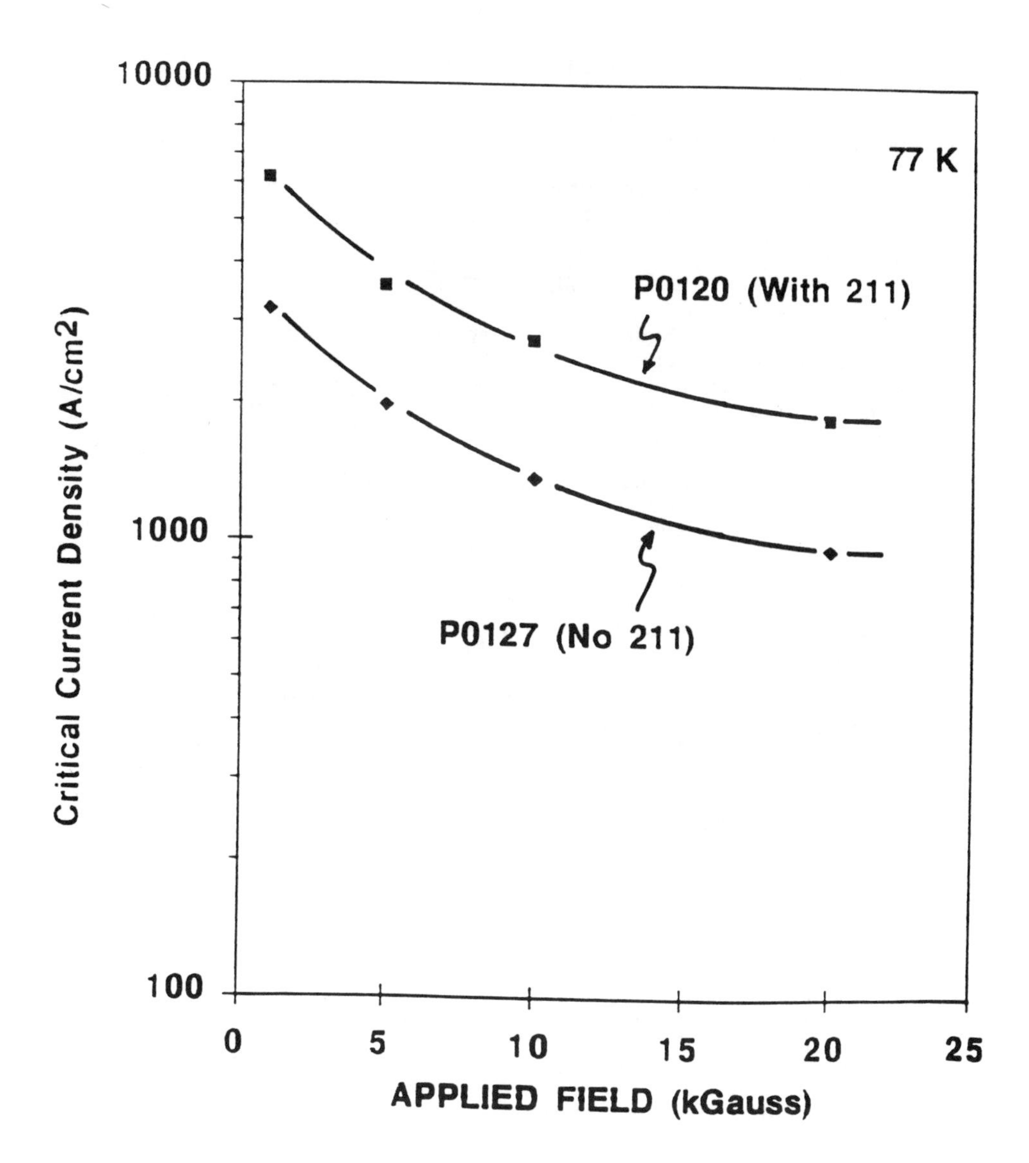

Figure 7: Critical current density versus applied field at 77 K for samples P0120 and P0127.

SUPERCONDUCTIVITY IN (Y,RE)-Ba-Ca-Sr-Cu-O SYSTEM

S. Arumugam and S. Natarajan
Department of Physics, Anna University, Madras-25, India

V.Ganesan, C.K.Subramaniam and R.Srinivasan
Department of Physics, I.I.T.,Madras-36, India.

ABSTRACT :

A new series of compounds in the family (Y,RE)-Ba-Ca-Sr-Cu-O (RE = Nd, Sm, Eu and Gd) with a general formula $(Y,RE)Ba_2Ca_3Sr_4Cu_5O_x$ have been synthesised using conventional solid state reaction technique. The materials were charecterised by X-ray diffraction (XRD), electrical resistivity, and magnetisation studies. The results of powder XRD shows that the samples were single phase. The XRD pattern could be indexed either to an orthorhombic cell of dimensions a = 5.470(5) A, b = 7.390(4) A and c = 14.56(1) A or to a near tetragonal cell of dimensions a = 5.45(3) A, b = 5.46(1) A and c = 14.60(3) A. The exact structural details can be confirmed only after obtaining sufficient data from electron or neutron diffraction measurements. From the DC magnetisation studies on a SQUID magnetometer these samples showed superconducting transitions in the range of temperatures 74 K to 52 K. The superconductivity in bulk were also confirmed by the measurements of electrical resistance in these samples. The details of preparation and characterisation will be discussed.

INTRODUCTION

After the discovery of superconductivity in the $La_{2-x}M_xCuO_4$ (M = Ba, Ca & Sr) system by Bednorz & Muller[1] and in $YBa_2Cu_3O_{7-X}$ (123) by Wu et al[2] there has been a lot of interest in preparing other high T_c superconducting materials containing copper - oxygen planes. The materials which have been found superconducting are $YBa_2Cu_4O_{8+y}$[3], $Y_2Ba_4Cu_7O_x$[4], $A_2Sr_2CaCu_2O_{8+x}$ (A = Bi or Tl)[5], $TlBa_2CaCu_2O_{7-y}$[6], $Pb_2Sr_2LnCu_3O_{8+y}$[7] etc., which are superconducting in the range 77 to 100 K. There are other compounds which have T_c above 100 K like $(Bi,Tl)_2Sr_2Ca_2Cu_3O_{10+y}$[8], $TlBa_2Ca_3Cu_4O_{11-y}$[9] and $TlBa_2Ca_4Cu_5O_{13-y}$[10] etc. There were interesting reports of increasing Cu content from 2 to 12 in the $YSrCaBaCu_xO_y$[11] and $GdBaCaCu_3O_x$[12]. The different compounds have T_c ranging from 30 to 60 K for the system reported in the reference [11] and 90K for the system reported in the reference [12]. All the above compounds are p - type superconductors. Tokura et al prepared an n-type superconductor in $Nd_{2-x}Ce_xCuO_4$ [13].

In this report we attempted to vary Ba, Ca, Sr and Cu content in (Y,RE)-Ba-Ca-Sr-Cu-O system to see if we could obtain a single phase superconducting compound at a composition different from what has been studied so far. We have prepared a new series of compounds $(Y,RE)Ba_2Ca_3Sr_4Cu_5O_x$ (12345) which appeared to be single phase from XRD and showed superconducting transitions in magnetic susceptibility measurements in the range 74 K - 52 K. The superconductivity in bulk were confirmed from the measurements of electrical resistance in these samples.

EXPERIMENTAL :

The samples were prepared by conventional solid state reaction technique. High purity $(Y,RE)_2O_3$, $BaCo_3$, $CaCo_3$, $SrCo_3$ and Cuo were used. After initial mixing with proper molal ratios, the compounds were transfered to a ball mill and mixed well for about 12 hrs. These powders were calcined at 950 C for 24 hours twice with intermitent grinding. Pellets were made out of these powders and sintered at 950 C for 24 hours. The pellets were then oxygen annealed at 900 C for 24 hours and 600 C for another 24 hours and furnace cooled to room temperature.

The DC electrical resistance measurements were performed using standard four probe technique. The field cooled magnetisation measurements were made on a MPMS SQUID magnetometer. The materials were characterised by powder XRD using CuK_α radiation.

RESULTS AND DISCUSSION :

Figure 1. shows XRD pattern of one of these samples Y(12345) . The XRD pattern could be indexed either to an orthorhombic cell of dimensions a = 5.470(5) A, b = 7.390(4) A, c = 14.56(1) A or to a near tetragonal cell of dimensions a = 5.45(3) A , b = 5.46(1) A and c = 14.60(3) A. The XRD pattern indicates that these samples were single phase. No impurity peaks were seen within the limits of accuracy of measurements. The exact structural details can be confirmed only after obtaining sufficient information from either electron or neutron diffraction measurements in these samples.

The electrical resistance was measured using standard four probe technique. The sample showed superconductivity with the onset of transition at 74 K and zero resistance at 63 K . Figure 2 shows the variation of DC electrical resistivity versus temperature for this sample. Figure 3 shows the DC magnetic susceptibility measurements made using MPMS SQUID magnetometer at a field of 20 Gauss for this sample as a function of temperature. The onset of diamagnetic response is at 74 K. The transition is also sharp with the susceptibility saturates at low temperatures.

DC magnetisation studies were carried out for the samples in which Yttrium was replaced by Rare Earths like Nd, Sm, Eu and Gd. These samples were also superconducting in the range 52-74 K. Figure 4 shows the DC magnetic susceptibility measurements for these samples at a field of 100 G using the MPMS SQUID magnetometer. XRD measurements showed that all these compounds were single phase having unit cell dimensions around the same as that of Y sample.

In conclusion the superconductivity in the newly found system 12345 was investigated. The effect of RE substitution decreases Tc in this system. The exact structural details can be confirmed only after obtaining sufficient data.

ACKNOWLEDGEMENTS :

Authors thank Program Management Board on Superconductivity for financial assistance. One of the authors (SA) would like to thank University Grants Commission, New Delhi, India for the award of research fellowship.

REFERENCES :

1. Bednorz J.G and Muller K.A.
   Z. Phys.**B64** 189 (1987).
2. Wu.M.K, Ashburn J.R, Tong.C.T, Hor.P.H, Meng.R.L, gao.l, Huang.Z.J, Wang.Y.Q and C.W. Chu
   Phys. Rev. Lett. **58** 908 (1987).
3. Donald E. Morris, Janice H. Nickel, John Y.T. Wei, Naggi G. Asmar, Jeffrey S. Scott, Ulrich M.Scheven, Charles T. Hultgren and Andrea G. Markelz.
   Phys. Rev.B. **39(10)** 7347 4 (1989).
   Fisher. P, Karpinski. J,Kaldis. E et al.,
   S. S. Commun. **69** 531 (1989).
5. Maeda.H, Tanaka.Y, Fukutumi.M et al.,
   Jpn. J. Appl. Phys. **27** L209 (1988).
   Tarascon.J.M, Mckinnon . W.R, Barboux .P et al
   Phys. Rev. B., **38** 8885 (1988).
   Kikuchi. M, Kajitani.T, Suzki. T et al.,
   Jpn. J. Appl. Phys. **28** L382 (1989).
6. Parkin S.S.P. et al Phys. Rev. Lett.**61** 750 (1988).
7. Gasnier. M, Ruault. M.O and Suryanarayanan. R.,
   S.S.Commun. **71** 485 (1989).
8. Sheng Z.Z. and Herman A.M. Nature 332 **55** (1988).
   Cava.R.J, Batlogg.B, Krajewski et al.,
   Nature **336** 211 (1988).
9. Ihara. H, Sugise.R, Hirabayashi. M et al.,
   Nature **334** 511 (1988).

Ling. J.K, Zhang. Y.l, Huang. J.Q, et al.,
Science in China, **A32** 827 (1989).

10. Kusuhara. H, Kotani.T, Takei. H and Tada. K
Jpn. J. Appl. Phys. **28** L1772 (1989).
11. Kurihama, T, Iwasaki. K, and Isumi T,
Jap. Jl. Appl. Phys **29(4)** 658 (1990).
12. Wang. X.J and bauuerle . D
Physica. C, **176** 507 (1991).
13. Tokura. Y, Takagi. H and Uchida. S.
Nature **337** 345 (1989).

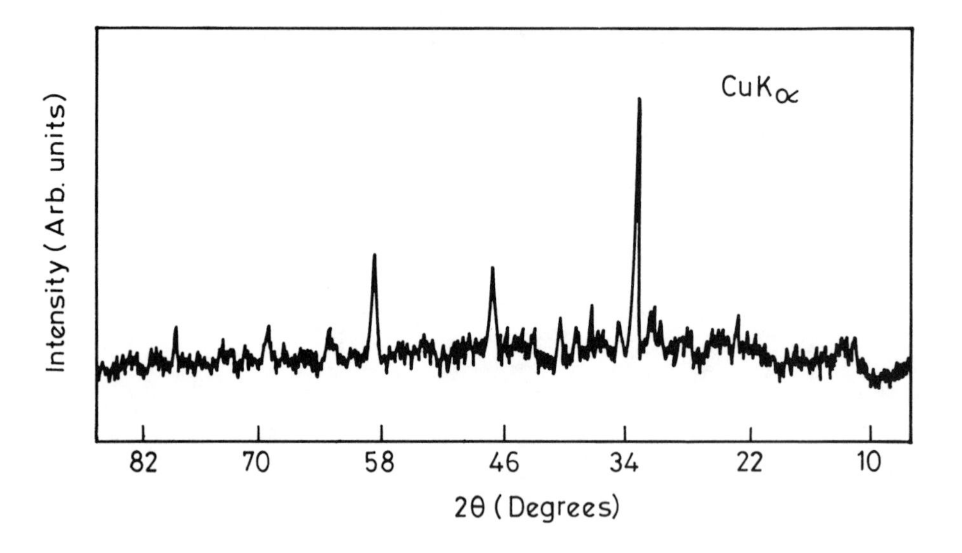

Figure 1. XRD pattern of $YBa_2Ca_3Sr_4Cu_5O_x$(12345).

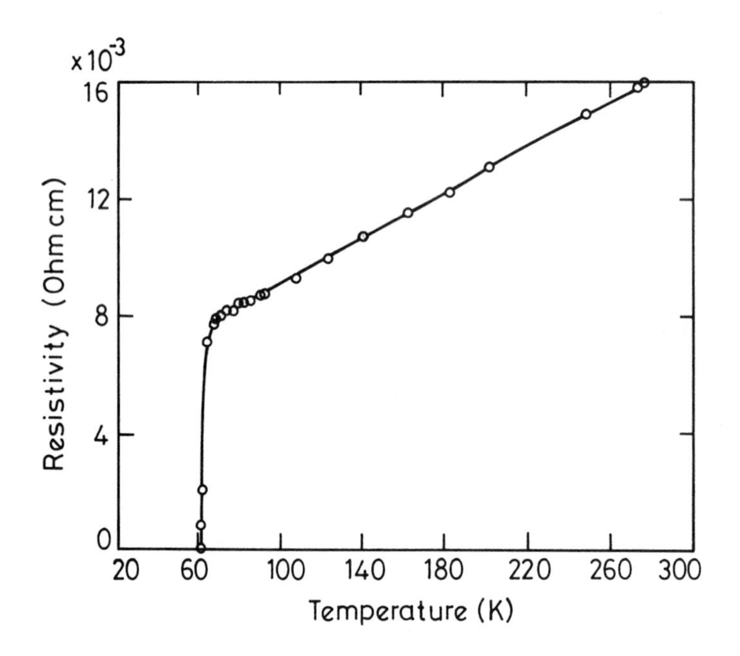

Figure 2. Variation of DC electrical resistivity with temperature for the sample Y(12345).

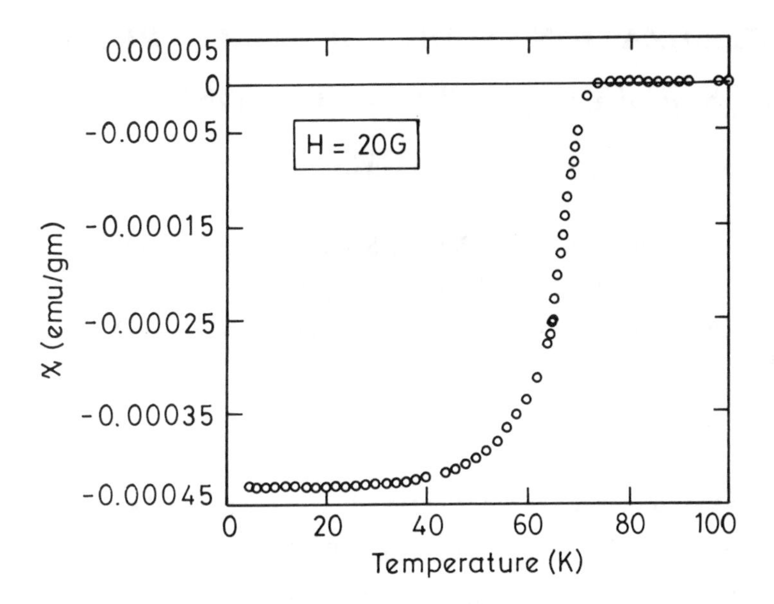

Figure 3. DC magnetic susceptibility versus temperature for the sample Y(12345).

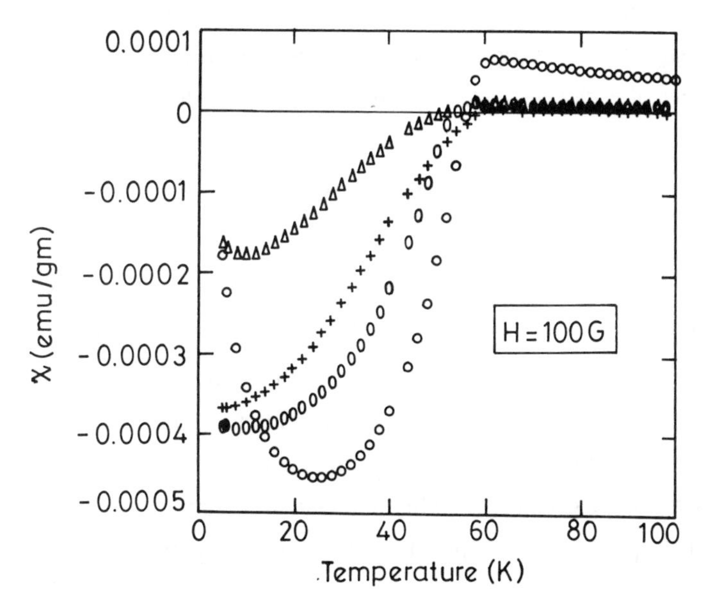

Figure 4. DC magnetic susceptibility measurements for the samples RE(12345).
[- Nd, - Sm, - Eu, - Gd].

OPTIMIZATION STUDIES OF HIGH CURRENT $Bi_2Sr_2CaCu_2O_x$ SUPERCONDUCTING TAPE SYSTEMS

T. Haugan, J. Ye, M. Pitsakis, S. Patel and D. T. Shaw
New York State Institute on Superconductivity,
State University of New York at Buffalo, Amherst, NY 14260 U.S.A.

Abstract

A review and comparison of different tape systems used to produce high critical current density (J_c), high critical current $Bi_2Sr_2CaCu_2O_x$ (2212) tapes is described. Four different tape configurations have been investigated: 2212/Ag "uncovered" single layer, Ag/2212/Ag "covered" single layer, [2212/Ag] x N multilayer tapes, and powder-in-(Ag tube) wires and tapes. Each type of system has advantages and disadvantages that affect the growth mechanisms and performance of the tapes in meeting engineering design criteria. Critical current densities, magnetic field properties, examples of growth and the advantages and disadvantages of each tape system are reviewed and updated.

Introduction

To be used for applications such as transmission lines, high field magnets, energy storage devices etc[1]... high transition temperature (T_c) superconducting wires and tapes will have to meet stringent design requirements in several areas such as 1) critical current density 2) magnetic field properties 3) cooling and winding stress/strain tolerance 4) growth process reliability and sensitivity 5) ability to scale up production 6) ability to maintain performance in long lengths of wires and 7) production cost profitability in units of [amp-feet/dollar].

One high T_c / substrate candidate being considered to meet these criteria is the 2212/Ag system. Interest in the 2212/Ag system began after initial reports that critical current densities (J_c's) of powder-in-(silver)tube wires measured at 4.2 K in magnetic fields greater than 20 T were higher than critical current densities of NbTi, Nb_3Sn, and $(Nb,Ta)_3Sn$ low temperature metallic superconductors commercially available.[2] The J_c of these powder-in-(Ag)tube wires at 77 K, however, was only in the range of 1000 to 2000 A/cm^2. Since that initial investigation, J_c's > 2000 A/cm^2 at 77 K for only one type of 2212/Ag tape system have been reported by only one research team.[3]

Herein, we present results of studies on four different 2212/Ag systems that could ultimately be used to make long lengths of wires. Each tape system has unique advantages and disadvantages that affect the growth processes of the 2212 phase and the ability of a final wire product made from each tape system to meet engineering design criteria. The critical current densities of these tape systems can (thus far) be consistently reproduced in a range from 4000 - 8000 A/cm^2 at 77 K in a self field, depending on the tape system. Critical currents at 77 K of the four systems vary from ~ 2 A to over 25 A. At 4.2 K, critical currents (and critical current densities) of the tapes are consistently 10 to 15 times higher than values achieved at 77 K.

Experimental

Of the four different tape systems studied, three were made by variations of brush-on coating and the last used powder-in-tube methodology. "Uncovered" single layer 2212/Ag tapes (0.5 cm wide and 2.0 cm long) were made by brush-on coating of powder-solvent-dispersant solutions onto 50 μm thick silver foil. Single layer "covered" Ag/2212/Ag tapes were made by brush-on coating the same powder solution onto one half side of a 25 μm thick foil. The extra half of silver foil was folded over to cover the brushed on powder layer after the dispersant was burned off by preheating the tape to ~ 500 ^oC. During processing at temperatures of ~ 885 ^oC and lower, the silver foil would grow soft and reactive enough so that the tape would be completely sealed after processing. Multilayer [2212/Ag] x N tapes were made similar to Ag/2212/Ag tapes however the surface area of coating was greater and the tapes were folded in jellyroll fashion to make the multilayer structure. Powder-in-tube wires were made by cold-drawing 7 mm diameter silver tubes bored out and packed with 2212 powder. After drawing, the powder-in-tube wires were cold-rolled to form tapes greater than or equal to ~ 60 μm in total thickness. Heat treatment temperature profiles used to process the tapes and wires were similar to heat treatment temperature profiles described previously.[4]

Critical current densities and transition temperatures (T_c's, defined as the temperature at which zero resistivity is reached upon cooling the sample) were determined with transport measurements using resistance-temperature systems calibrated weekly. Voltage taps were typically placed on the samples a distance of ~ 0.6 to 1.0 cm apart, with current leads attached near the ends of the samples. A criteria of 1 μv/cm was used to determine the critical current (I_c) and the cross-sectional superconductor area was used to calculate the critical current density (J_c).

Data from microbridge measurements are also included for comparison to earlier work.[4] Bridge sizes used for both studies were typically 0.3 to 0.5 mm in width and 50 to 100 μm long. Because the width of the bridge was usually longer than the length, the length of film that offered resistance to the critical current was roughly estimated to equal the width of the bridge.

Results

The highest, reproducibly obtainable, values of transition temperature and critical current density for each of the four 2212/Ag tape systems investigated are shown in Table I.

Table I. Review and Comparison of 2212/Ag Tape Systems (SUNY at Buffalo, effective 9/91)

Tape	Highest	77 K		4.2 K		77 K, 0.1 T
System	T_c (K)	I_c (A)	J_c (A/cm^2)	I_c (A)	J_c (A/cm^2)	J_c (A/cm^2)
2212/Ag	92	8	8300			2000
Ag/2212/Ag	90	7	5300	66	4.3×10^4	
[2212/Ag] x 3	86	20	2700	321	4.3×10^4	500
Pwd-in-tube	87	3	8200			
[2212/Ag] x 4 (*)	84	25	4000			
2212/Ag (**)	90	2	20,000	22	2.0×10^5	6500
2212/Ag (***)	90	3	35,000			7000

* Recent values

** Measured with microbridges, old values

*** Measured with microbridges, recent values

Analysis

Advantages and disadvantages of the four 2212/Ag systems investigated are summarized in this section.

A) 2212/Ag "Open" or "Uncovered" Single Layer Tapes:

Advantages:

1) Excellent texturing occurs at film/air and film/silver interfaces of "uncovered" 2212/Ag films. Especially, the flow of liquid material at the film/air surface may be enhance the overall flow of material more for this system than for any other type of 2212 tape system where the 2212 material is covered or enclosed with silver. Silver has a "sticking effect" for the Bi-oxides that can restrict the flow at silver/film interfaces and cause crystallization to occur, i.e. it is not uncommon to see needle-like crystals form at the edges of the films where molten Bi-oxide material has flowed out and crystalized on underlying pieces of silver. "Uncovered" films allow molten Bi-oxide material to flow with the least amount of restriction.
2) The oxygen content in "uncovered" films may be controlled more easily than for any other type of 2212/Ag growth system.
3) The highest T_c's can be consistently be obtained for "uncovered" films, which can possibly be related to the oxygen content of the films.[5]
4) "Uncovered" 2212/Ag tapes may be the easiest system of tape to irradiate for flux pinning experiments.
5) Increasing the overall flow of material promotes the growth of large aspect ratio grains that are nearly 100% dense and well connected in all directions. Weak links connecting grains in the perpendicular direction in Bi-oxide superconductors are thought to be the type of grain boundaries that have small magnetic field dependence (at 4.2 K- 35 K) and give the Bi-oxide superconductors their unique ability to maintain high critical current densities in high magnetic fields.[6] Large aspect ratio grains are

expected to improve the performance of long lengths of tapes by increasing the percentage of these favorable weak links, decreasing the average incidence angles of grain boundary junctions, and reducing the number and density of grain boundaries.

6) The density and coating uniformity of "uncovered" films are the highest of any 2212/Ag system tested, which should improve the ability of these tapes to consistently carry high critical current densities over extremely long lengths.

Disadvantages:

1) "Uncovered" 2212/Ag tapes must be coated with a protective material that will not cause undue stress to the film during winding and cooling processes.
2) The critical current carrying capability of a single layer tape is low, which will reduce the production profitability of this system in comparison to other systems tested.
3) Growth of single layer "uncovered" tapes is much more sensitive to changes in process parameters than growth of multilayer or thick film tapes.
4) There is a possibility of film separation from the silver layer during processing because of the thermal coefficient of expansion mismatch between the 2212 phase and silver. Ultimately, microcracks or cracks induced in the film caused by the mismatch could enhance or accelerate fatigue induced stress during winding and cooling processes.

B) Ag/2212/Ag "Covered" Single Layer Tapes:

Advantages:

1) A coated and protected tape is produced in final form that can be wound and used for applications.
2) During growth, escape of gases from unreacted carbonates or impurities in the powders can occur if the tapes are not completely sealed before processing.
3) Similarly, if the film is allowed to "breathe" before the silver melts

enough to completely seal the tape, the T_c can be slightly enhanced.

Disadvantages:

1) The single 2212 layer sandwiched between two silver layers seems to be more susceptible to cracking from growth processes than other systems tested, most probably because the ratio of silver thickness to film thickness is usually greatest for this system.
2) The "covered" single layer tape is also quite sensitive to processing conditions, much the same as the "uncovered" single layer tape.
3) The critical current carrying capability of "covered" 2212/Ag tapes is low, which affects the profitability of production.

C) [2212/Ag] x N Multilayer Tapes:

Advantages:

1) A coated and protected tape is produced ready for winding and testing.
2) The critical current carrying capability of multilayer tapes is high (e.g. 5-10 times higher than single layer tapes), which will increase the production profitability of this system.
3) Critical current densities of multilayer tapes are much less sensitive to changes in processing conditions than low current carrying tapes, a very important advantage when considering what type of tape system to use for production. The reproducibility of results and reliability of processing has, in general, been much greater for multilayer tapes than single layer tapes.
4) Only a slight decrease in critical current density has been observed for multilayer tapes containing up to four 2212/Ag layers (~ 80 - 100 μm superconductor thickness), in comparison to single layer tapes (~ 15 μm superconductor thickness) produced under similar annealing conditions. (Optimizing critical current densities of multilayer tapes using improved heat treatment programs has not yet been attempted, i.e. the value of J_c shown in Table I is expected to increase to nearly 7000 A/cm^2 from 4000 A/cm^2 by applying the latest improvements in heat treatment

temperature profiles.)

5) Multilayer tapes increase the number of film/silver interfaces which are critically important for texturing and alignment of the 2212 grains. Critical current densities of [2212/Ag] x N multilayer tapes are expected to be higher than critical current densities of (similar thickness) single layer tapes, provided the overall superconductor layer thickness is > 20 μm.

Disadvantages:

1) Depending on the methodology used to manufacture and process the multilayer tapes, gases from air or impurity phases in the powder layers can become trapped and cause bubbles that ruin the surrounding microstructures and degrade the critical current density.

D) Powder-in-Tube (Multifiliamentary) Wires and Tapes:

Advantages:

1) An important advantage powder-in-tube (multifiliamentary) wires and tapes have is their improved stress/strain properties.[7]
2) The use of multifiliamentary wires with filament sizes < 5 μm in diameter may help reduce the size of Bi-free defects that are an important intermediate phase in the growth of the 2212 phase from the melt. Controlling and reducing the size of the Bi-free defects may improve the growth process.
3) Multifiliamentary wires will have less c-axis orientation than tapes. This will improve the anisotropic performance of the wires in applied magnetic fields (at 77 K) but will reduce the overall performance in applied magnetic fields.

Disadvantages:

1) Critical current densities of each filament (and ultimately the entire wire) will be controlled by the worst section of the individual filaments. For Bi-oxide materials, growth of any large defects or outgassing from impurities in these microscopic filaments could cause catastrophic defects

that would completely ruin the current carrying capability of extremely long lengths or wires.

2) It will be difficult, if not impossible, to properly control the oxygen content of the 2212 phase while it is completely enclosed in silver. T_c's of powder-in-tube wires will probably always be lower than T_c's of "uncovered" 2212/Ag tapes.

3) Outgassing from impurities or unreacted carbonates in the powders will continue to cause problems for processing these wires. Also it isn't clear what will happen to air trapped in the wires packed with powders by mechanical methods.

Conclusions

Significant progress has been made in increasing the critical current densities of four $Bi_2Sr_2CaCu_2O_x$/Ag tape systems tested, each of which have advantages and disadvantages for meeting engineering design criteria in large-scale production. At 77 K, zero field critical current densities of the tape systems are consistently of the order of 4000 A/cm^2 to 8000 A/cm^2 and critical currents range from 2 to 25 A.

Acknowledgements

We would like to thank J. A. Rice and P. Haldar of Intermagnetics General Corporation, Guilderland, New York for critical current measurements of multilayer tapes at 4.2 K. This work was partially supported by the Electric Power Research Institute and the Martin Marietta Energy Systems Inc. - Oak Ridge National Laboratory, High Temperature Superconductor Pilot Center.

References

1. S. J. Dale, S. M. Wolf, T. R. Schneider, Electric Power Research Institute Report No. ER-6682 (1991).

2. K. Heine, J. Tenbrink, and M. Thöner, Appl. Phys. Lett. **55**, 2441 (1989).

3. J. Kase, K. Togano, H. Kumakura, D. R. Dietderich, N. Irisawa, T. Morimoto, and H. Maeda, Jpn. J. Appl. Phys. **29**, 1096 (1990).

4. J. Ye, S. Hwa, S. Patel and D. T. Shaw, in Superconductivity and Its Applications, AIP Conference Proceedings 219, edited by Y.H. Kao, P.Coppens and H.S. Kwok, (AIP, New York, 1991) p. 524.

5. N. Khare, S. Chaudhry, A.K. Gupta, V.S. Tomar and V.N. Ojha, Supercon. Sci. Tech. **3**, 514 (1990).

6. A. P. Malozemoff, presented at the Fifth Annual Conference on Superconductivity and Applications (Buffalo, NY Sept. 1991).

7. K. Sato, T. Hikata, H. Mukai, M. Ueyama, N. Shibuta, T. Kato, T. Masuda, M. Nagata, K.Iwata and T. Mitsui, IEEE Trans. Magn. **27**, 1231 (1991).

FABRICATION OF OXIDE SUPERCONDUCTING THICK FILMS BY A THERMAL SPRAYING METHOD

K. NUMATA, K. HOSHINO
Corporate R&D Center, Mitsui Kinzoku, 1333-2 Haraichi, Ageo, Saitama, 362 Japan

Y. YOSHIDA, K. INOUE, H. MAEDA
National Research Institute for Metals, 1-chome Sengen, Tsukuba, Ibaraki, 305 Japan

Y. KAMEKAWA, M. SUZUKI
R & D Center, Nippon Keiki Works, Ltd.
4-3-5 Goshogaoka, Moriya, Kitasoma-gun, Ibaraki, 302-01 Japan

ABSTRACT

$Bi_2Sr_2CaCu_2O_x$ thick film was fabricated by laminar flow thermal spray onto various types of substrates, such as; stainless steels and nickel-based alloys with buffer layers of yttrium stabilized zirconia (YSZ) and silver. After annealing the specimens, it was observed that the both buffer layer materials can effectively prevent the reaction between the oxide superconductor and the substrate, although the oxide superconductor sometimes peeled off from the YSZ layer. The highest critical temperature so far is 88 K by annealing the thick film in the low oxygen pressure atmosphere. This process preparing the thick film will be applied to the fabrication of a magnetic shield vessel for use in the biomagnetic measurements.

INTRODUCTION

Recently there is growing interest in the application of oxide superconductors as magnetic shielding for biomagnetic measurements, especially for the measurement of magnetic fields from the human brain. Hoshino et. al. fabricated rather large vessels (32 cm in diameter and 64 cm in length) by cold isostatic pressing and sintering. They measured the magnetic shielding factor S (S = external applied field / magnetic field inside the vessels) to be about 10^5 even in the low frequency range of 0.1-10 Hz [1,2]. Though the attenuation ratio of 10^5 is promising for magnetoencephalography, the vessels are not large enough to cover the whole human body. In addition the fabrication process has some problems in the apparatus and the cost of enlarging the vessels.

Tachikawa et. al. fabricated Y-Ba-Cu-O thick film with a critical temperature of 90 K by low-pressure plasma spraying [3]. Critical temperatures of 96 K and 107 K have been reported for the Bi-Sr-Ca-Cu-O films [4] and Bi-Pb-Sr-Ca-Cu-O films [5], respectively. However, it is difficult to fabricate a Bi-system thick film due to compositional change or decomposition caused by the high temperature of the plasma flow [6].

Since high deposition rate, controllable crystallinity and easiness of enlargement in this method are attractive, we initiated the research on the fabrication of very large vessels coated with Bi-system oxide superconducting thick film by a thermal spraying method.

In this paper, we report the synthesis of Bi-2212 powder suitable for thermal spraying and the superconducting characteristics of the sprayed thick film onto several kinds of substrates with the buffer layer of yttrium stabilized zirconia (YSZ) or silver..

EXPERIMENTAL

$Bi_2Sr_2CaCu_2O_x$ was synthesized by a solid state reaction among Bi_2O_3, $SrCO_3$, $CaCO_3$ and CuO powders. Appropriate amounts of the powders were mixed and calcined at 800 ℃ for 24 hours in air. The calcined powder was then pulverized and fired at 870 ℃ for 24 hours. A powder with particles of 53 - 105 μm in diameter was prepared by crushing and classifying the fired powder.

Figure 1 shows a schematic diagram of the laminar flow plasma spraying apparatus (APS 7000, Onoda Cement Co., Ltd.). This apparatus has the advantage that the feeding position, where the powder is fed into the flame, can be varied to control the composition and the crystallinity. Using the laminar flow plasma spraying, the powder was fed from particle nozzle into the plasma flow and sprayed onto various kinds of small substrates (4 mm x 30 mm), such as nickel, stainless steel (SUS-304, SUS-430), Inconel and Hasteloy (Inconel and Hasteloy are Ni-based alloy) with the buffer layer of yttrium stabilized zirconia (YSZ) or silver. When a substrate is being coated with the superconducting material, the material is sprayed from a fixed source whilst the substrate is moved. The plasma spray parameters for oxide superconductor are listed in Table 1. These specimens were then annealed at 870 - 890 ℃ in air. Then some of the specimens were annealed at 600 ℃ in low oxygen pressure atmosphere to improve the critical temperature [7]. The critical temperature was measured by the conventional four-probe technique. The microstructure was observed by scanning electron microscopy

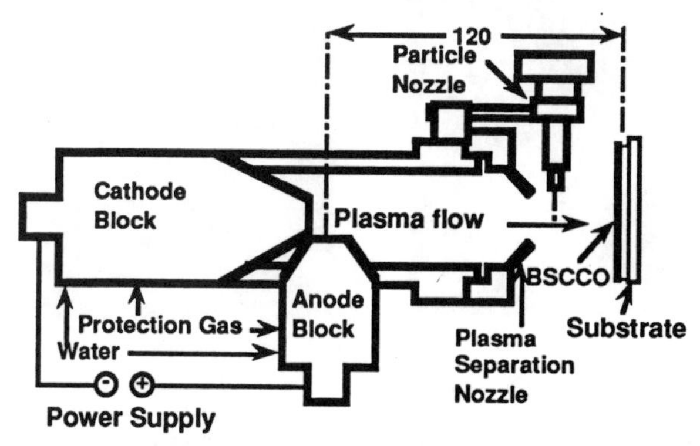

Fig.1 A schematic diagram of the laminar flow plasma spraying apparatus.

(SEM) and electron probe micro-analysis (EPMA).

Table I Process parameters for plasma spraying

Arc voltage	132 V
Arc current	80 A
Plasma gas	Air
Protection gas	Ar
Plasma separation gas	Air
Powder feed rate	6 g/min.
Carrier gas	Air
Spray distance	110 - 120 mm

RESULTS AND DISCUSSION

Figure 2 shows the SEM photograph of the classified powder with particles of 53 - 105 μm in diameter. This powder size was suitable for transporting from feedstock to the plasma flow by air flow.

Figure 3 shows the cross sectional view of the specimen annealed at 870 ℃ for 3 hours. In this specimen, the substrate was Inconel and the buffer layer was YSZ. The YSZ layer adhered well to the substrate and the oxide superconducting layer was also deposited to the YSZ buffer layer by thermal spraying. No apparent reaction in the either interface was observed. Similar views were observed in case of other substrates with both the YSZ and the silver buffer layer. However, it was sometimes observed that the oxide layer had come away from the YSZ layer after annealing. From thermal dilatation data by Emmen [8], thermal dilatation coefficient of $Bi_2Sr_2CaCu_2O_x$ is 1.1 x 10^{-5} K^{-1} (600 ℃). This value is closer to the coefficient of YSZ (9.6 x 10^{-6} K^{-1} (400 ℃)) than that of silver (2.1 x 10^{-5} K^{-1} (527 ℃)). So the mismatch of the

Fig. 2 SEM photograph of the classified oxide superconducting powder.

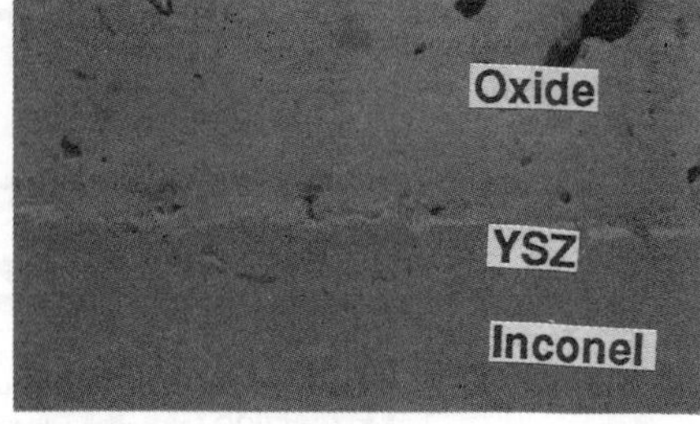

Fig. 3 SEM photograph of the cross sectional view of the specimen.

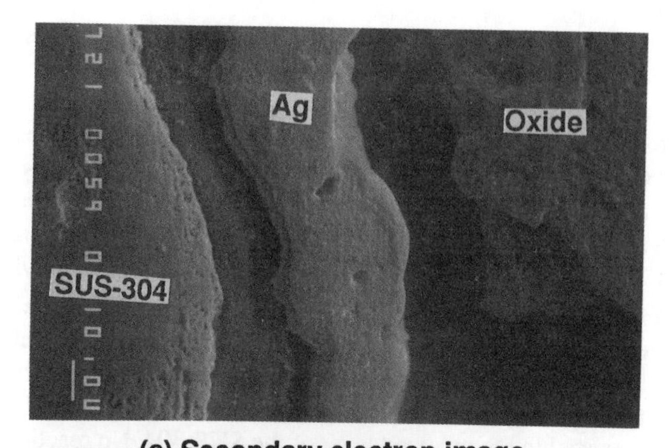

(a) Secondary electron image

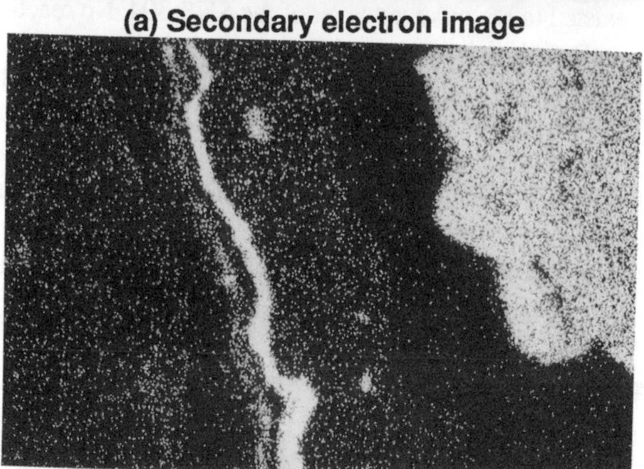

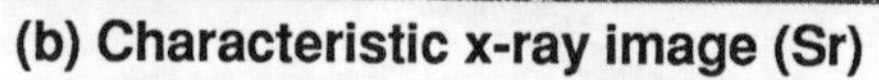

(b) Characteristic x-ray image (Sr)

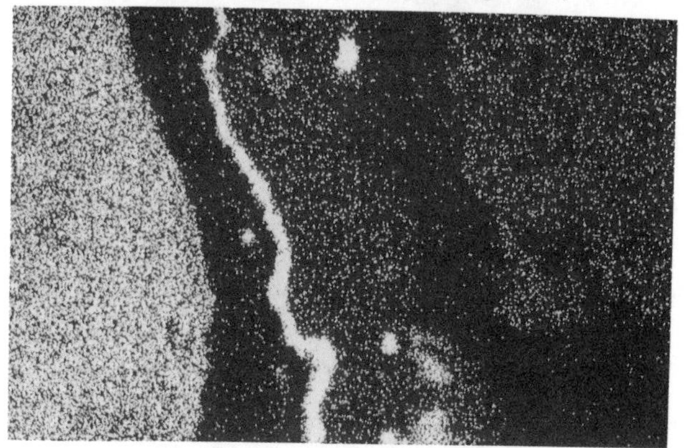

(c) Characteristic x-ray image (Cr) 10μm

Fig. 4. SEM photograph and characteristic x-ray photographs of a cross sectional view of the specimen. The substrate and the buffer layer are SUS-304 and silver, respectively.

thermal dilatation is not a serious reason for peeling off. Although there was no apparent reaction between oxide superconductor and YSZ in SEM photo, some reaction might occur in the interface during annealing.

It is noted that the oxide superconductor reacted to the substrate through the buffer layer of silver, when the thickness of the buffer layer was not sufficient. Fig. 4 shows the SEM photograph and the characteristic x-ray photographs of a cross sectional view of the specimen with the substrate of SUS-304 and thin buffer layer of silver annealed at 870 ℃ for 3 hours. The thickness of the buffer layer was 100 μm after spraying and about 50 μm after annealing. The buffer layer had come away from both the oxide layer and the substrate. After annealing, the buffer layer became thinner and some defects were observed in the layer. It was observed that strontium - chromium oxide compound was formed between the buffer layer and the substrate. It was also observed that calcium and silicon co-existed in the oxidized layer of the SUS-304. No reaction was observed when the substrate was nickel.

Figure 5 shows the transition curves of the specimens (substrate : SUS304, buffer : silver) annealed at 870 ℃ for 3 hours (curve (a)) and 890 ℃ for 15 min. (curve (b)). A current of 1 mA was used for the measurement. Zero resistance was obtained at 66 K for 870 ℃ annealing and 78 K for 890 ℃ annealing. In case of 870 ℃ annealing, the transition width is greater than 10 K. On the contrary, the transition width became narrower, about 6 K, when annealed at 890 ℃. This sharp transition may be due to a homogeneity of the oxide superconductor by partial melting.

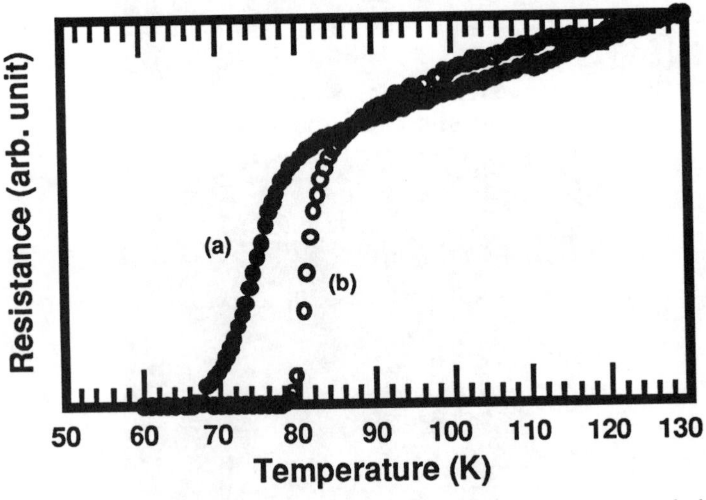

Fig. 5. Resistive transition curves of the sprayed specimens annealed at (a) 870 ℃ for 3 hours in air (b) 890 ℃ for 15 minutes in air

Figure 6 shows the transition curve of the specimen annealed at 600 ℃ for 3 hours in 100 ppm O_2 after annealing at 890 ℃ for 15 minutes in air. The critical temperature was improved to 88 K, higher than that of bulk specimen annealed at 600 ℃ in 200 ppm oxygen atmosphere reported by Groen [7]. This critical temperature should be high enough to use the thick film as magnetic shield at 77 K.

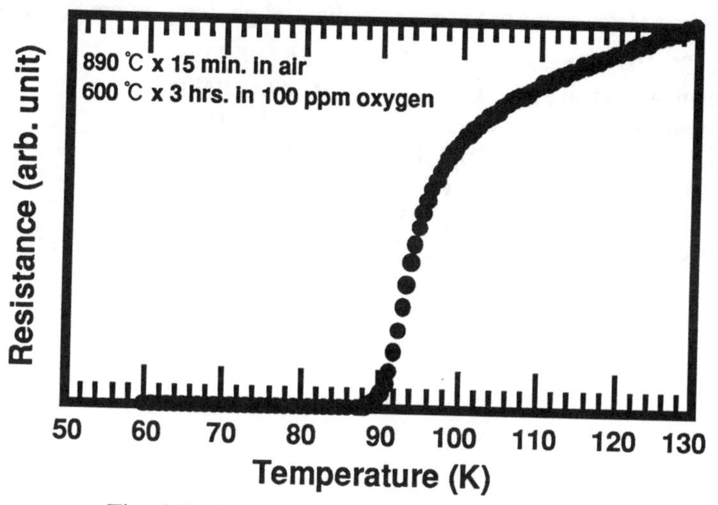

Fig. 6. Resistive transition curve of a specimen annealed at 600 ℃ for 3 hours in 100 ppm oxygen.

CONCLUSIONS

A critical temperature of 88 K was obtained by annealing at 890 ℃ for 15 minutes in air and 600 ℃ for 3 hours in 100 ppm oxygen atmosphere with the substrate of SUS-304 and the buffer layer of silver. Though both YSZ and silver can prevent the reaction as a buffer layer, silver will be a better material as the buffer layer. Because it was sometimes observed that the oxide layer had come away from the YSZ layer after annealing.

This process for preparing the thick film will be applied to the fabrication of a magnetic shielding vessel for the biomagnetic measurement. In order to fabricate large magnetic shielding vessel, the thermal dilatation and the magnetism of the substrate must be considered. Inconel alloy 625 is one of the most suitable substrates at the present time because of the nonmagnetism, the small thermal dilatation coefficient (1.4×10^{-5} K^{-1} (538 ℃)) and the high heat resistance. In the future, however, ferromagnetic stainless steel with smaller thermal dilatation coefficient will be able to be applied as the substrate for the magnetic shielding vessels, if the cylindrical substrate is large enough to spray the oxide superconducting materials onto the inner surface of the pipe.

REFERENCES

1. K. Hoshino et.al., IEEE Trans. Mag., 27, 2202 (1991).
2. A. Koike et. al., to be published in Proceedings of International Conference on Materials and Mechanisms of Superconductivity, High-Temperature Superconductors (M^2S-HTSC III), Kanazawa Japan, July 22 - 26 (1991).
3. K. Tachikawa et. al., Adv. Cryog. Eng., 36, 473 (1990).
4. A. Asthana et. al., Appl. Phys. Lett., 53, 799 (1988)
5. Y. Yoshida et. al., Jpn. J. Appl. Phys., 28, L639 (1989)
6. R. Suryanarayanan et. al., Physica C, 162, 139 (1989).
7. W.A. Groen et. al., Phisica C, 165, 55 (1990).
8. J.H.P.M. Emmen et. al., Physica C, 165, 293 (1990).

GROWTH OF THE HIGH-T_c PHASE, FLUX PINNING, AND CRITICAL CURRENT OF Bi-Sr-Ca-Cu-O SUPERCONDUCTORS

H.M. Shao, E.L. Chen, C.Y. Lee, M. Yang, L.W. Song, and Y.H. Kao

Department of Physics, State University of New York at Buffalo,

Amherst, NY 14260

Abstract

High-T_c superconductors of Bi-Sr-Ca-Cu-O were synthesized by means of repeated processes of repressing-sintering to yield a high content of the $Bi_2Sr_2Ca_2Cu_3O_{10}$ phase which exhibits consistently a transition temperature around 110K from ac susceptibility and a temperature at zero resistance above 100K from transport measurements. The critical current density was determined from magnetic hysteresis using a SQUID magnetometer. Comparison of the magnetic field dependence of critical current density is made for materials obtained with different processing conditions.

I. INTRODUCTION

Much effort has been directed towards the preparation of the high-T_c phase of the rare-earth-free superconductor Bi-Sr-Ca-Cu-O (BSCCO) system with the composition ratio Bi:Sr:Ca:Cu=2:2:2:3 (BSCCO-2223) which exhibits a transition temperature (T_c) around 110K. Several methods reported are known to be successful in achieving such a single-phase compound, e.g. by varying the nominal content of Cu or Ca in the mixture [1], and by partial substitution of Pb for Bi [2,3].

In the synthesis of BSCCO-2223 by solid state reaction, $SrCO_3$ and $CaCO_3$ raw materials are commonly used. In the reaction, the carbonates are believed to decompose first into oxides (SrO and CaO) which then interact with the other oxides (CuO and Bi_2O_3) present to form the desired perovskite structure of BSCCO. Since SrO and CaO have high melting points at 2430°C and 2578°C, respectively, and both have the NaCl structure which is quite different from that of the perovskite, it is desirable that the solid state reaction should take place at high temperatures. On the other hand, the melting point of Bi_2O_3 is only 817°C, it is also necessary to maintain the reaction for a reasonably long time at temperatures close to this melting point in order to achieve the required stoichiometry of 2223 and to prevent loss of Bi. These seemingly conflicting requirements can be satisfied by using precursor compounds of Sr-Ca-Cu-O and Bi-Pb-Ca-O synthesized with desired stoichiometry, and then subject these precursors to final sintering.

Although the values of T_c around 110K at the onset of superconductivity in bulk BSCCO-2223 can be consistently reproduced, the critical current density J_c in this compound is still quite low in comparison to that in the other well-known high-T_c superconductor Y-Ba-Cu-O (YBCO). The relatively low flux-pinning strength in BSCCO is a problem presently under more detailed investigations with different stoichiometry variations and heat treatment. The common bulk ceramic

BSCCO can only support a weak transport current at 77K and J_c decreases rapidly in a magnetic field [4]. Many factors can be attributed to the low values of transport J_c, e.g. mismatch between adjacent grains with different orientations, presence of impurity phases (such as BSCCO 2212 or 2201) or structural disturbance on the grain surface, and weak coupling between grains with a short coherence length [5]. For the interest of increasing the J_c values in bulk BSCCO for possible large-scale technical applications, it is desirable to explore various techniques in order to improve the surface morphology, alignment of grains, and material densification.

In this report we present our recent results on single phase BSCCO-2223 compounds obtained by synthesizing Sr-Ca-Cu-O and Bi-Pb-Ca-O precursors coupled with Pb- substitution. Different conditions of heat treatment, mechanical grinding, and repressing were used in the process. The magnetic-field dependence of the critical current density, i.e. J_c-H curves, were compared for three groups of BSCCO-2223 superconductors prepared under different processing conditions.

II. EXPERIMENTAL

High purity Bi_2O_3, $SrCO_3$, $CaCO_3$, CuO, and PbO were used to synthesize precursors of $(Sr,Ca)CuO_2$ with composition ratio Sr:Ca:Cu = 2:1:3 and $(Bi,Pb,Ca)_2O_{3-y}$ with composition ratio Bi:Pb:Ca = 1.6:0.4:1.0. The calcining conditions are 24 hours at 950°C and at 800°C, respectively. These precursors were repeatedly ground, pressed, and calcined at least twice before used for sintering the final BSCCO compound. Proper amounts of these two precursors were thoroughly ground and mixed, then pressed into the shape of pellets and sintered at 850°C for 100-200 hours, cooled to 500°C at a rate of 50°C/hour, followed by annealing in oxygen atmosphere for over four hours and then gradually cooled to room temperature.

In the process described above, mechanical grinding, repressing, and sintering were usually repeated several times in order to assure complete reaction and to improve the sample homogeneity. However, it is also known that mechanical grinding could lead to serious degradation of the high-T_c phase [6,7]. For this reason, we have prepared three different groups of samples with different mechanical treatments between the sintering processes. Group A specimens were obtained by sintering for 100 hours, subject to regrinding and repressing, followed by another sintering for 100 hours. Group B specimens were prepared in a processing sequence similar to group A but without regrinding. Group C specimens were prepared by continuous sintering of 200 hours after calcining without intermediate regrinding and repressing. Preparation and measurements of the specimens of each type were repeated several times, all yielded results of the same nature.

All the samples studied in the present work were characterized by powder x-ray diffraction, indicating single phase of BSCCO-2223. The magnetization curve and hysteresis were determined using a Quantum Design SQUID magnetometer. The ac magnetic susceptibility was measured by using a double-coil induction method with an ac magnetic field amplitude below 0.05 Oe. The temperature-dependent resistivity was measured by the standard four-terminal method.

III. RESULTS AND DISCUSSION

A typical resistive transition curve and x-ray diffraction pattern are shown in Fig. 1. Specimens of all three groups show virtually the same behavior with a transition temperature in the 105-107K range. The temperature dependence of ac magnetic susceptibility is shown in Fig. 2, with only some minor variations in the transition region for specimens of different groups. The main difference in samples of different groups can be clearly seen in the dc magnetization curves, as shown in Fig. 3. The field variation of the critical current density at 5K derived from the hysteresis curves is shown in Fig. 4.

The J_c-H curves in Fig. 4 show that critical current density of group-A specimens decreases rapidly with H while the group-B specimens exhibit a higher J_c and more gradual field dependence. On the other hand, group-C specimens all show a much lower value of J_c and weak dependence on magnetic field. These results demonstrate the strong influence on the critical current density due to intermediate grinding and repressing between sintering processes.

The group-A specimens suffered from grinding and repressing between sintering, in this process grinding is believed to have caused some damage to the connectivity between neighboring grains [7]. Although the final annealing might have recovered the intergranular coupling to some extent, the degradation is evidenced by the rather fast decrease of J_c in the low field region, a sign indicative of weakly coupled grains [5].

The group-B specimens show higher J_c values and weaker field dependence than those of group-A. Repressing without grinding between sintering is believed to have improved the material density as well as the intergrain alignment [8,9]. This is also supported by the powder XRD spectrum shown in Fig. 5. The gradual field dependence and higher values of J_c in the high field region suggest that intergranular coupling is indeed enhanced along with a possible increase in the intragrain flux pinning.

The group-C specimens show a low value of J_c even though the structure is still largely single phase BSCCO-2223 as found in XRD (Fig. 5). The degradation in J_c can be attributed to a decreased density due to thermal expansion as a result of long time sintering at a temperature around 850°C. The sample diameter can increase by as much as 50% during this process. The loosely connected grains in the ceramics result in a low value of J_c which is insensitive to magnetic field.

It should be remarked that although the specimens of group-B show a superior superconducting behavior, repeated processes of alternate repressing and sintering do not necessarily give rise to a linear increase in J_c. Repressing can only improve the intergranular coupling to some extent, but it would not have a strong effect on the flux pinning inside the grains.

The two step transition observed in our ac susceptibility measurements (Fig. 2) also suggest that there may exist a minor new phase with T_c higher than that of BSCCO-2223. This could be a consequence of partial Pb-substitution in the material. It seems possible that some grains may have a Pb-rich component residing on the surface layer, thereby showing a higher shielding diamagnetism.

IV. CONCLUSIONS

By using precursors of Sr-Ca-Cu-O and Bi-Pb-Ca-O with proper heat treatment, single phase bulk BSCCO-2223 superconductors can be consistently obtained. Although long time continuous sintering could improve the content of single phase material with a T_c around 110K, the critical current density suffers serious degradation due to loosely connected structure in the ceramics caused by thermal expansion. Repressing without regrinding between successive sinterings has been demonstrated to improve the intergranular coupling as evidenced by the superior behavior of J_c in magnetic field. The useful effects of repressing between sinterings also allow more alternatives in the processing of superconducting wires and tapes of BSCCO-2223 for large scale applications.

The present research is supported in part by DOE and NYSIS.

REFERENCES

1. Y.T. Huang et al., Appl. Phys. Lett. 56 (1990) 779.
2. N. Kijima et al., Jpn. J. Appl. Phys. 27 (1988) L1852.
3. T. Hatano et al. Croygenics 30 (1990) 611.
4. M. Matsuda et al., Jpn. J. Appl. Phys. 27 (1990) L1650.
5. H. Küpfer et al., Cryogenics 28 (1988) 650.
6. M. Awano et al., Jpn. J. Appl. Phys. 29 (1990) L254.
7. T. Kanai, T. Kamo, and S.P. Matsuda, Jpn. J. Appl. Phys. 29 (1990) L412.
8. P.S. Mukherjee et al., Solid State Comm. 74 (1990) 477.
9. C.J. Kim et al., J. Mat. Sci. Lett. 9 (1990) 774.

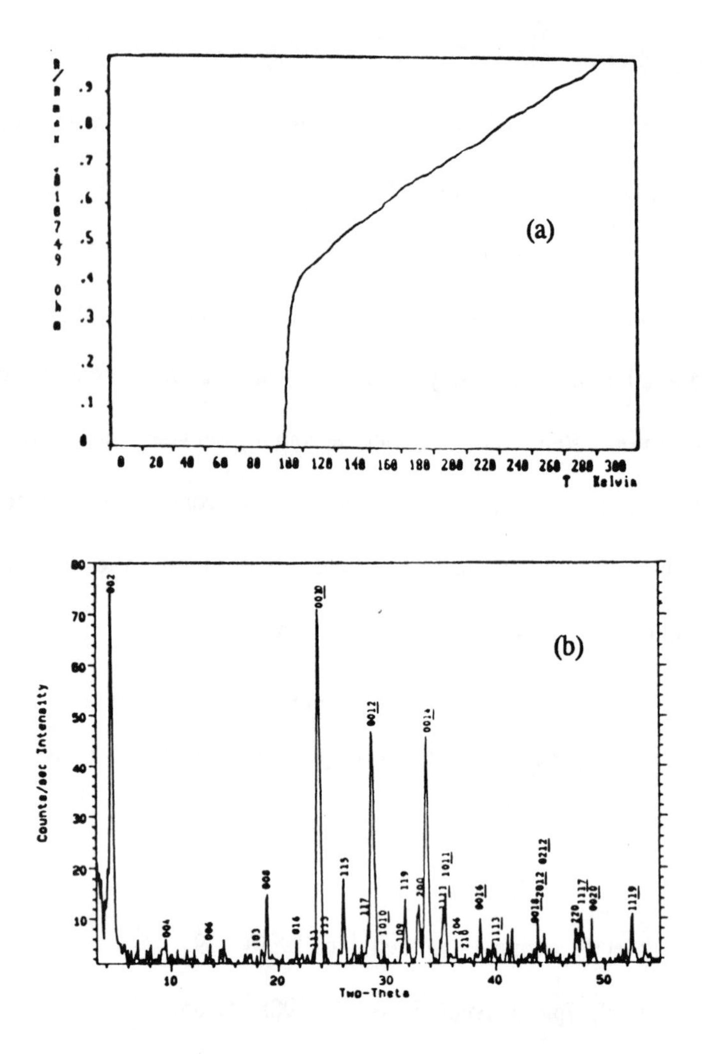

Fig. 1. A typical resistive transition curve (a) and x-ray diffraction pattern (b) for the specimens studied in the present experiment.

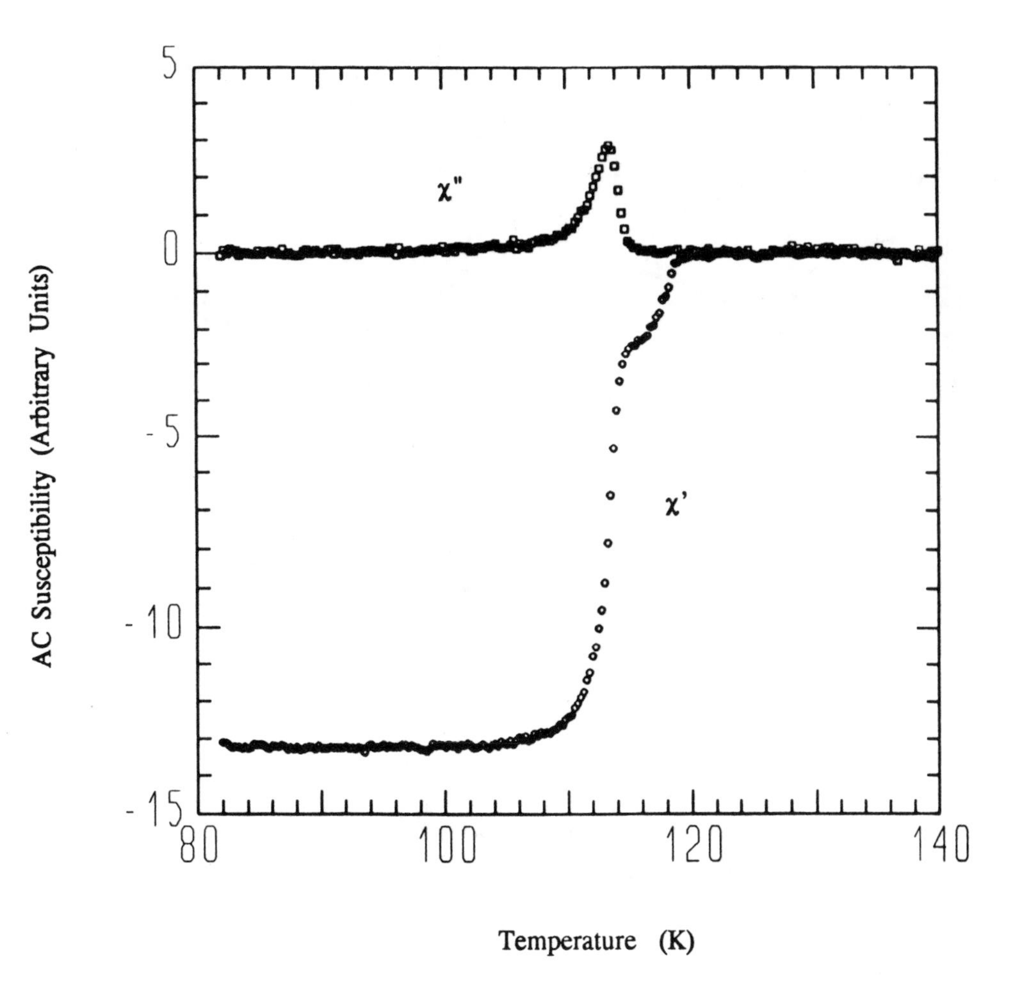

Fig. 2. A typical temperature dependence of ac magnetic susceptibility of the specimens studied in the present work.

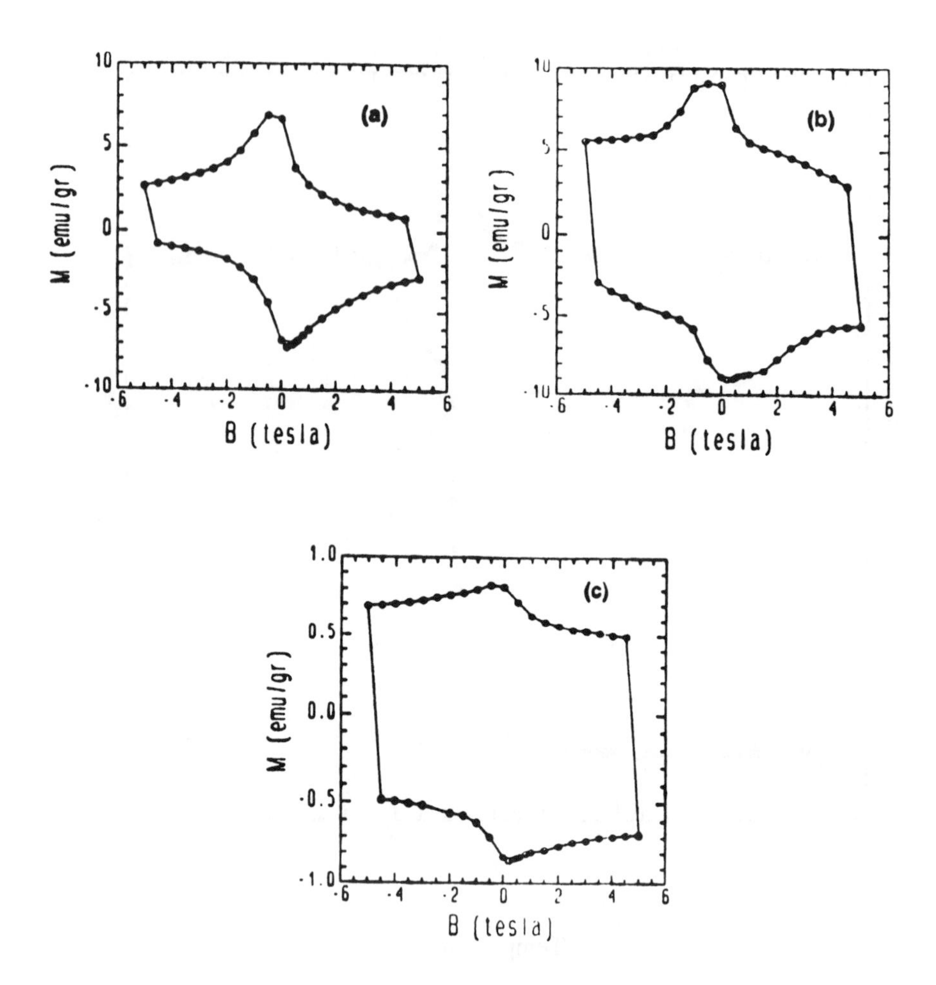

Fig. 3. Magnetic hysteresis curves obtained at 5K showing the difference in critical current density for specimens of group-A (a), group-B (b), and group-C (c). (Notice the change of scale in (c)).

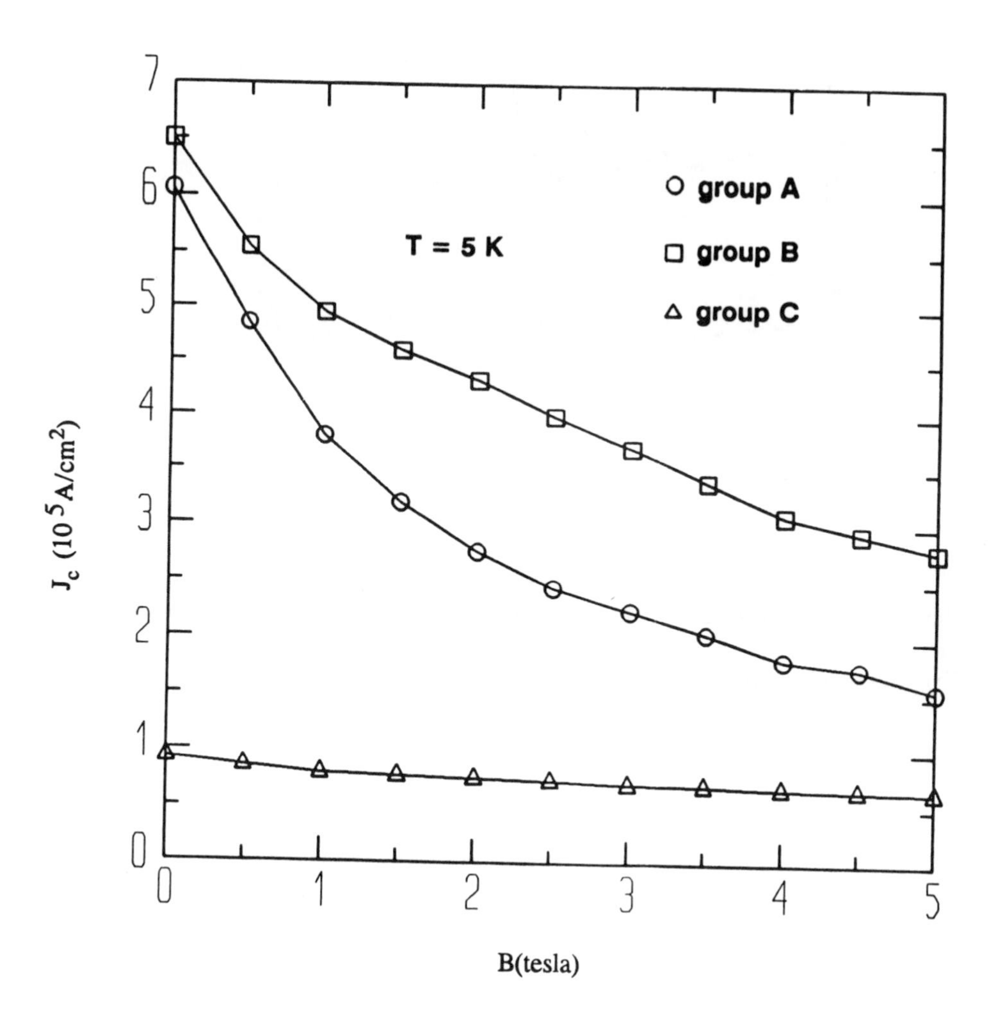

Fig. 4. Field dependence of critical current density derived from Fig. 3.

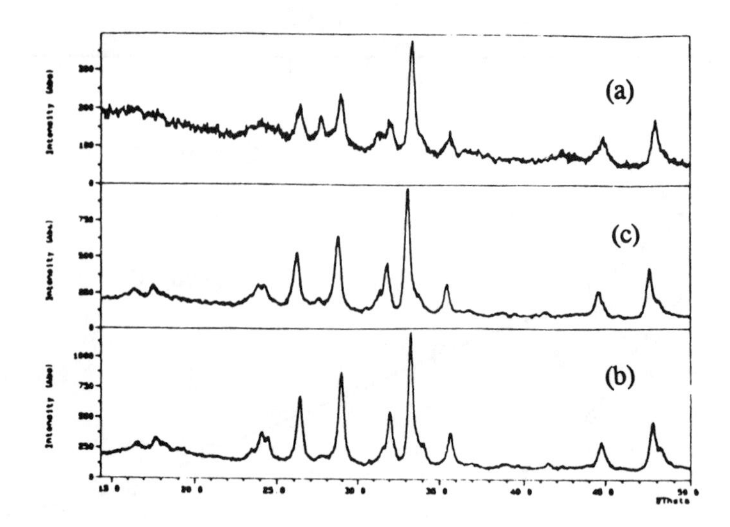

Fig. 5. X-ray powder diffraction patterns for specimens of group-A (a), group-B (b), and group-C (c).

A PRECURSOR METHOD FOR REACTING AND ALIGNING $Bi_2Sr_2Ca_2Cu_3O_{10}$

N.X.Tan, A.J.Bourdillon, and W.H.Tsai

Department of Materials Science & Engineering
State University of New York at Stony Brook
Stony Brook, NY 11794-2275, U.S.A.

ABSTRACT

Preferred orientation during crystal growth of Bi-Sr-Ca-Cu-O superconductors results in a peculiar platelet morphology. Aligned platelets can be formed by reaction between Pb-Bi-O and a ceramic precursor, such as $SrCaCu_2O_{4+y}$. The alignment is due to gravitational pull on the liquid phase during sintering. The processing, microstructure and superconducting transport properties of these aligned BSCCO materials have been characterized. Scanning electron microscopy shows that thick, fiber-textured, films grow in single domains. Zero resistivity at 100 K was observed in a textured specimen sintered for 100 hours. With differential thermal analysis, the flux action of lead, which accelerates the kinetics of $Bi_2Sr_2Ca_2Cu_3O_{10+y}$ formation, is understood.

I. INTRODUCTION

Diffusion texture results from crystal growth in a vertical plane due to the flow of liquid phase during sintering. A major effort has been devoted by many researchers to processing of aligned high-temperature superconducting material in order to reduce the resistive effects of intergranular weak links [1,2,3]. Much of this work has employed partial melting in the Y-Ba-Cu-O phase diagram [4]; but the success of this enterprise is limited by many reasons including processing difficulties associated with displacive phase transformations in $YBa_2Cu_3O_{7-x}$, high-angle boundaries between domains containing aligned micro-crystals, and the comparatively high (partial) melting temperature, i.e. higher than the melting temperature of Ag sheaths. In Bi-Sr-Ca-Cu-O (BSCCO) systems, two high-temperature superconducting phases are commonly observed, i.e. $Bi_2Sr_2Ca_2Cu_3O_{10+y}$ [2223] and $Bi_2Sr_2Ca_1Cu_2O_{8+y}$ [2212] with transition temperatures (T_c) at ~ 110 K and 80 K [5] respectively. [2223] can be stabilized by doping with Pb [5]. The BSCCO superconductor phases have a peculiar morphology - i.e. platelets along the crystallographic a-b planes. These platelets can be easily grown. Texture processing by mechanical stress has been reported [6,7] on BSCCO systems, e.g. by pressing and cold-rolling. In the former case, the BSCCO materials were repeatedly pressed in a uniaxial press, while in the latter case, the platelets of BSCCO were aligned by directional shear stress. Here we report a novel method which is used to align the BSCCO materials, i.e. by diffusion through the reaction of liquid phase Pb-Bi-O with a ceramic precursor.

II. EXPERIMENTAL PROCEDURE

The multi-step materials preparation is described below. Microstructure was studied on Scanning Electron Microscope (SEM). The secondary electron images (SEI) were recorded on an ISI SX 30 SEM and back-scattered electron images (BEI) on a JEOL JSM 840 SEM, equipped with Link Systems energy-dispersive spectrometer (EDS) for phase analysis with laboratory standards. The superconducting transition temperature (T_c) and critical current density (J_c) of the specimens were measured resistively by standard 4-probe methods.

(1) Fabrication of Pb-Bi-O Mixture and $SrCaCu_2O_x$ Precursor

The Pb-Bi-O mixture was made by mixing Bi_2O_3 and PbO in a molecular ratio of 1:3. The powder was then reacted, by sintering, at 630°C for 12 h in air, followed by natural furnace-cooling. Intermediate compounds are formed in accordance with the phase diagram [8].

The appropriate mixture of $SrCO_3$, $CaCO_3$ and CuO (all $\geq$ 99.99% pure), weighed according to the chemical composition $SrCaCu_2O_x$, was prepared by conventional ceramic powder procedures as follows: mixing, calcining at 900°C overnight, grinding, pelletizing in a uniaxial press (with 1/2" diameter), and finally sintering in air at 980°C for 12 h, followed by natural cooling in the furnace. The [0112] product is tough and dense (~ 30% shrinkage), with uniform single phase after the solid state reaction (figure 1).

(2) Diffusion texture

The Pb-Bi-O mixture was deposited on several [0112] substrates in each of various ways, before final sintering. These ways were selected to increase the thickness of the textured film and to elucidate the growth mechanism:

- A Pb-Bi-O slurry was made by suspension of Pb-Bi-O powder in methanol. Pb-Bi-O was deposited by sedimentation onto a 2 mm thick, sintered, [0112] substrate (specimen A);
- Pb-Bi-O powder was directly brushed onto a 5 mm thick, sintered, [0112] substrate (specimen B);
- A 3 mm thick recess was drilled into an **unsintered**, 6 mm thick [0112] pellet. The recess was filled with Pb-Bi-O powder before sintering (specimen C); and
- A recess was drilled into a 6 mm thick sintered pellet, and the hole was filled with Pb-Bi-O (specimen D).

Figure 1. BEI from polished surface of sintered [0112] precursor showing a uniform single phase .

Multiple steps of deposition and sintering were used to thicken the diffusion-textured film, except in the case of specimen B, where the brushed-on layer was comparatively thick.

Specimen A demonstrates the basic features and growth rate of the diffusion textured films. During sequential steps, the Pb-Bi-O was deposited on either side after turning the pellet upside down. After each deposition, the specimen was sintered at 840°C for 24 h. Figure 2 is a typical BEI of one fracture surface of the pellet, where the substrate and the diffusion-grown layer are separated by a flat interface boundary. The platelets, oriented perpendicular to the substrate pellet, form a single domain as shown in this figure. After each of five cycles of deposition and sintering, a part of the specimen was fractured and the film examined. Figure 3 shows the linear correlation between the thickness of the film with the summed weight of the deposited Pb-Bi-O mixture. This linear correlation shows that the film growth, under the conditions used, is not diffusion limited. Figure 4a shows a SEI of specimen A, with films on both sides of the [0112] substrate grown after 3 deposition cycles on each side. The superconducting layers are about 60 μm thick. The figure shows that texture growth can be produced from both sides of the pellet, and that the textured surfaces may be made to join at high angle boundaries. Figure 4 b is recorded at higher magnification from a region of the film where an uniform aligned microstructure is shown. To form the [2223] phase, a long sintering time is needed. After sintering for 100 h, a specimen was measured with zero resistivity at 100 K (figure 5). The J_c of this specimen is greater than 1000 A/cm^2, the measurement being limited by Ohmic heating at the contacts.

Figure 2. Typical BEI micrograph from specimen A shows flat interface between diffusion-grown superconducting layer and [0112] substrate.

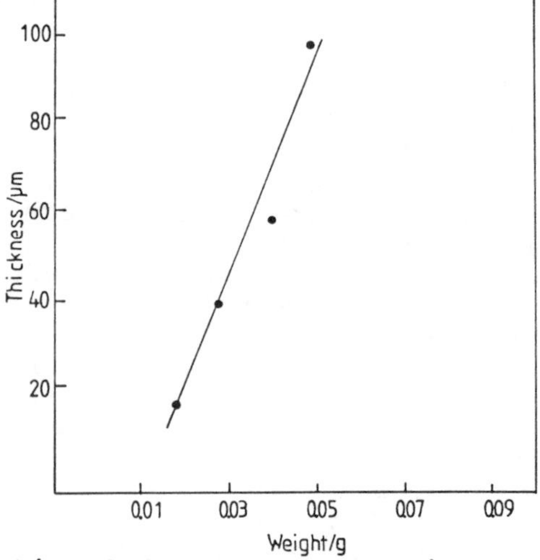

Figure 3. Correlation between summed weight of Pb-Bi-O deposited by sedimentation and thickness of grown superconducting layer, illustrated with linear fit.

Figure 4 a. SEI micrograph on the fractured cross-section of specimen A, after 3 deposition cycles of Pb-Bi-O on each of top and bottom sides.

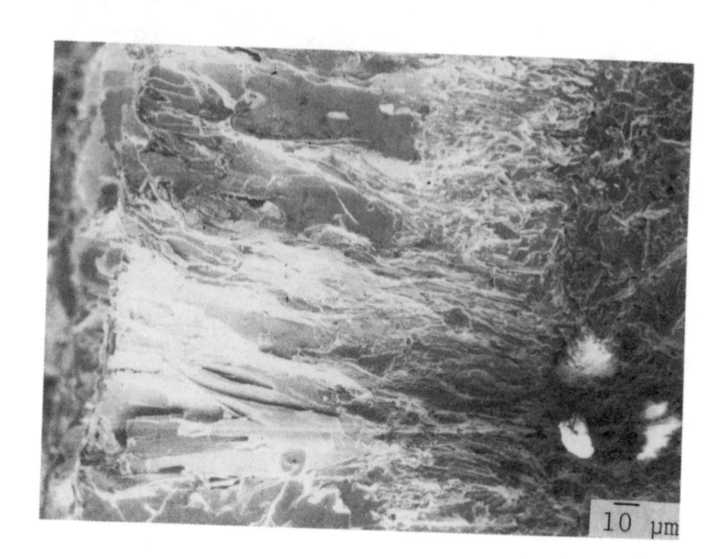

Figure 4 b. SEI micrograph showing microstructure detail in same specimen

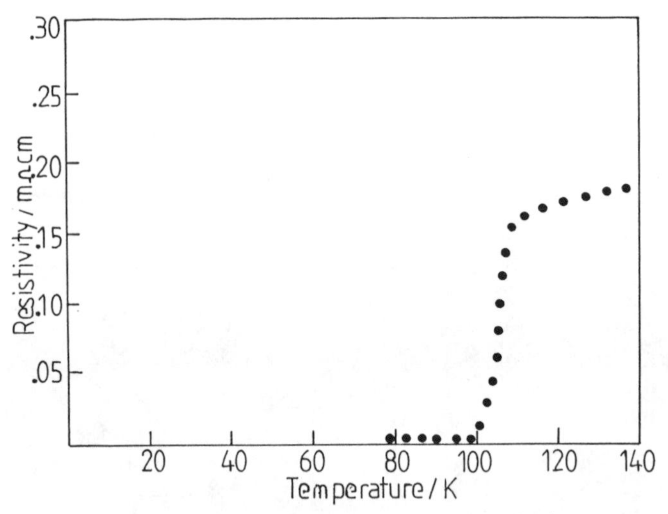

Figure 5. Temperature dependent resistivity of specimen A after sintering for 100 h.

Figure 6. SEI from fractured section of specimen B.

Specimen B was also sintered at 840°C for 24 h. Figure 6 is a SEM micrograph of this specimen after one deposition cycle. This layer is considerably thicker than those described in specimen A, which had multiple depositions. This is due to the larger mass of Pb-Bi-O placed on the substrate by brushing, which is a quicker process than the sedimentation described above. The interface is straight, as before. Even though the brushed-on layer is initially not as uniformly distributed as the sedimented layer, it appears that the melting which occurs before sintering evens out the surface.

Specimen C, which was made from pre-reacted [0112] but only pelletized without pre-sintering, had its conical cavity filled with the Pb-Bi-O mixture. The specimen was sintered at 840°C for 24 h, and the cycle repeated once. Figure 7 shows the fractured cross-section of specimen C. Rapid growth has occurred through the relatively porous substrate, right through to the bottom surface where excess liquid phase ran out during firing.

Figure 7. SEI from fractured section of specimen C. Reaction has proceeded right through to substrate base, at arrow.

In specimen D, by contrast, where reaction occurred in a dense sintered substrate, the film thickness around the conical hole was more uniform and the film did not grow through to the bottom. The film growth rate is thus shown to depend on the porosity of the substrate. This strongly suggests that the reaction kinetics depend on the mobility of the molten Pb-Bi-O.

The kinetics of the thermal reaction between Pb-Bi-O and [0112] precursor were studied by differential thermal analysis (DTA). The specimens for DTA were prepared by mixing [0112] and PbO-BiO in the ratio of $(Pb,Bi)SrCaCu_2$.

The specimen was heated at a rate of 10°C/min. Figure 8 shows the measurement which was reproduced several times.

III. Discussion

An understanding of the synthesis reactions of this multi-element system and of the DTA results can be obtained from published pseudobinary phase diagrams. The following discussion is summarized in table I. The role of Pb, which acts as a flux, is particularly important. The DTA of PB-containing BSCCO has marked differences from those of pure BSCCO [9]. The differences are greatest below 750°C owing to the Pb flux.

Table I

Explanation for differential thermal analysis shown in figure 8.

Temperature	DTA feature	Explanation
610°C	endo-thermic	Pb-Bi-O precursor eutectic
660	exo-	$PbCu_2O_4$ synthesis
680	endo-	Pb-Cu-O eutectic melt
710	exo-	$PbCa_2O_4$ synthesis
770	exo-	[2201] synthesis
820	endo-	Pb-Ca-O eutectic
850	exo-	[2212] and [2223] synthesis
860	endo-	[2223] melt
~900		[2212] melt
~950		[2201] melt

The first endothermic valley at 610°C (±15°C) occurs at a temperature a little lower than the published eutectic of $PbO-1.5Bi_2O_3$ [8] (635°C); but corresponds with our independent observations of melting of this mixture. At 660°C an exothermic peak occurs, apparently due to synthesis of the phase $PbCu_2O_4$ followed by endothermic eutectic melting at 680°C [10]. At 710°C, another exothermic reaction is observed, which corresponds to crystallization of $PbCa_2O_4$, stable in the binary Pb-Ca-O system up to 815°C [11]. These compounds are frequently observed in SEM as minority phases, for example in materials sintered at high temperatures above 900°C. These compounds appear as evidence for the solid state reaction, which occurs at the exothermic shoulder (figure 8) between 750 and 800°C:

$$2(Pb\text{-}3BiO_x)+6SrCaCu_2O_x \text{---}$$
$$3Bi_2Sr_2CuO_y+PbCu_2O_y+PbCa_2O_y+2Ca_2CuO_y+5CuO$$

These reaction products are frequently observed in analytical SEM as minority phases, for example in materials sintered at high temperatures above 900°C. $Bi_2Sr_2CuO_6$ [2201] is the low T_c phase (~ 6 K) in the BSCCO superconducting family [12]. $PbCa_2O_x$ undergoes partial melting at 815°C [11] corresponding to a pronounced endothermic valley. When the temperature exceeds 820°C, the Pb-Ca-Cu-O melt provides Ca and Cu which will accelerate the formation of [2212] and of [2223].

We have found that specimens sintered at temperatures above 860°C are depleted in [2223] and do not show zero resistance above 80 K. The endothermic valley at this temperature in the DTA suggests partial melting of [2223]. Moreover we have found that exaggerated grain growth occurs in this system at around 900°C, when sintering occurs in thermal gradients. Again the product is depleted in [2223], while the grain growth again suggests partial melting. At ~950°C the system melts completely.

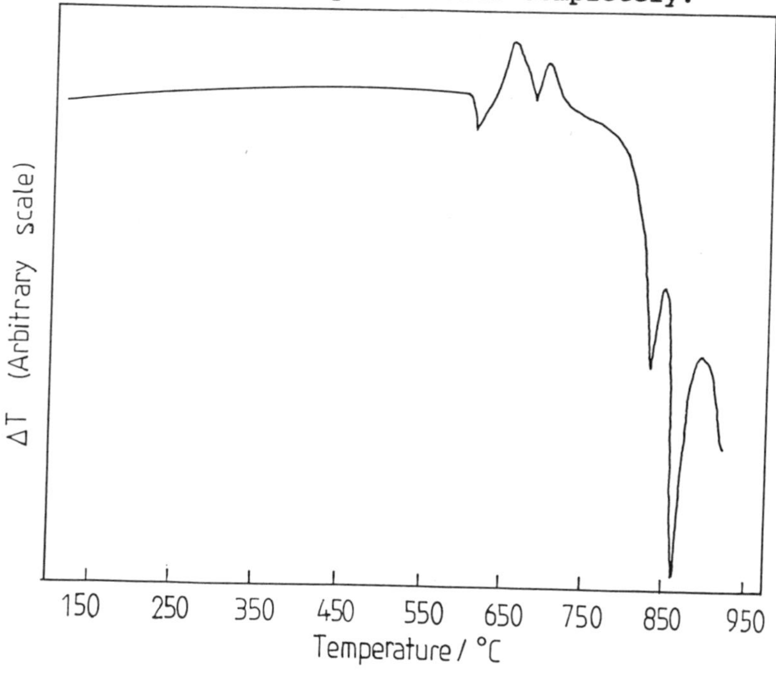

Figure 8. Differential thermal analysis from heating cycle of Pb-Bi-O with [0112].

This series of reactions is consistent with our studies of Pb dopants in BSCCO: formation of [2223] generally requires a long sintering time because of the slow diffusion kinetics of Ca and Cu into [2212]. The Pb dopant accelerates the formation of [2223] by liquid phase sintering, which accelerates the diffusion of Ca and Cu needed for the formation of [2223] [5].

The principles governing this diffusion texture growth appear to be as follows: molten $PbO\text{-}3BiO_{1.5}$ reacts with the [0112] substrate to form [2201], Pb-Cu-O, Pb-Ca-O, Ca-Cu-O and CuO. The superconducting [2212] phase is firstly formed. Crystal growth occurs downwards, along the direction of liquid (Bi-Pb-O, Pb-Cu-O and Pb-Ca-O) flow. This downward growth is similar to gravity aided texture growth through partial melting, previously observed in $YBa_2Cu_3O_{7-x}$ [3,13].

The experimental results from specimen A show a linear relation (figure 3) between the thickness of the diffusion-grown layer and the summed weight of deposited Pb-Bi-O. The relationship demonstrates that, under the experimental conditions employed, the thickness of the superconducting layer is controlled by available mass of Pb-Bi-O for reaction. By further repeated cycling, much thicker layers can be grown. This is also consistent with the result on specimen B, where a thicker superconducting layer is formed during only one deposition cycle. Substrate properties, including porosity, affect the texture processing. Thus a faster grain growth is observed in pre-sintered, porous [0112], as in specimen C, than in dense substrates. However the product is also more porous than films grown from pre-sintered precursors.

These diffusion-textured films have special properties owing to the orientation of the anisotropic platelets, i.e. with c-axis in the sheet plane. This orientation is normal to that commonly produced in thin film sputtering and evaporation techniques. The orientation of diffusion textured films suggests many potential applications, for example in shielding magnetic fields with field vectors parallel to the sheet plane, since the Bean model demands high values of J_c normal to this plane.

Acknowledgement: This work was carried out under grant proposal number 90F009 from the New York State Institute for Superconductivity.

REFERENCES

1. A.M.Wolsky, R.F.Giese and E.J.Daniels, Sci. Am. Feb. 45 (1989)

2. D.Dimos, P.Chaudhari, J.Mannhart and K.LeGoues, Phys. Rev. lett. 61 219 (1988)

3. A.J.Bourdillon, N.X.Tan, N.Savvides and J.Sharp, Mod. Phys. Lett. B3 1053 (1989)

4. S.Jin, T.H.Tiefel, R.C. Sherwood, R.B. van Dover, M.E.Davis, G.W.Kammlott and R.A.Fastnacht, Phys. Rev. B 37 7850 (1988)

5. H.K.Liu, S.X.Dou, A.J.Bourdillon, M.Kviz, N.X.Tan and C.C.Sorrell, Phys. Rev. B 40 5266 (1989)

6. T.Asano, Y.Tanaka, M.Fukutomi, K.Jikihara and H.Maeda, Jap. J. of Appl. Phys. 28 L595 (1989)

7. S.Jin, R.B.van Dover, T.H.Tiefel and J.E.Graebner, Appl. Phys. lett. (1991) (in press)

8. R.S.Roth, J.R.Dennis, H.F.McMurdie, Phase Diagrams for Ceramists, Vol. 6, Fig. 6298, (Am. Ceram. Soc. Inc., 1987) p. 40

9. P. Strobel, W.Korczak and T.Fournier, Physica C 161 167 (1989)

10. R.S.Roth, J.R.Dennis and H.F.McMurdie, Phase Diagrams for Ceramists, Vol. 1, Fig. 163, (Am. Ceram. Soc. Inc., 1964) p. 86

11. R.S.Roth, T.Negas and L.P.Cook, phase diagrams for ceramists, Vol. 5, fig. 5140, (Am. Ceram. Soc. Inc., 1981) p. 94

12. J.Q.Lin, C.Chen, D.Y.Yang, F.H.Li, Y.S.Yao, Z.Y.Ran, W.K.Wang and Z.X.Zhou, Z. Phys. B74 165 (1989)

13. N.X.Tan, and A.J.Bourdillon, Materials Lett. 9 339 (1990)

PREPARATION AND PROPERTIES OF Y-124 SUPERCONDUCTOR MADE BY A CHEMICAL PRECIPITATION METHOD *

Henry L. Laquer, *CryoPower Associates, Los Alamos NM, 87544-0478* [a]
J.R. Gaines Jr., Scot Brainard, Scott D. Hutson, Julie Pisanelli,
Superconductive Components Inc., Columbus OH, 43212 [b]
D.W. Cooke, E.R. Gray, Kevin C. Ott, Eric J. Peterson, J.F. Smith,
Los Alamos National Laboratory, 87545 [c]
U. Balachandran, M.T. Lanagan, R.B. Poeppel, *Argonne National Laboratory, 60439* [d]
J. Douglas Wolf, *University of Dayton Research Institute, 45469*
Frederic C. Laquer, *Dept. of Chemistry - University of Nebraska at Omaha, 68182*

ABSTRACT

We have prepared the thermodynamically stable *YBCO-124* high temperature superconductor in powder form by a chemical precipitation method and have characterized the material by a number of chemical and physical methods, including carbon content, x-ray diffraction, thermogravimetry, scanning electron microscopy, surface area, thermally stimulated luminescence, and the superconducting transition and magnetization curves. We have also started work on consolidating the powders by hot pressing.

1. INTRODUCTION

The only problem with bulk high temperature superconductors is their limited current carrying capacity. This deficiency is generally attributed to "weak links" at the grain boundaries and to the crystalline anisotropy of the grains. The obvious remedy is to remove and/or neutralize the grain boundaries by growing larger grains and by texturing the material, *i.e.* aligning the grains. The difficulty with this approach, however, is that the temperatures and pressures necessary to produce texturing can also lead to decomposition of the superconductor, with the creation of non-superconducting impurity phases, which then act as weak links. Clearly, it would be advantageous to use superconductors that are more stable.

The yttrium-barium-cuprate high temperature superconductor with the composition $YBa_2Cu_4O_8$ (*Y-124*) was first seen by Zandbergen *et al.* [1] as a defect in partially decomposed *Y-123* material and has since been synthesized in high pressure autoclaves by Karpinski *et al.* [2] and by Morris *et al.*, [3] Since *124* does not undergo the tetragonal-orthorhombic phase change and since it has a fixed oxygen content of exactly 8 atoms, it should be more stable than the widely studied *123*, with its variable oxygen stoichiometry of $(7-\delta)$.

The high pressure synthesis of *124* requires equipment that is not readily scaled to production quantities. Published pressure-temperature phase diagrams [4] [5] [6] give conflicting results for the more easily manageable and safer pressure region below 1 MPa, but it is clear that for each temperature there is a minimum permissible pressure to assure the *124* equilibrium composition, rather than *247* or *123*.

---

* Research sponsored by SDIO/IST and managed by DNA under Phase 1 SBIR Contract DNA001-89-C-0118.

[a] Additional technical support provided by USDOE under LANL Superconductivity Pilot Center Cooperative R&D Agreement CRDA-90-06.

[b] Additional technical support provided by USDOE under ANL Superconductivity Pilot Center Cooperative R&D Contract 85287.

[c] Work supported by USDOE, Office of Energy Management, Advanced Utility Concepts.

[d] Work supported by USDOE, Office of Energy Storage, Distribution, Conservation and Renewable Energy under Contract W-31-109-ENG-38.

A number of publications report "essentially" phase-pure *124*, or *Ca* substituted *124*. [7] [8] [9] [10] [11] [12] [13] All except one, [8] limit synthesis temperatures to 800 or 815 C and operate at or below atmospheric pressure. Most also use conventional mix-and-grind ceramic techniques with long annealing times to insure homogeneity.

The present investigation is based on material prepared by a chemical co-precipitation technique [14] and has used vacuum, [15] as well as "mild" pressure (up to 7 bar) calcinations in flowing oxygen to prepare *Y-124* in 10 to 100 g lots. Co-precipitation should permit closer control of composition and homogeneity than the more conventional techniques and the subsequent compound formation should take place at lower temperatures. A possible disadvantage of "chem-prep" materials may be their fine particle size and greater susceptibility to surface contamination.

2. POWDER PREPARATION

Superconductive Components, Inc. (SCI) has been manufacturing "standard" *Y-123* under a license to use the patented process developed at Sandia National Laboratory by Bunker *et al.* [14] The process mixes and precipitates in an ultrasonic horn aqueous solutions in tetramethylammoniumhydroxide (TMAOH) of the cations and of carbonate ion at controlled pH, and can be adapted to make *Y-124* precursor by simply changing the copper concentration. The subsequent calcination of the intimately mixed hydroxides and/or carbonates of *Y*, *Ba*, and *Cu* includes removal of residual TMAOH by pyrolysis. This should be done as quickly as possible, to avoid the reduction of *CuO* to *Cu* by the organic material, while avoiding melting and the possible segregation of the intimately mixed elements. On the other hand, there is the requirement to completely (and slowly) decompose all carbonates, at temperatures only slightly above the pyrolysis region.

Initial batches were processed by "flash" calcination, *i.e.* the precursor slurry was directly introduced into the hot furnace. Subsequent batches were either pressure or vacuum calcined at different heating rates and temperatures. Our best powders still contain some carbon and the optimization of the heating protocol remains to be worked out.

It should be noted that the exact nature of the intermediate phases and the kinetics of their conversions are largely unknown in the *123* as well as in the *124* (and *247*) systems. Most work on ternary phase diagrams has been done at temperatures above 850 C. Our results confirm the observations of Balachandran *et al.* [9] that, in order to form *124* rather than *123*, a short "nucleating" excursion to 800 C in 1 bar of O_2, followed by an ≈24 hr reaction at 750 C is necessary, and that a subsequent "finishing" treatment at 800 C (in O_2) will improve phase purity. However, it is not yet clear whether the oxygen-vacuum CO_2 removal is necessary for our sub-micron powders. Treatments above 800 C require elevated oxygen pressures.

3. MATERIALS CHARACTERIZATION

Our aim is to control purity to better than 0.5 volume percent. Quantitative materials characterization and quality control therefore become the focus of our endeavor. The available techniques fall into three groups:

a) Primarily chemical procedures that measure the elemental composition of a sample,

b) Standard physical, materials science, or metallurgical methods that show the structure and morphology at the macroscopic, microscopic and atomic levels, and

c) Specialized cryogenic techniques that define and elucidate the superconducting properties.

The equipment, expertise and manpower to apply all of these can only be mustered through the cooperative effort of a number of organizations.

3.1. CHEMICAL TECHNIQUES

An accurate knowledge of the chemical composition of the manufactured superconductor is necessary, but not sufficient, to determine its stoichiometry. Basic **Wet Chemistry** techniques involve only modest equipment costs, but are time consuming, if accuracies of 1 part in 1000 are attempted. For *YBCO*, copper has to be removed quantitatively by **Electrodeposition**, before yttrium and barium can be obtained by **EDTA** titration.

Quicker instrumental methods for elemental analysis, such as Atomic Absorption Spectroscopy (**AAS**), Inductively Coupled Plasma Atomic Emission Spectroscopy (**ICPAES**), or X-ray Fluorescence (**XRF**), only give accuracies of 3% at best, but are useful for trace impurities. **Oxygen** analysis by **Iodometry** [16] is, unfortunately, model dependent and indirect, since it involves separate titrations to differentiate between copper atoms of valence +2 and +3.

Carbon content, or residual carbonates, are easily measured to trace levels by the **LECO** method developed for the steel industry, as included in ASTM Standard E-350. [17] Carbon can also be seen with great sensitivity by Thermal Decomposition Mass Spectroscopy (**TDMS**) of a 50 mg heated sample through the appearance of CO and CO_2.

Insulating impurities within 1 μm of the surface can be detected with high sensitivity by Thermally Stimulated Luminescence (**TSL**), a technique developed at LANL. [18]

3.2. PHYSICAL MEASUREMENTS

ThermoGravimetric Analysis (**TGA**) stands out among standard techniques in producing highly informative results, both for analysis and for guiding the manufacture. The only limitation is on the magnitude of the gas (oxygen) pressure the instrument can tolerate. TGA clearly distinguishes between *123*, *124* and *247* and can furnish quantitative data for 2-phase mixtures. It also shows the desorption of moisture and the decomposition of carbonates.

X-Ray Diffraction (**XRD**) can only find impurities above the 5 to 10 % level. **Low Angle** XRD must be used to differentiate between similar compounds with large unit cells, such as *123*, *124* and *247*, where many of the strongest lines overlap. Precision Lattice Constant (**PLC**) analysis of the diffraction lines may indicate subtle distortions in the crystal structure, related to the manufacturing process or to oxygen deficiency.

Electron Microscopy, both Scanning (**SEM**) and Transmission (**TEM**), can show individual grains and other morphological features. *In situ* chemical analysis of small regions by x-ray fluorescence during SEM or TEM is also possible, but only for concentrations above 5 or 10%, and the TEM equipment can identify individual crystallites by electron diffraction.

The **BET** (Brunauer-Emmett-Teller) [19] surface area method indicates an **average** particle size, but other techniques are needed for the particle size distribution and shape.

3.3. CRYOPHYSICAL MEASUREMENTS

The measurements of the unique electrical and magnetic properties of superconductors furnish the essential check on the quality of our product and fall into three categories:

a) Low field (< 10 mT) magnetic susceptibility, as a function of temperature under dc, ac, or rf conditions,

b) Full field (0 to 5 T) magnetization, hysteresis and implicit intra-granular critical current densities, and

c) Transport critical current densities as a function of temperature and magnetic field.

Magnetometers -- The **MPMS** (Magnetic Property Measuring System), based on the SQUID has been the mainstay of our work, as it is in many studies on high temperature superconductors. Its resolution, precision and accuracy are outstanding, but, for other than the discovery of new superconductors, as much as four orders of magnitude in sensitivity may not be needed and, although fully automated, even a simple study takes hours. A **Vibrating Sample Magnetometer** (VSM) can make similar measurements at a faster rate and an **Integrating Coil** magnetometer could be assembled from components and would be faster yet.

AC Susceptibility and Loss equipment is available both in the low frequency (10 to 100 Hz) and radio frequency (MHz) range. It can provide information about structural heterogeneity, but is mostly used to qualitatively indicate the transition and its width.

Resistivity and Transport Critical Current Density -- For high current measurements, the temperature ranges accessible with common cryogenic liquids are limited. Few installations have such equipment available inside a high field magnet at other than liquid nitrogen and helium temperatures. One way around this limitation is to generate **Induced Persistent Currents** in hollow cylindrical structures. This avoids contact and lead heating problems and, therefore, only needs a variable-temperature flow-cryostat inside a high field magnet.

Surface Resistivity -- Unique information about weak link behavior is obtained from a technique, developed in part at LANL, where the change in the Q of a 3 GHz TM_{010} niobium cavity (with a flat 0.5 to 1 cm^2 specimen insert) is measured while the rf magnetic field is increased from 10^{-4} to 10^{-1} Oe. Similar measurements can be made over an extended temperature range in a 28 GHz copper cavity cooled by a refrigerator, but require samples of 3.8 cm diameter.

4. PROPERTIES OF POWDERS

In the following, we first discuss for each group of calcinations, the measurements that primarily guided the development of that group: the superconducting transition, XRD and SEM. We then cover TGA, BET, critical current density and other results, which cut across and unify the groups.

4.1. FLASH CALCINATIONS

The four flash calcined batches that were investigated in detail, were part of a larger set, prepared at SCI, where the gaseous environment (air or oxygen), the temperature (750, 800, 850 or 900 C), and the calcining times (4 or 10 hrs) were varied. The importance of the kinetics of the intermediate phase transformations is emphasized by the fact that none of these samples gave a clear indication of *124*.

Even so, the work served as an introduction to the capabilities and limitations of the various techniques. Onsets of the transition at 92 K indicated the presence of *Y-123*; a strong Curie-Weiss, $1/(T-\theta)$, paramagnetic rise in susceptibility with decreasing temperature suggested unreacted CuO or Cu_2O. X-Ray Diffraction showed $BaCO_3$ for all calcinations under 850 C, and CuO in all samples except the 900 C one. It also showed some form of *YBCO* in most samples, but we could not differentiate between *123*, *124*, and *247*.

4.2. PRESSURE CALCINATIONS

The main variables studied were processing time and temperature, with the pressure selected so that the reaction would occur in the *124* region of the phase diagram. We also varied heating rates, holding times and the important step of nucleating at elevated temperatures, prior to forming *124* at 750 C, as suggested in the ANL vacuum processing studies. The hazards of handling oxygen gas in a ceramic tube at elevated pressures and temperatures were reduced by installing a pressure switch to shut off the throttled oxygen supply in case of a tube failure.

The **first** sample (355-L) was taken directly to its reaction temperature of 825 C and held there for 14 hrs at an absolute pressure of 3 bar. Its **transition curves** still have an onset temperature of 92 K, indicating the presence of *123*, but also have a second major drop at 82 or 83 K, as expected for *124*. The transition is broad and shows no leveling or completion, even at 7 K. We tend to ascribe the broadening primarily to the extreme fineness of our powders, but the presence of *Y-247* may also be a contributing factor.

SEM shows platelets with major dimensions of 2 μm and less. We estimate the thickness of the platelets to be no more than 200 nm, which would give a large demagnetization factor.

X-Ray diffraction. The complete set of *Y-124* lines, for 2θ angles between 20 and 60°, were observed in this first batch. The insulating *Y-211* phase is also seen, plus some *CuO* and, possibly, Y_2O_3. There are "bumps" at 27.15 and 31.65°, that could be attributed to the [108] and [10.12] lines of *247*, but none that are unambiguously assignable to *123*. Nevertheless, the presence of *123* is quite likely, since the batch had not been given the nucleation treatment, and, indeed, strongly suggested by the by the previously mentioned 92 K onset and by the TGA data on mass loss as a function of temperature.

The **transition curves** for the **second** batch (355-N), Fig. 1, confirm the importance of a higher temperature nucleation, prior to 750 C reaction, to get rid of *123*. The batch was "finished" for 48 hrs at 830 C in oxygen at 5 bar. The transition has a clean onset at 83 or 82 K for all measuring fields. There is a hint of a paramagnetic rise in the 20 Oe curve below 20 K.

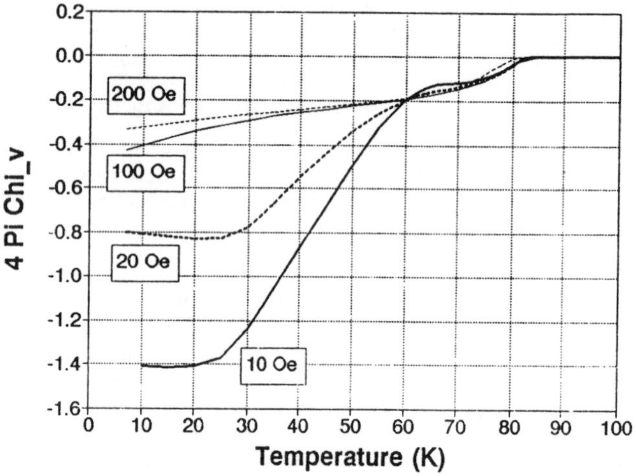

Fig. 1. Effect of Measuring Field on Apparent Width and Magnitude of Zero-Field-Cooled Transition of Pressure Calcined Material (355-N) - No demagnetization corrections.

X-Ray diffraction patterns of this and subsequent batches were done at twice the earlier resolution and over an extended 2θ range from 5 to 74°. We can now distinguish seven weaker lines, attributable to *247*, and identify Cu_2O and, possibly, Y_2O_3 and *CuO*, but no *123*. The presence of the high-temperature, reduced form of copper oxide, Cu_2O, is unexpected.

The **final** pressure calcined batch (355-O), was nucleated at 800 C, prior to the 750 C reaction and then finished at 840 C and 6 bar for 31 hours. It gave sharper and better resolved diffraction peaks than the samples finished at lower temperatures. It contains about the same amount of *211*, but only one third as much Cu_2O. Its transition was only measured on a cold pressed cylinder. It has an onset at 83 K and has not leveled off at 7 K.

4.3. VACUUM CALCINATIONS

Three vacuum calcinations were done by SCI in cooperation with and at the Argonne National Laboratory. The vacuum process was originally developed at ANL to improve the removal of CO_2 in ceramic process *123*.[15] It was subsequently adapted to *124*, pioneering the 800 C nucleation step, and eventually to our chem-prep materials. It is evident that the process for *124* has undergone successful improvements, with the earlier samples containing paramagnetic impurities and having an onset temperature above 90 K. Also a finishing treatment, typically 24 hours in flowing oxygen at atmospheric pressure and 800 C, is clearly needed to improve the phase purity of the *124* compound.

The **final** sample (430-B), given the full nucleation and finishing treatment at 800 C, has a clean transition, starting at 82 K. The transition is broad, shows no leveling and indicates a large demagnetization. All of this is reasonable for an assembly of very small platelets that are barely resolved at an SEM magnification of 30,000X, Fig. 2. These facts also correlate with the broader, less sharp x-ray lines, compared to samples that were finished at higher temperatures and pressures. There are clearly identifiable *247* lines, but no copper oxide.

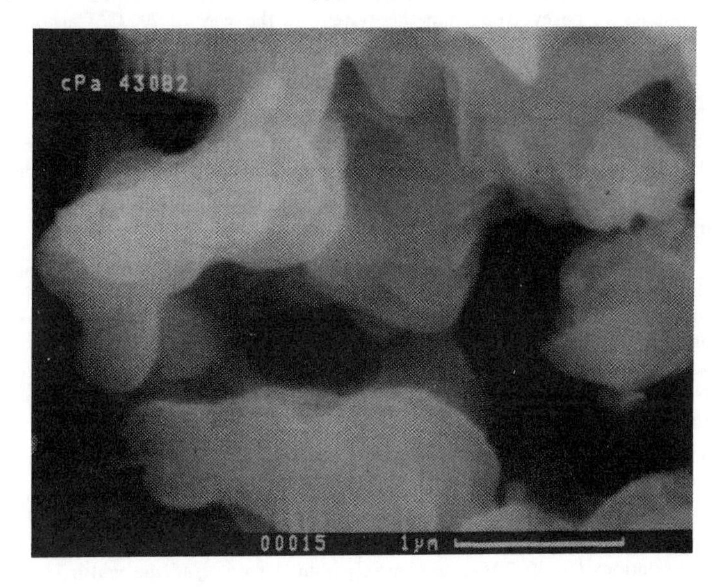

Fig. 2. SEM for Vacuum Processed Material (430-B) at Magnification of 30,000X.

4.4. THERMOGRAVIMETRY

Morris *et al.*[3] and Karpinski *et al.*[4] have pointed out that TGA differentiates between *123*, *124* and *247*, with great sensitivity. *123* starts to lose oxygen at 400 C and has lost 1.2 % of its mass at 800 C, even in an oxygen atmosphere, whereas pure *124* has lost only 0.07 % of its mass at that temperature and *247* is in between with a loss of 0.44 %. The 700 C losses are 0.9, 0.3 and 0.03 % for *123*, *247* and *124*, respectively.

From the mass loss TGA data of Fig. 3, we can estimate the *247* content, assuming that the absence of a 92 K onset can be used to exclude *123*. Depending on the chosen reference temperature, our best *124* samples still contain 10±5 % of *247*.

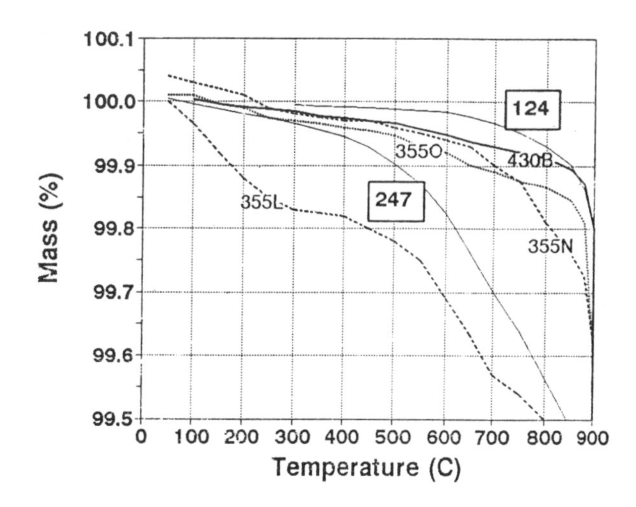

Fig. 3. Comparison of TGA Observations with Literature Data [4]. Note that only the early sample (355-L), which was not given the 800 C nucleation and still had a 92 K onset, is clearly outside the *247-124* region.

4.5. THERMAL DECOMPOSITION MASS SPECTROSCOPY

TDMS was run in batch mode on three samples. Moisture peaks between 150 and 250 C, but continues to be evolved from the system all the way to 700 C. Carbon monoxide and dioxide usually run parallel and seem to have double peaks at about 200 and 400 C. The areas under the curves scale roughly as the carbon obtained from combustion.

4.6. CARBON ANALYSES

Preliminary carbon analyses gave 0.05 % for one of our pressure calcined samples (355-O) and 0.03 % for two of the ANL vacuum calcined samples. These numbers correspond to $BaCO_3$ contents of 0.8 and 0.5 %, respectively.

4.7. SURFACE AREA

The BET surface area values obtained by a commercial laboratory, ranged from 4.5 m^2/gm for the pressure calcined 355-O to 90 m^2/gm for the vacuum calcined 430-B. The corresponding diameters of uniform spherical particles would be 220 and 12 nm. The former is in reasonable agreement with the SEM observations, but the values 12 or 16 nm (for 355-N) will need to be verified by alternative techniques that provide information on particle shape and size distribution. Nevertheless, there appears to be a trend in that the smallest surface areas go with the highest finishing temperatures.

4.8. MAGNETIZATION AND CRITICAL CURRENT DENSITIES

Magnetization-Hysteresis curves were obtained for most of the pressure and vacuum calcined powders at 10 and 40 K. The hysteretic width of the magnetization curve is commonly used to calculate an intragranular critical current density according to the Bean [20] critical state model and the relation: $J_c = 30\ \Delta M / 2r$, where J_c is in A/cm^2, ΔM the width in emu/cm^3 or Gauss, and $2r$ the particle diameter in cm.

Surprisingly, the hysteresis widths vary by less than a factor of two for the fine, pressure calcined powder (355-N), Fig. 4, a cylinder cold pressed from it, a coarser powder (355-O), and the extremely fine vacuum calcined powder (430-B). On the other hand, in view of the uncertainties in particle size and shape discussed in the preceding section, the intragranular current densities, indicated by the tabbed numbers on the right, could be in error by an order of magnitude in either direction. Nevertheless, the indicated critical current densities, although close to the depairing limit, are not unreasonable, especially if it should turn out that *247* or possible byproducts from the transformations between the different *YBCO* phases provide flux pinning sites, as suggested by Morris *et al.* [5] and by Jin *et al.* [21]

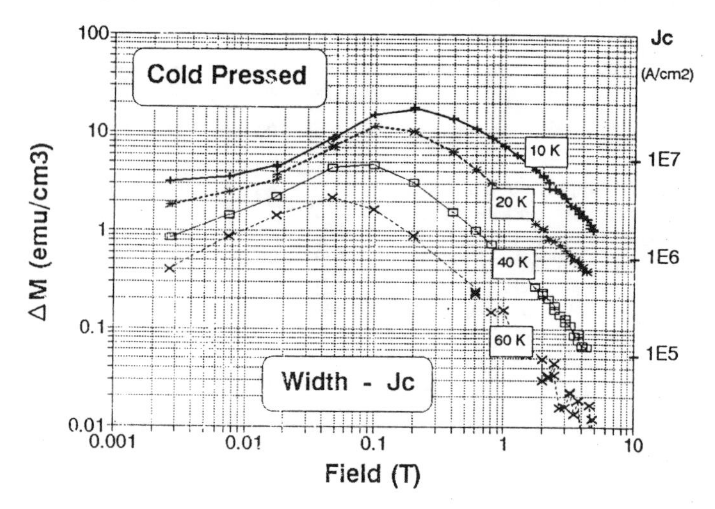

Fig. 4. Magnetization Width and Intragranular Critical Current Density as a Function of Field (0-5 T) and at 10, 20, 40 and 60 K for Cylinder Cold Pressed from Pressure-Calcined Powder (355-N). A particle diameter of 160nm is assumed to calculate J_c.

5. PROPERTIES OF CONSOLIDATED MATERIALS

Our work on consolidating the powders by hot pressing has been of a preliminary nature. Protection of the material from de-oxygenation is essential. The best run made at 810 C and 500 bar produced flakes with a slate-like morphology. The dc transition, measured with a 10 Oe field perpendicular to the plane of the flakes, shows a slow onset at 82 K, with the main drop below 50 K. The susceptibility at 7 K exhibits the largest negative value, or demagnetization, observed in our study. Clearly, the material is very anisotropic.

The rf transition at 14 MHz is complete at 60 K. The surface resistivity at 3 GHz and 4 K increases from 2 mΩ for an ac field of 10^{-4} Oe to 7 mΩ at 10^{-1} Oe. There still are weak links in the surface, but the material is competitive with the best bulk *123* seen to date. Similarly, light output during thermally stimulated luminescence (TSL) is quite low, implying the relative absence of insulating surface impurities.

6. SUMMARY

We believe that we have a method for producing *Y-124* powders well in hand and that we can usefully access a number of inter-disciplinary materials characterization and evaluation techniques to control and improve the quality of our superconductor. The manufacturing, forming and deforming of bulk shapes at elevated temperatures has just been started. Here the main challenge will be to maintain the correct oxygen pressure, so as to avoid decomposition and de-oxygenation, while allowing the grains to grow and deform.

ACKNOWLEDGMENTS

We would like to thank D.C. Larbalestier and M.P.Maley for enlightening discussions, W.L. Hults for an introduction to the technique of cold pressing fine powders, and R.D. Taylor, J.D. Thompson and J.O. Willis for magnetization measurements.

REFERENCES

1. H.W. Zandbergen, R. Gronsky, K. Wang, G.Thomas, *Nature* **331**, 596-9 (1988)
2. J. Karpinski, E. Kaldis, E. Jilek, S. Rusiecki, B. Bucher, *Nature* **336**, 660-2 (1988)
3. D.E. Morris, J.H. Nickel *et al.*, *Phys. Rev.* **B 39**, 7347-94 (1989)
4. J. Karpinski, S. Rusiecki, E. Kaldis, B. Bucher, E. Jilek, *Physica* **C 160**, 449-57 (1989)
5. D.E.Morris, A.G. Markelz, B. Fayn, J.H. Nickel, *Physica* **C 168**, 153-160 (1990)
6. T. Wada, N. Suzuki, ..., S. Tanaka, *Appl. Phys. Lett.* **57**, 81-3 (1990)
7. K. Kourtakis, M. Robbins, P.K. Gallagher, T. Tiefel, *J. Mater. Res.* **4**, 1289-91 (1989)
8. T. Miyatake, K. Yamaguchi, ..., S. Tanaka, *Physica* **C 160**, 541-44 (1989)
9. U. Balachandran, ..., R.B. Poeppel, *Physica* **C 165**, 335-39 (1990)
10. S. Jin, H.M. O'Bryan, ..., R.J. Cava, *Physica* **C 165**, 415-18 (1990)
11. D.M. Pooke, R.G. Buckley, M.R. Presland, J.L. Tallon, *Phys. Rev.* **B 41**, 6616-20 (1990)
12. T. Miyatake, S. Gotoh, N. Koshizuka, S. Tanaka, *Nature* **341**, 41-42 (1989)
13. R.G. Buckley J.L. Tallon, D.M. Pooke, M.R. Presland, *Physica* **C 165**, 391-3 (1990)
14. B.C. Bunker, J.A. Voigt, D.H. Doughty *et al.*, Ch. 7 in HTSC Materials W.E. Hatfield and J.H. Miller, Jr. eds, Marcel Dekker, NY (1988), pp. 121-9.
15. U. Balachandran, R.B. Poeppel *et al.*, *Mat. Lett.* **8**, 454-56 (1989)
16. E.H. Appleman *et al.*, Ox. content .., *Inorg. Chem.* **26**, 3237-39, (1987)
17. *Annual Book of ASTM Standards*, 1989, **03.05** p. 374.
18. D.W. Cooke, M.S. Jahan, J.L. Smith, M.A. Maez, *et al.*, *Appl. Phys. Lett.* **54**, 960-62 (1989)
19. S. Brunauer, P.H. Emett, E. Teller, *J. Am. Chem. Soc.* **60**, 309-19 (1938)
20. C.P. Bean, *Phys. Rev. Lett.* **8**, 250-3 (1962)
21. S. Jin, T.H. Tiefel, S. Nakahara, J.E. Graebner, *et al.*, *Appl. Phys. Lett.* **56**, 1287-89 (1990)

FUNDAMENTAL PROPERTIES

NEAR ROOM TEMPERATURE RESISTIVE TRANSITIONS IN SINGLE CRYSTALS OF $YBa_2Cu_3O_{7-x}$ WITH DEFECTS

J.T. Chen, L.E. Wenger, L-Q. Wang, and G-H. Chen
Department of Physics, Wayne State University, Detroit, MI 48202

ABSTRACT

Two thermally recycleable sharp resistance transitions - one just below and the other slightly above room temperature - have been observed in several defect-filled single crystals having a nominal composition of $YBa_2Cu_3O_{7-x}$. The occurrence as well as the stability of these room-temperature transitions require a long-term oxidation of the samples near room temperature and the maintenance of oxygen environment during thermal cyclings. From various electrical lead combinations in the electrical measurements and from the material analyses, we are led to the conclusion that these resistive transitions are associated with the localized defect regions in these "single-crystal" samples.

INTRODUCTION

It is well-known that the resistivity of a single-crystal $YBa_2Cu_3O_7$ is directional dependent, with the transition temperature, T_c, being the same for both perpendicular as well as parallel directions to the a-b plane of the single-crystal. Recently, we found a single crystal $YBa_2Cu_3O_{7-\delta}$ with a defect-filled surface layer having a T_c of 260 K in the direction parallel to the a-b plane while the T_c measured in the direction perpendicular to the a-b plane had the usual value of about 90 K. The results appear to be correlated to a similar phenomenon previously observed in mixed-phase ceramic samples which had a nominal composition of $Y_5Ba_6Cu_{11}O_x$ but mainly composed of preferentially oriented crystallites of $YBa_2Cu_3O_{7-\delta}$.[1,2] There are also other reports of higher temperature superconductivity-like anomalies in highly oriented YBaCuO films and ceramic samples.[3-6] To further understand the nature of this higher-T_c phenomenon and to identify its material origin, we have performed very detailed resistance versus temperature (R-T) and current -voltage (I-V) characteristic measurements on a minimum of six "single-crystal" samples. Some structure and composition analyses have also been done. A brief summary is given as follows:

(1) The resistive transition near 260 K has been observed in more than one "single-crystal" sample and it was reproducible over numerous thermal cycles. By using various combinations of electrical leads and contacts, the current-voltage characteristics have been measured to rule out that the observed transition phenomena arose from spurious signals caused by experimental artifacts. A weak magnetic-field dependent, nonlinear I-V curve showing a zero-voltage current has been observed, providing further evidence for superconductivity at this temperature.

(2) A very sharp resistive transition near 330 K was also observed in a minimum of three "single-crystal" samples. This observation may be related to a similar resistive transition previously observed in a slightly different cuprate ceramic sample in the same temperature range by Ihara *et al.*.[7] Due to the lack of a magnetic-field dependence study,we can not say whether this transition is related to superconductivity.

(3) Our "single-crystal" samples were not homogeneous throughout the entire samples. Instead, these samples have twins and/or stacking fault layers. One particular sample which exhibited anisotropic transitions with a T_c of 260 K in the parallel direction has a thin layer of material at the surface with a high density of defect

structures. Energy dispersive spectroscopy (EDS) results show that the surfaces of our "single-crystal" samples generally have a deficient concentration of Cu with an average ratio of 1:2:2.5 for Y:Ba:Cu.

EXPERIMENTAL INFORMATION

Sample Preparation The"single-crystal" samples studied in this report were isolated crystalline platelets, each typically 1mm x lmm x 0.1mm in size, and basically $YBa_2Cu_3O_7$ in composition according to x-ray diffraction studies. They were synthesized according to the recipe of Reference 8 with some modifications. The initial mixture with an atomic ratio of $Y_1Ba_4Cu_{10}O_x$ was heat treated according to the following procedure:

(1) Flux growth in air

Room temperature --6 hours to--> 800°C --8 hrs to--> 1010°C

--8 hrs at--> 1010°C --40 hrs to--> 970°C --48 hrs at--> 970°C

--100 hrs to--> 830°C --40 hrs to--> 300°C --7 days at--> 300°C --4 hrs to--> Room Temp.

(2) Annealing in O_2 at 500°C for 7 days

After the heat treatments described above, the samples were further oxidized near room temperature either by leaving them in air for a long time (nearly 18 months for one sample) or in pure oxygen gas under high pressure from a few days to weeks. Furthermore, it is important that a sample be kept in an oxygen environment during the measurements. One sample which did not show any resistive transition during the first few thermal cycles eventually developed a transition near room temperature after continuous thermal cycles in oxygen for more than one week. It was further found that rapid cooling is harmful to the sample.

Electrical Measurements For the resistance versus temperature (R-T) and the current-voltage (I-V) characteristics studies, four electrical leads were attached to a sample through four separate contacts by silver paste or gold epoxy. By using various combinations of two, three, and four contacts on the sample for the current and the voltage measurements, we determined the contact resistances as well as the sample resistance for different orientations. Three different measuring techniques including dc bias, ac bias, and lock-in phase sensitive detection schemes have been used. In the ac bias measurements[2], a very low frequency (mHz) current was used that permitted continuous monitoring of the voltage while the bias current was varied slowly between the two polarities. We have also studied the Seebeck voltage as a function of temperature near the transition to see if there was a correlation with its R-T. Some samples were studied for over two months continuously in order to obtain information on the growth and the stability of the samples in an oxygen gas.

EXPERIMENTAL RESULTS

Resistance Transition near 260 K As mentioned earlier, we have observed two drastically different resistive transition temperatures in a nominal $YBa_2Cu_3O_7$ single crystal depending upon the direction of the electrical current with respect to the crystalline axes. Figure 1 shows the impedance measured perpendicular to the flat

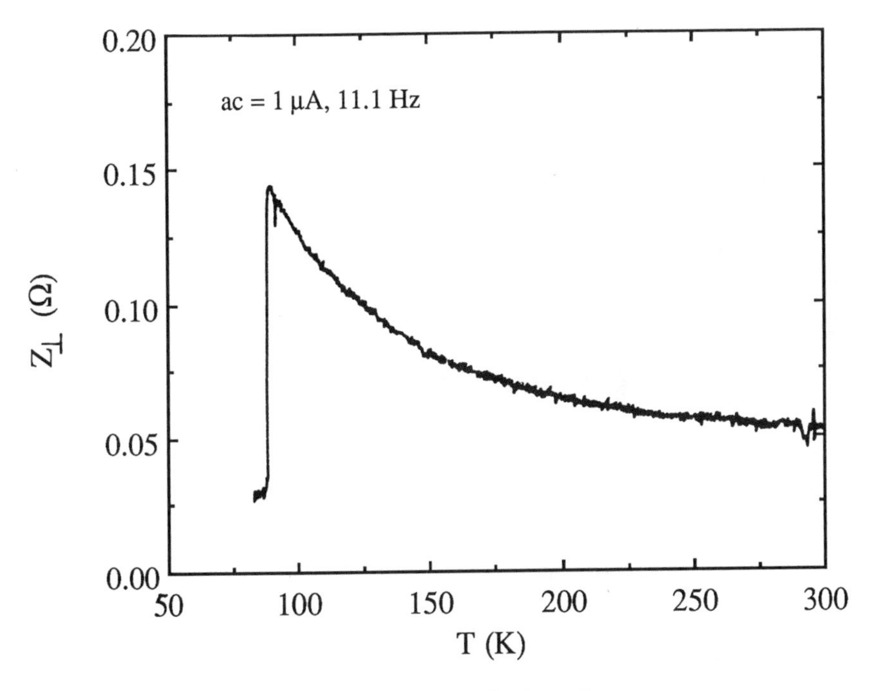

Fig. 1. The perpendicular (along c-axis) impedance vs temperature of a nominal $YBa_2Cu_3O_7$ single crystal. For explanation, see text.

platelet surface (along the c-axis) of the crystal by an ac lock-in technique with continuous phase adjustment. The signal, which is proportional to the amplitude of the ac voltage, has a negative-temperature-dependent coefficient down to 90 K and then exhibited the familiar resistive transition associated with the 90-K $YBa_2Cu_3O_7$ superconducting phase. This ac lock-in voltage did not actually reach zero below 86 K due to an unidentified ac background. However dc current-voltage (I-V) measurements verified that the sample was in a superconducting state, and if not, with an upper limit of 1 mΩ for the dc resistance at 84 K. By comparison, Figure 2(a) shows the similar measurement parallel to the platelet surface (along the a-b plane) on the same crystal. A sharp transition is evident in the temperature range 260 K to 275 K which is in the same temperature range as our previously reported zero-resistance transitions in several ceramic $Y_5Ba_6Cu_{11}O_y$ mixed-phase samples.[1,2] Again there was a nonvanishing, unidentified ac voltage below 260 K. The inset in Figure 2(b) is a magnified (20x) trace of the same measurement below 200 K which does not show any indication of the 90-K superconducting transition observed in the perpendicular direction. Subsequent dc current-voltage measurements indicated a similar upper limit of 1 mΩ for the dc resistance in this temperature region. To check whether the signal below the 260-K transition temperature is due to the ac background pick-up or to a nonvanishing resistance, the resistance versus temperature of the same sample was determined from its dc current-voltage characteristics. The resistance below 240 K, if any, is at least three orders of magnitude less than the value above the transition temperature. Since this sample was sealed in epoxy, it was not directly exposed to oxygen gas during the

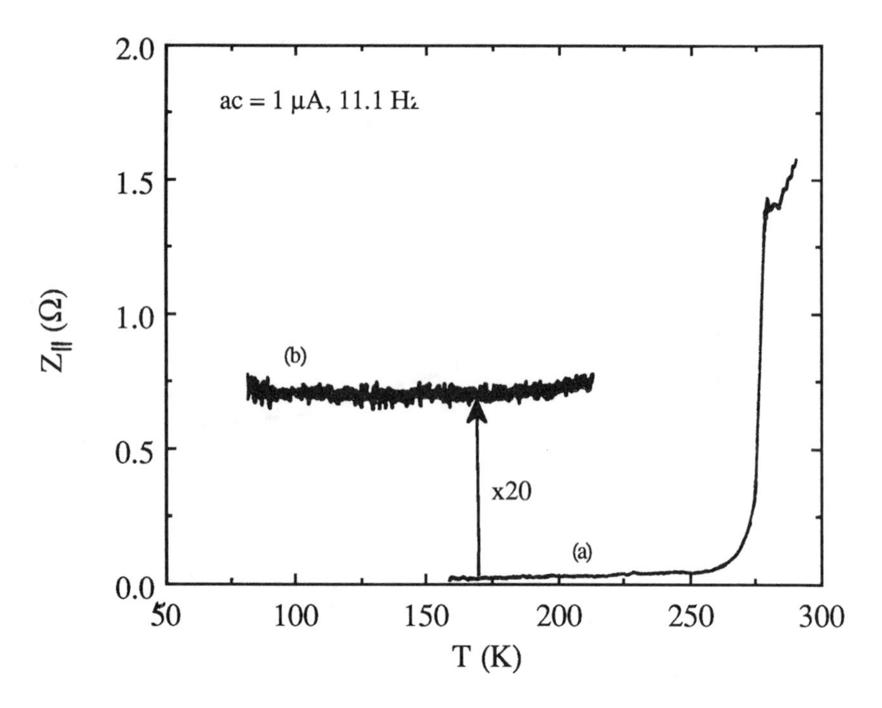

Fig. 2. (a) The parallel (in a-b plane) impedance vs temperature of the same sample. (b) The vertical sensitivity was increased by a factor of 20 in order to show the absence of an apparent transition near 90 K.

measurements. Perhaps due to this lack of oxygen exposure, the sample started to deteriorate in each successive thermal cycle after the tenth cycle with the normal-state resistance increasing from one-tenth of an ohm to several hundred ohms. However, the transition temperature remained the same even though the normal-state resistance had changed by three orders of magnitude. This indicates that the material has a well-defined transition temperature near 260 K. It has been previously found[9] that the Seebeck voltage of an oxide superconductor can also exhibit a sharp transition which is correlated with the superconducting T_c of the material. We have also observed the disappearance of the Seebeck voltage below 260 K in this sample as well.

Generally, dc characteristic measurements can yield more information than a "resistance" measurement by the lock-in technique. In a lock-in measurement, an alternating current may reduce the critical current I_o and may also pickup extraneous signals from the circuit. Furthermore, the lock-in technique measures dV/dI not V/I, and hence it measures the "true" resistance only for an ohmic device. It should be noted that measurements of I-V characteristics require a high degree of temperature stability and consequently are very difficult experiments. Nevertheless, some measurements were performed at selected temperatures to check for resistance and nonlinearity. Figure 3 shows the I-V curves of a sample having a zero-voltage current I_o and a magnetic-field dependence. The I_o in a magnetic field of 5.4 T is about 10 μA while I_o was greater than 30 μA for zero field.

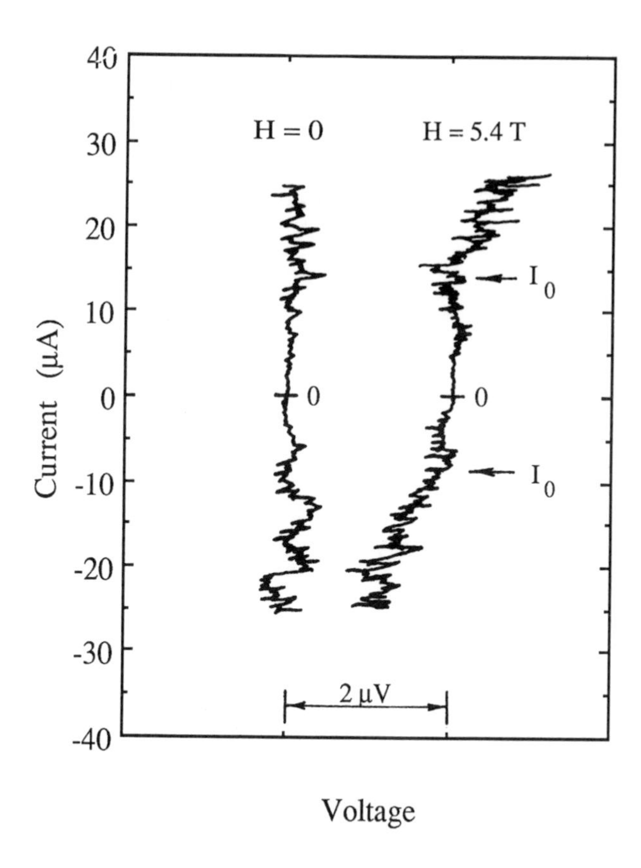

Fig. 3. The current-voltage characteristics of a "single-crystal" $YBa_2Cu_3O_{7-x}$ sample in magnetic fields of H=0 and H=5.4 tesla and at T=220K.

Resistance Transition near 330 K At least three "single-crystal" samples have shown a sharp resistance transition near 330 K. One example is shown in Fig. 4. The resistance of the samples above the transition temperature is ohmic in all directions as verified by dc I-V characteristic measurements using different combinations of electrical leads. However, below the transition temperature most but not all directions exhibit zero resistance. This indicates that the"zero resistance" paths are localized in certain parts of the sample. Nevertheless, the consistency in the transition temperature for a wide range of sample resistance values (from less than 0.1 Ω to a few kΩ) for several different samples suggests that the transition phenomenon is a real material property associated with the material. Also, contact resistance values did not affect the transition temperature. Although the contact resistance for one particular sample changed drastically from ohms to kilo-ohms after a few thermal cyclings, the resistive transition temperature did not change. To eliminate a possible artifact (such as separation of current and voltage paths) in a 4-contact arrangement, we have verified the transition by using a 2-contact current-voltage measurement. By removing the two contact resistances with a bridge circuit, the I-V characteristics near the transition show microbridge-type structures as seen in Fig. 5.

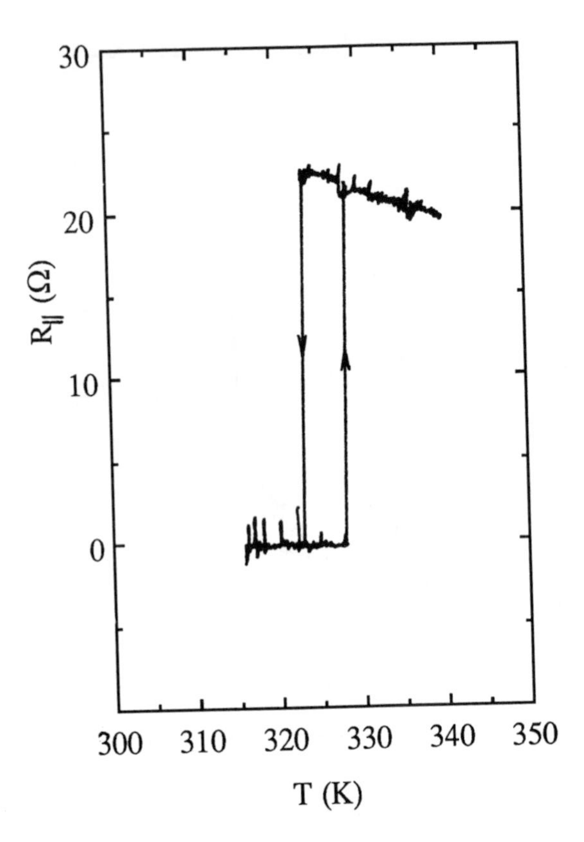

Fig. 4. These reproducible R vs T traces were obtained after 3 weeks of continuous thermal cycling in oxygen. The bias current was 10 μA. The hysteresis was not due to temperature error as the two traces overlap at both ends of the temperature.

Material Characterization In order to obtain additional information on the structure and composition, the single-crystal sample which had a transition near 260 K shown in Fig. 1 and 2 has been analyzed by high-resolution transmission electron microscopy. The result showed that there was a layer of about 1 μm in thickness on one surface containing numerous defects. In addition, EDS shows that the surface layer is Cu deficient. On average, the ratios of Y:Ba:Cu is approximately 1:2:2.5. By associating this structure information with the anisotropy shown in Figs. 1 and 2, we attribute the 260 K transition to the defect-filled surface layer and the 90-K transition to the previously established $YBa_2Cu_3O_7$.

DISCUSSION

Although our magnetization data are not shown, the magnetization on several ceramic samples have been measured with our Quantum Design SQUID magnetometer. The results consistently showed a weak diamagnetic deviation in the temperature range

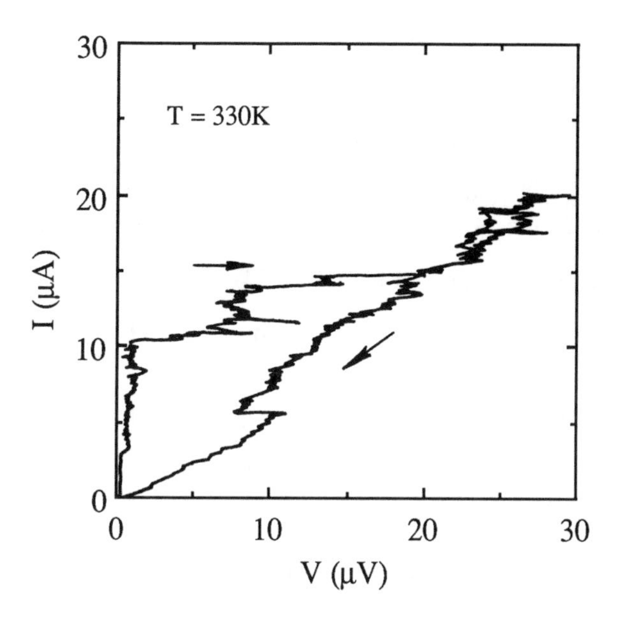

Fig. 5. The current-voltage characteristic near the transition temperature of the sample shown in Fig. 4.

of 240 to 270 K. This behavior is not inconsistent with the magnetic properties of thin superconducting layers in a mixed-phase material. Considering all the experimental facts, we are very optimistic that the transition in the 260 to 275 K is superconducting.

As for the material responsible for the 330 K transition, it is distributed more widely in a sample because the resistive transition appears in all directions. This transition was very sensitive to the oxygen environment and was affected by the thermal cyclings continuously for about two weeks. After one month of repeated thermal cycles between 300 K and 340 K in oxygen, the transition became stable and reproducible. We speculate that it is caused by the diffusion or adsorption of oxygen into defect regions, twin boundaries, or stacking faults which exist commonly in $YBa_2Cu_3O_{7-x}$. If the transition near 330 K is associated with superconductivity, it may be caused by the surface superconductivity [10-13] in the defect regions.[14,15] We are not certain that the transition near 330 K is superconducting since we have not obtained any magnetic-field dependence and it is difficult to identify surface superconductivity other than using a resistive transition for its detection. Surface superconductivity will not easily demonstrate a Meissner effect and has only a weak magnetic-field dependence even in the electrical measurements.[13] In any case, the transition near 330 K is also a real material property which must require further study in order to understand its nature.

ACKNOWLEDGEMENTS We thank C.H. Chen (AT&T) for his high resolution electron microscopy investigation of our samples and for giving us the useful structural information. This work is supported in part by the Electric Power Research Institute and the WSU Institute for Manufacturing Research.

REFERENCES

1. J. T. Chen, L-X. Qian, L-Q. Wang, L. E. Wenger, and E. M. Logothetis, *Mod. Phys. Lett. B* 3, 1197 (1989).
2. J. T. Chen, L- X. Qian, L - Q. Wang, L. E. Wenger, and E. M. Logothetis, in Superconductivity and Applications, edited by H. S. Kwok, Yi-Han Kao, and David T. Shaw, (Plenum, New York, 1989), p. 517.
3. R. Schönberger, H.H. Otto, B. Brunner, and K.F. Renk, *Physica C* 173, 159 (1991).
4. C.Y. Huang, L.J. Dries, P.H. Hor, R.L. Meng, C.W. Chu, and R.B. Frankel, Nature 328, 403 (1987).
5. V.N. Moorthy, S.K. Agarwal, B.V. Kumaraswamy, P.K. Dutta, V.P.S. Awana, P. Maruthikumar, and A.V. Narlikar, J. Phys.: Cond. Mat. 2, 8543 (1990).
6. P. Ayyab, P. Guptasarma, A.K. Rajarajan, L.C. Gupta, R. Vijayaraghavan,and M.S. Multani, J. Phys. C: Solid State Phys. 20, L673(1987).
7. H. Ihara, N. Terada, M. Jo, M. Hirabayashi, M. Tokumoto, Y. Kimura, T.Matsubara, and R. Sugise, *Jpn. J. Appl. Phys.* 26, L1413 (1987)
8. D. L. Kaiser, F. Holtzberg, B. A. Scott, and T. R. McGuire, *Appl. Phys. Lett.* 51, 1040 (1987).
9. J. T. Chen, C. J. McEwan, L. E. Wenger, and E. M. Logothetis, *Phys. Rev. B* 35, 7124 (1987).
10. V. L. Ginzburg and D. A. Kirzhnitz, *Soviet Phys.-JETP* 19, 269 (1964).
11. B. Weitzel and H. Micklitz, *Phys. Rev. Lett.* 66, 385 (1991).
12. M. Strongin, O. F. Kammerer, J. E. Crow, R. D. Parks, D. H. Douglass, and M A. Jensen, *Phys.Rev. Lett.* 21, 1320 (1968).
13. M. Strongin, A. Paskin, O. F. Kammerer, and M. Garber, *Phys. Rev. Lett.* 14, 362 (1965).
14. D. Wohlleben, J. F. Smith, F. M. Mueller, and S. P. Chen, *Physica C* 153 - 155, 586 (1988).
15. F. M. Mueller, S. P. Chen, M. L. Prueitt, J. F. Smith, J. L. Smith, and D. Wohlleben, *Phys. Rev. B* 37, 5837 (1988).

ANISOTROPIC SUPERCONDUCTING PROPERTIES FOR c–AXIS ALIGNED POWDERS OF THE (2223) COMPOUND $(Bi,Pb)_2Ca_2Sr_2Cu_3O_{10+\delta}$

H.C. Ku and J.B. Shi*
Department of Physics, National Tsing Hua University,
Hsinchu, Taiwan 30043, R.O.C.
*Institute of Electronics, National Chiao Tung University,
Hsinchu, Taiwan 30039, R.O.C.

ABSTRACT

Anisotropic superconducting properties of aligned powders of the bismuth–2223 $(Bi,Pb)_2Ca_2Sr_2Cu_3O_{10+\delta}$ compound are reported. The c–axis aligned powders embedded in epoxy are prepared in a 9.4 tesla applied magnetic field at room temperature. Temperature dependence of the anisotropy ratio $\chi_c(T)/\chi_{ab}(T)$ is derived from zero–field–cooled (ZFC) susceptibility measurements using small applied fields parallel or perpendicular to the orthorhombic c–axis. Large anisotropy ratios of 17.9 near T_c of 108 K and 9.8 at 5 K are observed. Field dependence of the anisotropy ratio $\chi_c(H)/\chi_{ab}(H)$ is derived from the initial magnetization $M_i(H)$ in the hysteresis measurements for **H**⊥c and **H**||c at 5 K. Anisotropic intragrain critical current densities J_c up to 4 kG are estimated. Anisotropic irreversibility lines $H_r(T)$ due to thermal fluctuation are determined from the merging point of ZFC and FC susceptibility measurements in various applied fields with **H**⊥c and **H**||c. The anisotropy ratio $H_r(\perp c)/H_r(\| c)$ decreases sharply from 13.6 at 100 K to a low value of 3.2 at 80 K and 3.1 at 70 K. The 3D–like power law $H_r = a\cdot(1 - T/T_c)^n$ is valid only in the low field region ($\lesssim$ 200 G) with n = 2.19 for **H**⊥c and 2.99 for **H**||c. In the higher field region up to 4 kG, the temperature dependence of $H_r(T)$ lines changes into a 2D–like exponential function $H_r = b\cdot\exp(-T/T_0)$ due to the breakdown of the interlayer and/or intralayer coupling of the conduction channel, with T_0 = 14.2 K for **H**⊥c and 13.7 K for **H**||c.

I. INTRODUCTION

Several high temperature superconducting phases prevail in the bismuth copper oxide family of the orthorhombic structure with the general formula $(Bi,Pb)_2Ca_2Sr_2Cu_3O_{2n+\delta}$ (2,n–1,2,n).[1-21] The maximum superconducting transition temperature T_c(max) is 32 K for the $Bi_2Sr_2CuO_{6+\delta}$–type (2021) structure,[1-2,4-13] 95 K for the $Bi_2CaSr_2Cu_2O_{8+\delta}$ (2122) structure and 110 K for the $Bi_2Ca_2Sr_2Cu_3O_{10+\delta}$ (2223) structure.[3,14-21] Excess oxygen atoms ($\delta > 0$) are commonly observed with the appearance of superstructure modulation along the orthorhombic b–axis.

The superconductivity of this family is closely related to the presence of hole as charge carriers in the quasi–two–dimensional Cu–O planes. These phases have units of CuO_x clusters where the elements are coordinated in different geometrical configurations. For the (2021) phase, CuO_6 forms an octahedral cluster; for the (2122) phase, there are two CuO_5 pyramidal clusters separated by Ca; and for the (2223) phase, in addition to two CuO_5 pyramids, there is a CuO_4 planar cluster which is separated from the CuO_5 pyramids by Ca atoms.

Many superconducting properties of the quasi–two–dimensional superconductors are highly anisotropic as expected. One of the most intriguing properties is the occurrence of the irreversibility line $H_r(T)$ due to thermal fluctuations in the vortex state region between lower critical field $H_{c1}(T)$ and upper critical field $H_{c2}(T)$. This irreversibility or "quasi de Almeida–Thouless" line was first observed in the $La_{2-x}Ba_xCuO_{4-y}$ superconductor where $H_r(T)$ can be fitted by a simple power law of $H_r(T) = a\cdot(1 - T/T_c)^n$ with $n \cong 1.5$.[22] The irreversibility line $H_r(T)$ was later observed in all high–T_c superconductors with the power n varying from 1.3 to 2.[23-31] However, in most cases the power law can be applied only in the low field (< 10^2–10^3 G) region, serious deviation from linearity in the logarithmic plot indicates that other forms of temperature dependence in higher field region are required. Recently, in the Bi–2223 $(Bi,Pb)_2Ca_2Sr_2Cu_3O_{10+\delta}$ bulk sample, an exponential law of the form $H_r(T) = b\cdot\exp(-T/T_0)$ for T < 80 K was reported.[28,30-1].

The effects of increasing disorder or other imperfections on the position of $H_r(T)$ in the H–T plane were also explored. After the 3–MeV proton irradiation which creates random local point defects, the irreversibility line of the $YBa_2Cu_3O_{7-x}$ single crystal remains unchanged.[32] However, a large shift was reported for the Bi–2122 $Bi_2CaSr_2Cu_2O_{8+\delta}$ single crystal after neutron irradiation.[33] A shift of $H_r(T)$ due to the thickness of $YBa_2Cu_3O_{7-x}$ thin film was also observed.[34]

The theoretical interpretations for the origin of the irreversibility line $H_r(T)$ due to strong thermal fluctuations for these quasi–two–dimensional high T_c superconductors are confusing. Various models ranging from the giant flux creep model to the vortex lattice (low random pinning) or vortex glass (strong random pinning) melting model were proposed.[35-38]

The anisotropic properties can be studied only using single phase samples of single crystal, c–axis oriented thin film or c–axis aligned microcrystalline powders. However, no good data were reported for the Bi–2223 compound due to the difficulty of single phase sample preparation. In this paper, we report the anisotropic superconducting properties of the highly oriented powders embedded in epoxy for the bismuth copper oxide $(Bi,Pb)_2Ca_2Sr_2Cu_3O_{10+\delta}$.

II. EXPERIMENTAL DETAILS

Bulk samples were synthesized using the solid–state reaction method. High–purity powders of Bi_2O_3 Pb_3O_4, $CaCO_3$, $SrCO_3$ and CuO were used with the ratio (Bi+Pb):(Ca+Sr):Cu = (1.85+0.15):(2.2+1.8):3 with excess PbO and CuO in order to preserve the entropy–stabilized metastable Bi–2223 phase with the nominal composition $(Bi_{1.85}Pb_{0.15})Ca_{2.2}Sr_{1.8}Cu_3O_{10+\delta}$. Well–mixed powders were calcined at 800 ^{0}C in air for 1 day with several intermediate regrindings. These powders were then pressed into pellets and sintered at 859 ^{0}C in air up to 3 days and then furnace–cooled.

For c–axis aligned powder samples, Farrel's method was employed.[36] Pellets were grounded to powders with average microcrystalline grain size of 1–10 μm, mixed with SPAR 5–minute epoxy/hardener in a 8 mm quartz holder with typical powder:epoxy ratio of 1:7, then aligned in a 9.4 T magnetic filed at room temperature as confirmed by the anisotropic normal state magnetic susceptibility. The block diagram of instrument is shown in Fig. 1. The degree of c–axis alignment is better than 90% as can be checked from the intensities of the orthorhombic (00l) lines from x–ray diffraction patterns.[29-31]

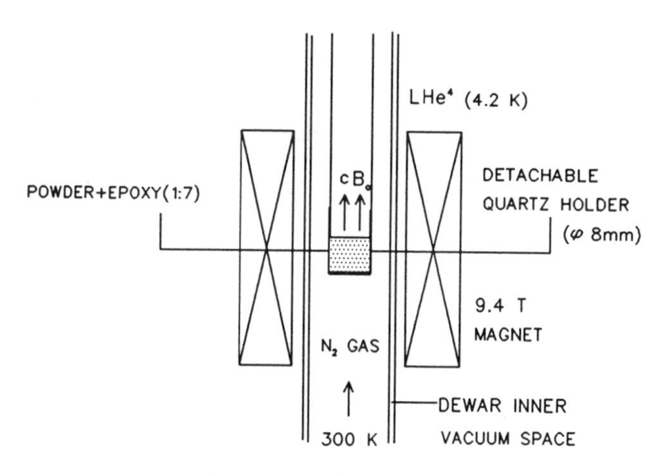

Fig. 1. Block diagram for powder alignment.

Superconducting data were obtained by using a Quantum Design MPMS SQUID magnetometer from 2 to 300 K. For zero–field cooled (ZFC) measurements, the "magnetic reset" option was used to quench the superconducting magnet and reduce the residual or remnant field to less than 1 G.

III. RESULTS AND DISCUSSION

For the Bi–2223 $Bi_2Ca_2Sr_2Cu_3O_{10+\delta}$ structure, sample with the composition $(Bi_{1.85}Pb_{0.15})Ca_{2.2}Sr_{1.8}Cu_3O_{10+\delta}$ was chosen. Excess PbO and CuO are necessary in order to ensure the prevention of the formation of Bi–2021 or Bi–2122 phases. X–ray diffraction patterns for randomly oriented powders and c–axis aligned powders embedded in epoxy can all be indexed with the (2223) orthorhombic structure with lattice parameters a = 5.410(5) Å, b = 5.413(5) Å and c = 37.07(3) Å. No (2021) or (2122)lines were observed.[29-31] For the c–axis aligned powder sample, only (00l) lines were observed which indicate excellent c–axis alignment.

The temperature dependence of magnetization M(T) for aligned powder samples, either field–cooled (FC) or zero–field–cooled (ZFC) in low applied fields parallel or perpendicular to c–axis, indicates a superconducting transition temperature of T_c = 108 K with excellent ZFC diamagnetic signal $-4\pi\chi_c \cong 0.77$ and the FC signal $\cong$ 0.55 for **H**||c using the x–ray derived density ρ_x = 6.20 g/cm^3.[29-31]

The temperature dependence of zero–field cooled (ZFC) anisotropic magnetic susceptibility ratio $\chi_c(T)/\chi_{ab}(T)$ for c–axis aligned powders in low applied fields of 8, 30 and 80 G are shown in Fig. 2. Large anisotropy ratio

χ_c/χ_{ab} of 9.8 was observed at low temperatures. The ratio in low applied field of 8 G increases sharply to a maximum value of 17.9 for temperatures near T_c of 108 K.

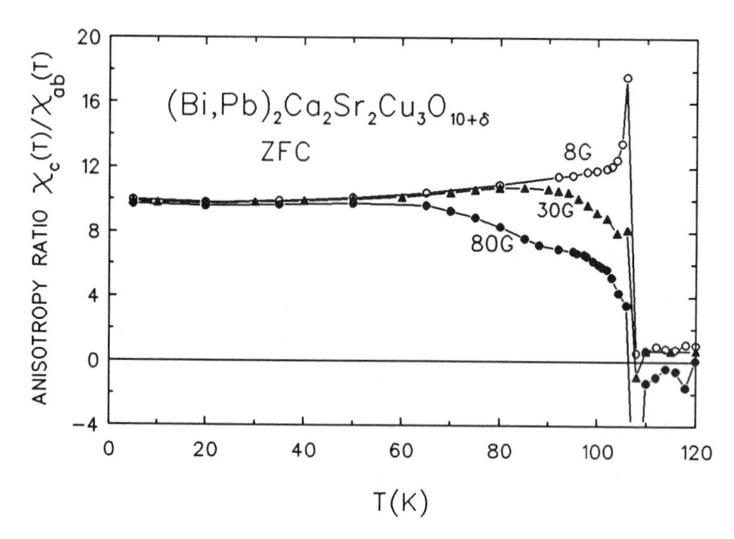

Fig. 2. Temperature dependence of zero–field cooled (ZFC) anisotropic magnetic susceptibility ratio $\chi_c(T)/\chi_{ab}(T)$ for aligned powders of $(Bi,Pb)_2Ca_2Sr_2Cu_3O_{10+\delta}$ in three low applied fields of 8, 30 and 80 G.

The field effect on anisotropy was obtained from the initial magnetization curve $M_i(H)$ and the magnetic hysteresis loop M(H) with applied magnetic field up to ± 5 kG parallel or perpendicular to c–axis as shown in Fig. 3. The field

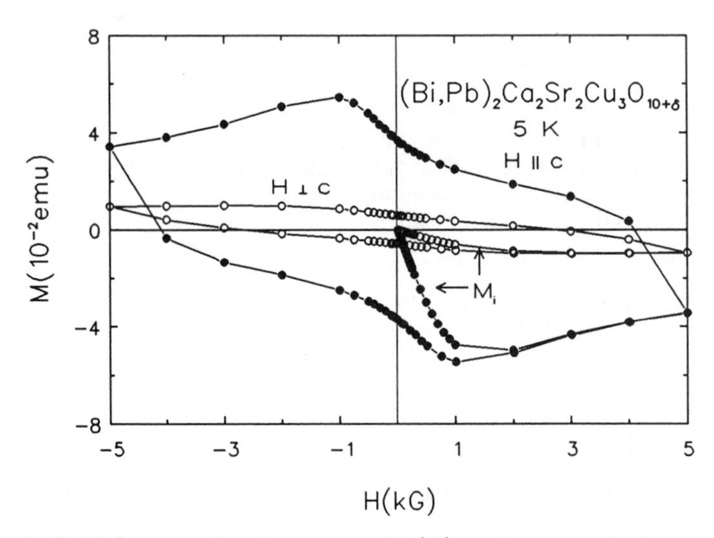

Fig. 3. Initial magnetization curve $M_i(H)$ and magnetic hysteresis loop M(H) for the aligned powder sample $(Bi,Pb)_2Ca_2Sr_2Cu_3O_{10+\delta}$ with applied field parallel or perpendicular to c–axis.

dependence of the anisotropy ratio $\chi_c(H)/\chi_{ab}(H)$ at 5 K extracted from the initial magnetization curve is shown in Fig. 4. The $\chi_c(H)/\chi_{ab}(H)$ ratio decreases from 9.8 in low field to 3.5 at 5 kG. The anisotropic field dependence of the intragrain critical current densities J_c at 5 K was estimated (Fig. 5) from the hysteresis loop using the simple Bean's formula $J_c = 30 \cdot \Delta M/d$, where ΔM (in emu/cm^3) is the magnetization difference between the increasing and decreasing field curves and d ($\cong$ 5 μm) is the average grain size of the powder observed from the scanning electron microscope (SEM).

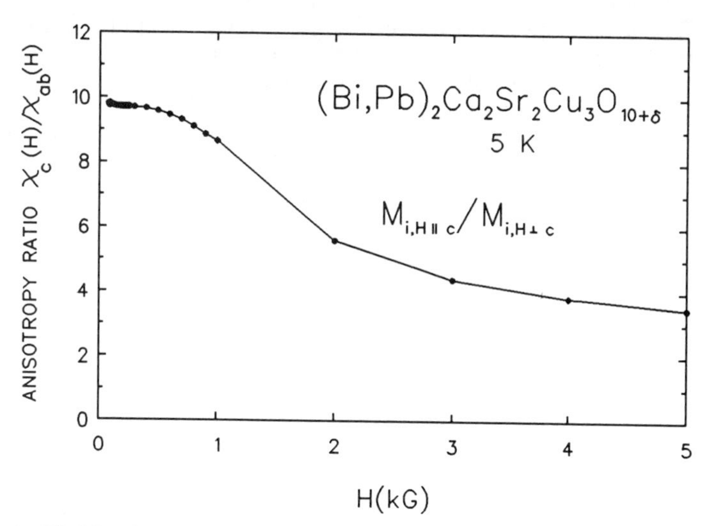

Fig. 4. Field dependence of the anisotropy ratio $\chi_c(H)/\chi_{ab}(H)$ of $(Bi,Pb)_2Ca_2Sr_2Cu_3O_{10+\delta}$ obtained from the initial magnetization curve $M_i(H)$.

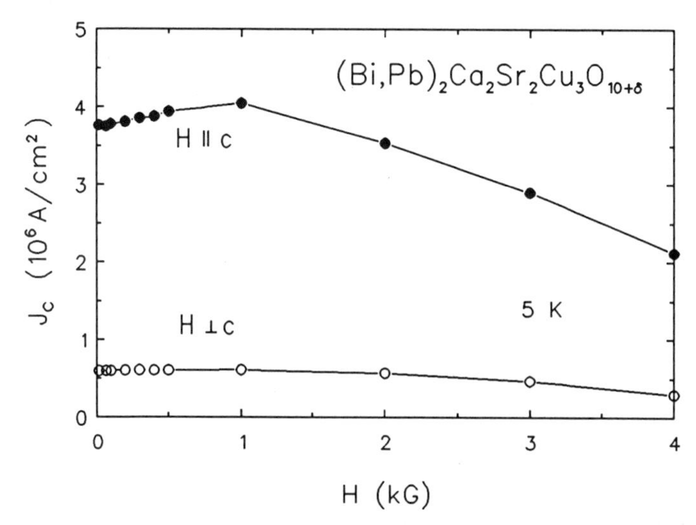

Fig. 5. Field dependence of the intragrain critical current density $J_c(H)$ at 5 K, obtained from magnetic hystersis loop M(H) with applied field parallel or perpendicular to c-axis.

The irreversibility temperature T_r for the aligned powder sample was obtained from the merging point of the field cooled (FC) and zero–field cooled (ZFC) curves of the temperature dependence of mass magnetic susceptibility χ_g. The irreversibility temperatures T_r's can be easily pinpointed using the ratio $\chi_g(ZFC)/\chi_g(FC)$, which decreases steadily to 1 when the field cooled and zero–field cooled curves merge together for $T \geq T_r$.

The irreversibility lines $H_r(T)$'s with applied field up to 4 kG parallel or perpendicular to c–axis were shown in Fig. 6. Dashed lines for the lower critical field H_{c1} and upper critical field H_{c2} are anisotropic in nature and are guides to the eyes. The anisotropy ratio $H_r(\perp c)/H_r(\|c)$ decreases rapidly from 13.6 at 100 K ($T_c = 108$ K) to 5.9 at 90 K, 3.2 at 80 K, and 3.1 at 70 K. The anisotropy ratio $H_r(\perp c)/H_r(\|c) > 1$ was also observed for the anisotropic upper critical field H_{c2} of all high–T_c superconductors due to the anisotropic coherence length ξ with $\xi_c < \xi_{ab}$. Since $H_r(T)$'s are closely related to pinning/depinning mechanisms and are sample dependent, in–depth studies between anisotropic properties of the irreversibility lines and anisotropic superconducting intrinsic parameters are necessary and are in progress.

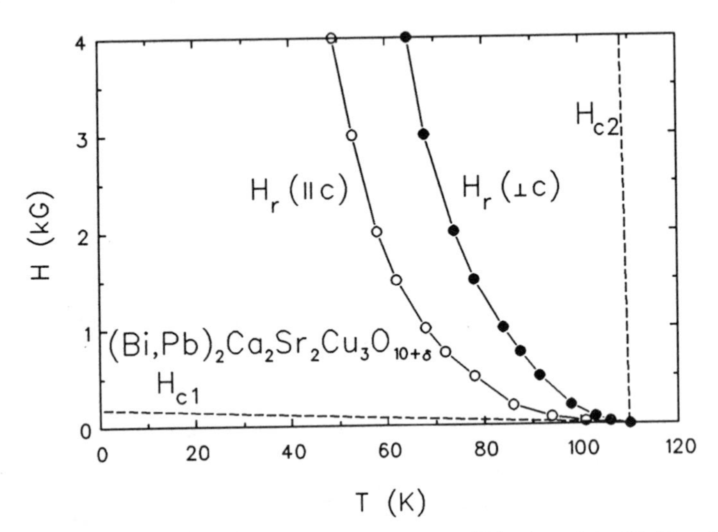

Fig. 6. Anisotropic irreversibility lines $H_r(T)$ in the H–T plane for $(Bi,Pb)_2Ca_2Sr_2Cu_3O_{10+\delta}$. Dashed lines are estimated to represent the lower critical field H_{c1} and upper critical field H_{c2}.

The temperature dependence of the irreversibility lines $H_r(\perp c)$ and $H_r(\|c)$ can be determined using a logarithmic plot of ln H_r versus ln $(1 - T/T_c)$ as shown in Fig. 7. The linear behavior in the low field region ($\leq$ 200 G) indicates that both lines can be nicely fitted by the simple power law

$$H_r = a \cdot (1 - T/T_c)^n$$

with n = 2.19, a = 3.68 T (36.8 kG) for **H**⊥c and n = 2.99, a = 2.23 T for **H**‖c. The power value n = 2.99 (**H**‖c) is the largest among values observed so far for all high–T_c superconductors. It is also larger than the value n = 1.5 predicated by the standard flux creep model and n = 2 by the vortex lattice melting model.[35,39] The large reversible region in the H–T plane indicates very low flux pinning for this c–axis aligned powder sample.

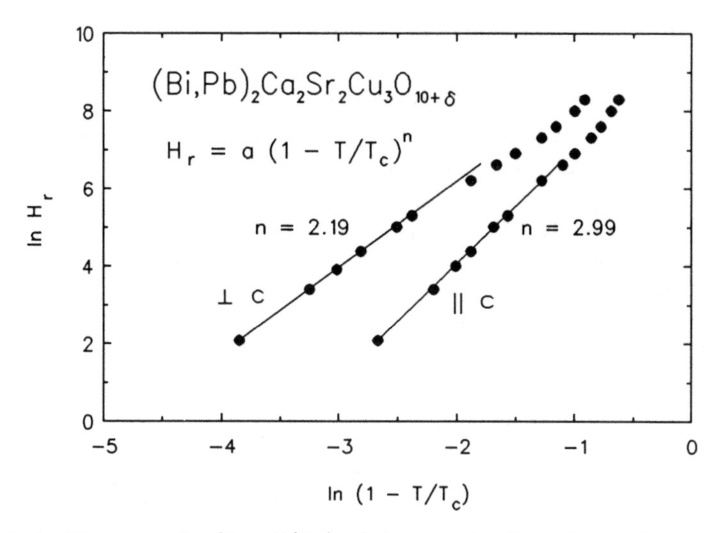

Fig. 7. ln H_r versus ln $(1 - T/T_c)$ of the c–axis aligned powder sample. A linear behavior was observed in the low field range up to 200 G.

In the higher field region (> 200 G), no simple power law can be found. However, using the ln H_r versus T plot (Fig. 8), the irreversibility lines can be fitted by the exponential function

$$H_r = b \cdot \exp(-T/T_0)$$

with $T_0 = 14.2$ K, b = 36.2 T for H⊥c and $T_0 = 13.7$ K, b = 14.0 T for H||c. The $H_r(\|c) = 14.0 \cdot \exp(-T/13.7)$ T for $H_r \geq 500$ G (T < 80 K) is compatible with the

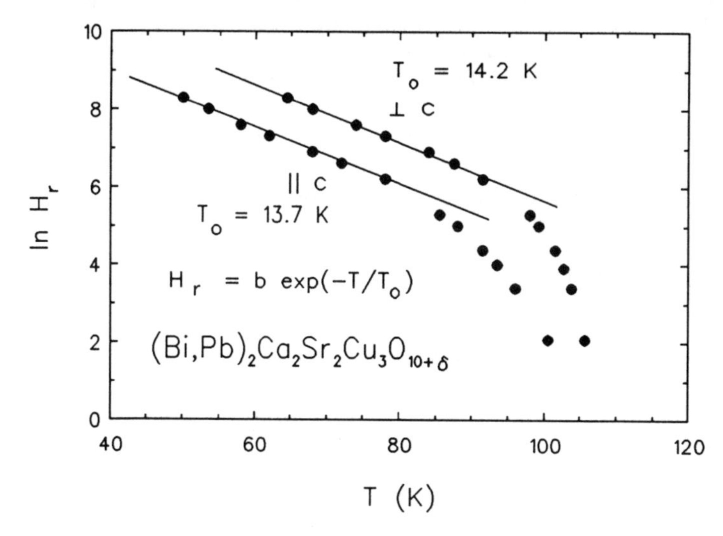

Fig. 8. ln H_r versus T of the c–axis aligned powder sample. An exponential dependence of the irreversibility lines for H > 200 G was observed.

$H_r = 14.7 \cdot \exp(-T/13.3)$ T, a value reported for bulk but possibly preferred oriented sample below 80 K.[28]

The Bi–2223 orthorhombic (a = 5.409 Å, b = 5.411 Å, c = 37.09 Å) phase has a structure in which conduction layers (with three Cu–O planes, CuO–Ca–CuO–Ca–CuO) of thickness d ≈ 9 Å alternate with charge reservoir layers (SrO–BiO–BiO–SrO) of thickness d' ≈ 9.5 Å. Interlayer and/or intralayer Josephson coupling are necessary due to small coherence length along the c–axis $\xi_c \approx 2$ Å. In the low field region, a 3D–like power law for $H_r(T)$ line is expected. With higher applied field in the vortex state region, interlayer and/or interlayer coupling will be broken and a crossover from the 3D–like to 2D–like exponential behavior is expected.[28]

The H–T phase diagram of this quasi–two–dimensional type II superconductor with strong thermal fluctuation indicates a possible phase transition along the anisotropic irreversibility lines $H_r(T)$ from the low temperature vortex glass phase to the vortex liquid phase for $T > T_r$.[36,37] The vortex glass long range ordered phase is due to the presence of random local pinning centers for sample with the complex nominal composition $(Bi_{1.85}Pb_{0.15})Ca_{2.2}Sr_{1.8}Cu_3O_{10+\delta}$. The vortex fluid phase is a fully disordered phase with only local pairing where the pairing field is strongly fluctuating with only a finite but large correlation length.

IV. CONCLUSIONS

The superconducting properties of the aligned powders sample of Bi–2223 with the nominal composition $(Bi_{1.85}Pb_{0.15})Ca_{2.2}Sr_{1.8}Cu_3O_{10+\delta}$ embedded in epoxy are reported. Temperature dependence of anisotropy ratio $\chi_c(T)/\chi_{ab}(T)$ in the superconducting state remains constant for temperatures below T_c and increases for lower applied field and decreases for higher field near T_c. A highly anisotropy ratio of $\chi_c/\chi_{ab} = 9.8$ is observed for n = 3 at low temperatures. The field dependence of the anisotropy ratio $\chi_c(H)/\chi_{ab}(H)$ at 5 K decreases from 9.8 in low field to 3.5 at 5 kG. Anisotropic intragrain critical current densities $J_c(H)$ were observed.

Anisotropic irreversibility lines $H_r(T)$'s were observed. The anisotropy ratio $H_r(\perp c)/H_r(\parallel c)$ decreases sharply from 13.6 at 100 K to a low value of 3.2 at 80 K and 3.1 at 70 K. The simple power law $H_r = a \cdot (1 - T/T_c)^n$ was observed only in the low field region ($\lesssim$ 200 G) with n = 2.19 for **H**⊥c and 2.99 for **H**||c. In the higher field region up to 4 kG, $H_r(T)$ lines can be fitted with the exponential law $H_r = b \cdot \exp(-T/T_0)$ with $T_0 = 14.2$ K for **H**⊥c and 13.7 K for **H**||c.

ACKNOWLEDGEMENT

The research was supported by the National Science Council of the Republic of China under Contract No. NSC79–0208–M007–73.

REFERENCES

1. C. Michel, M. Hervieu, M.M. Borel, A. Grandin, F. Deslandes, J. Provost, and B. Raveau, Z. Phys. **B68**, 421 (1987).
2. J. Akimitsu, A. Yamazaki, H. Sawa, and H. Fujiki, Jpn. J. Appl. Phys. **26**,

L2080 (1987).
3. H. Maeda, Y. Tanaka, M. Fukitomi, and T. Asano, Jpn. J. Appl. Phys. **27**, L209 (1988).
4. L.C. Porter, E. Appelman, M.A. Beno, C.S. Cariss, K.D. Carlson, H. Cohen, U. Geiser, R. J. Thorn, and J.M. Williams, Solid State Commum. **66**, 1105 (1988).
5. H. Sawa, H. Fujiki, K. Tomimoto, and J. Akimitsu, Jpn. J. Appl. Phys. **27**, L830 (1988).
6. G. van Tendeloo, H.W. Zandbergen, and S. Amenlinckx, Solid State Commun. **66**, 927 (1988).
7. J.M. Torardi, M.A. Subramanian, J.C. Calabrese, J. Gopalakrishnan, E.M. McCarron, K.J. Morrissey, T.R. Askew, R.B. Flippen, U. Chowdhry, and A. W. Sleight, Phys. Rev. **B38**, 225 (1988).
8. J.M. Tarascon, W.R. McKinnon, P. Barboux, D.M. Hwang, B.G. Bagley, L.H. Greene, G.W. Hull, Y. LePage, N. Stoffel, and M. Giroud, Phys. Rev. **B38**, 8885 (1988).
9. G.H. Hwang, J.B. Shi and H.C. Ku, Chin. J. Phys. **27**, 273 (1989).
10. H.C. Ku, J.B. Shi, G.H. Hwang, M.F. Tai, S.W. Hsu, D.C. Ling, M.J. Shieh, T.Y. Lin, and T.J. Watson–Yang, in **Superconductivity and Applications**, eds. P.T. Wu, H.C. Ku, W.H. Lee, and R.S. Liu, (Progress in High Temperature Superconductivity Vol. 19, World Scientific, 1989), p.96.
11. H.C. Ku, S.J. Sun, Y.T. Peng, G.H. Hwang, J.B. Shi, and B.S. Chiou, **Proc. of the ROC–Japan Symposium for High Temperature Superconductor** (Japan Interchange Association, 1990), p.25.
12. H.C. Ku, J.B. Shi, C.C. Lai, and S.J. Sun, Physica **B165&166**, 1153 (1990).
13. J.B. Shi, B.S. Chiou, Y.T. Peng, and H.C. Ku, Chin. J. Phys. **28**, 339 (1990).
14. H.M. Hazen, C.T. Prewitt, R.J. Angel, N.L.Ross, L.W. Finger, C.G. Hadidiacos, D.R. Veblen, P.J. Heaney, P.H. Hor, R.L. Meng, Y.Y. Sum, Y.Q. Wang, Y.Y. Xue, Z.J. Huang, L.Gao, J. Bechtold, and C.W. Chu, Phys. Rev. Lett. **60**, 1174 (1988).
15. S.A. Sunshine, T. Siegrist, L.F. Schneemeyer, D.W. Murphy, R.J. Cava, B. Batlogg, R.B. van Dover, R.M. Fleming, S.H. Glarum, S. Nakahara, R. Farrow, J.J. Krajewski, S.M. Zahurak, J.V. Waszczak, J.H. Marshall, P. Marsh, L.W. Rupp, Jr., and W.F. Peck, Phys Rev. **B38**, 893 (1988).
16. J.M. Taracon, Y. Le Page, P. Barboux, B.G. Bagley, L.H. Greene, W.R. McKinnon, G.W. Hull, M. Giroud, and D.M. Hwang, Phys. Rev. **B37**, 9382 (1988).
17. M.A. Subramanian, K.J. Morrissey, T.R. Askew, R.B. Flippen, U. Chowdhry, and A.W. Sleight, Science **239**, 1015 (1988).
18. G.S. Grader, E.M. Gyorgy, P.K. Gallagher, H.M. O'Bryan, D.W. Johnson, Jr., S. Sunshine, S.M. Zahurak, S.Jin, and R.C. Sherwood, Phys. Rev. **B38**, 38757 (1988).
19. M. Onoda, A. Yamamoto, E. Takayama–Muromachi, and S. Takekawa, Jpn J. Appl. Phys. **27**, L830 (1988)
20. N.H. Wang, C.M. Wang, H.C.I. Kao, D.C. Ling, H.C. Ku, and K.H. Lii, Jpn. J. Appl. Phys. **28**, L1505 (1989).
21. A. Maeda, M. Hase, I. Tsukada, K. Noda, S. Takebayashi, and K. Uchinokura, Phys. Rev. **B41**, 6418 (1990).
22. K.A. Muller, M. Takashige, and J.G. Bednorz, Phys. Rev. Lett. **58**, 408 (1987).
23. A.P. Malozemoff, in "**Physical Properties of High Temperature Superconductors I**", ed. D.M. Ginsberg (World Scientific 1989), Chap. 3 and

references cited therein.
24. H.C. Ku and J.B. Shi, in preparation (1991).
25. G.H. Hwang, T.H. Her, C.Y. Lin, and H.C. Ku, Physica **B165&166**, 1155 (1990)
26. C.C. Lai, T.Y. Lin, P.C. Li, P.C. Ho, C.Y. Hung, and H.C. Ku, Chin. J. Phys. **28**, 347 (1990).
27. G.H. Hwang, T.H. Her, and H.C. Ku, Chin. J. Phys. **28**, 453 (1990).
28. P. de Rango, B. Giordanengo, J.L. Genicon, P. Lejay, A. Sulpice, and R. Tournier, Physica **B165&166**, 1141 (1990).
29. J.B. Shi, B.S. Chiou, and H.C. Ku, Phys. Rev. **B43**, 13001 (1991).
30. J.B. Shi, P.L. Kuo, B.S. Chiou, and H.C. Ku, Physica **C**, in press (1991).
31. J.B. Shi, B.S. Chiou, P.L. Kuo, and H.C. Ku, Phys. Rev. B, submitted (1991).
32. L. Civale, A.D. Marwick, M.W. McElfresh, T.K. Worthington, A.P. Malozemoff, F.H. Holtzberg, J.R. Thompson, and M.A. Kirk, Phys. Rev.
33. W. Kritscha, F.M. Sauerzopf, H.W. Weber, G. W. Crabtree, Y.C. Chang, and P.Z. Jiang (unpublished).
34. L. Civale, T.K. Worthington, and A. Gupta, Phys. Rev. **B43**, 5425 (1991).
35. Y. Yeshurun and A.P. Malozemoff, Phys. Rev. Lett. **60**, 2202 (1988).
36. D.R. Nelson, Phys. Rev, Lett. **60**, 1973 (1988).
37. D.S. Fisher, M.P.A. Fisher, and D.A. Huse, Phys. Rev. **B43**, 130 (1991), and references cited therein.
38. Y. Xu and M. Suenaga, Phy. Rev. **B43**, 5516 (1991).
39. A. Houghton, R.A. Pelcovits, and A. Sudbo, Phy. Rev. **B40**, 6763 (1989).

CUPRATE SUPERCONDUCTORS WITH 1212-TYPE STRUCTURE

Z.Z.Sheng, Y.F.Li, and D.O.Pederson

Department of Physics, University of Arkansas, Fayetteville, AR 72701

ABSTRACT

The TlBaCaCuO system forms two series of superconducting compounds, double TlO-layer $Tl_2Ba_2Ca_{n-1}Cu_nO_{2n+4}$ and single TlO-layer $TlBa_2Ca_{n-1}Cu_nO_{2n+3}$. The member of n = 2 in the series with single TlO layer, $TlBa_2CaCu_2O_7$, has been often called 1212 phase. The 1212 structure can be widely subjected to elemental substitutions, forming a large family of 1212-type superconductors with a general formula of $M1M2_2M3Cu_2O_7$, where M1, M2, and M3 are metallic elements. In this paper, the existing 1212 phases are reviewed; formation and superconductivity of the 1212 phases are discussed in terms of the concept of mixed Cu valence; possible new 1212-type phases are predicted; some experimental results for the predicted 1212 phases are presented.

1. INTRODUCTION

The revolutionary discovery of the 30 K LaBaCuO by Bednorz and Muller[1] has led to a number of new cuprate superconducting systems, including YBaCuO, the first system with Tc above liquid nitrogen temperature[2], BiSrCaCuO, the first system with Tc above 100 K[3], and TlBaCaCuO, the system with the highest reproducible Tc above 120 K[4,5]. In the Tl-based system, two series of superconducting phases, double TlO-layer $Tl_2Ba_2Ca_{n-1}Cu_nO_{2n+4}$[6-8] and single TlO-layer $TlBa_2Ca_{n-1}$-Cu_nO_{2n+3}[9,10], were rapidly identified. The member of n = 2 in the series of single TlO-layer, $TlBa_2CaCu_2O_7$, has been widely called 1212 phase. The 1212 structure can be subjected to extensive elemental substitutions, forming a large family of 1212-type cuprate superconductors with a general formula of $M1M2_2M3M4_2O_7$ with M1, M2, and M3 = metallic elements and M4 = Cu. In this paper, we review the existing data for 1212 superconductors, and discuss the formation and superconductivity of the 1212 phase in terms of the concept of mixed Cu valence. We predict possible new 1212-type superconductors, and present some experimental results. We belive that the results for the 1212 structure is also applicable, to some extent, to other cuprate superconductors.

2. 1212-TYPE SUPERCONDUCTOR

An ideal 1212 structure has a tetragonal unit cell, space group P4/mmm. A 1212 unit cell consists of total 13 atoms: 6 metal atoms and 7 oxygen atoms. The six metal atoms occupy four different functional sites: one for M1, two for M2, one for M3, and two for M4. The seven O atoms occupy three different functional sites, one for O1, two for O2, and four for O3. Figure 1 shows the schematic drawing of the 1212 structure $M1M2_2M3M4_2O_7$ with M1 = Tl, M2 = Ba, M3 = Ca, and M4 = Cu. Figure 2 shows oxygen coordination of metallic ions in $TlBa_2CaCu_2O_7$. Tl is coordinated by six O (4 O1 and 2 O2), Ba by nine O (1 O1, 4 O2, and 4 O3), Ca by eight O (8 O3), and Cu by five O (1 O2 and 4 O3). Figure 3 shows metallic-atom coordination for O atoms in $TlBa_2YCu_2O_7$. It can be seen that each O atom is coordinated by six metallic atoms: O1 by four Tl and two Ba; O2 by one Tl, four Ba, and one Cu; O3 by two Ba, two Y, and two Cu.

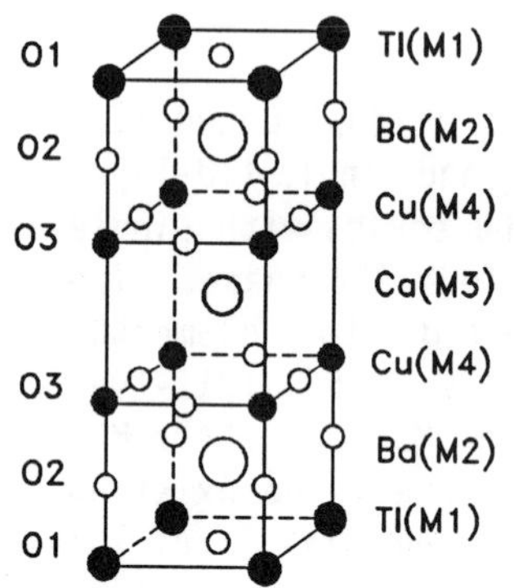

Figure 1 Schematic drawing of an ideal 1212 structure.

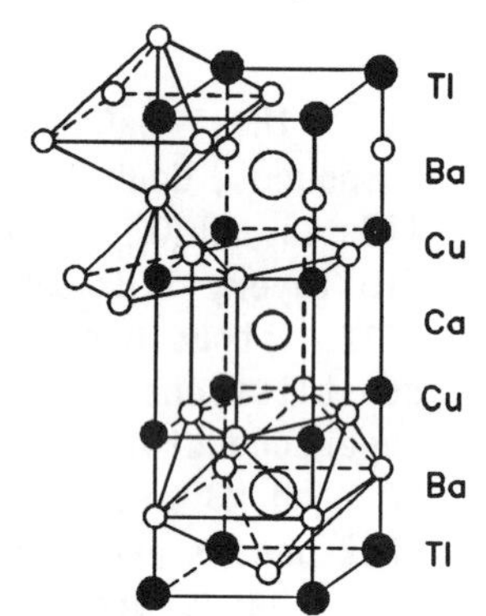

Figure 2 O-coordination of metallic ions in the 1212 structure $TlBa_2CaCu_2O_7$.

$YBa_2Cu_3O_7$, the so-called 123 structure, is the first superconductor with Tc above liquid nitrogen temperature[2,11], and has been investigated in the most detail. In fact, $YBa_2Cu_3O_7$ is an analogue and a special form of the 1212 structure. Fig. 4 shows a comparison between 123 phase $YBa_2Cu_3O_7$ and 1212 phase $TlBa_2YCu_2O_7$. The only difference is that Cu in the 123 structure plays the role of both M4 and M1. The TlO plane in the 1212 structure is replaced by the CuO chain in the 123 structure. This replacement decreases the symmetry of the structure, converting the tetragonal 1212 phase (space group P4/mmm) to the orthorhombic 123 phase (space group Pmmm). The 123 phase may be written as the form of

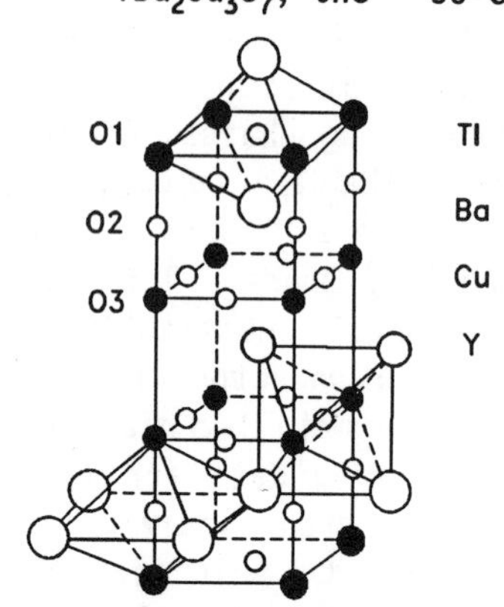

Figure 3 Metallic-ion coordination for oxygen in the 1212 structure.

1212 phase $CuBa_2YCu_2O_7$.

Various 1212-type phases have been synthesized. Table I summarizes the 1212 phases and their Tc's reported in the literature. The Tc's of the 1212 phases range from 75 K to 105 K. M1 can be Tl, Bi, Pb, and Cu (and Sr, Ca), M2 may be Ba, Sr, and La, and M3 is R, Ca, and Sr. In this paper, M4 is always Cu. According to the metallic atom M1, the 1212 phases may be classified as Tl-based, Pb-based, or Cu-based. It is not sure if there is a family of Bi-based 1212 superconductors since all Bi-containing 1212-type superconductors reported so far contain Tl.

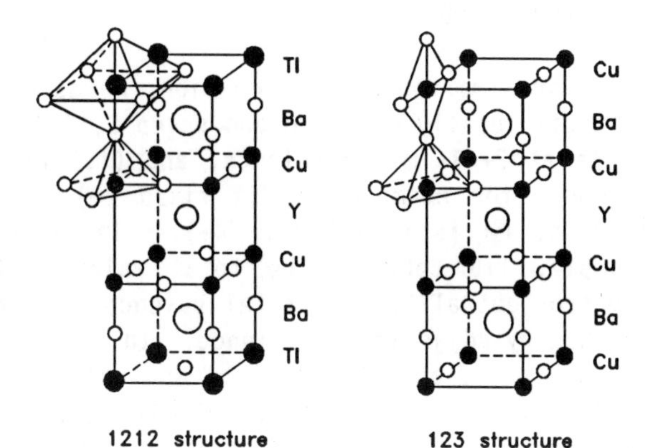

Figure 4 A comparison between 123 structure and 1212 structure.

Table I Existing 1212 phases $M1M2_2M3Cu_2O_7$ and their Tc.

M1	M2	M3	M4	Tc(K)	Ref.
Tl	Ba	Ca	Cu	85-100	9,10,12
Tl	Ba	Ca,R	Cu	105	13
Tl	Sr	Ca	Cu	70	14-17
Tl	Sr	Ca,R	Cu	90	18,19
Tl	Sr	Sr,R	Cu	90	20-25
Tl,Bi	Sr	Ca	Cu	90	26,27
Tl,Bi	Sr	Ca,R	Cu	105	28
Tl,Pb	Sr	Ca	Cu	75	17,29,30
Tl,Pb	Sr	Ca,R	Cu	105	31
Tl,Pb	Sr	Sr,R	Cu	105	23-25,32,33
Tl,Pb	Sr,La	Ca	Cu	105	34
Pb,Sr	Sr	Ca,R	Cu	90	35
Pb,Ca	Sr	Ca,R	Cu	90	36
Cu	Ba	R	Cu	90	2,11

3. MIXED Cu VALENCE

For simplicity of discussion, we assume that (i) 1212 structure consists ideally of six M and seven O, (ii) O is at 2- valence, (iii) Cu is at 2+/3+ mixed valence, and (iv) other M are either at fixed valence (for example, Sr is fixed 2+) or at variable valence (for example, Pb is variable 2+/4+). In an ideal 1212 structure, for example, $TlBa_2CaCu_2O_7$, the total valence of seven O is 14-. For electro-neutrality, the total valence of six M must be 14+. This is reached by adjusting Cu valence. In $TlBa_2CaCu_2O_7$, Tl is 3+, Ba is 2+, and Ca is 2+, and thus the average valence of Cu (avc) is (14-3-2x2-2)/2 = 2.5+. The concept of mixed $Cu^{2+/3+}$ valence can explain many experimental results to a certain extent, although it is somewhat oversimplified. Here we would like to point out the followings. First, the mixed $Cu^{2+/3+}$ valence is one of the guidelines that Bednorz and Muller used to discover the high Tc oxide superconductivity[1,37]. This fact suggests the intrinsic reasonability of the concept of mixed $Cu^{2+/3+}$ valence. Second, mixed $Cu^{2+/3+}$ valence has been determined experimentally in many cuprate superconductors using different methods. Third, mixed $Cu^{2+/3+}$ valence is also supported by other experimental results. For example, Figure 5 shows lattice parameters *a* and *c* vs y for $TlBa_2(Ca_{1-x}R_x)Cu_2O_7$ with R = Nd, Gd and Y[13]. With increasing x, the *c*-axis decreases, whereas the *a*-axis increases. That R^{3+} has ionic radius (1.11, 1.05 and 1.02 Å for Nd, Gd, and Y, respectively) smaller than that of Ca^{2+} (1.12 Å) can easily explain the decrease of *c*-axis with increasing y, but can not explain the increase of *a*-axis. The latter can be well explained by the concept of mixed $Cu^{2+/3+}$ valence. The *a*-axis of a cuprate superconductor depends strongly on the CuO framework. Substitution of R^{3+} for Ca^{2+} leads to the conversion of smaller Cu^{3+} to lager Cu^{2+}, expanding the CuO framework, and thus increasing *a*-axis.

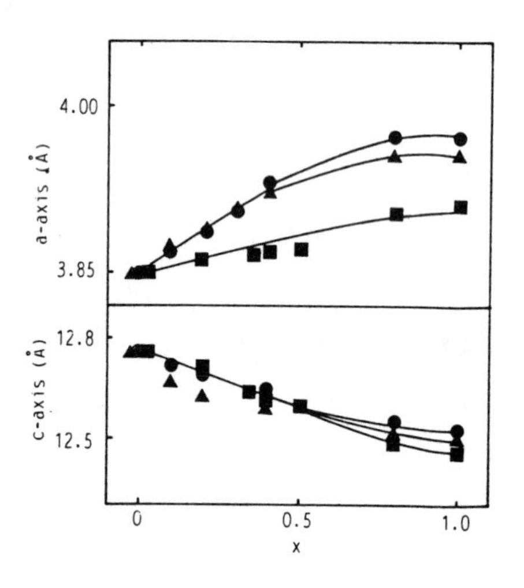

Figure 5 Lattice parameters *a* and *c* vs. y in $TlBa_2(Ca_{1-x}R_x)Cu_2O_7$ with R = Nd(●), Gd(▲), and Y (■)(from ref. 13).

4. STABILITY AND SUPERCONDUCTIVITY OF 1212 STRUCTURE

Several factors influence the stability of a crystal structure, so does the stability of the 1212 structure. The very important is the geometry factor. For example, $TlBa_2CaCu_2O_7$ is much more stable than $TlSr_2CaCu_2O_7$ although both have avc of 2.5+ due to that Sr^{2+} is much smaller than Ba^{2+}. $TlSr_2CaCu_2O_7$ and $TlSr_2SrCu_2O_7$ provide another example. Since Sr^{2+} is much larger than Ca^{2+}, $TlSr_2CaCu_2O_7$ exists[16,17], but $TlSr_2SrCu_2O_7$ does not exist[32,33]. The third example is $TlSr_2(Sr_{0.5}R_{0.5})Cu_2O_7$ and $TlSr_2(Sr_{0.5}Sc_{0.5})Cu_2O_7$. The former is stable but the latter is not stable since Sc^{3+} is too small for a stable 1212 structure[25]. For the cuprates with 1212-type structure, the average Cu valence is another important factor to influence the stability of the 1212 structure. The experimental results showed that a stable 1212 structure, the avc should be 2.0+ - 2.5+. When M2 = Sr, the 1212 phase with avc of 2.0+ is most stable. For example, $TlSr_2RCu_2O_7$ (with avc of 2.0+) is most stable[32], whereas $TlSr_2CaCu_2O_7$ (with avc of 2.5+) is not very stable.

When a 1212 structure, e.g., $TlSr_2RCu_2O_7$, has avc of 2.0+, it is stable but is not superconducting[32]. When R^{3+} in $TlSr_2RCu_2O_7$ is substituted partially by Ca^{2+} or Sr^{2+}, i.e. increasing the avc, $TlSr_2(R,Ca)Cu_2O_7$ or $TlSr_2(R,Sr)Cu_2O_7$ becomes superconducting[18,19,25,32]. On the other hand, $TlSr_2CaCu_2O_7$ has avc of 2.5+, and is superconducting, but not very stable[14-17]. The avc of $TlSr_2CaCu_2O_7$ can be decreased by substitutions, for example, Pb^{4+} for Tl^{3+}. Thus, $(Tl,Pb)Sr_2RCu_2O_7$ is stable and superconducting[17,19,30].

For a specific 1212 phase, there is an optimal avc at which Tc of the 1212 phase reaches maximum. Table II lists the optimal avc for some 1212 phases. Obviously, there are two groups of 1212 phases which have different optimal avc. The one with M2 = Sr has avc of 2.15-2.25, for example, the avc of $TlSr_2(Sr,R)Cu_2O_7$ is 2.20[32,33]. The another group

Table II Optimal avc for some Tl-based 1212 phases.

1212 phase	avc	Tc(K)	Ref.
$TlSr_2(Sr_{0.4}Pr_{0.6})Cu_2O_7$	2.20	90	32,33
$TlBa_2(Ca_{0.8}Y_{0.2})Cu_2O_7$	2.40	105	13
$(Tl_{0.5}Pb_{0.5})Sr_2CaCu_2O_7$	2.25	85	17,29,30
$(Tl_{0.5}Pb_{0.5})Sr_2(Ca_{0.8}Y_{0.2})Cu_2O_7$	2.15	105	31
$(Tl_{0.8}Pb_{0.2})Sr_2(Sr_{0.6}Pr_{0.4})Cu_2O_7$	2.20	100	32,33

with M2 = Ba has avc of about 2.4, for example, $TlBa_2(Ca,Y)Cu_2O_7$ has avc of 2.40[13]. Although the listed avc values are calculated from the ideal stoichiometry and not from the determined O content, we believe that this difference of the optimal avc is a reality even if the oxygen content is determined.

According to Tc, 1212 phase can be generally divided into two groups, one has Tc of 80-90 K, and another 100-105 K. For $TlSr_2CaCu_2O_7$, single substitution only reaches Tc around 90 K, and double substitution is necessary to increase Tc to 105 K. The only exception is the single Cr-substitution which does enhance Tc to 105 K[38]. One can adjust avc to a certain optimal value by changing the concentration of substitute(s), but one can not reach Tc of 105 K by a single element substitution (except Cr). This strongly indicates that hole concentration is not the only factor influencing Tc. This statement is also supported by the result that a small change in O content (thus average Cu valence, thus hole concentration) causes a large variation of Tc from 85 K to 105 K for $TlBa_2CaCu_2O_7$[12]. The same effect is observed for $Tl_2Ba_2CuO_6$[39].

5. POSSIBLE SUPERCONDUCTING 1212 PHASES

The above discussion showed that when avc = 2.0+ - 2.5+, the 1212 structure may be stable and superconducting. We may write a general form for the 1212 structure:

$$(M1,S1)(M2,S2)_2(M3,S3)Cu_2O_7 \qquad (1)$$

where M1, M2, and M3 are cations, and S1, S2, and S3 are substitutes. Average Cu valence of a 1212 phase may be adjusted to 2.0+ - 2.5+ in terms of changing metallic ions. Some possible 1212 phases are listed in Table III. Many of them have been synthesized (see, Table I), and

Table III Some possible 1212 phases $(M1,S1)(M2,S2)_2(M3,S3)Cu_2O_7$ from cation adjustment.

M1	S1	M2	S2	M3	S3
Tl	Pb,Bi	Ba,Sr	La	Ca,Sr	R,Cr
Tl	Cu,Hg	Ba,Sr	K,Rb	R	Ca,Sr
Pb	Tl,Bi,Cu	Ba,Sr	K,Rb	Ca,Sr	Na,K
Cu	Tl,Bi	Ba,Sr	La	R	Ce,Zr

we believe that more will be synthesized. It must be pointed out that possible 1212 phases are not limited to those in Table III, and that the principle is to adjust the average Cu valence to 2.0+ - 2.5+.

The average Cu valence can also be adjusted by substitution of an anion for O. Formula (1) may be written as a more general form

$$(M1,S1)(M2,S2)_2(M3,S3)Cu_2(O,X)_7 \qquad (2)$$

where X is an anion. We take $TlSr_2CaCu_2O_7$ as an example. $TlSr_2CaCu_2O_7$ has avc of 2.5+, too high to good stability and superconductivity of the 1212 phase. If O is partially substituted by F, the average Cu valence will decrease, and thus the stability and superconductivity are expected to be improved.

For simplicity, we have assumed the ideal stoichiometry. However, high Tc superconductors often show non-stoichiometry[40], so does the 1212 phase. $CuBa_2YCu_2O_{7-y}$ (i.e., $YBa_2Cu_3O_{7-y}$) is a typical example, where y is vacancy of O. O vacancy decreases the average valence of Cu. The avc of $CuBa_2YCu_2O_{7-y}$ is = (2.5-y)+. If y is 0.2, avc will decrease to 2.3+. The superconducting behavior of the O-vacant $CuBa_2YCu_2O_{7-y}$ is depressed. On the other hand, the avc of $TlSr_2CaCu_2O_7$ is 2.5+, too high for good stability and superconductivity of this 1212 phase. O-vacancy will decreases its avc, and thus, $TlSr_2CaCu_2O_{7-y}$ is expected to exhibit better stability and superconducting behavior. In this case, annealing in a non-oxidation atmosphere would improve the superconductivity. Metallic ions in a structure may also be deficient[40]. Deficiency of metallic ions increases avc. Therefore, in some cases, stoichiometry phase is not supercondcting, whereas cation deficient phase is superconducting. Detailed discussion for deficiency is out of the scope of this paper. Readers may refer to refs. 40 & 41.

6. EXPERIMENTAL RESULTS

There is no doubt that many new superconductors with 1212 structure will be synthesized. For example, we have synthesized Mo or W-substituted $TlSr_2CaCu_2O_7$[42], Cu-substituted $TlSr_2RCu_2O_7$[43], F-substituted $TlSr_2CaCu_2O_7$[43], and so on. Here we would like to present the results from Pb, La, and Y substitutions for $TlSr_2CaCu_2O_7$. As mentioned above, $TlSr_2CaCu_2O_7$ itself can not be synthesized easily in the pure form and shows Tc from 0 to 80 K[14-17]. If a metallic ion is substituted partially by a higher valence ion, the 1212 phase is stabilized and its superconducting behavior is improved. When the substitutes are Pb^{4+} (for Tl^{3+}), La^{3+}(for Sr^{2+}), and Y^{3+} (for Ca^{2+}), there are totally seven derivatives: three single-substitution derivatives $(Tl,Pb)Sr_2CaCu_2O_7$, $Tl(Sr,La)_2CaCu_2O_7$, and $TlSr_2(Ca,Y)Cu_2O_7$; three double-substitution derivatives $(Tl,Pb)(Sr,La)_2CaCu_2O_7$, $(Tl,Pb)Sr_2(Ca,Y)Cu_2O_7$, and $Tl(Sr,La)_2(Ca,Y)Cu_2O_7$; and one triple-substitution derivative (Tl,Pb)-$(Sr,La)_2(Ca,Y)Cu_2O_7$. To our knowledge, all the derivatives but $Tl(Sr,La)_2CaCu_2O_7$, $Tl(Sr,La)_2(Ca,Y)Cu_2O_7$, and $(Tl,Pb)(Sr,La)_2(Ca,Y)Cu_2O_7$ have been synthesized. The three remaining derivatives are expected to also be synthesized, and we have synthesized and characterized them. We prepared three series of nominal samples of $Tl(Sr_{2-y}La_y)CaCu_2O_v$ with y = 0.2, 0.3, 0.4, 0.5, 0.6, and 0.7, $Tl(Sr_{2-y}La_y)(Ca_{0.8}Y_{0.2})Cu_2O_v$ with y = 0.1, 0.2, 0.3, 0.4, 0.5, and 0.6, and $(Tl_{1-x}Pb_x)(Sr_{1.8}La_{0.2})(Ca_{0.8}Y_{0.2})Cu_2O_v$ with x = 0.0, 0.1, 0.2, 0.3, and 0.45. Powder x-ray diffraction analyses showed that all the samples are nearly pure 1212 phase. The diffraction patterns can be well indexed based on a tetragonal unit cell with $a \approx 3.80$ Å and $c \approx 12.0$ Å (space group P4/mmm). As an example, Figure 6 shows the indexed x-ray diffraction pattern for the sample $Tl(Sr_{2-y}La_y)(Ca_{0.8}Y_{0.2})Cu_2O_v$ of y = 0.4. All samples are superconducting. Figures 7 shows resistance-temperature curves for the serial samples $Tl(Sr_{2-y}La_y)(Ca_{0.8}Y_{0.2})Cu_2O_v$ with y = 0.1-0.6 as examples. The Tc for all three series of samples increases with the substitution of a higher valence ion for a lower valence ion (La^{3+} for Sr^{2+}, or Pb^{4+} for Tl^{3+}), and reaches a maximum at a certain value: y = 0.6 for $Tl(Sr_{2-y}La_y)CaCu_2O_7$, y = 0.5 for $Tl(Sr_{2-y}La_y)(Ca_{0.8}Y_{0.2})Cu_2O_7$, and x = 0.3 for $(Tl_{1-x}Pb_x)(Sr_{1.8}La_{0.2})$-$(Ca_{0.8}Y_{0.2})Cu_2O_7$. If an ideal stoichiometry is assumed, the average Cu valence of $(Tl_{1-x}Pb_x)(Sr_{2-y}La_y)(Ca_{1-z}Y_z)Cu_2O_7$ is 2.50-(x+y+z)/2. Therefore, the optimal average Cu valence for the above three series of samples is 2.20, 2.15, and 2.15, respectively. These results are well consistent with those reported for other derivatives. It is interesting that the optimal average Cu valence is 2.20 ± 0.05, the same for all derivatives. Table IV lists the optimal average Cu valence and the corresponding Tc for each derivative of $TlSr_2CaCu_2O_7$.

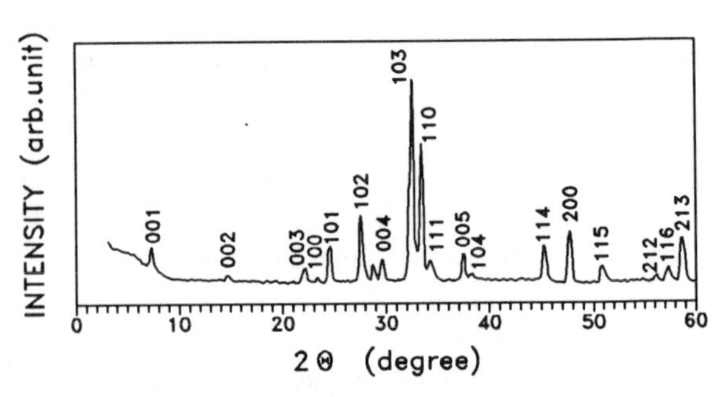

Figure 6 X-ray diffraction pattern for the sample $Tl(Sr_{2-y}La_y)(Ca_{0.8}Y_{0.2})Cu_2O_v$ of y=0.4.

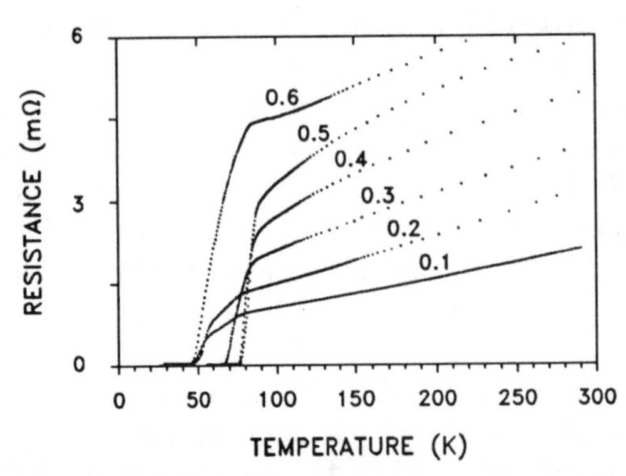

Figure 7 Resistance-temperature curves for samples $Tl(Sr_{2-y}La_y)(Ca_{0.8}Y_{0.2})Cu_2O_v$.

Table IV Tc and optimal average Cu valence for each derivative of $TlSr_2CaCu_2O_7$.

M1	M2	M3	M4	Optimal avc	Tc(K)	Ref.
Tl	Sr	Ca	Cu	(2.50)	0-70	14-17
Tl,Pb	Sr	Ca	Cu	2.20-2.25	85	17,29,30
Tl	Sr,La	Ca	Cu	2.20-2.25	90	this work
Tl	Sr	Ca,Y	Cu	2.20-2.25	90	18,19
Tl,Pb	Sr,La	Ca	Cu	2.15	105	34
Tl,Pb	Sr	Ca,Y	Cu	2.15	105	31
Tl	Sr,La	Ca	Cu	2.15-2.20	90	this work
Tl,Pb	Sr,La	Ca,Y	Cu	2.15-2.20	103	this work

7. HIGH Tc SUPERCONDUCTIVITY IN CUPRATES

So far no theory on high Tc superconductivity has been accepted. Much work, theoretical and experimental, needs to be done. Here we would like to point out the followings.

First, as shown in figure 3, all O are coordinated by six metallic ions. This fact strongly suggests that the six electrons ($2s^2 1p^4$) in the outside shell of O atom in the 1212 structure all (total 42 electrons) are involved in the formation of the structure. The total number of outside electrons of the metallic atoms in the 1212 phase (for example, $TlBa_2YCu_2O_7$) is also 42: 13 for Tl ($5d^{10}6s^2 6p^1$), 2 for Ba ($6s^2$), 3 for Y ($5s^2 4d^1$), and 11 for Cu ($4s^1 3d^{10}$). This may suggests that all outside electrons of the metallic atoms combine with the outside electrons of O atoms (forming electron pairs?) to contribute to the formation of the 1212 structure (and superconductivity ?).

Second, one considers hole concentration, or more accurately, mobile hole concentration. Many factors influence hole concentration, such as non-stoichiometric, non-mobile hole, etc. However, sometimes, mobile hole concentration is well consistent with Cu^{3+} concentration. For example, in a set of experiments on thermoelectronic power for $TlSr_2(Er_{1-y}Sr_y)Cu_2O_7$, the results showed that the thermoelectric power (S) for the 1212 phase of y = 0.2-0.5 is nearly temperature-independent in the temperature range from 150 to 300 K[44]. The temperature-independent thermopower invites the Hubbard model in the high temperature limit with $k_B T$ greater than the bandwidth, but much less than the on-site Coulomb interaction U, and the thermopower S can be written as[45]

$$S=\frac{k}{|e|}\ln\frac{1-c}{2c}, \qquad (1)$$

where c is the number of holes per Cu site. The hole concentration calculated from the thermopower data using Equation (1) is quantitatively consistent with the y/2 value (i.e. Cu^{3+} concentration)[44]. Table V lists thermopower at 290 K, calculated hole concentration (c), y/2, and Tc for $TlSr_2(Er_{1-y}Sr_y)Cu_2O_7$ of y = 0.2-0.5. This agreement not only explains why the concept of average valence of Cu can be fairly well applied to many experimental results and predict new 1212 superconductors, but also suggests that Hubbard model is one of candidates which may solve the problem of high Tc superconductivity.

Table V Thermoelectric-power (S) at 290 K, calculated hole concentration (c), y/2, and Tc for $TlSr_2(Er_{1-y}Sr_y)Cu_2O_7$ of y = 0.2-0.5. (from ref. 44).

y	S (μV/K)	c	y/2	Tc(zero R) K
0.2	110	0.12	0.10	<20
0.3	75	0.17	0.15	62
0.4	47	0.22	0.20	71
0.5	39	0.24	0.25	73

Acknowledgements: We would like to thank Michael Sorenson, David Ford, and Millicent Megginson for their assistance. This work was supported by a grant of Arkansas Energy Office.

REFERENCES

[1] J.G.Bednorz and K.A.Muller, Z. Phys. B 64, 189 (1986).
[2] M.K.Wu, J.R.Ashburn, C.T.Torng, P.H.Hor, R.L.Meng, L.Gao, Z.J.Huang, Y.Q.Wang, and C.W.Chu, Phys.Rev.Lett. 58, 908 (1987).
[3] H.Maeda, Y.Tanaka, M.Fukutomi, and T.Asano, Jpn.J.Appl.Phys.Lett. 27, L209 (1988).
[4] Z.Z.Sheng and A.M.Hermann, Nature 332, 138 (1988).
[5] Z.Z.Sheng, W.Kiehl, J.Bennett, A.El Ali, D.Marsh, G.D. Mooney, F.Arammash, J.Smith, D.Viar, and A.M.Hermann, Appl.Phys.Lett.

52 1738 (1988).
[6] R.M.Hazen, L.W.Finger, R.J.Angel, C.T.Prewitt, N.L.Ross, C.G.Hadidiacos, P.J.Heaney, D.R.Veblen, Z.Z.Sheng, A.El Ali, and A.M.Hermann, Phys.Rev.Lett. 60, 1657 (1988).
[7] C.C.Torardi, M.A.Subramanian, J.C.Calabrese, J.Gopalakrishnan, K.J.Morrissey, T.R.Askew, R.B.Flippen, U.Chowdhry, and A.M.Sleight, Science 240, 631 (1988).
[8] S.S.P.Parkin, V.Y.Lee, E.M.Engler, A.I.Nazzal, T.C.Huang, G.Gorman, R.Savoy, and R.Beyers, Phys.Rev.Lett. 60, 2539 (1988).
[9] S.S.P.Parkin, V.Y.Lee, A.I.Nazzal, R.Savoy, R.Beyers, and S.J.La Placa, Phys.Rev.Lett. 61, 750 (1988).
[10] R.Beyers, S.S.P.Parkin, V.Y.Lee, A.I.Nazzal, R.Savoy, G.Gorman, T.C.Huang, and S.J.La Placa, Appl.Phys.Lett. 53, 432 (1988).
[11] R.J.Cave, B.Batlogg, R.B. van Dover, D.W.Murphy, S.A.Sunshine, T.Siegrist, J.P. Remeika, A.Rietman, S.Zahurak, and G.P.Espinosa, Phys.Rev.Lett. 58, 1676 (1987).
[12] S.Nakajima, M.Kikuchi, Y.Syono, K.Nagase, T.Oku, N.Kobayashi, D.Shindo, and K.Hiraga, Physica C 170, 443 (1990).
[13] S.Nakajima, M.Kikuchi, Y.Syono, N.Kobayashi, and Y.Muto, Physica C 168, 57 (1990).
[14] Z.Z.Sheng, A.M.Hermann, D.C.Vier, S.Schultz, S.B.Oseroff, D.J.George, and R.M.Hazen, Phys.Rev.B 38, 7074 (1988).
[15] S.Maysuda, S.Takeuchi, A.Soeta, T.Suzuki, K.Alihara, and T.Kamo, Jpn.J.Appl.Phys. 27 2062 (1988).
[16] T.Doi, K.Usami, and T.Kamo, Jpn.J.Appl.Phys. 29, L57 (1990).
[17] C.Martin, J.Provost, D.Bourgault, B.Domenges, C.Michel, M.Hervieu, and B.Raveau, Physica C 157 , 460 (1989).
[17] Z.Z.Sheng L.Sheng, X.Fei, and A.M.Hermann, Phys.Rev.B 38 2918 (1988).
[18] Liang,J.K., Zhang,Y.L., Rao,G.H., Cheng,X.R., Xie.S.S., and Zhao,Z.X., Solid State Comm., 661 (1989).
[20] Z.Z.Sheng, C.Dong, Y.H.Lui, X.Fei, L.Sheng, J.H.Wang, and A.M.Hermann, Solid State Comm. 71, 739 (1989).
[21] A.K.Ganguli, V.Manivannan, A.K.Sood and C.N.R.Rao, Appl.Phys.Lett. 55, 2664 (1989).
[22] Z.Z.Sheng, L.A.Burchfield, and J.M.Meason, Mod.Phys.Lett. B 4, 967 (1990).
[23] Z.Z.Sheng, Y.Xin, J.M.Meason, and L.A.Burchfield, Physica C 172, 43 (1990).
[24] Z.Z.Sheng, Y.Xin, and J.M.Meason, Mod.Phys.Lett. B 4, 1281 (1990).
[25] Y.Xin, Z.Z.Sheng, D.X.Gu, and D.O.Pederson, Physica C 177, 183 (1991).

[26] P.Haldar, S.Sridhar, A.Roig-Janicki, W.Kennedy, D.H.Wu, C.Zahopoulos, and B.C.Giessen, J.Superconductivity. 1, 211 (1988).
[27] S.Li and M.Dreenblatt, Physica C 157, 365 (1989).
[28] Y.T.Hunag, R.S.Liu, W.N.Wang, and P.T.Wu, Physica C 162-164, 39 (1989).
[29] M.A.Subramanian, C.C.Torardi, J.Gopalakrishnan, P.L.Gai, J.C.Calabrese, T.R.Askew, R.B.Flippen, and A.M.Sleight Science 242, 249 (1988).
[30] J.C.Barry, Z.Iqubal, B.L.Ramakrishna, R.Sharma, H.Eckhardt, and F.Raidinger, Appl.Phys.Lett. 65, 5207 (1988).
[31] R.S.Liu, J.M.Liang, Y.T.Huang, W.N.Wang, S.F.Wu, H.S.Koo, P.T.Wu, and L.J.Chen, Physica C 162-164, 869 (1989).
[32] Z.Z.Sheng, Y.Xin, D.X.Gu, J.M.Meason, J.Bennet, D.Ford, and D.O.Pederson (to be published in Z.Phys.B).
[33] Z.Z.Sheng, Y.Xin, J.M.Meason, D.X.Gu, and D.O.Pederson, Supercond.Sci.Technol. 4, 212 (1991).
[34] M.R.Presland and J.L.Tallon, Physica C 177, 1 (1991).
[35] T.Rouillon, J.Provost, M.Hervieu, D.Groult, C.Michel, and B.Raveau, Physica C 159 201 (1989).
[36] T.Rouillon, A.Maignan, M.Hervieu, C.Michel, D.Groult, and B.Raveau, Physica C 171 7 (1990).
[37] J.G.Bednorz and A.K.Muller, Angew.Chem. 100, 759 (1988).
[38] Z.Z.Sheng, D.X.Gu, Y.Xin, D.O.Pederson, L.W.Finger, C.G.Hadidiacos, and R.M.Hazen, Mod.Phys.Lett.B 5, 635 (1991).
[39] Y.Kubo, Y.Shimakawa, T.Manako, T.Satoh, S.Iijima, T.Ichihashi, and H.Igarash, Physica C 162-164, 991 (1990).
[40] B.Raveau, C.Michel, M.Hervieu, J.Provost, and F.Studer, in Ealier and Recent Aspects of Superconductivity, ed. by J.G.Bednorz and K.A.Muller (Springer-Verlag, 1990), pp. 66-95.
[41] J.D.Jorgensen, Phys.Today 44, No.6, 34 (1991).
[42] Y.F.Li, Z.Z.Sheng, and D.O.Pederson (this proceedings).
[43] Z.Z.Sheng and Y.F.Li (unpublished data).
[44] Y.Xin, Y.F.Li, D.Ford, D.O.Pederson, and Z.Z.Sheng (to be published in Jpn.J.Appl.Phys.Lett.).
[45] P.M.Chaikin and G.Beni, Phys.Rev.B 13, 647 (1976).

THEORY OF TILTED VORTICES IN JOSEPHSON-COUPLED LAYERED SUPERCONDUCTORS.

D. FEINBERG
Laboratoire d'Etudes des Propriétés Electroniques des Solides (LEPES), Centre National de la Recherche Scientifique (CNRS), BP 166, 38042 Grenoble CEDEX 9, France.

ABSTRACT

Field penetration is studied for arbitrary directions in the London limit for Josephson-coupled layered superconductors, in the quasi-2D regime. In the absence of Josephson coupling, the normal field component creates an Abrikosov vortex lattice normal to the layers while the parallel component penetrates freely. For finite Josephson coupling there exists a region called the vortex line "nucleus", where the behaviour is essentially 2D, while it is 3D beyond. Its size parallel to the layers is given by the "Josephson bending length" $r_J = \Gamma d$, where $\Gamma = (M/m)^{1/2}$ is the anisotropy ratio and d the distance between layers. The consequence is a transition from a 3D vortex lattice to a quasi-2D vortex lattice when $H_\perp \geq \Phi_0/\pi r_J^2$. Furthermore, for flux lines making an angle $\theta > \theta_0 = \tan^{-1}\Gamma$ with the normal direction, vortex lines become staircase vortices with smooth kinks (of length r_J) separated by segments of Josephson vortices.

1. INTRODUCTION

In high-T_c materials, the electronic anisotropy can vary within several orders of magnitude. In an effective mass model with masses m within the layers and M across them, it is characterized by the parameter $\Gamma = (M/m)^{1/2}$. The discrete layer structure plays here a crucial role. At temperatures above a crossover value T* where the transverse coherence length $\xi_\perp(T^*)$ is of order the distance d between superconducting layers, one can apply the 3D anisotropic Ginzburg-Landau (3DAGL) or London (3DAL) approach[1]. Such a picture fails below $T^* = T_c(1 - \rho)$ with $\rho = 2\xi_{//}^2(0)/d^2\Gamma^2$, in the so-called quasi-2D regime. ρ is of order 10^{-1} in Y:123, 10^{-3} in Bi:2212, if Γ is measured by torque magnetometry[2], and even smaller in artificial multilayeres. The Lawrence-Doniach (LD) model assuming a Josephson coupling between superconducting layers is commonly used to describe this situation[1,3].

The present study is concerned with the general situation where the field is tilted with respect to the c-axis. We start from the situation with zero Josephson coupling[4-6]. In this limiting case screening currents exist only in the layers and generate a magnetization normal to the layers. Then the only interaction between 2D "pancake" vortices in different layers is the electromagnetic (dipolar) interaction which tends to align them along the c-axis. On the other hand, due to the absence of currents across the layers the parallel field component is unscreened. Thus in this case an hexagonal Abrikosov vortex lattice is formed along the normal to the layers, due to the normal field component only. A number of experiments have been interpreted in this way for very anisotropic materials[7]. We investigate this problem by considering a small Josephson coupling and show that the 2D character persists at distances smaller than $r_J = \Gamma d$ from 2D vortex cores forming vortex lines. The "Josephson bending length" r_J plays here a crucial role.

This paper is organized as follows: the LD model is recalled in Section 2. In Section 3 we prove in detail the decomposition property for zero Josephson coupling. The effects of a small Josephson coupling are analyzed in Section 4. The resulting vortex lattice structure is determined in Section 5.

2. THE LAWRENCE-DONIACH MODEL

Let us consider superconducting (S) layers, separated by insulating (I) layers. In the LD model[1,3], valid if $\rho < 1$, S layers are connected by Josephson tunneling. The free energy for extreme type-II superconductors can then be written in the London picture, assuming an uniform order parameter amplitude

$$F = \frac{\Phi_0^2 d}{32\pi^3 \lambda_{//}^2(T)} \sum_n \int dR \left\{ \left|\nabla_{//}\chi_n\right|^2 + \frac{2}{\Gamma^2 d^2}\left[1 - \cos(\chi_n - \chi_{n+1})\right] \right\} + \int \frac{h^2}{8\pi} d^2\mathbf{R}\, dZ \tag{2.1}$$

where $\Phi_0 = hc/2e$ is the flux quantum, $\chi_n(\mathbf{R}) = \phi_n(\mathbf{R}) - 2\pi/\Phi_0 \int_0^{\mathbf{R}} \mathbf{A}.\,d\mathbf{l}$ is the gauge-invariant phase. The XYZ frame stands with the Z-axis along the normal to the layers. One defines the parallel $\lambda_{//} = (mc^2/4\pi n_s e^2)^{1/2}$ and transverse (Josephson) $\lambda_\perp = (Mc^2/4\pi n_s e^2)^{1/2}$ penetration lengths for currents circulating in the XY plane and along the Z direction respectively. **A** is the vector potential and **h** = curl **A** the microscopic field. We shall hereafter use this London-Lawrence-Doniach (LLD) description.

3. THE CASE OF NO JOSEPHSON COUPLING

The case where no electronic coupling exists between the layers can be modeled by expression (2.1) with $M/m = \Gamma^2 = \infty$. Screening currents flow only in the layers and one has 2D vortices (or "pancake vortices") in the layers. Vortices in different layers are coupled through the magnetic (dipolar) field that they create throughout the full space. The London equations deriving from (2.1) with $\Gamma = \infty$ are

$$\nabla^2 \mathbf{A}_{//} = \frac{d}{\lambda_{//}^2} \sum_n \left(\mathbf{A}_{//} - \frac{\Phi_0}{2\pi} \nabla_{//} \phi_n \right) \delta(Z - nd) \quad ; \qquad \nabla^2 A_Z = 0 \tag{3.1}$$

$$\nabla_{//}^2 \phi_n = \frac{2\pi}{\phi_0} \nabla_{//} \mathbf{A}_{//}(nd) \tag{3.2}$$

In solving these equations[4-6] one usually assumes that the vector potential $\mathbf{A}$ is parallel to the layers, thus in the London gauge $\mathrm{div}\mathbf{A}_{//} = 0$ and the magnetic field $\mathbf{h}_1 = \mathrm{curl}\mathbf{A}$ is given by $\mathbf{h}_1 = \left(-\frac{\partial A_Y}{\partial Z}, \frac{\partial A_X}{\partial Z}, \frac{\partial A_Y}{\partial X} - \frac{\partial A_X}{\partial Y} \right)$. Due to cylindrical symmetry around each 2D vortex, the average of $\mathbf{h}_1$ giving the total flux is normal to the layers for any 2D vortex configuration. Therefore such a solution cannot take into account the presence of a parallel field component. To obtain a more general solution, one must introduce a normal component of $\mathbf{A}$. Choosing the gauge such as $\partial A_Z/\partial Z = 0$, then $\mathrm{div}\mathbf{A} = 0$ implies $\mathrm{div}\mathbf{A}_{//} = 0$, and $\mathbf{A}_{//}$ and A_Z are independent. The total field $\mathbf{h}$ can then be written as $\mathbf{h} = \mathbf{h}_1 + \mathbf{h}_2$, with $\mathbf{h}_2 = \left(\frac{\partial A_Z}{\partial Y}, -\frac{\partial A_Z}{\partial X}, 0 \right)$. Eqs. (3.1-2) then show that the normal induction is given by $B_\perp = \int h_Z d^2\mathbf{R} = \Phi_0 n_v$, where n_v is the density of 2D vortices per layer[5,6,9].

Therefore the free energy (2.1) can be decomposed as $F = F_\perp + F_{//}$, with

$$F_\perp = \frac{\Phi_0^2 d}{32\pi^3 \lambda_{//}^2(T)} \sum_n \int d\mathbf{R} \left| \nabla_{//} \chi_n \right|^2 + \int \frac{\mathbf{h}_1^2}{8\pi} d^2\mathbf{R}\, dZ \quad ; \qquad F_{//} = \int \frac{\mathbf{h}_2^2}{8\pi} d^2\mathbf{R}\, dZ$$

and the Gibbs energy density in an uniform field $\mathbf{H} = (\mathbf{H}_{//}, H_\perp)$ can be written as

$$G = \left(F_\perp - n_v \frac{\Phi_0 H_\perp}{4\pi}\right) + \left(F_{//} - \frac{\mathbf{h}_2 . \mathbf{H}_{//}}{4\pi}\right) \qquad (3.3)$$

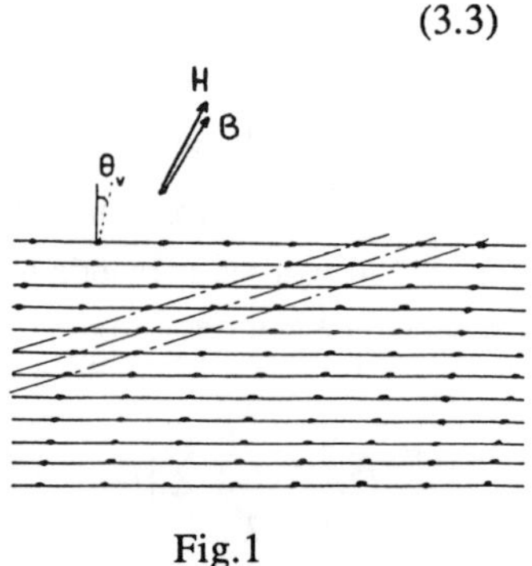

Fig.1

Minimization of G with respect to the parallel field $\mathbf{h}_2$ yields $\mathbf{h}_2 = \mathbf{H}_{//}$. The parallel field is therefore unscreened and independent of the 2D vortex configuration. The latter is determined at equilibrium by minimizing the first term in the r.h.s. of (3.3).The solution is an Abrikosov lattice of lines of 2D vortices. Due to the dipolar interactions between 2D vortices in different layers, these lines must be normal to the layers. This can be shown directly, but we shall need later on the free energy of a tilted vortex lattice.

Let us thus consider a perfect tilted lattice of 2D vortices, made of a 2D lattice in each layers, translated in the x-direction by a distance ℓ from layer to layer (Fig.1). The angle θ_v of the vortex lines with the layers is $\tan\theta_v = \ell/d$. Using Fourier transform and after some transformations, $F_\perp$ can be written as[5]

$$F_\perp = \frac{n_v^2}{8\pi\lambda_{//}^2} \sum_k \frac{\Phi_0^2}{k^2} \left[1 - \sum_n e^{ik_x n\ell} \frac{d \sinh kd}{2\lambda_{//}^2 k \sqrt{G_k^2 - 1}} \left(G_k - \sqrt{G_k^2 - 1}\right)^{|n|}\right] \qquad (3.4)$$

with $G_k = \cosh(kd) + \sinh(kd) / 2\lambda_{//}^2 k$. The summation is performed on the reciprocal vortex lattice. Let us consider the case of an intermediate normal component, defined as $H_{c1\perp} << H_\perp << H_{c2\perp}$. Let $a \approx (\Phi_0/B_\perp)^{1/2}$ be the average distance between 2D vortices in the layers, and replace $\phi_0 n_v$ by $B_\perp$. Performing the summation, one finds if $\ell < a$ (if $\ell >$ a one can redefine ℓ such as $\ell < a$, see Fig.1)

$$F_\perp = \frac{B_\perp^2}{8\pi} + \frac{\Phi_0 B_\perp}{16\pi^2 \lambda_{//}^2} \mathrm{Ln}\left[\frac{a}{\xi_{//}} \left(\frac{1 + (d/\lambda_{//})^2 \tan^2\theta_v \cos\theta_v}{1 + (a/\lambda_{//})^2 \cos\theta_v}\right)^{1/2}\right] \qquad (3.5)$$

Minimizing G (see Eq. (3.3)) with respect to θ_v and n_v yields $\theta_v = 0$, therefore an Abrikosov lattice normal to the layers, stabilized by the normal field component , with

$$B_{\perp} = H_{\perp} - \frac{\Phi_0}{4\pi \lambda_{//}^2} Ln \sqrt{\frac{\eta H_{c2\perp}}{H_{\perp}}} \tag{3.6}$$

(η is of order 1 for an hexagonal lattice) while $\mathbf{B}_{//} = \mathbf{H}_{//}$. The magnetization is normal to the layers and the parallel field is unscreened. In high fields $B_{\perp} \approx H_{\perp}$, **B** and **H** are nearly parallel, but are not parallel to the vortex line. Expression (3.6) can be compared to the 3DAL expressions[8] (for which $\theta \approx \theta_v$ where θ is the field angle)

$$B_{\perp} = H_{\perp} - \frac{\Phi_0}{4\pi \lambda_{//}^2} (1 + \Gamma^{-2} \tan^2\theta)^{-1/2} Ln \sqrt{\frac{\eta H_{c2}}{H}} \tag{3.7a}$$

$$B_{//} = H_{//} - \frac{\Phi_0}{4\pi \lambda_{//}^2} \Gamma^{-2} \tan\theta \, (1 + \Gamma^{-2} \tan^2\theta)^{-1/2} Ln \sqrt{\frac{\eta H_{c2}}{H}} \tag{3.7b}$$

In this latter result, the component parallel to the layers of the reversible magnetization is much smaller than the normal component, except very close to the layer direction, i.e. for $\tan\theta > \Gamma^2$. As a consequence, experiments with fields nearly parallel to the layers are required to distinguish between the 3DAL and 2D results.

Therefore a decomposition property holds for zero Josephson coupling: the normal component creates an Abrikosov lattice normal to the layers while the parallel component is unscreened. Kes et al. have assumed that this property holds for finite but small interlayer coupling and have explained in this way a number of experiments on Bi and Tl cuprates[7].

4. THE CORE STRUCTURE FOR SMALL COUPLING

Minimization of the functional (2.1) with respect to **A** and the phases ϕ_n leads to the London-Lawrence-Doniach (LLD) equations. In particular the phase equation reads

$$\nabla_{//}^2 \chi_n = \frac{1}{r_J^2} \left[\sin(\chi_n - \chi_{n-1}) - \sin(\chi_{n+1} - \chi_n) \right] \tag{4.1}$$

It involves a length $r_J = \Gamma d$, which characterizes the distance beyond which the Josephson coupling is able to "damp" large phase differences between neighbouring layers. It can be called "Josephson bending length". Basically, at a distance $r << r_J$ in the layers from a vortex core the LLD equations are nonlinear. On the contrary, if $r >> r_J$, the Josephson contributions can be expanded in small phase differences and the equations become the ordinary 3DAL equations.

Minimization with respect to the magnetic field yields in particular the equation for the normal field

$$h_Z - \lambda_{//}^2 \nabla^2 h_Z = \Phi_0 d \sum_{in} \delta(\mathbf{R} - \mathbf{R}_i) \, \delta(Z - nd) \tag{4.2}$$

which is linear, contrarily to the corresponding equations for h_X, h_Y (see Ref.9). Integrating (4.2) over the unit volume gives

$$B_{\perp} = \int h_Z d^2\mathbf{R} = \Phi_0 n_v \tag{4.3}$$

Therefore the result found in Section 3 in absence of Josephson coupling still holds for weak coupling (such as the LD analysis holds): the normal flux is simply determined by the density of 2D vortices in the layers. These vortices can be in any (ordered or disordered) configuration. They can form normal or tilted vortex lines (see below). Equation (4.3) shows that all properties (magnetization, transport) which essentially depend on the total number of 2D vortices will scale as $B_{\perp}$, as in many experiments[7]. The question of the orientation of the vortex lattice in very anisotropic layered superconductors is therefore subtle and the observed scaling with $B_{\perp}$ is not a definite proof of a decomposition.

Let us now consider two 2D vortices in adjacent layers, at points $(-\ell/2, 0, nd)$ and $(\ell/2, 0, (n+1)d)$ (Fig.2a). In the absence of Josephson coupling the phases in layers n and (n+1) are given by $\phi_n = \tan^{-1}(-Y/(X + \ell/2))$ and $\phi_{n+1} = \tan^{-1}(-Y/(X - \ell/2))$. Lines of constant phase difference $\phi = \phi_n - \phi_{n+1}$ are represented on Fig.3a.

Let us turn to the case of Josephson vortex, oriented along the X-direction, whose nucleus is situated between layers n and n+1 (Fig.2b where the hatched horizontal cylinder denotes the vortex nucleus). A good approximation for the phase consists in taking the phase function for an anisotropic vortex line centered at $Y_0 = 0$, $Z_0 = nd + d/2$. It yields $\phi = \tan^{-1}(Y/\Gamma(Z - Z_0))$. Thus in particular, in layers n and n+1,
$\phi_n = \tan^{-1}(-2Y/r_J)$ and $\phi_{n+1} = \tan^{-1}(2Y/r_J)$. Lines of constant ϕ are represented on Fig.3b.

Let us now compare the two cases, or Figs.3a and 3b. At equal distances between the two 2D vortices (X = 0), the two phase functions appear to coïncide when $\ell = r_J$. Therefore for $\ell > r_J$ a Josephson core (or better nucleus) naturally develops between 2D cores (Fig.2c) and one obtains the lines of constant phase represented on Fig.3c. In the other layers the matching of the two solutions is realized on a length of order r_J.

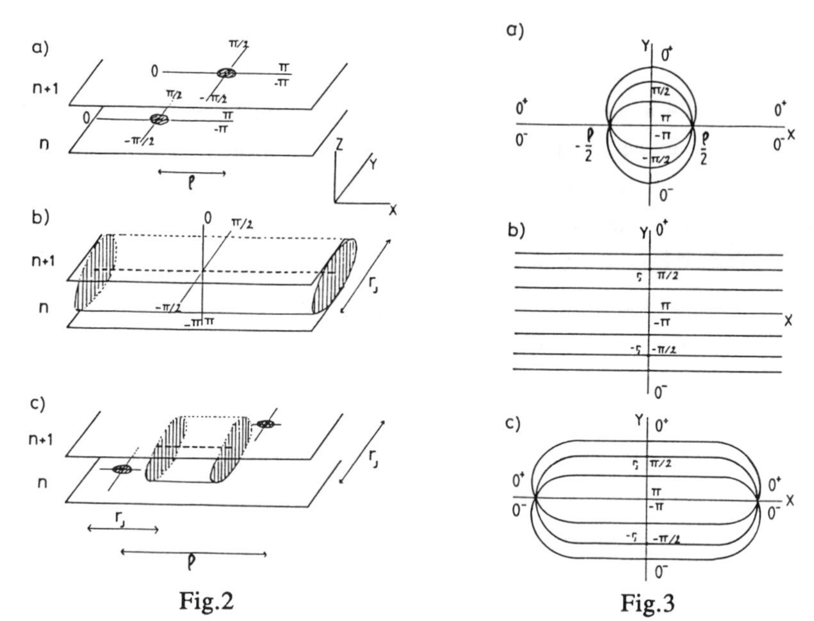

Fig.2 Fig.3

Therefore one is led to the following picture for a "vortex line" making an angle $\theta_v = \tan^{-1}(\ell/d)$ with the Z-direction, according to the value of θ_v:

i) $\theta_v < \theta_0 = \tan^{-1}\Gamma$. The 2D vortices are too close together to form Josephson vortices. Their juxtaposition forms effective straight "tilted" vortex lines (Fig.4a where hatched regions denote the vortex nuclei).

ii) $\theta_v > \theta_0$. The vortex nucleus has a "variquous" shape, made of steps of height d (2D vortices) and straight portions of Josephson vortices, of length $\ell - r_J$. Each of the steps extends on a length r_J along the layers. These vortices can be called "staircase vortices" (Fig.4b).

The finite length of the steps or "kinks" makes an important difference with other treatments[10-11] using the 3DAL theory with abrupt "kinks" on vortex lines without considering their intrinsic width r_J, which can be very large. To fix ideas, for Y:123, $r_J \approx 60Å$ is quite small, but for Bi(Tl):2212, $r_J \approx 600$-$1000Å$ is only slightly smaller than $\lambda_{//}$.

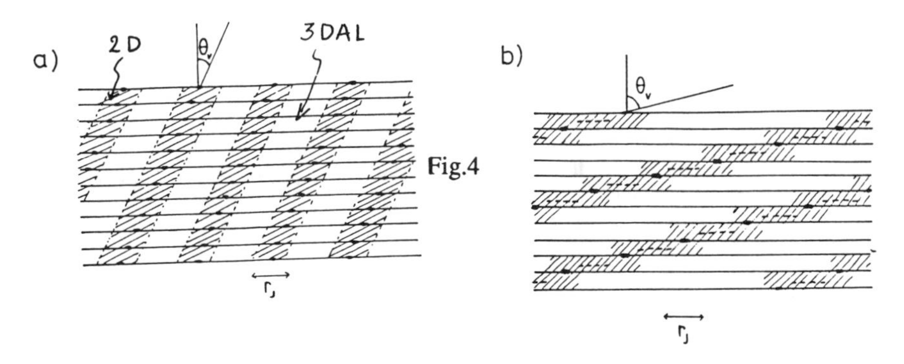

Fig.4

5. THE VORTEX LATTICE STRUCTURE

In Section 4 we have qualitatively elucidated the structure of the cores (nuclei) of tilted vortex lines in the LD model. The next step is to calculate the energy of a lattice of such lines. Two cases have to be considered, depending whether the distance a between 2D vortex cores in the same layer is larger or smaller than r_J. This involves a characteristic field $H_0 = \phi_0 / \pi r_J^2$.

a) $H_\perp < H_0$: the effective 3DAL lattice.

In this case the nucleus is a small region, and the phase is essentially continuous everywhere, except within the nucleus. One can therefore use the 3DAL solution, with a lower cutoff corresponding to the nucleus dimension r_J. This approximation is quite similar to that used in the usual London solution where the cutoff is associated to the GL core. Then one has to calculate the nucleus energy, which in the same analogy plays the role of the GL core energy. This can be done in perturbations in the Josephson coupling: the zeroth order term is obtained from Section 3, where the integrals are calculated between $R = \xi_{//}$ and $R = r_J$ from each 2D vortex core. The first order Josephson contribution is obtained by integrating the Josephson energy (see Eq. (1.1)) within the nucleus. Finally one must add the energy of the "true" GL core of the 2D vortices.

Then the Gibbs energy can be calculated. The total flux is the sum of a nucleus and an out-of-nucleus contributions. The result is expressed as a function of the 2D vortex density n_v, the tilt angle θ_v and the parallel field in the nucleus $\mathbf{h}_2$ (see Ref.9)

$$G = \frac{B_\perp^2}{8\pi}\left[1 + \tan^2\theta_v\,(1 - \frac{B_\perp}{H_0})\right] + \frac{B_\perp}{H_0}\frac{h_2^2}{8\pi}$$

$$+ \frac{\Phi_0 B_\perp}{16\pi^2\lambda_{//}^2}(1+\Gamma^{-2}\tan^2\theta_v)^{1/2}\ \mathrm{Ln}\sqrt{\frac{\eta H_0}{B_\perp}}$$

$$+ \frac{\Phi_0 B_\perp}{16\pi^2\lambda_{//}^2}\left[\mathrm{Ln}\frac{r_J}{\xi_{//}}(1 + \frac{\tan^2\theta_v}{2\Gamma^2}) - \frac{\tan\theta\,\tan\theta_v}{2\Gamma^2} + \alpha_c + \frac{d^2\tan^2\theta_v - r_J^2}{2\lambda_{//}^2}\cos\theta_v\right]$$

$$- \frac{B_\perp H_\perp}{4\pi} - \frac{B_\perp}{4\pi H_0}\mathbf{h}_2.\mathbf{H}_{//} - \frac{B_\perp H_{//}}{4\pi}\tan\theta_v\,(1 - \frac{B_\perp}{H_0}) \qquad (5.1)$$

α_c being a constant of the order of 0.5. One must minimize with respect to $\mathbf{h}_2$, $B_\perp$ and θ_v. This immediately gives $\mathbf{h}_2 \approx \mathbf{H}_{//}$: the parallel field is unscreened in the nucleus. Then one notices that the 3DAL contribution (third term) which favours tilted vortices is weak. But the terms in the nucleus contribution (fourth term) which favour the normal vortex lattice are still weaker. This shows that the tilted configuration is very stable up to $B_\perp \approx H_0$. Keeping the dominant terms in (5.1) indeed gives for $B_\perp < H_0$ the usual solution $B_\perp \approx H_\perp$ and $\theta_v \approx \theta$, the field angle.

b) Case $B_\perp > H_0$: quasi-2D vortex lattice.

In that case, the nuclei fill all the space and the Gibbs energy is given by the 2D result (3.3), modified by the perturbative Josephson energy

$$F_{NJ} = \frac{\Phi_0 B_\perp \tan^2\theta_v}{32\pi^2 \lambda_{//}^2 \Gamma^2}\left[\mathrm{Ln}\frac{a}{2\xi_{//}} - \frac{\tan\theta}{2\tan\theta_v}\right] \qquad (5.2)$$

and the core energy, identical to that of case a.

If $\theta_v < \theta_0$, there is a competition between the θ_v-dependent terms in the 2D free energy $F_\perp$ and the Josephson energy F_{NJ}: the former tends to make $\theta_v = 0$, due to the dipolar interaction, while the latter tends to tilt the vortex lines, due to the tilted field. However the latter effect is weak and is unable to give $\theta_v = \theta$ as in the 3D London case. It rather yields $\tan\theta_v \approx \tan\theta / 4\mathrm{Ln}(a/\xi_{//})$. The balance between the two above-mentioned terms involves a critical anisotropy $\Gamma_{cr} = (\lambda_{//}/d)^{1/2}$. If $\Gamma < \Gamma_{cr}$ the vortex lines acquire an orientation intermediate between the field direction and the normal to the layers. Such lines form what may be called a " tilted quasi-2D vortex lattice " (region II in Fig. 5). On the other hand, if the anisotropy is very large, i.e. $\Gamma > \Gamma_{cr}$, there are two possibilities: in very large fields, such as $H_\perp >> \phi_0\Gamma^2/\pi\lambda_{//}^2$, the lattice is again tilted, but if $H_\perp << \phi_0\Gamma^2/\pi\lambda_{//}^2$, θ_v is essentially zero: one has in that case a "normal quasi-2D vortex lattice" (region III in Fig.5). Region I corresponds to the effective 3DAL lattice.

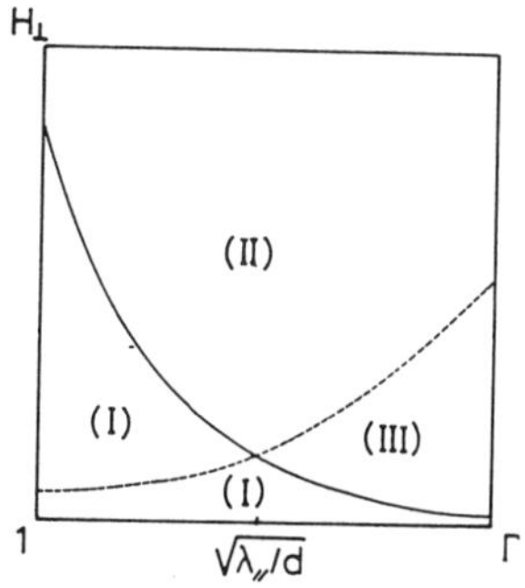

Fig.5

DISCUSSION

Let us summarize the main results. Around each 2D vortex core, in a nucleus of radius r_J, only the normal field component is screened, while the parallel component remains uniform. The Josephson currents, which favour a tilted vortex lattice, play a small role in this region where currents circulate essentially within the layers. The 2D currents create a dipolar interaction between 2D vortices which favours the normal orientation of the vortex lattice. When $B_\perp << H_0 = \phi_0/\pi r_J^2$, a large region exists between the vortex nuclei where the 3DAL analysis holds, which favours tilted vortex lines, oriented close to the field direction if $H >> H_{c1}$ (Fig 4a-b). On the opposite, if $B_\perp > H_0$, corresponding to overlap of the nuclei, vortices in the layers have essentially a 2D character. The Josephson coupling plays a small role and the direction of the vortex lattice is in general situated between the field direction and the Z direction (Fig.3). For large anisotropies (typically $\Gamma > 10$), and not too large fields, the vortex lattice becomes normal to the layers. The corresponding crossover field $H_{cr}(\theta) = \phi_0/\pi d^2\Gamma^2\cos\theta$ increases with the tilt angle. For instance, with $d \approx 10$Å and $\Gamma \approx 5$ (Y:123) one obtains for normal orientation the crossover field $H_0 \approx 22$T, while with $d \approx 15$Å and $\Gamma \approx 55$ (Bi:2212) one obtains $H_0 \approx 0.1$T. For directions close to the layers, one obtains high values $H = \Gamma H_0 \approx$ 120T (Y:123) and 5T (Bi:2212). Let us stress that the crossover involves a drastic change of the vortex lattice direction but not of the flux direction: at least for $H >> H_{c1}$, the direction of **B** remains close to that of **H**. More generally, the non alignment of the flux vector and the vortex lattice, due to imperfect screening of the parallel field component, is characteristic of Josephson-coupled systems.

Recently[12] decoration experiments on Bi:2212 crystals, beside the appearance of vortex chains, have shown the presence of a nearly isotropic Abrikosov lattice up $\theta \approx 85^o$. However this should not be taken as a direct proof of decomposition. In fact, Campbell et al.[13] have calculated the unit cell of the 3DAL lattice in the plane normal to flux lines. It yields a distorted hexagonal lattice, but once projecting it onto the layers it becomes distorted only in the extreme vicinity of the layer direction. For instance, for Bi:2212, $\theta_0 \approx 89^o$ and at $\theta = 85^o$ the distortion is less than 2%. Therefore, as in other experiments, the change from a 3DAL lattice to a lattice normal to the layers is difficult to detect with decoration experiments, unless the field direction is extremely close to the layer direction.

In very anisotropic high-T_c materials, the intrinsic effects of the layered structure on the vortex lattice properties are partly masked by the strong fluctuation effects. Therefore the existence of a field-induced crossover in the vortex lattice direction should be investigated at low temperatures in high-T_c materials, or in other systems as low-T_c multilayers.

The second result concerns the change in the nucleus structure beyond a critical tilt angle θ_0. Pieces of Josephson vortices form between 2D vortices and these staircase vortices become straight Josephson vortices when $\theta = 90^o$. The actual role of these

Josephson pieces needs a separate analysis. Complicated behaviour may occur, which could be responsible for the anomalies observed in transport properties in this range of field directions[14].

REFERENCES

1. L. N. Bulaevskii, Zh. Eksp. Teor. Fiz. 64, 2241 (1973)[Sov. Phys. JETP **37**, 1133 (1973)].

2. D. E. Farrell, C. M. Williams, S. A. Wolf, N. P. Bansal and V. G. Kogan, Phys. Rev. Lett. **61**, 2805 (1988); D. E. Farrell, S. Bonham, J. Foster, Y. C. Chang, P. Z. Ziang, K. G. Vandervoort, D. J. Lam and V. G. Kogan, Phys. Rev. Lett. **63**, 782 (1989).

3. W.E. Lawrence and S. Doniach, in *Proc. 12th Int. Conf. on Low Temperature Physics, Kyoto (1971)*, ed. E. Kanda (Keigaku, Tokyo, 1971) p 361.

4. K. B. Efetov, Zh. Eksp. Teor. Fiz. **76**, 1781 (1979) [Sov. Phys. Phys. JETP **49**, 905 (1979)].

5. A. Buzdin and D. Feinberg, J. Phys. France **51**, 1971 (1990).

6. J. R. Clem, Phys. Rev. **B43**, 7837 (1991).

7. P. H. Kes, J. Aarts, V. M. Vinokur and C. J. van der Beek, Phys. Rev. Lett. **64**, 1063 (1990).

8. V. G. Kogan, Phys. Rev. **B24**, 1572 (1981); V. G. Kogan, Phys. Rev. **B38**, 7049 (1988).

9. D. Feinberg, to appear.

10. L. N. Bulaevskii, Phys. Rev. **B44**, 910 (1991).

11. B. I. Ivlev, Yu. I. Ovchinnikov and V. L. Pokrovsky, Europhys. Lett. **13**, 187 (1990).

12. C. A. Bolle, P. L. Gammel, D. G. Grier, C. A. Murray, D. J. Bishop, D. B. Mitzi and A. Kapitulnik, Phys. Rev. Lett. **66**, 112 (1991).

13. L. J. Campbell, M. M. Doria and V. G. Kogan, Phys. Rev. **B38**, 2439 (1988).

14. Y. Iye, T. Tamegai and S. Nakamura, Physica **C174**, 227 (1991); Y. Iye, T. Terashima and Y. Bando, Physica **C177**, 393 (1991).

ANISOTROPY OF THE CRITICAL CURRENT IN BULK OXYGEN DEFICIENT AND Fe-DOPED YBCO MEASURED BY TORQUE

L. Fruchter, B. Janossy, H. Gu, I. Campbell
Laboratoire de Physique des Solides,
Bât. 510, Université Paris-Sud, 91405 Orsay Cédex, France.

R. Cabanel
L.C.R. Thomson-CSF, 91404 Orsay Cédex, France.

ABSTRACT

The angular dependence of the critical current was measured for single-crystal $Y_1Ba_2Cu_3O_{7-\delta}$ oriented grains with various oxygen deficiencies, using torque magnetometry. Strong discrepancies from transport measurements on thin films are outlined. An angular resolved contribution to the critical current of the twin boundaries is observed, which vanishes for $\delta \geq 0.15$. The $H\cos(\theta)$ scaling of the critical current, which was proposed for highly anisotropic superconductors, is not verified for the less anisotropic compounds. For Fe-doped samples, the critical current is dramatically altered: both pinning by twin planes and intrinsic pinning by CuO planes disappear.

INTRODUCTION

The angular dependence of the critical current in the high Tc superconductors has recently received a lot of attention, both on an experimental and a theoretical point of view[1,2,3,4]. Most of the discussions, however, rely on transport measurements on thin films. Indeed, measurements of the critical currents in bulk samples by means of the standard magnetization technique run into almost insuperable problems when faced to anisotropic materials. Simple evidence for twin plane pinning is hard to obtain for instance with such a technique[5,6]. However, it is desirable that bulk samples, which offer a wide range of different materials, are also investigated. In this paper, we show that the use of the torque technique provides a convenient way to overcome these difficulties. Our results differ strongly from the ones inferred from transport measurements on thin films.

SAMPLE PREPARATION AND CHARACTERIZATION

The YBCO samples with different oxygen contents are the same as those used in a previous study[7]. The fully oxidized compound was prepared by the conventional solid state reaction and oxidation process. Part of the material was then fully reduced under dynamic vacuum at 680 °C. The samples with intermediate oxygen contents were prepared by mixing the appropriate amounts of $Y_1Ba_2Cu_3O_7$ and Y_1Ba_2Cu-

$_3O_6$ and then allowing for oxygen equilibration at 650 °C. The fraction of the ground material sieved between a 45 µm and a 25 µm filter was embedded in epoxy resin and oriented in a 2T field at room temperature. Both the high viscosity and the short hardening time of the resin enabled the elaboration of samples with homogeneous concentration of superconducting material. The ac susceptibility superconducting transition and E.S.R. studies for such samples are reported elsewhere[7,8]. Transmission electron microscopy was carried out on the oriented samples, in order to study the twin patterns. Typical results are shown in figure 1. We observed a single twin set in most of the grains studied. The boundaries spacing was found roughly constant from δ=0 to δ=0.22. However, a rather large distribution for the twins density from one grain to another was found in the samples with δ less than 0.22.

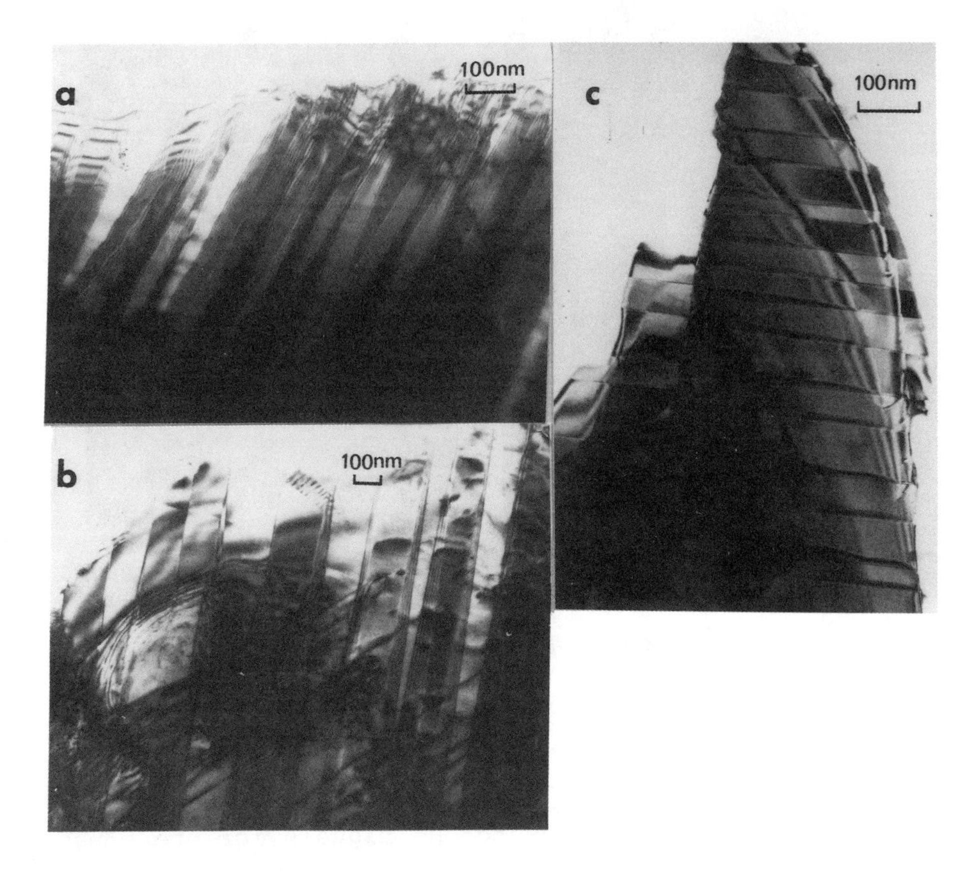

Figure 1: TEM observations. a) δ=0 b) δ=0.12 c) δ = 0.22

The iron-doped samples $Y_1Ba_2(Cu_{1-x}Fe_x)_3O_7$ with x=1,2,3 were prepared by mixing the appropriate amounts of carbonates and oxides and firing at 950 °C in the ambient atmosphere. The material was then ground and re-fired twice. Following this, a conventional growth in oxygen atmosphere was carried out. As reported elsewhere[9] and measured by us, the superconducting transition temperature (as given by the ac susceptibility transition) is constant within 1K for these doping levels. As expected, the twin patterns are dramatically altered by the introduction of Fe[9]. For increasing Fe content, the twin boundary spacing decreases and, for the higher concentration, a typical 'tweed' pattern is observed (fig. 2). Electron microscopy revealed that some of the grains in the resin could be misoriented. Such a problem is frequently encountered in substituted materials[10].

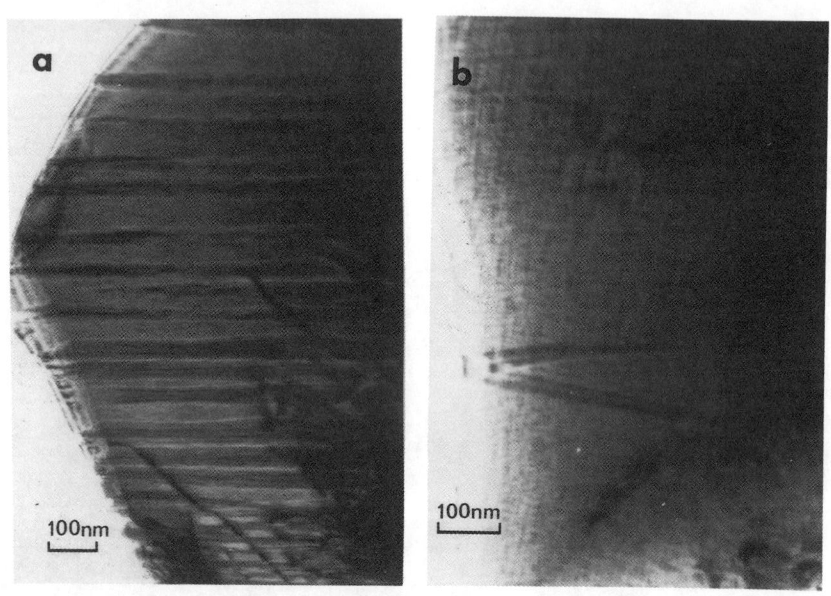

Figure 2: TEM observations of Fe-doped samples. a) x=0.01 showing twin pattern b) x=0.02 showing 'tweed' pattern.

CRITICAL CURRENT MEASUREMENTS

The pinning forces were measured from the irreversibility of the torque signal in the mixed state (rotational losses)[11]. For this purpose, a magnetic field, H, well above H_{c1} is applied to the sample and rotated in a plane containing the c axis. The irreversible torque is obtained from the difference of the torque signals for one sense of rotation and the other:

$$\Gamma_{irr}(\theta) = 1/2[\Gamma_+(\theta)-\Gamma_-(\theta)]$$

where θ is the angle between the **c** axis and the applied field. In the following, the angle between the applied field and the induction:

$$\theta_l = \sin^{-1}(4\pi\ \Gamma\ /\ H^2\ V)$$

was always small and θ could be identified with the angle between the induction and the **c** axis (V is the superconducting volume). For straight flux lines in the critical state, the irreversibility due to pinning is simply related to the macroscopic force per unit volume, F_p, and to the critical current

$$\Gamma_{irr}(\theta)/V=\eta\ F_p(\theta)\ R$$
$$J_c(\theta)=10\ \eta^{-1}\ (\Gamma_{irr}(\theta)/V)\ H^{-1}\ R^{-1}$$

This equation is similar to Bean's equation relating the critical current to the irreversible magnetization. Just as for hysteresis loops analysis, the radius R to be used is the one of the coupled superconducting domains, i.e. the grain radius in the case of single crystal grains.

PINNING BY TWIN PLANES

A typical irreversible torque signal is shown in fig.3.

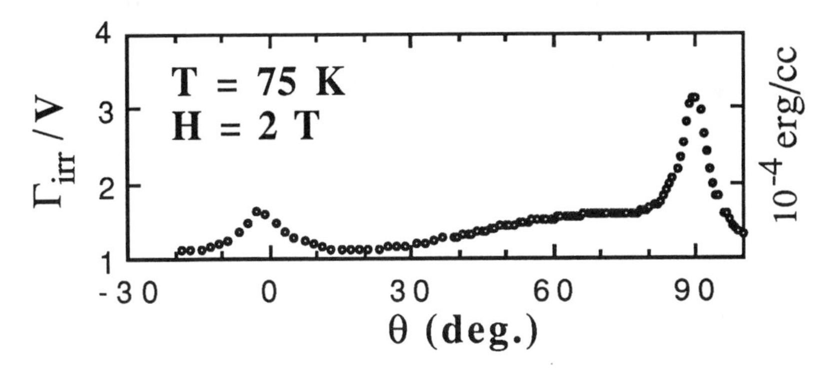

Figure 3: Irreversible torque signal for the fully oxidized compound.

It may be described as a background signal, with sharp enhancements when the field is aligned either with the **c** axis or the **ab** planes. We assign the peak along **c** to the pinning by twin boundaries. As demonstrated by torque measurements on single crystals[11,12], the enhancement is due to flux lines which are driven through the twin planes, while the gliding along the planes would be made easier. From the critical current enhancement along **c** with respect to the background signal, the temperature dependence of the specific contribution of the twin boundaries ($J_{c,TP}$) can be obtained (fig.4). We find that $J_{c,TP}(T)$ has a non-monotonous behavior, with a maximum at about 25 K. It has been suggested[13] that the decrease at the lower temperatures could be due to the competition with bulk pinning.

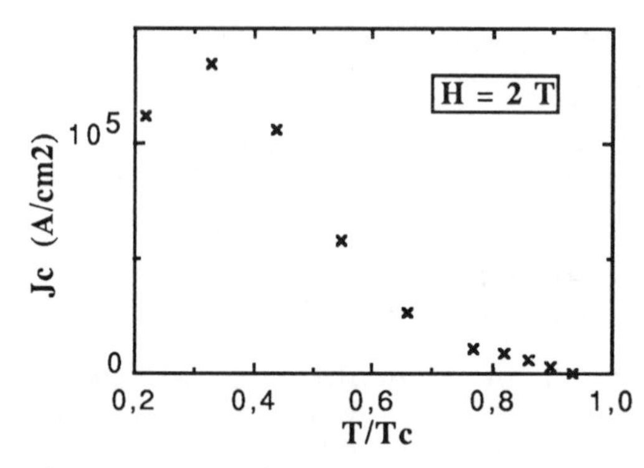

Figure 4: Twin planes contribution to Jc for the fully oxidized compound.

As demonstrated in figure 5, the angular resolved contribution of TP rapidly decreases with oxygen removal. For δ=0.22, the c peak is unobservable, although this sample shows well defined twin patterns. Two possibilities are likely to account for this phenomenon. First, as underlined by Tachiki et al[2], no angular resolved contribution of the TP to Jc should be observed for 2D flux lines in the case of decoupled superconducting planes. A contrario, our results show that, for the fully oxidized compound, flux lines are essentially 3D. As demonstrated in a previous paper, the coupling of the CuO planes gets weaker when the oxygen is removed[7]. Thus, the more pronounced 2D character of the oxygen deficient samples could account for the disappearance of the TP peak.

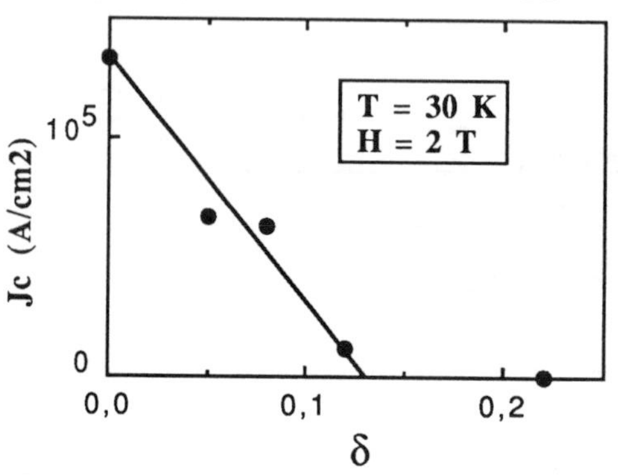

Figure 5: Variation of the TP contribution along c with oxygen content.

Alternatively, it was shown also[7] that the oxygen removal is associated with an increase of the London penetration depth in the planes, λ_{ab}. As a consequence, the flux lattice shear modulus, $C_{66} \propto \lambda_{ab}^{-2}$, should drop. In this case, diluted pinning centers (as compared to the flux lattice spacing) would be less effective to pin the flux lattice.

SCALING WITH TRANSVERSE FIELD COMPONENT

As suggested by Kes[1] and Tachiki[2], for highly anisotropic superconductors, the pinning of flux lines should be determined by the component of the field transverse to the planes only. Transport measurements of the critical current on YBCO2 and Bi-based[3]films have demonstrated the validity of this idea. Fig. 6 proides evidence that, for the fully oxidized compound, the scaling is limited here to $\theta \approx \pi/2$. This is in agreement with our previous conclusion that flux lines in this sample do not behave as 2D vortices.

Transport measurements show that this property can break down also in the case of YBCO thin films[4] However, there is no such data showing a sharp increase of the critical current when the field is applied along the twin planes. This could arise from structural properties of such films or from fundamental differences between transport and magnetic measurements of Jc.

For samples with large oxygen deficiency, our measurements show that the scaling property is retrieved to a better approximation. A typical example is given by the data in figure 7, obtained for a sample with oxygen stoichiometry 6.5.

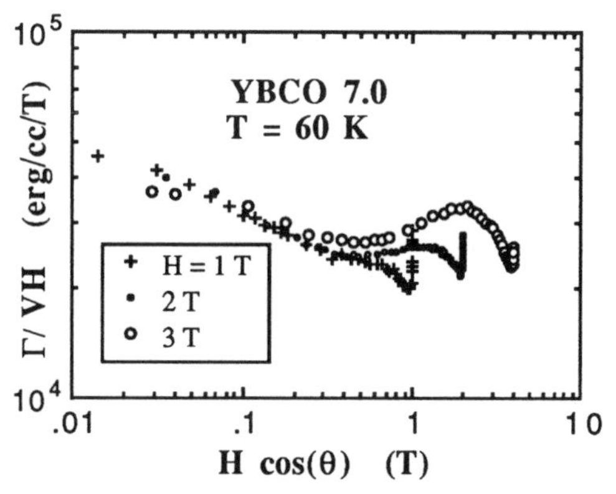

Figure 6: Attempt to scale the irreversible torque with $H_\perp$.

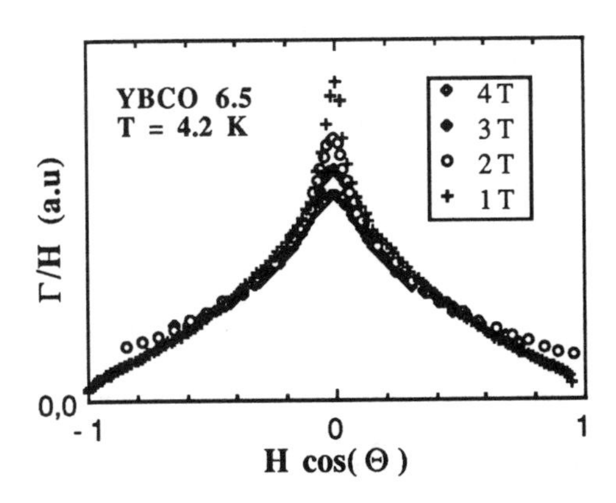

Figure 7: Scaling of the irreversible torque with transverse field component for a highly anisotropic YBCO.

For this compound, the anisotropy (ratio of the effective carrier masses across the planes and in the planes) was found to be about 900, while it is only about 30 for fully oxidized YBCO [7]

ANISOTROPY OF THE CRITICAL CURRENT IN Fe-DOPED YBCO

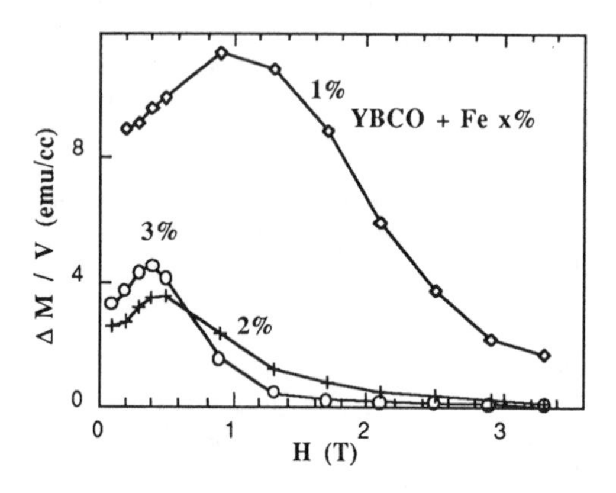

Figure 8: Irreversible magnetization of the oriented Fe-doped samples, for fields applied along the **c** axis (60 K).

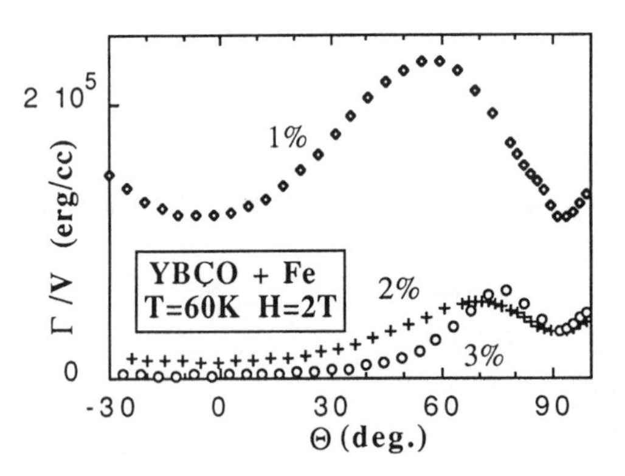

Figure 9: Irreversible torque for Fe-doped samples.

As shown in figure 2, Fe-doping at low concentrations results in significant variations of the twin pattern. For reduced twin spacing, as in the case of the 'tweed' pattern, one could expect enhanced pinning by twin boundaries. However, as shown by the analysis of magnetization curves (fig. 8), the in-plane critical current for fields applied along the **c** axis decreases sharply with doping. Equivalently, the critical field at which irreversibility sets in decreases.
The investigation of such samples by torque reveals dramatic changes in the angular dependence of the critical current (fig. 9). First, the **c** pinning enhancement is replaced by a minimum. Then, the signature of the intrinsic pinning between planes (**ab** peak) is also replaced by a minimum (some residue of it, however, can still be observed for the 1% doped sample). A possible interpretation which could account for both these remarkable features is that a denser twin pattern can provide a network of easy-gliding planes. In this way, it could become possible for the flux lines to cross the CuO planes without experiencing the core energy modulation. To check this idea, more would have to be known about the fundamental superconducting parameters of these samples. At the present moment, we have indications only that the anisotropy for such samples is comparable to the one of

the standard, fully oxidized YBCO. Finally, it should be noticed that, for these samples, the scaling property is well verified. Thus, the maximum of Jc in fig. 9 corresponds to the one in fig. 8.

CONCLUSION

As measured by torque, the angular dependence of the critical current in bulk YBCO can present important differences with the ones measured on thin films. Using two different parent materials, we have shown that some of the conclusions drawn from such data should be carefully put into relation with the sample structure. In particular, we have been able to show that twin planes can yield an angular resolved contribution to the critical current- which is expected in the 3D case -while most transport measurements would suggest 2D flux lines.

REFERENCES

1. P. Kes, J. Aarts, V. Vinokur, C. van der Beek, PRL **64** (1990), 1063.
2. M. Tachiki, S. Takahashi, Solid State Com. **72** (1989), 1083.
3. H. Raffy, S. Labdi, O. Laborde, P. Monceau, PRL **66** (1991), 2515.
4. B. Roas, L. Schultz, G. Saemann-Ischenko, PRL **64** (1990), 479.
5. L. Swartzendruber, A. Roitburd, D. Kaiser, F. Gayle, L. Benett, PRL **64** (1990), 483.
6. J. Liu, Y. Jia, R. Shelton, M. Fluss, PRL **66** (1991), 1354.
7. B. Janossy, D. Prost, S. Pekker, G. Oslanyi, to be pub. in Physica C.
8. A. Janossy, A. Rockenbauer, S. Pekker, G. Oslanyi, G. Faigel, L. Korecz, Physica C **171** (1990), 457.
9. R. Wördenweber, K. Heinemann, G. Sastry, H. Freyhardt, Cryogenics **29** (1990), 458.
10. H. Alloul, private com.

11. B. Janossy, R. Hergt, L. Fruchter,
Physica C 170 (1990), 22.

12. K. Fisher, L. Fruchter, R. Hergt, D. Linzen, G. Bruchlos, N. Cebotayev,
Mat. Lett. 11 (1991), 85.

13. G. Blatter, J. Rhyner, V. Vinokur,
preprint (1990).

LABORATORY ASPECTS OF TORQUE EXPERIMENTS ON HIGH-T_c MATERIALS

D.E. Farrell
CaseWestern Reserve University, Cleveland Ohio, 44106

ABSTRACT

A variety of torque-based experimental methods are finding increasing application in the study of high-T_c superconductivity. This paper discusses laboratory aspects of two of them, grain alignment and torque magnetometry.

INTRODUCTION

Torque-based experimental methods have been used extensively for nearly two hundred years in gravitational research,[1] and for a comparable period in the study of condensed matter.[2] On the practical side, torque-based research methods have contributed significantly to the steady improvement of permanent magnetic materials. Even if it has no permanent magnetic moment, an anisotropic material experiences a mechanical torque when placed in a magnetic field. First investigated systematically by Faraday,[2] these (relatively weak) torques have provided fundamental information, e.g., in electronic structure studies.[3] Torque methods were first extended to (conventional) superconducting materials by Genberg,[4] but difficulties were encountered due to the presence of extrinsic effects. In copper-oxide superconductors such effects are much smaller, and equilibrium torque methods are proving useful in high-T_c work. Non-equilibrium techniques are also finding application as a probe of both critical currents[5] and vortex transitions.[6]

The present paper is focussed on equilibrium torque methods. The first such method to find widespread application was the grain-alignment process, based on the anisotropic magnetic susceptibility in the normal state.[7] In addition, torque magnetometry[8] performed below T_c is now thought to provide the best available probe of superconducting anisotropy. It has also been suggested that detailed features of such experiments can provide dimensionality information.[9] In sharp contrast to early reports,[1,2] papers in the present era tend to contain few helpful laboratory details. The goal of the present paper is to provide such information for both the grain alignment process and torque magnetometry.

GRAIN ALIGNMENT

Many experimental techniques in condensed matter physics require larger crystals than are available in the case of high-T_c materials. The method of grain-alignment has provided a useful way of circumventing this difficulty. In practice, laboratory details control the degree of alignment achieved. This can be demonstrated by considering a simple model of the alignment process.

For high-T_c materials based on La,Y, Bi and Tl, the volume magnetic susceptibility measured with the field along the **c** axis is greater than the susceptibility measured with the field in the CuO planes. In cgs units, the difference, $\Delta\chi$, has been found[10] to be of the order of 10^{-6}. If a magnetic field is applied to a crystal of volume V, at some angle, θ, to the **c** axis, it experiences a mechanical torque whose sense is to rotate the **c** axis toward the direction of the magnetic field. The magnitude of the torque is $\Delta\chi\, V H^2 \sin\theta\cos\theta$. For a typical crystal of volume 10^{-9} cm^3, in a field of 10^5 Oe, the torque is ~ 10^{-5} dyne-cm, too small to produce any useful alignment in a ceramic or powdered sample.

Nevertheless, if a crystal is isolated from its neighbors and suspended in a suitable fluid medium, a small torque can produce a useful degree of alignment. Details of the angular rotation are influenced by the shape of the crystal, but the essential physics is exhibited by considering the case of a spherical crystal supported in a fluid medium of viscosity η. Suppose that the **c** axis initially makes an angle θ_o to the magnetic field. Equating the viscous drag on the sphere to the mechanical torque produced by the field gives a simple differential equation. In this way it can be shown that at some later time, t, the angle of the **c** axis to the field is given by:

$$\tan\theta(t) = e^{-t/\tau}\tan\theta_o \qquad (1)$$

$$\tau = \frac{6\eta}{\Delta\chi\, H^2}$$

An interesting feauture of this result is that the characteristic alignment time, τ, is *independent* of the sample volume. Small grains align just as rapidly as large ones. A commercial epoxy[11] with a viscosity of $\eta \sim 10^2$ cgs and a curing time of $\sim 4\times10^2$ secs was chosen as a suspending medium in the original work.[7] The characteristic alignment time is then of the order of $\sim 10^{-1}$ secs in a field of 10^5 Oe, very much shorter than the curing time. In fact, for fields greater than $\sim 10^4$ Oe, the misalignments that are typically encountered (~1 degree) must be attributed to factors that are ignored in the simple model. In other words, laboratory considerations control the degree of alignment achieved in practice.

One such consideration is the phenomenon of sedimentation. High-T_c materials have a larger density (~6 gm/cm^3) than that of epoxies (~ 1 gm/cm^3), so the grains will fall during the curing time. Stokes law provides an estimate of ~10^{-3} cm/sec for the sedimentation rate of a grain of volume 10^{-6} cm^3. With a sample of height ~1mm, and a curing time of ~10^4 secs, such grains will form a poorly aligned aggregate at the bottom of the container. An epoxy with a curing time of hours should only be used to align small grains.

As is clear from the model discussed above, the individual grains of the starting powder must all be single crystals. A given ceramic process tends to produce crystalline grains whose mean size, and distribution about that mean, are fairly well-defined. After grinding the material, the powder should therefore be passed through a pair of standard sieves that select an appropriate region of the grain-size distribution. There is evidence[12] that it is particularly difficult to prepare single-crystal powders of the bismuth and thallium based materials. In this connection, the standard rocking curve characterization of alignment should only be trusted if the misalignment is known to be limited to about a degree. If any doubt exists, some sort of pole-figure analysis is required.[13,14]

Random torques originating from convection currents, bubbles, and neighboring particles will also reduce the degree of alignment from that predicted by Eq.1. The curing process is strongly exothermic and the associated temperature increase can be large if a substantial volume of epoxy is used. For example, 25 cm^3 of resin and an equal volume of hardener mixed in an open 100cm^3 beaker increase in temperature by ~100°C (!) during the curing process. To minimise this temperature rise, and the associated convection currents, the sample should therefore be no larger than actually required. If samples more than a few cm^3 in volume are needed, an epoxy with a longer curing time should be used, together with a small-grained powder. Bubbles can be essentially eliminated by vacuum pumping the epoxy and hardener at a temperature of ~40°C for about an hour prior to use. Grain interactions are reduced by working with a small volume fraction of the superconductor. In a typical procedure, ~10^{-1}cm^3 of YBCO powder in ~1cm^3 of epoxy is aligned in a magnetic field of ~ 10^5 Oe. A variety of experiments show[15] that the majority of the grains in such a sample will lie within one degree of the field direction.

The grain-alignment technique has provided a useful way of producing samples that have many of the properties of very large single crystals. A number

($\sim 10^2$) of papers reporting work on grain-aligned high-T_c material have been published.[15] These have investigated about twenty different physical properties, including a recent experimental tour-de-force in which the de Hass-van Alphen effect was reported for the first time in YBCO.[16]

EQUILIBRIUM TORQUE MAGNETOMETRY

The torque on a material in a magnetic field **H** is given by **MxH**, where **M** is the magnetic moment, i.e., torque is a direct measure of the transverse magnetic moment. Equilibrium torque magnetometry was used to obtain the first quantitative evidence for the existence of a transverse magnetization of the Abrikosov flux lattice.[8] It has since been used in many studies of the superconducting anisotropy and dimensionality of high-T_c materials. In this section we first outline the experimental considerations that have motivated the use of torque methods. The design and operation of a specific instrument that has been used in the author's laboratory for a number of years will then be discussed.

In high-T_c materials, the Ginzburg-Landau parameter, κ, is much larger, and flux pinning generally much smaller, than in conventional superconductors. Both these characteristics facilitate torque measurements. The first factor ensures that demagnetization effects are generally negligible. Complications associated with demagnetization effects and pinning were the difficulties that thwarted all the early torque work on conventional materials. On the other hand, a large value of κ makes SQUID magnetization measurements difficult, because the magnetization itself becomes small. Reduced pinning also leads to severe difficulties with resistive H_{c2} measurements.[17]

Sensitivity is an important consideration. In principle, torque sensitivities of $\sim 10^{-9}$ dyne-cm are available.[1] In a field of 10^5 Oe this corresponds to a transverse magnetic moment sensitivity of $\sim 10^{-14}$ emu, about six orders of magnitude better than the moment sensitivity obtainable with SQUID magnetometry. With a higher sensitivity, smaller samples can be used, providing two useful benefits. A critical state description of flux pinning[18] indicates that the fraction of the total magnetization that is irreversible is directly proportional to a typical sample dimension. This implies that the magnetization of a small sample will be more reversible than that of a large one with the same defect density. In addition, small samples are more likely to be homogeneous than larger ones. In view of the materials problems associated with high-T_c superconductors, sensitivity considerations alone suggest the choice of torque over SQUID magnetometry for investigating superconducting anisotropy.

Commercially available torque instrumentation is designed for the study of materials with a permanent moment and is limited in sensitivity to $\sim 10^{-2}$ dyne-cm. The instrument described below achieves a sensitivity of $\sim 10^{-4}$ dyne-cm. Although much less than is possible in principle, this sensitivity has allowed the use of sample volumes less than 10^{-6} cm^3.

Design Considerations

The figure below illustrates the elements of the torque magnetometer designed by the author for high-T_c work:

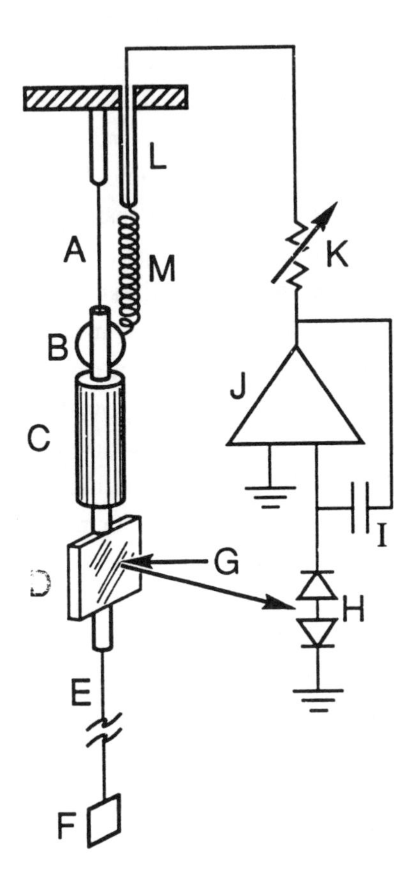

A. Tungsten suspension wire, 25μm in diameter, 2cm long.

B. Feedback coil, 50 turns.

C. Brass cylinder for frequency reduction and damping.

D. Mirror.

E. Quartz support rod: 1mm in diameter, 70 cm long.

F. Single crystal sample.

G. 1 mw Laser.

H. Photo-diodes.

I. Feedback capacitor.

J. 741 Op-Amp.

K. Sensitivity control.

L. Vacuum feed-through.

M. Helical electrical lead.

Fig.1. High-T_c Torque Magnetometer.

The design is based on that of an instrument developed for electronic structure studies by Condon and Marcus.[3] Torques experienced by the sample when a field is applied are coupled to a mirror at room temperature by a thin quartz rod. The assembly supporting the mirror is suspended by a thin tungsten wire, and the angular position of the mirror is sensed by an optoelectronic system. A feedback arrangement provides the torque necessary to maintain the sample at some specific angle to the laboratory. The current in the feedback loop is then a direct measure of the torque on the sample. As for the grain-alignment process, a statement of the procedure is straightforward. However, both the reliability and practicality of the procedure depend to an even greater extent on numerous laboratory details.

Building vibrations are generally the main source of noise for sensitive mechanical experiments. In gravitational studies, adequate noise reduction can only be achieved in an isolated location.[1] Fortunately, the requirements in the present case are less demanding, and a more convenient procedure is available. Vibrations couple mainly to the pendulum (swinging) mode of the suspension, whose frequency is ~2Hz. Energy in the pendulum mode is then transferred to the torsional mode via non-linearities in the suspension. The effect of vibrations can be therefore be reduced if the mode coupling is reduced. A useful measure of decoupling can be achieved by increasing the moment of inertia of the suspension so as to reduce the frequency of the torsional mode to well below that of the pendulum. The additional moment is provided by the brass cylinder of mass 40 gm which reduces the torsional frequency to 10^{-1}Hz. (The tungsten wire used has a torsional constant of 3.5 dyne-cm/rad and can support a load of ~150gm without breaking). This step alone reduces the angular noise to a few millidegrees. A further reduction is obtained by placing the brass cylinder in a field of ~10^3 Oe, provided by a permanent magnet mounted outside the vacuum chamber. With the associated eddy-current damping, no special anti-vibration mounting is required for the instrument. The angular noise observed in a busy laboratory environment is well below a millidegree.

The feedback arrangement employed in the Condon instrument[3] utilized a commercial device that is no longer available. It was replaced by the following scheme: Light from a helium-neon laser illuminates a pair of standard silicon photo-diodes located ~2 meters from the suspension mirror. The diodes are separated by ~1cm and connected in series so that no current is produced if they are equally illuminated. Any angular deflection produces a current, which is supplied to the virtual ground of a general-purpose (741) op-amp. Together with the capacitor and (sensitivity control) variable resistor shown in Fig.1, this comprises the entire feedback electronics. The circuit functions as an integrator and maintains the angle of the sample fixed with respect to the laboratory to within $\pm 10^{-3}$ degrees. A standard laboratory electromagnet allows the angle of the field to the **c** axis of the sample to be set with an angular uncertainty of $\pm 3 \times 10^{-2}$ degrees. For high angular resolution measurements, the two photo-diodes are mounted on a vernier translation stage. Keeping the field fixed and translating the diodes allows the angle of the sample to the field to be changed with a resolution of $\pm 10^{-3}$ degrees. (The null-deflection character of the measurement is sacrificed in this mode of operation. However, the increased angular resolution is useful for investigating highly anisotropic materials for field directions close to the CuO planes. The rapid change of torque with angle in this situation makes the null-deflection feauture less critical).

A final design problem arises in connection with the electrical leads that provide current for the feedback coil. Even if they are not under tension, these exert a torque on the suspension. As discussed below, the quantity actually recorded in an experiment is the *change* in feedback current associated with the appearance of superconductivity in the sample. Hence, if the lead torque remains constant, it will not effect the measurements. With the suspension locked in position by the feedback circuit, no changes can occur. Unfortunately, lock is sometimes lost during a measurement, allowing the suspension to suffer a temporary angular displacement away from the null position. This can slightly disturb the disposition of the leads and alter the torque they exert on the system. As will be clear from the discussion below, this does *not* alter the calibration of the instrument in any way. However it *can* introduce a zero offset and require a measurement cycle to be restarted. To minimise this problem, each lead is fabricated in the form of a 2mm diameter helix with ~50 turns, using very fine (50 gauge) copper wire. This precaution significantly reduces the background torque contributed by the leads and allows measurements to be resumed without interruption, except in cases of extreme angular displacement. The optical system is arranged so that if the feedback lock is lost, the laser spot immediately becomes visible to the experimenter. This facilitates taking appropriate corrective action, reducing the possibility of any extreme displacements occurring.

Laboratory Details

An adhesive is required in four of the operations that are discussed in this section. The commercial adhesive varnishes that are widely used in low-temperature physics can contain magnetic contaminants. By contrast, an extensive characterization effort failed to detect *any* metallic impurities in commercial epoxy resin at the ppm level.[19] The commercial epoxy[8] chosen for our grain-alignment work was therefore also used as the glue in all the operations discussed below.

When mounting a sample, magnetic components and contamination must be avoided. The CuO layers in the sample must be aligned in a vertical plane, and, because of a difficulty discussed below, the suspension should be disturbed as little as possible. It is also helpful to be able to perform the operation quickly. A procedure has been evolved that satisfies these requirements. It takes advantage of the fact that high-T_c crystals are available as thin flat plates, with the CuO layers lying in the plane of the plate. As a first step, the large surface of the experimental crystal is glued down to a 10^{-2} cm thick glass mounting plate. Glue is also applied to the other surfaces of the crystal to provide protective encapsulation against water vapor, etc. After curing, the plate is then supported, approximately in a vertical plane, with the top edge lying directly underneath the quartz rod. A small amount ($\sim 10^{-4}$ cm^3) of epoxy is applied to the bottom of the rod and the glass plate raised until contact is made. After about three minutes, the support is removed, allowing the plate to hang freely. The mass of the plate is sufficiently small ($\sim 10^{-2}$ gms) that the partially cured bond is usually able to support it. On the other hand, the same mass is very much greater than that of the sample and glue combined, so the gravitational torque quickly eliminates any vertical misalignment. After about seven minutes the bond is formed, but a complete cure would still take many hours. Nonetheless, the sample is cooled down after a further twenty minutes, i.e., without letting the curing process go to completion. In addition to saving time, it is then relatively easy to fracture the bond and remove the sample mounting plate without disturbing the suspension.

Tungsten is used for the suspension wire because the ratio of its tensile strength to torsion constant is higher than that of any other material except quartz. (Thin quartz fibers are troublesome because they tend to weaken on exposure to air). Unfortunately, tungsten wire is not easy to manipulate. It kinks much more readily than copper and will then break under small loads. It can not be welded into position and it will fracture if the breaking load is exceeded even briefly. This can occur all too easily if the instrument is accidentally jarred. The author's own experience is that the mean lifetime of a suspension is of the order of two months. The mounting procedure illustrated below was developed to cope with these problems. To make a suspension with a working length of 2 cm, scissors are used to cut out a ~6 cm length of wire. An end is then attached to a brass post, as shown in Fig.2. It is first tied down with thin (~40 gauge) copper wire and glued. The tungsten is then bent back, tied down, and again glued to the post. This process is repeated two more times, and the assembly allowed to cure for at least 24 hours. In contrast to the sample mounting procedure outlined above, a full curing time is essential. Otherwise, the wire tends to pull away from the post under load. The lower end of the wire is attached in a similar way to a second post. Set-screws allow a suspension unit made in this way to be readily removed from the instrument. A reserve supply allows replacement with little interruption to an experiment.

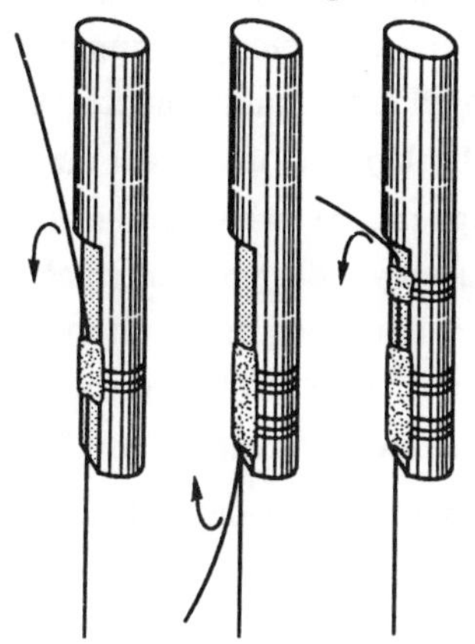

Fig.2 Steps in the suspension mounting procedure

Feedback current is supplied to a circular coil shown at the top of the brass cylinder in Fig.1, with the axis of the coil perpendicular to the field of the damping magnet. To calibrate the instrument, the torsional constant of the suspension is obtained from the measured period of free (undamped) oscillation and the calculated moment of inertia of the system. With the optoelectronic system removed, the angular displacement of the suspension for a given feedback current is measured, and the torsional constant used to relate this to a torque. Although the torsional constant appears as an intermediate step in the calibration, it is important to note that the calibration itself only depends on the magnetic flux in the feedback coil. It is *independent* of the torsional constant of the suspension. This is the primary advantage secured by the null-deflection aspect of the technique.

When a new suspension has been installed, or an existing one strongly disturbed (e.g., in the process of mounting a new sample), uni-directional drifts are observed. These can be as large as $\sim 10^{-2}$ dyne-cm/hour and are a peculiar property of tungsten wire.[20] Fortunately, the drift rate falls approximately exponentially with time; it is typically less than $\sim 10^{-4}$ dyne-cm/hour after a few days and remains constant over the period of an experiment (~1hour). Small drifts of this order may also arise from thermo-mechanical effects associated with falling levels of cryogenic fluids. For high-resolution work, they can be measured before and after an experimental run and subtracted out.

Helium exchange gas is required to establish thermal equilibrium between the sample and the adjacent thermometry. The pressure is reduced to 10^2 μm of Hg so as to avoid convective disturbances of the suspension. Commercial instrumentation is used for temperature control, as well as for standard data acquisition procedures. There are a number of sources of field-dependent background torques at the ~10^{-3} dyne-cm. level in a field of 10^4 Oe: Residual magnetic impurities in the sample mount and quartz rod, magnetic impurities introduced via dust, and interaction of the fringing field of the electromagnet with the feedback coil and with magnetic contamination in the mirror. Although the first two can not be estimated separately, the total background torque is allowed for by taking the difference $\tau(H,\theta,T<T_c)$ - $\tau(H,\theta,T>T_c)$. This procedure is used to obtain the temperature dependence of the torque at some fixed field and angle.

To obtain the *angular dependence* of the torque in the superconducting state, the above procedure is inconvenient because many large temperature steps are involved. (The time required to stabilize each temperature to ± 0.01K in the liquid nitrogen range is of the order of 30 minutes). Instead, the complete angle-dependent background is measured at the beginning of an experiment by raising the temperature above T_c and measuring the difference associated with the field. $\tau(H,\theta,T>T_c)$ - $\tau(H=0,\theta,T>T_c)$. The procedure is repeated below T_c and the torque associated with the superconductivity obtained from the double difference:

$$\tau(H,\theta,T<T_c) - \tau(H=0,\theta,T<T_c) - [\tau(H,\theta,T>T_c) - \tau(H=0,\theta,T>T_c)] \qquad (2)$$

Note that since the two zero-field terms in this expression are themselves equal, the procedure is formally equivalent to taking the simple difference discussed above.

Finally, it should be noted that the above type of instrument requires a horizontal magnetic field. The strength of such fields is presently limited to ~10^5 Oe. Higher fields are available in some superconducting and Bitter magnets, but these all employ a vertical field geometry. The capacitive technique pioneered by Naughten et al.[21] provides an interesting approach to torque magnetometry which can be employed with either field orientation.

AKNOWLEDGMENTS

The experimental efforts discussed here would not have come to fruition without the collaboration of many colleagues. The grain-alignment technique was developed by M.R. DeGuire, B.S. Chandrasekhar and the author, in active collaboration with D.K.Finnemore and his group. Torque magnetometry was motivated by the theoretical work of V.G. Kogan and by the remarkable single crystals grown by members of D.M. Ginsberg's group, particularly by J.P. Rice. Preparation of this paper was made possible by NASA grant No. NCC-123, and by National Science Foundation Grant No. NSF-DMR 89-13651.

REFERENCES

1. H. Cavendish, Phil. Trans. **88**, 469 (1798); P.H. Roll, R. Krotkov and R.H. Dicke, Ann. Phys. **34**, 442 (1964); J.K. Hoskins, R.D. Newman, R. Spero and J. Schultz, Phys. Rev. **D32**, 3084 (1985)
2. M. Faraday, Philosophical Transactions of the Royal Society, p19 (1849)
3. J.H. Condon and J.A. Marcus, Phys. Rev. **134A**, 446 (1964)
4. R.W. Genberg, J. Appl. Phys. **44**, 1288 (1973)
5. L. Fruchter and I.A. Cambell, Phys. Rev. **B40**, 5158 (1989)
6. D.E. Farrell, J.P. Rice and D.M. Ginsberg, Phys. Rev. Lett. **67**, 1165 (1991)
7. D.E. Farrell, B.S. Chandrasekhar, M.R. Deguire, M.M. Fang, V.G. Kogan, J.R. Clem, and D.K. Finnemore, Phys. Rev. **B36**, 4025 (1987)
8. D.E. Farrell, C.M. Williams, S.A. Wolf, N.P. Bansal, and V.G. Kogan, Phys. Rev. Lett. **61**, 2805 (1988)
9. D.E. Farrell, J.P. Rice, D.M. Ginsberg, and J.Z. Liu, Phys. Rev. Lett. **64**, 1573 (1990)
10. K. Fukuda, S. Shamoto, M. Sato, and K. Oka, Solid State. Comm. **65**, 1323, (1988)
11. "Master Mend" Epoxy. Manafactured by the Loctite Corporation, Cleveland, Ohio
12. D.E. Farrell, R.G. Beck, M.F. Booth, C.J. Allen, E.D. Bukowski, and D.M. Ginsberg, Phys. Rev. **B42**, 6758 (1990)
13. D.B. Knorr and J.D. Livingston, Supercon. Sci. Tech. **1**, 302 (1990)
14. G. Oszlanyi, G. Faigel, S. Pekker, and A. Janossy, Physica **C167**, 157 (1990)
15. A bibliography of grain-alignment papers is available from the author.
16. C.M. Fowler, B.L. Freeman, W.L. Hults, J.C. King, F.M. Mueller, and J.L Smith, to be published
17. A.P. Malozemoff, "Physical properties of high temperature superconductors " edited by D. Ginsberg, Vol.1, p71, World Scientific, Singapore. (1989)
18. C. Bean, Phys. Rev. Lett. **8**, 250 (1962)
19. W.W. Wright, Brit. Polym. Journ., **15**, 224 (1983)
20. V.B. Braginsky and A.B. Manukin, "Measurement of weak forces in Physics Experiments", University of Chicago Press, page 61, (1977)
21. R.J. Naughton, J.S. Brooks, L.Y. Chiang, R.V. Chamberlin and P.M. Chaikin, Phys. Rev. Lett. **55**, 969 (1985)

TUNNELING IN CUPRATE AND BISMUTHATE SUPERCONDUCTORS

J. F. Zasadzinski*, Q. Huang and N. Tralshawala
Illinois Institute of Technology, Chicago, IL. 60616

K. E. Gray
Materials Science Division,
Argonne National Laboratory, Argonne, IL. 60439

ABSTRACT

Tunneling measurements using a point-contact technique are reported for the following high temperature superconducting oxides: $Ba_{1-x}K_xBiO_3$(BKBO), $Nd_{2-x}Ce_xCuO_4$ (NCCO), $Bi_2Sr_2CaCu_2O_7$ (BSCCO) and $Tl_2Ba_2CaCu_2O_x$ (TBCCO). For the bismuthate, BKBO, ideal, S-I-N tunneling characteristics are observed using a Au tip. The normalized conductance is fitted to a BCS density of states and thermal smearing only proving there is no fundamental limitation in BKBO for device applications. For the cuprates, the normalized conductance displays BCS-like characteristics, but with a broadening larger than from thermal smearing. Energy gap values are presented for each material. For BKBO and NCCO the Eliashberg functions, $\alpha^2F(\omega)$, obtained from the tunneling are shown to be in good agreement with neutron scattering results. Proximity effect tunneling studies are reported for Au/BSCCO bilayers and show that the energy gap of BSCCO can be observed through Au layers up to 600 Å thick.

INTRODUCTION

Tunneling spectroscopy is a very powerful probe of the superconducting state[1] and, in ideal cases, provides a direct measure of the energy dependent gap function, $\Delta(\omega)$, and the electron-phonon spectral function, $\alpha^2F(\omega)$. In addition, a number of applications exist for thin-film tunnel junctions, including photon detectors, frequency standards and fast switches. Thus it is not surprising that a world-wide effort exists to form high quality

*Work supported by the National Science Foundation-Office of Science and Technology Centers (DMR 88-09854) and U.S. Department of Energy, Division of Basic Energy Sciences-Materials Sciences under contract #W-31-109-ENG-38.

junctions on high temperature superconductors (HTS) for fundamental studies and technological purposes. Unfortunately, the tunnel junctions obtained on cuprate superconductors have been far from ideal, showing BCS-like features, but with varying degrees of broadening and evidence of quasiparticle states inside the gap. In fact, the question of whether there exists a true energy gap in the cuprates remains an open one. Thus far, the bismuthate superconductor, $Ba_{1-x}K_xBiO_3$ (BKBO), is the only oxide with a Tc > 25K to display a well-defined energy gap[2] in a tunneling measurement.

Inroads into the problem of tunneling in HTS have been made by using a point-contact method, where the tip can be used to scrape, clean and even cleave the sample surface. We report here some of our best results on cuprate and bismuthate superconductors using this technique. Ideal junctions are found on BKBO with the normalized tunneling conductance fitted very well to a thermally smeared BCS density of states (dos). For $Bi_2Sr_2CaCu_2O_7$ (BSCCO)[3] and $Nd_{2-x}Ce_xCuO_4$ (NCCO)[4] we have observed a zero bias dos as low as 10%, which is a significant improvement over previously reported data. Structures have been found[4] in the high bias data of BKBO and NCCO which have the characteristics of phonon effects as found in conventional superconductors. We show that the $\alpha^2F(\omega)$ for BKBO and NCCO bear a close resemblance to the phonon density of states measured by neutron scattering, indicating that electron pairing is principally mediated by phonons.

To address the problem of broadening of the tunneling dos in cuprates, we have investigated a proximity effect approach whereby a thin layer of Au (200-600 Å) is deposited onto a freshly cleaved BSCCO crystal. The results show that the BSCCO energy gap is observed throught the Au layer, and for a 200 Å layer, the zero bias dos appears to be close to zero. This approach may thus be suitable to fabricate thin film juntions with low sub-gap currents and sharp onsets of current at the gap voltage, characteristics which are essential for device purposes.

ENERGY GAP MEASUREMENTS

In an ideal tunneling experiment, the dynamic conductance, $\sigma = dI/dV$, measured in the superconducting and normal state, leads directly to the quasiparticle dos[1], N(E), by the expression

$$N(E) = \sigma_s/\sigma_n = \mathrm{Re}\; E^2/(E^2-\Delta^2(E))^{1/2} \qquad \text{Eq. 1}$$

Here, $\Delta(E)$ is the complex gap parameter which, in conventional superconductors, has a strong energy dependence near energies which correspond to peaks in the phonon dos owing to the inelastic scattering of electrons by phonons. However for lower energies, $\Delta(E)$ is approximately a real constant, Δ, and Eq. 1 becomes the standard BCS dos. Finite temperature effects smear out N(E) by approximately 3.5 kT. An example of an ideal result is shown in Fig. 1 for a point-contact tunnel junction on the cubic bismuthate, BKBO (Tc=25-30K) at 4.2K using a Au tip.

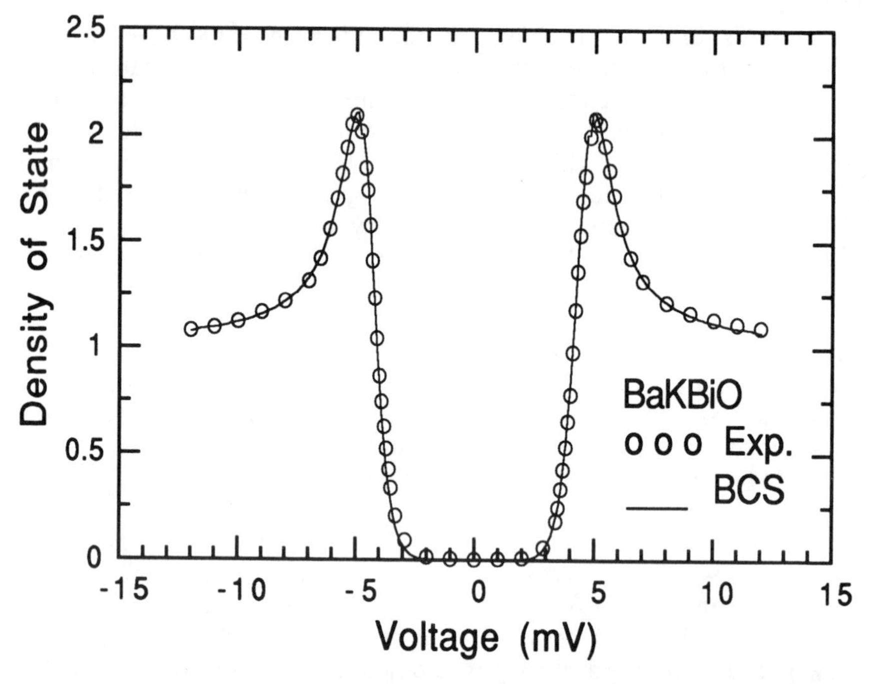

Figure 1. Fit of the experimental normalized conductance, open circles, with a BCS dos and thermal smearing of 4.2K (solid line). The gap parameter, Δ=4.5 meV.

The experimental data are fit quite well to the BCS dos with a single value of Δ=4.5 meV and thermal smearing only. Note the flat, near-zero dos for voltages, $eV < \Delta$, which proves that BKBO has a well-defined energy gap. Moving the Au tip from point to point on the polycrystalline sample, data similar to Fig. 1 were obtained although the value of Δ ranged from 3.6-4.6 meV. The different Δ values are likely due to variations of K concentration from grain to grain and associating the large(small) Δ values with the 10% (90%) points of the magnetic transition of this sample, we obtain $2\Delta/kT_c$ = 3.8-3.9. This indicates moderate coupling strength. Using a Nb tip, very low sub-gap conductance is found (<0.2%)[2] as well as a sharp current onset at the sum gap, proving there is no fundamental limitation in BKBO for device applications.

The electron-doped superconductor, NCCO, has a relatively low Tc~23K, but shares many of the properties of the higher T_c cuprates, including the quasi two-dimensional Cu-O planes. Generally high quality tunneling results were consistently obtained on single crystals of NCCO and a representative normalized conductance is shown in Fig. 2.

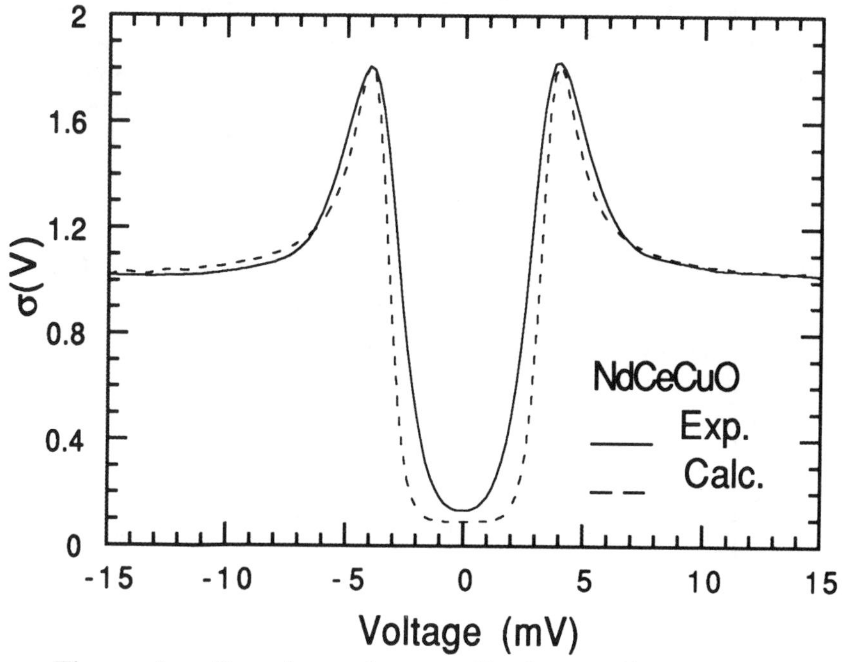

Figure 2. Experimental normalized tunneling conductance (solid line) for NCCO. Thermally smeared BCS dos (dashed line) includes a leakage conductance of 10%. The gap parameter, Δ=3.6 meV.

While the data diplay clear evidence of the BCS dos, there appears to be smearing of the structure in addition to that due to temperature effects and the normalized conductance at zero bias is not zero but 10-15%. However, the structure is much sharper than found in earlier studies.[5] We have fit the data with a thermally smeared BCS dos including a constant leakage conductance of 10% and while the fit is reasonably good, there are obvious differences in the two curves. Most notably, the shape of the data near zero-bias has a parabolic character while the BCS dos has the expected flat behavior indicative of a true energy gap.

The shape of the data indicates that the zero-bias dos is probably not due to leakage at all, but rather is manifestation of the overall broadening.

Despite the non-ideal fit of the NCCO data, the structures are sharp enough to extract the gap parameter, and a value Δ=3.5-3.6 meV was consistently obtained for over 15 junctions. Using the width of the magnetic transition, we obtain $2\Delta/kT_C$=3.5-4.1, indicating weak to moderate coupling strength.

Tunneling data on higher T_C cuprates such as BSCCO[3,6] (T_C=85K) and TBCCO[7,8] (T_C=120K) generally display much more broadening than we observe in NCCO. An example of one of our best tunneling curves for TBCCO is shown in Fig. 3. Important features to note include the reduced size of the normalized conductance peaks (1.4 compared to 2.1 in BKBO) and the zero-bias value of ~ 25%. These features are evidence of increased broadening of the BCS dos and thus we have attempted to fit the data using a phenomenological expression which adds an imaginary part, Γ, to the energy, E, in eq. 1. The value of Γ is thus a measure of the degree of broadening.

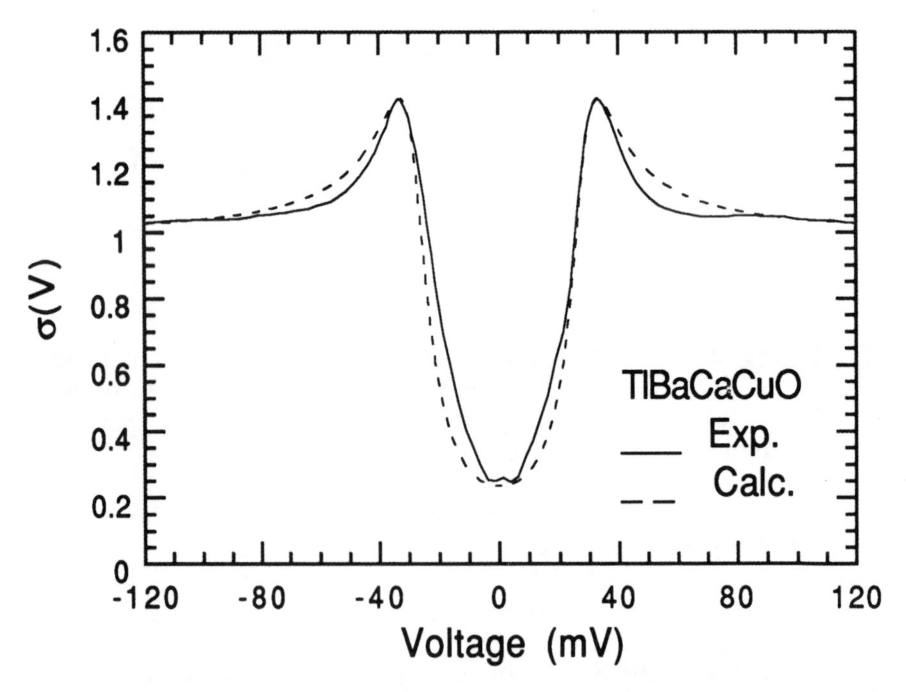

Figure 3. Experimental normalized conductance for TBCCO (solid line). The fit (dashed line) is a broadened BCS dos as explained in the text with Δ=28 meV and Γ=6.8 meV.

The fit of the TBCCO data is reasonably good, showing how the broadening of the BCS dos leads to reduced conductance peaks and a non-zero value of the zero-bias conductance. Another important feature is that the gap parameter for the fit in Fig. 3 (Δ=28 meV) is significantly less than the voltage of the conductance peaks (eV=$\pm$32 meV), and demonstrates that the extraction of the gap parameter directly from this voltage leads to an overestimate of the gap. It should be noted that for the cuprates with T_c>77K, a direct measurement of the normal state curve is very difficult using the point contact method. Thus we have generated the normal state data from high-bias superconducting data using a procedure described elsewhere.[3,8]

Another important HTS material is BSCCO and the general results for the normalized tunneling conductance in this compound are similar to that obtained for TBCCO, namely, broadened BCS-like features with zero-bias values in the range 20-50%. We have steadily improved the sharpness of the gap region tunneling data in BSCCO[3] using the point-contact method and one of our best results to date is shown in Fig. 4.

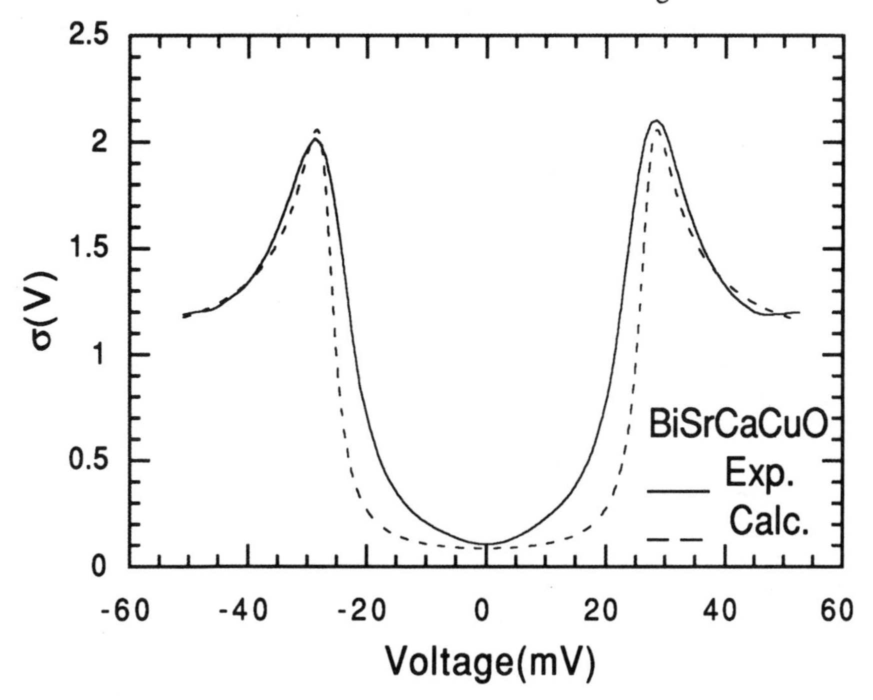

Figure 4. Experimental normalized conductance (solid line) for BSCCO compared to broadened BCS dos (dashed line) with parameters, Δ=27 meV, and Γ=2.3 meV.

What is most notable about the BSCCO data in Fig. 4 is the similar shape to the NCCO data of Fig. 2 for absolute voltages $eV<\Delta$. It appears that for those junctions which exhibit relatively sharp structure (zero bias values 10-15%), the normalized conductance has a parabolic-like shape near zero bias whereas the broadened BCS dos is flat. This indicates that neither thermal smearing plus leakage nor the phenomenological broadening introduced with Γ accurately describes the experimental data. Nevertheless, the gap parameters obtained with such fitting procedures are expected to be more accurate than obtained with constructs such as using the voltage of the conductance peak. The gap parameters and $2\Delta/kT_c$ values obtained from all the data sets are summarized in Table 1.

Table 1. Experimental Gap Parameters

Material	Δ (meV)	T_c (K)	$2\Delta/k_BT_c$
$Ba_{1-x}K_xBiO_3$	3.6~4.6	21.5~27.5	3.8~3.9
$Nd_{2-x}Ce_xCuO_{4-y}$	3.5~3.7	20~24	3.5~4.1
$Bi_2Sr_2CaCu_2O_x$	16~30	86~96	4.3~7.3
$Tl_2Ba_2CaCu_2O_x$	16~28	90~112	4.1~5.8

PHONON SPECTROSCOPY

The high quality tunneling data observed for BKBO and NCCO coupled with the ability to measure the normal state conductance in these relatively low Tc materials, makes tunneling spectroscopy a possibility. As mentioned in the Introduction, the high bias tunneling data for conventional superconductors such as Pb or Nb, deviate from the BCS dos by a few percent or less at energies which correspond to peaks in the phonon dos. We have shown that the point contact technique reproduces the phonon structures in Nb[9] observed with thin-film junctions. We have observed structures in the high-bias, normalized tunneling conductance data of BKBO and NCCO which are characteristic of phonon effects as seen in conventional superconductors. Using a modified version of the McMillan-Rowell inversion program[4] which allows for a thin proximity layer on the surface of the superconductor, we have inverted the tunneling data to obtain the electron-phonon spectral function, $\alpha^2F(\omega)$.

A comparison is shown in Figs. 5 and 6 between the $\alpha^2F(\omega)$

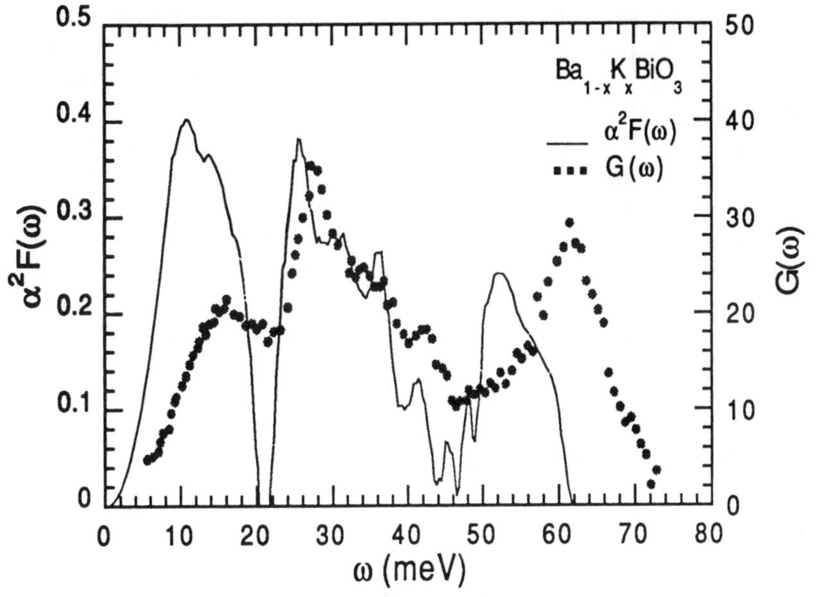

Figure 5. Comparison of $\alpha^2F(\omega)$ for BKBO (solid line) with the phonon G(ω) (solid circles) from neutron scattering.

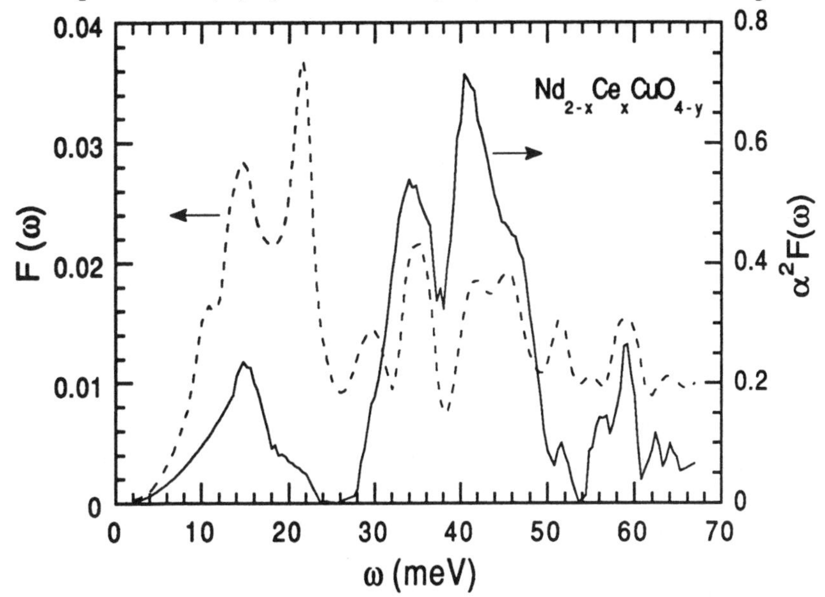

Figure 6. Comparison of $\alpha^2F(\omega)$ for NCCO with F(ω) (dashed line) from neutron scattering.

for BKBO and NCCO and the phonon density of states ($F(\omega)$ or $G(\omega)$) obtained by neutron scattering. For BKBO the phonon $G(\omega)$ is obtained from a polycrystalline pellet[10] and differs from $F(\omega)$ by scattering cross section terms, but the locations of peaks and valleys in the two functions should be similar. As is seen in Fig. 5, there is a good correlation of peaks and especially minima between $G(\omega)$ and $\alpha^2F(\omega)$ in BKBO.

For NCCO, measurements by Reichardt et al[11] on single crystals have yielded dispersion curves from which the $F(\omega)$ is generated. There is a remarkable agreement on the locations of peaks and valleys between $\alpha^2F(\omega)$ and $F(\omega)$ although the shapes of the two curves are different suggesting that α^2 has a strong energy dependence. The parameters, λ and μ^*, for two junctions each of BKBO and NCCO are given in Table 2, as are the calculated and measured T_c values. As is seen there is reasonable agreement between the measured Tc values and those calculated from the $\alpha^2F(\omega)$ functions, indicating that electron pairing is mediated principally by phonons in BKBO and NCCO.

Table 2. Measured and Calculated Parameters

Junction	Δ(meV)	λ	μ^*	ω_{log}(meV)	T_c (K)	T_c^{calc} (K)
BKBO #1	3.8	1.2±0.2	0.11±0.04	25.1	24.5±3	19
BKBO #2	4.5	1.3±0.3	0.04±0.04	13.7	24.5±3	20
NCCO #1	3.5	0.9±0.1	0.05±0.05	19.7	22.0±2	21
NCCO #2	3.6	1.0±0.1	0.08±0.04	25.1	22.0±2	19

PROXIMITY EFFECT STUDIES

The limitation of the point-contact technique is the difficulty in measuring the normal state conductance of a junction due to differential thermal expansion of the mechanical assembly during heating, and this is especially critical for HTS with T_c values above 77K. Thin film junctions on HTS are desireable for spectroscopy and, in addition, have device potential. For these reasons, we have investigated junctions on single crystals of BSCCO which have been cleaved in vacuum and immediately coated with a thin layer of Au to protect the surface. An important question is whether the underlying energy gap of the BSCCO is observable through the Au layer.

In Fig. 7 is shown a selected set of tunneling data for various Au thicknesses in the range 200-600 Å. Here a soft In tip was used so as not to perforate the Au layer and also the In-oxide surface of the tip provides a reasonably good tunnel barrier.

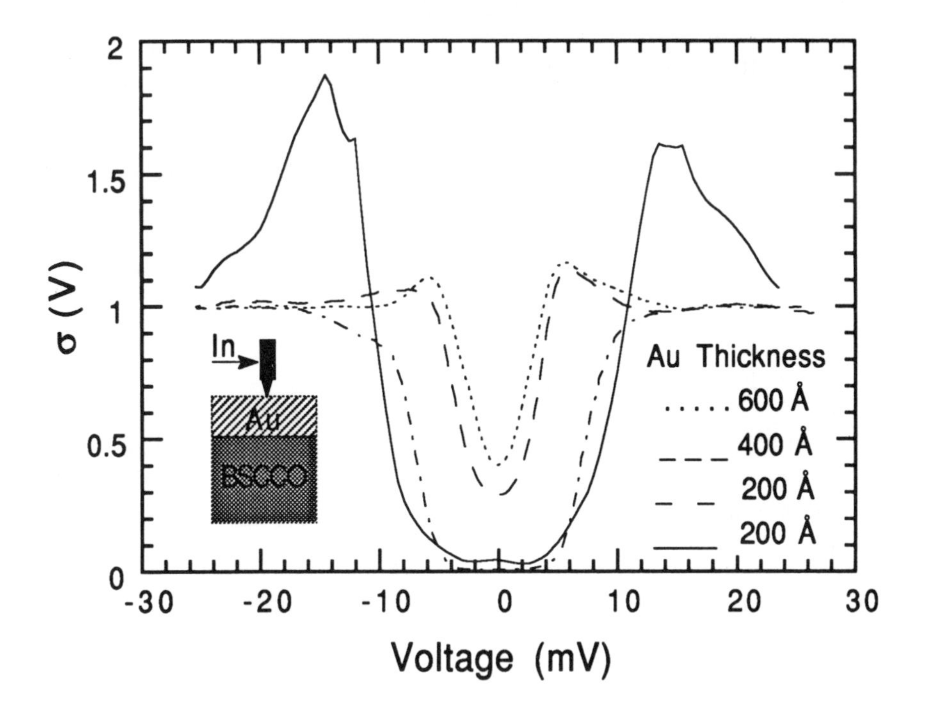

Figure 7. Normalized tunneling conductance for Au/BSCCO bilayers with various Au thicknesses. The tunneling geometry is shown in the inset.

The observed gap parameter of the Au/BSCCO bilayers decreases in a monotonic fashion with Au thickness, and even the 600 Å layers typically show a gap parameter of ~ 3 meV. This behavior can be explained using the proximity effect theory of Arnold.[1] Furthermore, two junctions with Au thicknesses of 200 Å showed a zero-bias dos close to zero, in contrast to point contact junctions formed directly on the BSCCO surface which always displayed at least 10-15% zero-bias dos. This indicates that the proximity effect approach has potential as a technique for fabricating thin-film junctions with low sub-gap conductance which is necessary for device applications.

ACKNOWLEDGEMENTS

We thank D.G. Hinks for providing the BKBO samples, J.Z. Liu for the BSCCO samples, D.M. Ginsberg for the TBCCO samples and R.L. Greene and J.L. Peng for the NCCO samples.

References

1. E.L. Wolf, Principles of Electron Tunneling Spectroscopy Ch. 2-5 (Oxford Univ. Press, New York, 1985).

2. Qiang Huang, J.F. Zasadzinski, K.E. Gray, D.R. Richards and D.G. Hinks, Appl. Phys. Lett. **57**, 2356 (1990).

3. Qiang Huang, J.F. Zasadzinski, K.E. Gray, J.Z. Liu, and H. Claus, Phys. Rev. **B40**, 9366-9369 (1989).

4. Q. Huang et al, Nature **347**, 369 (1990)

5. T. Ekino, and J. Akimitsu, Phys. Rev. **B40**, 7364-7367 (1989).

6. See for example, N. Miyakawa, D. Shimada, T. Kido and N. Tsuda, J. Phys. Soc. Jpn. **58**, 383 (1989)

7. I. Takeuchi, J.S. Tsai, Y. Shimakawa, T. Manako and Y. Kubo, Physica C **158**, 83 (1989).

8. Qiang Huang, J.F. Zasadzinski, K.E. Gray, E.D. Bukowski, and D.M. Ginsberg, Physica **C161**, 141-144 (1989).

9. Qiang Huang, J.F. Zasadzinski, and K.E. Gray Phys. Rev. **B42**, 7953 (1990).

10. C.K. Loong, P. Vashishta, R.K. Kalia, M.H. Degani, D.L. Price, J.D. Jorgensen, D.G. Hinks, B. Dabrowski, A.W. Mitchell, D.R. Richards and Y. Zheng, Phys.Rev. Letters **62**, 2628-2631 (1989).

11. W. Reichardt et al, Proceedings of the Conference on the Manifestations of Electron-Phonon Interaction in High Temperature Superconductors, Oaxtepec, Mexico, 1990.

LOCAL ENVIRONMENT AROUND THE OXYGEN ATOMS IN Y-Ba-Cu-O STUDIED BY X-RAY ABSORPTION FINE STRUCTURE (XAFS) SPECTROSCOPY

A. Krol, Y. L. Soo, S. W. Huang, Z. H. Ming, and Y. H. Kao
Department of Physics, SUNY at Buffalo, NY 14260, USA
N. Nücker, G. Roth, and J. Fink
Kernforschungszentrum Karlsruhe, Institut für NukleareFestkörperpysik
Postfach 3640, W-7500 Karlsruhe, Federal Republic of Germany
G. C. Smith
Brookhaven National Laboratory, Upton, NY 11973, USA

ABSTRACT

The local environment about oxygen atoms in the CuO_2 plane in $YBa_2Cu_3O_7$ single crystals has been investigated by orientation-dependent x-ray absorption fine structure (XAFS) spectroscopy at the O 1s edge using bulk-sensitive fluorescence-yield-detection method. Within experimental accuracy (± 0.03 Å) no structural change in local ordering about oxygen atoms in the CuO_2 plane at 90 K has been detected.

INTRODUCTION

Orientation dependent x-ray absorption fine structure (XAFS) spectroscopy at the O 1s edge is a useful tool to investigate local ordering around oxygen atoms. This method was applied to the $YBa_2Cu_3O_7$ (Y123) superconductors. Basic blocks of these systems are the CuO_2 planes, formed by Cu(2), O(2) and O(3) atoms, parallel to the ***a****- and* ***b***-axis, and the CuO chains along the ***b***-direction, formed by Cu(1), O(1), and the apex O(4) atoms. The O(2)-Cu(2) bond is oriented along ***a***-axis. By chosing the orientation of the x-ray electric $\boldsymbol{E}$-vector parallel to the $\boldsymbol{a}$-crystallographic axis of the single crystal one can selectivly probe bonds formed by O(2) atoms with its Cu(2) neighbor in the CuO_2 plane.

EXPERIMENTAL

Single crystals of Y123 were grown in a ZrO_2 crucible and oxidized to show a sharp superconducting transition at 91 K. They were heated in flowing oxygen to 400 °C and subjected to uniaxial stress while slowly cooling them down to form twin free samples.[1] The samples were glued to the sample holder using silver paint and oriented to permit measurements with the $\boldsymbol{E}$ vector of the x-rays parallel to the $\boldsymbol{a}$-axis of a single crystal. They were then mounted in a vacuum chamber near the soft-x-ray fluorescence detector. The soft-x-ray absorption at the O 1s edge was measured at the U 15 beam line at the NSLS at Brookhaven National Laboratory. During the experiment, the monochromator resolution was set at ~ 1.5 eV. The oxygen K_α fluorescence yield was monitored by means of a low pressure parallel plate avalanche chamber.[2] All measurements were performed in an energy range $500 \leq \hbar\omega \leq 780$ eV. The upper limit of the XAFS spectra was restricted by the interfering presence of the Ba M_V edge thus limiting the accessible energy- ergo k-range.

RESULTS

The spectra were first corrected for the thickess effect[3] and then normalized to the photoabsorption edge jump by standard procedures[4] using tables published by Henke et al.[5] Nonlinearity of the beam monitor which consisted of a mesh coverd with P46-Y2 phosphor was accounted for using data published by Yang et al.[6] A corrected spectrum obtained for $\boldsymbol{E}//\boldsymbol{a}$ at 90 K together with a cubic spline used to obtain XAFS is shown in Fig. 1. The XAFS interference function is defined as

$$\chi(\mathrm{k}) = \frac{\mu(\mathrm{k}) - \mu_{\mathrm{b}}(\mathrm{k})}{\mu(\mathrm{k_i})} \times \frac{\mu_{\mathrm{O}}(\mathrm{k_0})}{\mu_{\mathrm{O}}(\mathrm{k})} \tag{1}$$

Where k is the wavevector of the emitted photoelectron, k_0 and k_i are its values at the O 1s edge (assumed 532 eV) and at $E_i = 570$ eV. $\mu(k)$ is the measured absorption coefficient, $\mu_b(k)$ is simulated by cubic spline atomic-like absorption above the O 1s edge and $\mu_O(k)$ is the calculated photo-absorption due to oxygen only.[5] The obtained $\chi(k)$ function is shown in

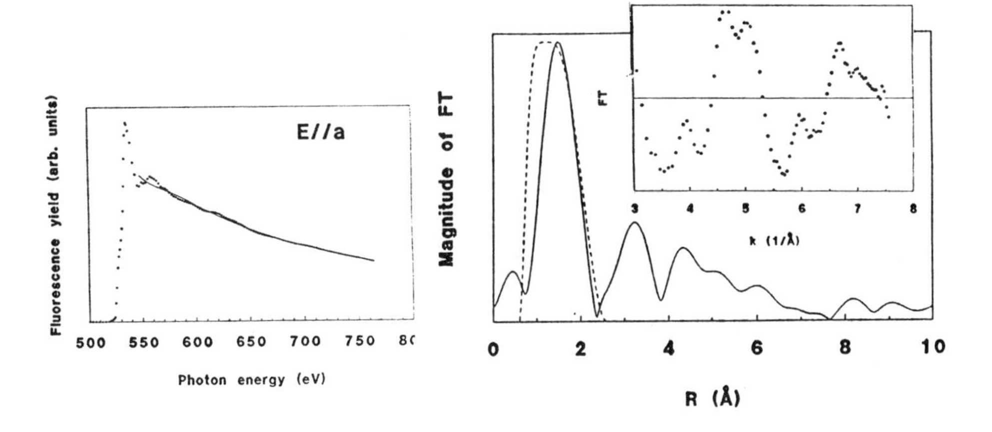

Fig. 1. Corrected FY spectrum transform obtained for ***E*//*a*** at 90 K: circles. $\mu_b(E)$: solid line.

Fig. 2. XAFS spectrum and magnitude of its Fourier transform. The window used to backtransform is also shown.

Fig. 2 together with the magnitude of its Fourier transform. Due to selection rules only some selected neighbors of the central oxygen atom contribute to the O 1s XAFS signal, see Table I. The effective coordination number N_i^* describes the contribution of the i-th shell to the XAFS signal and is given by

$$N_i^* = 3 \sum_{j=1}^{N_i} \cos^2\alpha_{ij} \tag{2}$$

The index j denotes the individual atoms in the i-th shell. α_{ij} is the angle between the vector $\mathbf{r}_{ij}$ from the central atom to the j-th atom in the i-th shell and the electric field vector $\boldsymbol{E}$ of the x-rays.

Table I. Effective contribution to XAFS at the O 1s in Y 123 for ***E//a***.
Crystal structure parameters were obtained by neutron diffraction.[7]

Absorbing atom	Backscattering atom	Relative disctance (Å)	Effective coordination N_i^*
O(2)	Cu(2)	1.93	11.84
O(3)	Y	2.38	7.8
O(2)	O(3)	2.73	11.76
O(3)	O(2)	2.73	11.76
O(4)	Ba	2.74	11.68
O(1)	Ba	2.88	5.28
O(3)	Ba	2.97	4.96
O(2)	O(4)	3.20	4.28
O(4)	O(2)	3.20	4.28

Based on the data gathered in the Table I the dominant peak in the magnitude of Fourier transform, between 0.8 and 2.4 Å is due to XAFS interference from O(2) – Cu(2), O(3) – Y and O(2,3) – O(3,2) pairs. The incorporation of Ba shells to the model XAFS function gave rise to a very poor fit. This means that contribution from O-Ba pairs to the considered peak is neglegible. In other words in this orientation the dominant contribution to the XAFS is given by neighbors of oxygen atoms in the CuO_2 plane (see also Figs. 3 and 4). In order to spectrally isolate the first three shells the XAFS spectrum was backtransformed using window shown in Fig. 2. The resulting $\chi_b(k)$ function is shown in Fig. 3 as open circles. A theoretical XAFS was then fitted and the result of this procedure is represented by solid line. Amplitude and phase functions were obtained from tables published by McKale et al.[8] with exception of the phase function for O-Cu pair which was taken from the table published by Stöhr.[9] Contributions due to individual shells are also shown in the same figure. In Fig. 4 the quality of the obtained fit is veryfied by comparison of the experimental magnitude and real part of Fourier transform together with the respective transforms of the fit. Contribution from individual shells are also shown. Parameters of the three shell model used to obtain the best fit are collected in Table II.

Table II. Parameters of the best least-squares curve-fit of the three shell model.

O - X pair	R(Å) fitted	N_i* fitted	$\sigma^2(10^{-3}$ Å$^2)$	σ^2 from references [a]	[b]	[c]	[d]
O(2)-Cu(2)	1.90 ± 0.03	13.6 ± 2.0	6.5 ± 2	2.0	6.5	6.3	1.7
O(3)-Y	2.36 ± 0.03	9.0 ± 1.4	2.0 ± 0.7	5.0	7.0	8.5	–
O(2,3)-O(3,2)	2.70 ± 0.03	27.0 ± 4	17.9 ± 5	–	15.0	–	–

a. From Ref. [10], polycrystalline samples at the Cu and Y K edges at 77 K.

b. From Ref. [11], polycrystalline samples at the O K edge at 110 K.

c. From Ref. [12], polycrystalline samples doped with Zn at the Cu and Y K edges at 300 K.

d. From Ref. [13], polycrystalline samples at the Cu K edge at 20 K.

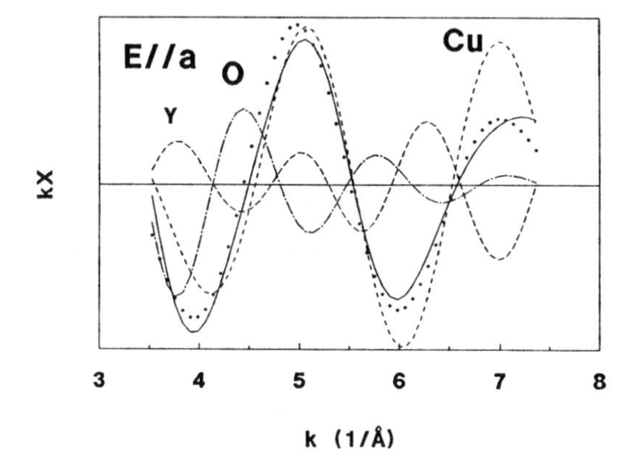

Fig. 3. Filtered XAFS spectrum (open circles) vs. fit (solid line). Contributions form different shells are also shown.

At this point we would like to note that the lenght of bonds is practically unchanged within error limits ($\sim \pm 0.03$Å) as compared to crystallographic data at room temperature. This was also observed by Gurman et al.[10] and Maruyama et al.[13] in polycrystalline samples. The amplitude reduction factor S_0^2 for small k ($3.5 \leq k \leq 8$ Å^{-1}) due to oxygen shake-up and shake-off

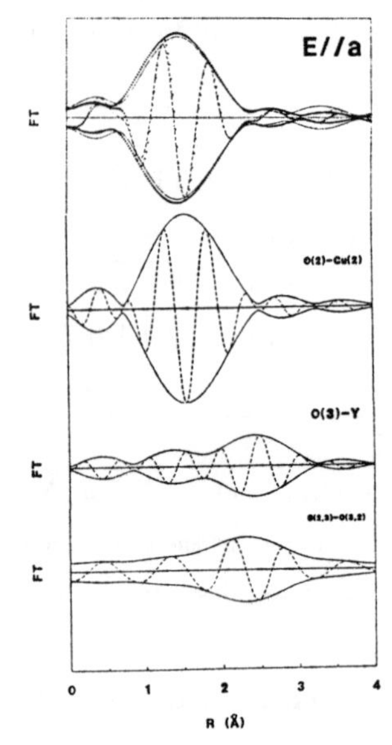

Fig. 4. Magnitude (open circles) and real part (crosses) of the filtered Fourier transform shown in Fig. 3 vs fit (solid line and dashed line, respectively). Contributions from different shells are also shown.

effects can be estimated as equal to 0.87 ± 0.15 which is close to the expected value.[14] The obtained Debay-Waller factors are close to results obtained for polycrystals. However, contrary to earlier investigations, due to application of orientation dependent XAFS to single Y123 crystal the data on $O(2) - Cu(2)$ bond are obtained in a direct way.

CONCLUSIONS

In summary, the O 1s XAFS spectroscopy at 90 K was applied to a single twin free crystal of $YBa_2Cu_3O_7$ high-T_c superconductor with the x-ray electric field ***E***-vector parallel to its ***a***-axis. In this orientation mainly neighbors of oxygen in the CuO_2 plane and specifically the $O(2) - Cu(2)$ bonds are probed. ***Within experimental accuracy ($\pm$ 0.03 Å) no structural***

change in local ordering about oxygen O(2) atoms in the CuO_2 plane at 90 K has been detected.

The work by the Buffalo group and at Brookhaven is supported by DOE.

REFERENCES

1. A. Zibold et al. Physica C **171**, 151 (1990). U. Welp et al. Physica C**161**, 1 (1989).
2. G. C. Smith, A. Krol and Y. H. Kao, Nucl. Instr. Meth. A **291**, 135 (1990).
3. P. A. Lee et al. , Rev. Modern Phys. **53**, 769 (1981).
4. D. E. Sayers and B. A. Bunker in *X-Ray Absorption* ed. D. C. Koningsberger and R. Prins, (Wiley 1988), p. 211.
5. B. L. Henke et al., Atomic Data Nucl. Data Tables **27**, 3 (1982).
6. B. X. Yang et al. Nucl. Instr. Meth. A**258**, 141 (1987).
7. J. E. Greedan, A.H. O'Reilly, and C. V. Stager, Phys. Rev. B **35**, 8770 (1987).
8. A. G. McKale et al., J. Amer. Chem. Soc. **110**, 3763 (1988).
9. J. Stöhr in *X-Ray Absorption* ed. D. C. Koningsberger and R. Prins, (Wiley 1988), p. 663.
10. S. J. Gurman et al., J. Phys. C **21**, L475 (1988).
11. L.Tröger et al., Solid St. Comm. **79**, 479 (1991).
12. H. Maeda et al., Physica C **157**, 483 (1989).
13. H. Maruyama et al., Physica C **160**, 524 (1989).
14. T. A. Carlson et al., Phys. Rev. **169**, 7 (1968).

LOW-FIELD IRREVERSIBILITY LINE IN BULK YBCO HIGH-T_c SUPERCONDUCTORS

C.Y. Lee, L.W. Song and Y.H. Kao
Department of Physics and Astronomy
State University of New York at Buffalo, Amherst, NY 14260

ABSTRACT

From the zero-field-cooled and field-cooled ac magnetic susceptibility measurements of bulk ceramic $YBa_2Cu_3O_y$ (YBCO) high-T_c superconductors, a low-field irreversibility line has been found which lies very close to $H_{c1}(T)$ in the H-T plane as a result of strong flux trapping. Although both the real and imaginary parts of the ac magnetic susceptibility in these ceramic samples are very sensitive to the ac field amplitude, no significant irreversible behavior could be observed in the absence of a dc magnetic field. By measuring the ac susceptibilities at constant temperature as a function of dc and ac fields, our results exhibit clear evidence for the existence of a field near H_{c1} that demarcates the transition from a reversible to an irreversible (hysteretic) behavior.

After the earlier work of Bednorz and Müller[1] on high-T_c superconductors, Yeshurun and Malozemoff[2] made an important observation and demonstrated that there exists an "irreversibility line" which can be expressed as

$$(1 - t) \alpha H^{\frac{2}{3}}$$

where $t = T/T_c$. This line demarcates two regions in the H-T plane corresponding to the respective reversible and irreversible magnetic behavior, such as that usually found in the dc magnetization measurements using SQUID. Above this line in the H-T plane, J_c becomes immeasurably small, and the flux lines are believed to be depinned as a result of thermal activation. On this basis, Malozemoff et al.[3] argued that the upper critical field $H_{c2}(T)$ should occur well above this irreversibility line. The physical significance of this irreversibility line was further discussed by Tinkham[4]. However, most of the previous results related to the irreversibility line were found in dc magnetization experiments while little has been reported on a similar study using ac techniques.

Inasmuch as there exists an irreversibility line in the high field region caused by thermally activated *depinning* of the flux lines, it is naturally conceivable that there should also exist another irreversibility line in the low field region corresponding to the onset of flux *pinning*. We shall call the latter a *low field irreversibility line* (LFIL) in contrast to the (high field) irreversibility line (HFIL) observed by Yeshurun and Malozemoff. By following a path in the H-T plane starting from the reversible Meissner state to the flux-flow region, the LFIL marks the crossover from reversible to irreversible behavior whereas the HFIL designates a transition from irreversible to reversible behavior.

In this note we report an investigation of the LFIL in bulk YBCO samples by means of ac magnetic susceptibility measurements. The samples used in this experiment were prepared by a solid state reaction technique. The samples were shaped into cylinders of approximately 4.3 mm in diameter and 16 mm long. With the sample axis parallel to the magnetic fields, the demagnetization factor was calculated to be ≈0.08. The average grain size determined by SEM was around 5-10 μm in all the samples studied.

Our ac susceptibility measurement system mainly consists of three coaxial cylindrical coils. Over the innermost primary coil of 250 turns, the secondary (pickup) coil of 1000 turns was wound and balanced to minimize the background signal at 77K without the sample. The outermost coil generates the dc magnetic field. The homogeneity of the longitudinal magnetic field in the sample has been calibrated experimentally with a Hall field probe, the result shows ≈±1% deviation at the ends of the sample with respect to the field at its center. Transverse field inhomogeneity was calculated from the coil geometry to be of the order of 0.1% across the sample area. All the coils were immersed in liquid nitrogen for higher stability. The entire cryostat was triply shielded by fully annealed μ-metal, attenuating the residual field at the sample position to ~1 mOe. The pickup coil is divided into two sections with two of the four terminals center-tapped. The other two ends were fed into a two-phase lock-in amplifier operated in differential mode. The phase angle of the reference signal was adjusted to nullify the out-of-phase susceptibility at 77 K with a small ac field (0.01 Oe). From the frequency dependence of ac susceptibilities, the eddy current loss was observable only for frequencies higher than 50 kHz.

Figure 1 shows ac susceptibilities measured at 2 kHz for several ac field amplitudes with no dc field superimposed. As it is well known, the ac susceptibilities are very sensitive to the ac field amplitude h_0 and the imaginary part normally shows two peaks corresponding to the inter- and intra-granular losses in the ceramic material when h_0 is sufficiently large (>0.1 Oe in the present case) to probe these regions. As h_0 is increased, the peak occurring at high temperature near T_c (termed P_1) shows a slight shift to the left of the figure while the peak at low temperature (termed P_2) exhibits a large shift in the same direction. The two nearly retraceable curves measured with $h_0 > 1$ Oe represent our data obtained under field-cooled (FC) and zero-field-cooled (ZFC) conditions. It can be seen that an ac magnetic field alone does not cause significant irreversible behavior. The small hysteresis shown in this figure can be attributed to weak flux trapping between superconducting grains when the external magnetic field exceeds the Josephson critical field in the intergranular region (~ 1 Oe).

Figure 2 shows FC and ZFC ac susceptibilities with a dc field of 30 Oe superimposed on the ac field of 0.01 Oe at 2 kHz. The susceptibilities become highly hysteretic in the presence of a dc field, and the onset of deviation between FC and ZFC traces occurs around P_1. The location of P_1 is believed to be determined mainly by losses in the intragranular region when the magnetic field fully penetrates the superconducting grains[5] in accordance with the critical state model[6]. This result therefore indicates that below a threshold field which penetrates into the superconducting grains, the observed hysteretic behavior is mainly caused by the penetration and trapping of flux (Josephson vortices) between grains. The magnetization of the intergranular region in a ceramic superconductor is shown to depend on its history.

On the basis of this observation, it can therefore be expected that when the dc magnetic field is increased beyond a threshold value H^*, which corresponds to the onset of field penetration into the superconducting grains, an irreversible behavior should appear and this will define the low field irreversibility line in the H-T plane.

Figure 3 shows the anticipated dc field dependence of ac susceptibilities measured at 88.9 K by using a 10 kHz ac field of 0.03 Oe with various ranges of dc field sweep. The sample was first heated to above T_c and then zero-field-cooled in order to eliminate any remnant fluxoids. It is shown from this plot that there indeed exists a threshold field H^* above which the susceptibility does not retrace itself in a reducing magnetic field. These results can be interpreted as follows. When $H < H^*$, magnetic field penetrates into the intergranular region and create Josephson vortices while the superconducting grains are shielded by a screening supercurrent along the boundaries. When $H > H^*$, the screening supercurrent is insufficient to shield the interior of grains and the field enters in the form of Abrikosov vortices, thereby causing the observed hysteresis as a function of dc field shown in Fig. 3. In general, there exists a surface barrier for flux entry[7], hence the interpretation of H^* can be complicated. It should be noted, however, that flux entry can occur more easily in granular superconductors because the surface roughness of the grains can lead to a local field enhancement in convex regions, thus the barrier is effectively lowered there[8]. In this case, H^* should be very close to the average lower critical field H_{c1G} pertaining to the grains. By measuring the temperature dependence of H^* with several samples and by comparison with the values of H_{c1} determined for single crystals[9-10], H^* has been found to be very close to H_{c1} for bulk YBCO[11].

In summary, the existence of a low field irreversibility line in high-T_c superconductors is ascertained by measurements of ac susceptibility as a function of dc magnetic field. Below a threshold dc field H^*, the magnetic field can penetrate into the intergranular region in the form of Josephson vortices but not into the interior of grains, the magnetic behavior of the bulk superconductor remains largely reversible. Above H^*, however, Abrikosov vortices are present inside the grains, flux pinning and energy losses in the grains are responsible for the observed irreversibility. By measuring the FC and ZFC ac susceptibilities as a function of dc magnetic field and temperature, the threshold field H^* for irreversibility was found to occur at the average lower critical field H_{c1} for the grains.

This research is supported in part by DOE.

REFERENCES

1. K.A. Müller, M. Takashige, and J.G. Bednorz, Phys. Rev. Lett. **58**, 1143 (1987).
2. Y. Yeshurun and A.P. Malozemoff, Phys. Rev. Lett. **60**, 2202 (1988).
3. A.P. Malozemoff et al., Phys. Rev. B **38**, 7203 (1988).
4. M. Tinkham, Phys. Rev. Lett. **61**, 1658 (1988).
5. Küpfer et al., Cryogenics **28**, 650 (1988).
6. C.P. Bean, Rev. Mod. Phys. **36**, 31 (1964).
7. C.P. Bean and J.D. Livingston, Phys. Rev. Lett. **12**, 14 (1964).
8. T. Van Duzer and C.W. Turner, *Principles of Superconductive Circuits and Devices*, pp. 342-345, New York, 1981.
9. S. Sridhar, Dong-Ho Wu, and W. Kennedy, Phys. Rev. Lett. **63**, 1873 (1989).
10. D.R. Harshman *et al.*, Phys. Rev. B **36**, 2386 (1987).
11. C.Y. Lee, L.W. Song, and Y.H. Kao, preprint.

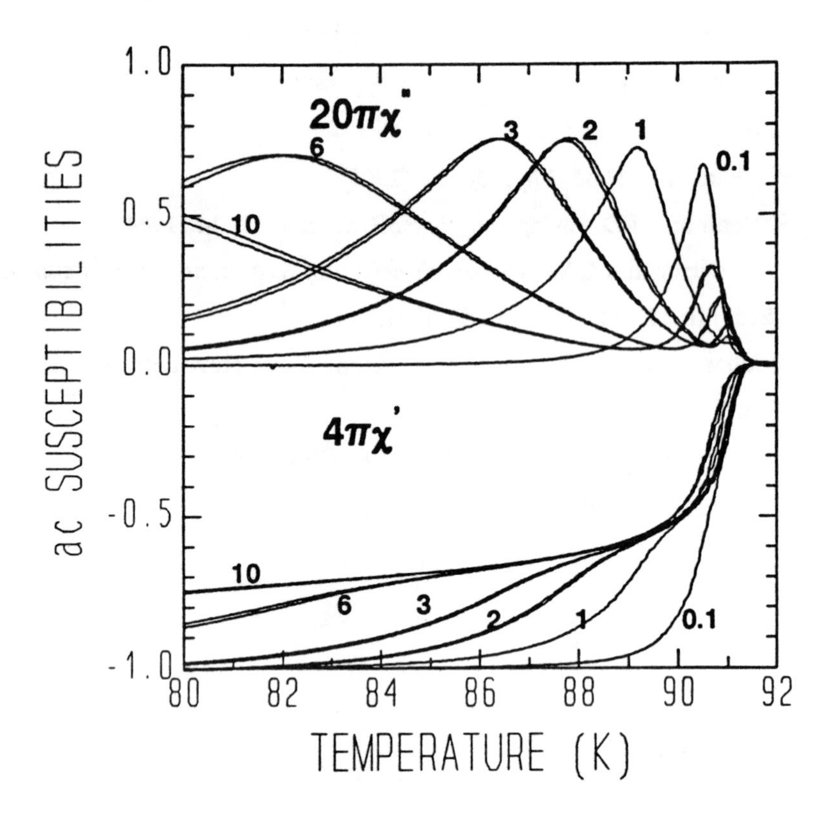

Fig.1 ac susceptibilities with various ac field amplitudes of 2 kHz.

No dc field was applied. ac field amplitudes are as labeled.

All the quantities are in Gaussian units.

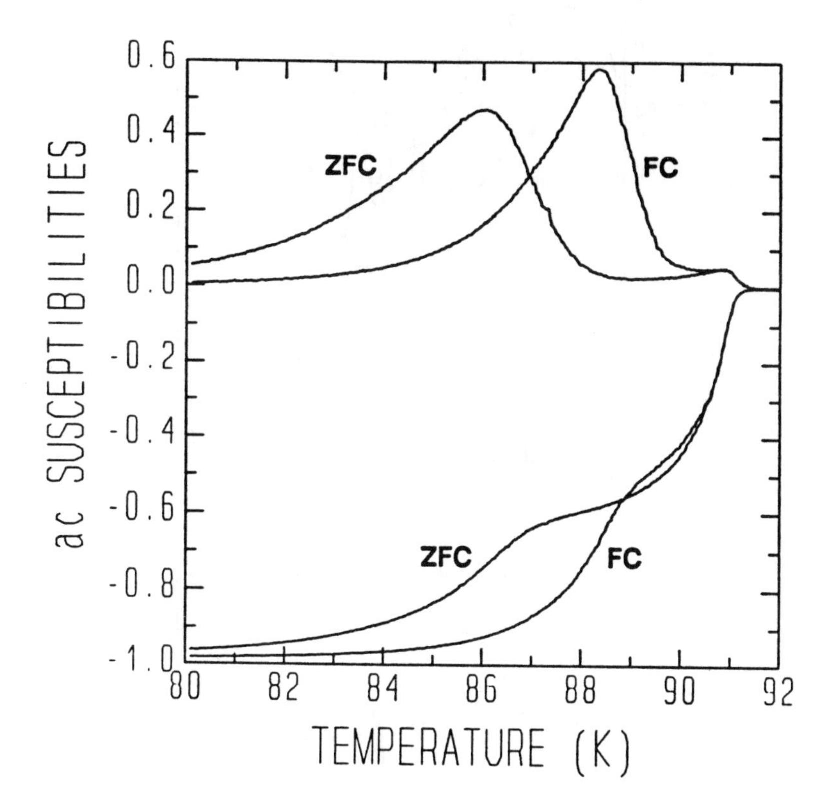

Fig.2 Field-cooled (FC) and zero-field-cooled (ZFC) ac susceptibilities by using 2 kHz ac field of 0.03 Oe with dc field of 30 Oe superimposed on it. The y-axis represents $4\pi\chi'$ and $20\pi\chi''$ for in-phase (the lower part) and out-of-phase (the upper part), respectively.

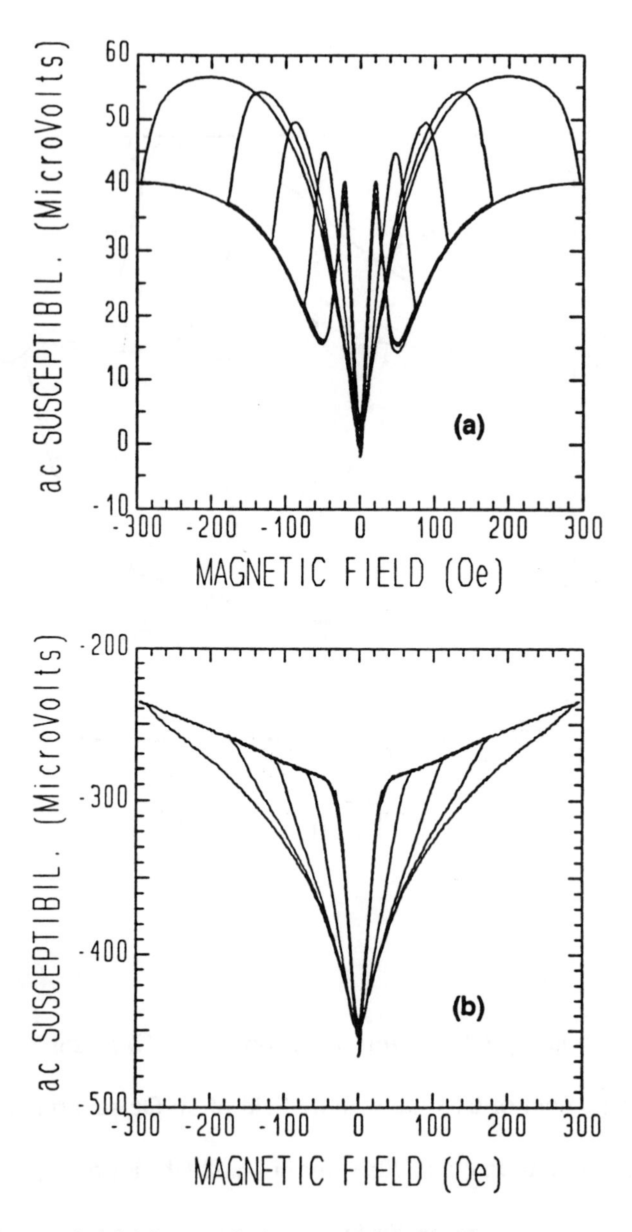

Fig. 3 Hysteresis of (a) out-of-phase and (b) in-phase ac susceptibilities with various dc field sweep ranges by using 10 kHz ac field of 0.03 Oe. Temperature is 88.9 K. Beyond a certain threshold value (around 30 Oe at this temperature), the behavior becomes hysteretic.

LOWER CRITICAL FIELD DETERMINATION IN THICK RODS OF HIGH-Jc MELT-PROCESSED YBCO

M. D. Sumption, and E. W. Collings
Battelle Memorial Institute,
Columbus, OH 43201

M. V. Parish
CPS Superconductor Corporation
Milford, MA 01757

Wang Jingrong
Northwest Institute for Nonferrous Metal Research
Baoji, China

W. J. Carr Jr.
Pittsburgh, PA 15235

ABSTRACT

The direct method of H_{c1} determination requires the detection of the first deviation from linearity of the initial magnetization curve. But in thick high-J_c samples such deviations from M *versus* H linearity turn out to be barely perceptible. They can, however, be detected if the sample radius, R_0, is relatively small, but then field penetration may influence the result. Attempts to observe small deviations from linearity of a few data points above H_{c1} will always be fraught with uncertainty. Accurate methods of H_{c1} determination should utilize an extensive set of data above that field in association with a theoretical expression to enable H_{c1} to be extracted from it by means of some mathematical fitting procedure. Two examples of the application of this philosophy to data sets representing the field-sweep-amplitude (H_m) dependence of hysteretic loss (Q_h), are presented. For *field-transverse* cylinders, the recently derived expression

$$Q_h = (10/3\pi^2 J_c R_0)\{(3/2)h_{c1}(H_m-h_{c1})^2 + (H_m-h_{c1})^3\}$$

is employed, in which h_{c1} is the *applied* field corresponding to the sample's actual H_{c1} (= $2h_{c1}$). Values of H_{c1} obtained for three samples of melt-processed YBCO at 77 K are: $DC1_{AB}$ (first method), 192 Oe, $DC1_{AB}$ (second method), 225 Oe; CPS3 and CPS4 (second method), 163 and 162 Oe, respectively. Both differences and similarities in superconducting property are to be expected since the DC1 and the CPS samples were made using slightly differing variants of the float-zone melting procedure, while the two batches of CPS fiber had been prepared by similar processess.

INTRODUCTION

In relatively thick superconducting wires, attempts to measure the lower critical field, H_{c1}, in terms of the first deviation from linearity of the initial magnetization curve tend to give uncertain results. For this and other reasons a method has been developed for obtaining H_{c1} from measurements of hysteresis loss. The method involves fitting the measured hysteresis loss data to a derived function of H_{c1}, the field-sweep amplitude (H_m), and some multiplier, $A(J_c,R_0)$. The lower limit of the H_m range is not particularly important, since little weight attaches to the very low field data; but the upper limit should be small compared to the penetration field, H*. The first task was to derive an expression to which the data could be fitted. In the Bean model it is well known that in the range of partial penetration the loss is proportional to H_m^3. Next, in the simple case of a slab[1], the effect of adding a Meissner current to the Bean model is to change the cubic term to $A(H_m-H_{c1})^3$, and to introduce a quadratic term, $A(3/2)H_{c1}(H_m-H_{c1})^2$. However a problem of great interest is the case of a *transverse* field applied to a *cylindrical* wire. It is known that an exact analytical expression for the partial-penetration loss cannot be obtained in this case, even in the pure Bean model. Nevertheless we have shown that to a reasonable approximation the general form of the slab expression is preserved (but with a numerically different prefactor), provided H_{c1} is corrected for demagnetization. Examples of the application of the new equation to melt-processed $YBa_2Cu_3O_x$ (YBCO) are presented.

EXPERIMENTAL

Measurements

The experimental methods used to study lower critical field, H_{c1}, were based on vibrating sample magnetometry. Samples typically 6 mm long were exposed to the transverse field of an electromagnet energized with a bipolar power supply by means of which the field could be cycled from about ±10 Oe to ±16 kOe. Prior to starting any series of magnetization (M) *versus* field (H) measurements, the sample was always cooled to the operating temperature (77 K in this case) in zero field. Two sets of measurements were made: (1) the initial magnetization curve of a virgin sample; (2) a series of hysteresis loops taken at field-amplitude (H_m) increments of about 15 Oe. With increasing H_m demagnetization was not necessary between steps.

Materials

Melt-processed YBCO samples were prepared and supplied by the Northwest Institute for Nonferrous Metals (NIN), Baoji, China, and CPS Superconductor (CPSS), Milford, MA.

NIN Material: The NIN sample, herein designated $DC1_{AB}$, had been prepared by applying the vertical floating zone technique to a cold-pressed rod of mixed Y_2BaCuO_5 (211 phase) and BaCuO powders[2]. It was subsequently shaped by hand to a crude cylinder (effective diam. 8.97×10^{-2}cm, length 0.685 cm).

CPSS Material: Two samples were prepared from CPSS material: (1) CPS3, a pair of monofilaments (each of diam. 3.60×10^{-2}cm, total length 2.053 cm); (2) CPS4, a bundle of five filaments (each of diam. 1.91×10^{-2}cm, bundle length 0.561 cm). In the CPSS process[3] a mixture of YBCO powder and thermoplastic binder is melt spun into a fiber which is sintered, transported through a short-hot-zone furnace, and slowly cooled in a thermal gradient thereby causing the growth of long, axis-orientated grains. The fiber was then Ag coated, re-sintered, and oxygen annealed to form the superconductor CPS3. To form the filament-bundle CPS4, five Ag coated fibers were imbedded in excess Ag slurry, sintered, and oxygen annealed.

THE LOWER CRITICAL FIELD -- INITIAL MAGNETIZATION METHODS

The Direct Method

Consider, for the purpose of illustration, the volume magnetization, M, of an infinite cylinder of radius R_0, in an axis-parallel magnetic field, H (in our actual experiments, the transverse orientation is employed). The sample is presumed to have been cooled in zero field. Then as H is increased, the initial Meissner state, within which $M = (-1/4\pi)H$, gives way at $H = H_{c1}$, to the Bean critical state[4], indicated in Fig. 1 by primed characters, within which

$$M' = (-1/4\pi)\{H'-H'(H'/H^{*\prime}) + (H'/3)(H'/H^{*\prime})^2\} \qquad (1)$$

Here $H^{*\prime}$ is the "penetration field", which in parallel-field geometry is equal to $(4\pi/10)J_cR_0$. The corresponding magnetization is $M^{*\prime}$.

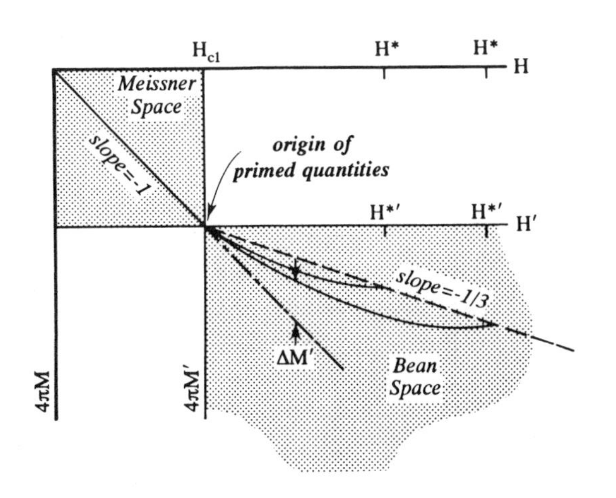

Fig. 1
Schematic representation of the variation of M with H within the "Meissner Space" and the "Bean Space" (i.e the Bean critical state[4]).

The magnetization enters the "Bean Space" still with the Meissner slope, -1/4π. M′ attains zero slope just at the penetration point $P^* \equiv (H^{*\prime}, M^{*\prime})$. It turns out that the locus of such points is a straight line of slope -1/12π. Since $H^{*\prime}$ is proportional to J_cR_0, P^* shifts to higher and higher fields as J_c and/or R_0 get larger. Thus for thick high-J_c cylinders it is difficult to discern in M(H) any departure from linearity as H passes through H_{c1}. This can be illustrated in the following way:

At fields beyond H_{c1}, the deviation of M′ from a linear extrapolation of the Meissner line is $\Delta M' = (-1/4\pi)H' - M'$ which, according to Eqn. (1), is approximately $(-1/4\pi)H'(H'/H^{*\prime})$. The fractional deviation is thus

$$\Delta M'/M = H'^2/HH^{*\prime}$$

$$= H'^2/\{H(4\pi/10)J_cR_0\} \qquad (2)$$

after the substitution $H^{*\prime} = (4\pi/10)J_cR_0$ has been made.

For a sample with $R_0 = 2\text{x}10^{-2}$ cm, $J_c = 10^5$ A/cm^2, and H_{c1} = 100 Oe, when H has risen to 200 Oe (H′ = 100 Oe) the fractional deviation of the magnetization from linearity is still only 2%. Indeed, as Fig. 2 amply illustrates, a departure of M(H) from linearity somewhere between 100 and 200 Oe is barely detectable.

Equation (2) suggests that for a material with a given J_c, successful application of the direct M(H) method would be possible only as $R_0 \rightarrow 0$. However at small sizes, such that R_0 becomes comparable with the magnetic penetration depth, λ, a new complication enters the picture. As discussed elsewhere[5], for small-radius

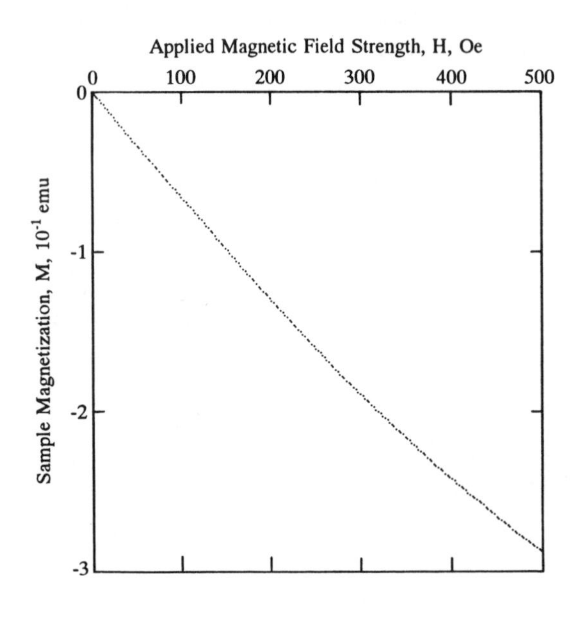

Fig. 2
Initial magnetization of sample $DC1_{AB}$ (vol., $4.325\text{x}10^{-3}\text{cm}^3$) at 77 K. h_{c1} is about 100 Oe.

field-parallel cylinders, the apparent H_{c1} becomes enhanced by a factor $1/\sqrt{\{1-(1/\zeta)\tanh\zeta\}}$, where $\zeta = R_0/\lambda$. Estimation of H_{c1} then requires a separate determination of λ.

The Amplified Deviation Method

The deviation from linearity can be amplified and so better detected by adopting a procedure first explained by Umezawa *et al.*[6]. In the Umezawa method the deviation of M(H) from a numerical extrapolation of the very low field initial magnetization line is amplified as large as possible and plotted *versus* H. As illustrated in Fig. 3, again for $DC1_{AB}$ at 77 K, this method allows linearity-departure to be more readily detected. Nevertheless, H_{c1} may still be only crudely estimable by this technique.

The difficulty with the initial magnetization methods lies in the fact that in both the Direct Method and the Umezawa Method attention is focused on the changes of M or ΔM that take place just at H_{c1} itself. Little if any of the information contained in M(H) data well beyond H_{c1} is being utilized. In the methods to be described, this deficiency is corrected. The functional form of a data set extending to well beyond H_{c1} is exploited to extract a value for H_{c1} itself.

THE LOWER CRITICAL FIELD -- HYSTERETIC METHODS

Introduction

In principle, when an AC modulation field of fixed amplitude is superimposed on a steadily increasing DC bias field, H, hysteretic loss is encountered as soon as H exceeds H_{c1}; likewise, in the absence of a bias field, when the amplitude, H_m, of

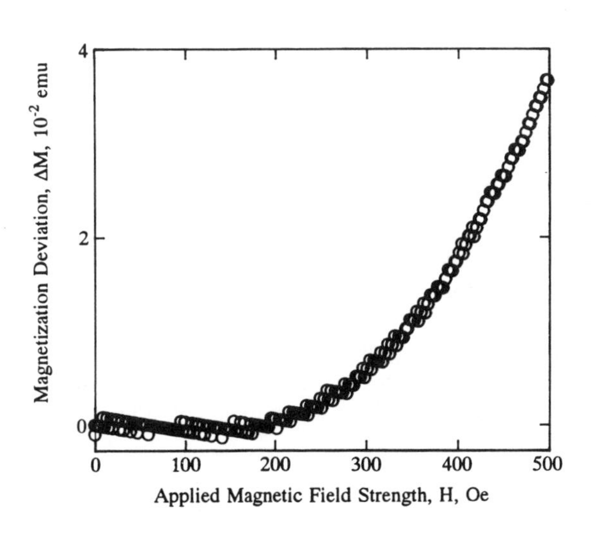

Fig. 3
For sample $DC1_{AB}$, deviation of M(H) from a computed extrapolation of the very low field slope. h_{c1} is about 100 Oe.

an oscillating field is steadily increased. The loss may be measured magnetically or calorimetrically. But if attention is focused just on the *onset* of the hysteretic loss, one encounters the same difficulties as those described above -- especially in the presence of parasitic surface-roughness loss. Both of these difficulties are minimized when the hysteretic loss, Q_h (energy per cycle), is fitted to a known function of H_{c1} and H_m, and H_{c1} is extracted analytically.

Some authors have relied on a simple extension of the Bean model (which in its original form ignores H_{c1} or assumes it to be zero) which for $H_m < H^*$ would suggest that Q_h is proportional to $(H_m - H_{c1})^3$. H_{c1} would then be obtained as the intercept of a plot of $\sqrt[3]{Q_h}$ *versus* H_m. But this approach is doomed to failure, since when introducing a nonzero H_{c1} one must at the same time also introduce the very significant (when $H_{c1}<H<<H^*$) additional losses associated with the Meissner surface current in the mixed state[1].

AC Hysteretic Loss Below H*

In the mixed state of a type-II superconductor, three types of current can flow: (1) a Meissner surface current; (2) a set of vortex currents; (3) in the irreversible (flux-pinned) case, the Bean vortex-density-gradient current -- also known as the bulk or body current. The body current density in the field-penetrated region in the Bean approximation is given by $\mathbf{J}_B = J_c(\mathbf{E}/E)$, where $\mathbf{E}/E$ is either the direction of an existing electric field or that of the last field that existed. For a slab, the integral of $\mathbf{J}_B.\mathbf{E}$ over the volume and around a closed field cycle gives a term, Q_{hB}, proportional to $(H_m-H_{c1})^3$. Since the Meissner surface current of density J_M is also acted on by the electric field it also contributes to the loss and provides a term, Q_{hM} proportional to $H_{c1}(H_m-H_{c1})^2$. The vortex term is neglected. An expression for $Q_{hM} + Q_{hB}$ has already been analytically derived for a slab.[1] But for a transverse-field cylinder, the case of present interest, an exact analytical expression cannot be obtained, even for Q_{hB}. Our non-analytical result for a slab is the same as that derived previously.[1] The new expression for the transverse-field cylinder is, to a reasonable approximation, given by

$$Q_h = Q_{hM} + Q_{hB}$$

$$\approx (10/3\pi^2 J_c R_0)\{(3/2)h_{c1}(H_m-h_{c1})^2 + (H_m-h_{c1})^3\} \qquad (3)$$

where h_{c1} is the applied field needed to supply a field H_{c1} at the "equator" of the cylinder (hence $h_{c1} = H_{c1}/2$). We refer to Eqn. (3) as the "2-3 Equation".

Experimentally, h_{c1} is determined by fitting an extensive (Q_h,H_m) data set to the 2-3 Equation -- hence the "2-3 Method". If parasitic surface loss is negligible -- as for a polished cylinder -- the data fit should also yield an average value for J_c, and has done so in one particular case.

Surface-barrier loss is occasionally mentioned in this context.[1] However, our M(H) loops contain no measurable evidence for a surface barrier. Besides, the 2-3 method itself would interpret surface-barrier loss as surface-roughness loss to which h_{c1}, by that method, is relatively insensitive.

RESULTS

H_{c1} by Direct Fit to the 2-3 Equation

Using commercially available curve-fitting software[7] an extensive (Q_h,H_m) data set, obtained from hysteresis measurements performed on $DC1_{AB}$, was fitted to the 2-3 Function . A two-stage procedure was employed: (1) Taking a "short" data set (H_m from about 10 Oe to a little greater than an estimated h_{c1}) h_{c1} was computed as the prefactor, $A \equiv 10/3\pi^2 J_c R_0$, was varied over a physically reasonable range. The procedure was next repeated for a "long" data set (upper limit of H_m, a little less than H*). The intersection of the two h_{c1} *versus* A curves yielded values for A (hence J_c) and h_{c1}. (2) Further refinement and confirmation of the h_{c1} so obtained was achieved by retaining the above A, and solving the 2-3 Function for h_{c1} with respect to a series of continually truncated data sets -- i.e as function of $H_{m,max}$ at fixed $H_{m,min} \sim 10$ Oe. A "good" value of A should provide only small variation of h_{c1} over a wide range of $H_{m,max}$. It was found necessary to exclude very large values of $H_{m,max}$, since as H_m approached H*, Q_h became dominated by Q_B and hence became relatively insensitive to h_{c1}. It was also noticed that h_{c1} was relatively insensitive to the choice of $H_{m,min}$ (usually $< h_{c1}$), since the corresponding values of Q_h are insignificantly small.

Some results for $DC1_{AB}$ at 77 K are presented in Fig. 4. In it are shown fits to the 2-3 Function over the H_m ranges of: (a) 9.8 to 881 Oe (h_{c1} = 96 Oe) and (b) 244 to 881 Oe (h_{c1} = 96 Oe, again). In spite of this agreement, several repetitions of the direct-fitting procedure led us to believe that the error associated with h_{c1} by that method is about ±10 Oe.

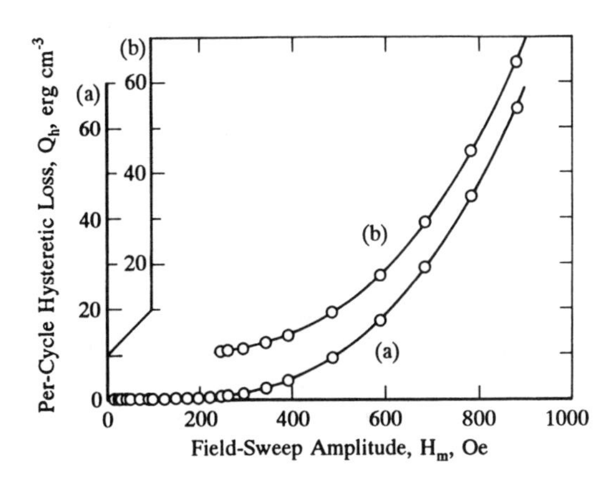

Fig. 4
Hysteresis loss for $DC1_{AB}$ at 77 K *versus* field sweep amplitude, H_m, fitted to Eqn. (3) from: (a) 9.8 to 881 Oe; (b) 244 to 881 Oe .

H_{c1} from Coefficient Ratios within the 2-3 Kernel

If parasitic surface-roughness loss is expected it is better to direct attention, not to the full 2-3 Function, but rather to the part within the brackets -- the so-called "2-3 Kernel", Q_{hK}, which is easily seen to be a three-term quantity

$$Q_{hK} = (1/2)h_{c1}^3 - (3/2)h_{c1}H_m^2 + H_m^3 \tag{4}$$

It follows that if the (Q_h,H_m) data are fitted to the cubic expression

$$y = A_0 + A_2x^2 + A_3x^3 \tag{5}$$

two independent expressions for h_{c1} ("h1" and "h2", say) are identifiable. They are: $h1 = -(2/3)(A_2/A_3)$ and $h2 = {}^3\sqrt{(2A_0/A_3)}$. The correct form of the 2-3 Kernel is guaranteed when h1 = h2; and of course they then both equal the sought-after h_{c1}.

To provide scope for h1 and h2 to vary, and hopefully to achieve equality under some conditions, Eqn. (5) is fitted to continually truncated (Q_h,H_m) data sets. The fits may be made according to Mode-I in which $H_{m,max}$ is varied at fixed $H_{m,min}$ or Mode-II in which $H_{m,min}$ is varied at fixed $H_{m,max}$. Mode-I is the better procedure since the fit is relatively insensitive to choice of $H_{m,min}$ with its relatively insignificant Q_h. With Mode-II, on the other hand, it is difficult to arbitrarily choose a satisfactory fixed $H_{m,max}$ since, as mentioned before, too high a value includes heavily weighted data that are satisfied rather well simply by H_m^3.

The results of applying the Mode-I Coefficient-Ratio method to $DC1_{AB}$, CPS3, and CPS4 are illustrated in Fig. 5. $DC1_{AB}$ yields three points of intersection with an average h_{c1} of 112.4 Oe. CPS3 and CPS4 yield one intersection point each at h_{c1}s of 81.5 and 80.9 Oe, respectively.

CONCLUDING SUMMARY

It has been pointed out that in thick high-J_c samples the deviation from M *versus* H linearity that is expected as H passes through H_{c1} is barely perceptible. Such a deviation can be detected if R_0 is relatively small, but then field penetration may influence the result. Attempts to detect small deviations from linearity of a few data points above H_{c1} will always be fraught with uncertainty. Accurate methods of H_{c1} determination should utilize an extensive set of data above that field in association with a theoretical expression to enable H_{c1} to be extracted from it by means of some mathematical fitting procedure. Two examples of the application of this philosophy to the measurement of 77-K H_{c1} in melt-processed YBCO were presented. The Direct-Fit method applied to $DC1_{AB}$ yielded 192±20 Oe. The Coefficient-Ratio method applied to three different samples yielded: $DC1_{AB}$, 225 Oe; CPS3, 163 Oe; CPS4, 162 Oe. Since DC1 and the CPS samples were made using slightly differing variants of the float-zone melting procedure it might be anticipated

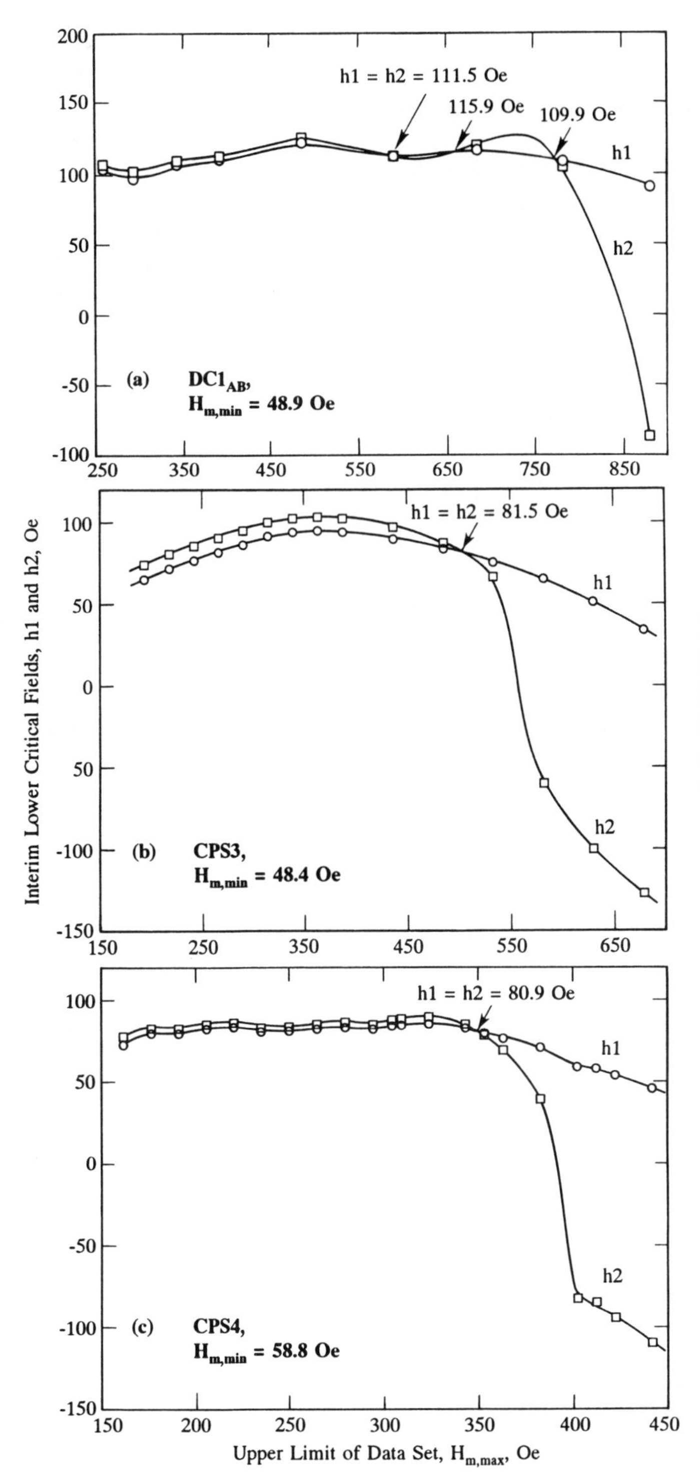

Fig. 5
Interim critical fields, h1 and h2, *versus* $H_{m,max}$ at fixed $H_{m,min}$ for: (a) $DC1_{AB}$, (b) CPS3, and (c) CPS4.

that the measured H_{c1}s for samples as anisotropic as YBCO would be different. It is reassuring to note that two different batches of CPS fiber, differing in diameter, but prepared by similar processes, yielded similar H_{c1}s.

ACKNOWLEDGEMENTS

The magnetization measurements were conducted by K. R. Marken, Jr. The research at Battelle was supported by the Electric Power Research Institute (EPRI) under Contract No. RP8009-06; sample preparation at CPSS was supported by DARPA under Contract No. N00014-88-C-0512.

REFERENCES

1. A. M. Campbell and J. E. Evetts, **Critical Currents in Superconductors**, Taylor and Francis Ltd., London; Barnes and Noble Books, New York, 1972, p. 75-76.

2. Zhou Lian, Zhang Pingxiang, Ji Ping, *et al.*, The properties of YBCO superconductors prepared by a new approach: the 'powder melting process', *Supercond. Sci. Technol.* **3**, 490-492 (1990).

3. J. D. Hodge, L. J. Klemptner, and M. V. Parish, Production of continuous filaments of melt-textured $YBa_2Cu_3O_{7-\delta}$, *J. Mater. Sci. Let.*, in press.

4. C. P. Bean, Magnetization of high-field superconductors, *Rev. Mod. Phys.* **36**, 31-39 (1964).

5. E. W. Collings, K. R. Marken, Jr., A. J. Markworth *et al.*, Critical field enhancement due to field penetration in fine-filament superconductors, *Adv. Cryo. Eng. (Materials)* **36**, 255-261 (1990).

6. A. Umezawa, G. W. Crabtree, J. L. Liu, *et al.*, Anisotropy of the lower critical field, magnetic penetration depth, and equilibrium shielding current in single crystal $YBa_2Cu_3O_{7-\delta}$, *Phys. Rev.* B**38**, 2843-2846 (1988).

7. MINSQ, copyright © 1988 MicroMath, Inc, Scientific Software.

METAL-INSULATOR TRANSITION IN DOPED $Bi_2Sr_2Ca_{1-x}Y_xCu_2O_8$

D. Mandrus, L. Forro*, C. Kendziora and L. Mihaly
Department of Physics, SUNY @ Stony Brook

ABSTRACT

The temperature dependence of the ab-plane electrical transport was investigated on a series of single crystal $Bi_2Sr_2Ca_{1-x}Y_xCu_2O_8$ samples with a wide range of yttrium content. The observed metal-insulator transition is discussed in terms of two dimensional localization. The driving force of the localization is the random impurity potential.

The cuprate-based metals, known for high temperature superconductivity, exhibit a transition to an insulating ground state upon appropriate doping. In principle, starting from a Fermi liquid, there are two routes to the insulating state: Mott localization, where the driving force is the electron-electron interaction [1], and Anderson localization, driven by the random potential of the impurities [2]. The details of the metal-insulator transition depend on the dimensionality of the electronic system. Doped high T_c materials may serve as good examples for two dimensional localization, since their electronic structure is quasi two-dimensional [3],[4].

Two-dimensional systems are special, for the resistivity is measured in terms of the sheet resistance, R□ (Ohms), without reference to a length scale. The dimensionless combination of R□/R*, with $R^*=h/e^2=25.8k\Omega$, was expected to play a crucial role in the separation of the metallic and insulating states [5], until it was shown that at T=0 even small disorder leads to localized electronic states and non-zero metallic conductivity is forbidden [6]. More recently, based on the experiments of Haviland et al. [7], Pang [8] and Fisher et al. [9] suggested that metallic conductance is possible at a sheet resistance of $R^*=c_0h/4e^2$, where c_0 is a universal number of order of unity. The latter authors, as well as Doniach and Inui [10], approach the transition as localization of bosons ("Cooper pairs") instead of fermions ("single electrons").

We performed a dc electronic transport study on single crystals of $Bi_{2.2}Sr_2Ca_{0.8-x}Y_xCu_2O_y$. According to the resistivity measurements, at low temperature the end members of the series, with nominal composition of $Bi_2Sr_2CaCu_2O_8$ and $Bi_2Sr_2YCu_2O_8$ are superconductor and insulator, respectively. We observed the suppression of the superconducting transition temperature as x was increased from zero to x=0.45. The Hall coefficient increases with x, corresponding to the decreasing carrier density in the Y doped samples. Above x=0.45 the samples are insulators at low temperature. The thermoelectric power of all samples approaches zero at low temperature, indicating that there is no energy gap or pseudogap at the Fermi energy. We interpret the results in terms of electron localization, due to the increase of impurity potential and the decrease of hole bandwidth upon doping. We argue that the structurally distinct copper oxide double layers (sandwiching the Ca(Y) layers) confine the electrons to two dimensions, and the localization is best characterized in

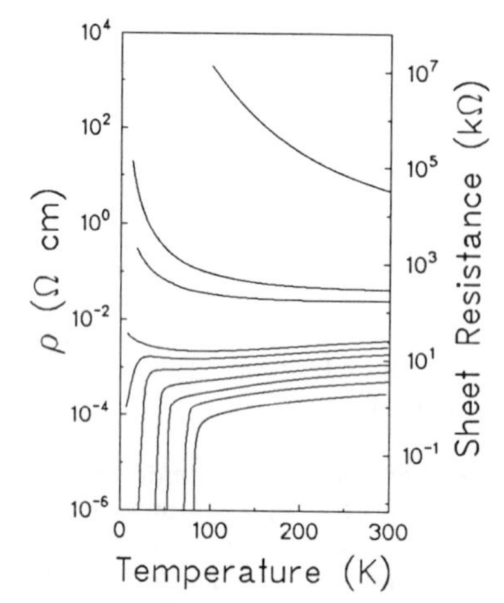

Fig. 1. Temperature dependence of the resistivity (left hand scale) of $Bi_{2.2}Sr_2Ca_{0.8-x}Y_xCu_2O_y$ samples. The right hand scale refers to the sheet resistance of the copper oxide double layers. The Y content x of the samples are (in the order of increasing resistivity): 0, 0.20, 0.34, 0.38, 0.42, 0.44, 0.47, 0.51, 0.55, 0.80.

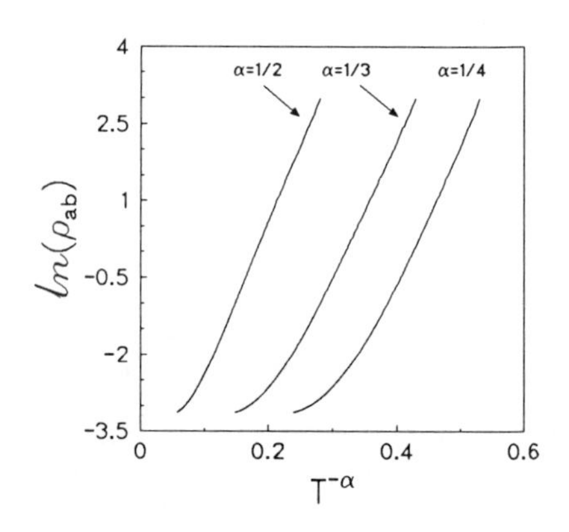

Fig. 2. Variable range hopping, $\rho=\rho_o\exp((T_o/T)^{\alpha})$, fits to the resistivity of the x=0.55 sample. The $\alpha=1/2$ exponent gives the best straight line fit with T_o=900K. (for this exponent 0.1 unit on the horizontal scale corresponds to 100K). We emphasize that Eq. (2) also fits the data over the same temperature range, with $\gamma=-1$ and Δ=50K.

terms of the sheet resistance of these sandwiches. We obtain $R^*=8\pm1$kΩ from our experiments. The behavior of $Bi_2Sr_2CuO_6$, La_2CuO_{4+y}, $La_{2-x}Sr_xCuO_4$ and $YBa_2Cu_3O_7$ (Refs. [11],[12],[13] and [14], respectively) has also been investigated from point of view of two dimensional localization.

The $Bi_{2.2}Sr_2Ca_{0.8-x}Y_xCu_2O_y$ crystals were grown from copper oxide rich melt, as described in a separate publication [15]. The x/2 = Y/Sr ratio of the samples has been determined by X-ray fluorescence with an estimated error of $\delta x=\pm0.01$. For the other elements in the chemical formula the estimated error is ±0.1. Each crystal was heat treated at $600C^0$ in air to ensure a reproducible oxygen content, but we did not determine y. The structure has been characterized by X-ray and electron diffraction. The c axis lattice parameter was found to depend on the Y content, in good agreement with the earlier studies [16]. The experimental details of the resistivity, Hall effect and thermopower measurements are discussed elsewhere [15].

The effect of Y doping is expected to be twofold. First, the introduction of Y^{3+} in place of Ca^{2+}, in the immediate neighborhood of the CuO layers, will lead to a random potential felt by the charge carriers. Second, the Y^{3+} ions take away holes from the conduction band, and reduce the carrier density. Accordingly, the Hall number ($n=1/R_He$) is found to decrease rapidly with doping [15].

In Figure 1, we show the resistivity vs. temperature for several samples of various yttrium content. The sheet resistance of the copper oxide double-layers was calculated by dividing the measured specific resistivity with the distance between the double layers, d=15Å. The samples are either insulators or superconductors at low temperature; the separatrix of the two sets of curves extrapolates to zero temperature at ρ=1200$\mu\Omega$cm or R□=8kΩ. The resemblance to the results of Haviland et al. [7] is striking. Note, however, that our results were obtained not on thin films, but on bulk single crystals.

The samples with superconducting ground state but close to 8kΩ sheet resistance exhibit a slight increase of resistance before the superconducting transition. This may be attributed to a weak localization [12]. The samples with an insulating ground state and room temperature sheet resistance less than 1MΩ can be compared to the variable range hopping formula [17],

$$\rho=\rho_o\exp\{(T_o/T)^{\alpha}\} \qquad (1)$$

where T_0 is an activation energy. For Anderson localization the exponent is α=1/(d+1), depending on the dimensionality d of the system. If interaction effects are important, a "soft" Coulomb gap may open at the Fermi energy and α=1/2, independent of d [18]. As illustrated in Figure 2, the best fit is obtained, over a large range of temperature and resistivity, with α=1/2 and T_o=900K. The α=1, or the three dimensional exponent, α=1/4, seems to be entirely out of question, but α=1/3 can not be excluded. Valles et al. [14] obtained α=1/2 for ion beam irradiated $YBa_2Cu_3O_{7-y}$, while Fiory et al. [11] got α=1/3 in $Bi_{2+x}Sr_{2-y}CuO_{6+z}$.

We caution that a weak (e.g. power law) temperature dependence of ρ_o (also allowed in the variable range hopping models) may change the fits dramatically. For example, the "fixed gap" expression,

$$\rho=AT^{\gamma}\exp\{(\Delta/k_BT)\} \qquad (2)$$

also fits the data below 100K with γ=-1 and Δ=50K. This formula was used for $Bi_2Sr_2CaCu_2O_8$ samples by Zettl and co-workers [19], [20], but they obtained γ>0. The common feature of this approach and Eq. (1) with α=1/2 is that they point to a gap or pseudogap in the density of states.

The absence of the gap in the density of states is evidenced by the temperature independent Hall coefficient of the high resistance samples. If Coulomb correlations are important, the Hall effect is expected to have a temperature dependence [17], in contrast to our experiment. Similar observations were made by Valles et al. on ion beam irradiated $YBa_2Cu_3O_7$ [18], while Preyer et al. found activated temperature dependence for La_2CuO_{4+y} [12]. Interestingly, for the metallic high T_c samples, when temperature dependence is not expected, the Hall effect varies with temperature [21].

In conclusion, we have shown that in the $Bi_{2.2}Sr_2Ca_{0.8-x}Y_xCu_2O_y$ material exhibits a superconductor-insulator transition upon increasing x. The temperature dependence of the resistivity of the high resistance samples excludes three dimensional variable range hopping (α=1/4 in Eq. (1)). We argue that the transition is two dimensional and the characteristic sheet resistance is R*=8±1kΩ. The resistivity data can be best fitted by assuming strong Coulomb correlations (α=1/2), but the thermopower and Hall effect measurements do not support this interpretation. Our discussion of the experimental results has relied on ideas developed for fermionic charge carriers. It remains to be seen if the apparent contradiction can be resolved if Cooper pairs survive in the large resistance samples, as suggested by several authors [8],[9],[10].

This work has been supported by the National Science Foundation Grant DMR 9016456. Useful discussions with P. Allen, G. Gruner and P. Chaikin are acknowledged.

REFERENCES

[1] N.F. Mott, Proc. Phys. Soc. A 62, 416 (1949)
[2] P. W. Anderson, Phys. Rev. 109 1492 (1958)
[3] S. Martin et al., Phys. Rev. Letters, 60, 2194 (1988)
[4] D.E. Farrell et al. Phys. Rev. Letters, 63 782 (1989)
[5] D.C Licciardello and D.J. Thouless, J. Phys. C 8 4157, (1975)
[6] E. Abrahams, P.W. Anderson, D.C. Liccardello and T.V. Ramakrishna, Phys. Rev. Letters 42 763 (1979);
P. A. Lee and T. V. Ramakrishnan, Rev. Mod. Phys. 57 287 (1985)
[7] D.B. Haviland, Y. Liu and A.M. Goldman, Phys. Rev. Letters 62, 2180 (1989)
[8] Tao Pang, Phys. Rev. Letters, 62, 2176 (1989)
[9] M.P.A. Fisher, G.Grinstein and S.M. Girvin, Phys. Rev. Letters 64, 587 (1990)
[10] S. Doniach and M. Inui Phys. Rev. B 41, 6668 (1990)
[11] A.T. Fiory, S. Martin, R.M. Fleming, L.F. Schneemeyer, J.V. Waszczak, Phys. Rev. B 41 2627 (1990)
[12] C.Y. Chen et al. Phys. Rev. Letters, 63, 2307 (1989)
N.W. Preyer et al, Phys. Rev. B 39, 11563 (1989)
[13] M. Suzuki, Phys. Rev. B 39, 2312 (1989)
[14] J. M. Valles, Jr., et al., Phys. Rev. B 39 11599 (1989)
[15] C. Kendziora et al., to be published
[16] A. Maeda, M. Hase, I. Isikuda, K. Noda, S. Takebayashi and K. Uchinokura, Phys. Rev. B 41, 6418 (1990)
[17] N.F. Mott and E.A. Davis, "Electronic Processes in no-crystalline materials" (Clarendon, Oxford, 1979)
[18] A.L. Efros and B. J. Shklovskii, J. Phys. C 8, L49, (1975)
[19] M.F. Crommie, A.Y. Liu, M.L. Cohen and A. Zettl, Phys. Rev. B 41 2526 (1990)
[20] G. Briceno and A. Zettl, Phys. Rev. B 40, 11352 (1989)
[21] L. Forro, D. Mandrus C. Kendziora L. Mihaly an R. Reeder, Phys. Rev. B 42 8704 (1990)
[22] B.L. Altshuler, D. Khmelnitzkii A.I. Larkin and P.A. Lee, Phys. Rev. B 22 5142 (1980)
[23] J. Clayhold et al. Phys. Rev. B 39, 7324 (1989)

ANISOTROPY IN SELECTED ELECTRICAL, MAGNETIC AND OPTICAL PROPERTIES OF $Pb_2Sr_2(Y/Ca)Cu_3O_8$

M. Reedyk, J.S. Xue, J.E. Greedan, C.V. Stager and T. Timusk
McMaster University, Hamilton, ON, Canada L8S 4M1

ABSTRACT

Measurements of anisotropy in the dc-resistivity, magneto-resistivity, low and intermediate field magnetization and mid infrared reflectance of single crystals of $Pb_2Sr_2(Y/Ca)Cu_3O_{8+\delta}$ which have a superconducting transition temperature of ≈ 76 K are reported. The ab-plane properties - a high resistivity, large magnetic penetration depth and low plasma edge - suggest that the in-plane carrier concentration is unusually small in this material given the relatively high value of T_c. The extent of the anisotropy between ab-plane and c-axis properties appears to be similar to that observed in $YBa_2Cu_3O_{7-\delta}$, and is thus less severe than in $Bi_2Sr_2CaCu_2O_8$.

INTRODUCTION

The highly anisotropic structure of the copper oxide superconductors leads to a significant anisotropy between ab-plane and c-axis properties. The common structural element in these systems is a series of one or more CuO_2 planes. The best known systems, $YBa_2Cu_3O_{7-\delta}$ and $Bi_2Sr_2CaCu_2O_8$ contain double CuO_2 layers. The a-c anisotropy in $Bi_2Sr_2CaCu_2O_8$ is found to be much greater than that in $YBa_2Cu_3O_{7-\delta}$.[1,2,3] This distinction in the extent of the anisotropy in these two materials likely stems from the differences in their intercalary structure. $Pb_2Sr_2(Y/Ca)Cu_3O_{8+\delta}$ is another system containing double CuO_2 planes.[4] A comparison with these two other systems is interesting because its interlayer structure contains elements in common with both materials. In $Pb_2Sr_2(Y/Ca)Cu_3O_{8+\delta}$ sets of CuO_2 bilayers are separated by a PbO-CuO_δ-PbO sequence, the two PbO layers being analogous to the double BiO sheets in $Bi_2Sr_2CaCu_2O_8$ and the CuO_δ layer comparable to an oxygen depleted chain layer such as in $YBa_2Cu_3O_{6+\delta}$. Investigations of the extent of the anisotropy in this system may thus make a contribution to our understanding of these materials in general.

DC RESISTIVITY

The sign of the temperature coefficient of the c-axis resistivity, ρ_c is still an issue in the cuprates, although a consensus seems to be forming that it may be an indication of sample quality. In high quality $YBa_2Cu_3O_{7-\delta}$ samples a positive temperature coefficient prevails. Our results in the $Pb_2Sr_2(Y/Ca)Cu_3O_{8+\delta}$ system

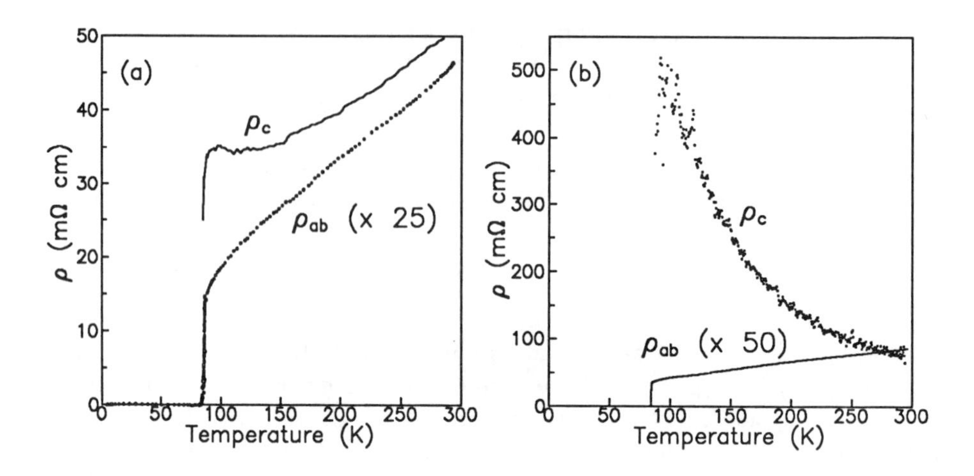

Figure 1 Anisotropic temperature dependence of the resistivity for a good (a) and poorer (b) quality $Pb_2Sr_2(Y/Ca)Cu_3O_{8+\delta}$ crystal.

support this view. Figures 1(a) and (b) show the anisotropic temperature dependence of the dc-resistivity for two different crystals of $Pb_2Sr_2(Y/Ca)Cu_3O_{8+\delta}$. The samples were synthesized via a NaCl-PbO flux method described previously.[5] The resistivity was measured in both the ac-plane and the pseudo-tetragonal ab-plane, such that the behaviour along the c-direction could be deconvolved from the data using the van der Pauw analysis for anisotropic samples.[6]

The crystal in (a) which shows a positive temperature coefficient for ρ_c is from a batch of crystals which consistently show superior properties crystallographically, magnetically and optically. The crystal in (b) which has a similar ab-plane resistivity, ρ_{ab}, to the crystal in (a) shows a negative temperature coefficient for ρ_c. The magnetically measured transition for crystals from this growth is, however, not nearly as sharp as it is for those of the batch of crystal (a). We attribute these differences primarily to oxygen content and crystallographic quality. For the remaining measurements discussed herein only samples of quality comparable to that of the crystal of figure 1(a) were used.

The extent of the anisotropy is described by the anisotropy ratio $\frac{\rho_c}{\rho_{ab}}$, which for crystal (a) is about 30 at room temperature and increases to approximately 50 near T_c. This is within the range of values typically reported for $YBa_2Cu_3O_{7-\delta}$,[7] and shows that the extent of the anisotropy in this system is closer to that of $YBa_2Cu_3O_{7-\delta}$ than to that of the highly anisotropic $Bi_2Sr_2CaCu_2O_8$ where the c-axis resistivity is a factor of $\approx 10^5$ larger than that in the ab-plane.[1]

The absolute magnitude of the ab-plane resistivity of $Pb_2Sr_2(Y/Ca)Cu_3O_{8+\delta}$ is close to an order of magnitude larger than in these other two systems. Based on optical reflectivity measurements which reveal a much lower effective optical carrier density[8] and on intermediate field magnetization measurements from

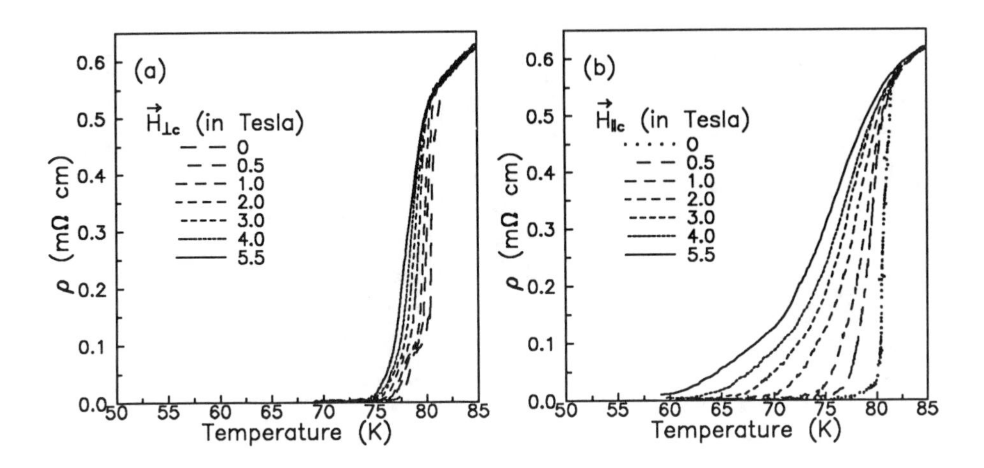

Figure 2 In-plane magnetoresistivity of $Pb_2Sr_2(Y/Ca)Cu_3O_{8+\delta}$ near T_c for the field oriented within the planes (a) and along the c axis (b).

which a considerably larger low temperature penetration depth is extracted[9] we conclude that the higher absolute resistivity reflects an intrinsically lower carrier concentration in this system.[10]

Figure 2 shows the effect of a magnetic field on the ab-plane superconducting transition. As in the other two systems[2] it broadens as opposed to undergoing a parallel shift to lower temperature. As expected the broadening is less severe for the field within the planes (a) than along the c axis (b). Once again the extent of the anisotropy observed is more typical of that in $YBa_2Cu_3O_{7-\delta}$ than in $Bi_2Sr_2CaCu_2O_8$, where especially for $H \parallel c$ the broadening is much more pronounced.[2]

Because the crystal dimensions are small, and in relation contacts are large, there is always some uncertainty associated with these results. It is thus perhaps more reliable to examine anisotropy in magnetization derived properties. The next section discusses low and intermediate field magnetization measurements from which the temperature dependence of the lower critical field and penetration depth near T_c is extracted.

MAGNETIZATION

Intermediate field magnetization measurements were carried out using a Quantum Design SQUID magnetometer with the c-axis of the crystal oriented both perpendicular and parallel to the field direction. For each orientation measurements were made as a function of temperature for fields of 0.1, 0.4, 0.7 and 1.0 to 5.5 T in steps of 0.5 Tesla (T) in the vicinity of T_c where the magnetization was reversible. The field dependence of the magnetization was then determined

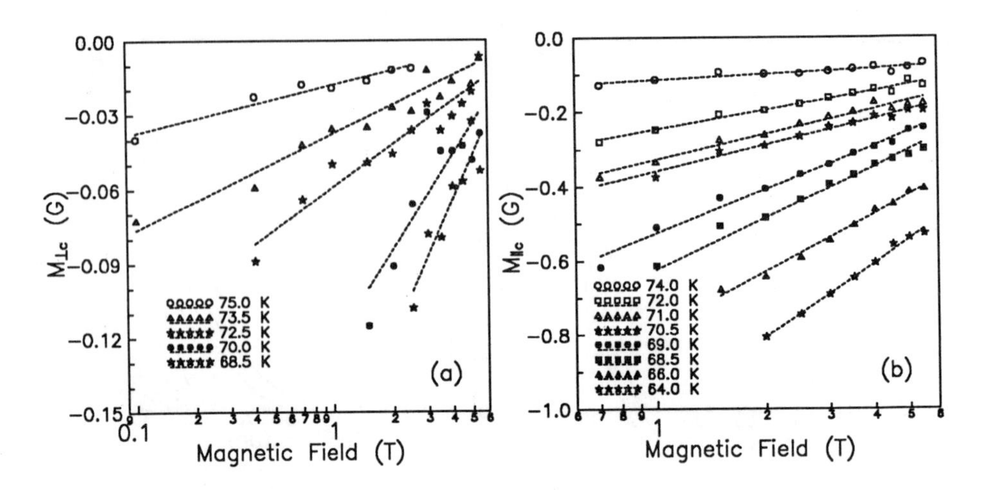

Figure 3 Magnetic field dependence of the magnetization of $Pb_2Sr_2(Y/Ca)Cu_3O_{8+\delta}$ at various temperatures for $H \perp c$ (a) and $H \parallel c$ (b). The dashed curves are least squares fits to the data.

from these results and is plotted in figures 3(a) and (b) for $H \perp c$ and $H \parallel c$ respectively.

The slopes of the best fit lines shown in figure 3 are given according to Kogan *et. al.* by

$$\frac{dM_{\perp c}}{d\ln H} = \frac{\Phi}{32\pi^2\lambda_{ab}\lambda_c} \quad \text{and} \quad \frac{dM_{\parallel c}}{d\ln H} = \frac{\Phi}{32\pi^2\lambda_{ab}^2}, \tag{1}$$

for (a) and (b) respectively.[11] Here Φ is the flux quantum and λ_{ab} and λ_c are the ab-plane and c-axis magnetic penetration depth respectively. The temperature dependence of λ_{ab} and λ_c can then be extracted from the measured slopes and is shown in figure 4. The zero temperature penetration depth can be estimated by comparing with the expected behaviour for various models. The temperature dependence of λ_{ab} is found to be much better described by the BCS (clean or dirty) model than the two-fluid model.[9] Since these materials are generally believed to be in the clean limit we show in figure 4 the best fit to the BCS clean limit, which yields a value of 2575 Å for $\lambda_{ab}(0)$. The c-axis penetration depth data are more scattered and limited in extent and thus rather than predict $\lambda_c(0)$ by fitting to theory we make use of the experimentally determined mass ratio. Since λ^2 is proportional to the effective mass, m,[12] the effective mass anisotropy parameter $\gamma = (m_c/m_{ab})^{\frac{1}{2}} = \lambda_c/\lambda_{ab}$ can be obtained by taking the average ratio from figure 4. For $\lambda_{ab}(0) = 2575\mathring{A}$ and $\gamma = 2.5$ we obtain 6425 Å for $\lambda_c(0)$.

The ab-plane penetration depth in $YBa_2Cu_3O_{7-\delta}$, (typically 1200-1500 Å), is considerably smaller than the value obtained for $Pb_2Sr_2(Y/Ca)Cu_3O_{8+\delta}$. The lower critical field, H_{c1}, is related to the penetration depth via the London

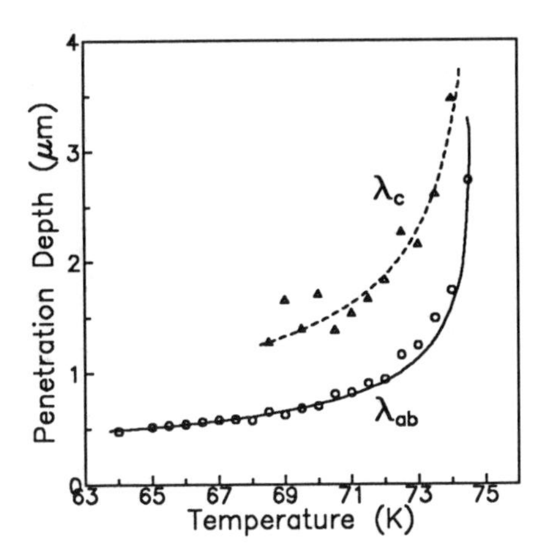

Figure 4 Temperature dependence of λ_{ab} and λ_c in the vicinity of T_c. The solid curve drawn through the data for λ_{ab} is a fit to the BCS clean limit prediction for a $\lambda_{ab}(0)$ of 2575 Å. The dashed curve through the λ_c data is a guide to the eye.

formula:[12]

$$H_{c1} = \frac{\Phi}{4\pi\lambda^2}[\ln(\frac{\lambda}{\xi}) + 0.5], \tag{2}$$

where ξ is the coherence length. A larger penetration depth should therefore lead to a noticeably smaller value for H_{c1}.

We have attempted to evaluate the lower critical field in $Pb_2Sr_2(Y/Ca)Cu_3O_{8+\delta}$ using a method developed by Naito *et. al.*.[13] By definition, for $H \leq H_{c1}$, the equilibrium flux density, $B = H + 4\pi M$ is zero so that M varies linearly with H. For $H > H_{c1}$, $H + 4\pi M (\equiv 4\pi\Delta M)$ represents the deviation of $M(H)$ from perfect diamagnetism. In Bean's critical state model the equilibrium flux density for a sample with a slab-like geometry in a field just above H_{c1} is found to be proportional to $(H + 4\pi M)^{\frac{1}{2}}$.[14] Thus a plot of $(\Delta M)^{\frac{1}{2}}$ as a function of magnetic field strength should be zero until H reaches H_{c1} whereupon it should begin to rise linearly with increasing field.

Figure 5(a) shows a typical low field magnetization curve. The best fit through the lowest field points is shown. The deviation of $M(H)$ from perfect diamagnetism, ΔM, is obtained by subtracting this curve from the experimental data. $(\Delta M)^{\frac{1}{2}}$ is plotted as a function of demagnetization-corrected field strength, H_{eff} in (b). Note that the expected behaviour is observed, $(\Delta M)^{\frac{1}{2}}$ is zero (scatter in the data has been rectified to enable evaluation of the square root) up to a threshold value of H_{eff} at which point it begins to rise. H_{c1} is obtained by extrapolating the higher field linear regime to the horizontal axis. Figure 6 shows the temperature dependence of the lower critical field for the two orientations of interest obtained using this procedure, the uncertainty reflecting different choices

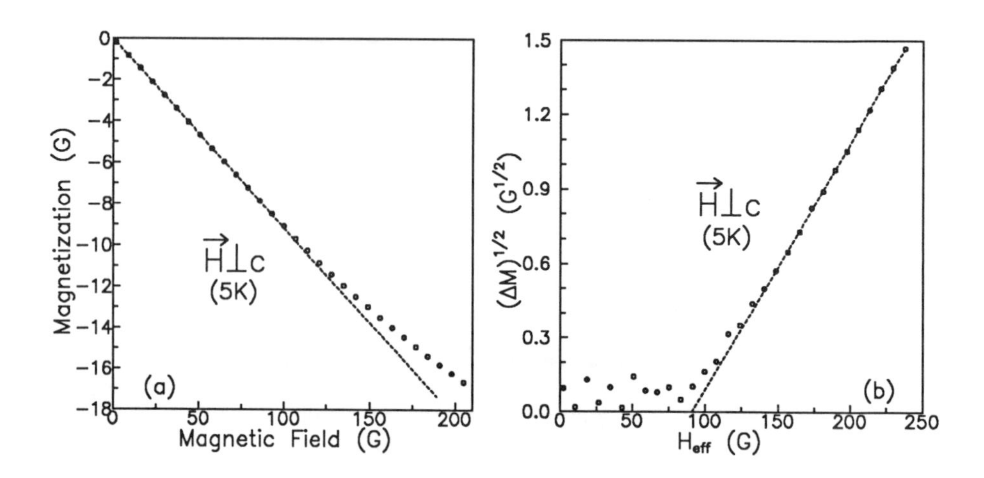

Figure 5 Low field magnetization of $Pb_2Sr_2(Y/Ca)Cu_3O_{8+\delta}$ at 5 K for $H \perp c$. (a) M vs H. The dashed curves represent a least squares fit to the very low field data. (b) $(\Delta M)^{\frac{1}{2}}$ vs the demagnetization corrected field, H_{eff}. The dashed curves represent a least squares fit to the highest field data.

for the range of points used in the least squares fitting.

The values of $H_{c1}(0)$ extrapolated from these results are 95 ± 10 G and 505 ± 20 G respectively for H perpendicular and parallel to the c axis of the crystal respectively, leading to an anisotropy ratio of $H_{c1}^{\|}/H_{c1}^{\perp}$ of approximately 5. These values of $H_{c1}(0)$ are not noticably smaller than those for $YBa_2Cu_3O_{7-\delta}$.[15] The source of the discrepancy between these results and what is expected according to the penetration depth measurements most likely lies mainly in the determination of H_{c1}, (in particular $H_{c1}^{\|}$), where difficulties such as surface pinning, and imperfect demagnetization and model geometries can cause large uncertainty in the values obtained.

MID INFRARED REFLECTANCE

In order to investigate the optical anisotropy, room temperature polarized mid infrared reflectance measurements were carried out using a Spectra Tech infrared microscope. The results are shown in figure 7. The reflectance for the electric field vector polarized in the ab-plane ($E \perp c$) shows metallic behaviour, while that for $E \parallel c$ exhibits an insulating character. This same trend has been observed for $La_{2-x}Sr_xCuO_{4-\delta}$ and $Bi_2Sr_2CaCu_2O_8$ (references 16 and 17). $YBa_2Cu_3O_{7-\delta}$ is somewhat different in that the c-axis reflectance shows a more metallic-like character.[17,18]

The ab-plane reflectance of $Pb_2Sr_2(Y/Ca)Cu_3O_{8+\delta}$ is distinguishable from that of the other two systems containing double CuO_2 planes in the mid infrared. In these other materials the plasma edge occurs *beyond* the spectral region shown

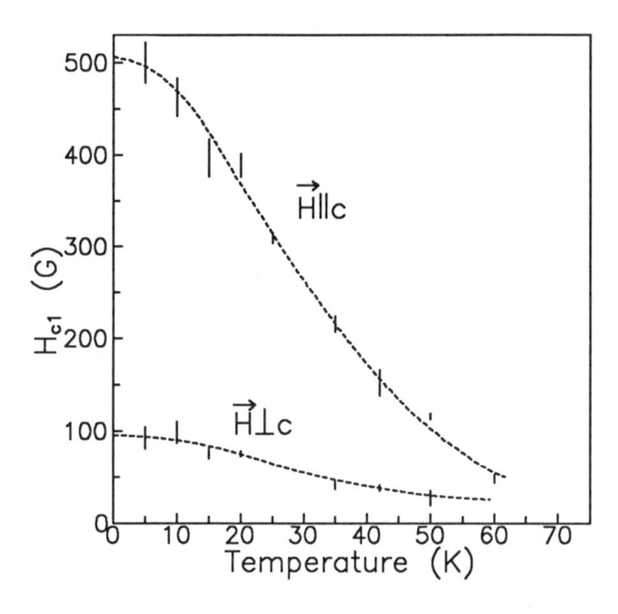

Figure 6 Temperature dependence of $H_{c1}^{\perp}$ and $H_{c1}^{\parallel}$. The dashed curves are a guide tc the eye.

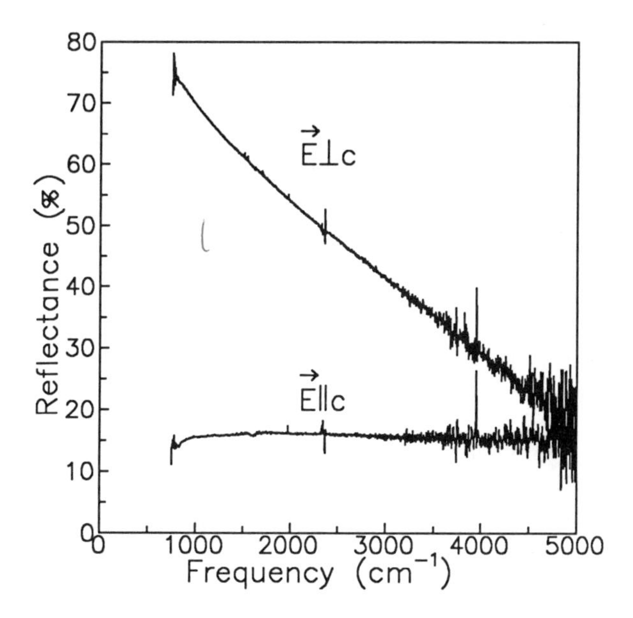

Figure 7 Mid infrared reflectance of $Pb_2Sr_2(Y/Ca)Cu_3O_{8+\delta}$ for $E \perp c$ and $E \parallel c$.

in figure 7, so that the reflectance is still above 60 % throughout this range. In $Pb_2Sr_2(Y/Ca)Cu_3O_{8+\delta}$ the essentially linearly decreasing reflectance seen in the frequency range shown *is* the plasma edge. In fact, the position of the plasma edge in $Pb_2Sr_2(Y/Ca)Cu_3O_{8+\delta}$ is even lower than that of $La_{2-x}Sr_xCuO_{4-\delta}$,[16] which is unexpected in light of the comparatively high T_c ($\approx$ 76 K). This shift of the plasma edge to lower frequency once again points to a significantly lower carrier density in this system.

CONCLUDING REMARKS

These studies of selected properties of $Pb_2Sr_2(Y/Ca)Cu_3O_{8+\delta}$ have provided some insight into this member of the copper-oxide superconductor family. The high resistivity, large magnetic penetration depth and low plasma edge suggest that the carrier concentration is unusually small in this material given that the transition temperature is not that much lower than it is in other materials with CuO_2 bi-layers. The extent of the anisotropy in this system appears to be closer to that of $YBa_2Cu_3O_{7-\delta}$ than to that of the highly anisotropic $Bi_2Sr_2CaCu_2O_8$. The intermediate field magnetization measurements likely yield the most reliable estimate of the effective mass anisotropy: $m_c/m_{ab} \approx 6$.

ACKNOWLEDGEMENTS

We thank E.J. Nicol for providing the theoretical $\lambda(T)$ curve shown in figure 4. This work was supported by the Natural Science and Engineering Research Council of Canada (NSERC), including infrastructure support for the Institute for Materials Research.

REFERENCES

1. S. Martin, A.T. Fiory, R.M. Fleming, L.F. Schneemeyer, and J.V. Waszczak, Phys. Rev. Lett. **60**, 2194 (1988).

2. T.T.M. Palstra, B.Batlogg, R.B. van Dover, L.F. Schneemeyer, and J.V. Waszczak, Phys. Rev. B **41**, 6621, (1990).

3. D.E. Farrell, S. Bonham, J. Foster, Y.C. Chang, P.Z. Jiang, K.G. Vandervoort, D.J. Lam, and V.G. Kogan, Phys. Rev. Lett. **63**, 782, (1989); T.T.M. Palstra, B. Batlogg, L.F. Schneemeyer, R.B. van Dover; J.V. Waszczak, Phys. Rev. B **38**, 5102 (1988), and M. Touminen, A.M. Goldman, Y.Z. Chang, and P.Z. Jiang, Phys. Rev. B **42**, 412, (1990).

4. R.J. Cava, B. Batlogg, J.J. Krajewski, L.W. Rupp, L.F. Schneemeyer, T. Siegrist, R.B. van Dover, P. Marsh, W.F. Peck, Jr., P.K. Gallagher, S.H. Glarum, J.H. Marshall, R.C. Farrow, J.V. Waszczak, R. Hull, and P. Trevor,

Nature **336**, 211, (1988).

5. J.S. Xue, M. Reedyk, Y.P. Lin, C.V. Stager, and J.E. Greedan, Physica **C166**, 29, (1990) and J.S. Xue, M. Reedyk, A. Dabkowski, H. Dabkowska, J.E. Greedan, and C.H. Chen, (Journal of Crystal Growth, to be published).

6. J. Hornstra and L.J. van der Pauw, J. Electron. Control **7**, 169, (1959).

7. T.A. Friedmann, M.W. Rabin, J. Giapintzakis, J.P. Rice, and D.M. Ginsberg, Phys. Rev. B **42**, 6217, (1990).

8. M. Reedyk, T. Timusk, J.S. Xue, and J.E. Greedan, (unpublished).

9. M. Reedyk, C.V. Stager, T. Timusk, J.S. Xue, and J.E. Greedan, Phys. Rev. B **44**, 4539, (1991).

10. The relationship between the dc resistivity, ρ, the plasma frequency, ω_p, the London penetration depth, λ, and the carrier concentration, n, can be summarized as follows:

$$\omega_p^2 = \frac{60\Gamma}{\rho} = \frac{1}{4\pi^2\lambda^2} = \frac{4\pi ne^2}{m}(\frac{1}{2\pi c})^2.$$

Here ω_p and Γ, (the scattering rate), are in cm^{-1}, ρ is in Ω cm, λ in cm, n in cm^{-3}, m, (the effective mass of the carriers), in g, c is the speed of light (3×10^{10} cm/s), and e, (the electronic charge) is 4.8×10^{-10} esu. The factor of $(\frac{1}{2\pi c})$ converts ω_p in s^{-1} to ω_p in cm^{-1}, while $\frac{60\Gamma}{\rho}$ arises from $4\pi\sigma = \frac{\omega_p^2}{\Gamma}$ (all are in cm^{-1}, σ is the dc conductivity) as follows:

$$\omega_p^2 = 4\pi\sigma(\mathrm{cm}^{-1})\Gamma = 4\pi\frac{\sigma(\mathrm{s}^{-1})}{2\pi c}\Gamma = 4\pi\frac{(9 \times 10^{11})\sigma(\Omega^{-1}\mathrm{cm}^{-1})}{2\pi c}\Gamma$$

$$\approx 60\sigma(\Omega^{-1}\mathrm{cm}^{-1})\Gamma = \frac{60\Gamma}{\rho(\Omega\mathrm{cm})}.$$

11. V.G. Kogan, M.M. Fang, and S. Mitra, Phys. Rev. B **38**, 11958, (1988).

12. M. Tinkham, Introduction to Superconductivity (Robert E. Krieger Publishing Company, Malabar, Florida, 1980).

13. M. Naito, A. Matsuda, K. Kitazawa, S. Kambe, I. Tanaka, and H. Kojima, Phys. Rev. B **41**, 4823, (1990).

14. C.P. Bean, Rev. Mod. Phys. **36**, 31, (1964).

15. L. Krusin-Elbaum, A.P. Malozemoff, Y. Yeshurun, D.C. Cronemeyer, and F. Holtzberg, Phys. Rev. B **39**, 2936, (1989) and E.W. Sheidt, C. Hucho, K. Lüders, and V. Müller, Solid State Commun. **71**, 505, (1989).

16. S. Uchida, T. Ido, H. Takagi, T. Arima, Y. Tokura, and S. Tajima, Phys. Rev. B **43**, 7942, (1991).

17. J.H. Kim, I. Bozovic, D.B. Mitzi, A. Kapitulnik, and J.S. Harris, Jr., Phys. Rev. B **41**, 7251, (1990).

18. I. Bozovik, K. Char, S.J.B. Yoo, A. Kapitulnik, M.R. Beasley, T.H. Geballe, Z.Z. Wang, S. Hagen, N.P. Ong, D.E. Aspnes, and M.K. Kelly, Phys. Rev. B **38**, 5077, (1988).

GRAIN BOUNDARY MICROSTRUCTURE AND TRANSPORT CRITICAL CURRENT DENSITY IN $YBa_2Cu_3O_x$

Donglu Shi, Yufei Gao, A. C. Biondo[1], J. G. Chen, K. L. Merkle, and K. C. Goretta[1]

Materials Science Division, [1]Materials and Components Technology Division, Argonne National Laboratory, Argonne, IL 60439

ABSTRACT

It is proposed that the transport critical current (J_c) density for high–T_c thin films, bicrystals, and bulk ceramics is determined by magnetic field penetration into the grain boundaries; the grain orientation may not in all cases be an important factor. The parameter $(\lambda_L/\lambda_J)^2$ can characterize the strength of the grain boundary coupling, which depends mainly on the crystal coherence and the structure of the boundary.

INTRODUCTION

The grain boundaries in bulk samples of high–T_c superconductors generally exhibit so–called weak–link effects as a result of short coherence lengths. These weak links severely limit the transport J_c at very low fields of a few hundred Gauss (1–7). To study the transport current behavior in association with the weak–link effect at grain boundaries, researchers have focused on the correlation between the transport J_c and grain boundary structure and orientation.

Dimos et al. (8) developed a method to study the angular dependence of the critical current density across grain boundaries, J_{gb}. By growing epitaxial $YBa_2Cu_3O_x$ (123) thin films on $SrTiO_3$ bicrystals, they made films with single grain boundaries. These boundaries were oriented at different angles, θ. They found near 4.2 K that the ratio of J_{gb}/J_g, where J_g was the J_c measured within a single grain, was reduced 50 times as θ increased from 0° to 35°. Based on these results, they concluded that the strong angular dependence of J_{gb} was associated with the distortion of vortices by grain boundary dislocations. Their conclusion also clearly implied that materials with high–angle grain boundaries would not be capable of carrying high transport J_c.

However, Babcock et al. reported recently (9) that weak–link behavior was not observed in bicrystals with high–angle grain boundaries. J_{gb} of bicrystals with θ = 3°, 14°, and 90° was measured up to 7 T. Surprisingly, all of these crystals exhibited J_c behavior that was controlled by flux–pinning (i.e., weak–link effects were not observed). These striking experimental data clearly indicate that high–angle grain boundaries are not necessarily connected with Josephson junction weak links.

It may be that the major influence on the transport J_c in polycrystalline high–T_c superconductors is not the gross orientation of the grain boundaries, but rather the degree to which the grain boundaries can be penetrated by an external applied field. In this paper, the physical picture of the Josephson junction (10) is used to explain the flux–pinning–dominated transport J_c in $YBa_2Cu_3O_x$ samples that contain high–angle grain boundaries.

BACKGROUND

One can assume that each grain boundary in $YBa_2Cu_3O_x$ is a Josephson junction with a junction length, L, of the grain size and a junction thickness, t. The coupling strength, F, of the grain boundaries is related by the ratio of the London penetration depth to the Josephson penetration depth, $F = (\lambda_L/\lambda_J)^2$, as suggested by Hylton and Beasley in a laminar model (11). In the strong–coupling limit (7), where the barrier disappears, the expression can be written as

$$F = (\lambda_L/\lambda_J)^2 = J_{gb}/J_g . \quad (1)$$

Thus, one can see that grain boundaries exhibit weak coupling when $(\lambda_L/\lambda_J)^2 << 1$.

It is noted that the occurrence of a Josephson junction is dependent only on the ratio of the London penetration depth to the Josephson penetration depth; it is not directly related to the gross orientations of the grains. This point is illustrated schematically in Figure 1. Figure 1a shows a low–angle grain boundary with the applied field along the c axis. Because the London penetration depth along the ab plane is the same on both sides of the grains, one expects an evenly distributed field profile as indicated. If the the grain boundary is weakly coupled, as has been observed for conventionally sintered samples, the boundary area is uniformly penetrated by the applied field, and $(\lambda_L/\lambda_J)^2 << 1$. In contrast, if the grain boundary is strongly coupled, as is often the case for melt–textured bars or flux–grown bicrystals, high crystal coherency at the boundary makes it difficult for the field to penetrate the boundary. Thus, $(\lambda_L/\lambda_J)^2$ would approach or exceed 1 for this strongly coupled grain boundary.

Thus, the coupling strength has not only to do with grain orientation, but also with the nature of the boundary itself. For example, weak–link effects have been observed in magnetically aligned samples with low–angle boundaries (12). A similar situation is represented in Fig. 1b for c–axis–oriented grains disposed by a tilt angle θ. The field penetration is symmetrical on both sides of the boundary, since the values of λ_{ab} are the same in both grains. Again, in principle, both weak and strong coupling could exist for this grain orientation. The field penetration becomes uneven as both grains are misoriented by 90°, as shown in Fig. 1c. According to previous measurements (13), $\lambda_{ab} \approx 1500$ Å and $\lambda_c \approx 5000$ Å. Thus, for the configuration shown in Fig. 1c, the field penetration on the right side ($H \| c$) is more severe than that on the left side ($H \| ab$). It is noted here is that the situations shown in Fig. 1a and Fig. 1b can occur for the

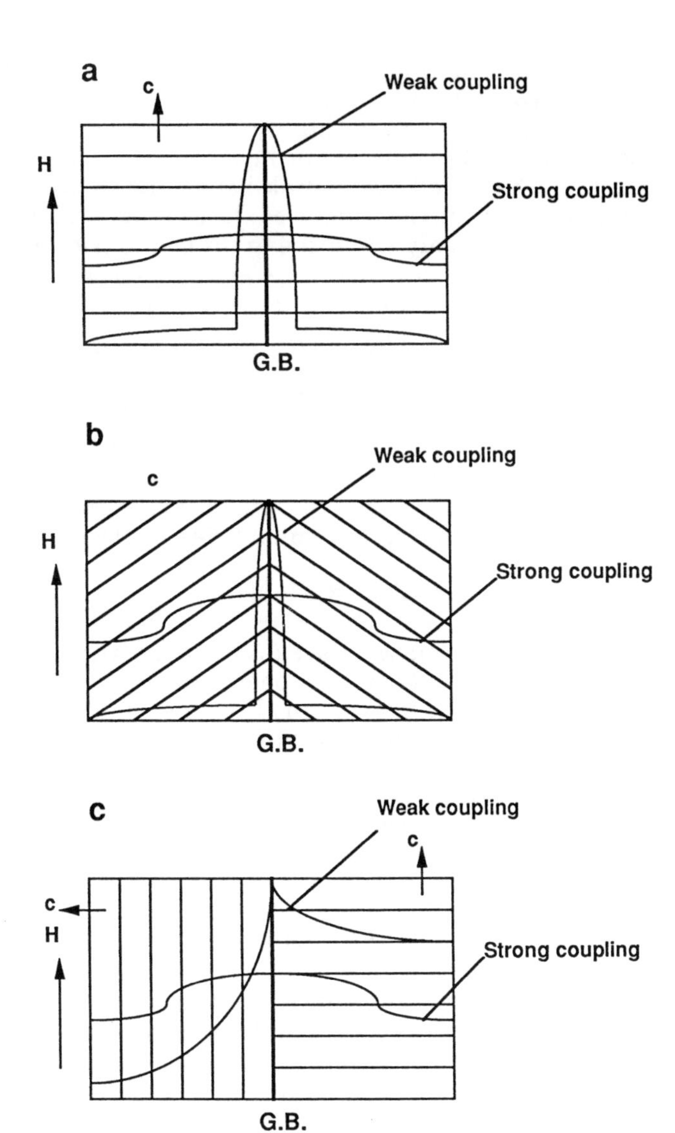

Fig. 1. Schematic diagrams showing the field penetration at grain boundaries with different orientations.

orientation shown in Fig. 1c, depending on the nature of the grain boundary structure.

Previous studies, which have generally emphasized the path for the transport of electrons at the grain boundaries, have failed to explain satisfactorily the complex nature of superconducting critical currents. Although the superconductor is highly anisotropic, the current will redistribute across high–angle grain boundaries if they are strongly coupled. On the other hand, weak–link behavior can still exist at low–angle grain boundaries if they are weakly coupled. The transport J_c will be severely reduced only when the grain boundaries are penetrated by an applied field. High transport of supercurrent is determined, to a large degree, by the resistance of the grain boundary area to the penetration of the applied field.

It has been well reported that transport J_c exhibits flux–pinning–dominated behavior in melt–processed samples, including flux–grown bicrystals (9), zone–melted rods (14), and melt–textured bars (15). Although these materials tend to contain mainly low–angle grain boundaries (14), substantial amounts of high–angle grain boundaries have been observed in these samples, which may provide evidence for strongly–coupled high–angle grain boundaries in these materials.

EXPERIMENTS

The melt–processed materials examined in these studies were produced by a technique that has been published (15). The bars contained large regions of well–aligned grains, but the regions often meet with large extents of misorientation (Fig. 2), as was observed by scanning electron microscopy (SEM). These bars had transport J_c values of 4.4×10^4 A/cm^2 at 77 K in fields greater than 1 T (Fig. 3). The current work focused on the properties of these specimens and transmission electron microscopy (TEM) of some of the grain boundaries.

DISCUSSION

Although microscopy data have not indicated the essential structural differences between strongly and weakly coupled grain boundaries in terms of field penetration, certain possibilities can be ruled out in terms of the formation of grain boundaries through different processing methods. Grain boundaries must be treated differently for liquid–processed and solid–state–sintered samples. It is reasonable to consider the high–angle grain boundaries to be of the following three types (16): (a) amorphous layer present (AGB); (b) transition structure (TGB), and (c) perfect grains up to the boundary (PGB), as for a coincident–site lattice. When there exists a layer with no well–defined lattice structure with relation to the grains, the properties of the boundary will be largely independent of the orientation, and the boundary may be considered as AGB. However, if the grain boundary is structurally related to the adjacent grains and

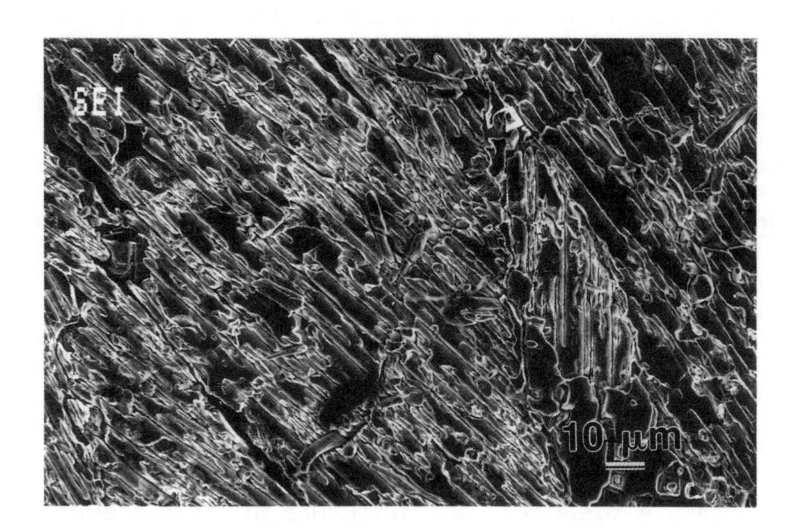

Fig. 2. SEM photograph of a melt–textured 123 bar.

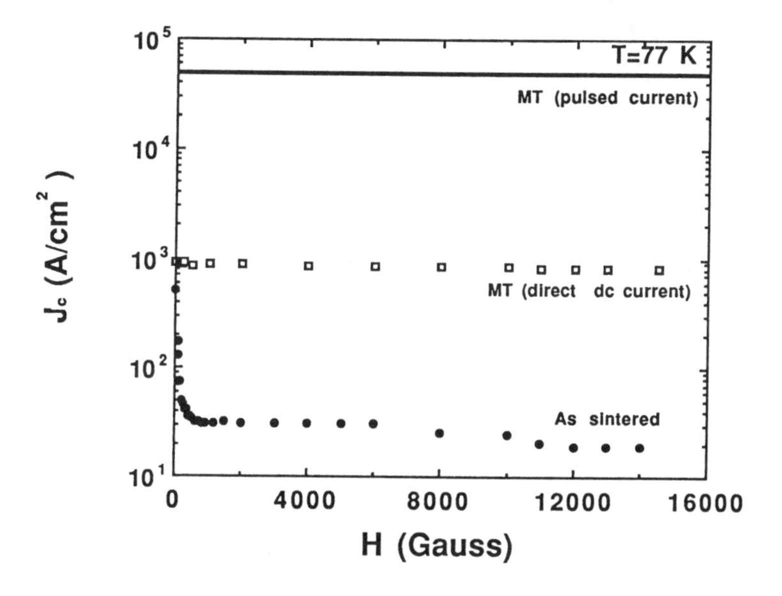

Fig. 3. Transport J_c values at 77 K for a melt–textured 123 bar.

the properties of the layer depend on the difference of the orientation, TGB may be an appropriate term to describe the boundary. The term PGB describes the distinct interface between two perfect crystals with negligible structural distortion at the edges of the crystals.

Amorphous–type or highly disordered grain boundaries are likely to be formed in sintered samples. Bulk samples are formed through diffusional processes at temperatures of ≈900–1000°C. Sintering by solid–state diffusion is slow, and grains have little chance to reorient. In addition, second phases can be present. Grain boundaries can possess a high degree of disorder and be glassy in nature. This type of boundary can be penetrated easily by an applied field, even for low–angle boundaries; thus, $F << 1$.

In contrast, grains are formed fundamentally differently in melt–texturing (17–19). Diffusion is rapid in a liquid. In addition, the duration of the growth process is typically significantly longer (for our studies, 3–5 days) than conventional sintering times (less than 1 day). Therefore, the grain boundaries can migrate and and are likely to assume low–energy configurations. Low–angle and low–energy boundaries are preferred, and texturing in large regions results. These grain boundaries tend to be types PGB and TGB (20).

Transmission electron microscopy was performed on many specimens. Films and melt–textured bars exhibited substantial misorientation and faceting of the grains, whereas the sintered specimens exhibited no facets on the high–angle boundaries (Fig. 4). Most grain boundaries in the melt–textured bars were of the low–angle type (Fig. 5). It is noted that the high–resolution images of the boundaries in the films and melt–textured bars indicated high coherence between the grains. It was observed that the high–angle grain boundaries of the melt–processed samples contained many facets, whereas such faceting appeared to be minimal in the sintered form (21). The crystal coherence and good connectivity at grain boundaries, including high–angle boundaries, were generally much higher for the melt–processed materials than for sintered materials. The high crystal coherency may have been achieved by a facet structure or by the orientations of the grains themselves. These types of grain boundaries are strongly coupled and thus are not easily penetrated by the applied field.

The degree of field penetration was examined by magnetization measurements. It has been shown (22) that the peak position in the flux creep–rate (dM/dlnt) versus field (or temperature) plot corresponds to the full penetration field, H^*. Figure 6 shows the field penetration, H^*, versus temperature for a sintered and a melt–textured sample with low– and high–angle grain boundaries. Note that the melt–textured sample contained large domains that were separated by high–angle grain boundaries (Fig. 2) and that the field was applied randomly to the sample. The sintered sample exhibited considerably lower H^* values than did the melt–textured sample.

It is clear that H^* is closely related to the field penetration at weakly linked areas and that H^* can serve as an indirect measure of the

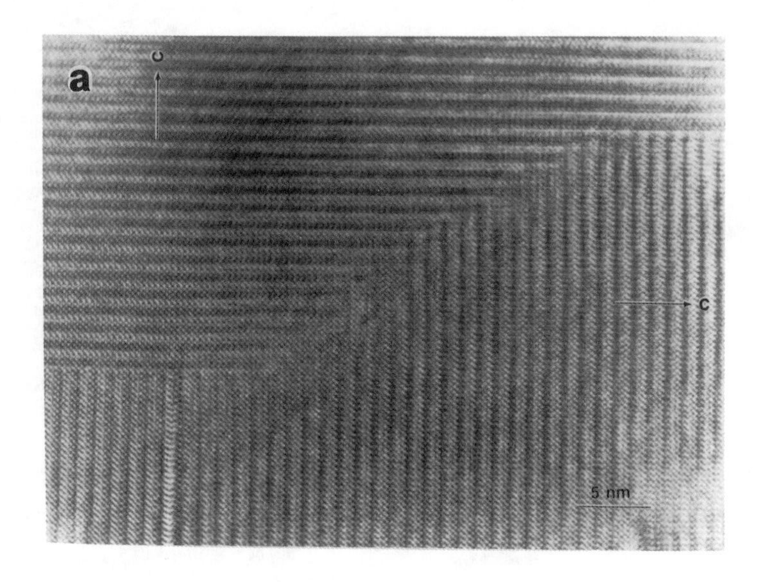

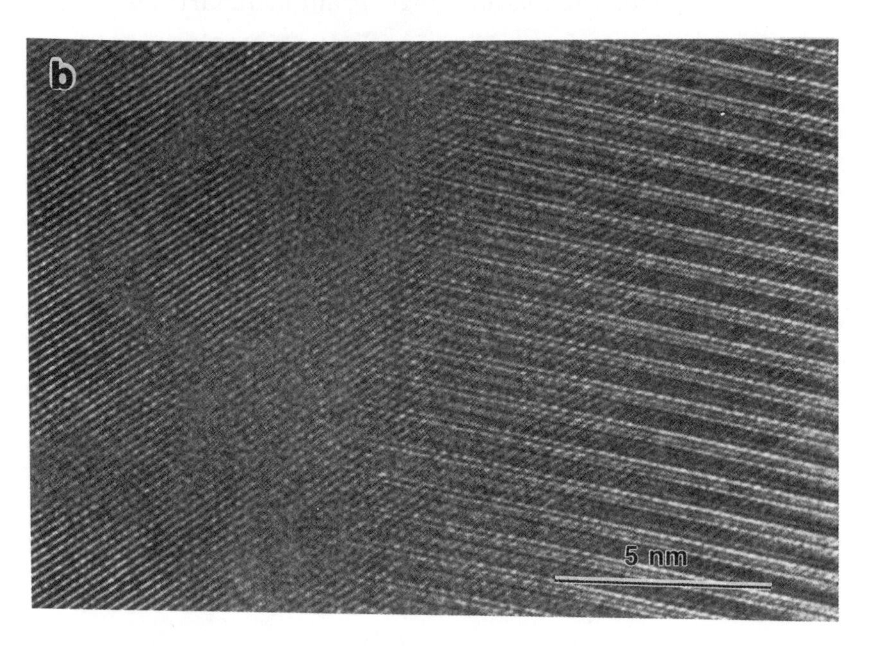

Fig. 4. High–resolution TEM images of (a) high–angle boundary in a 123 thin film and (b) boundary in sintered polycrystalline 123.

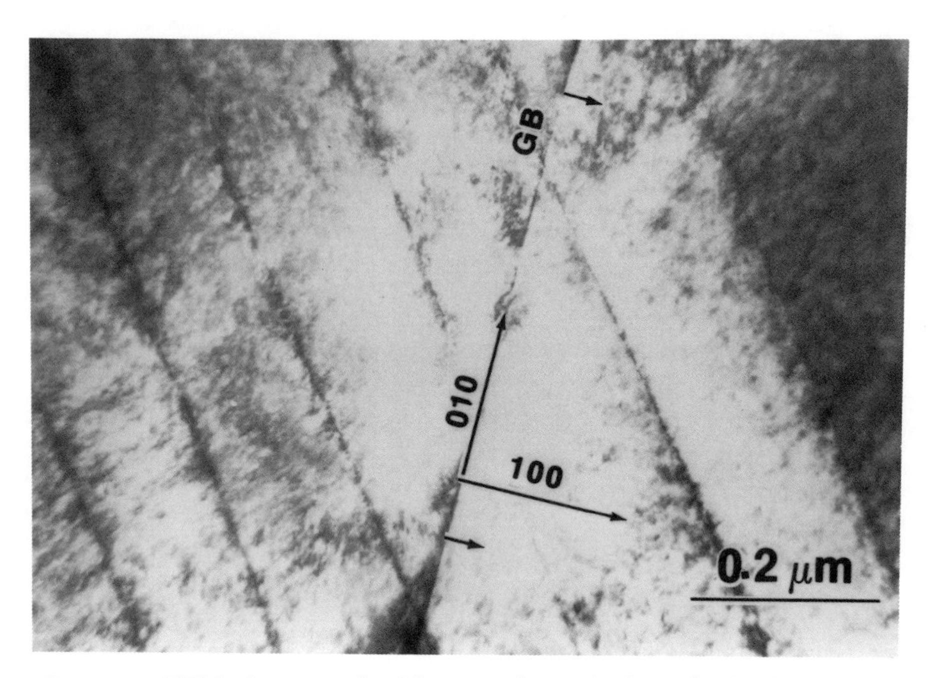

Fig. 5. TEM photograph of low–angle grain boundaries in 123.

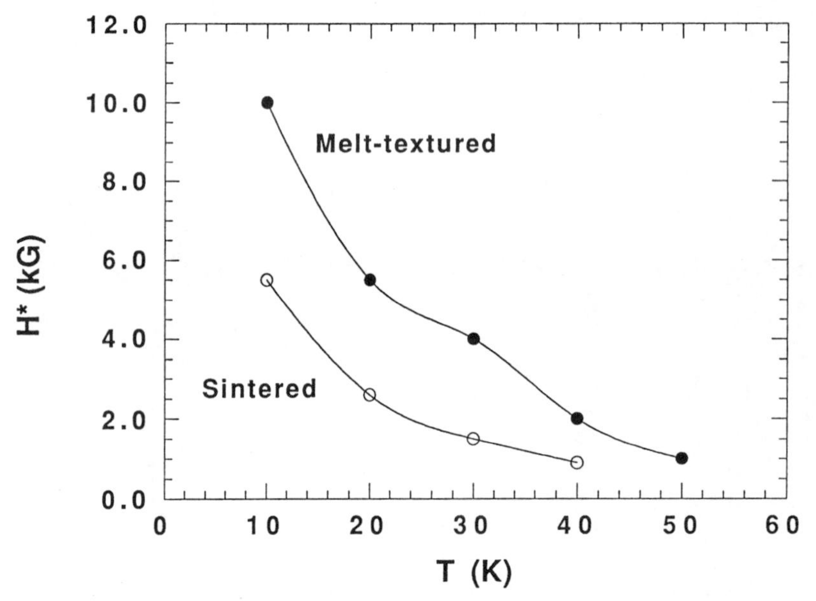

Fig. 6. Full field penetration, H*, versus temperature for a sintered sample and a melt–textured sample of 123. Note that the sizes of both samples are ≈ 2 mm.

coupling parameter, F. As calculated by Hylton and Beasley (11), F = 0.00063 and 8.6 for a ceramic sample and a single crystal, respectively. Magnetization measurements provide strong evidence for the difference of field penetration in weakly and strongly coupled samples, which indicates that the coupling strength is related to the connectivity of the crystals. The observations to date suggest that faceting of the grain boundaries of melt–textured samples may be a mechanism by which highly coherent boundaries are formed.

SUMMARY

Transport J_c is significantly reduced only when the grain boundaries are penetrated by the applied field and $(\lambda_L/\lambda_J)^2 << 1$. The degree to which a boundary area is penetrated by an applied magnetic field is related to the crystal coherence and connectivity at the boundary area, i.e., the boundary structure.

ACKNOWLEDGMENTS

D.S. is grateful to S. Salem–Sugui for valuable discussions during this research. This work was supported by the U.S. Department of Energy, Basic Energy Sciences–Materials Sciences, under Contract W–31–109–Eng–38; and by the National Science Foundation (DMR 88–09854) through the Science and Technology Center for Superconductivity.

REFERENCES

1. D. Shi, D. W. Capone II, G. T. Goudey, J. P. Singh, N. J. Zaluzec, and K. C. Goretta, Mater. Lett. **6**, 217 (1988).
2. D. Shi, J. G. Chen, M. Xu, A. L. Cornelius, U. Balachandran, and K. C. Goretta, Supercond. Sci. Technol. **3**, 222 (1990).
3. R. J. Cava, B. Batlogg, R. B. Van Dover, D. W. Murphy, S. Sunshine, T. Siegrist, J. P. Remeika, E. A. Rietman, S. Zahurak, and G. P. Espinosa, Phys. Rev. Lett. **58**, 1676 (1987).
4. J. W. Ekin, Adv. Ceram. Mater. **2**, 586 (1987).
5. R. L. Peterson and J. W. Ekin, Phys. Rev. B **37**, 9848 (1988).
6. R. L. Peterson and J. W. Ekin, Physica C **157**, 325 (1989).
7. R. B. Stephens, Cryogenics **29**, 399 (1989).
8. D. Dimos, P. Chaudhari, J. Mannhart, and F. K. LeGoues, Phys. Rev. Lett. **61**, 219 (1988).
9. S. E. Babcock, X. Y. Cai, D. L. Kaiser, and D. C. Larbalestier, Nature **347**, 167 (1990).
10. T. Van Duzer and C. W. Turner, "Principles of Superconductive Devices and Circuits" (Elsevier North Holland, Amsterdam, 1981) p. 139.
11. T. L. Hylton and M. R. Beasley, Phys. Rev. B. **39**, 9042 (1989).
12. J. E. Tkaczyk, K. W. Lay, and H. R. Hart, in "Superconductivity and Applications" ed. H. S. Kwok, Y.–H. Kao, and D. T. Shaw (Plenum,

New York, 1990) p. 557; J. D. Livingston, A. R. Gaddipati, and R. H. Arendt, private communication with the author.
13. L. Krusin–Elbaum, A. P. Malozemoff, Y. Yeshurun, D. C. Cronemeyer, and F. Holtzberg, Phys. Rev. B **39**, 2936 (1989).
14. D. Shi, H. Krishnan, J. M. Hong, D. Miller, P. J. McGinn, W. H. Chen, M. Xu, J. G. Chen, M. M. Fang, U. Welp, M. T. Lanagan, K. C. Goretta, J. T. Dusek, J. J. Picciolo, and U. Balachandran, J. Appl. Phys. **68**, 228 (1990).
15. D. Shi, M. M. Fang, J. Akujieze, M. Xu, J. G. Chen, and C. Segre, Appl. Phys. Lett. **57**, 2606 (1990).
16. P. G. Shewmon, in "Physical Metallurgy," ed. B. Chalmers (John Wiley and Sons, New York, 1959), p. 110.
17. M. Murakami, M. Morita, and N. Koyama, Jpn. J. Appl. Phys. **28**, L1754 (1989).
18. S. Jin, T. H. Tiefel, R. C. Sherwood, M. E. Davis, R. B. Van Dover, G. W. Kammlott, R. A. Fastnacht, and H. D. Keith, Appl. Phys. Lett. **52**, 2074 (1988).
19. P. J. McGinn, M. Black, and A. Valenzuela, Physica C **156**, 57 (1988).
20. Y. Zhu, H. Zhang, H. Wang, and M. Suenaga, preprint (1991).
21. Y. Gao, G. Bai, D. J. Lam, and K. L. Merkle, Physica C **173**, 487 (1991).
22. D. Shi, M. Xu, A. Umezawa, and R. F. Fox, Phys. Rev. B **42**, 2062 (1990).

POINT CONTACT TUNNELING STUDY OF THE HIGH TRANSITION TEMPERATURE SUPERCONDUCTOR $BI_2SR_2CACU_2O_8$

M.H. Jiang, C. A. Ventrice, K. J. Scoles, S. Tyagi, N. J. DiNardo and A. Rothwarf.

Ben Franklin Superconductivity Center,
and Drexel University, Philadelphia, PA 19104

Abstract

The point-contact (Pt-Ir probe) tunneling measurements on high-T_c super -conducting $Bi_2Sr_2CaCu_2O_8$ single crystals ($T_c = 82^o$ K) have been performed from 4.2 K to 77 K. A high conductance region around the origin, which is attributed to to the junction of the pure metallic-bridge type of point-contact tunneling, has been observed. There are also several current steps beyond the high conductance region in the I-V curves, corresponding to conductance troughs, which are either zero or negative, in the dI/dV characteristics of the tunneling spectra. These are pronounced structures indicative of bound states in tunneling spectroscopy. A superconducting energy gap of $\Delta = 26 \pm 2$ me V has been identified from tunneling spectra taken at T = 4.2 K and as a function of temperature. Current drops with some regularity in voltage intervals between drops, occur at high applied voltage (> 80mV). This resonance tunneling is the result of the effects of quantization due to the layered-structure of the crystal.

A model based on the superlattice-like nature of the crystal has been developed to interpret the tunneling observation of the bound states and the energy gap. When a normal-metal probe is brought into contact with the superconducting crystal, the region of the first CaCuO slab under that contact will be most likely reduced to a low T_c region with a smaller energy gap Δ_1, or even completely reduced to the normal state, due to the work function difference between the normal metal and the superconductor and/or the proximity effect. However, the second CaCuO slab remains in the intrinsic superconducting state with energy gap $\Delta_2 > \Delta_1$. Thus, when the energy of a tunneling quasiparticle is in the range $\Delta_1 > E > 0$, Andreev reflection takes place, giving rise to the high conductance near the origin (up to $V = \Delta_1/e$). If the energy of the quasiparticle is greater than Δ_1 but less than Δ_2 (or even slightly greater than Δ_2), quasiparticle reflections take place at the boundary of the first and second CaCuO slabs, resulting in the virtual bound states observed in the tunneling spectra. The drops in current at higher voltages result from a combination of resonant scattering and pair breaking.

Based upon such a model and experimental data, the energy levels of the bound states have been estimated to be "$n\varepsilon_0$" with n an integer and energy spacing $\varepsilon_0 = 9.1 \pm 0.3$ meV. The Fermi velocity of quasiparticles in the $Bi_2Sr_2CaCu_2O_8$ single crystal can be calculated from the energy spacing ε_0 and the lattice constant of the crystal, and the value of $v_F \approx 1.2 \times 10^6$ cm•sec^{-1} has been obtained. Hence the coherence length, has been evaluated to be around 1.0 Å using this value of the Fermi velocity. The values of the superconducting energy gap, Δ, the Fermi velocity, v_F, and the coherence length,obtained, are in good agreement with the results reported in the literature.

Introduction

Despite many "tunneling" measurements (1) on high Tc materials, using conventional geometries(2), break junctions(3), and point contacts(4), unambiguous interpretations of the features seen are still lacking. Differences amongst the above methods, which include a surface layer of unknown composition and thickness in conventional geometries(2), reconstructed surface layers in break junctions, and damage caused by the probe tip in point contact studies(4), make simple comparisons of features difficult. We have previously reported(4) point contact studies of $Bi_2Sr_2CaCu_2O_8$ single crystals using a scanning tunneling microscope system, in which a series of steps in the current-voltage curves were seen at voltages below ~70 mV. These steps were interpreted as being due to geometrical resonances, associated with quasiparticles injected into a region of the superconductor in which the gap had been reduced, by a combination of the proximity effect and space charge effects. Reflection of the quasiparticles resulted in a range of voltage over which the current was flat. The equation derived from Andreev reflection considerations enabled us to deduce values for Δ, v_F, and the coherence length, of $\Delta = 26$ meV, $v_F \sim 1.0 \times 10^6$ cm/sec and $\zeta \sim 1$ A°, by assuming a value of 15 A° for the relevant thickness of the layer in which reflections occurred.

In our experimental results, in addition to the steps, which are characterized by flat regions in the I-V and zero values of the conductance G = dI/dV, at higher voltages a series of giant saw tooth features appear in the I-V, with negative conductance spikes. It is these features that we wish to address in this paper. In contrast to most tunneling theory(5), which considers almost exclusively the conductance, and the position in energy of various maxima and minima in G, we are in addition interested in the current-voltage curve itself, and the magnitudes of the features seen. In particular, we are interested in what causes the widths of the steps and negative conductance regions, and the magnitude of the current drops in the saw tooth structure.

Experimental Procedures

Our experimental approach has been described previously (4). It consists of a Pt-Ir tip brought into contact with the Bi-Sr-Ca-Cu-O sample, using the STM design of Fein et al [6]. The current-voltage curve is obtained by a computer controlled system, while the conductance is obtained by numerically taking the derivative of the I-V curve.

Experimental Results and Discussion

A variety of curves can be obtained[4,7] depending upon how the tip is brought into the sample. For the purpose of this paper we wish to concentrate on the results taken at 4° K and illustrated in Figure 1, which shows both the steps at low voltage, and the higher voltage saw tooth structure. These features are repeatable, with some shifts in shape occurring from sample to sample, and after moving the tip. The features wash out with increasing temperature, and are gone slightly above 50° K.

The features of interest are indicated in Figure 1 and Table 1 gives the relevant data, broken into two series, based upon the model to be discussed in section 4. V_n is the voltage at the beginning of a step, or the voltage at the peak of the saw tooth current. δV_n is the width of the step or the width of the negative conductance region. δI_n is the drop in current from the peak of the saw tooth to its minimum. The $G_B\delta V$ values represent the rise in current above the peak that would have been reached at the position of the minimum in the absence of the current drop. G_B is the background conductance. M is a multiplication factor given by

$$M = \frac{\delta I + G_B \delta V}{G_B \delta V} = 1 + \frac{\delta I}{G_B \delta V} = 1 + \frac{G_P}{G_B} \qquad (1)$$

where G_P is the peak negative conductance at the position of the saw tooth current drop. M represents the multiplier that the incoming particles provide for the drop in current. The problems presented by the data are the following:

i) If incoming quasiparticles satisfy a resonance condition and are reflected out of the system, then they won't contribute to the current, and a flat region or step could occur. However, if there is a sinusoidal function a drop in current should occur not just a step. Hence what is the origin of the step, and what controls its width?

ii) For the saw tooth structure to occur with its drop in current, not only must the incident particles be reflected, but they must also impact on other carriers and cause them to tunnel back. Again we are faced with questions such as what determines the voltage of the peak, and the values of δV and δI?
In the next section we discuss a possible model for the observed structures.

Model

In our previous paper[4] we proposed a model for the lower voltage features, that involved a resonant condition for quasiparticle reflection. Basically carriers injected into the first Cu-O plane near the Pt-Ir normal metal tip could be reflected by both the physical BiO barrier layer and an the unperturbed gap of the second CuO plane. Figure 2 shows this model schematically. By building a wave packet of carriers bouncing back and forth in the first layer a resonance condition is found

$$\text{when } E_n = qV_n = \left[\Delta^2 + \left(\frac{n\pi\hbar v_F}{2d_1}\right)^2\right]^{1/2} = \left[\Delta^2 + (\varepsilon_0 n)^2\right]^{1/2} \qquad (2)$$

Plots of the square of this relation for several samples using low voltage values yield $\Delta \sim 26$ meV and values of $\frac{n\pi\hbar v_F}{2d_1}$ of ~10 meV. Using a value of $d_1 = 15$ A° enables a value for v_F of ~ 1.5×10^6 cm/sec to be deduced. Using the BCS relation for the coherence length $\zeta = \hbar v_F/\pi\Delta(0)$ then yield $\zeta \sim 1$ A°.

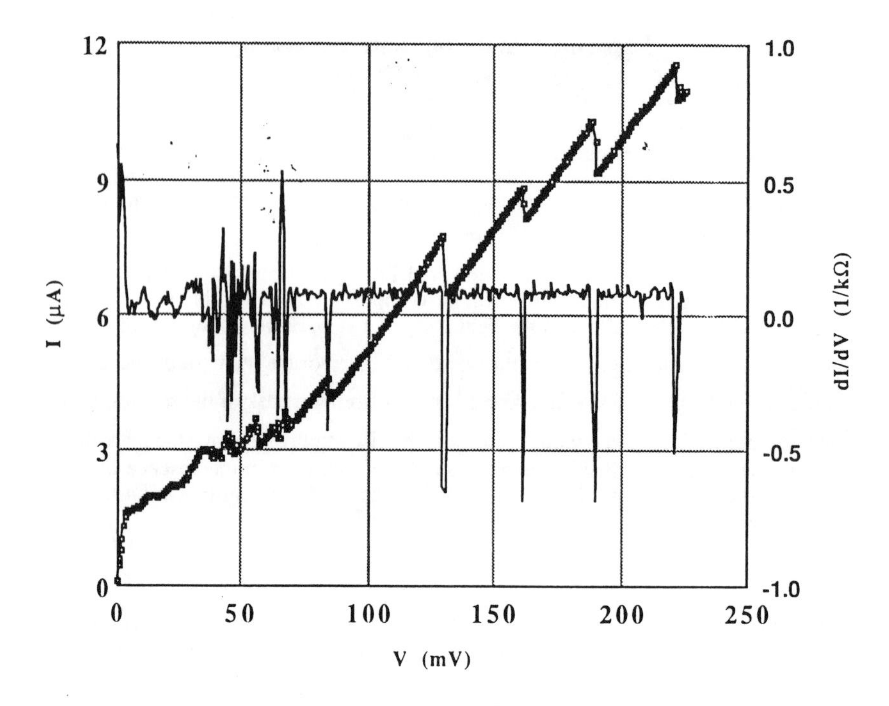

Fig. 1 Point contact tunneling spectra of Bi-Sr-Ca-Cu-O (2212) single crystal showing steps and saw tooth structure.

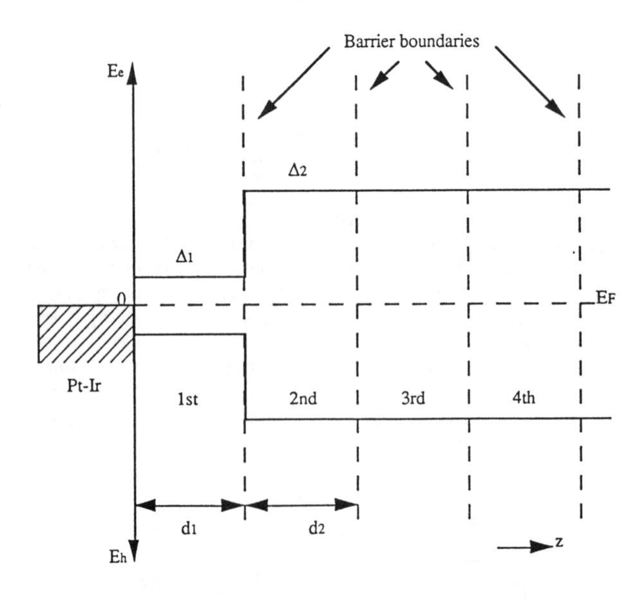

Fig. 2 Schematic representation of gap barrier at first layer and internal barriers.

Table 1 Relevant data from Figure 1. G_B = 80 uS

Series 1

n	V_n (mV)	δV_n (mV)	δI_n (uA)	$G_B\ \delta V$(uA)	G_p (uS)	M
1	11	~5	0	0.4	0	1
2	21	~5	0	0.4	0	1
3	30	~5	0	0.4	0	1
4	45	~1	~0.4	0.1	~400	6
5	53	~1	~0.6	0.1	~300	5
6	61	~1	~0.5	0.1	~400	6
7	69	~1	~0.5	0.1	~400	6

Series 2

n	V_n (mV)	δV_n (mV)	δI_n (uA)	$G_B\ \delta V$(uA)	G_p (uS)	M
1	41	~1	0.2	0.1	~200	~5
2	69	~1	0.4	0.1	~400	~5
3	84	~1	0.4	0.1	~450	~6
4	129.9	1.5	1.0	0.1	~650	~9
5	161.2	0.9	0.6	0.1	~700	~9
6	188.5	2.0	1.4	0.15	~700	~9
7	221.2	2.0	1.0	0.15	~500	~7

If we look at the features in Figure 1 there are clearly distinct structures at low voltages, below ~ 40 mV, which show the steps and minimum conductance of zero, and at high voltages, above ~ 70 mV, with saw tooth structures and negative conductance. In the region from 40 to 70 mV there is clutter with both positive and negative conductance peaks that indicates a possible close coincidence of structure from two distinct types of series. In particular, distinguishing which values in this range belong to each series, determines the deduced value of Δ, and how well the data fits Equation 2.

For both series 1 and series 2 we have used Equation 2 and fitted least squares deviations curves for various combinations of values in the two series. The results are shown in Figure 3 with the corresponding numbers. The fits for series 1 imply a low value of Δ characteristic of the layer adjacent to the probe tip, while those for series 2 point to higher values of Δ characteristic of the bulk material.

In Figure 3a we show a least squares fit to the last four points in series 1, (using 45mV point) that yield from the $(qV_n)^2$ vs n^2) curve values of $\Delta = 26.86$ meV, and from the slope $\varepsilon_0 = 9.10$ meV with a coefficient of determination of 0.99966.

For series 2 using the 41mV and 69mV points, Figure 3b shows at least square fit with $\Delta = 19.03$ meV $E_0 = 31.47$ meV with a coefficient of determination of 0.99621.

However, if we use all the points in series 1, Δ's from 8 to 12 meV can be fitted depending upon whether the 41 or 45 mV point is used, while for series 2, including the 45mV point instead of 41mV reduces Δ to 15.59 meV. The coefficients of determination remain at 0.99. Hence identifying which points correspond to a given series can be crucial in determining the value of Δ deduced from Equation 2.

To explain the high voltage saw tooth structure we note that resonant effects have been seen in normal superconducting systems in the past[(5)], with thicknesses sampled by tunneling depending on scattering processes and coherence lengths. However, effects from reflection at the back of films many microns thick have been reported[(8)]. Since the $Bi_2Sr_2CaCu_2O_8$ is a layered compound, with multiple interior barriers as illustrated in Figure 2, we propose that at specified voltages, corresponding to specific k vectors in the c direction, that quasiparticles that may have penetrated many layers into the superconductor, but remain unthermalized, are reflected from the internal barriers, and bounce back and forth within the layers forming standing waves. It is the resonating quasiparticles that have sufficient energy to break pairs, and impart c-direction momentum, to allow the broken pair members to reverse tunnel, and cause a drop in current. Furthermore, these reverse tunneling quasiparticles, in passing through the first layer with its lowered gap, can break additional pairs. Thus, with sufficient energy, an incident quasiparticle can break a pair forming three quasiparticles, which each in turn can break a pair producing up to nine reverse tunneling quasiparticles, and perhaps even higher multiples if the incident energy is high enough.

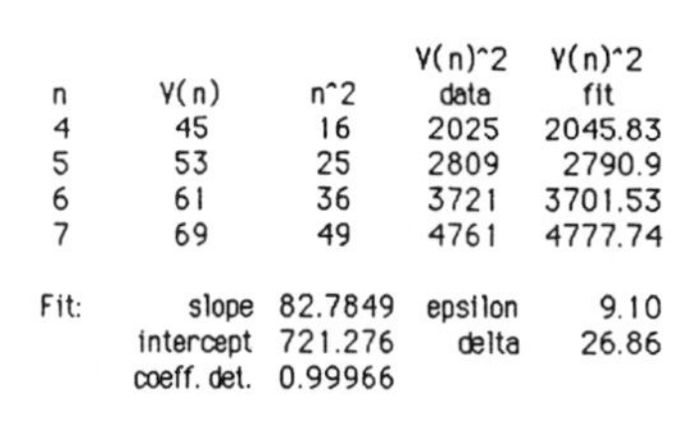

n	V(n)	n^2	V(n)^2 data	V(n)^2 fit
4	45	16	2025	2045.83
5	53	25	2809	2790.9
6	61	36	3721	3701.53
7	69	49	4761	4777.74

Fit:				
	slope	82.7849	epsilon	9.10
	intercept	721.276	delta	26.86
	coeff. det.	0.99966		

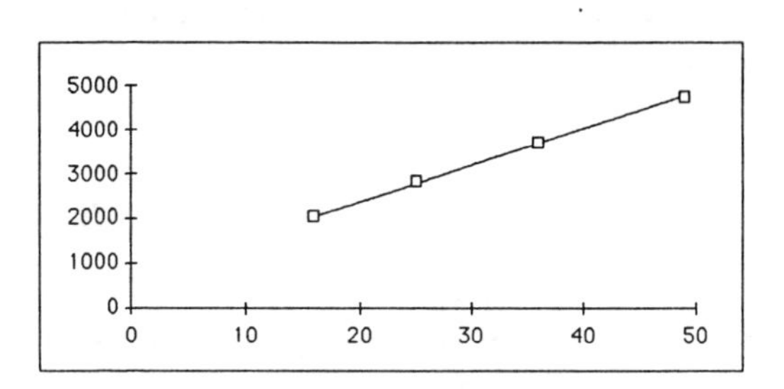

Fig. 3a Least square fit of last four points of series 1.

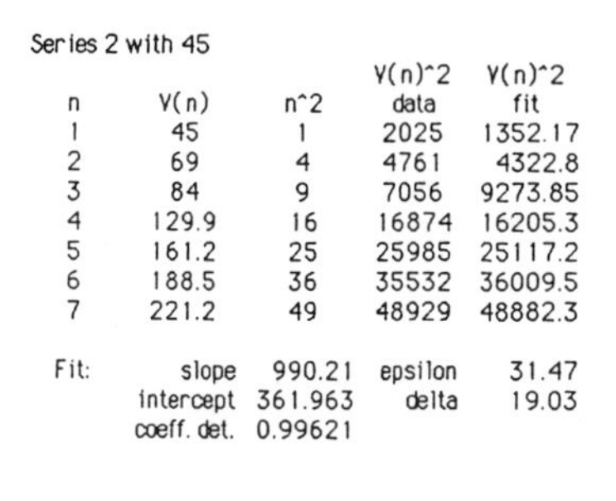

Series 2 with 45

n	V(n)	n^2	V(n)^2 data	V(n)^2 fit
1	45	1	2025	1352.17
2	69	4	4761	4322.8
3	84	9	7056	9273.85
4	129.9	16	16874	16205.3
5	161.2	25	25985	25117.2
6	188.5	36	35532	36009.5
7	221.2	49	48929	48882.3

Fit:				
	slope	990.21	epsilon	31.47
	intercept	361.963	delta	19.03
	coeff. det.	0.99621		

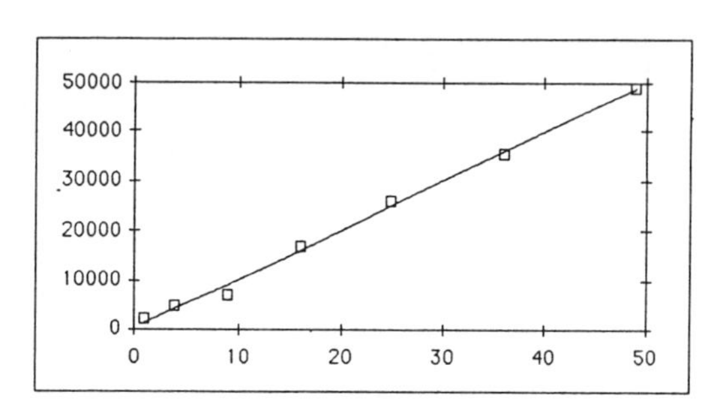

Fig. 3b Least square fit to series 2.

The agreement of this picture with the results listed in Table 1, is **i)** the M factor is unity for incident energies below 40 meV since these energies lie below the 3 Δ, threshold. **ii)** the maximum M value in Table 1 is 9, corresponding to the cascade process mentioned above. Below we indicate a mathematical model for the two series listed in Table 1.

The origin of Equation 2 is that in a Fermi liquid, excitations with momentum k measured from a full Fermi surface, have energy $\in$ given by:

$$\in = \frac{\hbar^2}{2m}(k^2 - k_F^2) \cong \frac{\hbar^2 k_F}{m}(k - k_F) \tag{3}$$

With $E_F = (\hbar^2/2m)\,k_F^2 >> \in$, $\in$ can be written as

$$\in = \hbar v_F (k - k_F) \tag{4}$$

On reflection a change in momentum $\hbar\Delta K = 2\hbar(k - k_F)$ occurs. For constructive interference $\Delta Kd = n\,2\pi$, with n an integer is needed. Since the energy of a quasiparticle in a superconductor is $E = (\Delta^2 + \in^2)^{1/2}$ using Equation (4) and the interference condition, one obtains Equation 2.

Series 1 in Table 1 corresponds to Equation (2). To obtain series 2, we invoke the picture presented above, of the injected quasiparticle bouncing back and forth between the interior barriers of a single unit cell, and apply the condition that when the quasiparticle breaks a pair it shares its energy and c-axis momentum equally with the two created quasiparticles. We can further require for additional pair breaking that the resulting quasiparticles also satisfy the resonance condition. This yields the condition $\frac{\Delta Kd}{3} = n'\,2\pi$ where n' is an interger. This condition results in recovering Equation (2) but with a factor of 3 multiplying n. This factor of 3 accounts for the major difference in slopes seen in Figure 3 for the plots of series 1 and series 2. A further factor results from a possible difference between d_1 which is assumed for series 1, and d_2 for series 2. Recall that d_1 is the distance between the normal metal tip and the first internal barrier, while d_2 is the distance between two internal barriers within a unit cell.

Further corrections, related to the large difference in magnitude of the Fermi energies[5] between the Pt-Ir tip and the superconductor, as well as their k_F vectors[5], and effective masses, should be considered if accurate fits are sought.

The width of the steps in voltage which are about 5 mV, and of the saw tooth structures which are less than 2 mV, may arise from the finite radius of curvature of the tunneling tip at the tunneling site, for the steps, and/or slight variations in the d spacing of the layers due to the pressure of the tunneling tip. A detailed model of these has not been attempted.

Conclusions

We have presented current-voltage curves obtained from a Pt-Ir point contact on single crystal $Bi_2Sr_2CaCu_2O_8$, that reveal a series of steps at low voltages, and saw tooth structures at high voltages. We interpret the features as arising from interference effects due to both internal barriers and a step in the pair potential between the layer in contact with the probe tip, which has a reduced gap, and the deeper unperturbed layers with normal gaps. Two series of features have been distinguished. The higher voltage saw tooth features result from resonant scattering of the injected quasiparticles in the deeper layers, with pairs being broken and the newly formed quasiparticles reverse tunneling to produce the drops seen in the current. We have defined a multiplication factor M, to numerically compare the current drop seen to a calclated quantity, and find reasonable agreement. Both series of features can be fitted to the same expression, from which a gap parameter, Fermi velocity, and coherence length can be obtained.

Acknowledgements

This work was supported in part by the U.S. Army Research Office through the Scientific Services Program administered by Battelle under Delivery Orders 1195 and 1604, and by the Ben Franklin Superconductivity Center a consortium of Drexel University, Temple University, and the University of Pennsylvania, with financial support from GE Astrospace, ITT Defense Inc., and the Ben Franklin Partnership program of the Commonwealth of Pennsylvania.

References

1. J.R. Kirtley, Int. J. Mod. Phys. 4, 201-237 (1990).

2. LH. Greene, J. Lesueur, W.L. Feldmann and A. Inam in High Temperature Superconductivity: Physical Properties, Microscopic Theory and Mechanisms. ed. J. Ashkenazi, S.E. Barns, F. Zuo, G.C. Vezzoli and B.M. Klein. Plenum Press, New York, 1991.

3. D. Mandrus, L. Forro, D. Koller, and L. Mihaly, Nature 351, 460 (1991).

4. M.H. Jiang, C.A. Ventrice Jr., K.J. Scoles, S. Tyagi, N.J. DiNardo, and A. Rothwarf, Physica C (to be published).

5. E.L. Wolf, Principles of Tunneling Spectroscopy Oxford University Press, Oxford 1989 Chapter 5.

6. A.P. Fein, J.R. Kirtley, and R.M. Feenstra Rev. Sci. Instrum. 58, 1806 (1987).

7. M.H. Jiang, Ph.D. dissertation Drexel University 1991.

8. W.J. Tomash, Phys. Rev. Lett. 15, 672 (1965).

BREAK-JUNCTION TUNNELING AND INFRARED SPECTROSCOPY ON $Bi_2Sr_2CaCu_2O_8$

L. Forro, D. Mandrus, C. Kendziora, and L. Mihaly
Department of Physics, SUNY @ Stony Brook

ABSTRACT

Single crystals of the high temperature superconductor of nominal composition $Bi_2Sr_2CaCu_2O_8$ (T_c=84K) have been studied by break junction tunneling and infrared transmission spectroscopy. At low temperatures the results are consistent with a pseudo-gap in the density of states with 2Δ = 84meV. The gap decreases with increasing temperature, but it does not follow BCS temperature dependence. At high temperatures the tunneling spectra are smooth and featureless.

Superconductor-Insulator-Superconductor (SIS) tunneling and infrared spectroscopy played a crucial role in the exploration of the physical properties of superconductors *[1,2]*. What makes these methods extremely useful is that the natural energy scale of the measurement can be matched to the relevant coupling and thermal energies in the metal. The differential conductance of the various tunneling junctions usually exhibits a sharp peak associated with the energy gap in the single particle density of states *[1]*. Infrared transmission, explored in great detail by Tinkham and others, also shows clear signs of the gap in the superconducting state *[2]*. These methods have been refined to utmost accuracy in order to study phonon effects and other details of the superconducting state *[3]*.

Great efforts have been made to apply these techniques to high temperature superconductors, but up to now there is no general agreement even on simple questions like the value of the energy gap. Tunneling data are ambiguous *[4]*, and there is an ongoing controversy about the interpretation of infrared reflectivity measurements *[5,6]*. Infrared transmission measurements do not show evidence of a sharp BCS type onset in the optical conductivity *[7]*, but the results are consistent with an energy gap in the range of 2Δ = 50mV-80mV (8-12 k_BT_c) *[8]*. Early tunneling measurements, mostly on ceramic samples, resulted in much smaller gaps *[4]*. Gaps from careful SIN tunneling measurements on $Bi_2Sr_2CaCu_2O_8$ by Miyakawa et al. *[9]* were in the range of 42-44mV, while Lee et al. obtained 2Δ=50mV *[10]*. In two recent studies, closely related to ours, Briceno and Zettl reported 2Δ=45mV for ab plane tunneling in $Bi_2Sr_2CaCu_2O_8$ break junctions and Ponomarev et al. obtained energy gaps in the range of 35-50mV on BiSrCaPbCuO break junctions *[11]*.

In this article we report SIS tunneling results on junctions made by breaking ultrathin $Bi_2Sr_2CaCu_2O_8$ single crystals. The superconducting transition temperature of the samples was T_c=84K. We report the temperature dependence of tunneling spectra (differential conductance, dI/dV, *vs.* voltage) between temperatures of 10K and 160K. The spectra are

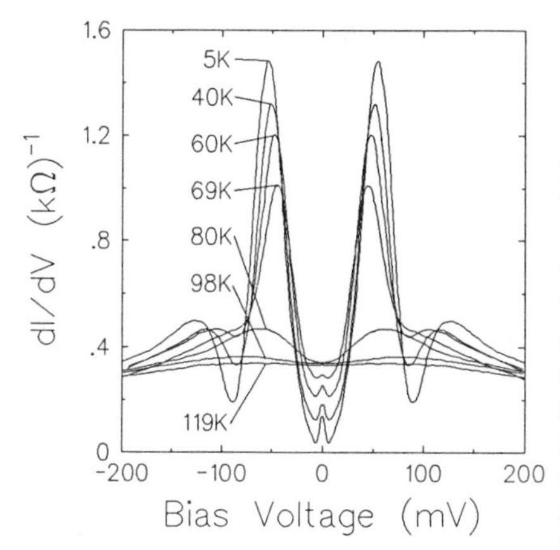

Fig. 1 Differential conductance (dI/dV) vs voltage at various temperatures. Note that a zero bias supression of dI/dV is observable even above T_c. The small zero bias peak at low temperatures is due to Josephson tunneling.

suppressed, probably to zero, in the superconducting ground state. Most of our junctions were "good" in this sense. For two samples we were able to mechanically "tune" the junction in the cold state, and repeatedly reduce and increase the R_N by a factor of 30, while keeping $R_0/R_N>10$. Occasionally, we found the "V shaped" or "parabolic" differential conductance curves reported in earlier tunneling measurements *[14],[15]*, but only if $R_0/R_N<10$. We suspect that the differences may originate from the different junction geometry: while we were mostly probing electron transfer parallel to the CuO planes, other measurements may be looking at the tunneling perpendicular to the planes.

In Figure 1 we show differential conductance *vs.* voltage at selected temperatures. For SIS tunneling between BCS superconductors the low temperature differential conductance is zero up to $V=2\Delta/e$, then it exhibits a sharp peak. The rounding of the measured curves can not be attributed to BCS type "finite temperature" effects. At low temperatures the differential conductance peaks at $V=\pm55$mV. The temperature dependence of the characteristic parameters of the tunneling spectra are plotted in Fig. 2. The zero bias depression in the differential conductance curves diminishes rapidly as the temperature approaches T_c (T_c was independently measured before breaking the sample). There is, however, a residual gap-like structure above T_c. It remains to be seen if this is due to inclusions of a 2223 phase, or it is an intrinsic property of the two dimensional CuO layers.

Low temperature tunneling spectra were recorded on at least ten other samples with very similar results. Variations occur in the magnitude of "pair tunneling" current I_c (*i.e.* zero bias anomaly of the differential conductance), in the height of the peak in the differential conductance and in the junction resistance R_N. We found that the high resistance junctions have smaller pair tunneling currents; the $V_c=I_cR_N$ product decreased with the increasing junction resistance. One expects $V_c=\alpha\Delta$, where α is a number of order of unity *[16]*. However, for high R_N junctions the Josephson energy ($E=\hbar I_c/2e=(2k\Omega/R_N)\alpha e\Delta$) decreases, and the junction is more susceptible to electronic and thermal noise.

The infrared transmission study was performed on similarly prepared samples at the National Synchrotron Light Source in Brookhaven. The thin single crystals of area 0.7mm² were laid on sample holders with an aperture of 0.6 mm diameter. The thickness of the samples were accurately determined by X-ray transmission measurements and it was found to be

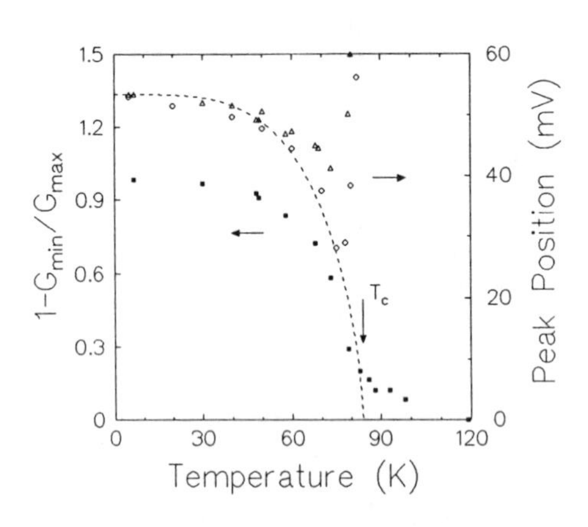

Fig. 2. Temperature dependence of the parameters of the tunneling spectra. The structure in the differential conductance G=dI/dV is characterized by the quantity 1-Gmin/Gmax (full squares), where Gmin and Gmax are the minimum and peak values of G. Open triangles represent the voltage belonging to the peak position. Open circles correspond to the peak position measured with another sample. The dashed line indicates the weak coupling BCS temperature dependence of the gap, fitted to the low temperature value of the peak position with a vertical scale factor.

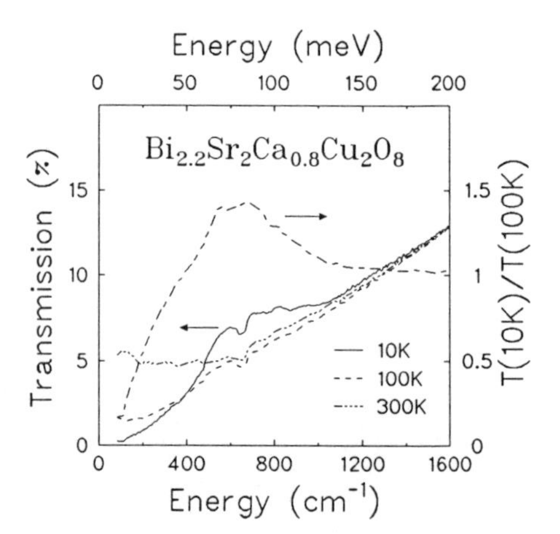

Fig. 3. Optical transmission of $Bi_2Sr_2CaCu_2O_8$ single crystals with infrared radiation polarized perpendicular to the c axis. The dashed line shows the ratio of transmissions at 25K and 100K.

uniform to within 50Å. The samples were cooled in an optical cryostat. The measurements were made in two frequency ranges: $100cm^{-1}$ - $750cm^{-1}$ in a Nicolet 20F spectrometer *[7]* and $400cm^{-1}$ - $3000cm^{-1}$ in a Nicolet 60SX spectrometer. Similar studies have been performed earlier by Romero et al. *[8]*.

The optical transmission spectrum of a typical sample is plotted in Fig. 3. Although the low temperature spectra do not show the pronounced peak characteristic of lead or tin in the superconducting state [2], the ratio to the normal state (T=100K) transmission exhibits a maximum around 700 cm^{-1}. The detailed analysis of the infrared transmission and reflection data is rather involved *[5-8]*. There is no doubt, however, that spectra similar to those presented in Fig 3. for low temperatures (and the corresponding infrared reflectance spectra) can be obtained only if the real part of the optical conductivity, σ_1, has an increase at frequencies below $700cm^{-1}$. This increase has been alternatively associated with mid-infrared oscillators *[5]* or to the onset of σ_1 at the gap frequency 2Δ *[6]*. In accordance with the earlier measurements, we found that the onset frequency does not have a strong temperature dependence.

In many ways the experimental results presented above speak for themselves in their strong resemblance to the classic measurements on BCS superconductors *[1,2]*. In the framework of the BCS theory, there is an important qualitative difference between the infrared transmission and

symmetric and they exhibit a strong anomaly on an energy scale extending to 100mV. We interpret the tunneling spectra in terms of a pseudo-gap Δ in the single electron density of states. We compare the results to infrared transmission measurements, on the same material, in the frequency range of $100cm^{-1}$ $1600cm^{-1}$. We conclude that at low temperatures the gap parameter 2Δ is certainly not less than 7 k_BT_c and may be as high as 12 k_BT_c. The gap does not reach zero at the critical temperature, but electronic states fill in the low energy states as the transition temperature is approached from below.

The sample preparation and characterization by electron microscopy and dc transport measurement are described elsewhere *[12]*. The typical specimens are cleaved to 1000Å thickness with an area of $0.5mm^2$. The copper oxide layers are parallel to the cleavage planes. Transmission electron microscopy on similarly prepared samples indicate that the crystals are not twinned. Each sample undergoes heat treatment in air at 600C in order to set an equilibrium oxygen content; after the heat treatment the superconducting transition temperature is 84K.

For the tunneling measurements a "break junction" device was built *[13]*. The samples were cooled in vacuum, and the superconducting transition was monitored by resistance measurement. The transition temperatures obtained in the device were identical with those from the regular cryostat measurements. When low temperatures (around 10K) were reached, the sample was broken by bending its support. In about 50 percent of the attempts SIS junctions were produced, the insulator being the vacuum between the surfaces of the broken sample. The junctions were connected to a combined dc and ac voltage source and the current, the voltage and the differential conductance were recorded *[13]*. The large-bias junction resistances were in the range of R_N=500Ω–50kΩ. For the samples heated above T_c, this resistance was equal to the normal state junction resistance. At low temperature all samples had zero bias (pair tunneling) current. We found that a 4kG magnetic field, applied parallel to the c axis of the sample, strongly suppresses the critical current I_c of the pair tunneling; for fields in the other directions I_c was less sensitive. From this result we conclude that the break is perpendicular to the copper oxide planes and the junction is likely to probe the "in plane" wavefunction of the electrons. The cross section of the junction is estimated to be about 1000Åx1000Å. (Note that the typical sample thickness is t=30c where c is the lattice spacing perpendicular to the CuO planes.) The comparison of measurements of slow (1mV/sec) and fast ($4x10^5$mV/sec) sweep rates did not show any evidence for heating effects.

Junctions were selected for detailed study by looking at the the zero bias/large bias resistance ratio, R_0/R_N (when calculating R_0 the pair tunneling contribution to the current was subtracted). A junction was considered "good" if it had an R_0/R_N of at least 10. This ratio is by no means comparable to the best junctions obtained from lead or aluminum *[1]*, but it is much better than the values reported for high T_c materials *[4,9-11,14]*. In fact the large resistance ratio alone allows us to conclude that the density of states at the Fermi level is strongly

the tunneling measurements: in contrast to tunneling, the optical conductivity is influenced by the coherence factors. Consequently, the SIS tunneling conductance has a Dirac δ divergence at $V=2\Delta/e$, while σ_1 has a continuous onset at $\hbar\omega=2\Delta$. The major optical features in the optical spectra usually appear at frequencies higher than the gap frequency; *e.g.* in the measurement of Ginsberg and Tinkham the transmission peaks at $h\omega=2.6\Delta$ *[2]*. This general argument may help to reconcile the large difference in the energies belonging to the peaks in the tunneling spectrum (55mV) and in the optical transmission ratio (80mV). This interpretation assumes that the density of states has a true gap, the rounding of the tunneling conductance curves is due to "dirt effects". The $2\Delta/k_BT_c$ ratio will be around 7-8, similar to the value suggested by Schlesinger et al. for another high temperature superconductor, $YBa_2Cu_3O_7$ *[6]*.

The inspection of the data reveals major deviations from the BCS model, however. The most pronounced one is the finite conductance at $V\neq0$ in the low temperature tunneling spectra. We noticed that at low voltages $dI/dV \propto V^2$: if plotted against V^2, the differential conductance is a straight line up to V=40mV. By making a few simple approximations, one can deduce the density of states, g(E), from the low temperature data. If the tunneling matrix element is independent of the momentum, energy and bias voltage, then the tunneling current is given by $I \sim \int g(E-E_F-V/2)g(E-E_F+V/2)dE$, where the integral is from $-V/2$ to $+V/2$, representing the cut-off due to the low temperature Fermi functions. For $g(E-E_F)=g(E_F-E)$ (density of states function symmetric around E_F), $dI/dV \propto V^2$ leads to $g(E-E_F)\propto|E-E_F|$ *i.e.* the density of states is linear close to the Fermi energy.

We were able to reproduce the major features of the differential conductance, including the minima at V=±80mV, with a $g(|E-E_F|)$ linearly increasing from zero to $2.2g_o$ between zero and 42mV, linearly decreasing from $2.2g_o$ to g_o between 42mV and 60mV and constant g_o above that. It is remarkable that the maximum in the density of states is at V=84mV (corresponding to $\Delta = E-E_F=V/2=42mV$), while the differential conductance peaks at 55mV. In the absence of a microscopic theory we can not calculate the optical transmission, but it is very likely that the linear density of states leads to a non-singular, smooth $\sigma_1(\omega)$ function. We want to emphasize the near coincidence of the optical transmission peak and the value of the pseudo-gap, both of them being at $2\Delta=12k_BT_c$.

Another non-BCS feature is apparent if we consider the area under the differential conductance curves (*i.e.* the IV curves). In the BCS theory the area under the tunneling conductance curves, calculated up to a fixed voltage of a few times Δ, is independent of the temperature *[17]*. In our measurements the area is larger at low temperatures. Although we can not entirely exclude that the difference is due to experimental artifacts, we suggest two other possibilities: a./ at the superconducting transition electronic states are imported from energies $|E-E_F|$ much larger than Δ, or b./ in the normal state the charge carrying quasi-particles cannot tunnel. The latter case would occur if the electrons in the normal state are heavily "dressed" with other excitations - the extreme case is when the elementary excitations are holons and spinons, and they have to pair up before tunneling *[18]*.

In summary, we have shown that break junction tunneling and infrared transmission spectroscopy point to an energy gap or pseudogap on the scale of $7k_BT_c$ to $12k_BT_c$, depending on the interpretation of the results. The magnitude of this parameter is weakly temperature dependent. The density of states, evaluated from the tunneling conductance at low temperatures, is linear near the Fermi surface. The results indicate that the number of states accessible for tunneling around the Fermi surface is different in the normal and superconducting states.

Acknowledgements. We are indebted for discussions with P.B. Allen, M. Gurvitch and L. Carr. We thank G.P. Williams for valuable contributions to the infrared studies. This work has been supported by NSF Grant DMR 9016456.

REFERENCES

[1] I. Giaver, Phys. Rev. Letters 5 464 (1960) .
[2] D.M. Ginsberg, M. Tinkham, Phys. Rev. 119 990 (1960).
[3] W.L.McMillan and J.M. Rowell in "Superconductivity" ed. R.D.Parks p.561 (M.Dekker, New York, 1969).
[4] For a recent review see M. Yu Kupriyanov and K.K.Likharev, Usp. Fiz. Nauk. 160, 49 (1990) {Sov. Phys. Usp., 33 340 (1990)}
[5] K. Kamaras et al. Phys. Rev. Letters 64 84 (1990)
[6] Z. Schlesinger et al. Phys. Rev. Letters 65 801 (1990)
[7] L. Forro et al. Phys. Rev. Letters, 65, 1941 (1990)
[8] D. Romero et al., to be published .
[9] N. Miyakawa, D. Shimada, T. Kido and N. Tsuda, J. Phys. Soc. Jpn. 58 383 (1989) and references therein.
[10] M. Lee, D.B Mitzi, A. Kapitulnik, M.R. Beasley, Phys. Rev. B 39 801 (1989)
[11] Ponomarev et al., Proceeedings of LT 19
G. Briceno and A. Zettl, Solid State Commun. 70, 1055 (1989)
[12] L. Forro, D. Mandrus, B. Keszei and L. Mihaly, J. Appl. Phys., 68 4876 (1990)
[13] The device is similar to that described by J. Moreland and J.W. Elkin J. Appl. Phys., 58, 3888 (1985). We used a simple screw mechanism to bend the sample holder. A detailed report on the device, and the the IV curves of various junctions investigated will be published elsewhere.
[14] M. Gurvitch et al. Phys. Rev. Letters, 63 1008 (1989)
[15] J. Moreland et al. Phys. Rev. B 35, 8856 (1987)
[16] V. Ambegaokar and A. Baratoff, Phys. Rev. Letters, 10 486 (1963); see also P.W. Anderson, in Lectures at Rovello Spring School (1963)
[17] At the phase transition the electronic states are rearranged within an energy range of a few Δ and the total number of states does not change.
[18] We thank P.W. Anderson for pointing out this possibility.

STUDIES OF Y-Ba-Cu-O SINGLE CRYSTALS BY X-RAY ABSORPTION SPECTROSCOPY

A. Krol, Z. H. Ming, and Y. H. Kao
Department of Physics, SUNY at Buffalo, NY 14260, USA

N. Nücker, G. Roth, and J. Fink
Kernforschungszentrum Karlsruhe, Institut für NukleareFestkörperpysik
Postfach 3640, W-7500 Karlsruhe, Federal Republic of Germany

G. C. Smith
Brookhaven National Laboratory, Upton, NY 11973, USA

A. Erband G. Müller-Vogt
Universität Karlsruhe, Institut für Kristallchemie
W-7500 Karsruhe, Federal Republic of Germany

J. Karpinski and E. Kaldis
Labor für Festkörperphysik, ETH Zürich, Hönggerberg/HPF,
CH 8093 Zürich, Switzerland

K. Schönmann
Walter-Meißner-Institut, W-8046 Garching, Federal Republic of Germany

ABSTRACT

The symmetry and density of unoccupied states of $YBa_2Cu_3O_7$ and $YBa_2Cu_4O_8$ have been investigated by orientation dependent x-ray absorption spectroscopy on the O 1s edge using a bulk-sensitive fluorescence-yield-detection method. It has been found that the O 2p holes are distributed equally between the CuO_2 planes and CuO chains and that the partial density of unoccupied O 2p states in the CuO_2 planes are identical in both systems investigated. The upper Hubbard band has been observed in the planes but not in the chains in both systems.

INTRODUCTION

Orientation dependent x-ray absorption spectroscopy (XAS) at the O 1s edge is a useful tool to measure the partial unoccupied DOS and its symmetry at the O sites. This method was applied to the $YBa_2Cu_3O_7$ (Y123) and $YBa_2Cu_4O_8$ (Y124) superconductors. In these systems besides the CuO_2 planes formed by Cu(2) and O(2,3) atoms there are also chains along the ***b***-direction formed by Cu(1), O(1), and the apex O(4) atoms. In Y124 there is an additional CuO chain in antiregister with the chain which exists in Y123. As a result the O(1) atoms in Y124 are coordinated with 3 Cu(1) atoms while in Y123 they are coordinated only with 2 Cu(1) atoms. The holes are formed upon doping the respective parent compound with the additional O(1) atom which oxidizes the Cu(1) atom from Cu^+ to Cu^{++} and therefore contributes more positive charge. Created holes are shared by the CuO_2 planes and the CuO chains. The goal of this work was to obtain information on the ratio of the number of holes in planes and that in chains. This problem has been studied in a previous investigation of orientation-dependent O 1s absorption edges of twinned Y123 single crystals using EELS[1] in transmission. The EELS experiments gave evidence for about two times as much holes in O $2p_{x,y}$ orbitals compared to those in O $2p_z$ orbitals. However twinning present in these crystals hindered reliable evaluation of the O 2p distribution between the CuO chains and the CuO_2 planes.

EXPERIMENTAL

Single crystals of Y123 were grown in a ZrO_2 crucible and oxidized to show a sharp superconducting transition at 91 K. They were heated in flowing oxygen to 400 °C and subjected to uniaxial stress while slowly cooling them down to form twin free samples.[2] The Y124 single crystals were grown[3] using high oxygen pressure (2.7 Kbar) at 1170 °C; the measured T_C was 80 K. Y123 and Y124 crystals were glued to the sample holder using silver paint and oriented to permit measurements with the ***E*** vector of the x-rays parallel to the ***a***-axis and to the ***b***-axis. They were then mounted in a vacuum chamber near the soft-x-ray fluorescence detector. The soft-x-ray absorption at the O 1s edge was measured at the AT&T Bell Laboratories Dragon high resolution beam line at the NSLS at Brookhaven National Laboratory.[4] During the experiment, the monochromator resolution was set at 250 meV. The oxygen K_α fluorescence yield was monitored by means of a low pressure parallel plate avalanche chamber.[5] All measurements were performed in an energy range $520 \leq \hbar\omega \leq 570$ eV. The spectra were normalized to the integrated area between 534 and 560 eV after subtracting an energy-independent background.

RESULTS

In Fig. 1 we show a typical O 1s absorption spectrum of a ***c***-axis oriented Y123 film on YSZ substrate obtained by laser ablation.[6] The ***E*** vector was parallel to the CuO_2 plane.

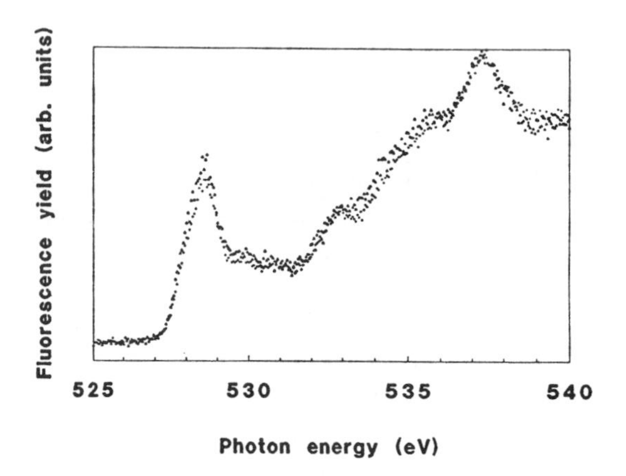

Fig. 1. Half of the sum of the O 1s absorption spectrum for ***E//a*** and ***E//b*** (triangles) compared to a spectrum of a ***c***-axis oriented $YBa_2Cu_3O_7$ prepared by laser ablation (crosses).

According to previous measurements the first peak at 528.8 eV is due to hole states formed upon doping. The rise in spectral weight above 531 eV is due to Ba 5d and Y 4d states hybridized with O 2p states. In Fig. 1 we also show a second O 1s spectrum which is composed of spectra obtained with untwinned Y123 crystals for ***E*//*a*** and ***E*//*b***. The perfect agreement of the two spectra shown in Fig. 1 demonstrates the reliability of our data and normalization procedure used.

In Fig. 2 we show the O 1s edges in the pre-edge region (527-531 eV) for Y123 and Y124 as measured for ***E*//*a*** and ***E*//*b*** together with the difference spectra between the two directions (***E*//*b*** – ***E*//*a***).

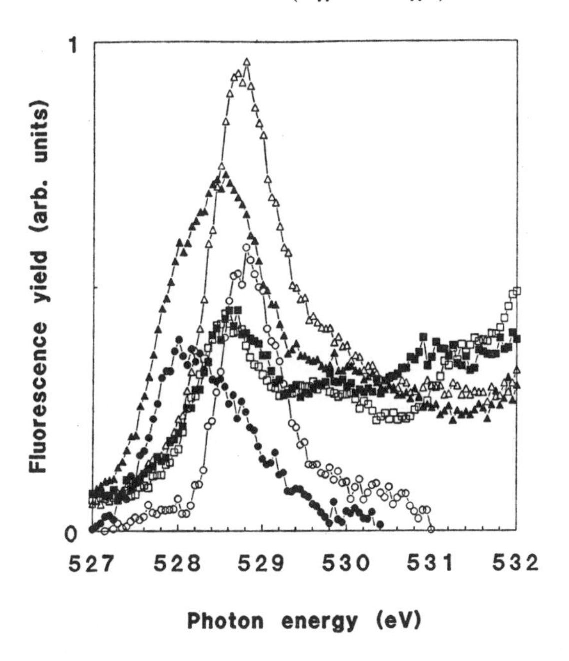

Fig. 2. O 1s x-ray absorption spectra for $YBa_2Cu_3O_7$ ***E*//*a*:** filled squares, ***E*//*b*:** filled triangles, difference spectrum (***E*//*b*** – ***E*//*a***): filled circles, and $YBa_2Cu_4O_8$ ***E*//*a*:** open squares, ***E*//*b*:** open triangles, difference spectrum (***E*//*b*** – ***E*//*a***): open circles.

For ***E*//*a*** the absorption edges for Y123 and Y124 as measured are almost identical with two maxima at 528.6 and 529.7 eV; however, the threshold observed in Y124 is slightly shifted towards higher energy as compared with Y123. For ***E//b*** the spectral weight near 528.6 eV is almost twice as high as for ***E//a***. The pre-edge peak of the difference spectrum (***E//b*** – ***E//a***) is slightly shifted to lower energies as compared to the spectrum ***E//a***, while for Y124 the opposite shift is observed.

In order to understand these spectra it is necessary to consider the orbital character of the unoccupied states as derived from band structure calculations.[7-12] According to these calculations the unoccupied partial density of states on O sites near the Fermi level E_F in Y123 and Y124 is predominantly caused by holes in orbitals which are σ-bonded to Cu atoms. In Y123 there is an additional contribution from a narrow dpπ band formed by O(1,4) and Cu(1) atoms. Since the Fermi level is right at the top of this band, the contribution to the total unoccupied DOS is small although it contributes considerably to the density of states at E_F. In Y124 this band is completely filled and no contributions from π-bonded O2p orbitals are expected for the unoccupied DOS. Therefore, according to band structure calculations the unoccupied partial DOS at O sites is predominantly formed in O orbitals σ-bonded to Cu atoms.

Based on this result, the XAS spectrum for ***E//a*** is mainly due to O(2) atoms and that for ***E//b*** is caused by O(3) and O(1) atoms. The difference spectrum is caused *solely* by O(1) atoms when assuming an isotropic hole-distribution in the planes. In Y123 the experimental O 1s binding energies as determined from the absorption edge of O(1) and O(4)[14] are found to be smaller by 0.4 and 0.7 eV, respectively, when compared to those of O(2,3) atoms. This is in the range of calculated values.[15,16,18] In Y124 the observed O 1s binding energies of the O(2) and O(3) atoms are the same within 0.1 eV and the O 1s binding energy of the O(1) is about 1.1 eV larger than in Y123. As it was mentioned before this latter finding can be explained by the fact that in Y124 the O(1) atom is coordinated with three Cu atoms while in Y123 there are only two nearest-neighbor Cu atoms. The large shift ($\sim$ 3 eV) of the threshold for the unoccupied DOS at the O(1) atoms in Y124 to higher energies was predicted by the band structure calculations in the LDA approximation.[13] These calculations gave rise to the similar O 2p hole DOS at the O(2,3) atoms in both the investigated structures.

The second peak at ~530 eV is observed for ***E//a*** and ***E//b*** polarization in Y123 and Y124. Zaanen et al.[10] proposed that this feature was due to van Hove singularity of the top of (1D) chain band probed by the O(1) $2p_y$ states. However, the mere fact that we observed this peak for ***E//a*** rules out this explanation since transition to the chain band are for this orientation dipole-forbidden. In previous investigations[14] this peak has been observed in $YBa_2Cu_3O_6$ as well. According to recent studies[15,16] on $La_{2-x}Sr_xCuO_4$, this peak can be ascribed to transitions into the conduction band (upper Hubbard band) which is predominantly formed by Cu 3d states but has some admixture of O 2p states. At higher dopant concentration, this peak is reduced due to a reduction of correlation effects and due to hybridization of the valence and the conduction band in the final state.[17] It is interesting to note that this second peak is not observed for O(1) in the difference spectrum ***E//b*** – ***E//a*** nor for O(4) in the EELS spectra.[18] We tentatively interpret the absence of the conduction band peak in the chains in terms of a stronger Cu-O hybridization in the chains compared to that in the planes which was derived from band structure calculations as well.[11] However, it might also be possible that the charge transfer gap is smaller for the Cu-O chains as compared to that for the CuO_2 planes.

From the experimental ratios for the in-plane and out-of-plane absorption $I_{a,b}/I_c$ = 2.0 (2.6) measured by EELS on twinned samples and the present XAS result I_b/I_a = 2.0 (2.0) on untwinned crystals for Y123 (Y124) one can easily calculate the partial hole concentration on different sites when taking the sum of the holes to be one. The evaluation gives nearly the same hole concentration at O(2), O(3) and O(4) atoms in Y123 and Y124 . The holes present on O(1) sites in Y123 are found to be evenly distributed between two O(1) sites in Y124. It is interesting to note the equidistribution of the O 2p holes between chains and planes in both investigated systems.

CONCLUSIONS

In summary, from orientation-dependent O 1s XAS spectra of

twin-free Y123 and Y124 single crystals together with a previous EELS study, it has been found that the O 2p holes are distributed equally between the CuO_2 planes and CuO chains and that the partial density of unoccupied O 2p states in the CuO_2 planes are almost identical in both investigated systems. The measured unoccupied DOS in the chain is different in Y123 and Y124. The upper Hubbard band is observed in planes but not in chains in both systems.

The work by the Buffalo group and at Brookhaven is supported by DOE.

REFERENCES

1. F. J. Himpsel et al. Phys. Rev. B, **38**, 11946 (1988).
2. A. Zibold et al. Physica C **171**, 151 (1990). U. Welp et al. Physica C **161**, 1 (1989).
3. S. Karpinski et al. Physica C **160**, 449 (1989).
4. C. T. Chen and F. Sette, Rev. Sci. Instrum. **60**, 1616 (1989).
5. G. C. Smith, A. Krol and Y. H. Kao, Nucl. Instr. Meth. A **291**, 135 (1990).
6. E. Narumi et al., Appl. Phys. Lett. **5b**, 2684 (1990).
7. S. Massida et al., Phys. Lett. A **122**, 198 (1987).
8. J. Yu et al., Phys. Lett. A **122**, 203 (1987).
9. W. M. Temmerman et al., J. Phys. F **17**, L 135 (1987).
10. J. Zaanen, M. Aluani, and O. Jepsen, Phys. Rev. B, **40**, 837 (1989).
11. J. Yu, K.T. Park, and A. J. Freeman, Physica C **172**, 467 (1991).
12. W.Y Ching et al., Phys. Rev. B, **43**, 6159 (1991).
13. K.T. Park, J. Yu and A. J. Freeman, private communication.
14. N. Nücker et al. Phys. Rev. B, **39**, 6619 (1989).
15. H. Romberg et al., Phys. Rev. B, **42**, 8768 (1990).
16. C.T. Chen et al., Phys. Rev. Lett. **66**, 104 (1991).
17. H. Eskes and G. A. Sawatzky, Phys. Rev. B, **43**, 119 (1991).
18. N. Nücker et al, in *Studies of High Temperature Superconductors*, ed. by A. Narlikar, Nova Science Publishers **6**, 145 (1990).

STRUCTURE AND TRANSPORT PROPERTIES OF YBCO THIN FILMS AND SINGLE CRYSTALS

V.M. Pan, A.L. Kasatkin, M.A. Kuznetsov, V.S. Flis, V.F. Taborov, V.F.Soloviev, V.L. Svetchnikov, G.G. Kaminsky, V.G. Prokhorov, V.I. Matsui,Y.A. Petrukhno, S.K. Yushchenko & V.S. Mel'nikov.
Institute of Metal Physics , Kiev 252142, Ukraine, USSR

ABSTRACT

Mechanisms of nucleation and growth of YBCO thin films and their real microstructures are investigated and discussed regarding to effective pinning sites and critical current densities. Edge dislocations with Burgers vector b = a[100] on low angle tilt grain boundaries appeared to be served the strongest pins providing critical current densities $J_c(77K) > 5x10^6$ A/cm^2. Edge dislocation density in thin YBCO films grown at high substrate temperatures by island mechanism exceeds $10^{11} cm^{-2}$ saving quasi - single - crystalline film structure with average misorientation angle between islands in ab - plane about 4°.

Direct measurements of transport critical current density and resistive state for YBCO single crystals have been performed using thin microbridges. Critical current densities more than 10^5 A/cm^2 at 82K have been observed.It was shown flux flow phenomena are essentially different for thin films and single crystals.

Peculiarities of resistive state for thin films in magnetic fields H || c appeared to be connected with vortex flow along one - dimensional easy slip channels, that is, planar defects (low angle grain boundaries as one suggests) near which superconducting order parameter is locally suppressed because of proximity effect. It was confirmed, at least at high enough temperatures, $T < T_c$.

INTRODUCTION

Transport critical current densities J_c at 77K and $J_c(H)$ needed for the most of practical high current and/or high field applications of HTSC materials so far were achieved only in their thin films.

Neither in textured wires and tapes, nor even in single crystals nobody could manage to reach $J_c(77K) \geq 10^4$ A/cm^2 and $J_c(77K,\ 5T\ H\ ||\ c) \geq 10^3$ A/cm^2 by measurements of transport critical current densities.

It is a consequence of two main circumstances: (i) Grain boundaries in HTSC materials play dramatic sinister role resulting in suppression of supercurrent. Dimos et al [1,2] have shown clean grain boundaries with misorientation angles more than $\vartheta \geq 10 - 11^\circ$ become Josephson - like weak links.

Ones undertake a number of efforts (theoretically and experimentally) to find out ways to overcome this principal obstacle to develop continuous HTSC material conductors with high J_c(77K) and enough high J_c(77K, 5T H || c), since it is would be impossible to avoid occurrence of grain boundaries in continuous conductors.

Babcock et al [3] have shown that 90 - degree high angle boundary in YBCO bicrystals can be transparent for non - Josephson supercurrent, that is, does not serve as a weak link. After this work many researchers started to study atomic structure of special grain boundaries, low energy grain boundaries with Near Coincident Site Lattice approach [4,5] in HTSC materials. However this direction of investigation so far developed separately from experimental measurements of real critical current densities across different kind of single grain boundary, as Dimos et al [2] did.

(ii) To achieve high levels of J_c and J_c(H) it is necessary to provide an effective pinning of Abrikosov vortices. In single crystals YBCO the highest measurable value J_c(77K) at transport dc current measurement does not exceed $J_c \leq 10^4$ A/cm^2. One consider as the most probable pinning centers in single crystals point defects and twin domain boundaries. However recent results [6,7] strongly contradict these suggestions. Since we studied only H || c geometry the intrinsic pinning did not consider at all.

Much higher J_c(77K) take place in thin YBCO films - up to 8 10^6 A/cm^2 [8]. Recently the highest levels of J_c were found out for thin YBCO films having spiral - like island morphology with screw growth dislocations. Authors [9 - 11] supposed that screw dislocations of growth with high density ($n_d \simeq 10^9$ cm^{-2}) which are observed by use of scanning tunnel microscope (STM) may serve as an effective pins and therefore the high value of J_c(77K) can be easily explained.

In present work the experimental evidences were obtained:

(i) Transport critical current densities for YBCO single crystals at temperature range just below T_c appeared to be much higher than it was measured earlier: namely $J_c(77K) \geq 2\ 10^5$ A/cm^2.

(ii) Thin YBCO films with quasi - single - crystalline spiral, island - like morphology originated by growth mechanism, with the highest J_c have had in their microstructure low angle tilt boundaries. Mean azimuthal misorientation angle in a-b plane for these boundaries is approximately 4°. Edge dislocations with density as high as $n_d \propto 10^{11}$ cm^{-2} seem to be forming such low angle tilt boundaries. One may suggest these grain boundary edge dislocations with Burgers vector b = a[100] obviously are served as the most effective pins when H || c. Such model is discussed in this work.

SAMPLE PREPARATION

Single crystals were grown by spontaneous crystallisation method in zirconia crucibles. After annealing in pure oxygen atmosphere current and potential leads were connected by Ag paste. Contact resistance 0.1Ω was typical one. Reduction of the cross section area of the specimen was achieved by chemical etching. We succeeded in producing uniform microbridges with cross section 20x20μm and length 2mm.

The $YBa_2Cu_3O_7$ films were deposited by pulse laser ablation. The film thicknesses was d $\propto$ 2000 $\div$ 3000 Å. The superconducting transition temperature was T_c = 89 $\div$ 93 K, the transition width was ΔT_c = 1.0 $\div$ 1.5 K, the residual resistivity were ρ_o = 100 $\div$ 200 $\mu\Omega$cm, the resistivity ratio was $\gamma = \rho_{300}/\rho_o \propto 2.6 \div 3.0$. The samples were prepared by mask technique to achieve the following sample shape: 0.3 $\div$ 1.0 mm film width, 2 $\div$ 10 mm distance between potential probes. The silver contacts were vacuum deposited. The films were quasi - single - crystalline. They had low angle boundaries between islands or regions of coherent scattering.

EXPERIMENTAL

To avoid the influence of thermal electricity, contact potentials, and electric hindrances, the resistive measurements of $YBa_2Cu_3O_7$ films were performed by pulse technique. The resistive transition curves were measured with lock - in device using quasi - direct current. The sample was supplied with rectangular bipolar current pulses, which were integrated and stored in computer memory after relaxation of transition processes. The difference signal was monitored by X - Y recorder. The pulse duration was about 2.3 ms and the sequence frequency was 220 Hz. The I - V curves were measured similarly, but to lower the sample heating the current pulse duration was significantly shorter about 36 μs and the sequence frequency was approximately 870 Hz.

For single crystals measurements pulse technique was used as well. The pulse duration for single crystal was about 2 μs and the sequence frequency was 50 Hz. Since pulse technique is not wide - spread in such measurements for single crystal (at least in microsecond range) the parallel dc and pulse measurements of J_c in high field area were carried out. We have found out, that J_c values determined by pulse method exceed ones of dc measurements by a factor of 2 - 3. Taking into account different voltage criterion (10 and 1 μV) and some heating in the dc case, it is not surprising.

STRUCTURE

It is well known that there are three film growth mechanisms [12]. They are Volmer - Weber, Frank - van der Merve and Krastanov - Stranski ones. The first one occurs in the case of weak bonding of a substrate with deposited film and incoherent matching of crystal lattices. On the contrary the second one occurs in the case of strong bonding between substrate and deposited matter provided that lattices match coherently. At last the third mechanism is an intermediate one with partially coherent coupling.

As for morphology of deposited films the first mechanism results in three - dimensional island nucleation and growth. For the second one the two - dimensional nucleation of pseudomorphous layers on the substrate surface is the case. During the further growth elastic strains relax by generation of misfit dislocations [13]. The two - dimensional nucleation and growth of the film with following transition to three dimensional island growth is the third mechanism (Krastanov - Stranski).

Norton and Carter [12] suggested that all three mechanisms may occur during the deposition of $YBa_2Cu_3O_7$ films.

The scanning electron microscopy and X - ray studies of $YBa_2Cu_3O_7$ films deposited onto single crystalline (100) YSZ substrates have shown the following.

A typical island morphology of the film surface with the mean island (grain) size from ≤ 0.5 μm up to 4.0 μm is observed in the substrate temperature range $T_s \simeq 730 - 800^{\circ}C$. Thus, supposing that each island corresponds to one screw growth dislocation, as it was done in[9 - 11],the density of screw dislocations ranges from 10^9 to $6\ 10^6\ cm^{-2}$ (Fig. 1). The X - ray studies of the film structure show that despite of the island morphology the films are quasi - single - crystalline ones with the mean mozaicity level of 4° (misorientation angle in a-b plane). As for misorientation of the c - axis, it does not exceed $0.4 \div 0.5^{o}$.

Recently the dislocation structure of a low angle [001] tilt boundary with 3.5^{o} misorientation angle has been studied by high - resolution electron microscopy [14]. It has been shown that such boundaries consist of edge dislocation array with the Burgers vector b = a [100]. These dislocations are spaced at $\langle D \rangle \simeq 60$ Å. The size of dislocation core is about 20 Å. It is important to point out that lattices of the regions in a low angle boundary separated from each other with dislocation cores match well and there is a considerable number of almost undisturbed Cu - O rows providing the supercurrent flow through the boundary. The bulk (non - Josephson) supercurrent should be locked when the dislocation cores would form continuous array, i.e. when $\langle D \rangle$ would be about 20 Å. It is clear that this value accurately corresponds to the critical misorientation angle of

10 - 11°, which has been found in [1 - 2].

When the substrate temperature changes from 730 to 780°C, the density of screw dislocations ranges from 10^{9} to 1.6 10^{7} cm^{-2}, while j_c reduces less than an order of magnitude remaining about 10^{6} A/cm^2 for the film prepared at T_s = 780° C [15]. As for the edge dislocations mentioned above, its concentration reduces less than an order of magnitude (as j_c does) from 10^{11} cm^{-2} down to 1.5 10^{10} cm^{-2}. Therefore, it could be suggested that the edge dislocations indeed provide the most effective flux pinning.

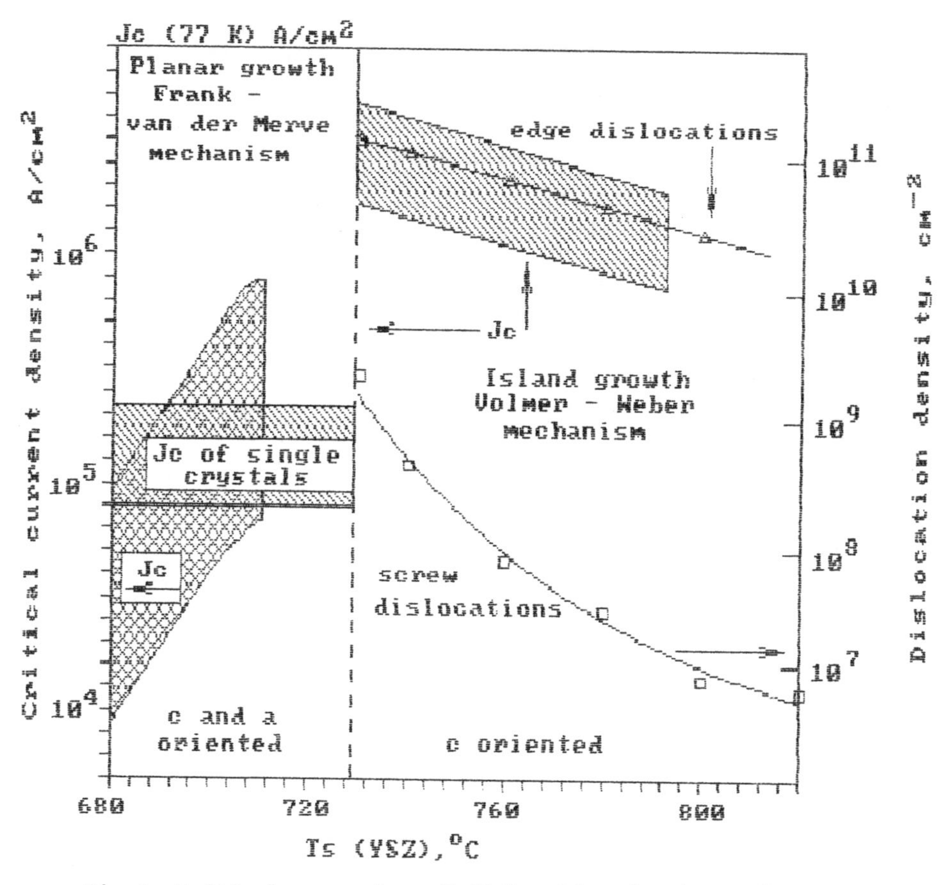

Fig.1. Critical currents and dislocation density of $YBa_2Cu_3O_7$ films and single crystals [19].

Accordingly HREM data [14] the effective size of dislocation core is about 20 Å and the ξ_{ab} = 15 Å [16]. So, in this case the pinning mechanism connected with the condensation energy gain will be very effective (core interaction pinning). If the film was grown by Volmer - Weber mechanism, the low angle boundaries and dislocation arrays can be assumed to penetrate through the whole film normally to the substrate. Therefore, when the field is parallel to the c - axis, the vortices are strongly pinned at dislocation cores.

When the substrate temperature drops lower than 730°C the film surface morphology, seen with SEM, principally changes. The surface become flat. The growth mechanism seems to change to planar quasi - two - dimensional (Frank - van der Merve).

X - ray diffractions and topography investigations show that misorientation angles between coherent scattering regions are significantly less in epitaxially grown by planar mechanism thin films than that in films deposited at higher substrate temperatures, T_s > 730° C. The value of azimuthal misorientation is near 10 - 20′, and since low angle boundaries in a - b plane suppose to be formed, as in [14], by edge dislocations with b = a[100] Burgers vector the dislocation spacing will be ≤ 1000 Å, i. e. for 1.5 order of magnitude more than in previous case. Mean edge dislocations concentration will depend from size of regions of coherent scattering in a - b plane, which is not yet determined precisely, however, one can expect that it should be about 0.2 - 0.5 μm. In this case it is worth to suppose that J_c(77K) as well as concentration of dislocations will be by 1.5 order of magnitude less, i. e. near ∝ 10^5 A/cm^2.

The specific for the films grown at T_s < 710 - 720°C is that the precipitations, often regularly cut, are observed at the surface. Additional studies show that such inclusions are mostly a - oriented and are the vertexes of pyramids or columns, which grow from the substrate through the whole film. The density of stacking faults along [001] direction detected by X - ray diffraction measuring of $\Delta c/c$ value increases since T_s reduces.

Such pyramids or columns of a - oriented crystals can be seen in patterns obtained by computer simulation of the film nucleation and growth (will be published elsewhere).

Such films have the highest T_{Conset} = 95 K, but some lower j_c(77K) ≃ 2 10^5 A/cm^2. With further reduction of the substrate temperature down to 680°C j_c drops down to 1 ÷ 2 10^4. The structure at the same time changes to Vidmanstaedt - like of a - oriented precipitations in c - oriented matrix.

As to single crystals, X - ray investigations show that their structure is similar to that of thin films grown by planar mechanism. They also have regions of coherent scattering of 0.7 - 0.8 μm in size divided from each others by low angle boundaries with mean misorientation angle in a - b plane about $\propto$ 50 - 60′ .

TRANSPORT PROPERTIES

The I - V curve measurements revealed the existence of non - linear parts of I - V curve in the region of $J \geq J_c$, usual for flux creep, which followed by the flat parts of I - V curve that corresponds to the flux flow regime [17,18].

In Fig.2 the field dependence of flux flow resistance $\rho_f(B)$ is shown. Its behavior at low fields seems to be rather unusual. The dependence $\rho_f \propto B^{1/2}$ is observed, whereas the Bardeen - Stephen theory predicts a linear dependence $\rho_f(B)$ [17,18].

Meanwhile in single crystals the $\rho_f(B)$ dependence shown in Fig. 3 has partially flat rather than square root character and $\rho_f(B)$ is linearly depends form field in the low field region [19].

Such difference in $\rho_f(B)$ dependencies for thin films and single crystals as well as difference in J_c values (up to 1.5 order of magnitude), to our mind, is due to the structural peculiarities of thin films and single crystals, in particular due to difference of low angle boundaries edge dislocation density.

DISCUSSION

Real defects structure of the crystal lattice of HTSC materials, particularly occurrence of edge dislocations, forming low angle block or grain (more precisely - regions of coherent X - ray scattering) boundaries in films and single crystals of HTSC under investigation may be a very essential for determination the character of vortex motion and transport properties of single crystals and thin films of HTSC.

This difference of transport properties of single crystals and screw dislocation mediated grown films can be explained by substantially different concentration of edge dislocations in the both cases.

For the films under investigation the dislocations concentration was high enough to consider them as the main pinning centers for vortices located along low angle grain boundaries. Assuming that the local suppression of order parameter $\Delta(r)$ occurs in the neighborhood of dislocation core (and respectively it causes the condensation energy reduction) due to the proximity effect, one can expect that boundaries oriented perpendicular to the transport current (along Lorentz force) form the so called easy slip channels (ESC) of vortices [17,18], meanwhile boundaries parallel to the transport current (perpendicular to the Lorentz force) form the lock - in layers, preventing vortices to leave grains.

Supposing, that vortices are pinned on dislocations by core - pinning mechanism (the dimension of dislocation core $\propto$ 20 Å $\propto$ ξ_{ab})

it is possible to estimate the value of critical current density (neglecting of the flux creep):

$$J_{co} \propto \frac{\phi_o c}{64 \pi^2 \lambda^2 \delta} \frac{\langle \Delta^2(r) \rangle}{\Delta_o^2} \quad (1)$$

were δ - is the effective radius of pinning potential, λ - penetration depth, Δ_o - order parameter far from boundaries. The $\frac{\langle \Delta^2(r) \rangle}{\Delta_o^2}$ multiplier describes the change of condensation energy on ESC (averaging is done along the ESC). Assuming that $\delta \simeq \xi_{ab} \simeq 15$ Å, $\lambda \simeq 1500$ Å, $\frac{\langle \Delta^2(r) \rangle}{\Delta_o^2} \propto 1$, one can evaluate $J_{co} \propto 10^8$ A/cm^2. When misorientation angle increases, i.e. when the concentration of dislocation increases, the $\langle \Delta^2(r) \rangle$ decreases, resulting in reduction of the critical current and in transformation of the grain boundary

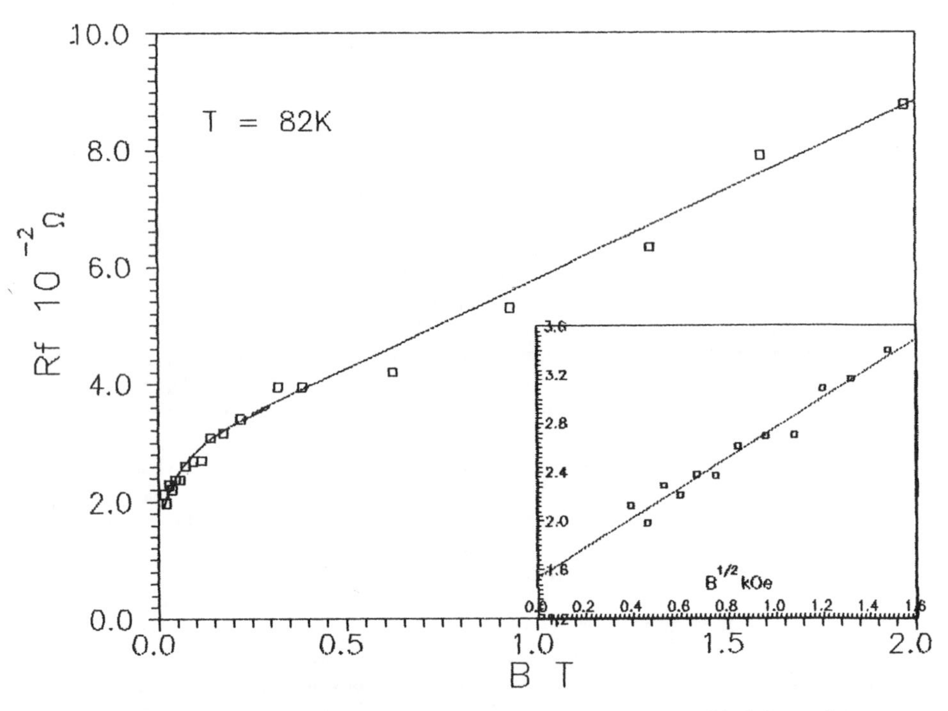

Fig.2. ρ_f (B) dependence for thin films at low field region.

into Josephson contact of SNS type at high enough misorientation angles.

The similar dependence of transport properties vs azimuthal misorientation angle in a - b plane was experimentally observed in the works [1,2,20].

The fact that vortices are preferably move along grain boundaries at $J > J_c$ was confirmed by unusual square root $\rho_f(B)$ dependencies in low field region, shown in Fig. 2 [17,18].

Movement of vortices along distinguished one - dimensional easy slip channels rather than through whole volume of film, as it was shown in [17,18], results in the fact that instead of common Bardeen - Stephen [21,22] field dependence $\rho_f \propto \rho_n \dfrac{B}{B_{c2}}$, one can expect $\rho_f \propto \dfrac{1}{L} \dfrac{(\phi_0 B)^{1/2}}{B_{c2}}$, were L - mean distance between channels (grain size).

In single crystals the edge dislocations densities at boundaries of coherent scattered regions are significantly less ($\propto$ 1.0 - 1.5 orders of magnitude). Because of this reason J_c values seem to be not so high ($\propto 10^5$ A/cm^2 at 77K) comparatively to those

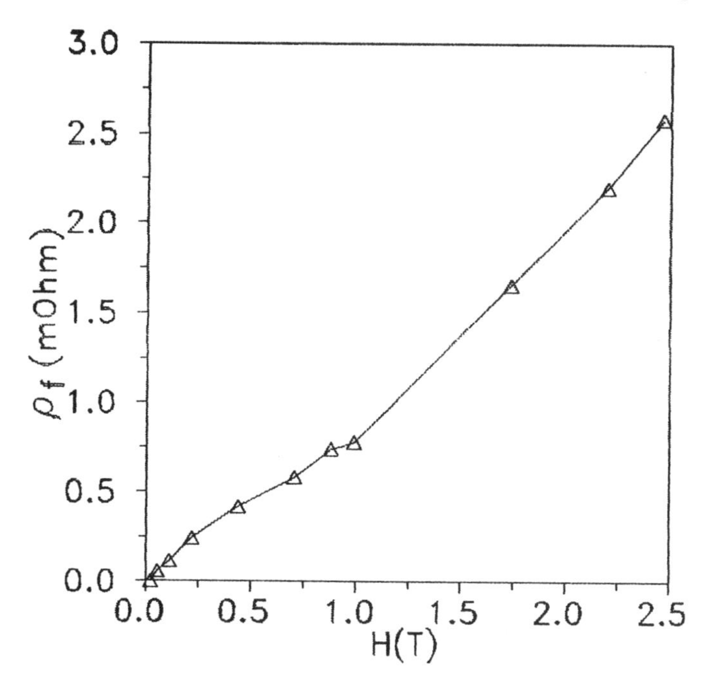

Fig.3. $\rho_f(B)$ dependence for single crystals.

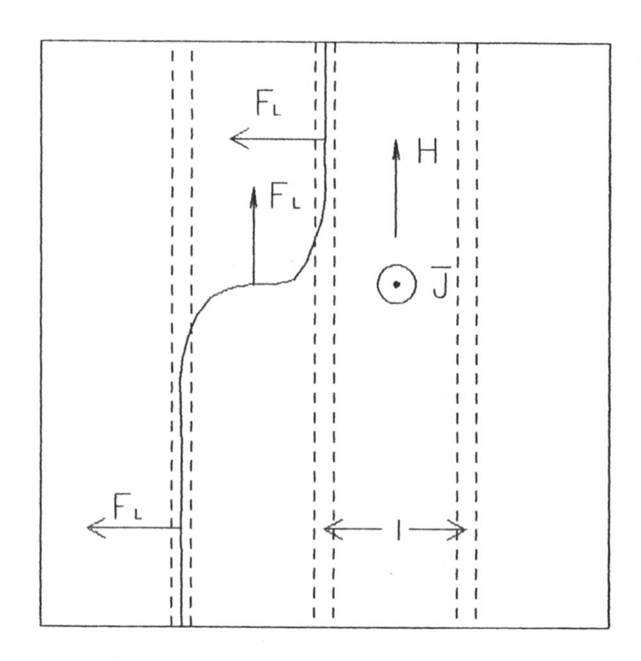

Fig.4 Possible mechanism of vortex motion along the chain of dislocations.

in screw dislocation mediated grown thin films, and $\rho_f(B)$ dependence has common behavior as it should be in modified Bardeen - Stephen rule (Fig. 3). Thus, it seems that in this case the vortices movement under the Lorentz force occurs in the whole crystals volume rather than along boundaries.

It is worth to notice that vortex can easily move along chain of dislocations by creation a kink which move along dislocations axis as shown in Fig. 4, and that process doesn't need a large energy of activation. Energy of activation for such process at $J \to 0$ can be estimated as:

$$U \propto \frac{\phi_o^2}{16\pi^2\lambda_{ab}^2}\left[\frac{\xi_{ab}}{4} + \ln\varkappa\left(\frac{m_{ab}}{m_c}\right)^{1/2} l\right]\frac{\langle\Delta^2(r)\rangle}{\Delta_o^2} \qquad (2)$$

where the first term corresponds to the energy of nucleation, and second one - to the energy of kink origination, taking into account anisotropy [23]. For specific values of $\varkappa \propto 100$, $l \propto 100$ Å, $m_{ab}/m_c \propto 1/25$, one can receive typical estimations of activation energy $U_o \propto 10^{-2} \div 10^{-1}$ eV.

The approach presented in this section gives the opportunity to explain transport properties of films without suggestion about extremely high density of points defects, as it was done in work [16].

CONCLUSION

The real defect structure of HTSC material can strongly effect the character of vortex motion and transport properties of superconductor under magnetic field. Particularly, the edge dislocation arrays, forming low angle boundaries with significantly different spacings between dislocation cores may be responsible for the difference in the transport properties of $YBa_2Cu_3O_7$ single crystals and thin films as it is shown in this work. Analysis of experimental data allows us to suppose that in thin films the vortex motion occurs preferably along the low - angle grain boundaries, which form the network of so called easy slip channels (at least at high enough temperatures) whereas in single crystals such channels are not formed and vortex motion under the Lorentz force takes place in whole volume of the crystal. The difference in the critical current density values for this two cases is caused by the difference in concentrationof edge dislocations ,which serve as the most effective pins in the case of H || c.

1. D. Dimos, P. Chaudhari, J. Mannhart and F.K. Le Goues, Phys. Rev. Lett. 61, 219 (1988).
2. D. Dimos, P. Chaudhari, J. Mannhart, Phys. Rev. B41, 4038 (1990).
3. S. E. Babcock, X. Y. Cai, D. L. Kaiser, D. C. Larbalestier, Nature, 347, 167 (1990)
4. T. S. Ravi, D. M. Hwang, R. Ramesh, S. W. Chan, L. Nazar, C. Y. Chen, A. Inam, T. Venkatesan, Phys. Rev., B42, 10141 (1990)
5. K. Jagannadham and J. Narayan, Materials Science and Engineering, B8, 1, (1991)
6. B. M. Lairson, J. L. Vargas, S. K. Streiffer, D. C. Larbalestier and J. C. Bravman, Critical Currents and Magnetization Decay in Oxygen Deficient $YBa_2Cu_3O_{7-\delta}$ Thin Films, Preprint, to be published in Physica C, (1991).
7. L. I. Swartzendruber, A. Raitfurd, D. L. Kaiser et al., Phys. Rev. Lett., 64, 483, (1990)
8. V. M. Pan, V. G. Prokhorov, G. G. Kaminsky et al. in High - Temperature Superconductors Material Aspects, vol. 1 (Ed. H. C. Freyhardt, R. Flukiger, M. Peuckert, DGMI, Verlag, FRG, 1991) p. 51 - 58.
9. Ch. Gerber, D. Anselmetti, J.D. Bednorz, J. Mannhart and D.G. Schlom, Nature, 350, 279 (1991).
10. M. Hawley, I. D. Raistrick, J. G. Beery and R. J. Houlton, Science, 251, 1587, (1991)
11. H.P. Lang, T. Frey and H. J. Guntherodt, Europhys. Lett., 15, 667 (1991).
12. M. G. Norton and C. B. Carter, J. Cryst. Growth, 110, 641 (1991).
13. C. L. Jia, B. Kabius, H. Soltner, U. Poppe and K. Urban, Physica C, 172, 81 - 89, (1990).
14. Y. Gao, K. L. Merkle, G. Bai, H. L. M. Chang, and D. J. Lam,

Physica C, 174, 1/3, 1 - 10, (1991).
15. R. Broom, A. Catana, T. Frey, Ch. Gerber, H. - J. Guntherodt, H. P. Lang, J. Mannhart and K. A. Muller, Submitted to Z. Phys. B (1991).
16. T. L. Hylton and M. R. Beasley, Phys. Rev. B, 41, 16, 11669 - 11672 (1990).
17. V. G. Prokhorov, A. L. Kasatkin, G. G. Kaminski, M. A. Kuznetsov, V. I. Matsui, V. M. Pan, Fiz. Nizk. Temp. (Sov. Low Temp. Phys.), 17, 4, pp. 467 - 475, (1991).
18. V. M. Pan, G. G. Kaminsky, A. L. Kasatkin, M. A. Kuznetsov, V. G. Prokhorov, C. G. Tretiachenko IEEE Trans. on Magn., 27, 2, (1991).
Supercond. Sci. Techn. 4, 18, 594 - 96, (1991).
19. V. M. Pan, V. F. Taborov, V. F. Soloviev, Supercond. Sci. and Technol, Proc. 6th Int. Workshop on Critical Currents 8 - 11 July , Cambridge, England, (1991).
20. J. Mannhart J. of Superconductivity 3, 3, 281 - 285, (1990).
21. J. Bardeen, M. J. Stephen Phys. Rev. 140, p. A1197, (1965).
22. L. P. Gor'kov, N. B. Kopnin Usp. Fiz. Nauk (Sov.) 116, p. 413, (1975).
23. V.G. Kogan Phys. Rev. B, 24, 3, p. 1572 - 1575, (1981).

COLLAPSE OF CIRCULAR CURRENT IN HIGH-T_c RINGS

L.M.Fisher, V.V.Borisovskii, N.V.Il'in, I.F.Voloshin
All-Union Electrical Engineering Institute, Krasnokazarmennaya Str.12, 111250, Moscow, USSR

I.V.Baltaga, N.M.Makarov, V.A.Yampol'skii
Institute for Radiotechnic and Electronics Ukrainian Academy of Sciences, academician Proscura str.,12, Kharkov 25, USSR

ABSTRACT

The collapse of supercurrent in high-T_c ceramic rings was studied experimentally and theoretically. Under a longitudinal ac magnetic field the region of flow of current I circulating through the ring is found out to contract, and upon the amplitude of ac magnetic field H reaches the threshold value H_p the current I disappears. The role of dissipative processes in this phenomenon is analyzed. It is shown that the collapse of current circulating through a ring goes in two stages. At the first stage ac field presses the current into the sample depth. The current does not depend on ac field H at this stage practically. At the second stage a great dissipation of current energy takes place.

INTRODUCTION

High-T_c materials are known to be essentially non-linear mediums.So, an electromagnetic field distribution in high-T_c ceramics is described by modified equation of a critical state which is true in a wide range of amplitudes and frequencies[1-4]:

$$\mathrm{curl}\ \mathbf{B} = (4\pi\mu/c)\mathbf{j}_c(B)\cdot\mathbf{E}/E. \quad (1)$$

Here $\mathbf{B}$ is the magnetic induction, $\mathbf{E}$ is the electric field, $\mathbf{j}_c(B)$ is the critical density of intergranular currents, μ is the magnetic permeability of ceramics without weak links between grains (only intragranular currents are taken into account), c is the speed of light. The investigations on non-linear electrodynamics of superconductors have a long prehistory, however, the non-linear properties of Eq.1 has not been studied practically. The first attempt in this field made a year ago gave an unexpected result. The new phenomenon - the collapse of a transport current was observed in a cylindrical ceramic specimen placed in an external ac field [4]. Let us suppose that the value of transport current I through it is less than the critical one I_c. That means that the current flows near the surface of the cylinder and there exists a region of its cross section where current does not flow. The phenomenon of collapse consists in the next: the region of current I flow removes to the cylinder axis if ac magnetic field is switched on. The compression of current with the increase of the amplitude of ac field goes on until the current fills the full core of sample i.e. the current tube transforms into the cylinder, After that the voltage drop between the cylinder-ends appears and the superconductor transits to the resistive state.

It is remarkable that the phenomenon described takes place at forceless configuration when the external ac magnetic field $H(t)$ is directed along transport current (parallel to the cylinder axis). The compression of current goes on not because of Lorenz force as it usually takes place in the pinch effect but owing to specific non-linear properties of the critical state of hard superconductors. Obviously, the azimuthal electric field E_φ is generated in a space region where a longitudinal ac magnetic field penetrates. (In the region where the longitudinal ac magnetic field $H_z(t)$ penetrates there exists the azimuthal electric field E_φ caused by the former.) The longitudinal electric field E_z in the sample is absent in stationary regime. As it follows from Eq.1 in outer layers, the φ-component of the current screening the sample core from the ac field may flow and here z-component of current equals zero. If the transport current I is fixed the location of its flow is inevitably displaced to the cylinder axis .

The collapse of transport current was predicted and observed in [4-5], however, the mechanism of transport current displacement remains practically unstudied . Besides, only the situations when a current I was supported by an external source were investigated earlier. In this case a transport current remains invariable as in superconducting as well as in resistive state. The final stage of collapse consisted in appearing of voltage drop between the cylinder-ends [4].

We think that more convenient systems to study the dynamics of collapse are those where a current source is absent. In this case the compression of a transport current has to follow by the irreversible relaxation processes, which role is not obvious a priori. In this paper we have used the toroidal rings as the object of investigation.

RESULTS AND DISCUSSION

1. The typical sizes of rings prepared from yttrium ceramics were: $a \simeq 4$ mm, $R \simeq 0.6$ mm (Fig.1). The circular current was induced by switching off an external magnetic field. For that purpose the long solenoid was placed coaxically inside the ring. The electric current was switched on through it when the ring temperature was higher than Tc. After cooling the ring down to the liquid nitrogen temperature the current through the solenoid was switched off. As a result, the current in the ring raised. Its value is defined from the condition of the conservation of the magnetic flux through the ring. It is necessary to remark, that the induced current has a non-uniform distribution on the cross section of the toroid. The current occupies the surface region (marked region in Fig.1). The density j of this current equals the critical one j_c. Outside this region j=0. It is clear that the size of region 1 is defined by the value of full induced current I.

An azimuthal ac magnetic field having the form

$$H(t) = H_m \cos(\omega t) \quad (2)$$

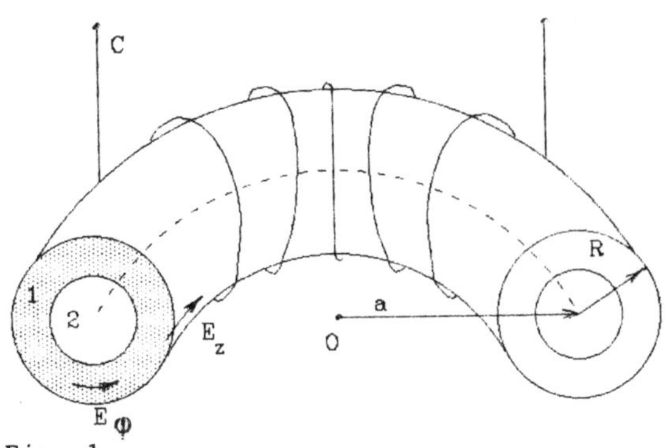

Fig. 1.

was created by the coil C (Fig. 1). The collapse of the circular current was studied by measuring the static magnetic moment M of a ring. It is clear that the value of M for the thin ring with $R \ll a$ is proportional to the circular current, i.e. $M \backsim I$. The measurement of the magnetic moment was carried out by the vibrating magnetometer with oscillating pick-up coils. The ring investigated was placed near the center of symmetry of pick-up coils.

2. The record of the function $M(H_m)$ in the conditions, when the maximum current was induced in the ring, is presented by curve 1 in Fig.2. If the amplitude H_m increases the magnetic moment of the ring decreases monotonously and tends to zero at the finite amplitude $H_m = H_p$. The value H_p corresponds to the "penetration" amplitude at which ac magnetic field fills the ring volume completely. One should pay attention that the circular current can be partly suppressed also by a static magnetic field (the dashed line in Fig.2). This is due to the sensitivity of the critical current density to a magnetic induction. Note that the dependence of M on the static magnetic field essentially differs from the dependence of M on ac field. While the function $M(H_m)$ tends to zero at H_p, the magnitude of $M(H_{dc})$ differs from zero at any dc magnetic field. The shape of curve $M(H_m)$ and H_p value do not depend practically on the frequency $\omega/2\pi$ in the range 10-10^3 Hz.

The curve $M(H_m)$ can be obtained by the simple theoretical calculations based on the critical state model. In the region $r < r_o$, where the ac magnetic field B_z does not penetrate, Eq. 1 for the component of magnetic induction B_φ connected with the circular current may be written as follows

$$B_\varphi' + B_\varphi/r = (4\pi/c)\mu j_c(B_\varphi). \tag{3}$$

Here r is a radial coordinate in the cross section of ring, B_z' means a differential coefficient on r. At the point r_0 the value B_φ is equal to

$$B_\varphi(r_0) = 2I\mu/(r_0 c). \tag{4}$$

In the region $r > r_0$ ac magnetic field B_z exists but the circular current j_z directed perpendicular to the cross section of ring is absent. The component B_φ decreases as $1/r$ in this region:

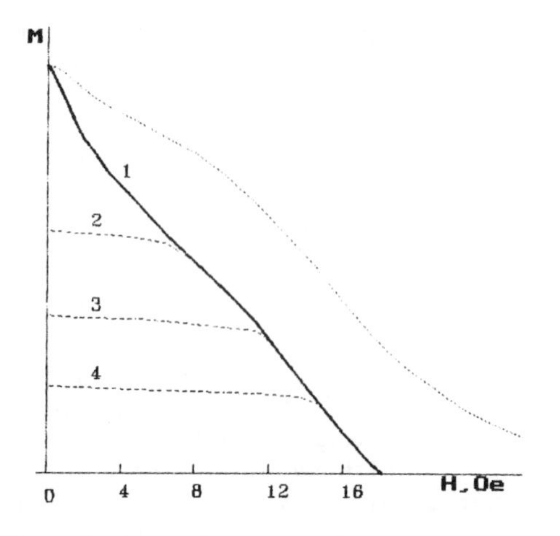

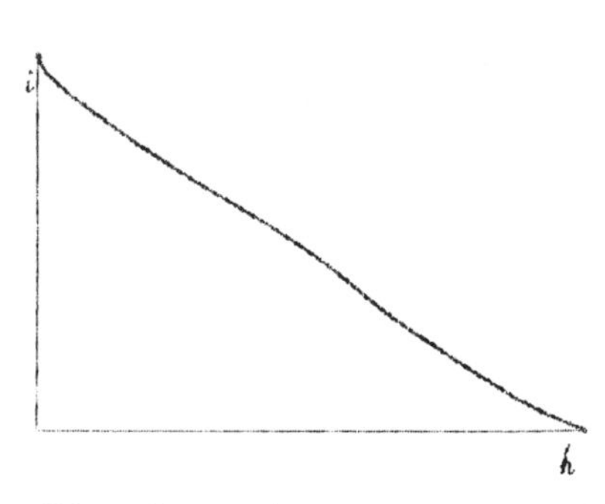

Fig. 3. Calculated dependence of the current through the ring on ac field.

Fig. 2. Dependencies of moment M on magnetic ac field (curves 1 - 4) and dc field (dashed line). The initial current through the ring $I=I_c$ for cases presented by N 1 and dashed curves, for the rest curves $I < I_c$.

$$B_\varphi(r) = 2I\mu/(rc), \qquad r_0 < r < R. \tag{5}$$

The distribution of magnetic induction B_z at the moment t = 0 is described by equation

$$B_z' = [4\pi/c)\mu j_c(B_z^2 + (2I/rc)^2]^{1/2}. \tag{6}$$

The boundary condition may be written in the form

$$B_z(R) = \mu H, \qquad B_z(r_0) = 0. \tag{7}$$

The system of equations and boundary conditions (3)-(7) give relation between I and H. It is possible to obtain this relation by the analytical methods at an arbitrary dependence of $I(B)$ in two limiting cases of low $I \ll I_c$ and high $I_c - I \ll I_c$ value of I:

$$I(H) = \begin{cases} I_c - HRc/2, & I_c - I \ll I_c; \quad H \ll H_p \\ c^2/(16\pi)\cdot\left[j_c(0)/j_c^{\,2}(\mu H_p)\right](H_p-H)^2, & I \ll I_c;\ H_p-H \ll H_p. \end{cases} \tag{8}$$

To obtain the function $I(H)$ in full range of H from zero to H_p, let us take the empirical dependence $j_c(B)$ in a form

$$j_c(B) = j_0/[1+(B/B^*)^{7/4}], \tag{9}$$

taken from [6]. Here B^* is the characteristic scale where the function

$j_c(B)$ changes noticeably, j_0 is the critical current density at $B = 0$.

The system (3)-(9) was solved numerically. The form of the curve $i(h)$ in dimensionless parameters $i = I/I_c$ and $h = H/H_p$ depends on single parameter α:

$$\alpha = 4\pi j_0 R/(cB^*), \qquad (10)$$

which defines, as a matter of fact, the sensitivity of the critical current density to the self magnetic field of current. The plot of the function $i(h)$ is presented in Fig.3 for $\alpha = 100$.

Let us pay attention that the curve in Fig. 3 as well as the curve 1 in Fig.2 have the inflection points, where the derivatives $\partial^2 i/\partial h^2$ and $\partial^2 M/\partial H^2$ are equal to zero. The existence of such points is connected with a competition of two causes having influence on the form of $i(H)$. The first consists in the sensitivity of the critical current density to a magnetic field and the second is connected with non-linear decrease with H (square-law in the case $I \ll I_c$) of the sample cross section occupied by the current. If the function $j_c(B)$ constant and only the geometry is essential we obtain the dependence of $i(h)$ in the form

$$i(h) = (1 - h)^2, \qquad (11)$$

which has no the points of inflection. If we take into account the dependence of j_c on B but do not pay attention to the geometrical factor (as in the case of plate), we shall have the curve $i(h)$ which curvature changes sign when field h varies in the region $0 < h < 1$. In the case of cylindrical geometry the existence of the inflection points depends on the value of parameter α. In model considered (Eq.9) the points of inflection on curve $i(h)$ appears at $\alpha > 18$.

3. If the current I induced in ring is less than its critical value I_c, the magnetic moment of the ring decreases with magnetic field amplitude H_m as one can see from Fig.2 (curves 2-4). The magnetic moment does not change at low amplitude. When the curve $M(H_m)$ reaches a curve 1 magnetic moment begins to decrease noticeably . Note that the starting points of abrupt decreasing of curves 2-4 in Fig.2 lay on curve 1 practically. These facts permit us to propose the next picture of the collapse of circular current. At the beginning circular current is displaced to the core of ring under the influence of external ac field; its value does not change practically at this stage. As a result, we observe a small variation of the magnetic moment of current. After the circular current occupies the full core of sample (at $H = H_M$) the second stage of collapse starts. At this stage the strong dissipation of current energy arises. It is clear that the beginning of the second stage (the magnitude of H_M) depends on the initial value of circular current I. The law of decrease of $M(H_m)$ for the second stage of collapse is almost the same for all curves (1, 2, 3, 4).

A special analysis is necessary to understand the role of

dissipative process at the first stage of collapse. It is clear the transit of current into deeper layers of superconductor may be realized owing to an appearance of induced electric field E_z there. This field causes Joule losses $jE = j_c E$. To describe this stage of collapse we employ a simple model. Imagine that the ring is divided into two parts (see Fig. 1). The region 1 is a closed toroidal tube where a circular current I_{1c} flows at initial moment. The region 2 is the toroid where current I_{1c} is displaced under the electromagnetic wave. In the displacement process current in the first region diminishes from I_{1c} to zero and the current in the second region increases from zero to I_{2max}. The value of I_{2max} is less than I_{1c} due to the dissipative process. The system of equations describing this process may be written in the form

$$2\pi a E_z(I_1) + (L_1/c^2)\partial I_1/\partial t + (L_{12}/c^2)\partial I_2/\partial t = 0,$$
$$(L_2/c^2)\partial I_2/\partial t + (L_{21}/c^2)\partial I_1/\partial t = 0. \tag{12}$$

Here L_1 and $L_2 \cong (2\pi a)\ln(a/r)$ are the inductances of contours presented by the regions 1 and 2 correspondingly, $L_{12} = L_{21} \cong L_1$ is the coefficient of the mutual inductance between the regions 1 and 2, E_z is the electric field existing in the region 1, directed parallel to axis of toroid and arising due to the penetration of ac field. This field may be evaluated by the next way. As it follows from Eq.1, we can obtain

$$I_1(t) \cong I_{c1}E_z(I_1)/E. \tag{13}$$

The field **E** in the region 1 has two components- the unknown component E_z and ac component

$$E_\varphi \cong H\omega R/c, \text{i.e. } E = [(H\omega R/c)^2 + E_z^2(I_1)]^{1/2} \tag{14}$$

Using Eq.13 and Eq.14 we find

$$E_z(I_1) = (H\omega R/c)I_1/\,(I_{1c}^2 - I_1^2)^{1/2}. \tag{15}$$

So, the process of the circular current displacement from region 1 to region 2 is described by the system of Eqs.12 and 15. The physical picture of it is following. The effective resistance $2\pi a E_z(I_1)/I_1$ arises in the contour 1 under the influence of ac field and leads to the relaxation of current I_1. Due to the mutual inductance the current I_2 excites in the region 2. Excluding the parameters I_2 and E_z from the system Eqs.12 and 15 we obtain the following equation for current I_1:

$$2\pi a(H\omega R/c)I_1/[I_{1c}^2 - I_1^2]^{1/2} + (1/c^2)(L_1 - L_{12}^2/L_2)\partial I_1/\partial t = 0. \quad (16)$$

The effective inductance of the contour 1 is approximately equal to

$$L_1 - L_{12}^2/L_2 \cong 2\pi R. \quad (17)$$

Eq.16 permits to establish the several conclusions about the first stage of the circular current collapse.

1) If $I_1 \ll I_{1c}$ and $H \cong H_p$ the current I_1 relaxes exponentially

$$I_1(t) \cong I_1\exp(-t/\tau). \quad (18)$$

The value of τ is short and may be evaluated as

$$\tau \cong I_{1c}/H_p\omega ac \sim (1/\omega)\cdot(R/a). \quad (19)$$

The period τ is compared with the period of ac field and the current relaxation in region 1 occurs very short.

2) During the relaxation time τ the dissipation energy may be evaluated as

$$Q \cong 2\pi a\cdot\pi R^2 \mathbf{jE}\tau \cong 2(\pi R)^2 aj_c(H_p\omega R/c)(R/\omega a) \cong (8\pi^3R^5j_c^2)/c^2. \quad (20)$$

This energy is in $(a/R)\ln(a/R)$ times less than the current energy $\mathcal{E}$ at $t = 0$:

$$\mathcal{E} = (1/c^2)L_1I^2 \cong (2\pi^3R^4a)/c^2j_c^2\ln(a/R) \cong Q\cdot(a/R)\ln(a/R) \gg Q. \quad (21)$$

Consequently, the main part of current disappearing in region 1 passes into region 2 and only a little part of total energy of current has dissipated in collapse. Both these conclusions agree with the measurements.

4. At the second stage of collapse when the whole cross-section of sample is occupied by current, the circular current must dissipate. The dissipative process is described by equations similar to Eq.12:

$$2\pi a\cdot E_z(I_1) + (L_1/c^2)\partial I_1/\partial t + L_{12}\partial I_2/\partial t = 0,$$
$$2\pi a\cdot E_z(I_2) + (L_2/c^2)\partial I_2/\partial t + L_{12}\partial I_1/\partial t = 0 \quad (22)$$

Electric field $E_z(I_1)$, as at first stage, is defined by Eq.15. However, the electric field E_z at the region 2 is not equal to zero now. The current 1 passing from the region 1 increases the critical current flowing in the region 2 up to the value higher than the critical one. On this reason the critical state forms in the region 2.

The electric field $E_z(I_2)$ has a non-linear dependence on the supercritical current $\Delta I_2 = I_2 - I_{2c}$:

$$E_z(I_2) \cong A\cdot\Delta I_2^{\,2}, \qquad A = \text{const}. \tag{23}$$

Eq.23 describes the squared approximation of the usual I-V plot for ceramic samples at $\Delta I_2 < I_{2c}$.

Analyzing Eqs.22, 15 and 23 we can understand the process of collapse at the second stage. At a short time $\tau \cong \omega^{-1}R/a$ an essential redistribution of current between the regions 1 and 2 takes place. The current flowing in the region 2 increases by the value ΔI_0 equal to decreasing part of current I_1, i.e. the total circular current $I = I_1 + I_2$ in the ring does not change practically. Besides, during the period τ the electric fields $E_z(I_1)$ and $E_z(I_2)$ equalized :

$$E_z(I_1) = E_z(I_2). \tag{24}$$

The total current I_1+I_2 starts to decrease after this moment only. The current I_1 drops from value $I_{1c}-\Delta I_o$ to zero and current I_2 drops from value $I_{2c}+\Delta I_o$ to I_{2c}. During the relaxation process of I_1 and I_2 Eq.24 remains valid.

As follows from Eqs.24 and 15, 23 the relaxation of the current I_1 is quicker than I_2. As a result, the full current $I(t) = I_1(t) + I_2(t) = I_1(t) + I_{2c} + \Delta I_2(t)$, if t tends to infinity, is defined by the current in the second region:

$$I(t) = I_{2c} + \Delta I_2(t). \tag{25}$$

Using Eqs. 22, 15, 23 and 24 we obtain the transcendental equation describing the relaxation of current ΔI_2:

$$(\Delta I_2)^{-1} - (\Delta I_c)^{-1} + (Ac^2/\omega)\ln(\Delta I_0/\Delta I_2) = [c^2A/\ln(a/R)]t. \tag{26}$$

The function $\Delta I_2(t)$ is proportional to t^{-1} when t tends to infinity. Consequently, the relaxation of total current and magnetic moment of a ring is described by power function. According to Eq.26 the relaxation rate depends on frequency of ac field. This rate is greater at high frequency.

The proposed description of collapse on its second stage agrees with experimental results. Fig.4 shows the relaxation of $M(t)$ after switching on an external magnetic field for three frequencies. At initial state the full cross-section of the ring was occupied by circular current I_c. The relaxation time in all cases occurs long enough and reaches hundreds seconds. The relaxation time decreases with increase of frequency.

5. The collapse of circular current may be caused not only by alternating magnetic field but by sharp switching on an external static field also. Fig. 5 demonstrates decreasing of the ring magnetic moment after switching on a static magnetic field H. Note, that decreasing of $M(t)$ is deeper in this case than after slow

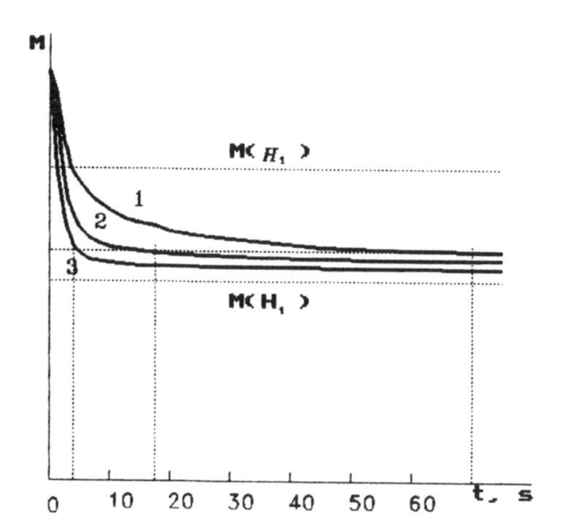

Fig. 4. Relaxation of magnetic moment of ring under ac magnetic field $H_1 < Hp$ at various $\omega/2\pi$ = 20 Hz (curve 1), 200 Hz (2) and 2000 Hz (3). Horizontal lines, denoted by M(H) and M(H), show "limiting" levels under dc field (upper line) and under ac field (lower) at $H_1 = H_1$.

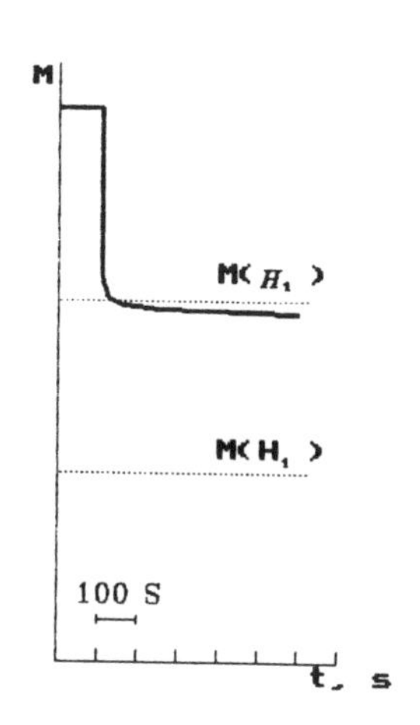

Fig 5. Relaxation of M under dc magnetic field H .

increasing of H up to the same value. The reason is connected with the induced electric field (appearing in the first case) which causes the current relaxation more effective than the influence of magnetic field on the critical current density.

CONCLUSION

The main results obtained in this paper are as follows. After investigations of dynamics of current collapse in a ring with a circular current we found out that there are two different stages of collapse. At the first stage the current flowing in the outer layers is displaced into a volume of sample under influence of external ac magnetic field. The effect of the mutual inductance between two space regions plays the main role here. The current is displaced from region 1 and induced in region 2. The dissipation process is not essential at this stage. The second stage begins from the moment when the displacement current occupies the full cross-section of samples. At this stage the current density of circular current becomes higher than the critical one and the dissipative process is very essential. The relaxation of the circular current is described by power function of time.

REFERENCES

1. C. P. Bean, Phys. Rev. Lett. 8, 250 (1962)
2. H. London, Phys. Lett. 6, 162 (1963)
3. H. Dersch, G. Blatter, Phys. Rev. B 38, 11391 (1988)
4. I. V. Baltaga, N. M. Makarov, V. A. Yampol'skii, L. M. Fisher, N. V. Il'in, and I. F. Voloshin, Phys. Lett. A 148, 213 (1990)
5. I. F. Voloshin, N. V. Il'in, N. M. Makarov, L. M. Fisher, and V. A. Yampol'skii, Pis'ma JETP, 53, 109 (1991)
6. L. M. Fisher, N. V. Il'in, O. I. Lyubimov, N. M. Makarov, I. F. Voloshin, and V. A. Yampol'skii, Sol. State Comm., 76, 141 (1990)

THE UNIVERSAL MAGNETIC FIELD DEPENDENCE OF THE CRITICAL CURRENT DENSITY IN HIGH-T_c CERAMICS

L.M. Fisher, V.S. Gorbachev, N.V. Il'in, N.M. Makarov,
I.F. Voloshin, V.A. Yampol'skii
All-Union Electrical Engineering Institute,
Krasnokazarmennaya St., 12, Moscow, 111250, USSR
R.L. Snyder, J.A.T. Taylor, V.W.R. Amarakoon, M.A. Rodriguez,
S.T. Misture, D.P. Matheis, A.M.M. Barus, J.G. Fagan
New York State College of Ceramics at Alfred University
Alfred, New York 14802 USA

Abstract

We present a new contactless method for determining the local critical current density $J_c(B)$ and the effective magnetic permeability μ of ceramic superconductors. Using this method we have carried out systematic investigation of these parameters in ceramic samples with different microstructures. We find that the dependence of $J_c(B)$ can be described by a single universal function over a wide region of magnetic field. At low fields this function has a plateau and at higher field it varies as $B^{-3/2}$. The universal behavior of $J_c(B)$ breaks down as the field exceeds the first critical magnetic field of grains.

1. Introduction

Earlier [1,2,3], a new contactless method was developed to determine the local critical current density in ceramic samples using the measured surface impedance. This method was founded ideologically by Campbell [4] based on Bean's critical state model [5]. The main equation of this model connects the magnetic field distribution **H** inside the superconductor with the local critical current density $J_c(H)$:

$$curl\ \mathbf{H} = (4\pi/c) \cdot \mathbf{J}_c(H)\ , \qquad (1)$$

where c is the speed of light. This formula correctly describes the critical state in normal superconductors. Ceramics are essentially heterogeneous material consisting of a large number of grains connected by a network of weak links. To describe the current carrying capability it is necessary to take into account the intergranular as well as intragranular currents. The transport current of ceramics is governed by intergranular currents. The parameter J_c results from averaging this current over the bulk volume which includes a large number of grains whose volume is small in comparison with the entire specimen volume. On the other hand, the intragranular currents define an effective magnetic permeability μ of the ceramic medium. The critical state model was modified in [9] to apply to such non- uniform material as ceramics. According to this model the distribution of the magnetic induction $\mathbf{B}$ is described by the equation:

$$curl\ \mathbf{B} = (4\pi/c) \cdot \mu \mathbf{J}_c(B) \ , \tag{2}$$

The authors of [5] have pointed out the necessity to take into account the dependence of the critical current density on on B. The strong dedependence of J_c (B) is connected with the sensitivity of Josephson intergranular current to magnetic field. As was shown in [2], the function $J_c(B)$ varies as $B^{-3/2}$ in moderate magnetic fields. The same result was obtained in [7]. Such behavior agrees with the results obtained for the model of a single- sphere Josephson junction . In the region of low magnetic fields, the dependence of $J_c(B)$ becomes weaker . In particular, the function $J_c(B)$ at $B \cong 0$ does not change with field. Similar effects of B on J_c have been observed in a number of previous studies [7,9,10]. In fact, if the magnetic flux through the junction does not exceed the magnetic flux quantum Φ_0, the maximum undissipative current of the junction remains nearly constant, and J_c is independent of B. Although many papers have been devoted to this theme, the majority of the published studies were made by the well known four-point method which yields the average critical current density. It is important to note that the four points method ignores the self-field of the transport current. The method proposed in [2] has no such drawbacks. However, the results obtained in [2] have some essential limitations; we could determine only the product $\mu \cdot J_c(B)$, and our theory applied to cylindrical samples only. Here we consider a modified method which allows us to measure the local critical current $J_c(B)$ and magnetic permeability μ separately and may be used for samples having the form of a cylinder as well as a plate. Using this method we have measured J_c vs. B for ceramic samples with different microstructures. In spite of noticeable variations of ceramic structure, critical current density, and μ, we have observed similar behavior of function $J_c(B)$. Moreover, at fields less than the first critical field of ceramics grains the dependence of $J_c(B)$ for different samples may be described by single universal function.

2. Experimental Procedure

We have investigated several groups of $YBa_2Cu_3O_{7-\delta}$ samples prepared by different techniques. The main group (samples 1 - 18) were fabricated by solid state reaction. The second group (samples 19 - 25) was prepared via a sol-gel process. The samples differ in grain size, grain size dispersion, and other structural parameters. The function $J_c(B)$ is determined by measuring the surface impedance of samples. The measurements were carried out at the frequency f $= \omega/2\pi = 1300$ Hz, at liquid nitrogen temperature and in the magnetic field region 0-400 Oe. The magnitude of ac probe field was less than 1 Oe.

3. Theory of the Measurement Method

Consider the penetration of a magnetic field into specimens having two different geometries, a plate and a cylinder. These sample geometries are the most convenient for experimental research and some results have been published in [1,2,3] for cylindrical samples. To conduct the measurement, the sample is placed in a magnetic field having the form

$$\mathcal{H} = H + h\cos\omega t \, . \tag{3}$$

The field $\mathcal{H}$ is directed parallel to the plate surface or, for cylindrical samples, along the cylinder axis. Let one of the following conditions be fulfilled

$$h \ll B^*/\mu \quad or \quad h \ll H \, . \tag{4}$$

Here B^* is the flux where the local critical current density changes noticeably. Under these circumstances we can neglect the influence of ac probe field on J_c. It is well known that the response of a sample to an external field H(t) is described by the surface impedance. The surface impedance $\mathcal{Z} = \mathcal{R} - i\chi$ for a plate is determined by the relation

$$\mathcal{Z} = (8\pi/c) \cdot E_\omega(\mathrm{x}=0)/h \quad where \tag{5}$$

$$E_\omega = (\omega/2) \cdot \int_0^{2\pi/\omega} E(t) \cdot \exp(i\omega t)dt \, . \tag{6}$$

(The magnetic field is oriented along the z-axis, the electric field E is oriented along the y-axis, the x-axis is perpendicular to the plate, and the points x = 0 and x = d correspond to the surfaces of the plate). In the case of a cylinder the surface impedance is defined by Eqs. (5)- (6) where the electric field E has only the ϕ-component which is taken at the surface of cylinder where $r = R$. The surface resistance describes the losses in a specimen while χ is connected with the ac field penetration depth. For the full description we also need in the Maxwell's equations:

$$curl \;\; \mathbf{E} = -(1/c) \cdot \partial \mathbf{B}/\partial t \, . \tag{7}$$

The boundary conditions for the ac part of induction **b** in both cases may be written in the form:

$$\mathbf{b}_{|surface} = \mu h \cdot \cos\omega t \, . \tag{8}$$

The solution to Eqs. (2) and (5)-(8) permits us to obtain the following relations for the surface impedance for both geometries:

$$Z = \mu \cdot \chi_n \cdot F_1(h/h_p) \quad for\ a\ plate \tag{9}$$
$$Z = \mu \cdot \chi_n \cdot F_2(h/h_p) \quad for\ a\ cylinder \tag{10}$$

where χ_n is the reactance of a sample in a normal state

$$\chi_n = 2\pi\omega d/c^2 \quad for\ a\ plate \tag{11}$$
$$\chi_n = 2\pi\omega R/c^2 \quad for\ a\ cylinder\ . \tag{12}$$

The functions F_1 and F_2 will be presented later. The penetration field h_p is the ac field at which the ac magnetic induction exceeds zero at some point in the sample volume. The field h_p is defined by the following relations

$$h_p = (4\pi/c) \cdot J_c(\mu H) \cdot d/2 \quad for\ a\ plate \tag{13}$$
$$h_p = (4\pi/c) \cdot J_c(\mu H) \cdot R \quad for\ a\ cylinder\ . \tag{14}$$

Notice that the penetration field in (13) and (14) depends on the magnetic field H due to $J_c(H)$. Since $J_c(H)$ decreases rapidly as H increases, h_p will show similar behavior.

It is clear that Eqs. (9),(13) and (10),(14) determine the connection between J_c, μ, and $Z(\mu H)$. Equations (10) and (14) were considered in [1,3] for cylindrical samples. Unfortunately, each of the pairs of equations determines only one functional connection between J_c, μ, and $Z(\mu H)$. We can not determine the two independent quantities, J_c and μ, having only a single equation.

According to Eq. (2), J_c and μ affect the magnetic field distribution differently. While μ effects a change of the induction B on a sample surface, the derivative $\partial B/\partial x$ is defined by the product $\mu \cdot J_c$. The role of $B(x)$ is illustrated in Fig. 1, for a ceramic plate having $J_c=$ const.

To find an additional equation to determine μ and J_c separately, let us consider the behavior of the function $\mathcal{R}(H)$ in detail. When $h < h_p$ the surface resistance increases with h. However, when $h > h_p$ the induced electric fields penetrating from opposite sides of the plate are compensated. As a result, the surface resistance $\mathcal{R}$ decreases. The surface resistance as a function of H must therefore have a maximum near $h = h_p(H)$. This simple physical consideration is confirmed by direct calculation. If the magnetic field H is small enough to satisfy $h < h_p$, the functions F_1 and F_2 in Eqs. (9) and (10) can be written in the form

$$F_1(x) = (2/3\pi)x \cdot (1 - (3\pi i/4)) \tag{15}$$
$$F_2(x) = (4/3\pi)x \cdot \{(1 - 3\pi i/4) - (x/2)(1 - 15\pi i/32)\}\ , \tag{16}$$

for $x = h/h_p < 1$. When $h > h_p$, simple formulae exist only for the real parts of F_1 and F_2.

$$Re \;\; F_1(x) = (2/\pi x) \cdot (1 - 2/3x) \tag{17}$$
$$Re \;\; F_2(x) = (4/3\pi x) \cdot (1 - 1/2x) \, , \tag{18}$$

for $x > 1$. Analyzing these formulae, it is simple to establish that the maximum of $F(x)$ occurs at x = 4/3, or $h = 4h_p(H)/3$ for plates. For cylinders the maximum $F(x)$ occurs at $x = 1$ or $h = h_p(H)$. The maximum values of the surface resistance are described for a plate and a cylinder by

$$Re \;\; Z_{max} = R_{max} = 3(\mu\omega/c^2) \cdot d/2 \tag{19}$$
$$= (3/4\pi)\mu\chi_n$$
$$\mathcal{R}_{max} = (4/3)(\mu\omega R)/c^2 = (2/3\pi)\mu\chi_n . \tag{20}$$

Equations (9), (11), (13), (15), (17), and (19) allow μ and J_c to be determined separately for plates. The same treatment applies to Eqs. (10), (12), (14), (16), (18), and (20) for cylinders.

To achieve the maximum value of the surface resistance $\mathcal{R}$ in an experiment, the dependence of $\mathcal{R}$ on a static magnetic field H must be determined. In reality, increasing H decreases h_p and, as a result, h_p may become comparable to h at some value of H. Under these conditions $\mathcal{R}(H)$ will be at a maximum, while its position depends on the magnitude of h. Analyzing the formulae (13)-(18), it is clear that increasing h leads to a displacement of the maximum position to the region of higher magnetic fields. If we calibrate the plot of $\mathcal{R}(H)$ in units of χ_n we have sufficient information to determine μ and the function $J_c(\mu H)$.

A list of the necessary equations is included to simplify data analysis. These equations are derived from the above for the critical state parameters of high-T_c samples in terms of the measured surface resistance $\mathcal{R}(H)$.

For a plate

$$\mu = (4\pi/3) \cdot \mathcal{R}_{max}/\chi_n \, , \tag{21}$$
$$J_c(\mu H) = (4ch/9\pi d) \cdot \mathcal{R}_{max}/\mathcal{R}(H) \quad H < H_1 \, , \tag{22}$$
$$J_c(\mu H) = (3ch/8\pi d) \cdot \left\{1 \pm [1 - \mathcal{R}(H)/\mathcal{R}_{max}]^{1/2}\right\} \quad H > H_1 \, . \tag{23}$$

Here H_1 is the field H where $h = h_p(H)$. The plus sign in (23) is used in the region $H < H_m$ and the minus sign corresponds to fields $H > H_m$, where H_m is the field when $\mathcal{R}(H)$ is a maximum. A simple way to define the value of H_1 is by the relation $\mathcal{R}(H_1) = (8/9) \cdot \mathcal{R}_{max}$

For a cylinder

$$\mu = (3\pi/2) \cdot \mathcal{R}_{max}/\chi_n \, , \tag{24}$$

$$J_c(\mu H) = (ch/4\pi R) \cdot [\mathcal{R}_{max}/R(H)] \times \left\{1 + [1 - \mathcal{R}(H)/\mathcal{R}_{max}]^{1/2}\right\} \quad H < H_m \, , \tag{25}$$

$$J_c(\mu H) = (ch/4\pi R) \times \left\{1 - [1 - \mathcal{R}(H)/\mathcal{R}_{max}]^{1/2}\right\} \quad H > H_m \, . \tag{26}$$

Note that in the case of cylinders the field H_1 coincides with H_m.

The proposed method is valid independent of any relation between the ceramic grain size α and London penetration depth λ_L. However, in a forthcoming study we shall investigate ceramics with grain size $\alpha > \lambda_L$.

4. Results and Discussion

4.1 The Effective Magnetic Permeability, μ.

Figure 2 shows the dependence of the surface resistance $\mathcal{R}$ on applied field H for sample 19, a plate having thickness $d = 1$mm, which was fabricated by the sol-gel method. Note that $\mathcal{R}$ is normalized to to the normal-state surface reactance, χ_n. The function $\mathcal{R}(H)$ changes slowly at low magnetic field then increases, passes through maximum, and decreases weakly at high fields. Such behavior of the function $\mathcal{R}(H)$ is expected from the previous considerations. Using Eq. (21) and the measured maximum value of $\mathcal{R}(H)$, μ is calculated as 0.6. As mentioned earlier, μ is governed by the sample volume which is occupied by weak links plus the fraction of the grain volume which is penetrated by the magnetic field to the London penetration depth. Naturally, this parameter depends strongly on the ceramic microstructure.

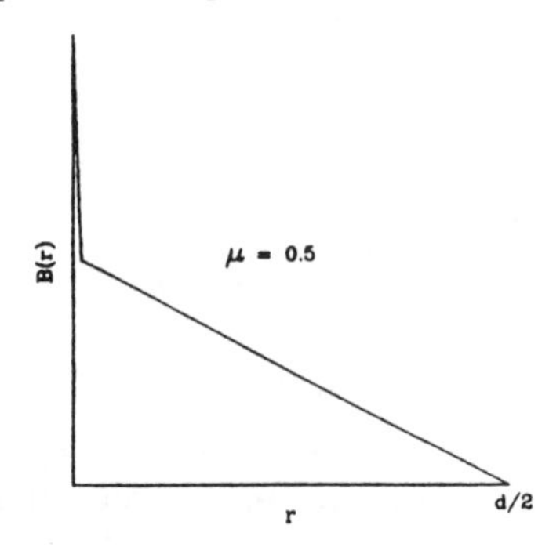

Fig. 1: The distribution of magnetic induction inside a ceramic plate having $\mu = 0.5$.

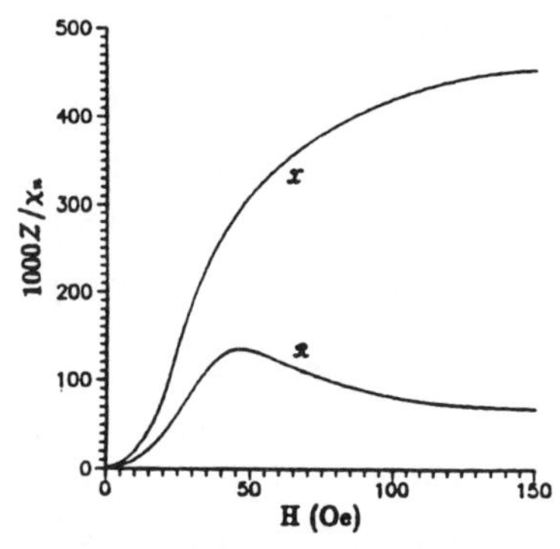

Fig. 2: The dependence of $\mathcal{R}(H)$ for a sample prepared by the sol-gel method. T = 77K, $d = 1$mm.

We have measured μ for samples prepared by the different methods described in Sec. 2. The value of μ for these samples varies over the range of $0.05 < \mu < 0.95$. The maximum value of μ was achieved for samples prepared by solid state reaction and sintered at 870°C.

The observed variations in μ are easily explained on the basis of grain size. The grains are generally about equal to the London penetration depth λ_L. Increasing the sintering temperature results in grain growth and consequently the volume of the sample which is penetrated by the magnetic field decreases. This results in a lower value of μ. The minimum value of $\mu = 0.05$ is observed for samples prepared by melt growth technology. We have also investigated the influence of silver doping into these ceramics and found that this addition leads to grain growth and the associated decrease in μ.

4.2 Critical Current Density

Using the surface resistance data and Eqs. (22) and (23) we have obtained the critical current density as a function of magnetic induction. Some results for sample 9 are presented in Fig. 3, which clearly shows a strong dependence of J_c on B. To understand this dependence, we examined it at low and high magnetic fields separately. The dependence of $J_c(B)/J_c(0)$ for samples 1 - 3 are shown in Fig. 4 on a double logarithmic scale. Figure 4 shows that the critical current density is approximately constant in the region of low magnetic field. Such behavior may be described on the basis of a Josephsen junction. The dependence of the maximum undissipative current I on B for a single plane Josephson junction is given by

$$I_c = I_0 \cdot |\sin(\pi\Phi/\Phi_0)|/(\pi\Phi/\Phi_0) , \tag{27}$$

where Φ_0 is a magnetic flux quantum and Φ is the magnetic flux through the junction. At low magnetic fields where $\Phi/\Phi_0 \ll 1$ the critical current I_c does not change significantly from its maximum value of I_0. I_c deviates from I_0 at $\Phi \geq \Phi_0$ which occurs in such Josephson media as High-T_c ceramics. For these materials the magnetic flux Φ is related to the grain size α by $\Phi = B \cdot 2\lambda_L \cdot \alpha$. A larger grain size results in a larger Φ, and, as shown by Eq. (27), the critical current deviates from its maximum value at lower B. Figure 4 shows this dependence of J_c (B) on grain size. Evaluations of the junction size obtained from the condition $\Phi \simeq \Phi_0$ agree with direct observations.

As shown in Fig. 3, the critical current density begins to decrease sharply in the next magnetic field region. As was established in [2], the dependence of J_c on B is described by power function $J_c \cong B^{-\alpha}$ in this region. To our surprise, all of the samples show this behavior despite their different microstructures. As we can see from Fig. 4 the exponent for every sample is approximately $\alpha = -3/2$. We propose that this type of dependence is connected with the character of the contacts between ceramic grains. To confirm this theory we have measured

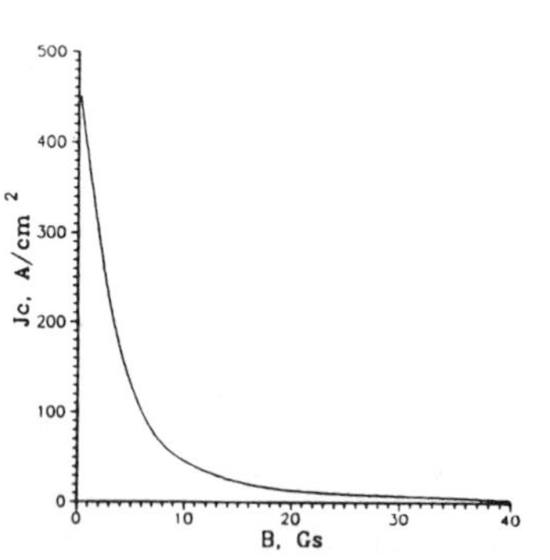

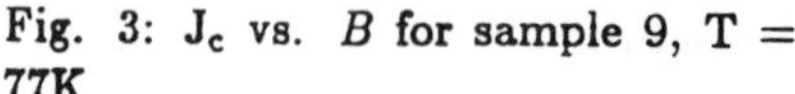

Fig. 3: J_c vs. B for sample 9, T = 77K.

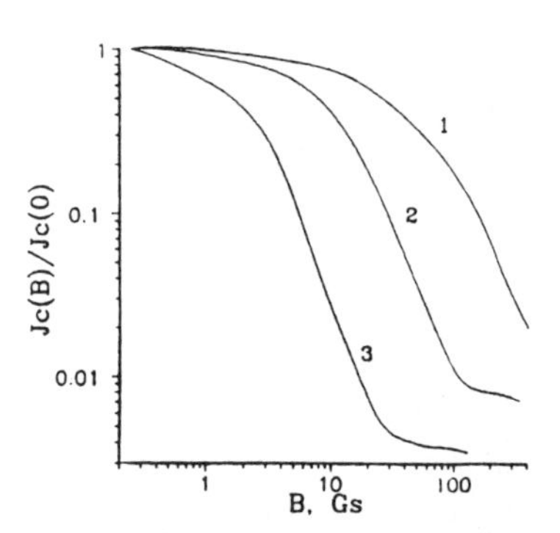

Fig. 4: $J_c(B)$ for samples 1-3 sintered at τ = 870, 930, and 960°C, respectively. T = 77K.

the temperature dependence of the critical current density. Some results of these measurements are shown in Fig. 5. Here we have plotted the local critical current density vs. temperature on a log-log scale for the different samples. This plot shows that the dependence of J_c on $(1 - T/T_c)$ is described by a power function having an exponent = 2 for the samples with a heat treatment temperature less than 930°C. $J_c \cong (1 - T/T_c)^{3/2}$ for the samples with $\tau > 930$°C. This shows that the contact character changes with heat treatment temperature, τ. It is well known that the first function corresponds to that of an S-N-S contact type, or S-I-S type for superconductors with a short coherence length [11,12]. In the second case the dependence of $J_c\ (T)$ corresponds to an S-N-I-N-S contact type. Taking into account all of our measurements, we may conclude that the function $J_c(B)$ does not change even when the contact character changes radically.

These results agree with those of Peterson [13], which show that the overall current is defined by the maximum current of a single junction and depends only on the junction geometry [13]. To understand the observed dependence of J_c on B we can model the ceramic grains as spheres with Josephson contacts between them [7,13]. In the framework of this model the diffraction pattern is described by the formula

$$I_c(H) = 2I_0 \cdot |J_1(\pi\Phi/\Phi_0)|/(\pi\Phi/\Phi_0)\ , \tag{28}$$

where J_1 is a Bessel function. After averaging over different grain sizes for the condition $\Phi \gg \Phi_0$, Eq. (28) leads to the dependence $I_c(H) \cong B^{-3/2}$, which is confirmed by our experiments.

At low magnetic field J_c does not depend on H until a magnetic flux quantum can penetrate into a single Josephson junction. Due to the grain size distribution of ceramics, the vortices begin to penetrate different junctions at different magnetic fields. The most prominent change in J_c with B will therefore occur in samples having only small variations in grain size. This behavior was observed in those samples (1, 4-7) which had narrow grain size distributions. Conversely, for samples with large variations in grain size, there is only a weak dependence of J_c (B). As the magnetic field increases, the critical current of each contact, as described by Eq. (28), decreases; the dependence of $J_c(B)$ becomes stronger and tends to the universal law $J_c \cong B^{-3/2}$.

4.3 Universal Dependence of J_c on Magnetic Induction

From the above considerations we conclude that the behavior of $J_c(B)$ is analogous to the current-carrying properties of a single Josephson junction. The magnitude of $J_c(B)$ is defined by the average parameter I_0, as given in Eq. (28). The magnitude of the variations in $J_c(B)$ are on the order of B^*, which is the average magnetic flux through the junctions, and is equal to Φ_0. Thus, we expect that $J_c(B)$ for any ceramic medium can be scaled to a single universal curve. After a simple analysis we have established that this supposition is correct. The results for several samples are given in Fig. 6. These results are the same as those given in Fig. 4, presented in dimensionless units as $J_c/J_c(0)$ and B/B^*. It is clearly seen that all experimental points lie near the theoretical curve, which was obtained by averaging the critical current dependence from Eq. (28) for a single Josephson junction. Naturally, the scaling parameter B^* is different for the various samples. It is 9.5 Oe, 3.1 Oe and 1.0 Oe for samples 7, 8, and 9, respectively. Thus, we have established the universal behavior of the critical current dependence on magnetic induction in a wide range of magnetic fields. Of course, the universal behavior of $J_c(B)$ is not absolute.The essential deviation can take place at low field region for samples with a wide grain size distribution. Apparently this is the reason of the observed convergence between the calculated curve and the experimental results for sample 9 in Fig. 6.

4.4. Critical Current in High Magnetic Field

The universal behavior of $J_c(B)$ occurs as long as the magnetic field does not exceed the critical value H_{c1g}, the field at which Abricosov vortices begin to penetrate the grains. At fields larger than H_{c1g}, $J_c(B)$ tends to a constant asymptotic value of J_a [6]. Such an effect is observed in Fig. 4 for samples 2 and 3. We did not find the plateau on the curve for sample 1 which is characterized by the finest grains.

The value J_a may be evaluated from a simple physical consideration

$$J_a \simeq J_c(0) \cdot (\lambda_L/\alpha)^{3/2} . \tag{29}$$

This formula may be obtained if we substitute the parameter $\Phi = H_{c1g} \cdot 2\lambda_L \cdot \alpha$ and use the asymptotic expression for the Bessel function at large values of its

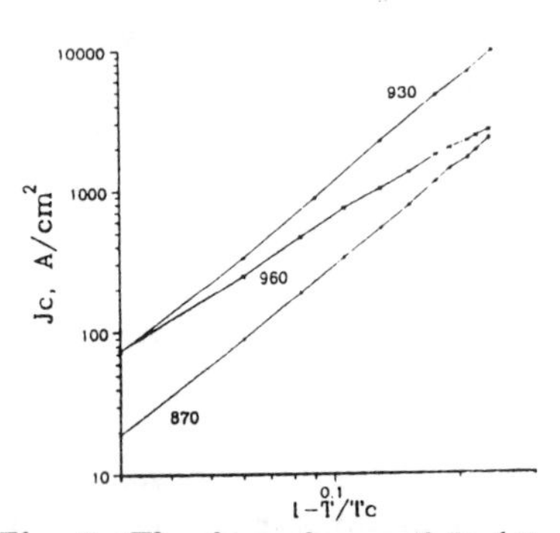

Fig. 5: The dependence of J_c (B=0) on T for samples 4-6 sintered at the temperatures indicated.

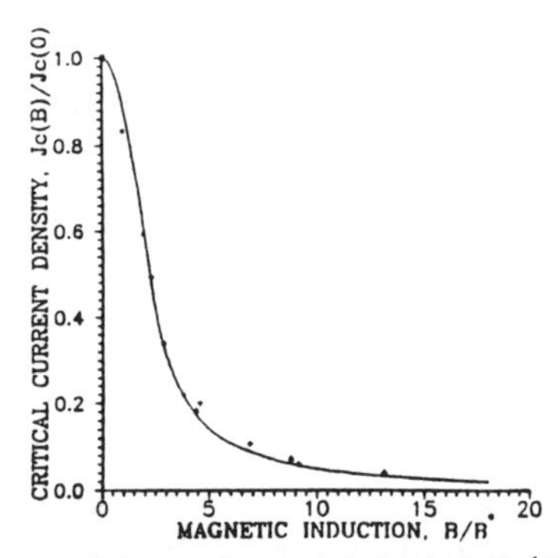

Fig. 6: The universal function $J_c(B)$ in dimensionless units. The solid curve is calculated.

argument. The experimental results agree with our estimation from Eq. (29). The calculated values of $J_a/J_c(0)$ for samples 2 and 3, where α was extracted from measured values of B^*, correlate with the direct measurement of $J_a/J_c(0)$. We note, for these values of the field, that the dependence of $J_c(B)$ in the general case is more complex. In particular, for fields near H_{c1g} the current may increase due to peak-effect [14,15,16].Nevertheless, Eq.(29)givesthecorrectorderforJ_a. At higher fields the situation is very complex due to the interaction between Josephson and Abricosov vortices. The behavior of $J_c(B)$ in these fields has been examined in a number of papers [10,16], so we will not pursue this subject.

The existence of a high-field plateau on the $J_c(B)$ plot may change the form of the surface resistance dependence on H. As mentioned previously, the position of the maximum in $\mathcal{R}(H)$ is defined by the relation

$$h = (4/3) \cdot h_p = (8\pi d/3c) \cdot J_c(B) \ . \tag{30}$$

Since the function $J_c(B)$ exceeds the value J_a everywhere, Eq. (30) can not be satisfied at $h < 8\pi d J_a/3c$. This means that the maximum in $\mathcal{R}(H)$ does not exist at low h. This effect is demonstrated in Fig. 7 for sample 2. The maximum in $\mathcal{R}(H)$ for this sample is observed only at $h > h_{min} = 0.14$ Oe. The calculated value of h_{min} coincides exactly with the experimental one.

In the above considerations we did not account for the dissipative process in the grains. This process begins to give a noticeable contribution to the surface resistance at fields where $\mathcal{R}(H)$ tends to plateau. The losses in the grains lead to the approximately linear increase of $\mathcal{R}(\mathcal{H})$ in this field region (see Fig. 7) which is connected with the critical state in the grains. Using this experimental result, we conclude that the intragranular critical current density changes with magnetic field in accordance with the Kim - Anderson law.

4.5. Surface Impedance in the Resistive State

Above we considered the behavior of the surface impedance of high-T_C materials as a function of external magnetic field. Here we investigate the behavior of Z as a function of the transport current I flowing through the sample. We may expect that the dependence of Z on current I will be similar to the function $Z(H)$. The self magnetic field of transport current will play the same role as an external magnetic field H. Some results of these investigations were discussed in [2]. Here we consider the problem of ac magnetic field penetration into a sample when the transport current I flows through the sample. Because the self current is proportional to I, increasing I leads to the suppression of the critical current density of the superconductor. The surface impedance may be described by Eqs. (9)-(16) after replacing the external magnetic field H by the self field of the current. As a result, the surface impedance proves to be a smooth function of transport current.

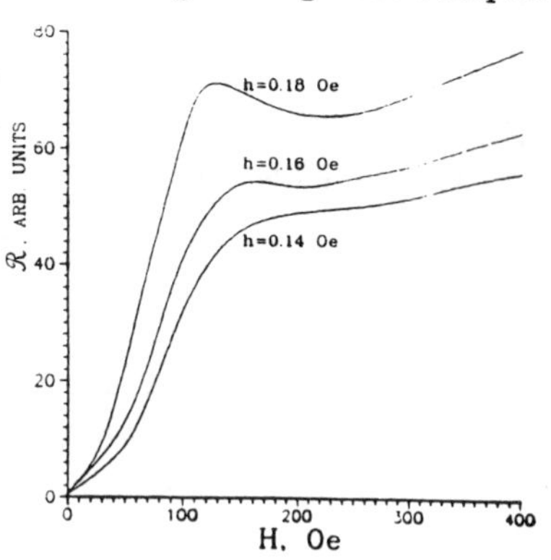

Fig. 7: The function $\mathcal{R}(H)$ for sample 7 obtained at different ac field amplitudes (h). T = 77K.

However, such behavior of $Z(I)$ occurs only when the current I is less than the critical one. The situation changes radically at $I > I_c$, when a sample reverts to the resistive state. The formation of a longitudinal electric field E_z decreases the current which screens an external ac field. For a cylinder the screening current may be written in a form:

$$J_\phi = J_c \cdot E_\phi / (E_\phi^2 + E_z^2)^{1/2} \tag{31}$$

where E is the azimuthal component of electric field. This dependence of the screening current shows that when I exceeds I_c, the ac penetration depth and surface impedance increase sharply. We have noted above that the surface resistance has a maximum when the penetration depth becomes equal to the size of the sample. This maximum is easily observed experimentally. The sharp increase of surface impedance near $I = I_c$ permits us to define the critical current value I with high precision. These results, of course, may be compared with the results of the contactless method for J_c measurement described earlier.

Acknowledgements

This work has been done in the framework of contract between Alfred University (USA) and State Committee for Science and Technology (USSR) and

National Programs for High-T_c Superconductivity. It is supported by Council on the Problem of High-T_c Superconductivity (USSR), project No. 90067.

References

[1] I.F. Voloshin, N.M. Makarov, L.M. Fisher, and V.A. Yampol'skii, JETP Lett. **51**, 255 (1990).

[2] N.V. Il'in, O.I. Ljubimov, N.M. Makarov, I.F. Voloshin, and V.A. Yampol'skii, Solid State Commun. **76**, 141 (1990).

[3] L.M. Fisher, N.V. Il'in, O.I. Ljubimov, N.M. Makarov, I.F. Voloshin, V.A. Yampol'skii, *AIP Conference Proceedings* "Superconductivity and its Application," Buffalo, N.Y. p. 153 (1990).

[4] A.M. Campbell, J. Phys. C **2**, 1492 (1969).

[5] C.P. Bean, Phys. Rev. Lett. **8**, 250 (1962), Rev. Mod. Phys. **36**, 31 (1964).

[6] H. Dersch and G. Blatter, Phys. Rev. B **38**, 11391 (1988).

[7] R.L. Peterson and J.M. Ekin, Physica C **157**, 325 (1989).

[8] A. Barone and G. Paterno, *Physics and Applications of the Josephson Effect*, (Wiley, New York, 1982).

[9] T.K. Dey, *et al.*, Solid State Commun. **68**, 635 (1988).

[10] L.M. Fisher, N.V. Il'in, N.A. Podlevskikh, S.I. Zakharchenko, Solid State Commun. **73**, 687 (1990).

[11] K.K. Likharev, Pis'ma ZTP **2**, 26 (1976) (Russian).

[12] C. Deutscher, IBM J. Res. Dev. **33**, 29, (1989).

[13] R.L. Peterson and J.W. Ekin, Phys. Rev. B **42**, 8014 (1990).

[14] E.G. Zwartz, B.A. Judd, E. Batalla and L.S. Wright, J. Low Temp. Phys. **74**, 277 (1959).

[15] J.R. Thompson, J. Brynestad, *et al.*, Phys. Rev. B **39**, 6652 (1989).

[16] M.V. Fistul', JETP Lett. **49**, 113 (1990).

APPLICATIONS

HIGH-TEMPERATURE SUPERCONDUCTIVE DELAY LINES AND FILTERS

W. G. Lyons, R. S. Withers,* J. M. Hamm, Alfredo C. Anderson, and D. E. Oates
Lincoln Laboratory, Massachusetts Institute of Technology
Lexington, Massachusetts 02173

P. M. Mankiewich and M. L. O'Malley
AT&T Bell Laboratories
Holmdel, New Jersey 07733

R. R. Bonetti and A. E. Williams
COMSAT Laboratories
Clarksburg, Maryland 20871

N. Newman†
Conductus
Sunnyvale, California 94086

ABSTRACT

Passive microwave devices are expected to be among the first devices made from high-temperature superconductors to be inserted into system applications outside of a laboratory environment. The deposition and patterning of $YBa_2Cu_3O_{7-x}$ (YBCO) thin films has progressed sufficiently that a variety of high-quality passive microwave devices can now be fabricated on $LaAlO_3$ substrates. Demonstrated microwave devices include tapped-delay-line transversal filters with multigigahertz bandwidths, untapped delay lines, and narrowband (2% or less) microstrip filters. Potential system applications at operating temperatures of 77 K and below are discussed. These devices are a proof of principle that YBCO superconductive passive microwave devices can be developed to a level that will make system insertion a reality. The YBCO microwave devices described here demonstrate the potential performance advantages that superconducting thin-film devices offer for high-frequency, wide-bandwidth, high-performance analog signal processing systems.

INTRODUCTION

The low conductor loss of superconducting materials makes them attractive for a number of passive microwave devices. Superconductive passive microwave devices can be divided into two general classes: transmission-line-based delay lines and resonator-based filters. Long delay lines take advantage of the low propagation loss and dispersionless nature of a thin-film superconductor. Resonator-based devices take advantage of the large conductor quality factor (Q) that can be obtained using thin-film

*Current address: Conductus, Inc., Sunnyvale, California 94086
†Current address: University of California, Dept. of Materials Science, Berkeley, California 94720

superconductors. These devices are particularly suited to high-frequency, wide-bandwidth analog signal processing applications.[1] In addition, microwave applications are especially attractive for high-temperature superconductors since these systems would usually be in nonlaboratory environments where power and weight are major considerations. Moderate cryogenic temperatures (e.g., 60 - 100 K) can be obtained far more readily than temperatures near the boiling point of liquid helium, which are required for traditional superconductors and have precluded most microwave applications of superconductors in the past. Furthermore, many semiconductor devices used in microwave systems perform at cryogenic temperatures with less noise, higher efficiency, and lower loss.

Recent advances in thin-film deposition techniques for high-temperature copper-oxide superconductors have made possible the demonstration of surface resistances well below that of cooled copper at 77 K. Figure 1 illustrates the current state of the art in thin films of YBCO by plotting surface resistance as a function of frequency for YBCO.[2] For comparison, curves are shown for copper at 77 K, superconducting niobium at 4.2 K, and superconducting Nb_3Sn at 4.2 K. Films of YBCO can be chemically unstable, and it is significant that the best results on patterned films agree well with the best results for unpatterned films, indicating that no degradation occurred in the patterned films. This is especially important at the edges of the signal lines where currents peak. To indicate the potential performance of YBCO that can be expected with further improvements in film quality, theoretical curves obtained using the standard two-fluid model for the surface resistance of YBCO are also plotted.[3] The parameters used were λ_o =150 nm, T_c = 93 K, and $\sigma_n = 2 \times 10^6\ \Omega^{-1}m^{-1}$. It should be noted that the Siemens/Wuppertal data point at 77 K is within a factor of two of the prediction of the two-fluid model if the actual T_c of that film is used in the model. This lends some credence to the use of such a simple model.

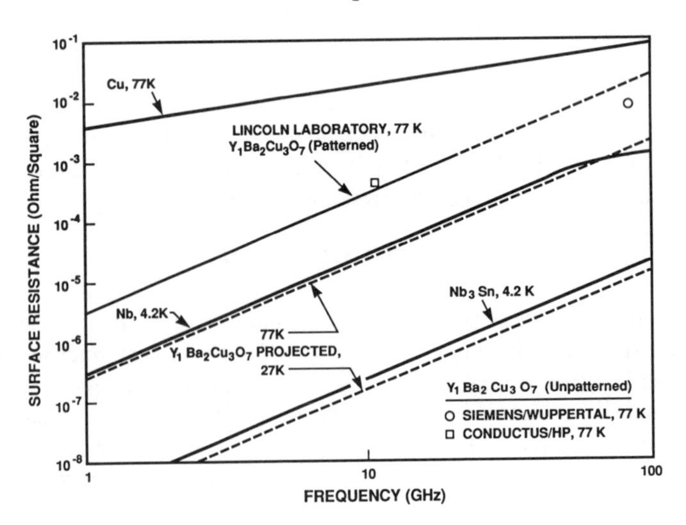

Fig. 1. Surface resistance plotted as a function of frequency for cooled copper and for superconducting niobium, Nb_3Sn, and YBCO. Theoretical values for YBCO are based on the simple two-fluid model.

Another advantage of superconductors is the ability to trade some of the low loss for increased circuit density. This is illustrated in Fig. 2. By reducing the thickness of the substrate, conductors can be placed much closer together because the coupling between lines has been greatly reduced. However, to maintain a reasonable characteristic impedance the linewidths of the conductor must be significantly reduced, which requires a lower surface resistance for the same performance. Currently, the deposition of high-temperature superconductors is limited to dielectric substrates on the order of 20 mils thick. Techniques for thinning substrates, similar to work done with niobium on silicon substrates,[1] and techniques for depositing compatible dielectrics are currently a focus of research. Work summarized here will be limited to YBCO deposited on 20-mil-thick $LaAlO_3$.

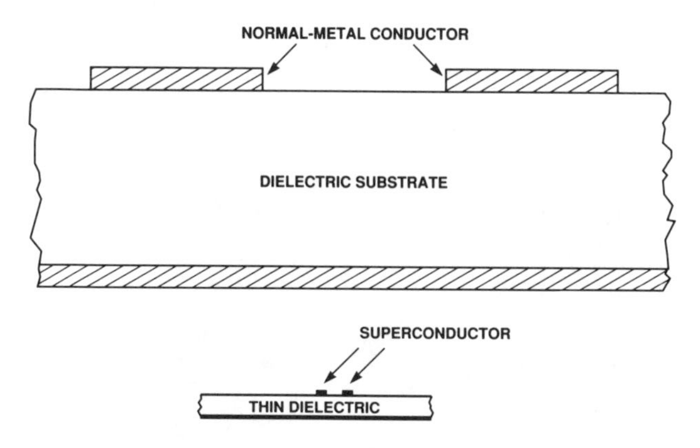

Fig. 2. Illustration of the increased circuit density made possible through the use of superconductors.

THIN FILM DEVICE FABRICATION

Thin films of YBCO were deposited by two different techniques. One method is an *ex-situ* process in which Y, Ba, and Cu are deposited with the correct stoichiometry by coevaporation of BaF_2, Y, and Cu followed by a postdeposition annealing, typically at 850 °C, in flowing O_2 containing water vapor.[4] The second method is an *in-situ*-growth process using off-axis single-target sputtering with substrate temperatures typically between 650 and 750 °C.[5,6] Both techniques can be used to deposit YBCO uniformly over large areas of at least 2-in. diameter. Typical YBCO film thicknesses used in this work were 200-300 nm. The postanneal growth technique can be readily applied to the deposition of films on both sides of a substrate. The off-axis sputtering is currently limited to single-sided deposition because silver paste is used to thermally anchor the substrate to a heater block. The off-axis sputtering technique has provided films with the lowest microwave surface resistance and the best power handling capability measured to date in our laboratory for YBCO.[2]

Patterning of the YBCO signal line was accomplished with standard AZ1470 photoresist and a spray etch of 0.25% H_2PO_4, which prevented the formation of a residual film typically found with other wet etching methods.[7,8] Undercutting of

1-2 μm is typical for this etch and these YBCO thicknesses. A trilayer resist (PMMA/Ti/AZ1470) and liftoff process were used to pattern 1.5-μm-thick silver contacts on the signal line. The trilayer resist was used so that the chemically sensitive YBCO was not exposed to AZ photoresist developer. The silver contacts were annealed in O_2 at 400 °C to form ohmic contacts, reduce the contact resistance,[9] improve the mechanical robustness of the contacts, and allow wire bonding to be performed. Final packaging was performed using ultrasonic wedge bonding of aluminum ribbon directly to the annealed silver contacts. This contact technique reliably produces a very robust and low-resistance contact. A 4-μm-thick, electron-beam-evaporated silver film was used either alone or over a YBCO film to form the device ground planes. Figure 3 shows the contact region of a packaged device with an aluminum ribbon wedge-bonded to an annealed silver contact on top of a YBCO signal line. The underlying substrate material is $LaAlO_3$. Note the excellent pattern definition obtained with the wet etch for the YBCO signal line.

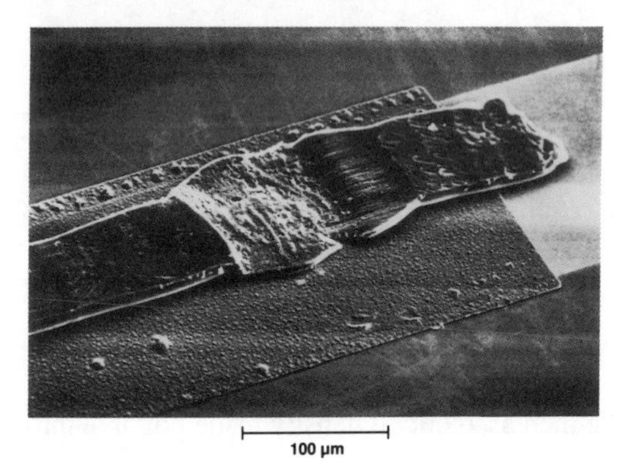

Fig. 3. Photograph of an aluminum ribbon wedge-bonded to an annealed silver contact on a YBCO signal line.

RESONATOR-BASED NARROWBAND FILTERS

Single stripline resonators have been used to measure the surface resistance of superconducting films (see ref. 2, for example) and may eventually be used for the stabilization of low-noise oscillators for high-performance radars.[10] Narrowband filters based on multiple-resonator structures are an even broader application area. The conductor quality factor of normal-metal microstrip lines is too low to provide the low insertion loss and sharp filter skirts needed for channelizing microwave communication bands on communication satellites. For this application, less than 3% bandwidth filters are needed, which in turn require high-Q resonators. As a result, dielectric-loaded cavity filters are used in present systems in order to obtain the requisite Q's. Figure 4 compares a four-pole superconducting microstrip filter with a six-pole, dielectric-loaded, dual-mode cavity filter that is currently flown on communication satellites. Superconducting filters potentially could replace these dielectric-loaded cavity filters with a significant reduction in the weight of a frequency multiplexer consisting of dozens of filters. In addition, the potential exists to produce superconducting filters with much sharper filter skirts than current-generation filters so that communication bands can be utilized more efficiently. Figure 5 illustrates a typical channelization scheme using ten 1%-bandwidth filters to channelize frequencies from 9 to 10 GHz. It is clear from the channelization scheme shown in Fig. 5 that low insertion loss, flat passband characteristics, and sharp filter skirts are the crucial criteria for channelizing filters.

Fig. 4. A photograph of a four-pole superconducting microstrip filter and a six-pole dielectric-loaded dual-mode cavity filter.

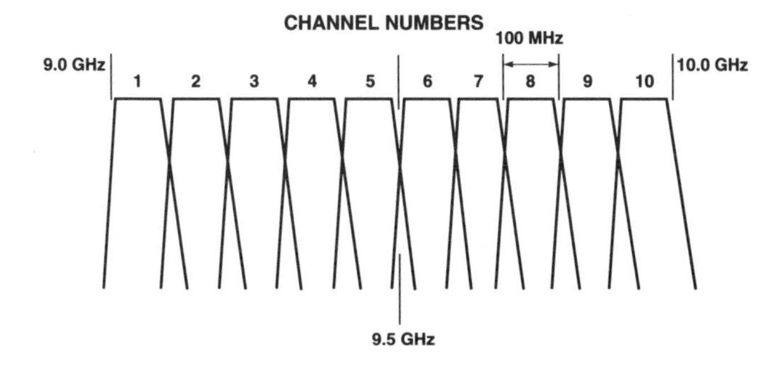

Fig. 5. Typical channelization scheme using ten 1%-bandwidth filters to channelize the band from 9 to 10 GHz.

High-temperature superconductive resonator-based filters were first fabricated at Lincoln Laboratory using postannealed YBCO.[7,11] These filters had YBCO signal lines and were demonstrated with either a normal silver ground plane or a superconducting YBCO ground plane (requiring double-sided coverage of a $LaAlO_3$ substrate). Figure 6 shows the transmission characteristic of an all-YBCO 2%-bandwidth four-pole filter measured at 77 K. Thin films of YBCO were deposited on both sides of a 1.3×2.3-cm, 425-μm-thick $LaAlO_3$ substrate. This filter had an insertion loss of 0.3 dB at 77 K. The same filter design with a YBCO signal line and a silver ground plane had 0.4 dB of insertion loss at 77 K. An all-gold version of the filter had 2.8 dB of insertion loss at 77 K. The conductor Q required for a certain level of filter performance and the current distribution in the signal line and ground plane for the microstrip layout chosen to implement that filter determine whether a superconducting ground plane is required. This is represented schematically in Fig. 7. Losses are much higher in the signal line, where the current densities are much larger and the current tends to peak at the edges of the line. Current distributions have been calculated accurately for stripline structures,[12] where the loss in the signal line is about 10 to 20 times larger than the loss in the ground plane for the typical structure used. Calculations are currently being done for microstrip configurations.[13] From Fig. 1 it is clear that the surface resistance of YBCO approaches the surface resistance of a normal metal at higher frequencies ($R_s \propto f^2$ for a superconductor). Thus, a normal ground plane will be the limiting loss in a filter structure at frequencies below which the ratio of the superconducting to normal metal surface resistances is equal to the ratio of loss in the ground plane to loss in the signal line for a structure made completely of the same material. This means that a superconducting ground plane is required only at lower frequencies.

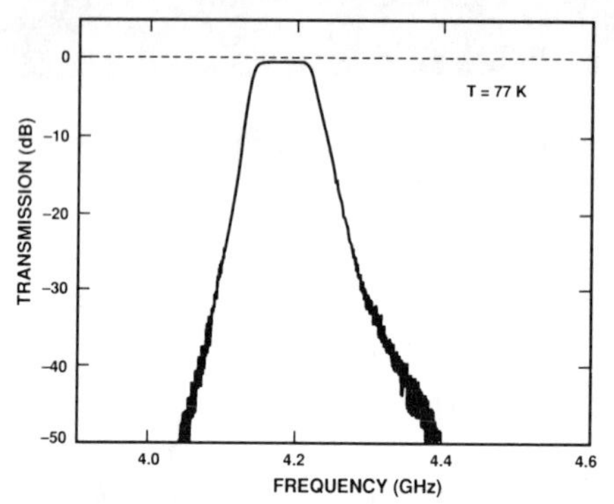

Fig. 6. Transmission versus frequency measured at 77 K for an all-YBCO four-pole, 2%-bandwidth microstrip filter.

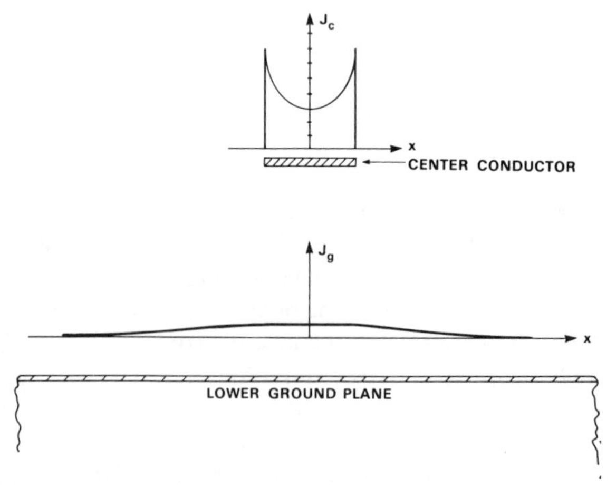

Fig. 7. Schematic representation of the current distribution in the signal line and ground plane of a microstrip transmission line.

The work reported here was begun in conjunction with the Navy's desire to use high-temperature superconductive devices on board an orbiting satellite. A satellite experiment will take place in 1993 and is called the first High-Temperature Superconductivity Space Experiment (HTSSE I).[14] A number of other groups associated with this program have also reported work on resonator-based filter structures.[15-19] This body of work demonstrates that high-performance devices can be made from the high-T_c materials and has made clear a number of technical difficulties and considerations. The low loss of the superconductor makes matching to a 50-Ω impedance challenging. Any reflections generated by a mismatch at some frequency will not be attenuated and will result in a filter passband with high insertion loss at that frequency. The very flat, low-insertion-loss passband characteristic shown in Fig. 6 is indicative of a good match to 50 Ω.

Because of inaccuracies in microwave computer-aided design (CAD) routines for substrates with large dielectric constants ($\varepsilon_r \approx 24$ for $LaAlO_3$), our design of the bandpass filters was accomplished in a semiempirical fashion.[20] A first iteration was designed using an existing CAD package. Line-to-line couplings were then measured on this design and used to adjust the CAD design to more closely match the actual filter parameters. Figure 8 shows a packaged four-pole filter that operated at 4.8 GHz with a 1.3% bandwidth. This filter was a redesigned version of an earlier filter, the frequency response of which is shown in Fig. 6. The redesigned filter had a reduced sensitivity to variations in linewidth and substrate dielectric constant.[21,22] Figure 9 shows the measured filter response for this redesign using several different conductors at temperatures of 300 and 77 K. The YBCO filter, fabricated on $LaAlO_3$, had less than 1 dB of insertion loss at 77 K compared with 5.9 dB for the silver filter at 77 K and 10.6 dB for the gold filter at 300 K.

Fig. 8. Photograph of a four-pole YBCO filter (with silver ground plane) in a space-qualified microwave package. The package is hermetically sealed with an indium gasket.

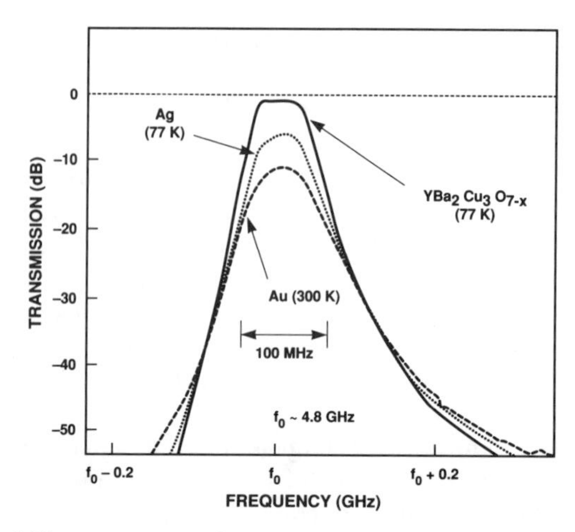

Fig. 9. Measured filter response of the 1%-bandwidth, four-pole filter design shown in Fig. 8. The design was fabricated from gold (measured at 300 K), silver (measured at 77 K), and YBCO (measured at 77 K).

The surface resistance of a superconductor is nonlinear at high fields, and this effect can be characterized by measuring the intermodulation distortion generated in a device.[2,23] This is a two-tone measurement (f_1, f_2) where the third-order intermodulation terms ($2f_1$ - f_2 and $2f_2$ - f_1) are the first and the largest intermodulation products that will fall back into the bandpass of a narrowband filter. The two tones at f_1 and f_2 are applied with equal amplitudes and can be located anywhere within the passband of the filter. Measurements indicate that the third-order intermodulation products at $2f_1$ - f_2 and $2f_2$ - f_1 are equal in amplitude as long as the equal-amplitude fundamental tones f_1 and f_2 are well within the passband of the filter. Figure 10 shows the results at 4.2 K on the filter shown in Fig. 8, which was fabricated from an *in-situ*-grown, off-axis-sputtered YBCO film on $LaAlO_3$. The output power for the third-order products is proportional to the cube of the input power and shows a slope of 3 on the log-log plot, while the output power vs input power for the fundamental tones has a slope of 1. The amplitude of one fundamental tone (f_1) and the amplitude of one intermodulation product ($2f_1$ - f_2) are plotted in Fig. 10. The input power at which the two lines intersect is the third-order intercept. The measured data are shown uncorrected for the loss in the probe. A 5-dB correction for the loss in the probe gives a third-order intercept of approximately +35 dBm at 4.2 K. This implies that at normal operating levels of -10 to 0 dBm the intermodulation products will be more than 75 dB below the main signal at 4.2 K. The third-order intercept is approximately + 25 dBm at 77 K, and the intermodulation products are more than 50 dB below the linear term at an operating level of 0 dBm.[23] This is adequate for most applications. The good power handling capability of this YBCO film is only obtained with a high-quality film. A variation of over four orders of magnitude has been observed in the third-order intercept points, indicating the wide differences in power handling capability of YBCO films that have similar properties at lower power levels. Recall that the energy stored in a resonator is approximately Q times the input energy. Thus, these high-Q superconducting resonators contain very high fields at only moderate input powers. Noise generated by nonlinearites and other mechanisms in the YBCO film was also measured[2] and found to be negligible for standard filter applications.

For channelization applications, filter skirts determine the channelization efficiency. The definition of a filter skirt is somewhat arbitrary, but a common choice is the width in frequency of the transition from the -3-dB passband edge to the -60-dB passband edge. Figure 11 shows the simulated performance of 1four-, six-, eight-, and ten-pole Chebyshev filters with 1% bandwidths centered at 4 GHz assuming a Q of 5000 for YBCO resonator sections operated at 77 K. The filter skirts for the four- and six-pole filters are very broad and would require very wide guard bands in any channelization scheme, resulting in a very inefficient use of the microwave band. These wide guardbands are unacceptable in many applications. We define channelization efficiency as the ratio of the -3-dB passband to the -60-dB passband. The required guardbands are equal to the -60-dB passband minus the -3-dB passband. The channelization efficiencies of the filters whose calculated response are shown in Fig. 11 are as follows: 21% for the four-pole filter, 40% for the six-pole filter, 55% for the eight-pole filter, and 64% for the ten-pole filter. Channelization efficiency is obviously a function of the filter bandwidth as well.

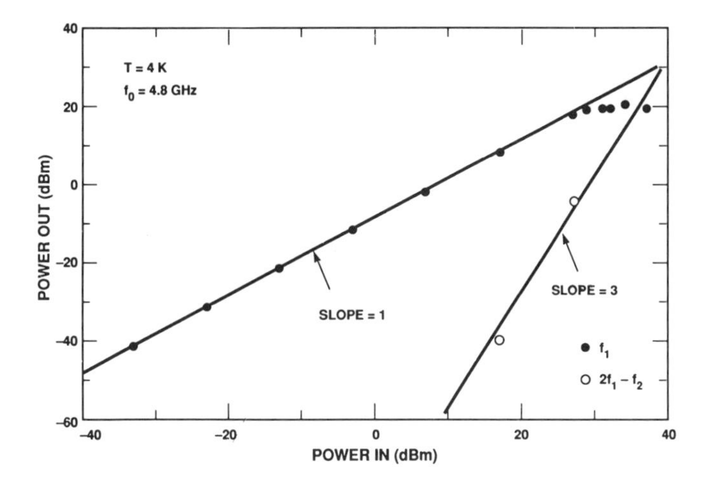

Fig. 10. Two-tone intermodulation measurements at 4.2 K on the YBCO filter shown in Fig. 8. Two fundamental tones at f_1 and f_2 (with $f_1 > f_2$) are applied to the input of the filter. The fundamental tones are equal in amplitude and were centered at 4.8 GHz with a separation of 1 MHz. The output power for the fundamental tone at f_1 and for the $2f_1$ - f_2 third-order intermodulation product are shown as a function of the input power at f_1.

For example, the Intelsat VII communications satellite will avoid wide guardbands by utilizing bulky dielectric-loaded cavity filters with an eight-pole elliptic response (equivalent to a ten-pole Chebyshev response). Figure 12 shows the simulated response of a ten-pole microstrip Chebyshev filter with a 1% bandwidth. Note the less than 2 dB of insertion loss for YBCO (77 K) and the 33 dB of insertion loss for gold (300 K). The 3-dB passband of the gold filter has been significantly reduced compared to the passband of an ideal, low-loss filter. Similarly, when normalized to a small passband insertion loss, the filter skirts of the gold filter have been greatly broadened compared to the ideal, low-loss filter characteristic. Such complex filter designs as the ten-pole filter shown here are required for channelization applications.

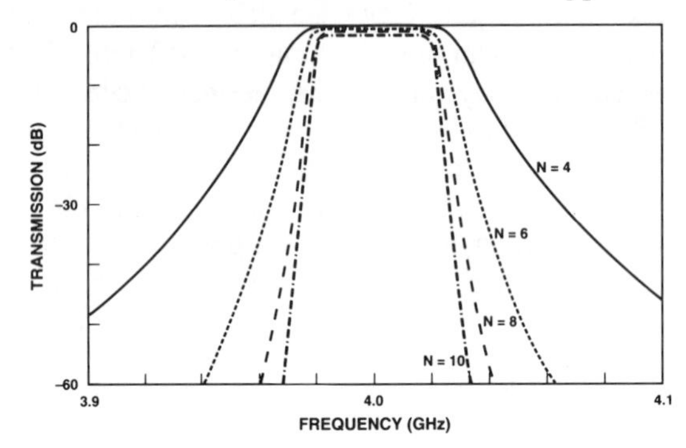

Fig. 11. Projected performance at 77 K of YBCO Chebyshev-response 1%-bandwidth filters of order N (number of poles). Note the greatly improved channelization efficiency provided by the ten-pole filter compared to the four- or six-pole filters.

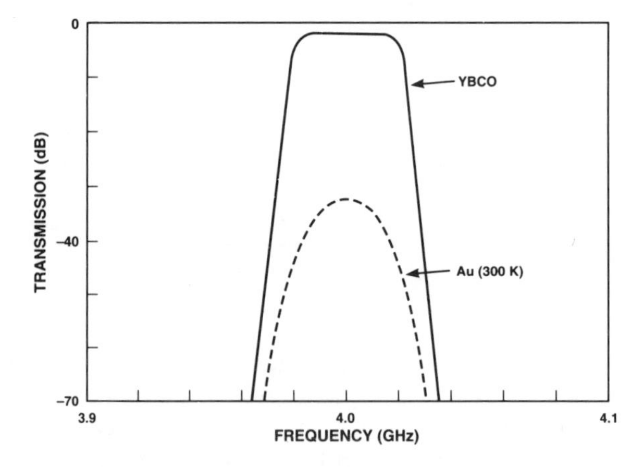

Fig. 12. Projected performance at 77 K of YBCO 1%-bandwidth, ten-pole Chebyshev filter compared to the same filter implemented with room temperature gold.

A severe materials constraint currently limits the complexity of resonator-based narrowband filters, however. Figure 13 highlights the matrix of twins typically found in $LaAlO_3$. Measurements of various resonator sections of filters on $LaAlO_3$ substrates indicate that the dielectric constant varies spatially on the order of 2%, apparently because of the twin structure. Cavity measurements also indicate a variation in the dielectric constant of $LaAlO_3$.[24] It is clear that the twin boundaries modify the optical index of refraction (the optical dielectric constant) because we can see them, so it is reasonable to assume that they also modify the microwave dielectric constant. The twins represent a random variation in the dielectric constant if this is true. Figure 14 shows that the twins move considerably at the YBCO growth temperatures. Figure 15 illustrates that mild stress is enough to generate twins. Thus, a new substrate material is needed for high-T_c microwave applications. Several candidate substrate materials are MgO and sapphire coated with a thin-film buffer material.

Fig. 13. Photograph of a six-pole YBCO filter (YBCO signal line and silver ground plane). The photograph was taken to highlight the twin structure in the pseudo-(100)-oriented $LaAlO_3$ substrate.

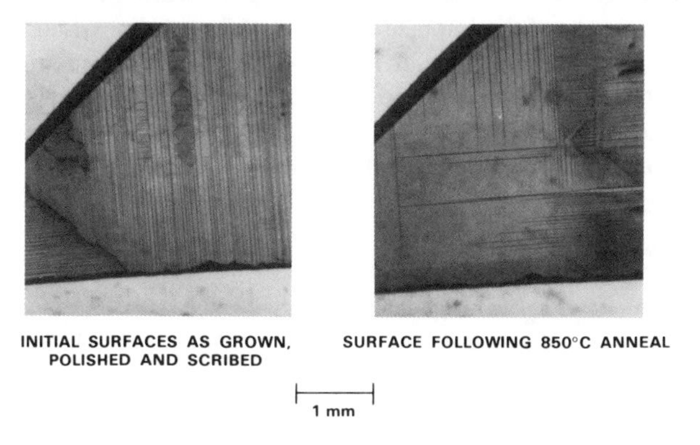

Fig. 14. Photograph of the twin motion in $LaAlO_3$ that occurs at the growth temperature of YBCO thin films.

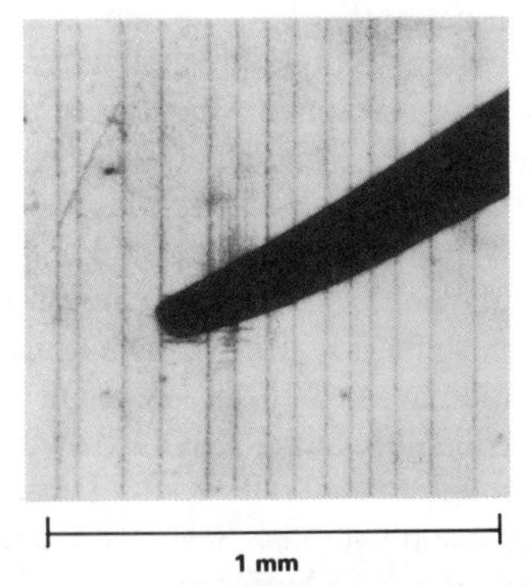

Fig. 15. Photograph of twin boundaries formed in a $LaAlO_3$ substrate by the pressure generated at a tweezer tip (horizontal lines just above and below tweezer tip). Twin boundaries running the length of the photograph were present before pressure was applied.

TAPPED-DELAY-LINE CHIRP FILTERS

Most analog signal processing structures are variations of the transversal filter. A tapped delay line can be used to implement matched filtering, correlation, and Fourier transformation. Compact transversal filters in the form of tapped delay lines have previously been fabricated from superconducting niobium to operate with multigigahertz signal processing bandwidths.[25,26] One example of a fixed-tap-weight transversal filter is a linearly frequency modulated delay line or chirp filter. Figure 16 shows the architecture of a chirp filter made using superconducting striplines with a cascaded series of backward-wave couplers, in this case forming a down-chirp, i.e., the instantaneous frequency of the impulse response decreases linearly in time. Each coupler has a peak response at the frequency for which it is 1/4 wavelength long. By making the length of the couplers proportional to the reciprocal of the length down the line, the filter has a local resonant frequency that is a linear function of delay. Design of the stripline chirp filters is based on the theory of coupled modes.[27]

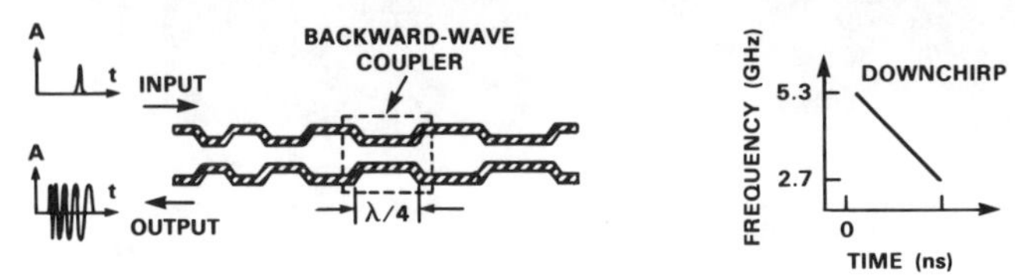

Fig. 16. Superconductive chirp filter architecture using the down-chirp input and output ports. The instantaneous frequency response to an impulse function input is also shown for a down-chirp filter.

Several untapped YBCO delay lines have been reported in the literature.[8,21,28-30] For example, one untapped delay line consisted of a short coplanar transmission line with lengths of various characteristic impedances designed to implement a low-pass filter.[30] Care must be taken in the design of long delay lines. The utility of the delay lines will be rather limited if the lines in the meander pattern are overcoupled, resulting in a transmission response with bandstop regions throughout the bandwidth. A 7-ns-long YBCO microstrip delay line fabricated at Lincoln Laboratory[8] was also demonstrated in conjunction with a GaAs sample-and-hold cell. The pair of devices was operated at 77 K to simulate a pretrigger function on a high performance oscilloscope.[21] The power dependence of the surface resistance is much less of an issue in these delay lines since energy is not stored but rather continually flows down the transmission line. This contrasts sharply with devices that store energy in high-Q superconducting resonators where the nonlinear surface resistance at high power levels is a crucial issue.[23]

Success with the YBCO microstrip line[8,21] led to the YBCO chirp-filter design shown in Fig. 17 (along with the stripline package and upper ground plane).[8] This device was fabricated on a 50-mm-diameter, 500-μm-thick $LaAlO_3$ substrate and used 40-Ω, 120-μm-wide signal lines in the tapped portion of the device. A Klopfenstein-taper impedance transformer from the narrow 50-Ω input lines to the much wider 40-Ω lines greatly reduced the length of narrow line required. Chirp filters were initially fabricated with 8 ns of dispersive delay[8] and more recently with 11.5 ns.[21,22] The latest devices consist of 48 backward-wave couplers and have a bandwidth of 2.7 GHz, giving a time-bandwidth product of 30, a center frequency of 4.2 GHz, and a designed insertion loss of 5 dB. The time-bandwidth product represents the number of information cycles gathered coherently in the filter, determining the signal processing gain. The output of this filter is produced by the backward-wave coupling of a signal to a second line. The 5 dB of insertion loss actually represents fairly strong coupling between the lines. The time-domain step-function response of a flat-weighted YBCO chirp filter is shown in Fig. 18. The response is fairly insensitive to temperature from 4.2 K up to 77 K, and in fact the propagation loss is immeasurable (< 0.2 dB) at temperatures below 77 K. Because the filter is flat weighted with respect to frequency, the taps are more strongly coupled at the lower frequencies where the number of taps acting together coherently, N_{eff}, is smaller than at the higher frequencies. Figures 19 and 20 show the frequency response of a flat and Hamming-weighted YBCO chirp filter, respectively. The agreement with design, which assumes lossless transmission lines, is quite good. The effects of kinetic inductance are obvious at 88.5 K.

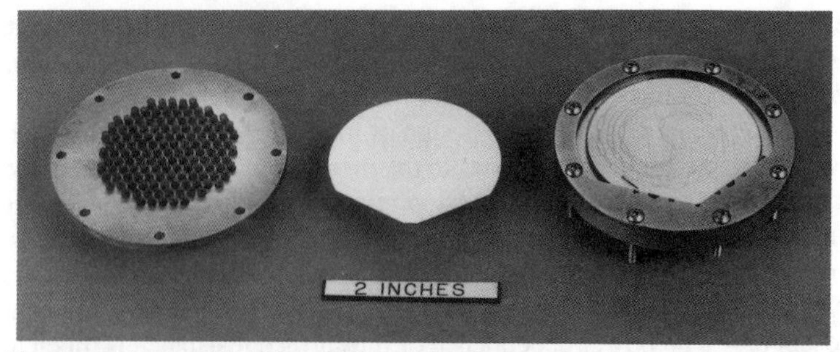

Fig. 17. Photograph of a YBCO stripline chirp filter showing the stripline package, YBCO signal line, and silver upper ground plane.

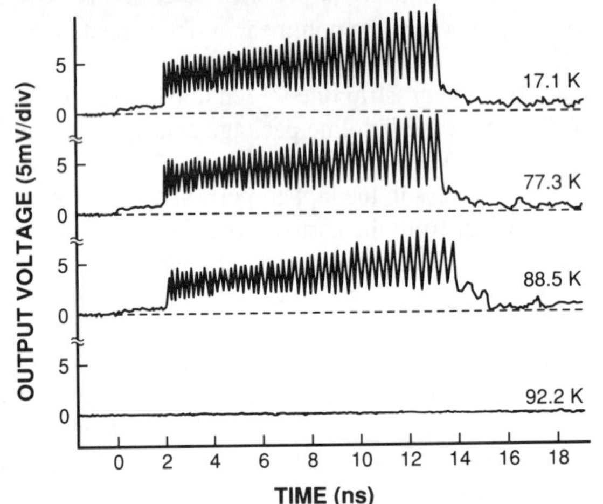

Fig. 18. Time-domain response to a step-function input for a YBCO chirp filter at various temperatures. The filter was operated in the down-chirp mode. Note the additional delay caused by kinetic inductance at 88.5 K.

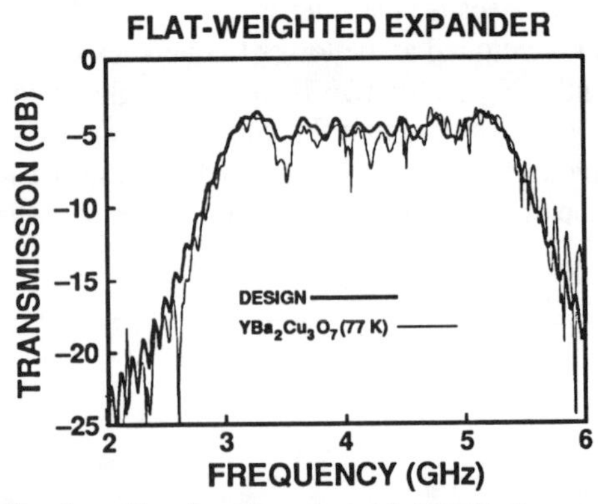

Fig. 19. Designed (lossless lines) and measured (77 K) frequency-domain performance of a flat-weighted YBCO chirp filter. Design parameters were 5-dB insertion loss, 11.5-ns dispersion, 2.7-GHz bandwidth, and 4.2-GHz center frequency.

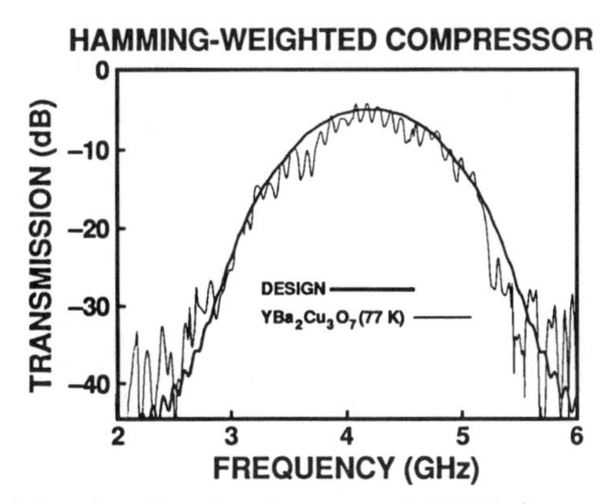

Fig. 20. Designed (lossless lines) and measured (77 K) frequency-domain performance of a Hamming-weighted YBCO chirp filter. Design parameters were the same as for the device in Fig. 19 with the addition of the Hamming weighting.

We have recently demonstrated the operation of a matched pair of YBCO chirp filters at 77 K.[31] A matched filter pair consists of two chirp filters, which have impulse responses that are time reversed with respect to one another. A single chirp filter, such as that shown in Fig. 16, can be used to generate both up-chirps and down-chirps. An up-chirp and a down-chirp represent time reversed signals if the bandwidths and time delays are the same. The compressed pulse response is a good measure of the quality of a matched filter pair. One chirp filter (the expander) is used to generate a chirp waveform by passing an impulse through it. The chirp waveform is then passed through the other chirp filter (the compressor) having an impulse response that is the time-reverse of the first filter. The matched signal energy is compressed coherently into a quasi-impulse, or compressed pulse, with a width that is inversely proportional to the bandwidth of the chirp filters. Since the input to the compressor is the time-reverse of its impulse response, the output is effectively the autocorrelation of the input signal. By varying the weighting of the backward-wave couplers, the sidelobe level of the autocorrelation can be tailored. A Hamming-weighted compressor is used in this YBCO chirp filter pair to substantially reduce the sidelobe level. Figure 21 shows the compressed pulse generated using a flat-weighted YBCO chirp filter as the expander and a Hamming-weighted chirp filter as the compressor. The compressed pulse response is quite good, although the relative sidelobe level of -23 dB is higher than the designed level of -32 dB. The total patterned YBCO transmission line length for each filter is 0.8 m.

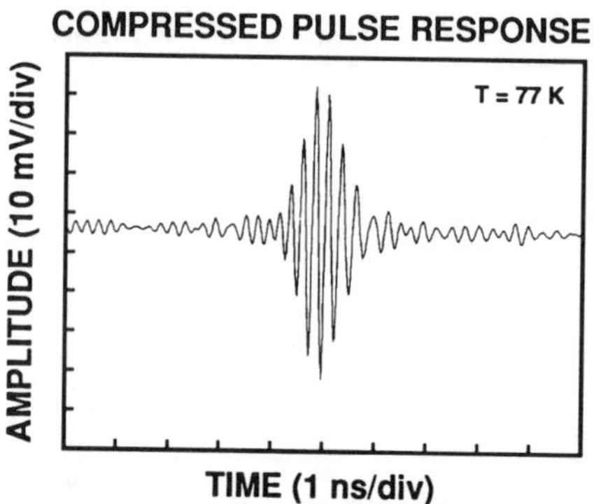

Fig. 21. Measured compressed pulse response (77 K) using flat-weighted YBCO chirp filter shown in Fig. 19 as the expander and Hamming-weighted chirp filter shown in Fig. 20 as the compressor.

Chirp filters can also be used to perform real-time Fourier transforms. Spectral analysis using the chirp transform is illustrated in Fig. 22. The input to the system is a number of narrowband tones. The dashed line in the figure will allow us to follow a single tone through the chirp-transform system. The input signal is mixed with a chirp waveform, generated by propagating an impulse through a chirp filter. Here an up-chirp is used and the sum frequency from the mixer is selected. This results in an up-chirp for each tone present at the input since the mixer has merely shifted the up-chirp in frequency by each input tone. The second chirp filter (compressor) has a time-reversed impulse response compared to the first, i.e., the frequency-delay slope is the negative of the first, so each up-chirp is transformed to a pulse at the output. Pulses corresponding to the higher input frequencies emerge first. The time of arrival of the pulses at the output is linearly related to the frequency of the input signal.

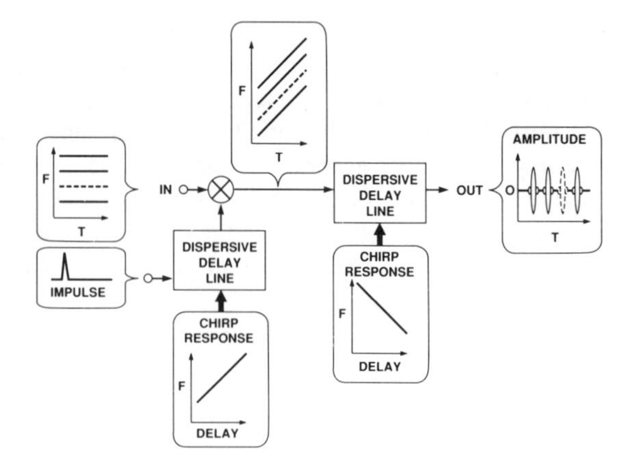

Fig. 22. Schematic of a chirp-transform spectrum analyzer.

DISCUSSION

Dramatic progress has been made in the development of high-temperature superconductive passive microwave devices. A variety of delay line and filter structures have been demonstrated using YBCO with operation temperatures as high as 77 K. Techniques have been developed to grow high-quality thin films of YBCO, pattern devices without degrading the superconducting material, and package the devices in a robust fashion that will survive a satellite environment. Improvements are still required, however, in a variety of materials-related areas.

In order to move forward with more aggressive filter designs, an alternative twin-free, low-loss substrate material must be obtained. Random variations in the dielectric constant caused by twinning or some other spatially varying effect are completely unacceptable for complex filter designs. Sapphire is a possible choice, but the anisotropic dielectric constant of sapphire is a difficult design challenge for many microwave device structures. Perovskites such as $LaAlO_3$ typically have high dielectric constants, which can be a problem because most computer-aided design routines are not valid for dielectric constants greater than about 18. Substrates significantly thinner than 20 mils must also be used in order to reduce the line-to-line spacing and increase the length of delay lines. Resonator-based filters are very sensitive to linewidth variations. More complex filter designs will require a very accurate linewidth control. The chemical sensitivity of YBCO continues to be a problem even with the best quality films. Damage at the edge of the patterned signal line must be avoided in any patterning process, and passivation methods for the films need to be examined.

Film quality must be improved in order to completely eliminate defects in the large-area (2-in. diameter) YBCO films. Current film quality often results in one or two small defects in a delay line that is about 1 m long. These defects must be bridged using annealed silver contacts in order to obtain a working delay line.[21] Wide linewidths were used in the delay lines reported here to reduce the sensitivity to defects in the film. Much longer YBCO delay lines will require significantly lower residual surface resistance. The residual surface resistance, which determines the surface resistance at lower temperature, is sufficiently large in YBCO that it is not yet possible to tell if the 77 K surface resistance will be lower for higher-quality YBCO films. In general, YBCO deposition techniques will continue to improve and new dielectric materials or combinations of dielectrics will be examined.

CONCLUSIONS

Two classes of superconductive passive microwave devices have been demonstrated: long delay lines and resonator-based filters. Both classes of devices have a significant performance advantage over normal metal versions of the same devices. Examples of the most recently demonstrated YBCO devices include tapped-delay-line stripline chirp filters and resonator-based narrowband microstrip filters.

Microstrip narrowband filters have also been demonstrated at 4.8 GHz and 77 K with bandwidths as narrow as 1%. Performance worthy of select system applications was obtained from a four-pole Chebyshev filter. Good power handling capability, low noise, and a good match to 50 Ω were obtained in this YBCO filter at 77 K. Stripline chirp filters were fabricated from YBCO and operated at 77 K with a 2.7-GHz bandwidth, a 4.2-GHz center frequency, and 11.5 ns of dispersive delay giving a time-bandwidth product of 30. No other device is capable of processing this bandwidth in a real-time fashion, and the chirp-transform performance is equivalent to a digital computer operating at a rate of 10^{11} operations per second. Two chirp filters were also operated as a matched filter pair to produce a compressed pulse at 77 K. The total transmission length of two cascaded matched filters was 28 ns.

All of these superconductive passive microwave devices represent preliminary versions of devices that, when fully developed, will far exceed the performance of normal metal versions of the same devices to such an extent that the devices effectively cannot be built with normal metal conductors. The devices presented here clearly demonstrate that high-performance passive microwave devices can be fabricated from high-temperature superconducting thin films. Operation of superconductive devices is now possible at temperatures up to 77 K, opening up new opportunities for high-frequency, wide-bandwidth analog signal processing systems.

ACKNOWLEDGMENTS

The authors wish to thank A.P. Denneno for mask generation, R.S. Tompkins for device packaging, G.L. Fitch for computer programming, R.L. Slattery for assistance with the microwave package fabrication, and E.M. Macedo for conventional thin-film depositions. We acknowledge the technical guidance of R.W. Ralston and T.C.L.G. Sollner, and the secretarial assistance of J.F. Clark and A.M. LaRose. This work was supported, in part, by the Department of the Navy, the Department of the Air Force and, under the auspices of the Consortium for Superconducting Electronics, the Defense Advanced Research Projects Agency.

REFERENCES

1. R. S. Withers and R. W. Ralston, Proc. IEEE 77, 1247 (1989).
2. D. E. Oates, A. C. Anderson, D. M. Sheen, and S. M. Ali, IEEE Trans. Microwave Theory Tech. 39, 1522 (1991).
3. W. G. Lyons and R. S. Withers, Microwave J. 33 (11), 85 (1990); and references therein.
4. P. M. Mankiewich, J. H. Scofield, W. J. Skocpol, R. E. Howard, A. H. Dayem, and E. Good, Appl. Phys. Lett. 51, 1753 (1987).
5. A. C. Westerheim, L. S. Yu-Jahnes, and A. C. Anderson, IEEE Trans. Magn. 27, 1001 (1991).
6. N. Newman, B. F. Cole, S. M. Garrison, K. Char, and R. C. Taber, IEEE Trans Magn. 27, 1276 (1991).
7. W. G. Lyons, R. R. Bonetti, A. E. Williams, P. M. Mankiewich, M. L. O'Malley, J. M. Hamm, A. C. Anderson, R. S. Withers, A. Meulenberg, and R. E. Howard, IEEE Trans. Magn. 27, 2537 (1991).
8. W. G. Lyons, R. S. Withers, J. M. Hamm, A. C. Anderson, P. M. Mankiewich, M. L. O'Malley, and R. E. Howard, IEEE Trans. Magn. 27, 2932 (1991).

9. J. W. Ekin, T. M. Larson, N. F. Bergren, A. J. Nelson, A. B. Swartzlander, L. L. Kazmerski, A. J. Panson, and B. A. Blankenship, Appl. Phys. Lett. 52, 1819 (1991).
10. D. E. Oates, A. C. Anderson, and B. H. Shih, in 1989 IEEE MTT-S International Microwave Symposium Digest (IEEE, New York, 1989), Vol. 2, pp. 627-630.
11. W. G. Lyons, R. R. Bonetti, A. E. Williams, P. M. Mankiewich, A. Meulenberg, M. L. O'Malley, A. C. Anderson, R. S. Withers, and R. E. Howard, in Solid State Research Report, MIT Lincoln Laboratory, Lexington, MA, 1990:1, pp. 64-70.
12. D. M. Sheen, S. M. Ali, D. E. Oates, R. S. Withers, and J. A. Kong, IEEE Trans. Appl. Superconduct. 1, 108 (1991).
13. S. M. Ali (private communication).
14. J. C. Ritter, M. Nisenoff, G. Price, and S. A. Wolf, IEEE Trans. Magn. 27, 2533 (1991).
15. H. S. Newman, D. B. Chrisey, J. S. Horwitz, B. D. Weaver, and M E. Reeves, IEEE Trans. Magn. 27, 2540 (1991).
16. D. B. Rensch, J. Y. Josefowicz, P. Macdonald, C. W. Nieh, W. Hoefer, and F. Krajenbrink, IEEE Trans. Magn. 27, 2553 (1991).
17. M. S. Schmidt, R. J. Forse, R. B. Hammond, M. M. Eddy, and W. L. Olson, IEEE Trans. Microwave Theory Tech. 39, 1475 (1991).
18. D. Kalokitis, A. Fathy, V. Pendrick, E. Belohoubek, A . Findikoglu, A. Inam, X. X. Xi, T. Venkatesan, and J. B. Barner, Appl. Phys. Lett. 58, 537 (1991).
19. S. H. Talisa, M. A. Janocko, C. Moskowitz, J. Talvacchio, J. F. Billing, R. Brown, D. C. Buck, C. K. Jones, B. R. McAvoy, G. R. Wagner, and D. H. Watt, IEEE Trans. Microwave Theory Tech. 39, 1448 (1991).
20. R. R. Bonetti and A. E. Williams, in 1990 IEEE MTT-S International Microwave Symposium Digest (IEEE, New York, 1990), Vol. 1, pp. 273-276.
21. W. G. Lyons, R. S. Withers, J. M. Hamm, R. H. Mathews, B. J. Clifton, P. M. Mankiewich, M. L. O'Malley, and N. Newman, to be published in Proceedings of the 1991 Picosecond Electronics and Optoelectronics Conference (Optical Society of America, Washington, DC, 1991).
22. W. G. Lyons, R. S. Withers, J. M. Hamm, A. C. Anderson, P. M. Mankiewich, M. L. O'Malley, R. E. Howard, and N. Newman, in 1991 IEEE MTT-S International Microwave Symposium Digest (IEEE, New York, 1991), Vol. 3, pp. 1227-1230.
23. D. E. Oates, W. G. Lyons, and A. C. Anderson, in Proceedings of the 45th Annual Symposium on Frequency Control (IEEE, New York, 1991), pp. 460-466.
24. D. Reagor and F. Garzon, Appl. Phys. Lett. 58, 2741 (1991).
25. R. S. Withers, A. C. Anderson, J. B. Green, and S. A. Reible, IEEE Trans. Magn. MAG-21, 186 (1985).
26. M. S. DiIorio, R. S. Withers, and A. C. Anderson, IEEE Trans. Microwave Theory Tech. 37, 706 (1989).
27. R. S. Withers and P. V. Wright, in Proceedings of the 37th Annual Symposium on Frequency Control (IEEE, New York, 1983), pp. 81-86.

28. L. A. Hornak, M. Hatamian, S. K. Tewksbury, E. G. Burkhardt, R. E. Howard, P. M. Mankiewich, B. L. Straughn, and C. D. Brandle, in 1989 IEEE International Microwave Symposium Digest (IEEE, New York, 1989), Vol. 2, pp. 623-626.
29. E. K. Track, G. K. G. Hohenwarter, L. R. Madhavrao, R. Patt, R. E. Drake, and M. Radparvar, IEEE Trans. Magn. 27, 2936 (1991).
30. W. Chew, A. L. Riley, D. L. Rascoe, B. D. Hunt, M. C. Foote, T. W. Cooley, and L. J. Bajuk, IEEE Trans. Microwave Theory Tech. 39, 1455 (1991).
31. W. G. Lyons, et al. (to be published).

PROSPECTS FOR HIGH-T_c SUPERCONDUCTING OPTOELECTRONICS

Roman Sobolewski*
Department of Electrical Engineering and Laboratory for Laser Energetics
University of Rochester, Rochester, NY 14627

ABSTRACT

Two possible approaches for the development of a complete optoelectronic system with the elements based on high-temperature superconducting (HTS) films are discussed. The first approach consists of manufacturing the devices made of conventional electro-optic materials and containing HTS transmission lines and electrodes. The second, more futuristic approach, is to exploit contrasting properties of the oxygen-poor and oxygen-rich HTS phases to fabricate novel, *monolithic* devices. In this latter case, a laser writing process is implemented to define superconducting and nonsuperconducting regions in the same, epitaxial HTS film. Several practical devices, such as high-speed interconnects, high-frequency traveling-wave optical modulators, picosecond electrical pulse generators, sensitive photodetectors, and a novel HTS charging-effect transistor are proposed. All the devices can operate in the 30–80 K temperature range, where refrigeration is cheap and the parameters of semiconducting (e.g., GaAs) devices are optimal.

1. INTRODUCTION

Present-day optoelectronics is a mature area of engineering, with a proven technology based on $LiNbO_3$, and many practical devices, such as, e.g., directional couplers or modulators. Unfortunately, the $LiNbO_3$ elements are limited to sub-GHz frequencies and cannot be easily integrated with standard electronic materials such as Si. The other optoelectronic materials that attract the most attention today are $A_{III}B_V$-type semiconductors, mostly GaAs, InP, and related compounds. Their main advantages are attractive physical (especially high-speed) properties and advanced technology for their growth and processing. In addition, the highest performance electronic devices, such as high electron mobility transistors or permeable base transistors, are made of these materials. Individual $A_{III}B_V$ devices (e.g., ultrafast switches with the opening time of 1 ps or less, or high-speed waveguide modulators) have already been demonstrated in the laboratory. At the same time, semiconducting laser diodes have found a variety of practical applications, ranging from integrated optoelectronics to compact disk players. Despite these impressive achievements, the semiconductor optoelectronic is not without problems. Device fabrication and integration with other electronic elements is not a trivial task, and the technology itself is extremely expensive. In addition, a given device is operational only in a limited range of wavelengths.

Applications of metallic superconductors in optoelectronics have been very limited and have primarily consisted of far-infrared bolometric detectors and fast opening switches, based on the light-induced superconducting-to-normal transition. The superconducting electrodes and transmission lines have not been implemented in the semiconductor-based optoelectronics, since it would require operation of the devices at liquid helium temperatures, where semiconducting elements usually do not work properly. The high cost of helium refrigeration was also a factor in preventing a wide use of superconductors in electronics and optoelectronics.

The discovery of high-temperature superconductors[1,2] (HTS), which superconduct above the liquid nitrogen temperature has opened completely new opportunities in applying superconductors in optoelectronics. The HTS materials are perovskites and their crystalline structure is very similar to several optoelectronic materials, enabling epitaxial growth of HTS films on electro-optic substrates. The HTS crystals exhibit interesting nonlinear optical properties, which can be "tuned" by controlling their oxygen content. The contrasting electrical and optical properties of the oxygen-poor and oxygen-rich HTS phases can be exploited to make novel devices.

In this presentation we argue that an entire optoelectronic technology based on HTS is feasible. In the next section, we review major properties of HTS crystals, demonstrating that they indeed represent a new class of solid-state materials. We also briefly discuss their basic optical properties. The main concept of an optoelectronic system based on HTS materials is introduced in Sec. 3, where we discuss two possible scenarios for its development. In Sec. 3.1 a conservative approach is presented. We discuss HTS interconnects and propose to implement superconducting electrodes in the state-of-the-art, $LiNbO_3$-based optoelectronic devices, such as, e.g., traveling-wave modulators. The second and more ambitious approach (Secs. 3.2–3.5) is to develop fully *monolithic* devices, where the entire optoelectronic element is made in a HTS film. We review possible designs of several specific devices, such as an all-HTS electro-optical modulator, optical-to-electrical transducer, photoconducting pulse generator, and charging-effect transistor. Finally, conclusions and projections for future investigations are presented in Sec. 4.

2. HIGH-T_c PEROVSKITES AS A NEW CLASS OF SOLID-STATE MATERIALS

All HTS materials are so-called oxygen-deficient perovskites and their basic crystalline structure is similar to that of $CaTiO_3$ (parent mineral for the perovskite family).[2] Similarly to the other perovskites, HTS electrical and optical properties are very sensitive to small variations of their crystallographic structure and atomic composition in the elementary cell. The most important is the oxygen doping, which plays a similar role to the impurity doping in semiconductors, leading to a strong dependence of the material's electrical properties.[3] The $YBa_2Cu_3O_{7-y}$ (YBCO) crystals with $0 < y \leq 0.1$ exhibit metallic-type transport properties at room temperature and show a superconducting transition at $T_c \simeq 91$ K, while those with $0.5 < y < 1$ are nonsuperconducting, with a semiconducting, thermally activated transport. As it was found by Veal *et al.*,[4] at the intermediate oxygen levels ($0.1 < y \leq 0.5$), YBCO films are superconducting with T_c given by a simple, although empirical, relation:

$$T_c \propto \sqrt{N(E_F)} \quad , \tag{1}$$

where $N(E_F)$ is the material free-carrier density of states at the Fermi energy, and is directly related to oxygen doping. Recent experiments revealed that the oxygen content in YBCO can be reversibly varied between y equal to 1 and 0 by annealing the sample in a controlled ambient atmosphere.

The level of oxygen doping, and resulting free-carrier concentration in the conduction band have also a pronounced effect on the material's optical properties.[5] The existing experimental data are very scattered and not precise. Nevertheless, Thomas *et al.*[6] observed that at $T << T_c$ the best, fully oxygenated YBCO films are perfectly reflecting at far-infrared frequencies up to above 150 cm^{-1} (4.5 THz), which the authors interpreted as the signature of the superconductor energy gap. At the same time, partially oxygen depleted samples (with depressed T_c and energy gap) exhibited a certain degree of far-infrared loss. The dielectric properties of HTS are also dependent on oxygen content, and vary from typically metallic ($\varepsilon' \simeq 0$ at room temperature and about -1×10^5 just below T_c—Ref. 7) for films with $0 < y \leq 0.1$ up to that similar to ferroelectrics ($\varepsilon' \simeq 700$ at 300 K and $\simeq 30$ at 4.3 K for a YBCO film—Ref. 8). Finally, recent spectroellipsometric studies of YBCO ceramics, conducted by Aspnes and Kelly,[9] have demonstrated a very large difference in the dielectric function of oxygen-poor and oxygen-rich samples in the visible light range.

Nonlinear optical materials are the foundation for any optoelectronic applications. Below the transition temperature, the HTS materials are not likely to display useful nonlinear optical properties, as the electromagnetic field does not penetrate the volume of the superconductor. However, above T_c and/or in the oxygen-depleted HTS phase, an interesting photovoltaic effect has recently been discovered in YBCO samples.[10–14] The effect implies that the HTS films can have electro-optic properties. The origin of this photovoltaic effect is unknown. Although, the most convincing argument was given by Scott,[12] who suggested that thin YBCO films possess the *mm*2 symmetry, instead of the *mmm* symmetry, commonly accepted for the bulk YBCO.[2] This lower symmetry makes a series of electro-optic effects possible.

Another set of recent experiments demonstrated the existence of an ultrafast optical response in HTS materials excited with <100 fs pulses at 2 eV.[15–18] In particular, the change in optical transmission or reflection of HTS thin films above and below T_c have been studied. Although these experiments suffered from a lack of wavelength tunability, which rendered the interpretation difficult, they suggested to us that it should be possible to switch the HTS film from the superconducting state to the normal state in a few hundred femtoseconds and back again to the superconducting state in less than 5 ps. Thus, several types of optoelectronic devices, such as ultrafast pulse generators and photodetectors, become possible.

3. SUPERCONDUCTING OPTOELECTRONICS

3.1. Optoelectronic devices with HTS electrodes

It has been experimentally demonstrated by us[19] as well as by others[20] that low-T_c superconducting transmission lines exhibit substantially lower signal distortion up to sub-THz frequencies, as compared to the normal-metal lines operational at the same temperature. The superconducting lines can also be accurately (quantitatively) modeled, using the simulation program developed by Whitaker *et al.*[21] The program is based on the frequency-domain full-wave analysis of the propagation process and includes the Mathis-Bardeen theory to describe the conductivity of superconducting electrodes.

Implementation of the HTS materials in high-frequency transmission lines offers a further, two-fold improvement: HTS lines can operate up to THz frequencies, because of their, apparently, large gap, and in the temperature range 30–80 K, where the parameters of semiconducting devices are optimal. Very recent measurements of the picosecond pulse propagation on YBCO coplanar transmission lines,[22–24] and microwave studies of high-quality YBCO single crystals[25] have clearly supported this view. Figure 1 presents the first experimentally measured picosecond electrical transient propagated over a distance of about 4 mm on the 15-µm coplanar YBCO transmission line, grown *in-situ* on the MgO substrate.[22] The result shows that a HTS line can indeed support distortion-free transport of picosecond signals over a substantial distance.

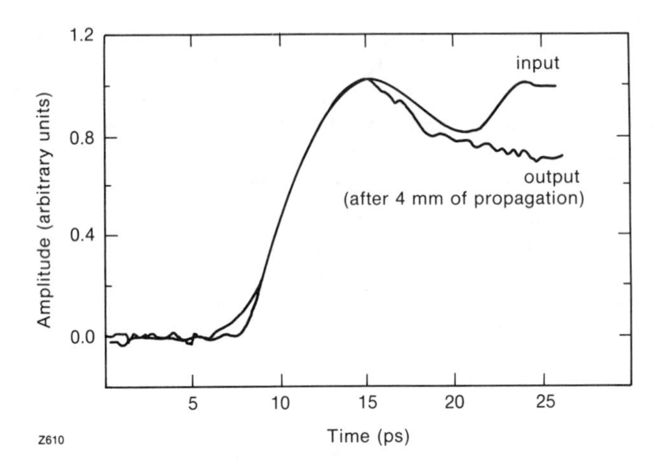

Fig. 1. Picosecond electrical transients measured before and after propagation of 4 mm on the coplanar YBCO transmission line. Temperature was 2 K. (From Ref. 22.)

The above measurements were later confirmed by Nuss *et al.*[24] in a substantially refined experimental setup. The characteristics of 1-THz-bandwidth electrical pulses propagated on epitaxial YBCO-on-$LaAlO_3$ coplanar transmission lines were investigated, and directly compared to the properties of gold lines deposited on the same substrate. Nuss's findings showed that the current upper frequency limit for practical (about 1 cm long) HTS interconnects is in the sub-THz frequency range,[26] which makes them very useful for intrachip and system-level communication.

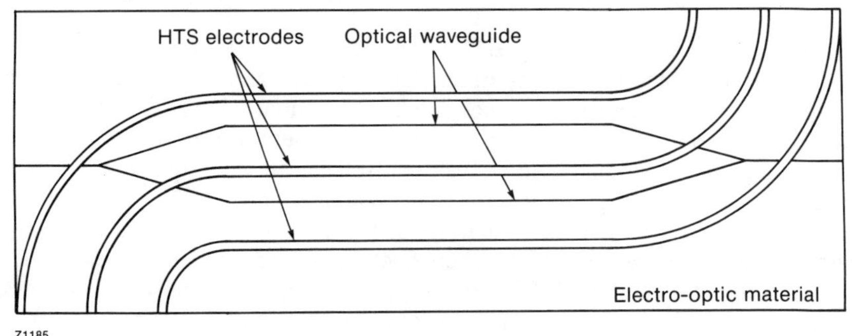

Fig. 2. Top view of the Mach-Zehnder-type HTS modulator.

Few-micrometer-wide HTS coplanar lines with an aspect ratio of about 1000 are ideal not only for medium-range communication, but also in electro-optical (E-O) modulators. This latter concept is illustrated in Fig. 2, which presents the top view of the Mach-Zehnder traveling-wave modulator. The HTS lines are fabricated directly on the E-O material with the diffused optical waveguide. Low geometrical dispersion of a coplanar waveguide, combined with the negligible dispersion and loss of HTS electrodes, guarantees the maximum bandwidth and the low modulation voltage of the device.

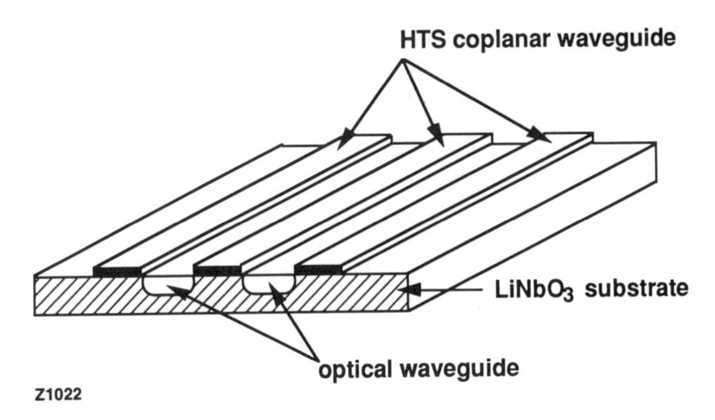

Fig. 3. Cross-sectional view of the YBCO-on-$LiNbO_3$ modulator.

The cross-sectional view of a practical YBCO-on-$LiNbO_3$ modulator is shown in Fig. 3. $LiNbO_3$ is known to have excellent E-O properties, low microwave losses, and is suitable for fabrication of nearly epitaxial YBCO films.[27,28] Thus, the only factors, which will limit the performance of the modulator shown in Fig. 3, are velocity mismatch between the optical and electrical signals, and the millimeter-wave dielectric losses of the $LiNbO_3$ substrate. The effect of the velocity mismatch can be minimized by using the zig-zag optical waveguide configuration to slow down the optical beam. The other option is to introduce air gaps into the transmission line to lower its effective dielectric permittivity and speed-up the electrical signal. We expect that the described above modulator, operational at 77 K, will have the bandwidth of above 100 GHz, what represents at least a ten-fold improvement over the existing devices.

3.2. Prospects for monolithic devices

Despite the enormous number of papers on HTS, published since their discovery in 1986, only a few reported work of relevance to optoelectronics. Nevertheless, the results, briefly reviewed by us in Sec. 2, are encouraging enough to propose several specific device-concepts for HTS optoelectronics. The proposed devices are fully monolithic, and their fabrication process is based on the laser writing technique, described in detail in Refs. 29–31. This method allows to form in the same HTS film, oxygen depleted regions next to the enriched ones, in a manner similar to *n* and *p* diffusion regions in semiconductors. As we mentioned in Sec. 2, the YBCO electrical and optical properties are very sensitive to the material's oxygen content. Heating the material in a controlled atmosphere allows oxygen to diffuse in or out of the sample, depending on the ambient concentration. The works of Dye *et al.*[30] and Shen *et al.*[31] have shown the heating can be done locally with a laser. Thus, an oxygen depleted, semiconducting HTS film can be patterned with imbedded superconducting lines or vice versa. The laser beam does not damage the material structure and, using a well-focused beam from the Ar-ion laser, the features as small as <3 μm can be patterned.[32] We regard the laser writing technique as a potentially very powerful tool to fabricate monolithic HTS structures, or even integrated circuits.

3.3. All-HTS traveling-wave modulator

In the Sec. 3.1, we presented the $LiNbO_3$ modulator with HTS superconducting transmission lines. Here we move one step forward and introduce a monolithic traveling-wave HTS modulator. The configuration (Mach-Zehnder-type) of the modulator is shown in Fig. 4, and is similar to that in Figs. 2 and 3. However, in this case, the oxygen-rich transmission line is laser drawn in a semiconducting HTS film [Fig. 4(a)], simultaneously forming two optical waveguides in the oxygen-poor regions between the coplanar waveguide electrodes. The opposite design, based on the fully oxygenated film [Fig. 4(b)] with laser written (oxygen-depleted) optical waveguides is also feasible. In this latter case the rest of the film naturally forms a superconducting coplanar waveguide. In order to properly operate the modulator, the permittivity of the substrate material has to be much smaller than that of the oxygen-poor film, assuring that both the electrical field and light are confined inside the HTS material. The substrate must also guarantee the high quality, epitaxial growth of the film. Currently, only the $LaAlO_3$, $SrLaAlO_4$, and $CaNdAlO_4$ single crystals fulfill the above requirements.[33]

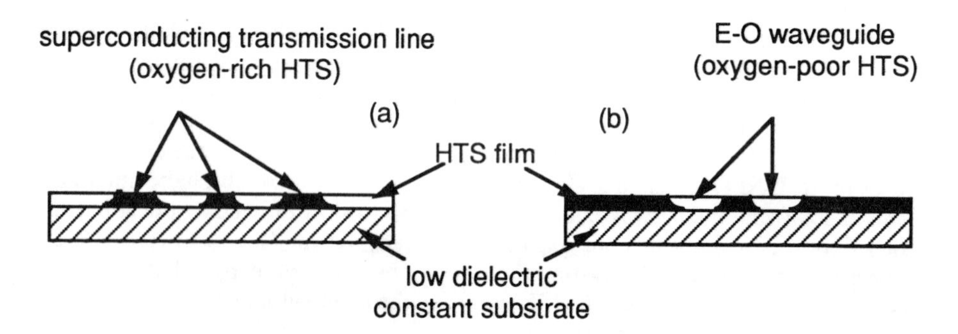

Fig. 4. Cross section of the all-HTS traveling-wave modulator.

Obviously, the ultimate performance of our all-HTS modulator will depend on the E-O properties of the oxygen-deficient HTS phase, which are still poorly characterized. The device optimal wavelength, corresponding to the minimum light dissipation in the E-O waveguide, must also be experimentally determined.

3.4. Ultrafast optical-to-electrical transducer and pulse generator

In a complete optoelectronic system, it is necessary to also have the devices in which an optical signal can modulate electrical signals. Figure 5 presents a device, which principle of operation is based on the observation that in HTS films the superconducting-to-normal transition, triggered by femtosecond optical pulses, can occur in less than 1 ps. The transducer, shown in Fig. 5, is simply a current-biased ($I < I_c$) superconducting constriction ("bridge") at the center of the transmission line, which upon illumination with a femtosecond optical pulse switches from the superconducting state into the normal (resistive) state, triggering an ultrafast electrical transient. The generated signal propagates along the line and can be coupled to any electrical device on the wafer. The main advantage of using the HTS material is that unlike metallic superconductors, which go from being a perfect conductor to a very good conductor, the HTS become poor conductors as they rise above their critical current (or equivalently their T_c). In addition, the oxygen-rich HTS material is black, assuring a high level of absorption of light. The above properties assure much higher (as compared to low-T_c materials) optical responsitivity of the HTS transducers, making them also sensitive photodetectors.

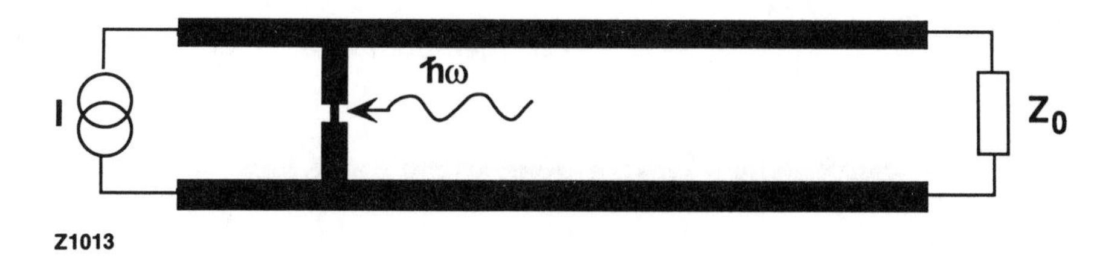

Fig. 5. Transmission line configuration of the light-activated HTS transducer.

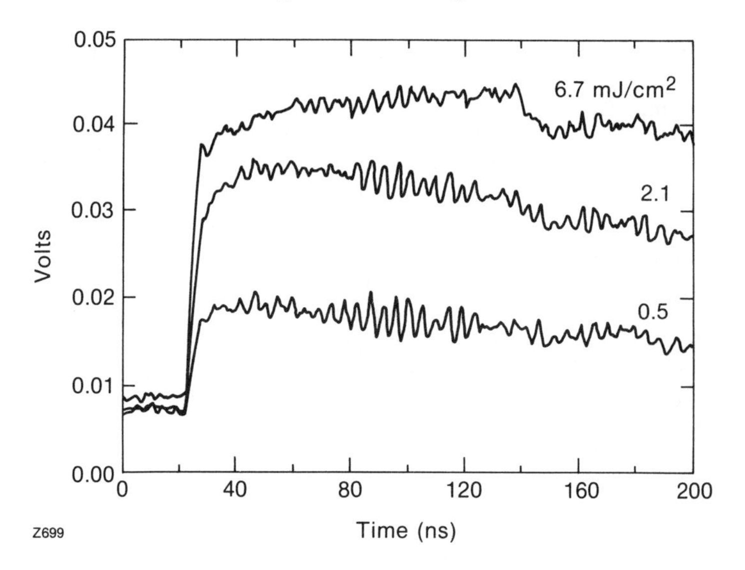

Fig. 6. Transient response of the YBCO thin-film stripe to a 150-ps optical pulse, for several values of laser fluence. Temperature was 17 K. (From Ref. 34.)

Figure 6 presents results of our preliminary studies[34] of the optically induced superconducting-to-normal transition in a granular YBCO film irradiated by infrared

(λ = 1.06 μm) 150-ps-long pulses. We note that although the overall response is predominantly bolometric (thermal), the initial fast rise of the pulse (~2-ns rise time is oscilloscope limited) is an indication of nonequilibrium, hot electron transport. We expect that in the future transducers, the bolometric contribution can be minimized by appropriate illumination conditions, in which the energy in an optical pulse will drive only a portion of a superconducting film into the normal state, creating the intermediate state, earlier observed in low-T_c superconductors.[35]

Another way optical pulses can interact with HTS materials follows from the ability to selectively control the level of oxygen doping in the film. In the example shown in Fig. 7, an oxygen-poor, semiconducting region is inserted ("laser written") into the HTS transmission line. The region is voltage biased and, since it was found that the semiconducting HTS phase has photoconductive properties,[36] upon femtosecond optical pulse irradiation it should inject a high-speed electrical transient into the HTS line, in a similar way to the GaAs photoconductive switch used in our picosecond sampling experiments.[22,23]

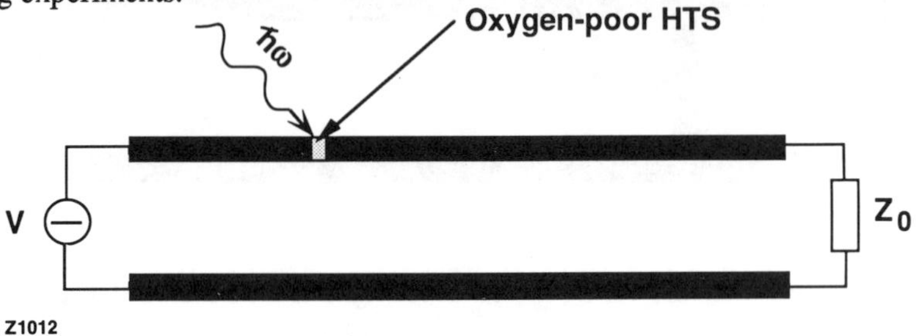

Fig. 7. Transmission line configuration of the HTS photoconducting transducer.

3.5. HTS charging-effect transistor

Charging (field and polarization) effects in superconducting thin films have been the subject of some research in the past.[37] As early as in 1960 Glover and Sherill[38] discovered that application of a constant electric field to the surface of a superconductor resulted in a shift in its T_c. Unfortunately, the observed effect was very small (only 10^{-4} degree for the electric field as high as 3×10^7 V/m).

It has been recently suggested by Kechiantz,[39] that in the case of HTS crystals, the charging effect leads to substantial modification of the material's $N(E_F)$, and results in a measurable difference between T_c at the surface, T_{cs}, and in the bulk, T_{c0}. The surface T_c modulation, ΔT_c, is the largest for partially oxygen-depleted films, which are characterized by the low value of $N(E_F)$ and suppressed T_{c0}. Indeed, for the parameters typical for a YBCO film, ΔT_c, according to Ref. 39, is at least of the order of several degrees.

Figure 8 shows the configuration of a new MOS-type HTS transistor. The structure is completely monolithic, and consists of the oxygen-depleted channel ($T_{c0} \approx$ 60 K), insulated metallic gate (G), and fully oxygenated HTS source (S) and drain (D) regions. The device operating temperature, T, is above T_{c0}, so the channel is initially in the normal (semi-insulating) state, while S and D are superconducting. Application of the negative voltage to the G electrode enhances the concentration of holes at the dielectric-HTS interface, leading to $T_{cs} > T_{c0}$, and formation of a two-dimensional superconducting layer at the channel surface, if the $T_{cs} > T > T_{c0}$ condition is fulfilled.

Successful operation of the proposed transistor is based on the possibility of the rapid, gate-voltage enforced switching of the transistor's channel between the normal (highly resistive) and the superconducting state. From the circuit point of view, the "off" state corresponds to the channel in the normal state, while the "on" state is the superconducting state of the channel. Unbiased transistor is in the "off" state.

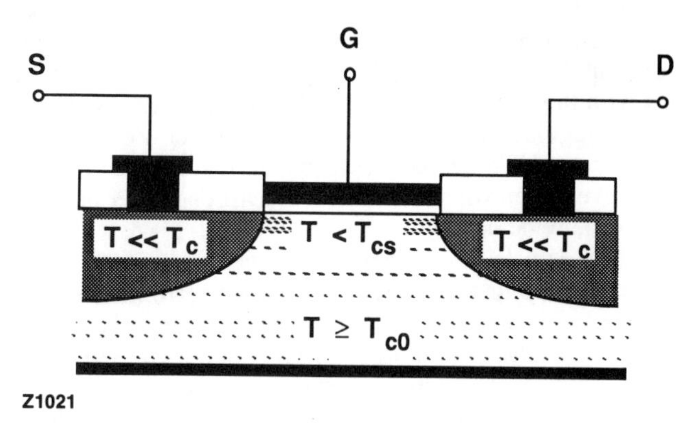

Fig. 8. Structure of a HTS charging-effect transistor. A negative voltage applied to the gate attracts holes to the surface of the HTS channel, establishing a superconducting connection between the source and the drain. Normal metal electrodes are labeled in black.

The proposed device can be built using the laser writing process. Both the approach which assumes writing of the S and D regions in the oxygen depleted HTS film, as well as the opposite one, based on the removal of oxygen from the channel are feasible. However, the former one seems to be more convenient, since it should result in desired S and D diffusion profiles (schematically shown in Fig. 8). The insulating layer under G must be thick enough to prevent any flow of the tunneling current across the gate. It can be fabricated as a lattice-matched dielectric film (e.g., a sputtered $LaAlO_3$ layer). One can also use a ferroelectric film (e.g., $BaTiO_3$ or $SrTiO_3$) to enhance the magnitude of the induced electric charge via the polarization effect. The G electrode does not have to be superconducting, and it can be fabricated together with the S and D metallization.

Despite apparent similarities with low-T_c superconducting FET transistors (see e.g., Ref. 37), the proposed device is very different, since it is not based on the proximity effect. Unlike Josephson FET structures,[37] which rely on the gate enhanced proximity-link between the source and drain, the superconducting channel in our transistor is created by the induced charge at the surface of the HTS material remaining in the normal state. Thus, no restrictions on the gate width (besides standard circuit requirements) apply in our case.

4. CONCLUSIONS

We have demonstrated the potential of HTS materials in optoelectronics. The HTS crystals represent a new class of solid-state materials, exhibiting many very interesting and potentially useful electronic, optical, and electro-optic properties, and, in our opinion, can form the basis for a complete optoelectronic technology. The main concept is based on the observation that we can merge the material's superconducting and optical properties in a single monolithic device (or equivalently in a film-substrate hybrid).

For near-term applications we propose hybrid structures, based on the "classical" optoelectronic devices with HTS transmission lines. Pursuing the hybrid structures, we have been partially motivated by an impressive progress observed in high-speed interconnects, operational at the 50–70 K range, and characterized by negligible distortion of signals with the bandwidth up to 100 GHz. We proposed implementation of the HTS coplanar waveguide into the $LiNbO_3$-based, Mach-Zehnder type traveling-wave modulator. The hybrid HTS devices represent an incremental progress, although we expect them to achieve the maximal bandwidth at least one order of magnitude larger than that of room temperature devices.

In contrast to hybrid devices, the monolithic devices are quite futuristic now, since prospects for their practical applications and their ultimate performance are still difficult to assess. Nevertheless, what we have proposed does not merely describe physical experiments—we regard it as the beginning of a new technology. Several devices, such as an all-HTS traveling-wave modulator, ultrafast transducers, and charging-effect transistor have been proposed. All devices are completely monolithic and can be fabricated using laser patterning and writing techniques.

ACKNOWLEDGEMENTS

The author is indebted to P. M. Fauchet for many valuable discussions. This work was partially supported by the Army Research Office grant DAAL03-91-G-0318. Additional support was provided by the Laser Fusion Feasibility Project at the Laboratory for Laser Energetics, which is sponsored by the New York State Energy Research and Development Authority and the University of Rochester.

REFERENCES

* Also at the Institute of Physics, Polish Academy of Sciences, PL-02668 Warszawa, Poland.
1. J. G. Bednorz and K. A. Müller, Z. Phys. B, Condens. Matter **64**, 189 (1986).
2. For a general review see, e.g., *High Temperature Superconductivity*, edited by J. W. Lynn (Springer-Verlag, Heidelberg, 1990).
3. See, e.g., A. Santoro, in *High Temperature Superconductivity*, edited by J. W. Lynn (Springer-Verlag, Heidelberg, 1990), p. 84.
4. B. W. Veal, J. Z. Liu, A. P. Paulikas, K. Vandervoort, H. Claus, J. C. Campuzano, C. Olson, A.-B. Yang, R. Liu, C. Gu, R. S. List, A. J. Arko, and R. Bartlett, Physica C **158**, 276 (1989).
5. See, e.g., "Special Issue on Quantum Electronic Applications and Optical Studies of High-T_C Superconductors," edited by H. J. Rosen and T. K. Gustafson, IEEE J. Quantum Electron. **QE-25**, 2357 (1989).
6. G. A. Thomas, J. Orenstein, D. H. Rapkine, M. Capizzi, A. J. Mills, R. N. Bhatt, L. F. Schneemeyer, and J. V. Waszczak, Phys. Rev. Lett. **61**, 1313 (1988).

7. A. A. Volkov, B. P. Gorshunov, G. V. Kozlov, I. O. Sirotynsky, E. V. Pechen, and R. Sobolewski, "Submillimeter BWO-Spectroscopy of High-T_c Superconducting Films," to appear in *Proc. of German-Soviet Bilateral Seminar on HTS* (Springer-Verlag, Heidelberg, 1991).
8. L. R. Testardi, W. G. Moulton, H. Mathias, H. K. Ng, and C. M. Rey, Phys. Rev. B **37**, 2324 (1988).
9. D. E. Aspnes and M. K. Kelly, IEEE J. Quantum Electron. **QE-25**, 2378 (1989).
10. C. L. Chang, A. Kleinhammes, W. G. Moulton, and L. R. Testardi, Phys. Rev. B **41**, 1564 (1990).
11. K. L. Tate, R. D. Johnson, C. L. Chang, E. F. Hilinski, and S. C. Foster, J. Appl. Phys. **67**, 4375 (1990).
12. J. F. Scott, Appl. Phys. Lett. **56**, 1914 (1990).
13. K. L. Tate, E. F. Hilinski, and S. C. Foster, Appl. Phys. Lett. **57**, 2407 (1990).
14. H. S. Kwok, J. P. Zheng, and S. Y. Dong, IEEE Trans. Magn. **MAG-27**, 1288 (1991).
15. D. W. Face, S. D. Brorson, A. Kazeroonian, J. S. Moodera, T. K. Cheng, G. L. Doll, M. S. Dresselhaus, G. Dresselhaus, E. P. Ippen, T. Venkatesan, X. D. Wu, and A. Inam, IEEE Trans. Magn. **MAG-27**, 1556 (1991).
16. J. M. Chwalek, C. Uher, J. F. Whitaker, G. A. Mourou, J. Agostinelli, and M. Lelental, Appl. Phys. Lett. **57**, 1696 (1990).
17. S. G. Han, Z. V. Vardeny, K. S. Wong, O. G. Symko, and G. Koren, Phys. Rev. Lett. **65**, 2708 (1990).
18. G. L. Eesley, J. Heremans, M. S. Meyer, G. L. Doll, and S. H. Liou, Phys. Rev. Lett. **65**, 3445 (1990).
19. T. Y. Hsiang, J. F. Whitaker, R. Sobolewski, D. R. Dykaar, and G. A. Mourou, Appl. Phys. Lett. **51**, 1551 (1987).
20. W. J. Gallagher, C.-C. Chi, I. N. Duling III, D. Grischkowsky, N. J. Halas, M. B. Ketchen, and A. W. Kleinsasser, Appl. Phys. Lett. **50**, 350 (1987).
21. J. F. Whitaker, R. Sobolewski, D. R. Dykaar, T. Y. Hsiang, and G. A. Mourou, IEEE Trans. Microwave Theory Tech. **MTT-36**, 277 (1987).
22. D. R. Dykaar, R. Sobolewski, J. M. Chwalek, J. F. Whitaker, T. Y. Hsiang, G. A. Mourou, D. K. Lathrop, S. E. Russek, and R. A. Buhrman, Appl. Phys. Lett. **52**, 1444 (1988).
23. J. M. Chwalek, D. R. Dykaar, J. F. Whitaker, R. Sobolewski, S. Gupta, T. Y. Hsiang, and G. A. Mourou, IEEE Trans. Magn. **MAG-25**, 814 (1989).
24. M. C. Nuss, P. M. Mankiewicz, R. E. Howard, B. L. Straughn, T. E. Harwey, C. D. Brandle, G. W. Berkstresser, K. W. Goossen, and P. R. Smith, Appl. Phys. Lett. **54**, 2265 (1989).
25. D.-H. Wu, W. L. Kennedy, C. Zahopoulos, and S. Sridhar, Appl. Phys. Lett. **55**, 696 (1989).
26. M. C. Nuss, K. W. Goossen, P. M. Mankiewicz, R. E. Howard, B. L. Straughn, G. W. Berkstresser, and C. D. Brandle, IEEE Electron Device Lett. **EDL-11**, 200 (1990).
27. A. Hohler, D. Guggi, H. Neeb, and C. Heiden, Appl. Phys. Lett. **54**, 1066 (1989).
28. S. G. Lee, G. Koren, A. Gupta, A. Segmuller, and C.-C. Chi, Appl. Phys. Lett. **55**, 1261 (1989).

29. R. Krchnavek, S. W. Chan, C. T. Rogers, F. DeRosa, M. Kelly, P. F. Miceli, and S. J. Allen, J. Appl. Phys. **65**, 1802 (1989).
30. R. C. Dye, R. E. Muenchausen, N. S. Nogar, A. Mukherjee, and S. R. J. Brueck, Appl. Phys. Lett. **57**, 1149 (1990).
31. Y. Q. Shen, T. Freltoft, and P. Vase, Appl. Phys. Lett. **59**, 1365 (1991).
32. R. Sobolewski and W. Xiong (unpublished).
33. R. Sobolewski, P. Gierlowski, W. Kula, S. Zarembinski, S. J. Lewandowski, M. Berkowski, A. Pajaczkowska, B. P. Gorshunov, D. B. Lyudmirsky, and O. I. Sirotinsky, IEEE Trans. Magn. **MAG-27**, 876 (1991).
34. W. R. Donaldson, A. M. Kadin, P. H. Ballentine, and R. Sobolewski, Appl. Phys. Lett. **54**, 2470 (1989).
35. R. Sobolewski, D. P. Butler, T. Y. Hsiang, C. V. Stancampiano, and G. A. Mourou, Phys. Rev. B **33**, 4604 (1986).
36. T. Venkatesan, X. Wu, A. Inam, C. C. Chang, M. S. Hegde, and B. Dutta, IEEE J. Quantum Electron. **QE-25**, 2388 (1989), and references therein.
37. For a general review see, e.g., A. W. Kleinsasser and W. J. Gallagher, in *Superconducting Devices*, edited by S. T. Ruggiero and D. A. Rudman (Academic Press, Boston, 1990), p. 325, and references therein.
38. R. E. Glover III and M. D. Sherrill, Phys. Rev. Lett. **5**, 248 (1960).
39. A. M. Kechiantz, in *Progress in High-Temperature Superconductivity*, edited by W. Gorzkowski, M. Gutowski, A. Reich, and H. Szymczak (World Scientific, Singapore, 1990), Vol. 24, p. 556.

THIN FILMS OF $YBa_2Cu_3O_{7-x}$ (YBCO) FOR MICROWAVE APPLICATIONS

L. (Rao) Madhavrao, E. K. Track, R. E. Drake, R. Patt, and M. Radparvar
HYPRES Inc., 175 Clearbrook Road, Elmsford, NY 10523

Abstract

The ability to fabricate YBCO films on large areas and on both sides of the substrate is of interest for microwave component applications such as resonators, filters and delay lines. We have fabricated thin films of YBCO on both sides of two inch diameter and 10 mils thick $LaAlO_3$ substrates. The films are deposited by rf magnetron sputtering of BaF_2, CuO and Y_2O_3 targets followed by post-annealing in oxygen. The films show superconducting transition temperatures of 90 K and a critical current density of one million amperes per square cm at 70 K. The AC Susceptibility transition temperature is 88 K with a width <0.5 K (at 100 kHz., 300 mGauss). The microstructural, transport and rf properties of these films are presented. Microstrip delay lines have been fabricated using these films. The fabrication and characteristics of the delay lines are presented and discussed.

Introduction

Since the discovery of high-temperature superconductivity significant effort has been dedicated to the fabrication of passive microwave devices as these are perceived to lead to the first systems applications of high-temperature superconductors (HTS)[1]. In order to realize the improvements in size, design, practicality and performance offered by HTS it is necessary to first reliably demonstrate HTS films with superior rf properties over large areas. The performance of the devices is further enhanced by a superconducting ground plane on the back side of the substrate. Therefore, deposition techniques that enable *both sides* of a *large* substrate to be coated with HTS films are pertinent for these applications. While continued efforts[2] are being made to extend in-situ techniques to accommodate these requirements (i.e, large area and double-sided), we have chosen an approach that is easily manufacturable and extendable to areas as large as 16 square inches - sequential multilayer rf magnetron sputtering and post anneal. We have so far fabricated YBCO films on both sides of two inch diameter, 10 mil $LaAlO_3$ substrates using the above technique. HTS films on 3 inch diameter $LaAlO_3$ will be attempted next as these substrates are now available. The crystal structure, transport and rf properties of the films are discussed. Fabrication and characterization of microstrip delay lines are also presented.

Fabrication

The films are deposited in a cryopumped vacuum system containing three targets BaF_2, Y_2O_3 and CuO each with a diameter of 6.5 inches. Film deposition is believed to be uniform over at least 4 inches, and has been proven to be uniform in deposition over 3-inch substrates. It is important to note that, since no heating is used at this stage, commercially available systems capable of uniform deposition over 150 square inch areas may be substituted at this stage. The composition of the film is determined by Energy Dispersive X-ray Spectroscopy using a bulk YBCO as standard. After deposition, the films are annealed in a 4-inch diameter quartz tube in the presence of steam. The optimum anneal consists of a ramp up to 850 °C in about 1/2 hour during which wet oxygen is introduced. The cool down is done under dry oxygen. It is our experience that the adhesion of the film is sensitive to the substrate quality, annealing atmosphere and the size of the substrate. Films on two-inch diameter substrates have been found many times to have lost adhesion to the substrate in some areas. The adhesion problem becomes more intense for double sided films on two-inch diameter substrates, since occasionally only one side comes out with loss of adhesion in some areas while the other side does not show any loss of adhesion. These problems with annealing of large area films point to the fact that the stresses developed in the film may contribute to loss of adhesion in addition to factors such as substrate quality, annealing atmosphere, etc.

Properties

All the properties discussed below are for YBCO films on $LaAlO_3$ substrates.

Crystal Structure

X-ray diffraction in the normal θ-2θ scan shows the films to be oriented with their c-axis perpendicular to the substrate (Fig. 1). Minor impurity peaks due to $Y_2Cu_2O_5$ are also present. The c-axis lattice parameter varies from 11.66 Å to 11.67 Å. The rocking curve FWHM of the (005) peak falls in the range of 0.7 -0.8 ^{0} compared to that of the substrate peak that shows a FWHM of 0.6°. The scanning electron micrograph of the surface of the films shows large c-axis plates and does not show any basket-weave pattern seen in a-axis oriented films. We have fabricated films more than 4000 Å thick, which do not show any basket-weave structure under the SEM. In contrast to our sputtered films, co-evaporated and post-annealed films[3] show a predominance of a-axis grains on the surface, when the film thickness becomes greater than 2000 Å, thus leading to degradation of the superconducting properties for thicker films.

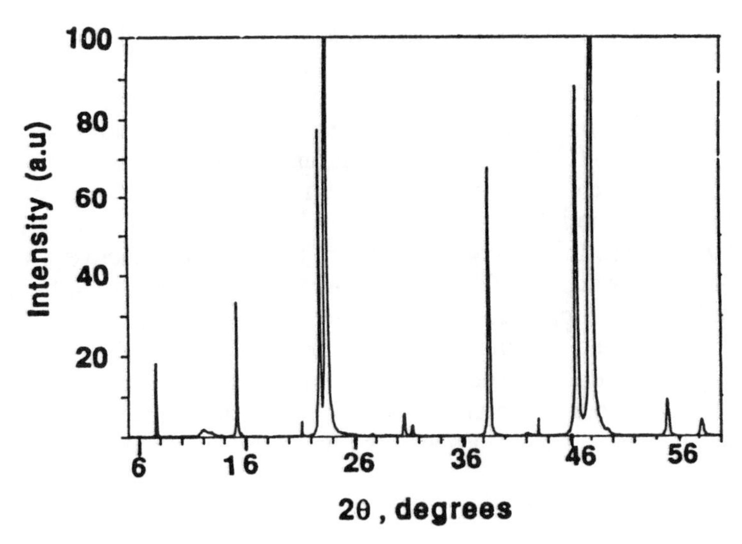

Fig. 1. X-ray diffraction pattern for a 3000 Å YBCO film on $LaAlO_3$.

Electrical transport (dc and low frequency)

A four probe configuration on a bridge pattern is used for determining the resistive transition temperature. The bridge is typically 150 μm wide and 500 μm in length. Silver paste is applied at the contact pads. Fig. 2 shows the variation of resistance with temperature.

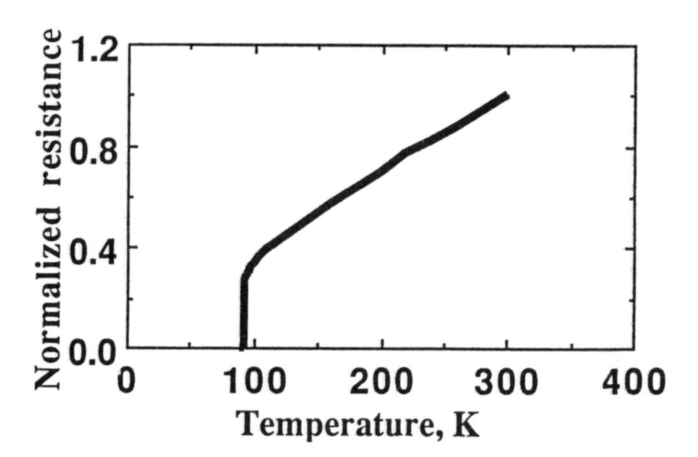

Fig. 2. Variation of the DC resistance of a YBCO film with temperature.

The zero resistance transition temperature is 90 K and the resistance extrapolates close to zero although not to zero as observed for in-situ films. This may be due to the presence of defects in the post-annealed films. The critical current density (J_c) of the film measured by direct transport method at 77 K is typically 0.3-0.5 MA/cm^2 although one particular film gave a current density of 1.6 MA/cm^2. The variation of current density with temperature is shown in Fig 3. for both the high current density film (●) and for a typical film (o). The AC susceptibility (Fig 4.) transition (also called inductive T_c) is 88 K with a transition width of <0.5 K. The susceptibility is measured by a set of mutual inductance copper coils of diameter about 1.5 mm. The drive coil generates an H-field of amplitude 300 mGauss and frequency 100 kHz perpendicular to the film. The pick-up coil on the other side of the substrate is connected to a lock-in amplifier with an input transformer. The inductive transition onset usually coincides with the full resistive transition.

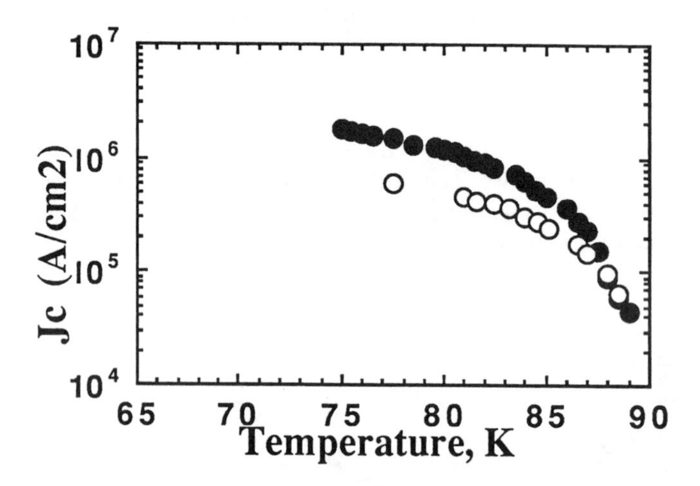

Fig 3. Variation of critical current density with temperature of our typical YBCO film (0), and a film that showed high current density (●)

Surface resistance

The surface resistance of the films (R_s) is measured by an endwall cavity method. The YBCO film replaces the bottom of a magnetically coupled cylindrical copper cavity. The cavity is immersed in liquid nitrogen and kept just below the surface of the liquid. The cavity resonance at 35.6 GHz is measured using a network analyzer. The quality factor Q for a copper film as

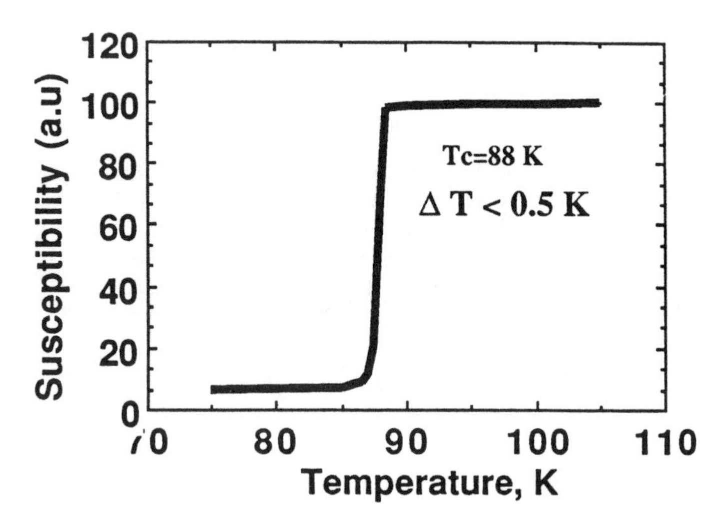

Fig 4. AC Susceptibility of a YBCO film as a function of temperature.

endwall at 77 K is typically 7900. For a YBCO film on $LaAlO_3$ as endwall we have measured a Q equal to that of copper for many films at 77 K and 35.5 GHz. The same Q values are obtained for double sided YBCO films too. Below 77 K Q increases further. We conclude that the surface resistance of our YBCO films is less than or equal to the surface resistance of copper at 77 K and 35.5 GHz. Using an f^2 dependence for surface resistance, we conclude that the surface resistance is $\leq 2\ m\Omega$ at 10 GHz and 77 K. The surface resistance of the YBCO films is very sensitive to composition of the film. Deviations in Ba and Y composition away from stoichiometry results in high R_s, whereas the R_s does not change significantly with excess copper. Likewise, the R_s is not affected by the presence of minor amounts of **insulator** impurities such as $BaCuO_2$ and $Y_2Cu_2O_5$. However, whenever a-axis grains are seen in the SEM or the (013) (110) peaks are detected in the x-ray patterns, the R_s is high.

For a better estimation of surface resistance ring resonators were fabricated from the YBCO films. In preliminary attempts, the resonators are fabricated from 1/2" x 1/2" $LaAlO_3$, 20 mils thick. The YBCO ring resonators have copper ground plane. The ring has a diameter of 43 mils and a linewidth of 6 mils. It is capacitively coupled at the input and output through 8 mils-gaps between the 50 Ω microstrip launches and the ring. Fig. 5. shows the resonance spectra for YBCO and Cu ring resonators. The insertion loss and the Q values at different resonant frequencies for YBCO and Cu are compared in Table I. YBCO resonators show a Q better than copper for a wide range of frequencies up to 20 GHz. However, extraction the surface resistance from the measured

Q values requires simulations, which have not been carried out yet.

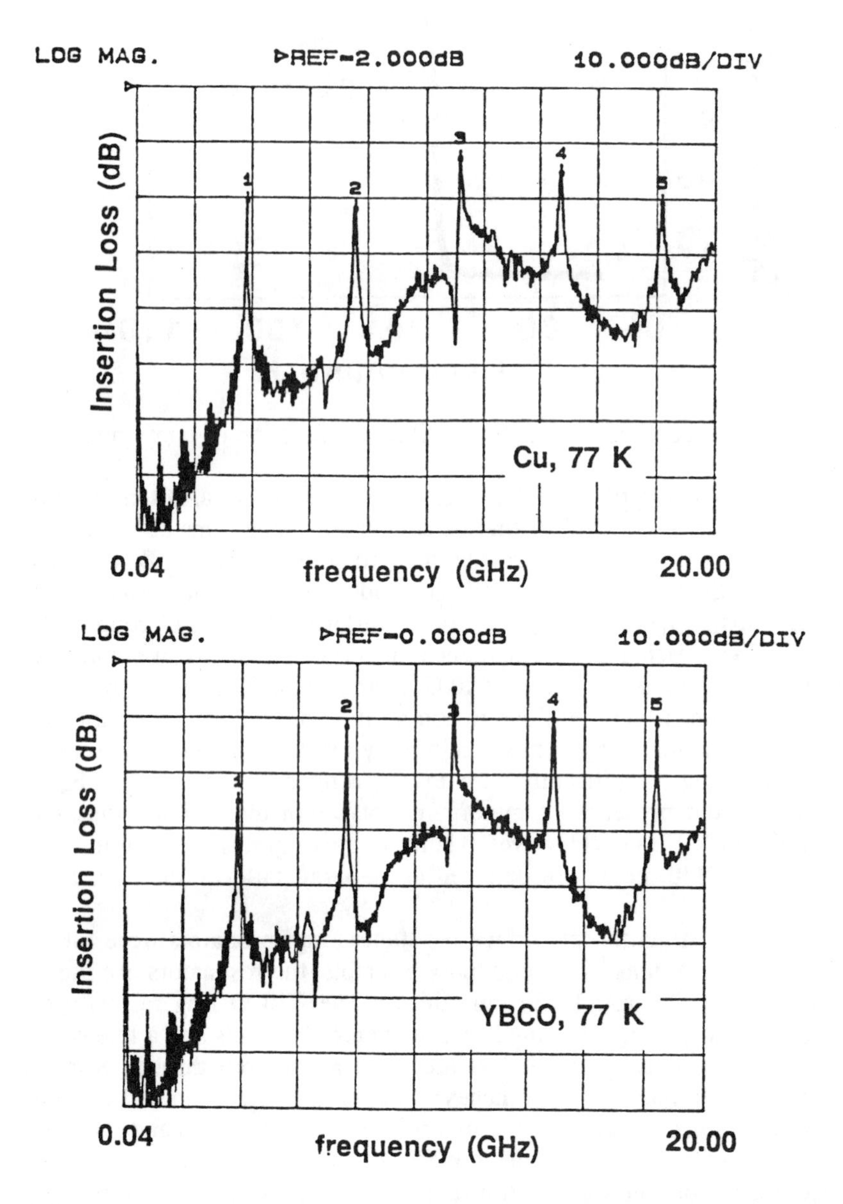

Fig. 5. Measured transmission loss .vs. frequency for Copper and YBCO ring resonators at 77 K.

Table I. Comparison between YBCO and Cu ring resonators at 77 K

f(GHz)	Q		Insertion loss (dB)	
	Cu	YBCO	Cu	YBCO
3.9	168	390	-18	-24
37.7	296	596	-19	-11
11.4	217	398	-10	-4
14.9	162	436	-13	-10
18.2	256	681	-18	-11

The table below summarizes the film properties that have been measured on our YBCO films on $LaAlO_3$.

X-ray diffraction: oriented films with c-axis perpendicular to substrate c-axis lattice parameter = 11.66 - 11.67 Å. Minor impurities such as $Y_2Cu_2O_5$. Rocking curve FWHM = 0.7 - 0.8^0.

Electrical Properties: Resistive T_c = 90 K, Inductive T_c, 10% drop = 88.5 K, 90% drop = 88 K, J_c(77 K) = 0.3-0.5 MA/cm^2 (typical).

Surface resistance: $R_s \leq 2$ mΩ at 77 K and 10 GHz.

Microwave Devices

Delay lines

HTS delay lines are particularly taxing on the film properties since they require large patterned film areas. In addition to the materials issues, design considerations become important since standard microwave design programs are not directly applicable due in part to the high dielectric constant of $LaAlO_3$ (ϵ=25). We are involved in the design, fabrication, and measured characteristics of two different delay lines.

1 - A 7 nanoseconds delay line[4,5] in an inverted microstrip configuration where the $LaAlO_3$ substrate carrying the YBCO line is inverted over a bulk polished copper ground plane and separated from it by a Kapton dielectric. The total line length is 60 cms, wound in a regular meander shape on the 1 square inch

$LaAlO_3$ substrate. The length of each line segment is approximately 3/4 inch. The YBCO is patterned by photolithography and ion milling. The thickness of the Kapton dielectric (in this case 4 mils) is adjusted to obtain a 50 Ω characteristic impedance for the line. Silver is evaporated over the contact areas to provide a low resistance contact to the coaxial connectors. No special efforts have been expended yet to optimize the coaxial-to-microstrip transition as this would require an extensive optimization process. As a result, some ripple in the measured characteristics is obtained due to the reflection of the signal at the input and output. For comparison to normal metals and to a good elemental superconductor, the same design is implemented using thick (6 microns) gold films and using niobium films. Transmission loss data is shown in Fig. 6. The 7 nanoseconds electrical length is obtained from phase vs.frequency measurements carried out using a Wiltron 360 network analyzer.

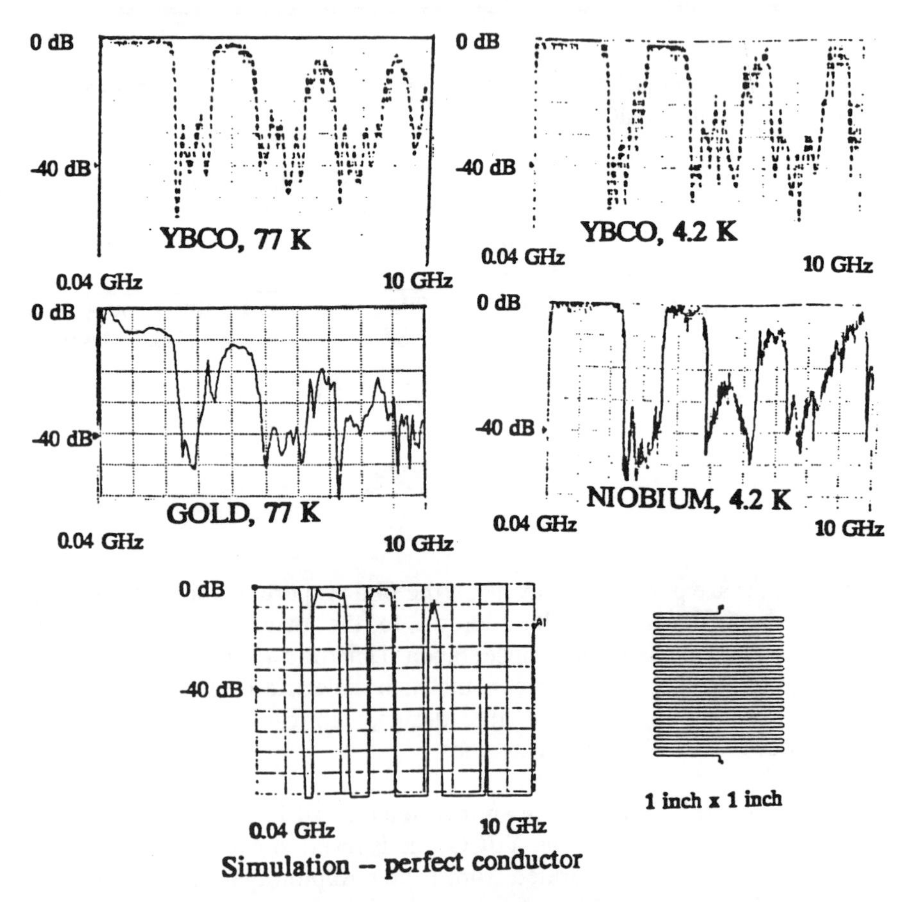

Fig. 6. 7 nanoseconds delay lines, narrow band (0-2 GHz).

As seen in the data, the practical bandwidth for the line is approximately 2 GHz. Above this frequency, there is a succession of stop-bands and pass-bands generated by the regular structure of the meander geometry. This is verified by the simulations carried out using the HP85150 MDS program which reproduce qualitatively the features observed. In fact, in the 4-4.5 GHz passband, the phase-frequency relationship is linear and also yields 7 nanoseconds of delay. The transmission loss for the YBCO line is comparable to that of the niobium line and both are about 10-15 dB lower than for the gold line. A superconducting ground plane may be substituted for the copper ground plane and should result in even lower transmission loss.

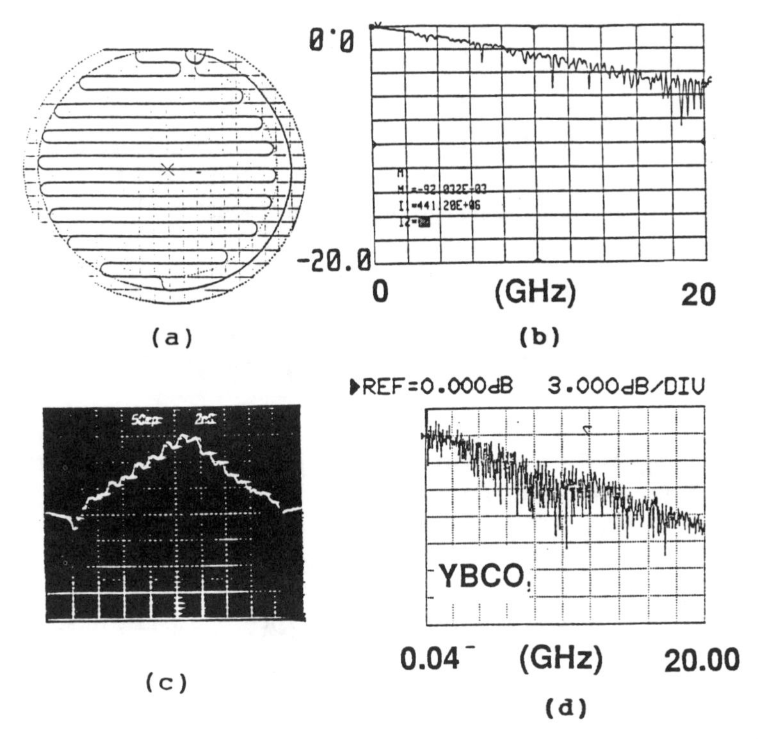

Fig. 7. 8 nanoseconds delay line, wide band(0-20 GHz). a) mask layout, b)simulation, c) TDR delay, and 4) measured transmission loss for YBCO at 11.5 K

2 - The second design was approached with the intent of increasing the bandwidth of the line. A 2-inch $LaAlO_3$ substrate of thickness 10-mils is used. A conventional microstrip design is applied with the substrate constituting the dielectric. The linewidth of 80 μm for the meander line yields a 50 Ω characteristic impedance. The line spacing is ten times the substrate thickness,

i.e. 100 mils. Total line length is 64 cms. Here again no optimization of the coax-to-microstrip transition is carried out. Simulations of the line show no stop bands to 20 GHz, but small amplitude resonances modulate the response. These may be tolerated if they do not cause significant dispersion. The design was implemented using niobium nitride films and a preliminary run with YBCO films was carried out. The data is shown in Fig. 7. For NbN at 11.5 K, an insertion loss of about 6 dB is measured at 20 GHz, whereas the YBCO line yields a loss of about 11 dB at 20 GHz. This particular double sided film on both sides showed a higher sheet resistance than our normal films. Measurements on our typical films are still in progress. Both NbN and YBCO delay lines show a linear phase-frequency relationship yielding an electrical length of 8 nanoseconds.

Acknowledgements

This work was funded in part by U. S. Air Force Contract Number F33615-90-C-1456. We are grateful to Adam Rachlin of the ETDL Army Lab for providing the design, masks, and the package of the ring resonator.

References

1. See for example Special Issue on Microwave Applications of Superconductivity in *IEEE Trans. Microwave Theory And Techniques* **39**, 1445 (1991).

2. S. R. Foltyn, R. E. Muenchausen, R. C. Dye et al., *Appl. Phys. Lett* **59**, 1374 (1991); J. Talvacchio, M. G. Forrester, J. R. Gavaler, and T. T. Braggins, *IEEE Trans. Mag.* 27, 978 (1991).

3. D. B. Rensch, J. Y. Josefowicz, P. Macdonald et al., *IEEE Trans. Magn.* 27, 2553 (1991); M. P. Siegal, J. M. Phillips, Y.F. Hsieh, and J. H. Marshall, Physica C (In Press).

4. E. K. Track, G. K. G. Hohenwarter, L. (Rao) Madhavrao, R. Patt, R. E. Drake, and M. Radparvar, *IEEE Trans. Magn.* 27, 2936 (1991).

5. L. (Rao) Madhavrao, E. K. Track, R. E. Drake, R. Patt, G. K. G. Hohenwarter, and M. Radparvar, *IEEE Trans. Magn.* 27, 1402 (1991).

THICK AND THIN FILM Y-BA-CU-O INFRARED DETECTORS

M. Fardmanesh, M. Ihsan, A. Rothwarf, K. Scoles, and K. Pourrezaei
Ben Franklin Superconductivity Center, and
Drexel University
Electrical and Computer Engineering Department
Philadelphia, PA 19104

ABSTRACT

The infrared response of Y-Ba-Cu-O superconducting thick and thin films has been studied. Infrared detectors were fabricated using screen printing and planar magnetron sputtering techniques, respectively. The films were deposited on polycrystalline or (001) crystalline MgO substrates.

IR-response vs. temperature, dc bias current, and chopping frequency were measured using an LED with 0.84 μm wavelength output. The IR-response of the films shows a peak at the maximum points of the slope of the resistance vs. temperature curve of the samples. A bolometric response, sharply peaked near the transition temperature was obtained for the films with narrow (sharp) resistive transitions. The sensitivity of the IR-response shows a dependence on the bias current.

The frequency response of the thick and thin films was measured, and the response has shown a dependence on the thermal time constants of both the superconducting film and the substrate. The plot of IR-response vs. chopping frequency for the thick films showed two breakpoints at about 400 Hz and 800 Hz, while the plot of the thin film, from 10 Hz to 2 KHz, showed only one breakpoint at 400 Hz.

INTRODUCTION

A very promising application of high-Tc superconductive films is in broad band optical detectors. IR-response of these materials can be classified into two main categories: equilibrium or bolometric response, and non-bolometric response. In the first case, the temperature dependence of equilibrium properties of the material is used. The bolometric response of these materials has many advantages due to the essentially flat response over a broad range of wavelengths, and simple absolute calibration.

In the bolometric response, the flux of photons can directly or indirectly heat the superconductor and change the temperature of the material. This change in the temperature, which can be local, will be reflected in the I-V curve characteristic of the superconductor. For this type of detection, the superconducting detector is biased where the temperature derivative of the resistance, dR/dT, is large. A superconductive bolometric detector can exhibit good performance in terms of high sensitivity and low noise at low temperatures, but it's sensitivity is inherently obtained by sacrificing the speed of the response. In this type of response (bolometric) there is a trade-off between response magnitude and response speed that is determined by the thermal mass and thermal conductance of the bolometer. For bolometric response, the induced voltage change, ΔV, due to a small temperature variation of the detector, ΔT, will be equal to I (dR/dT) ΔT, where I is the dc bias current across the detector as shown in figure 1.

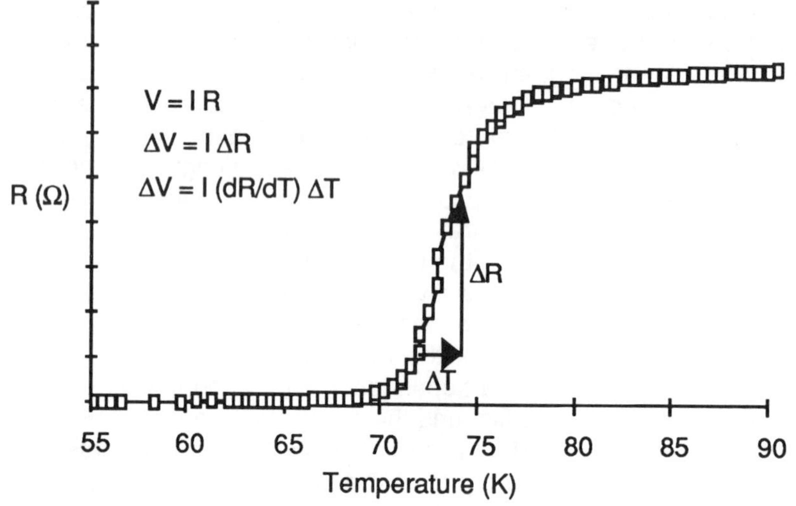

Figure 1. Bolometric mode of IR detector operation. Absorbed energy produces a temperature shift which can be observed as a change in the voltage drop across the detector.

Important parameters of a bolometer are the temperature derivative of the resistance (dR/dT), heat capacity and the effective thermal conductivity of the materials used.

Calculations for responsivity (r_V) of high-temperature bolometers, done by different groups,[1,2] lead to a responsivity given by

$$r_V = (dR/dT)I/K(1 + \omega^2\tau^2)^{1/2}, \quad (1)$$

where I is the current (normally dc) passed through the detector, ω is the chopping frequency of the radiation, and τ is the thermal time response equal to C/K of the bolometer, where C is the specific heat and K is the thermal conductivity.

In this work, the Y-Ba-Cu-O compound with 123 and 124 stoichiometries have been used for fabrication of thick film superconductive infrared detectors using hybrid technology. Thin film detectors were made by off-axis planar magnetron sputtering. To improve the properties of the material, fabricated films have been post annealed in an argon ambient at high temperatures and in an oxygen ambient at lower temperatures (T<600°C) resulting granular superconductive films with a 123 phase.[3] The thick film samples are screen printed on polycrystalline MgO substrates. The film geometry was a serpentine pattern, with a line of 250 μm width and 82.6 mm length, 0.89 mm pitch and 30-35 μm in thickness. Electrical contacts were made using silver epoxy to attach fine copper wires to sputtered gold pads of 850 Å thickness. The response of the samples under different experimental conditions is presented in this work, and a correlation between the IR-response of the fabricated thick films and thin films is given.

EXPERIMENTAL SETUP

A setup has been developed for measurements of the infrared-response of the fabricated Y-Ba-Cu-O films. Two types of IR-sources can be considered for a setup such as this; a diode-source and high power light-source. A dc or ac infrared light source can be mounted on an optical bench with the other components for filtering, chopping, and collimation. An IR-transparent window is required to admit the radiation to the vacuum chamber. A block diagram of a setup compatible with both types of sources is shown in figure 2. Using a Macintosh Plus computer as a controller, measurements of resistance vs. temperature and IR-response vs. temperature can be done simultaneously.

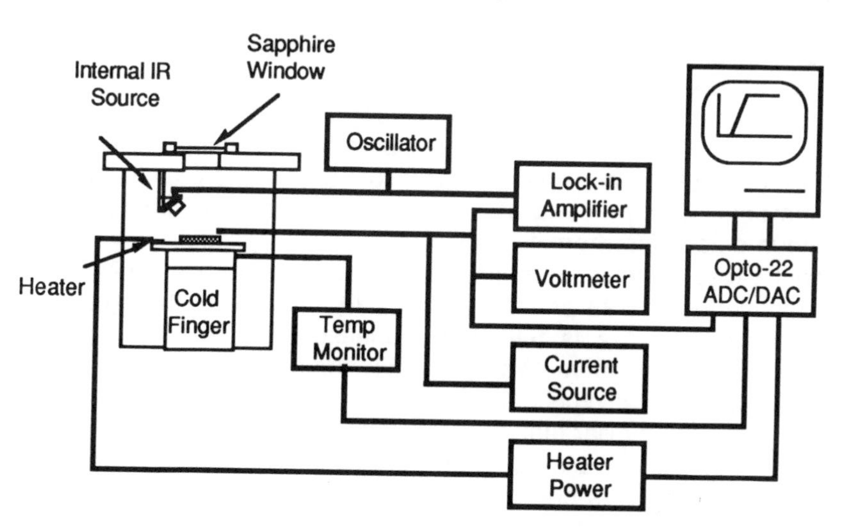

Figure 2. Schematic of IR detection apparatus

The cryopump used (Cryo-Torr 100, CTI-Cryogenics) has two cooling stages, at 80 K and 15 K. By removing the the 80 K and 15 K condensing arrays from the pump, the samples can be mounted directly on top of the 15 K second-stage.

A GaAlAs light emitting diode (HFE4020, Honeywell) with 0.84 µm peak wave length, has been used as the internal IR source of the setup. The diode can have a maximum power output of 6 mW. The diode is driven by an adjustable oscillator (Model 171, Wavetek) that can supply an output current up to 200 mA. The oscillator's output signal was connected to the reference input of the lock-in amplifier (Model 5204 Lock-In Analyzer, EG&G Princeton Applied Research). The distance between the diode and the sample is 0.3 to 1.5 cm.

Calibration of our temperature sensor, done after the data shown in this paper was taken, showed that our data is approximately 9 K too low at liquid nitrogen temperatures. The presence of the shift was confirmed by dipping samples directly into liquid nitrogen. Work is underway to correct this problem.

RESULTS AND DISCUSSION

The bolometric response of films (thick and thin) vs. temperature, bias current and chopping frequency were measured. To measure the IR-response, samples were connected in a 4-probe pattern. Measurements were done using pulsed light from the GaAlAs LED.

Thick film IR detectors were fabricated by screen printing pastes manufactured from 124-based powders on polycrystalline MgO substrates. At sufficiently high annealing temperatures, about 850 °C in air, the 124 phase of YBCO converts to the 123 phase plus a CuO species. The conversion of 124 material to 123 + CuO at high temperatures can give 123 material with higher critical currents,[4] permitting a higher bias current for the bolometer. The CuO species, considered to be uniformly distributed in the bulk, can behave as pinning centers for fluxons and prevent flux flow in the material. Our process of thin film and thick film fabrication is discussed in detail elsewhere.[3]

The IR-response vs. temperature measurement on a 124-based sample (converted to 123 phase) is shown in figure 3. In the figure, the results of the IR-response vs. temperature measurement are superimposed on its R vs. T transition curve. We note that in the response there is a small peak at 60 K to 65 K, which is below the region of its transition-temperature curve ($T<T_{c(zero)}$). A similar feature was not observed in the IR-response of the sample at 100 µA dc-bias current. This response, which shows a dependence on the dc-bias current through the sample, is attributed to a Kosterlitz-Thouless phase transition.

Multiple peaks in IR-response have also been observed and reported by other groups.[5,6] Black body photoresponse in granular YBCO films have shown that at a sufficiently high bias current there are two major peaks of response.[7] The response voltage vs. temperature curves obtained experimentally have shown a peak near the $T_{c(onset)}$ and another peak below $T_{c(zero)}$ of the material.[8-10]

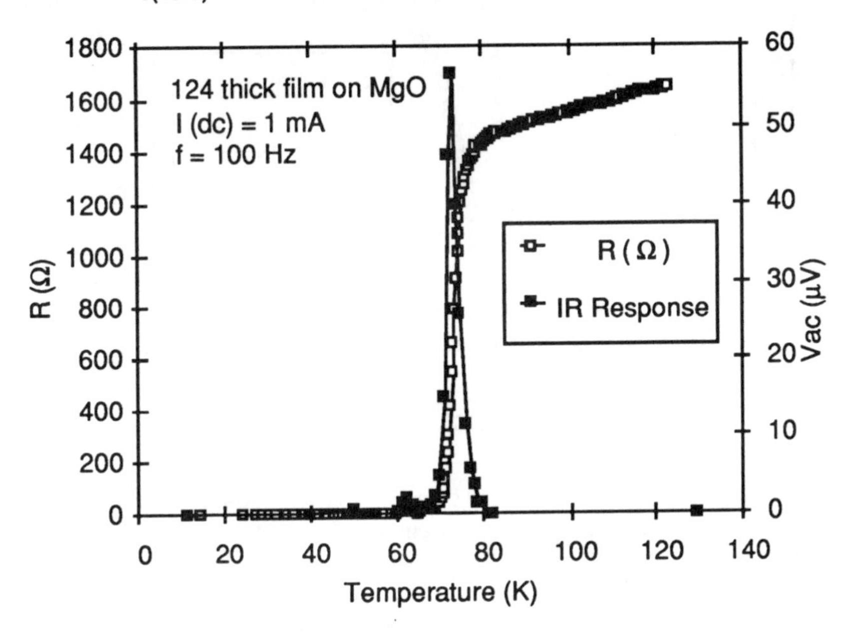

Figure 3. Resistance vs temperature and detector response for a 123 thick film on MgO, formed by heating a 124 paste in argon above its decomposition temperature. Maximum optical power = 2.5 mW/cm^2. Note that corrected temperatures would shift curves ≈ 9 K higher.

IR-response vs. temperature of the thin film sample at 50 Hz chopping frequency is shown superimposed on its R vs T result in figure 4. As shown in the figure, there are two peaks in the response that are due to changes in the slope of the R vs. T result. There is a somewhat higher response near $T_{c(onset)}$ of the sample that is due to the sharper transition of R vs. T giving a higher $\Delta R/\Delta T$.

Figure 5 compares the observed IR-response from a thin film with the direct derivative of the measured resistance vs temperature data. The match of the temperatures of the peaks confirms the bolometric behavior, as described by equation 1. The magnitude of the IR-response of the film is low due to the low normal resistance and consequently low $\Delta R/\Delta T$ of its active area.

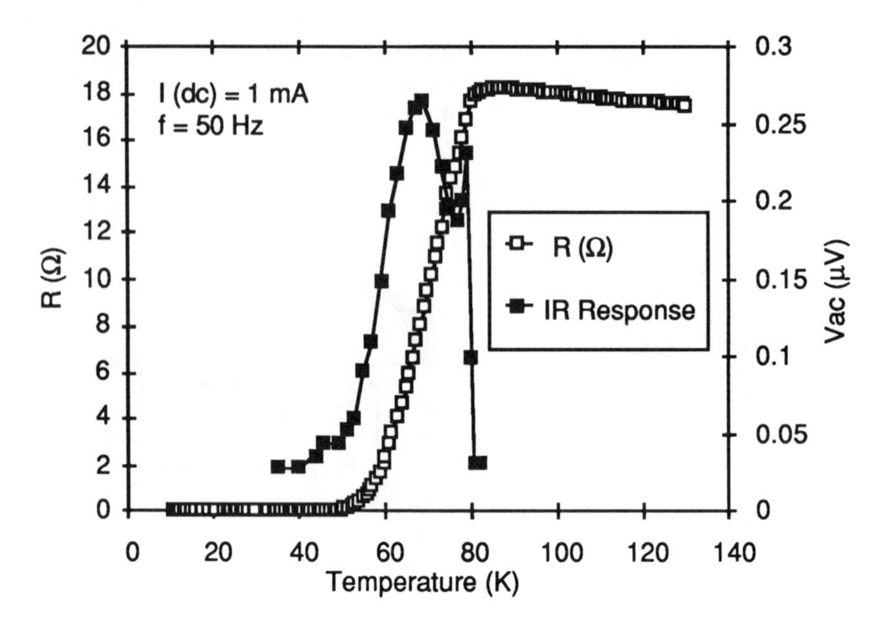

Figure 4. Resistance vs temperature and detector response for a 123 thin film on (001) MgO. The film is ≈ 1 μm thick, and has a 0.4 cm by 0.5 cm active area. Maximum optical power = 1.43 mW/cm^2. Note that corrected temperatures would shift curves ≈ 9 K higher.

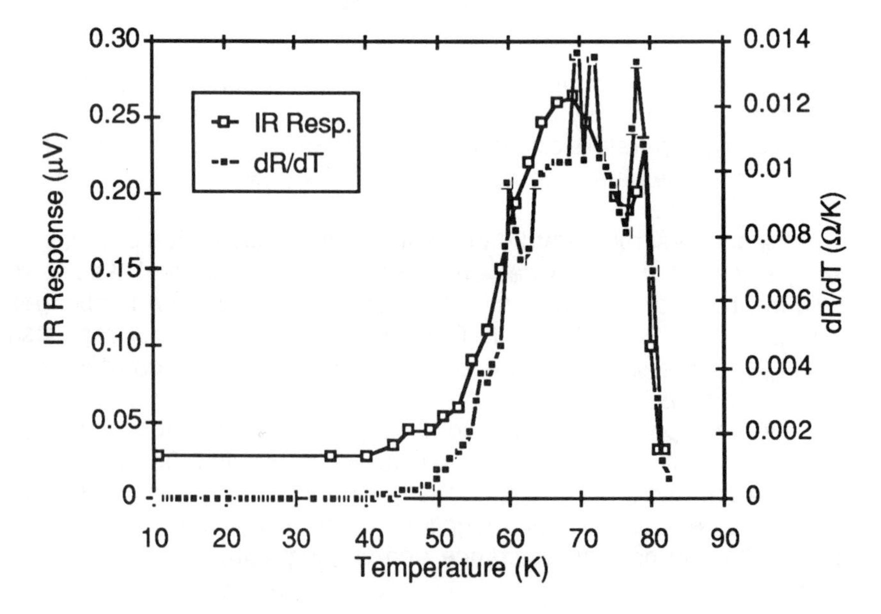

Figure 5. IR response of a thin film (see figure 4) compared to the derivative of its resistance vs temperature data. Note that corrected temperatures would shift curves ≈ 9 K higher.

The magnitude of the response can be increased by increasing the dc-bias current through the detector (Equation 1). This increase is limited by two main factors. One factor is the J_C of the sample. Currents higher than J_C at grain boundaries (weak links), can drive the weak links to the normal state, affecting the transition of the sample. The other factor is power dissipation in the sample (self-heating effect) that can heat up the material, especially at the grain boundaries where the resistance is higher.

To study the effect of the bias current on the responsivity of the samples, IR-response measurement vs dc bias current were done on a sputtered thin film. The sample was patterned after annealing using photolithography as explained elsewhere.[11] The film was deposited on single crystal MgO (001), and had 1 μm thickness, 1 x 10^{-5} cm^2 cross sectional area, and 0.4 cm^2 active area. The magnitude of the response of the film, divided by the bias current, is shown in figure 6. The experiment was done for bias currents up to 25 mA, limited by the heat dissipation in the sample resulting in the increase of resistance at the bias point.

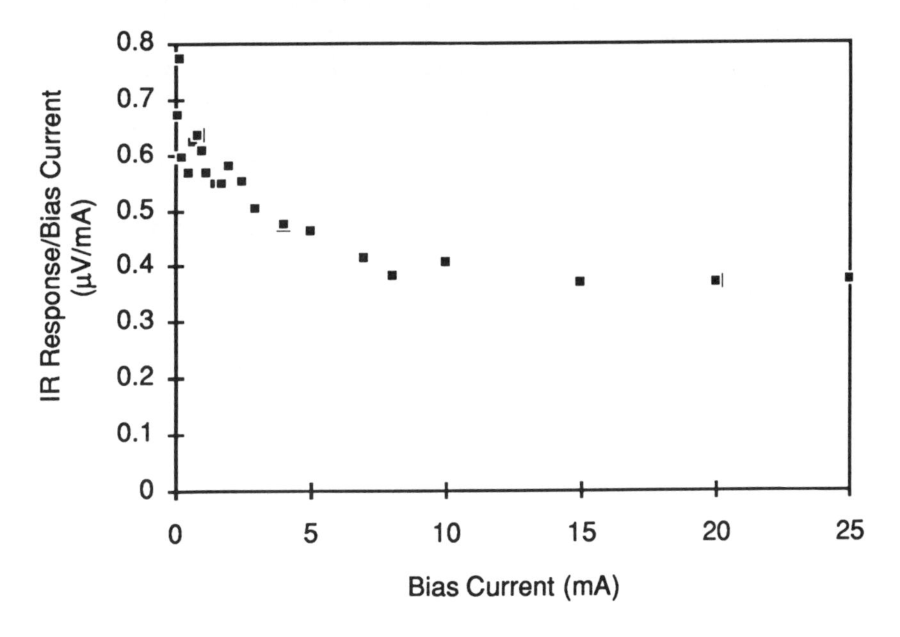

Figure 6. The current-normalized responsivity of a thin film IR detector at various dc bias currents. Temperature =70 K, chopping frequency = 50 Hz, maximum optical power = 1.71 mW/cm^2

As shown in figure 6, the current-normalized responsivity of the detector drops with increase of the bias current up to currents near 8 mA (equal to 800 A/cm^2 current density). This effect is attributed to the granular structure of the film (observed by SEM). The superconductive material at the grain boundaries can be partially driven to the normal state, resulting in a decrease in the sensitivity of the film.

It has been observed in some samples that the increase of current at low temperatures ($T<T_{c(zero)}$) drives the superconducting sample to a normal state irreversibly, resulting in a very high resistivity. This irreversible transition of samples from the superconducting state to the normal state (at $T < T_{c(zero)}$) by current is attributed to the high power dissipation at the weak links. To prevent this irreversible process, high currents through the sample at low temperatures have been avoided.

To measure IR-response vs. frequency of the samples, the frequency of the oscillator was ramped while the temperature of the sample was constant. To do the measurement, a constant current was passed through the sample and, at chosen temperatures, the magnitude of IR-response at different chopping frequencies was plotted. This measurement was done at the temperatures at which the samples had their maximum IR-response, as identified by the IR-response vs. temperature measurements.

An IR-response vs. frequency measurement was done on a thick film prepared from a 124-based paste (figure 7). This sample was annealed using a step profile, in argon/oxygen atmosphere,[3] giving a smooth and sharp transition with about 1.5 KΩ resistance at the $T_{c(onset)}$, the same as the sample shown in figure 3, and exhibiting the same form of IR response.

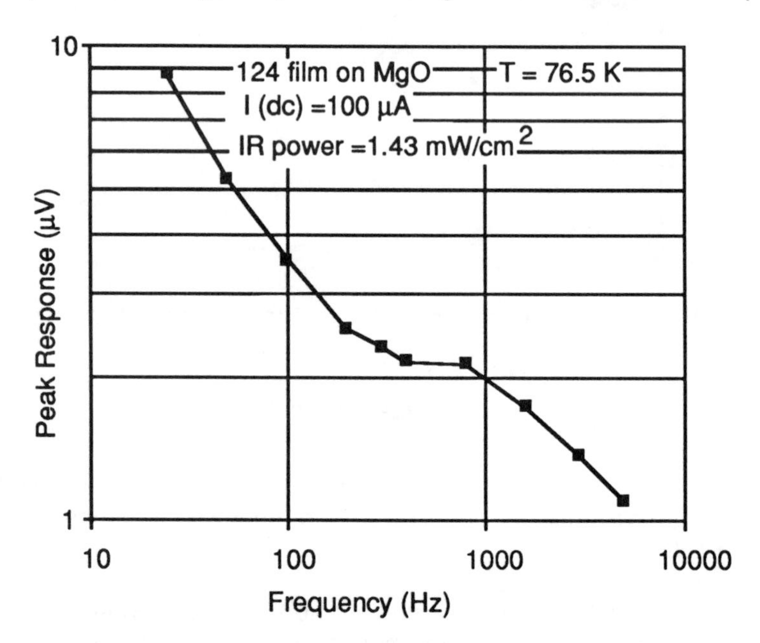

Figure 7. Peak response vs frequency for a thick film IR detector on polycrystalline MgO

As shown in the figure, there are two breakpoints in the response. The first breakpoint, near 400 Hz, is attributed to the thermal time response, τ (C/K), of the substrate. The second breakpoint, near 800 Hz, is attributed to the thermal time response of the superconducting film. Considering this assumption, the second breakpoint should be observed at higher chopping frequency for thinner films.

IR-response vs. frequency of the thin film sample was done at T=69 K, which was the temperature of the maximum response due to the maximum dR/dT in the R vs. T curve. The experiment was done for different chopping frequencies ranging from 10 Hz to 2000 Hz. Figure 8 shows the logarithmic plot of magnitude of the response vs. chopping frequency of the sample up to 1 kHz. A constant dc-bias current of 1 mA was passed through the sample.

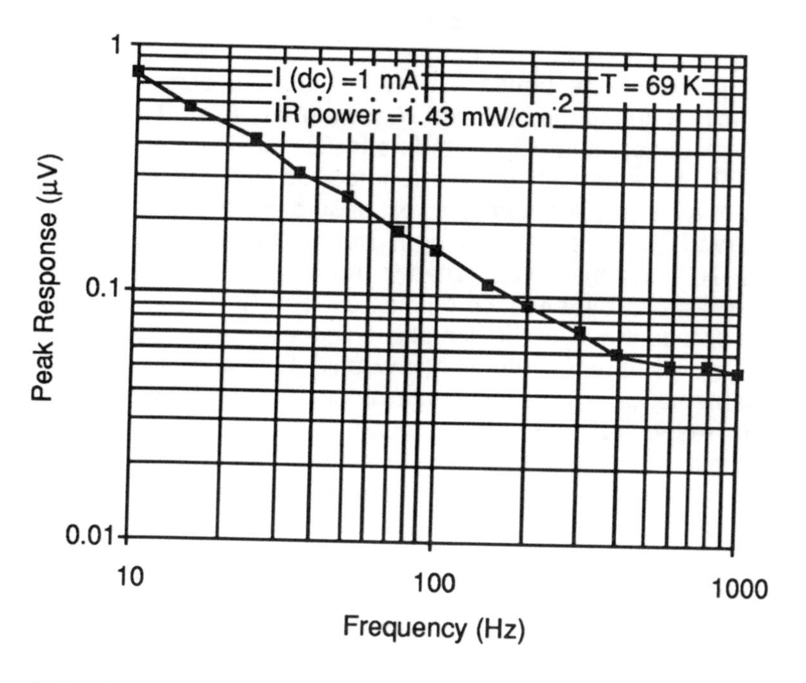

Figure 8. Peak response vs frequency for a thin film detector on (001) MgO.

As shown in figure 8, there is only one breakpoint in the response between 10 Hz and 1 kHz chopping frequency. The break point is at 400 Hz and is again attributed to the thermal time response of the substrate. Both thick and thin film samples are made on MgO substrates with nearly 0.5 mm thicknesses. The IR-response of the thin film was flat up to the maximum measured frequency, 2 kHz. This flat response of the film up to somewhat higher frequencies is attributed to the low thermal time response of the superconducting thin film, which had nearly 1 μm thickness. Due to low signal levels, measurements could not be made at frequencies high enough to observe the film-dependent breakpoint on the thin films.

The linearity in the IR-response vs. frequency in the logarithmic plots of the samples, obtained for both thick and thin films, shows a power relation between the magnitudes of the IR-response and the chopping frequency. Considering the response to be limited by the thermal time response of the substrates of the samples, this response will be similar to the form of the response of a bolometer at high frequencies, discussed earlier.

CONCLUSIONS

We have prepared thick and thin film Y-Ba-Cu-O IR detectors and show that they exhibit a bolometric response at their transition temperature. The IR-response vs. temperature of some films showed multiple maxima due to the changes in slope of the R vs. T curve. A single bolometric response, sharply peaked near the transition temperature, was obtained for the films with narrow (sharp) resistive transitions. A small response, relative to the bolometric response at the transition of the films, was observed at temperatures right below the $T_{c(zero)}$ under some conditions. This response is attributed to the Kosterlitz-Thouless effect, and has shown a strong dependence on bias current. Currents which are large but do not exceed the J_C of the samples shift the superconductive transition and $T_{c(zero)}$ to lower temperatures and wash out the effect.

Measurement of IR-response vs bias current has shown that the peak bolometric response does not increase linearly with the current. The current-normalized response drops to almost half of its magnitude with increasing current up to high current densities. This drop is attributed to driving the material at the grain boundaries to their normal state. At higher currents this process is limited by the size of the grains, resulting in a flat current-normalized IR-response.

The plot of IR-response vs. chopping frequency of the thick films showed two breakpoints at about 400 Hz and 800 Hz, where the plot of the thin film from 10 Hz to 2000 Hz, showed only one breakpoint at 400 Hz. These results show that there are two thermal time constants relevant to the response, one due to the film properties and the other due to the substrate. At low chopping frequencies the thermal delay time due to the substrate dominates the form of the response. Due to low signal levels, measurements could not be made at frequencies high enough to observe the film-dependent breakpoint on the thin films.

ACKNOWLEDGEMENTS

The authors would like to thank Dr. Robert Solomon and co-workers at Temple University for preparation of precursors used in the fabrication of the 124 powders, High Tc Superconco for 123 powders and sputtering targets, and ITT Defense Inc., GE Astrospace, and the Ben Franklin Partnership of the Commonwealth of Pennsylvania for financial support.

REFERENCES

1. A. I. Braginski, M. G. Forrester, and J. Talvacchio, ISEC' 89, June 1989.
2. S. A. Wolf, U. Strom, J. C. Culbertson, Solid State Technology, **187**, April, 1990.
3. M. Fardmanesh, K. Scoles, A. Rothwarf, E. Sanchez, E. Brosha, and P. Davies, NYSIS Conference, Buffalo, NY, September, 1991.
4. K. Char, A. D. Kent, M. R. Beasley, and T. H. Geballe, Appl. Phys. Lett. **51**, 1370, 1987.
5. M. G. Forrester, J. Talvacchio, and A. I. Braginski, Proc. GACIAC Workshop on High-Tc Superconductivity, May 1989.
6. E. Zeldov, N. M. Amer, G. Koren, A. Gupa, Phys. Rev. B, **39**, 9712, 1989.
7. M. Leung, U. Strom, et al., Appl. Phys. Lett., **51**, 2046, 1987.
8. W. S. Brocklesby, C. E. Rice, et al., Appl. Phys. Lett., **54**, 1175, 1989.
9. Y. Enomoto, T. Murakami, and M. Suzuki, Physica C, **153**, 1592, 1988.
10. A. Frenkel, et al., Appl. Phys. Lett., **54**(16), 1594, 1989.
11. M. Ihsan, M. Fardmanesh, A. Rothwarf, and K. Pourrezaei, NYSIS Conference, Buffalo, NY, September, 1991.

APPLICATION CONSIDERATIONS FOR HTSC POWER TRANSMISSION CABLES

John S. Engelhardt
Underground Systems Inc., Armonk, N.Y., 10504

Donald Von Dollen and Ralph Samm
Electric Power Research Institute, Palo Alto, Ca., 94304

ABSTRACT

The Electric Power transmission infrastructure is described to establish a background to examine the potential for applications of underground transmission cables using HTSC materials. Near-term and farther out scenarios are identified that could become niches for this type of cable system. Two design concepts for AC cables are presented and discussed in relation to historical systems that were proposed with LTSC materials.

INTRODUCTION

This paper examines the potential application of HTSC (High **Temperature** SuperConductor) materials to the transmission of electric power via underground superconducting cable systems. We begin with a general exploration of the power transmission infrastructure in the US, both overhead and underground, with an eye toward possible scenarios which would favor HTSC cable systems over conventional alternatives. We then look at practical HTSC conductor and cable system configurations which could compete with conventional cable technology and perhaps lower the cost of the underground alternative as societal pressures against overhead transmission lines continue to grow.

We hope this paper will clarify for the scientific community some of the issues that impact the potential use of HTSC materials in power transmission cables, and where such use is likely to begin.

HISTORICAL PERSPECTIVE

In the sixties and early seventies, when much work was done around the world on LTSC (Low Temperature SuperConductor) power transmission systems, there was a clear need by the utility industry for power transmission systems with much higher capacities than what was available with conventional High Voltage and Extra High Voltage cable technology.[1,2] This was due to the sustained strong growth of electric power usage, and the apparent need to build very large power generation complexes to keep pace with this growth. Generation parks were envisioned that would need to be located remotely from urban industrial and business centers and would require major transmission facilities to transport this bulk power to load centers. Underground transmission would be required to penetrate the rapidly developing suburban surroundings where aesthetic considerations were combining with huge right-of-way land costs to prohibit further overhead transmission construction.[3]

Several developing technologies were competing for the high capacity market thought to be right around the corner. Included in this list were heavily forced-cooled High-Pressure-Fluid-Filled (HPFF) pipe-type cables, compressed-gas (SF_6) insulated transmission lines (GITL's), cryo-resistive cables operating at liquid hydrogen or liquid nitrogen temperatures, and LTSC cables based on the A-15 series of metallic superconductors and cooled by liquid or gaseous Helium.[4] However, following the oil crisis of 1973-1974, the US became energy-conscious. Load growth dropped abruptly by a factor of two, causing utilities to put the brakes on major generation additions. Growth in the eighties was reduced again by half, thanks to the 1979 oil embargo and the general slow-down of US industry. With this stabilization of the utility structure, the need for very large capacity underground systems all but disappeared.[5] In the mid-seventies a few heavily forced-cooled HPFF 345 KV cable systems were built into the New York City area, and a few short GITL lines were built, mostly as bus ties within substations, but the anticipated need for very high power, and possibly very long cable systems such as promised by cryo-resistive and superconducting cable concepts, simply evaporated.

As we enter the last decade of the twentieth century, power transmission technology has reached a mature state. The utility planner has a broad arsenal of tools from which to chose, for virtually any type of transmission requirement he can foresee, using well-established designs with many alternatives to suit particular scenarios. 500 and 765 KV underground cable designs have been qualified for more than a decade,[6,7] but have yet to be required. The only 500 kv cables in the US were installed at Grand Coulee Dam in the mid seventies, and were supplied by the British and the Japanese. The eighties saw R&D expenditures concentrating on the improvement of the utilization of existing transmission systems through computerized real-time monitoring and rating technology, lower cost alternatives for small systems, and cost improvements in the operation and maintenance of transmission equipment.

The most significant development in the past several decades in underground transmission has been the commercialization of PPP, a laminated paper-polypropylene tape that replaces paper tape as the dielectric of High Pressure Fluid Filled cables.[8] PPP has lower dielectric loss and higher dielectric strength than paper. This results in smaller cables with more capacity for the same amount of copper conductor, which lowers the final cost of the system. PPP development began in the sixties, and was funded by EPRI since the early seventies, when EPRI was founded, through its commercialization in 1988, an evolutionary process that took more than twenty years.

The discovery of a perovskite that became superconducting at 30 K in 1986 by Bednorz and Muller of IBM, and the ensuing work on this family of copper oxides by Chu in Houston and Wu in Huntsville that resulted in the 1987 discovery of Yttrium-Barium-Copper oxide which superconducts above 77 K, the magic liquid Nitrogen temperature, rekindled the dying embers of the dream of lossless power transmission via superconducting cables. As these new materials make the transition from scientific curiosities to practical engineering materials, EPRI has been evaluating the opportunities that might result vis-a-vis power transmission cable applications in their STI Project (Superconducting Transmission Initiative) RP 7911-10. This paper briefly presents some of the key points of this ongoing work.

POWER TRANSMISSION TECHNOLOGY IN THE US

An Overview

It is difficult to drive very far in the United States, particularly on the Interstate System, without becoming aware of the presence of overhead transmission lines. In fact, transmission lines criss-cross the entire continent and serve as pathways for the transfer of electric energy from where its available in excess to where its in short supply. Complex interplay between and within major regional power pools is one of the reasons that the electric supply in North America is as stable and efficient as it is.

Power transmission is loosely defined as the transfer of electric energy from a source to a load over conductors that carry relatively large currents while being maintained at a high voltage. The energy transferred is the product of the current and the voltage. AC transmission is characterized by voltages of 69 KV to 765 KV while DC transmission is not so clearly defined but generally will be in the range of 100 KV to 600 KV. AC circuits operating at lower voltages are considered distribution class circuits.

The textbook power system will consist of a generator located in a remote area, a transformer to raise the voltage and lower the current output of the generator, a transmission line or cable to transfer the power to the load which will be in a developed area, a substation to receive the transmitted power and transform it from high voltage to a lower distribution voltage and send it out over many distribution lines to customers in the area. At each customer point, a final transformation will lower the voltage again to the service level. This system can be viewed as having two levels; the bulk power part of the system which generates and transports electrical energy, and the distribution system which delivers the energy to the customer. These two functions come together at the Utility substation. Superconducting cables may be used for both functions, but only underground transmission will be considered here.

The concept of networking, wherein the energy is given multiple paths to the customer, was first applied to distribution, but quickly spread to the bulk power side of the system as well. The modern power system contains many generators and many substations, all interconnected by a network of transmission lines. Each Utility is considered to be an independent power system; however, early in their evolution it became clear that interties between utilities would strengthen the systems and give flexibility to operations. Interconnections soon grew into defined power pools, which in turn have evolved into nine major power regions for the United States and Canada collectively known as the North American Electric reliability Council, or NERC, which was founded in 1968. The nine NERC regions are shown in Fig 1.[9]

Table I identifies each of these power regions, their 1990 peak demand, the circuit miles of EHV transmission lines (230-765 KV) in each region, a rough estimate of the percentage of these circuit miles that are underground (at 230 or 345 KV), and the 10 year estimated growth of EHV circuit miles of transmission for the period of 1991 to 2000.[10]

The first superconducting cable systems proposed in the early sixties were targeting 100,000 megawatts per circuit (100 GW). By the seventies the targets were in the 5,000 to 10,000 MW range. By the early eighties the level was at 1000 MW. Today we believe the first interesting applications may be at levels as low as a few

Figure 1.
MAP OF NORTH AMERICAN ELECTRIC RELIABILTY COUNCIL

hundred megawatts. Given the reluctance of a Utility to entrust too much of its capacity to one transmission facility, the peak demand levels shown in Table I give some insight to the magnitude of capacity contained within each region, and the size of a transmission facility that might be viable when one also looks at the thousands of circuit miles of EHV transmission already in place.

Transmission system growth is tied to load and capacity growth. Load grows steadily, but capacity is a quantum phenomena, increasing in blocks as new generators are added to the system. One function of the newer EHV transmission systems is to smooth the steps of capacity increases by "distributing" the new power over several load centers. Thus, utility planners study their regions as to the "adequacy" of existing transmission links to bring power to the major load centers, especially at times when one or two facilities may be suddenly lost. When possible shortfalls are projected, plans are made for additional lines or upgrades of existing lines. One potential market for superconducting cables will involve these regional expansions.

Table 2 gives a breakdown of the circuit miles of transmission by voltage class in the US only, and an indication of the range of capacity in megawatts of both overhead and underground circuits. It can be observed that the bulk of the transmission in the US (68%) takes place in the 69-161 KV range where the capacities are low, 200 megawatts or less. Superconducting cables have never been thought of as viable for systems of such low capacity.

Considering all transmission in the US, 69 KV and above, there are about 433,000 circuit miles of which 2500 circuit miles (0.6%) are underground pipe-type

Table 1
NERC COUNCIL STATISTICS

Region	Full Name	1990 Peak Load (GW)	EHV Circuit Miles	UG (%)	10 Year Growth (%)
ECAR	East Central Area Reliability Coordination Agreement	73	14,500	0.1	1.9
ERCOT	Electric Reliability Council of Texas	42	7,100	0	15
MAAC	Mid-Atlantic Area Council	39	6,500	0.1	9.5
MAIN	Mid-America Interconnected Network	37	5,130	0.1	3.8
MAPP	Mid-Continent Area Power Pool	27	18,500	0	1.9
NPCC	Northeast Power Coordinating Council	87	32,850**	3 (US)	1.4
SERC	Southeastern Electric Reliability Council	121	26,730	0.1	9.3
SPP	Southwest Power Pool	49	10,800	0	9
WSCC	Western Systems Coordinating Council	100	63,400	0.1	9
	TOTALS	575	185,510	0.2%	10%

** only 20% of these circuit miles are in the US

Table 2
CIRCUIT MILES AND CAPACITY

Voltage (kV)	Circuit (miles)	Typical Capacity Range - MW	
		OH	UG
69	110,000	20-150	40-100
115-138	185,600	100-300	50-200
230	64,800	150-600	100-400
345	45,738	400-1400	400-650
500	23,436	600-2000	N/A
765	3,654	2000-6000	N/A

cable systems. Approximately 400 circuit miles (16%) of this underground plant are EHV (230 KV or 345 KV) pipe-type cables, which leaves 84% of the underground pipe cables in the low capacity category, 161 KV or less. Roughly 600 circuit miles (24%) are on the Con Edison System in New York City. Of these 600 miles, 200 are 345 KV circuits; the rest are 69 and 138 KV circuits. Con Edison's 600 miles comprise about half of the underground cable miles in the NPCC.

The transmission growth projected for the next ten years is estimated at 10%. This is about half the expected load growth which is in the 20% range (2%/yr), and includes a significant amount of replacement of aging facilities. Table I shows the growth to be highest in the South and West in regions that have little underground transmission. The NPCC region shows the highest percentage of underground but the lowest projected growth rate. If HTSC cables are intended to compete with conventional undergound cables which are traditionally concentrated in the NorthEast, there may not be sufficient activity to justify the R&D effort. Fortunately we believe that other potential niches will appear that do, in fact, justify considerable R&D effort.

Capacity Limitations

Fig. 2 plots the power-current relationship for the common voltage levels in use to illustrate the power capacity behavior for the two different types of transmission considered here. Notice that the current carrying capacity of overhead lines increases with voltage while the opposite is true for underground cables. Overhead lines use larger conductors at higher voltages to control corona and radio noise, and at EHV levels each phase will have more than one conductor to lower the electric stress in the air at the conductor surface.

There are several factors that limit the capacity of a transmission circuit. The ultimate limit is the thermal capacity, which is that loading at which the ohmic and dielectric losses cause the conductor or dielectric to reach the maximum temperature that can be physically tolerated. These losses must be dissipated to the surroundings, either the air or the earth as the case may be.

Figure 2.
POWER vs CURRENT FOR US TRANSMISSION VOLTAGES

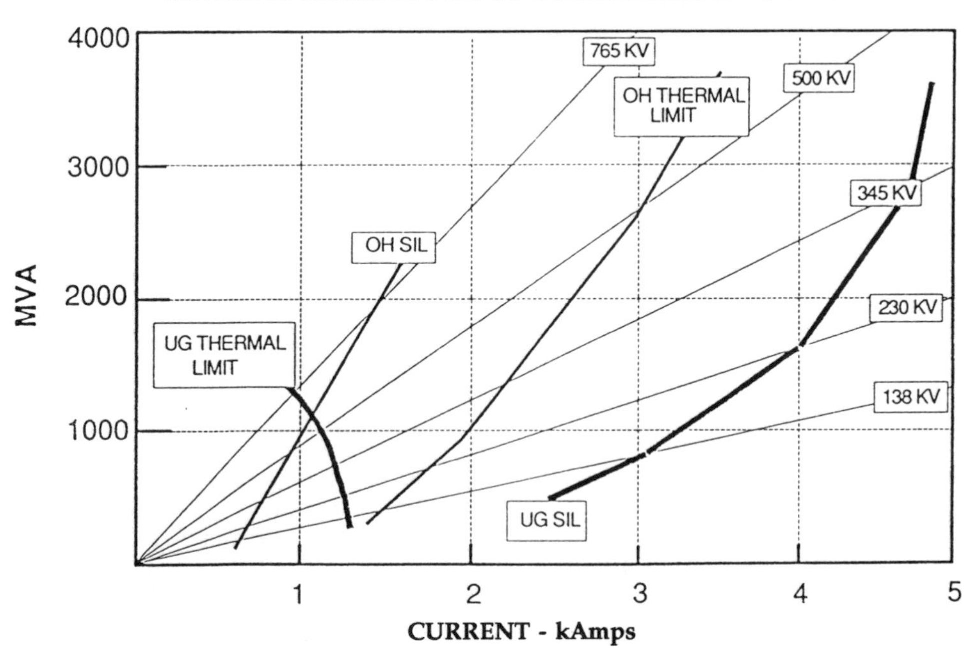

The thermal limit of an overhead conductor is usually established by sag; as the conductor heats it expands and the midpoint of each span drops closer to the ground. Sag violation can be hazardous in addition to lowering the dielectric strength of the circuit. A second concern is "aging" of the conductor by a phenomena called "creep", which is a tendency for slow irreversible extension of the conductor under stress which is accelerated by higher temperatures. Short-time excursions to very high temperatures will cause annealing and severe loss of strength.

Underground pipe-type cables are insulated with oil-impregnated paper or PPP which have well defined temperature limitations based on the thermal decomposition of cellulose. The operating temperature rise of the cable's conductor is caused by the ohmic losses in the conductor, shield and pipe due to the current, plus the dielectric losses that occur in the oil-impregnated paper or PPP insulation.

The "thermal" limit of a HTSC conductor will be a function of the HTSC characteristics. AC losses will vary as H^3/J_c and the cryogenic system will have a design limit for its capacity to control the temperature of the HTSC material. H_{C2}, the upper critical field, will limit the value of H that the HTSC material can tolerate and stay in the superconducting state.. Clearly, the higher the values of J_c, the critical current density, and H_{c2} that can be achieved, the higher will be the inherent capability of the conductor assembly.

Other aspects of the transmission system can also limit the loading, especially for lines longer than about 50 miles. System stability is one important factor that has several aspects.[11] The most obvious stability criteria for a large power system is that the sudden loss of the transmission line must not cause the system to become unstable. The power must have alternate paths to the load that are able to handle an increased loading without damage while the generation and loading adjust themselves to the new configuration. It was the loss of a heavily loaded transmission line out of the Niagara Falls Hydro complex that caused the 1965 Northeast Blackout. Alternate paths were unable to absorb the power demand without exceeding their own capabilities so that when their loading increased, they also tripped out due to overload. The hydro generators were unable to reduce their output fast enough to avoid speeding up, which caused an increase in frequency that triggered even more line and generator trips throughout the interconnected area which spanned most of the Northeast US. Thus, for overhead lines it is common for the normal loading to be established by load flow studies such that any first contingency problem on the system does not result in overloads that exceed the thermal limit of the line being studied.

The loss of the transmission line as a system component is similar to the loss of any other component in the system, be it a generator, transformer, circuit breaker, or any component whose failure would suddenly interrupt power flow. For this reason, the maximum loading of all components tend to be equalized. At the present time (1991) the maximum unit size of generators, transformers, transmission lines, and auxiliary equipment is in the 600 to 1000 MW range. Generators in particular reached a maximum size of about 1350 MW in the early eighties. More recent orders are for smaller units. Transformers have serious physical size problems that limit their capabilities to the 750 MW range. If larger units are desired, they will have to be manufactured on site. Superconducting transmission lines with high power ratings go against this trend. Power capabilities more in the range of other equipment will have more opportunity for application within the present framework of US transmission systems. However, the sheer size of the industry makes a return to the growth in unit size inevitable. The uncertainty is when, not if, this will happen. Therefore, we shouldn't abandon the quest for high capacity superconducting cables, but their development should follow smaller systems which might find immediate application opportunities.

One application for high capacity transmission lines is the transfer of bulk power between regions. These transfers are motivated by economic generation which can be prearranged in relatively large blocks with established contingency plans in the event of the loss of the transfer. This type of transfer is common for 500 and 765 KV overhead lines that have excess thermal capacity. Asynchronous DC connections are used for similar bulk transfers between frequency-independent power systems. Transfers in the range of 1000 to 3000 MW are possible in this category, and these levels can be expected to grow in the future. These interconnections could be a possible niche for medium power AC or DC HTSC cable systems which could provide this function at more economical sub-EHV voltage levels in environmentally sensitive or urban areas.

A more subtle stability criteria involves the dynamic stability of the power system and its ability to maintain control of the bus voltage at distribution substations. Voltage control considerations dictate loading limits for lines in the 50 to 200 mile range. A transmission circuit has two characteristic reactances that determine its behavior in, and effect on, the power system. The first is its inductive series reactance. The energy stored in this reactance is proportional to the square of the current in the line. The second is the shunt capacitive reactance whose energy is proportional to the square of the voltage on the line. The root of the ratio of the line's inductance to its capacitance is its characteristic impedance, Z_0, also known as its surge impedance. Z_0 corresponds to the specific load resistance at which the transmission line neither demands or contributes reactive energy. When a transmission circuit is loaded with its surge impedance, the power level is known as the surge impedance load, or SIL. Looked at another way, at surge impedance loading the series inductive energy exactly balances the shunt capacitive energy. If the load drops below the SIL, the transmission system is capacitive, while above its surge level it is inductive.

Table 3
SURGE IMPEDANCE AND LOAD LIMITS

Voltage (kV)	Zo - ohms		SIL - MW		Thermal Limits - MW			
	OH	UG	OH	UG	OH	%SIL	UG	%SIL
69	250	24	19	200	200	1050	150	75
115-161	260	27	73	705	350	480	320	45
230	270	33	196	1603	600	306	460	29
345	280	41	425	2900	1400	330	650	22
500	290	50	862	5000	2800	325	930	19
765	300	61	1950	9600	7000	360	1300	13

Table 3 summarizes the relations between surge impedance and load limits for overhead and underground systems. It tabulates the typical surge impedances, SIL's and thermal limits, as well as the ratios of thermal rating to SIL. Overhead lines have a characteristic impedance of 250 to 400 ohms which depends on conductor size more than voltage level but tends to increase with voltage.

Z_0 for conventional pipe-type cables is in the range of 20 to 70 ohms, increasing significantly with voltage. The thermal limits of EHV overhead lines tend to be about three times SIL. Conventional underground cables are thermally limited to a fraction of their SIL. This fraction also varies inversely with the voltage due to increasing dielectric loss and thermal resistance at higher voltages. As voltages increase, the thermal limit of a single overhead line becomes much larger than its underground counterpart. One common need for underground transmission is to bring the power of a long overhead line the last few miles into an urban load center. At EHV levels, this may require two or more underground circuits to match the capacity of one overhead line. An HTSC cable should be able to achieve parity with an overhead line for all voltages. It could then perform this function with significant economical advantage.

Power systems control distribution voltage with automatic tap changing transformers. These transformers allow a +/- 5% or +/- 10% range of adjustment of voltage. To maintain a constant secondary voltage in the substation at the load end of the transmission line, the magnitude of the sending end voltage will vary with load. It will be higher than the receiving end when loading is above the SIL but will fall below the receiving end when the loading is light. The working range of voltage for heavy loading is the minimum acceptable value at the receiving end to the maximum allowable at the sending end. For light loads, these limits become swapped. As the loading varies, transformers automatically adjust their tap settings to keep the distribution voltage constant. Fig. 3 shows a typical line loading curve as a multiple of SIL for long transmission lines based on the above limit of voltage difference between the two ends of the line.[11] For lines longer than 50 miles this stability limit drops well below the thermal limit. The usual concern is for loads above surge impedance. The voltage drop is due to the series reactance of the transmission line and is not sensitive to the line's resistance, or losses. Compensation is typically provided with series capacitors, however, during periods of very light loading the sending end voltage must be reduced in addition to switching out compensating capacitors. In extreme cases the operator may be required to take the line out of service and force the power to use alternate paths to the load.

The power angle of a transmission link, which is the phase difference between source and load, is another important parameter that limits the stable loadings of lines longer that 200 miles, and shorter lines as well during transient and fault disturbances. Space does not permit a full discussion of this critical aspect of power system operation. It is intended only to emphasize that thermal capacity is a minor concern for the bulk of the overhead transmission applications in the US. Surge impedance characteristics dominate the design criteria for longer high power transmission lines. While it is naive to expect that HTSC cables can economically compete with overhead lines, they can provide a viable alternative with much lower surge impedance than the overhead option. This may prove to be an enabling technology for some grandiose power supply schemes that aren't viable with conventional transmission technology.

One consequence of the difference in surge impedance between overhead and

Figure 3.
LOAD LIMITATIONS FOR LONG TRANSMISSION LINES

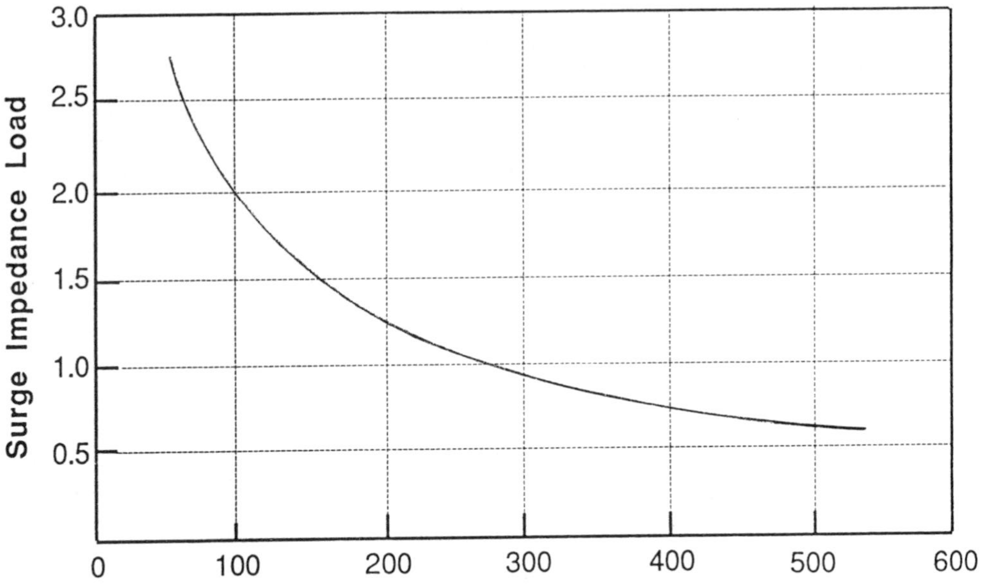

underground systems is that the reactance of the underground system is essentially constant because loading is thermally limited to levels well below the surge impedance load. Thus, long underground transmission circuits can be compensated with fixed shunt reactors at one or both ends that balance the capacitance of the cables and require no further control. It must be noted, however, that the longest underground AC line in the US is 25 miles. The capacitive reactance is not large enough to justify the cost of fixed compensation for underground cables that are less than about 15 miles long. Many utilities welcome the capacitive reactance to help balance overhead lines and loads that tend to be inductive.

Underground cables have an inherent problem with long lengths. Because their thermal capacity is well below their surge impedance loading, they get no help from their series reactance. The cable must carry the charging current demanded by its shunt capacitance, in addition to that required by the load. While the charging current is distributed, and is in quadrature with the load current, the ends of the circuit must carry the full amount, which reduces the circuit's capability to carry useful load current. If the charging current, which increases directly with length, reaches the total capacity of the cable, the circuit will be useless. The best arrangement is to supply the charging demand equally from both ends, which is why long underground lines are routinely equipped with shunt reactors at both ends. However, at lengths in the range of 25 to 50 miles, depending on voltage and conductor size, the cable charging demand will exceed its thermal capacity and a simple system is not possible. To overcome this limitation, shunt reactors must be placed in the middle of the circuit, essentially turning a single line into two or more lines. Unfortunately, breaking into a cable system in the middle is an expensive matter. Reactors are physically large and costly, and cable terminations are very expensive. A substation will be required that involves real estate, esthetics, auxiliary equipment, etc., all of which add to the cost of this necessity.

HTSC superconducting cables will have surge impedance characteristics similar to conventional HPFF cable systems. The fact that their ohmic losses may be significantly reduced has little bearing on their insertion into the power system. The loss component of the series impedance is very small compared to the reactance and it is common practice to ignore it in system studies using both overhead conductors and conventional underground cables. Dielectric stresses and losses will be similar to conventional systems because conventional dielectric materials and design will be used. The big difference will be a large increase in "thermal" capacity, limited now by the ability of the superconductors to carry relatively high currents in moderately high fields. The consequence of this one point will be the shift of the underground (UG) thermal limit curve in Figure 2 to the right, perhaps all the way past the underground SIL curve. This will give the Utility industry the ability to build longer underground circuits with higher capacity and lower compensation costs than conventional cables. These cables will clearly be able to match and exceed an overhead line's thermal capability for short and medium length applications, and because of their lower surge impedance they will open the door to higher AC transfers per circuit over medium to long distances when underground cables are required and DC is not justified.

Given this overview of the transmission function, where can we expect that superconducting cables might find a place? Before addressing this question we will review some design concepts of HTSC cable systems.

SUPERCONDUCTING TRANSMISSION SYSTEMS

Development work on LTSC cables in the sixties and seventies considered many alternative design concepts, some of which were truly bizarre. All designs were driven by two factors that may not apply to HTSC-based cable systems. First, the refrigeration energy cost for cooling cables to temperatures below 10 K placed the utmost demands on the minimization of losses, and secondly, system capacity targets were set at very high levels in anticipation of continued growth and economies of scale in vogue in the late sixties and early seventies. Cable designs were required that could not use conventional cable concepts and materials because of the necessity to reduce ohmic and dielectric losses to the absolute minimum possible. DC cables designs were straight-forward as this type of service produces no ohmic loss and insignificant dielectric dissipation. The loss problem for DC cables was one of controlling the heat in-leak. DC concepts had fewer restraints and those working to develop LTSC DC cables proposed several different concepts. all of which were viable to some degree. AC cable designers, however, quickly found they had few choices as to fundamental design.

High voltage cables must perform two functions; they must have a conductor that can carry a useable current and they must insulate that conductor from other phase conductors and ground for the voltages required by the system. The power transmitted is the product of voltage and current; the ability to do one without the other is of no value.

Conventional power transmission cables are comprised of a conductor assembly surrounded by a dielectric system which in turn is enclosed in a grounded metallic member that serves as the dielectric shield, and a system enclosure which contains the operating pressure necessary for proper dielectric function and protects the cable from the earth environment. The pressure medium is a dielectric fluid or Nitrogen gas. The conductor material is copper or aluminum strands, depending on the economic trade-offs at the time of manufacture. The dielectric is vacuum-dried oil-impregnated paper or PPP. The enclosure may be a steel pipe holding all three phases, or lead or aluminum sheaths over each phase. The latter is known as a self-contained cable, quite common in Europe but not popular in the US.

Ohmic losses occur in the conductor when the cable carries current. Ohmic losses also occur in the dielectric shields and enclosure due to induced circulating currents, eddy currents and hysteresis caused by the AC magnetic fields created by the currents in the three conductors. Dielectric losses occur in the dielectric due to the applied voltage. These losses appear as heat and cause the cable temperature to rise until the heat dissipation to the surroundings balances the heat generated by the losses. Thermal degradation of the cellulose in the dielectric limits the temperature at which the system can operate and there-by limits the power transfer capability of the system.

The concept of a superconducting cable is to replace the metallic conductor with a superconductor that can carry a larger current with lower ohmic loss, or zero ohmic loss in the DC case. The complication is the necessity to maintain the temperature of the superconductor at a cryogenic level. This requires a channel for the flow of the cryogen and a cryostat to thermally insulate the superconductor and cryogen from the surroundings. Fortunately the superconductor can support a very large current density so little material is needed for the conductor function. This leaves space for the cryogen channel and cryostat and allows a superconducting cable to be comparable in size to, or smaller than a conventional cable, an important conclusion because more than half the cost of conventional underground installations is the digging of the trench to house the system.

Early LTSC cable designers had to contend with the refrigeration cost of cooling helium at or near its liquefaction temperature at a penalty of about 500 watts per watt of heat removed. HTSC cable designers will work with liquid nitrogen and pay only about 10 watts per watt of heat removed. This luxury gives them considerable flexibility in the many choices they face as to possible cable configurations.

LTSC designers quickly found that the magnetic fields normal to the surface of the superconductor seriously degraded the superconductor's performance, and losses in the shields and enclosures due to eddy currents and circulating currents were unacceptable. Their only option was to make the outer shielding of the cable also superconducting. In this coaxial form the outer superconductor will experience an induced current equal and opposite to the current in the inner conductor which confines the magnetic fields to the space between the two and eliminates the driving force for eddy currents and circulating currents in the outer metallic components. The truly coaxial cable has the lowest inductance possible for a given dielectric spacing, which results in the lowest surge impedance obtainable. The cryogenic aspects of the cable are more complex because the entire cable must be maintained at the cryogenic level. The dielectric must therefore be able to function at the low temperature and immersed in the cryogen, and dielectric losses must be removed by the cooling system. Thus, all AC LTSC cable designs have required cryogenic dielectrics and two coaxial superconductors for each phase. Cost and space considerations favored the placement of the three phases in one large cryostat. The rigid design developed at Union Carbide[12], and the flexible design demonstrated at Brookhaven[13] shared this overall arrangement, as did most cable concepts put forth around the world. Figure 4 shows the flexible cable system proposed by Brookhaven. Professor Klaudy of Austria and Kabelmetal of Germany used a separate flexible cryostat for each phase, taking advantage of their proprietary expertise with corrugated tube technology.[14]

Figure 4.
BROOKHAVEN LTSC CABLE SYSTEM

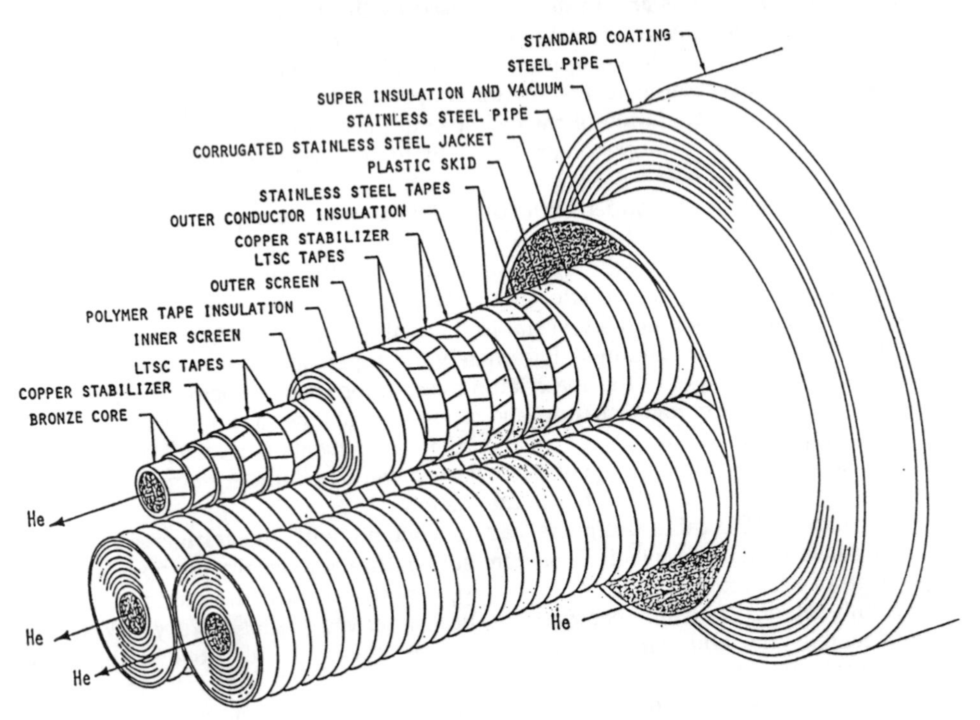

HTSC TRANSMISSION DESIGN CONCEPTS

We have carefully reviewed the early work on LTSC cables in search of similarities that may be applicable to HTSC designs. While the coaxial, fully cryogenic cable may be necessary for very high current applications of the future, the lower power levels in use today and for the near future open the door to simplification and subsequent cost reductions that take advantage of the reduced refrigeration penalty afforded by a liquid nitrogen cryogen.

As the first consequence of a smaller current carrying requirement, one asks if the superconducting outer shield is still mandatory or would it be possible to revert to conventional shield structures and tolerate the additional losses? The answer is yes; with careful shield optimization the shield loss can be reduced to an acceptable level. The next question is what about magnetic field interactions? The effect of normal fields on the surface of a superconductor is severe, but at lower currents and fields, it too may be tolerable. The lack of phase-to-phase magnetic isolation raises the inductance and surge impedance and reduces the surge impedance load, but at lower power levels a lower SIL is desirable. The most significant consequence affects the dielectric. If the outer shield is not superconducting, there is no reason for the dielectric to operate at cryogenic temperature. This means that a very simple design put forth for LTSC DC cables in the early seventies could be applied to HTSC AC cables. This concept placed the conductor inside a small flexible cryostat which was covered with a dielectric that operates at room temperature.[15]

One strong lesson learned during the development of LTSC cable systems was that a new dielectric design for a high voltage cable system and its subsequent qualification is not a trivial task. Many researchers found that the introduction of a new dielectric system was far more difficult than qualifying a radically new superconducting conductor assembly. Dielectric concerns can probably be blamed for the lack of acceptance of LTSC concepts. The Kabelmetal system was the only system to be demonstrated by connection into a Utility's operating transmission line. It used conventional paper impregnated with supercritical helium for its dielectric.[14] The use of a conventional dielectric system operating at room temperature eliminates this formidable hurdle and greatly increases the probability of an early successful system demonstration. We have therefore focused on the room temperature dielectric cable concept as our first choice for immediate development, recognizing, of course, that fully coaxial, cryogenic dielectric systems may also be of interest for future higher power levels.

Room-Temperature Dielectric Cable Concept

Figure 5 illustrates a room temperature dielectric, pipe-type cable system built around an HTSC superconducting conductor assembly which provides, in one structure, the cryogen channel, the superconductor, and the cryostat. The cable dielectric, its shielding and the steel pipe enclosure are conventional in every respect. Comparison of this assembly with a standard pipe-type cable reveals that only the conductor has been changed. If the cryostat assembly can be made in the diameter range of normal copper or aluminum conductors, which is three to six centimeters, the physical size and appearance of the cables will be identical to conventional practice and they will be easily assimilated by the cable industry/utility infrastructure.

The room temperature dielectric cable concept has many advantages in comparison to the cryogenic dielectric cable beyond the use of an established dielectric system. Only the heat in-leak and the AC and eddy current losses in the superconductor assembly load the refrigerator. Dielectric losses and ohmic losses in the cable shields and enclosure are conducted to the earth; they do not have to be removed by the refrigeration system. This reduces the overall system energy requirements, reduces the size of the refrigeration packages and lowers the flow requirements which extends the distance between cooling plants for a given channel diameter. Thermal contraction phenomena are limited to the former, the HTSC tapes,

Figure 5.
ROOM TEMPERATURE DIELECTRIC HTSC CABLE SYSTEM

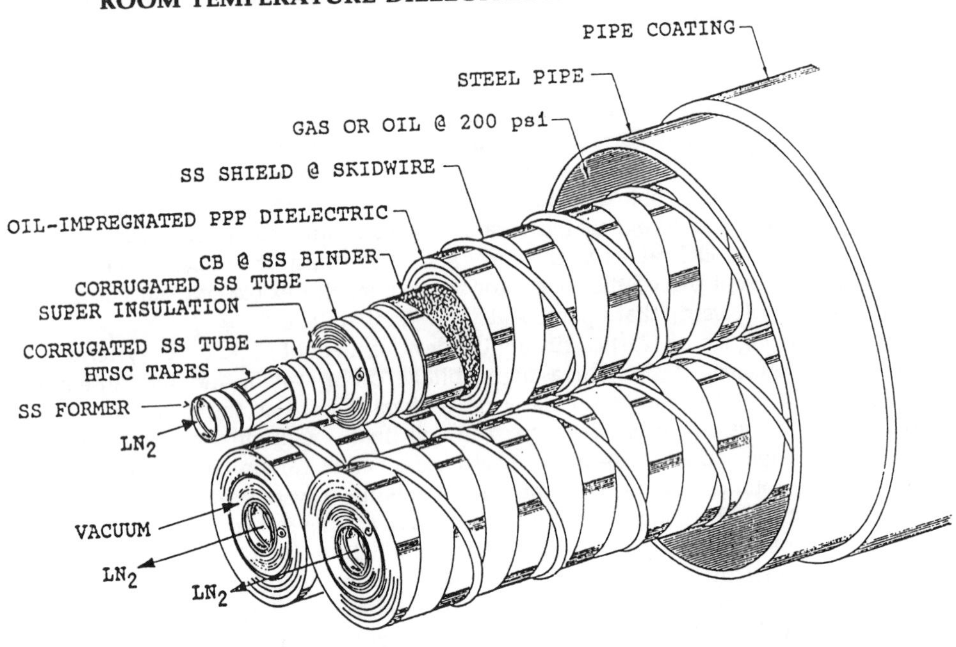

wires, or coating, and the inner wall of the cryostat. The remainder of the cable components operate at a temperature slightly above the earth ambient. The mass of material that must be cooled is greatly reduced, which simplifies service procedures and reduces down time, an important consideration for large transmission systems. Installation does not entail any special operations except for the splicing of the conductor assembly. The terminations will also be essentially conventional since the thermal insulation is not exposed to the voltage gradient. Perhaps the most important aspect of this design concept is it's ability to be retrofitted into existing pipes. This capability presents utility planners with a totally new option for the upgrading of existing urban facilities and a low cost alternative for bulk transfer through urban areas.

The key and only new component of the room temperature dielectric system is the conductor assembly, which forms the entire conductor system. Once manufactured, this system can be fully characterized and tested for compliance with its intended service. When its performance has been confirmed, it enters the conventional cable manufacturing process to be insulated, factory tested for dielectric integrity (at room temperature), shipped and installed. At all stages in the life of this system, the conductor function and the dielectric function are separated. Voltage testing can be done without cooling the system, and current testing can be done any time, even before installation in the pipe. With this concept, the challenge facing HTSC proponents is to develop only the conductor assembly and cryostat. A "new" dielectric system is not required.

The limitation of the room temperature dielectric HTSC cable installed in a conventional steel pipe will be its current carrying capability. Eddy current and hysteretic losses in the steel will limit the current that each conductor can carry to the 2000 to 2500 ampere range, decreasing with increasing pipe size. However, as can be seen in Figure 2, this capability more than doubles the range of conventional underground circuits, and extends the thermal capability beyond that of single circuit overhead lines except at the highest voltage levels. New installations can benefit by the use of a non-magnetic stainless steel pipe to achieve current ratings in the 3500 ampere range which is close to the SIL for 138 and 230 KV voltages. Thus, this concept can provide the ability to build very long underground circuits that operate at or near surge impedance loading with power transfer capabilities in the 1000 to 1500 MVA range at lower voltages (138 or 230 KV) than overhead lines.

Cryogenic Dielectric Cable Concept

For cable systems that carry more current than the room temperature dielectric system can accommodate, a coaxial cryogenic cable configuration will be required. Figure 6 illustrates a preferred assembly for such a system, derived from the earlier LTSC designs. This particular design employs an extruded polyethylene dielectric which also provides the enclosure for the cryogen channel, and a flexible corrugated inner cryostat wall, which gives this concept the ability to also be retrofitted into

Figure 6.
CRYOGENIC DIELECTRIC COAXIAL HTSC CABLE STYSTEM

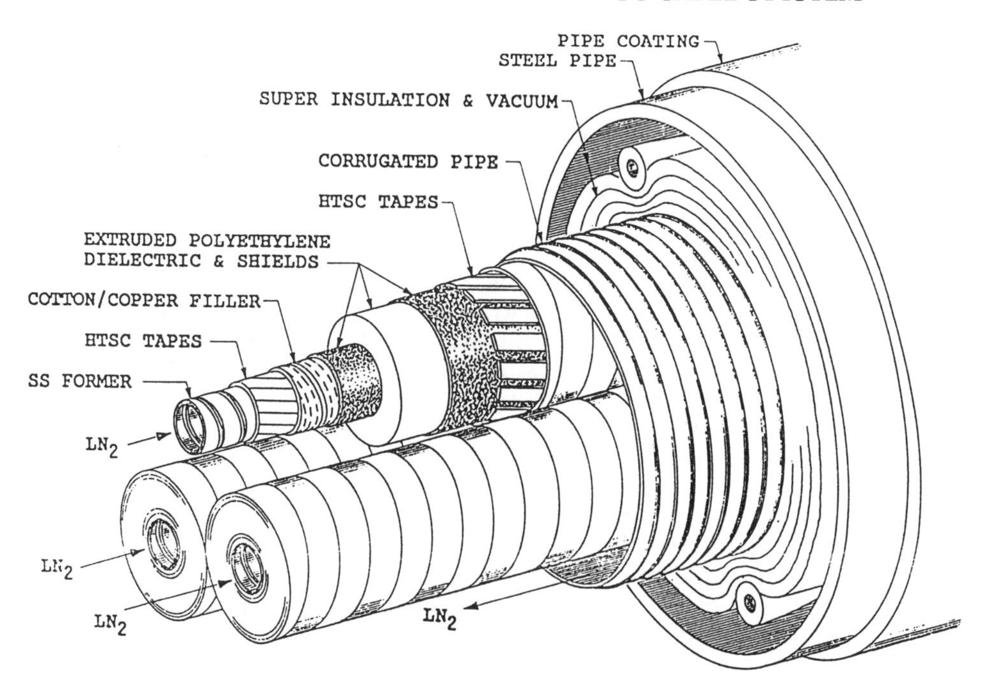

existing steel pipes. Of course, a conventional rigid cryostat as shown on Figure 4 could be used for new installations if a more robust outer covering and skid wires were added to each cable. The current capability of this structure is limited only by the magnetic field and current density characteristics of the HTSC elements.

The cryogenic dielectric of choice at this point in time is an extruded polyethylene rather than the cryogen-impregnated plastic tapes that were proposed in the seventies. Extruded dielectrics for transmission cables have evolved into a mature technology over the last twenty years, while plastic tapes have yet to be successfully demonstrated and adopted by the cable industry. EPRI funded a study of the use of extruded polyethylene at cryogenic temperatures in the early eighties which demonstrated their viability for operation in liquid nitrogen.[16] More recent work in Japan has confirmed the excellent properties of this material at liquid nitrogen temperatures and similar HTSC cable designs are being developed.[17]

The cooling of extruded polyethylene to 77 Kelvin actually improves its dielectric behavior relative to temperature-limited operation in conventional cable applications. This extends the voltage levels that are possible for HTSC systems to the EHV range and promises to provide an underground alternative transmission system with the highest power delivery capability that can be imagined in the present scope of utility applications.

NEAR-TERM APPLICATIONS FOR HTSC CABLE SYSTEMS

The primary advantages of an HTSC cable over conventional cables is the ability to carry more power per circuit with a lower energy cost per megawatt delivered. The former can lead to an economic advantage where space is at a premium or more than one conventional underground circuit is required to satisfy the capacity requirement and a single superconducting circuit does not violate reliability criteria. Two scenarios seem to fall into this category. The first would take advantage of the room temperature dielectric concept to retrofit HTSC cables into existing HPFF pipes. This would be beneficial in urban areas where, because of crowded streets and congested underground corridors, it is difficult and expensive to install new pipes. There are hundreds of miles of cables that have been installed in pipes over the past 40 years in all of the nation's larger cities. As urban load densities shift over time, some of these cable systems have reached their load limits, and many are approaching the end of their expected 40 year life. EPRI's development of PPP which reduces the insulation thickness has enabled Utilities to pull out old paper insulated cables and replace them with higher voltage PPP insulated cables that double the power capacity of the existing pipe. This results in substantial savings in construction costs as well as operation and maintenance savings. In 1989 Con Edison upgraded an old 69 KV paper cable system in lower Manhatten with a 138 KV PPP cable system and doubled the pipe's capacity. By combining the advantage of PPP insulation with an HTSC conductor assembly, an existing pipe's capacity might be increased three to six times. Such a large increase in capacity may not be required in all urban areas due to load growth alone because the power density in many crowded areas is already high. However, significant urban renewal projects that replace low rise slums with high rise, state-of-the-art population centers pose this type of problem to the host Utility. The World Trade Center in New York City is one example of this type of growth. There are now five 138 KV pipe cables feeding that neighborhood substation.

Another opportunity for HTSC cables retrofitted into existing pipes is in the area of inter- and intra-regional bulk transfers motivated by shifting economical generation locations as new units come on line and old, uneconomical units are retired. The major interconnections are EHV overhead networks that necessarily skirt urban areas. Some Utilities like Boston Edison have had to install phase angle regulators to prevent bulk load transfers from taking the low impedance direct underground route through the urban center, which overloaded the pipe cables feeding power into the city, and force the bulk power to go around the area on the overhead system. As the economic motivation for bulk power movement increases, as it is, the overhead routes are going to saturate. By replacing one or more underground cables with HTSC cables, the Utility could increase their pass-through capability and their local power density without jeopardizing service to the urban area or building new overhead facilities around the area in sensitive, now suburban environments.

The second scenario will be the undergrounding of sections of overhead lines, either as they approach populated load centers or in remote sections that are forced out of the sky for non-technical reasons. This application will favor EHV circuits where conventional cables don't come close to matching the thermal capacity of the overhead line and the circuit length is not great enough to limit the overhead line loading to levels that could be handled by one conventional circuit. The drawback to the possible application of fully coaxial cryogenic dielectric cable designs for this application is the EHV voltage level. The development and qualification of a cryogenic dielectric capable of EHV levels is many years away. However, as mentioned above, the room temperature dielectric capability is available and qualified now, all the way to 765 KV using PPP. Again, the only missing component is the conductor assembly. Therefore, once a conductor assembly is developed, this scenario could be satisfied by a potentially competitive single circuit HTSC room temperature dielectric cable system.

The reduction in energy used (lost) by an HTSC cable is another potential advantage. Unfortunately, this characteristic is not the panacea that has often been claimed. It does, however, have some merit for circuits that are to be operated at very high load factors, that is, loaded all the time. Our studies show that an AC HTSC cable can achieve a per unit energy savings of about 50% when loading conditions are most favorable. A cryogenic cable system requires a baseline amount of energy to be functional, even before it starts to deliver useable power. While the losses in the transmission circuit are dramatically reduced, the energy consumed by the refrigeration plant to maintain the cryogenic temperature negate much of the savings. Even "lossless" DC HTSC cables will consume considerable energy to balance the heat in-leak. For practical designs of HTSC DC cables, their energy consumption will about equal a conventional cable alternative which is loaded near its thermal limit, making HTSC dc cables attractive only for loads that are higher than the capability of conventional designs which is about 2000 amperes.

For base-loaded systems of medium length, say 10 to 50 miles in areas hostile to overhead construction, the HTSC cable can enjoy a considerable advantage due to it's lower energy costs. The cost of losses for conventional cables when capitalized over the life of the system can approach 25% of the installed cost. In this context even the 50% reduction in energy consumption per megawatt delivered adds to a sizeable cost advantage for the HTSC cable.

LONG-RANG POSSIBILITIES FOR HTSC CABLE AN ENABLING TECHNOLOGY?

Given the present low level of growth in the utility sector and the adequacy of the present transmission industry to meet the needs of utilities over the near-term, it is difficult to envision how superconducting cables may impact the future. However, consider the actual numbers. The present demand of North America is 575,000 megawatts. At present rates this demand will double in about 50 years. That means that the utility industry has to find and build 575,000 megawatts of energy and the means to transport it to load centers within this 50 year time frame. Given the realities of the regulatory process, the long delays in obtaining site approvals, public intervention and the like, the magnitude of this task begins to emerge. Assuming that energy sources and sites for generation can be found, the means for absorbing this additional load into the transmission grid are not obvious. It is unlikely that the public will tolerate even a 50% increase in the density of overhead transmission lines around the country. Thus, transmission considerations are going to impact generation sighting decisions and cause further delay in the process. Delays are going to cause shortfalls, and shortfalls will require reaction. Larger projects will become mandatory to stay even with growth. An unexpected upturn in the US economy could place utilities at risk, unable to construct the required facilities in the available time.

HTSC cable systems could have a major impact on the above scenario. Since they will easily perform the same function as the overhead line, and may offer some operational advantages because of their lower surge impedance and higher loading capability, the transmission function may default to the underground HTSC system. These systems will be environmentally acceptable, will speed the permit process, and may "enable" an otherwise unacceptable power source option.

The potential for very long, very high power DC lines is also on the horizon, and HTSC cables may be the system of choice because of their potential for very high capacity and low loss. The use of conventional DC cables would consume roughly 10% of the energy they transmit for distances on the order of 1000 miles. HTSC cables could reduce this penalty to the 1 to 2% range.

REFERENCES

1. "The Electric Utilities Industry Research and Development Goals Through the Year 2000", Report of the R&D Goals Task Force to the ERC, June 1971, ERC Pub. No. 1-71

2. Proc. of Conference on Research for the Electric Power Industry, Washington, D.C., December 1972, IEEE Pub. No. 72 CHO 726-0 PWR

3. "Underground Power Transmission: A study for the Electric Research Council", by Arthur D. Little, Inc, October 1971, ERC Pub. No. 1-72

4. Greenwood, A., Tanaka, T., Advanced Power Cable Technology, CRC Press, Inc., Boca Raton, Florida, 1983. See Volume II, Chapter 2

5. Barthold, L., "Need for High Power Systems and Generating Center Capacities by 2000-2010", Workshop Proceedings: Public Policy Aspects of High-Capacity Electric

Power Transmission, EPRI WS-79-164, September 1979

6. Engelhardt, J.S., Blodgett, R.B., "Preliminary Qualification of a HPOF Pipe Cable System for Service at 765 KV", IEEE Trans. on Power Apparatus and Systems, Vol. PAS-94, No. 5, September/October 1975

7. Development of Low-Loss 765 kV Pipe-type Cable, EPRI Project RP 7812-1, Final Report EL-2196, EPRI, Palo Alto, Ca., January 1982

8. Allam, E.M., McKean, A.L., Teti, F.A., "Optimized PPP-Insulated Pipe Type Cable System for the Commercial Voltage Range", IEEE Paper 86 T&D 569-8, September 1986

9. "Forecast", Transmission & Distribution, January 1990

10. "T&D in 1991: A Year of Growth", ibid., January 1991

11. Stoll, H.G., Least-Cost Electric Utility Planning, John Wiley & Sons, Inc., 1989

12. Superconducting Cable System Program (Phase II), Interim Report, Part I, Union Carbide Corp., EPRI Project RP 7807-1, EPRI, Palo Alto, Ca., 1973

13. Forsyth, E.B., "The Brookhaven Superconducting Underground Power Transmission System", Electronics & Power, May 1984

14. Klaudy, P.A., "Practical Conclusions from Field Trials of a Superconducting Cable", IEEE Transactions on Magnetics, Vol. MAG-19, No. 3, May 1983

15. Klaudy, P.A., Titisee, 1972

16. Development of Cross-linked polyethylene Insulated Cable for Cryogenic Operation, CTL, EPRI Project RP 7892-1, Final Report EL-3907, EPRI, Palo Alto, Ca., February 1985

17. Kosaki, M., et al, "Development and Test of Extruded Polyethylene Insulated Superconducting Cable", Proceedings of 2nd Int. Conf. on Properties and Applications of Dielectric Materials, Beijing, China, 1988

NOVEL CONCEPT FOR A SPACE POWER DISTRIBUTION BUSBAR USING HTS MATERIALS AND PASSIVE COOLING

Martin A Shimko
Christopher J. Crowley
Peter N. Wallis
Creare Inc., Hanover, NH

ABSTRACT

This paper presents the performance, defines the range of application, and shows the feasibility of using high temperature superconducting (HTS) materials with passive heat rejection for space power transmission. A conceptual design for the busbar is presented, and mass and resistive energy losses are estimated for various missions, power levels, and current types (AC and DC). All applications display a large increase in power transmission efficiency, while mass comparisons show the passively cooled HTS busbar mass ranges from 12% of the mass of a copper busbar at geosynchronous orbit (GEO) and beyond, to 38% at a 1000 km earth orbit (LEO).

The design of the HTS conductor is novel, consisting of interleaved HTS strip conductors (HTS plus substrate) separated by dielectric insulating material. Appropriate HTS materials are presently available in long length (>100 m) with current densities (>1000 amp/cm^2) and critical temperatures (95 K) which make the passively cooled busbar feasible. An original numerical model for the conductor/radiator assembly is described which includes the effects of solar insolation, reflected and IR thermal loads from the earth, and internally generated losses in the HTS. Completely passive operation at low earth orbits (LEO) of 1000 km is enabled by a novel asymmetric design for a directional radiator that includes a unique back-to-back busbar configuration that does not require active pointing. The design includes copper conductor downleads employing the same passive cooling scheme.

BUSBAR APPLICATION AND DESIGN ALTERNATIVES

Future space systems such as scientific laboratories, the Space Station, the mission to Mars, planetary exploration, Lunar bases, etc. will have much greater baseload power requirements (10s to 100s of kW) than existing systems. The power sources will be connected to the loads through high power transmission lines or power buses. At these power levels the power bus becomes a significant power distribution component in terms of weight, reliability, and efficiency. Alternatives previously considered for these buses have included conventional busbars, low temperature superconductors, and high purity metal conductors.

Buses made from conventional conductors (e.g. copper at ambient temperatures) need to operate at low-current densities to minimize energy losses. This can be accomplished by either using very high voltages or large cross-sectional areas. High-voltage conductors cannot be directly exposed to the space plasma environment because of arcing and discharge to ground requiring enclosures filled with SF_6 dielectric that must survive meteoroid impact. In addition, conventional low-voltage power conditioning equipment cannot be used. If bus cross-sectional area is increased instead of voltage, conductor mass per unit length becomes unattractive, even if resistive losses are allowed to rise to 2 percent of the source power (compared with 0.1 percent losses for superconductors).

The use of metallic superconductors or high purity metal conductors (which operate at conventional voltages) is constrained by the cryogenic operating temperatures which require active cooling using a cryocooler. Large penalties in mass are incurred due to the insulation and cryocooler equipment. Additionally, there is the power requirement for providing refrigeration and the added complexity and reliability considerations associated with long-life cryocoolers that are not presently available.

HTS BUSBAR DESIGN CONCEPT

Coupling of a HTS busbar with directional radiators was initially suggested by NASA LeRC[1]. The busbar concept analyzed in this paper integrates these two innovative approaches:

- a power transmission bus made from ceramic superconductor materials having high critical temperature characteristics (applying recent discoveries in HTS technology)
- passive radiation heat transfer to deep space to maintain the superconductor bus at the cryogenic operating temperature (70 K to 90 K).

In our design concept, the conductor and radiator are integral and continuous with one another. The basic structure of this innovative approach places the HTS conductor at the base of a linear cone or "trough" in a directional radiator. The busbar is shielded from incident radiation by the cone walls - it "sees" primarily cold space and rejects heat by radiation heat transfer to that cold sink. The downleads use the same passive cooling concept and make a gradual transition from normal to HTS conductor. The application of this busbar design in a 100 meter power transmission line is shown in Figure 1.

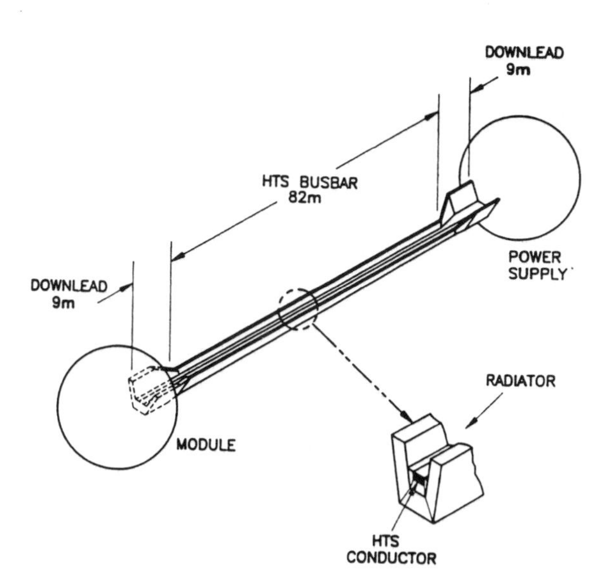

Figure 1. HTS BUSBAR APPLICATION

CONDUCTOR CONFIGURATION AND AVAILABILITY

The novel conductor configuration uses interleaved conductor strips of opposite current direction. This configuration minimizes the magnetically induced stress between conductors. In addition, the selection of an appropriate conductor thickness reduces the strength of the magnetic field that is generated (and its effect on J_c), satisfies the adiabatic stability criteria against flux jumps[2], and reduces the self-field energy loss which dominates AC operation.

The actual conductor assembly is composed of:

- 2 sets of HTS conductors for bi-directional current flow
- A metallic substrate for support, fabrication requirements, and electrical connection
- A dielectric material between opposing current conductors

The individual lamination can be either a continuous strip or a row of fibers surrounded by a substrate. Figure 2 shows a typical configuration and size for a 10 kW, 28 Volt DC application using long-length YBCO ($YBa_2Cu_3O_{7-\delta}$) superconductor material. The size is compared with that required when using conventional copper conductor material.

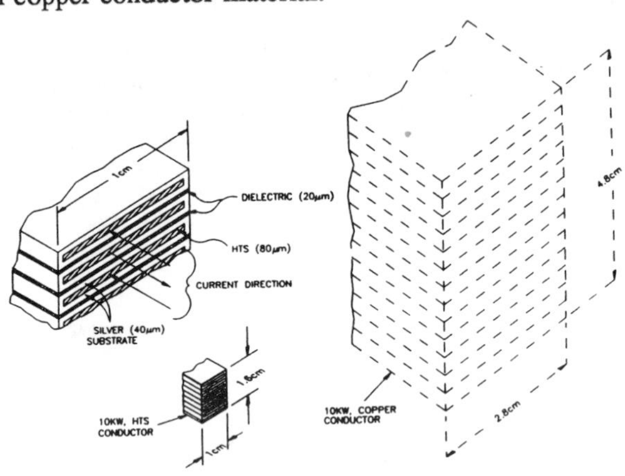

Figure 2. TYPICAL CONDUCTOR CONFIGURATION

The critical criterion of the three criteria on conductor dimension mentioned above is the one limiting the magnetic field (β) affecting the conductor. A magnetic field strength limit of less than 0.001 Tesla was used based on published β vs. J_c data for YBCO. For a strip configuration this criterion is

$$\beta \leq \mu_o J_c 2a \tag{1}$$

while for a row of fibers it becomes

$$\beta = \mu_o J_c \pi d^2/4s \tag{2}$$

where

d = fiber diameter (m)
s = center-to-center fiber spacing (m)

J_c = critical current density (A/m²)
a = half width of slab (m)
μ_o = permeability of vacuum ($2\pi x 10^{-7}$ H/m)
The resulting maximum dimensions are
- 80 μm thickness for strip
- 204 μm diameter for fibers, where s = 2d

Long-length wire and strip in the required sizes are both presently available. HTS electrical performance data from vendors shows high current densities (1000 to 3500 A/cm²) at acceptably low flux-flow loss conditions. Both YBCO and BISCCO HTS materials are presently available for prototype testing. Because of its immediate availability at lengths greater than 100 meters, published performance data, and the conservative nature of the critical current densities and critical temperatures, YBCO fiber properties were used in the busbar/radiator modelling.

HTS BUSBAR PERFORMANCE MODEL

Figure 3 shows the configuration of the symmetric busbar design used in the initial modelling. Multi-layer insulation (MLI) is used to thermally isolate the conductor from the inner cone base and protect the inner cone from radiative heating of the outer shell. To minimize heat leak, four stainless steel wires (10 mil in diameter) support the MLI at 0.3 m intervals and maintain the lateral position of the cone, conductor, and shell. Short ribs (not pictured) will also be placed periodically on the cone and shell to enhance stiffness.

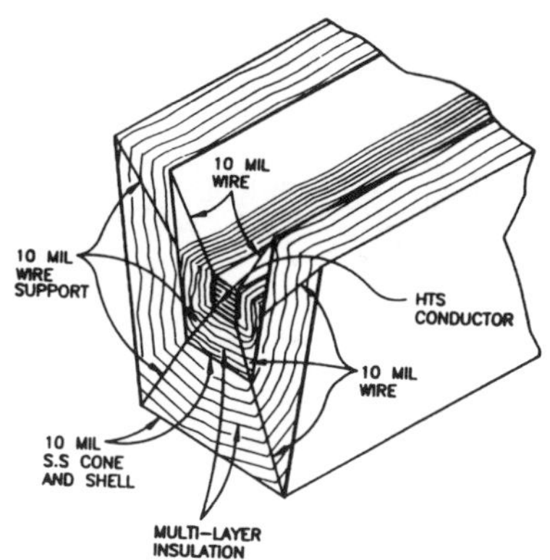

Figure 3. TYPICAL SYMMETRIC BUSBAR CROSS-SECTION

An original analytical model was written to describe the heat transfer performance and to calculate the mass of the busbar. The important elements included in the model are
- A specular model for radiation heat transfer,
- The conductor geometry,
- The cone geometry,
- External heat fluxes,
- Internal heat generation,
- Geometric view factors for radiation heat transfer, and
- Masses of elements in the busbar.

The numerical model was used to size and configure the conductor and radiator components for minimum conductor temperature as a function of orbit, power, and current type. Operating temperature, resistive losses, and the mass of the busbar are presented as a function of altitude (orbit).

The model uses a specular definition for reflected radiation, consistent with highly polished cone and shell surfaces and previous directional radiator analyses[3]. The model solves the energy balance equation for the conductor, the cone, and the outer shell. The view factors between deep space, the cone, and the conductor are calculated using Hottel's crossed-string method[4]. The input parameters required to define these equations are:

- Geometric Configuration
- Orientation Relative to the Earth and Sun
- Earth Orbit Altitude
- Material Emissivity
- Thermal Conductance
- Internal Energy Generation

External Energy Sources. Solar flux is assumed to strike the entire surface of the outer shell. Therefore, the combined thermal conductance of the MLI layers and support wires and material emissivity controls the energy input to the conductor from the shell. Reflected solar flux and infrared flux from the earth entering the radiator opening is calculated based on the orientation shown in Figure 4. The model configures the symmetric cone walls to reflect incident radiation back into space in two reflections and uses published methods for determining the view factor between the planet and spacecraft[5]. External energy input is minimized by setting $\psi = 0$ in Figure 4 (pointing north-south).

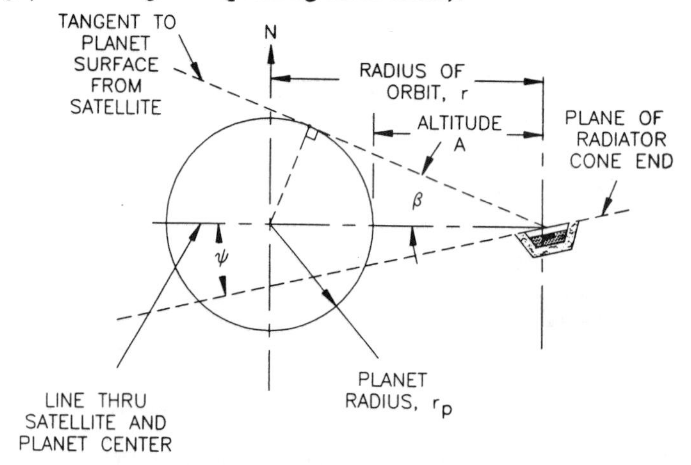

Figure 4. GEOMETRY OF BUSBAR ORBIT FOR REFLECTED SOLAR AND IR FLUXES

Internal Energy Generation. Four potential sources of internal losses have been identified for AC application.

- Self-field losses
- Eddy-current losses
- Dielectric (capacitance discharge) losses
- HTS Flux Flow (finite electric field) losses

Of these, the most important is the self-field loss. Eddy-current and dielectric losses can be made negligible relative to self-field losses by suitable design and material selection[6]. Flux flow losses can be reduced to arbitrarily small values by minor increases in conductor cross-section due to the steep flux flow loss vs. current density relationship typical of HTS materials. In DC applications, flux flow loss dominates and is therefore a less challenging application.

The model for the AC self-field losses in a slab of HTS has been derived from first principles by Kang[6]. This model agrees with the equations presented by Hull and Myers[1] in their preliminary evaluation of high temperature superconductors for space applications. The self-field losses can be described by

$$Q_v = \frac{2}{3}\mu_o f h^2 J_c^2 \tag{3}$$

where

Q_v	is the volumetric energy generation (W/m^3)
f	is the frequency of the voltage waveform (s^{-1})
J_c	is the current density (amp/m^2)
h	is the height of the conductor slab (m)
μ_o	is the permeability of free space ($4\pi x10^7$ henry/m)

When put in terms of the required energy flux from the top surface of a conductor composed of a stack of N laminations (current flow in both directions) this becomes:

$$Q_{sf} = \frac{4}{3} N \mu_o f h^3 J_c^2 \tag{4}$$

A baseline application of 10 kW at 110V, 400 Hz current was used in the performance modelling as a bounding case. Table I summarizes the relative contribution to the internal energy generation in terms of required surface flux for this case.

Table I. SUMMARY OF INTERNAL LOSSES IN HTS CONDUCTOR

Loss Mechanism	Value for Baseline Geometry (W/m^2)
Self-Field	0.305
HTS Flux Flow (Finite Electric Field)	0.143
Capacitance Discharge Across Dielectric	0.058
Eddy Current in Silver Substrate	0.010

Summary of Energy Sources. Figure 5 shows the contributions of all of the energy inputs to the busbar from the various sources for the baseline case. Internal loss generation is a small effect in the design of the HTS busbar. At GEO, the contribution of the solar flux on the outer surface is ten times larger than the internal losses, while the small contribution of radiation from the earth is about the same size as the internal losses. At lower altitudes, the reflected radiation from the earth becomes significant.

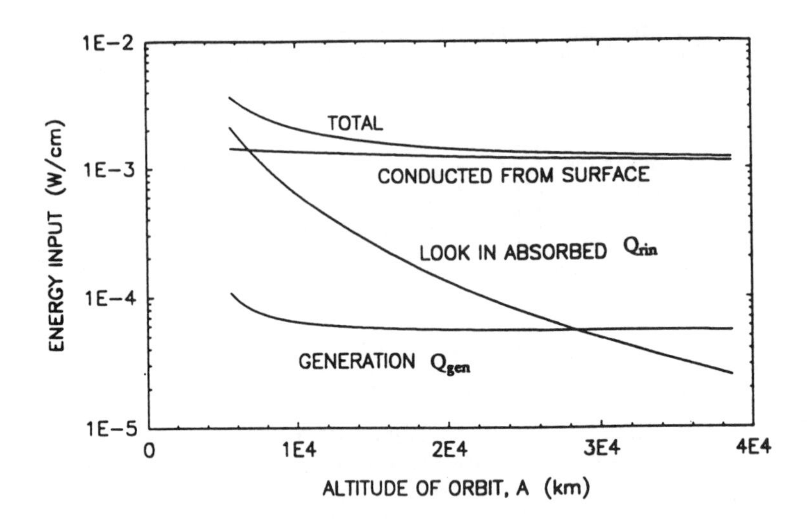

Figure 5. CONTRIBUTIONS OF ENERGY INPUT TO BUSBAR

BUSBAR PERFORMANCE AND APPLICATION RANGE

A preliminary round of performance modelling using the symmetric "trough" design was performed. The severe case of equatorial earth orbit was chosen for a 110 V, 400 Hz AC busbar to encompass most potential missions. A nominal 10 kW power level was chosen, since busbar weight scales linearly with power. A conservative limit of 80 K (0.87 of T_c for YBCO) was chosen for the conductor HTS material. The results of this analysis demonstrated an inherent limit of this configuration for low-earth orbit.

An asymmetric radiator was therefore designed that differentiates the "sun" and "earth" sides in low-earth orbit. Figure 6 shows this design and indicates several other features resulting from the initial analysis. In this design the cone end of the busbar lies in the plane through the equator, just as in the symmetric design. The cone wall nearest to the earth is at the same angle as the symmetric design and the conductor is located on that wall. The second cone wall is at an angle of 90° relative to the first cone wall. It reflects any incident radiation entering the cone from the earth side into space in just one reflection, thus minimizing the energy input from that source. Operating temperature and mass per unit length using this model are shown in Figure 7. These results demonstrate that the asymmetric busbar operates below 80 K down to an altitude of 1000 km with a busbar mass of only 20 g/cm.

The model assumes that direct exposure of the radiator interior to solar insolation is avoided. To accomplish this using a single busbar, either the radiator or a shield must be actively positioned. Figure 8 depicts an alternate approach that is completely passive (i.e., requires no active radiator repositioning in orbit) placing two trough shaped busbars back-to-back and perpendicular to the earth's surface. In this configuration, one side is always positioned with its back side sufficiently oriented away from the sun to prevent direct solar exposure of the radiator interior. This "side" then carries the load current and has a constant radiative load from earth shine dependent only on orbit altitude, planet radius, and reflective characteristics. It should be noted that no active current switching system is required as current will naturally select the conductor in the superconducting state from a pair of busbars wired in parallel.

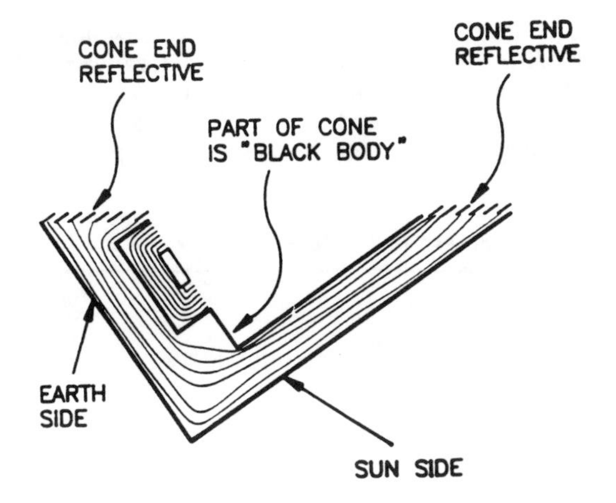

Figure 6. ASYMMETRIC BUSBAR DESIGN

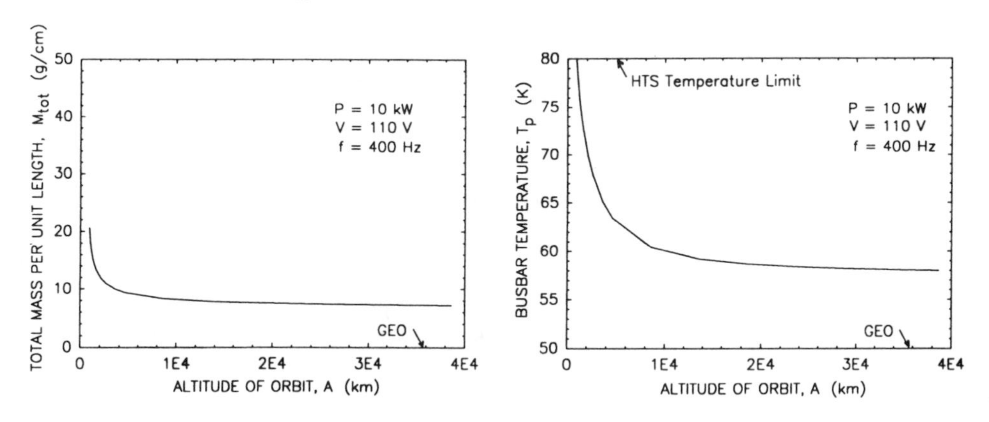

Figure 7. ASYMMETRIC BUSBAR PERFORMANCE ESTIMATE

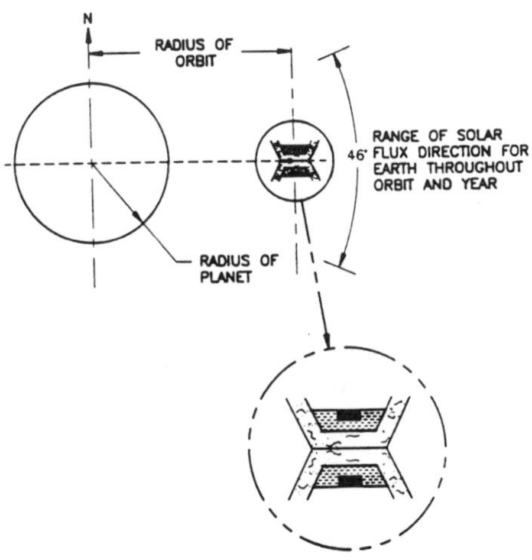

Figure 8. DUAL PASSIVE BUSBAR

The numerical model was also applied to the downlead design. Electrical resistive characteristics for high-purity copper conductors were substituted for HTS properties. Figure 9 shows the results of this initial analysis for deep space or synchronous orbit showing a downlead length of 9 meters with resistive losses of only 0.07% of the delivered power for 10 kW busbar operating at 110 VAC and 400 Hz.

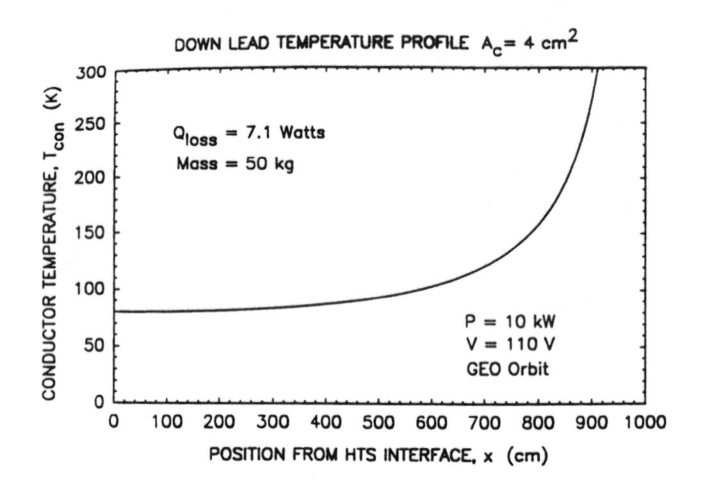

Figure 9. PRELIMINARY DOWNLEAD ANALYSIS RESULT

CONCLUSIONS

The passively cooled HTS busbar will provide high-reliability power transmission that does not depend on mechanical cooling systems and can be built with presently available HTS materials and configurations. The potential applications for HTS busbars span the range of space missions that require power levels in excess of 10 kW or use a baseload nuclear reactor. The space station, defense space platforms, Mars missions, and lunar bases would all benefit from reduced launch and maneuvering masses, as well as increased power system efficiencies. In addition, the novel HTS conductor and asymmetric radiator concepts developed here have general applications for the use of HTS and for spacecraft thermal management.

This work indicates that passive cooling of the HTS busbar to 80 K -- well below the critical temperature -- is possible at orbits ranging from low earth orbits (LEO) of 1000 km to geosynchronous orbit (GEO) 35,000 km and beyond. The mass scales linearly with power and non-linearly with altitude. In addition, preliminary analysis indicates that passively cooled downleads are feasible. The benefits of the HTS busbar relative to viable alternatives (i.e. conventional copper conductors) are summarized as follows:

Low-loss power transmission over long distances. HTS technology reduces resistive power losses at least an order of magnitude for all applications, from significant levels of 1% to 2% of the system power to negligible levels of less than 0.1%.

Low voltage operation integrates with existing space qualified equipment. By allowing the power transmission to take place at conventional voltage levels (28 Volts DC or 110 Volt, 400 Hz AC, for example) existing power conditioning equipment can be used.

Light-weight bus for reduced launch cost. A superconducting bus with a current density in the range of current technology (e.g. 10^3 A/cm^2) would be much better than a conventional copper bus on the basis of conductor mass alone, having only 12% (deep space) to 38% (LEO) of the mass. In addition, the increased efficiency translates into an additional system weight savings.

These benefits are based on very conservative estimates of HTS performance; increased benefits in mass reduction and an increased range of missions will occur as critical temperatures and current densities of HTS materials move even higher.

REFERENCES

1. Hull, J.R. and Myers, I.T.; *High Temperature Superconductors for Space Power Transmission Lines*; Superconductivity Advances and Applications - 1989, D.N. Palmer, ed., New York, NY: ASME, 1990, pp. 49-54.
2. Wilson, M.N.; *Practical Superconducting Materials;* Superconductor Materials Science: Metallurgy, Fabrication and Applications, S. Fourner and B.B. Schwarz, ed.s, NATO Advanced Studies Institute Series, V.68, New York, NY: Plenum Press, 1989, pp. 63-131.
3. Annable, R.V.,; *Radiant Cooling;* Applied Optics, V.9(1), Jan. 1970, pp. 185-193.
4. Hottel, H.C. and Sarofim, A.F., eds.; Radiative Transfer; New York, NY: McGraw-Hill, 1967.
5. Hottel, H.C.; *Radiant Heat Transmission;* Heat Transmission, 3rd ed., W.H. McAdams, ed., New York, NY: McGraw-Hill, 1954.
6. Kang, S.S. and Rothe, P.H.; High Temperature Superconductors for EM Launch Creare Technical Memorandum, TM-1330, Mar. 1989.

AC LOSSES IN HTSC CONDUCTOR ELEMENTS

S.A. Boggs
Underground Systems, Inc., Armonk, NY 10504

E.W. Collings
Battelle Memorial Institute, Columbus, OH 43201

M.V. Parish
CPS Superconductor Corp., Milford, MA 01757

ABSTRACT

AC self field losses and DC "dynamic resistivity" in HTSC conductor elements have been measured. The data are discussed with reference to the relevant theory. For two very different sample types, the J_c calculated on the basis of the AC self field loss is about an order of magnitude lower than the Jc based on the 1 $\mu V/cm$ DC voltage drop criterion often employed for HTSC materials. This is attributed to the onset of resistive behaviour at currents much lower than the reported J_c based on the 1 $\mu V/cm$ criterion. The J_c determined from measurements of the dynamic resistivity of a melt-textured YBCO fiber is within 15% of that determined by magnetic measurements.

INTRODUCTION

The research reported below was carried out in support of an EPRI-sponsored program to develop a technology base for high temperature superconducting power cable (RP7911-10) for both AC and DC applications. The samples YBCO samples on which the work is based were provided by CPS Superconductor with the support of DARPA Contract N00014-88-C-0512.

The primary cable concept of interest for both AC and DC applications is a room temperature dielectric cable in which only the conductor is within the cryostat, the dielectric being placed over the cryostat. This design has the advantages of avoiding a superconducting ground return well off the neutral bending axis and retrofittability to existing cable pipes for the purpose of circuit upgrading. As 70% of cable system capital costs can be associated with installation of the cable pipe, retrofittability is a powerful economic incentive which could well drive the first installations. This cable configuration has numerous subtle aspects related to the minimization of various types of ambient temperature losses which go beyond the scope of the present topic.

As is well known, a superconductor is not a perfect conductor but rather a material in a novel thermodynamic state which permits lossless DC conduction. However, DC conduction in the presence of an AC magnetic field and AC conduction are not lossless. If a Type II superconductor can be operated below H_{c1}, where it is in the Meissner state and conduction takes place in a very thin surface layer, AC conduction can, in principle, be lossless. However even in this case, losses occur which are related to extrinsic conductor surface conditions. H_{c1} for high temperature superconductors appears to be on the order of 100 Gauss[1]. For a conductor current of 1500 A_{rms} (900 MW at 345 kV), a conductor diameter of 90 mm would maintain the peak magnetic field below H_{c1}. As cable taping machines cannot tape much beyond 100 mm diameter, a conductor diameter of 90 mm is clearly impractical.

Given that a cable conductor will operate with a self field above H_{c1}, the magnitude of AC conductor losses becomes very important to cable design, as these losses must be removed by the refrigeration system at an energy cost on the order of 10 W of energy for every W of heat removed. As heat in-leak through the cryostat is likely to be in the range of 0.5 W/phase-m, an AC conductor dissipation of a few tenths of a W/phase-m is quite acceptable.

RELEVANT THEORY

Experiments concerning AC losses have been carried out in two contexts, (i) a DC conduction current exposed to an external transverse AC magnetic field, (ii) an AC conduction current in self field, and (iii) an AC conduction current exposed to an external, transverse, in-phase magnetic field. The first two conditions will be discussed in the present paper.

Dynamic Resistance during DC Conduction

The loss in a fully penetrated cylindrical Type II superconductor conducting a DC current while exposed to a transverse AC magnetic field is discussed by Carr[2] who shows that as the conduction current, I_m, approaches 0, the dynamic resistivity, ρ_{dyn}, goes to

$$\rho_{dyn}\,(I_m \rightarrow 0) = \frac{\pi}{4}\frac{R_o}{J_c}\left|\frac{dH_A}{dt}\right| \tag{1}$$

while as I_m approaches I_c, the critical current,

$$\rho_{dyn}\,(I_m \rightarrow I_c) = \frac{R_o}{J_c}\left|\frac{dH_A}{dt}\right| \tag{2}$$

so that the dynamic resistivity should increase by only 25% as the current is increased from a small value to the critical current.

AC Loss under Self Field Conditions

The formula for AC loss under AC self field conduction can be derived from Maxwell's equations with the assumption that the current in the conductor changes only at a moving boundary within the conductor[3]. The loss is caused by the periodic penetration of the magnetic field into the conductor in a (thermodynamically) irreversible manner. The resulting loss is given by (in MKS units)

$$Q = 2\,A\,f\,I_c\,J_c\,[\,i\,(2 - i) + 2\,(1 - i)\ln(\,1 - i\,)\,]\,x\,10^{-7}\,\frac{W}{m} \tag{3}$$

where "A" is the conductor cross sectional area, "f" is the frequency and "i" is the ratio of operating current density, J_m, to critical current density, J_c which is equal to the ratio of the operating current, I_m, to the critical current, I_c. For a solid cylindrical superconductor, this can be expanded to

$$Q = \frac{2f}{3}\frac{I_m^3}{I_c}\left\{\frac{I_m}{2I_c} + 1\right\}x\,10^{-7}\,\frac{W}{m} \tag{4}$$

where the expansion has been carried to second order. Note that the formula is independent of the conductor cross section and depends only on the critical current and operating current for the conductor. For a planar superconductor, the formula can be expressed as

$$Q = \frac{2\pi f w t}{3}\left\{\frac{\eta_m^3}{\eta_c}\right\}x\,10^{-7}\,\frac{W}{m} \tag{5}$$

where "f" is the frequency, "w" is the conductor width, "t" is the conductor thickness ($t << w$), and η_m and η_c are the conduction and critical currents per unit sheet width, respectively.

In each of these equations we see that to first order, the AC losses are proportional to the conduction current cubed and inversely to the critical current. This provides an obvious test to determine if the data conform to the theory. If data go as the current cubed, the critical current and critical current density can be backed out of the fit of the theory to the data.

DC DYNAMIC RESISTIVITY

Experimental Configuration

The sample was placed on a carrier which was placed within a stainless steel cryostat which passed through the bore of two orthogonal, metre-long saddle magnet coils which provided about 80-cm of uniform field transverse to the sample. LN_2 was pumped through the cryostat bore which contained the sample. The magnet and sample were placed within a bath of LN_2, which cooled the magnet to 77 K and reduced the magnet resistance from about 28 Ω to about 3.5 Ω, a value ideally for a stereo audio power amplifier which has been modified to provide a current (rather than voltage) proportional to the input voltage. Reactive compensation was used to null the inductance of the magnet coil so that the amplifier was driving a resistive impedance. Since the capacitive compensation eliminates the DC continuity which is necessary to maintain a zero DC quiescent voltage out of the current source, an operational amplifier-based feedback circuit was used to produce an overall amplifier circuit which acted as a voltage amplifier up to about 1 Hz and a current amplifier above 1 Hz. This produced both the desired AC current (independent of any impedance fluctuations of the magnet) and DC voltage stability in the driving circuit.

For DC currents up to about 10 A, the sample was driven by the other channel of the current source amplifier operating as a DC current supply. Alternatively, a Kepco power op amp was employed. For very high currents, a Sorrenson 115 A power supply was employed.

Experimental Results and Analysis

Figure 1 shows a graph of 51 Hz dynamic resistivity vs DC transport current for an early non-melt textured conductor tape made of a 17 conductor elements braided together with a total superconductor cross section of 1 mm^2. The data clearly indicate qualitative agreement with the above theory. The sample appears to be fully penetrated at only 17 Gauss, and the resistivity increases only slightly over an order of magnitude increase in conduction current, in agreement with the above theory. However, as seen in Figure 2, the resistivity does not increases in proportion to the magnetic field except between 42 and 170 G. The reason is probably related to the weak-link nature and low J_c of this material, which is not melt textured.

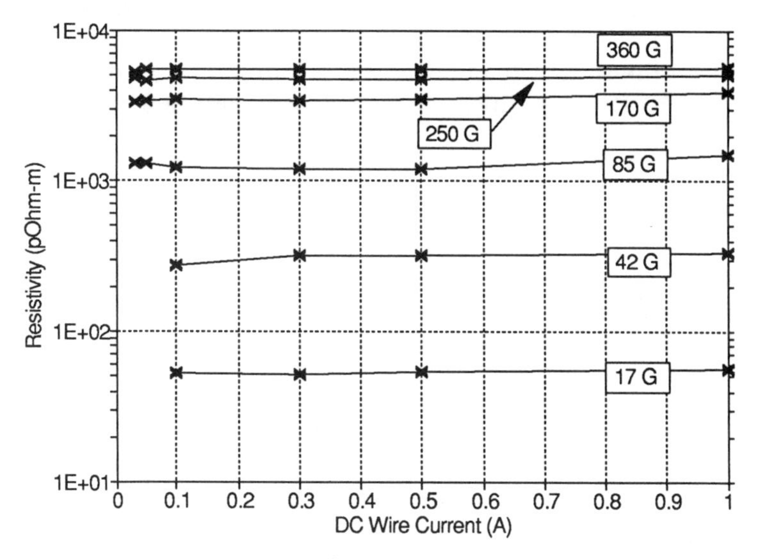

Figure 1. Dynamic resistivity for a YBCO conductor consisting of 17 YBCO fibers woven into a tape and coated with a thin layer of silver. The sample appears to be penetrated fully at 17 G, and the dynamic resistivity shows the expected behaviour as a function of current.

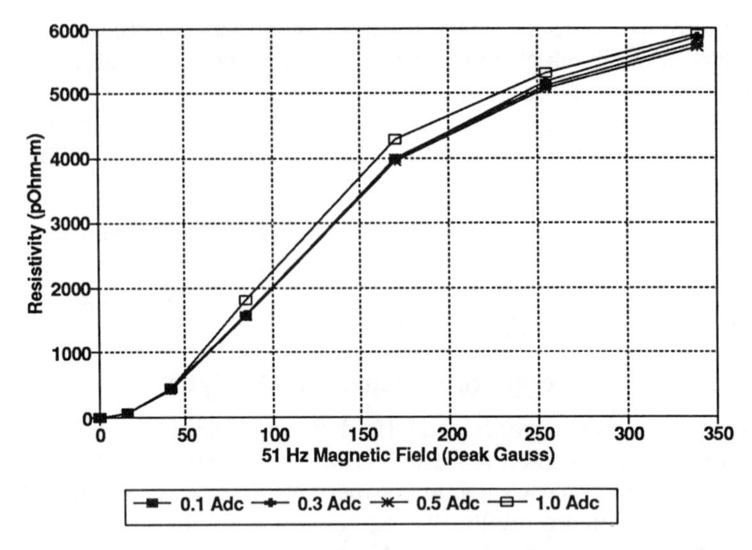

Figure 2. Dynamic resistivity for the sample of Figure 1 as a function of AC magnetic field for various DC conduction currents. The resistivity does not increase linearly with the field amplitude as predicted by the theory. This is likely the result of the weak-linked nature of the material and its low J_c.

Figure 3 shows the dynamic resistivity for a 0.38 mm diameter melt-textured YBCO fiber. The sample appears to be penetrated fully only at 850 G_{peak}. Given the resistivity of about 3.6 pOhm-m at a field of 850 G and a frequency of 59 Hz, the critical current density can be computed from the formula

$$\rho_{dyn}\,(I_m \to 0) = \frac{\pi}{4}\frac{R_o}{J_c}\left|\frac{dH_A}{dt}\right| \tag{6}$$

Here $\left|\frac{dH_A}{dt}\right|$ is $4\,H_{\mathrm{peak}}\,f$. Then

$$J_c = \frac{\pi\, H_{\mathrm{peak}}\, f\, R_o}{\rho_{dyn}} = 8.3\mathrm{x}10^4\ \frac{\mathrm{A}}{\mathrm{cm}^2} \tag{7}$$

which differs by about 15% from the magnetic J_c determined for this sample (9.62×10^4 A/cm^2)[1], which is reasonable agreement.

Thus the dynamic resistivity for this sample appears to agree with theoretical expectations. The most significant deviations are in the variation of the dynamic resistivity with magnetic field amplitude and frequency for the non-melt textured sample. However, the complex geometrical configuration of this sample and its weak linked nature make further analysis difficult.

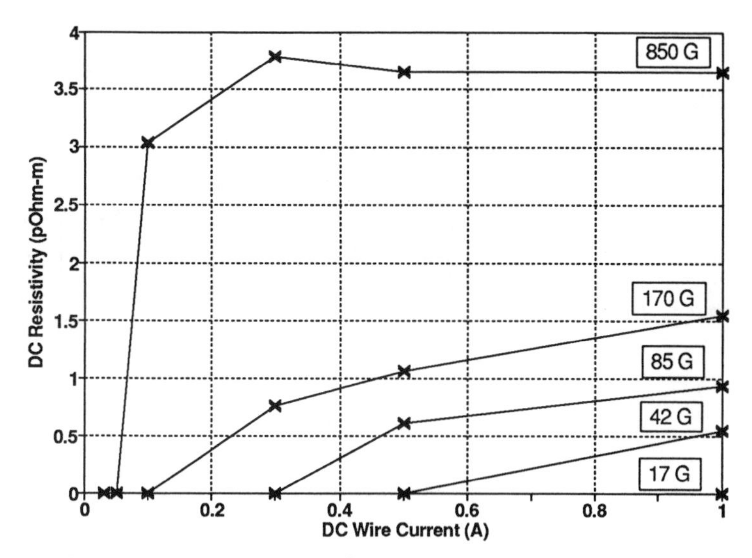

Figure 3. Dynamic resistivity for a 0.38 mm diameter melt-textured YBCO fiber. The fiber appears to be penetrated fully only at 850 G_{peak} applied magnetic field. Calculation of Jc from the 850 G dynamic resistivity results in a value which differs from the measured magnetic J_c by about 15%.

AC SELF FIELD LOSSES

Experimental Configuration

The experimental configuration for AC self field loss measurements was similar to that described above. The sample was normally placed within the magnet, as the additional loss caused by the application of an AC magnetic field was also of interest. The sample current was provided by the second channel of the stereo current amplifier. Up to about 10 A, the amplifier output could be used directly. Above 10 A and to about 80 A, the current amplifier was used to drive a 10:1 distribution step-down transformer, which could provide up to 80 A_{rms} at up to 6 V_{peak}. The transformer added a small amount (0.5%) of second harmonic distortion, and the oscillator driving the current source added a slightly greater amount of third harmonic distortion.

Figure 4 shows a schematic of the measurement configuration. The signal was detected by a transformer-coupled preamplifier with an equivalent input noise of 0.15 $nV/\sqrt{Hz}$, equivalent to the noise of a 1.4 Ω resistor at 300 K. However, if the sample to be measured was not sufficiently well stabilized to preclude sample and stabilizer failure, a 28 Ω fuse was employed in series with the transformer of the preamplifier in the hope that this would save the transformer in the event of sample failure. In this case, the fuse noise dominated the preamplifier input noise.

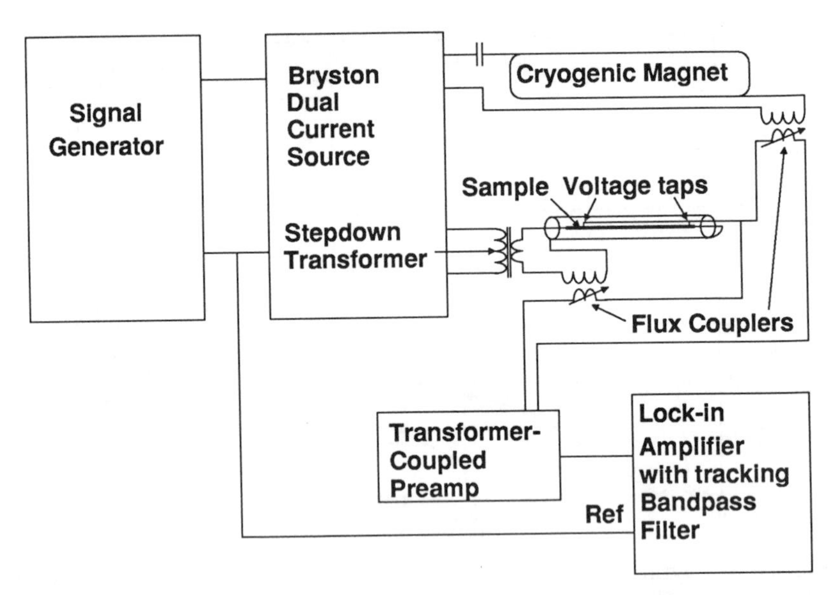

Figure 4. Configuration for measurement of AC power dissipation in both self field and with an applied external field. The sample (and magnet) current is passed through a toroid which contains two pick-up coils. These coils are used to generate an inductive voltage equal and opposite to that generated by the flux from the sample self field or magnet field which "cuts" the sample measurement loop.

Since stability and $1/f$ noise are not problems with AC measurements, sensitivity is generally limited only by noise and detection bandwidth. With the above parameters and a bandwidth in the range of 1 Hz, measurements down to a few nV could reasonably be expected. However for the sample discussed below, the inductive voltage produced as a result of the magnetic field generated by the sample cutting the loop formed by the voltage taps and the sample is expected to be in the range of $\omega \, x \, 10^{-9}$ V/cm or about 377 nV/cm at 60 Hz. Such a large interfering signal has two negative effect, viz., (i) it will overload the lock-in amplifier so that a small resistive signal cannot be measured and (ii) even if it does not overload the lock-in amplifier, any small error in the phase of the reference signal would cause a portion of the inductive signal to be indicated as resistive. For this reason, the inductive component must be eliminated if accurate measurements of the resistive component are to be made. This was achieved by passing the sample current through a toroid wound with #18 AWG wire on a non-magnetic, insulating core. Two small coils (one about 5 times smaller than the other) were placed on rotating shafts which penetrate the toroid so that the flux in the toroid passes through the coils. The coil shafts are driven by a combination 36:1, 6:1 reducer, so that very fine adjustment of the coil rotational position is possible. Further, the entire assembly was constructed so that it can operate with the toroid in LN_2 and with

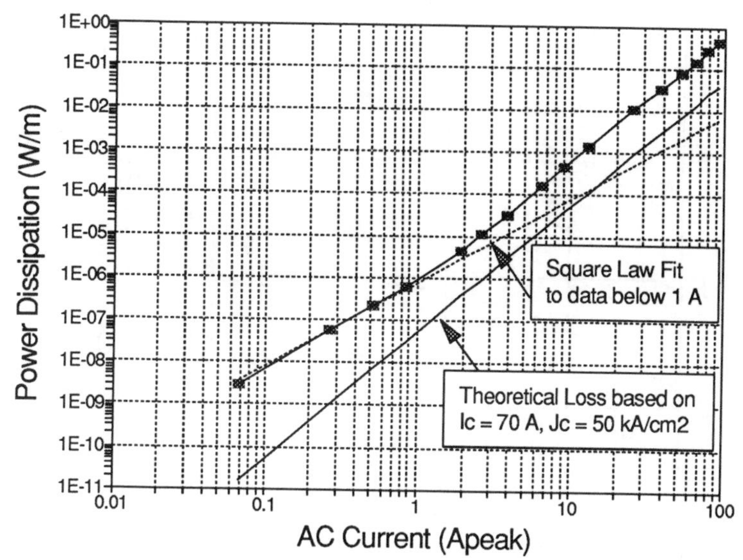

Figure 5. Self field power dissipation at 50.5 Hz for a bundle of five, 100 μm radius melt-textured YBCO fibers bound with silver. The solid line is the theoretical power dissipation based on an I_c of 70 A, which is a lower limit for the DC I_c based on the 1 $\mu V/cm$ criterion. The average I_c computed from the measured power dissipation is 5.8 A. The data have been corrected for eddy current losses in the silver by subtracting the extrapolation of the square law fit.

the reducing adjustments available to the operator. LN_2 cooling was necessary to reduce the resistance of the #18 wire so that currents up to 80 A could be carried without excessive voltage drop or power dissipation. The two coils are wired in series with the voltage taps, so that the coils can be rotated to produce an inductive voltage which is equal and opposite to that generated by the flux from the sample cutting the measurement loop. Adjustment of the coil rotational positions (for the coarse and fine adjustment coils) is made using a dual phase lock-in amplifier to null the inductive voltage. If the magnet is used to impose an external AC field, a second compensation toroid and is used with its coils also wired in series with the voltage taps. With this system of inductive voltage compensation, AC voltage drops to about 10 nV can be measured reproducibly.

AC self field loss measurements were carried out on a sample consisting of five, 0.1 mm radius melt-textured YBCO fibers bound together with silver. The overall sample diameter was about 2 mm, and the distance between voltage taps was about 11 cm. Figure 5 shows self field power dissipation vs conduction current for this sample. Below a conduction current of 1 A, the power dissipation clearly goes as the square of the current. The complex and

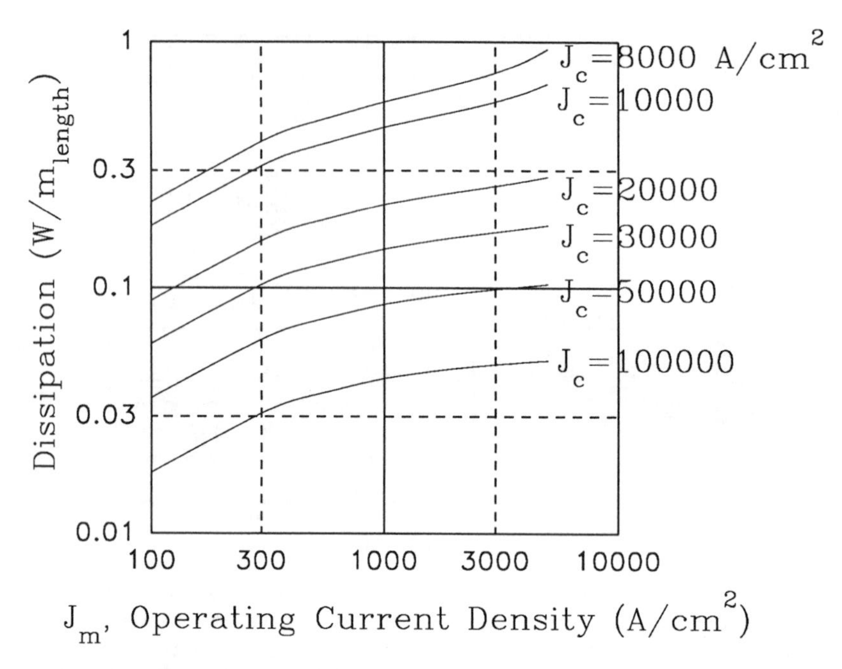

Figure 6. Theoretical self-field power dissipation for a 1500 A_{rms} AC cable conductor (e.g., 900 MW at 345 kV) as a function of operating current density and material critical current density. If the HTSC material conforms to this theoretical prediction, a J_c of 20,000 A/cm^2 is adequate, and J_c's of this magnitude are now being achieved in flexible conductor elements. However, if the DC J_c must be an order of magnitude greater to achieve this level of loss, substantial technical challenges remain.

somewhat irregular sample configuration precludes a rigorous calculation of the eddy current losses in the silver; however, an estimate of these losses is in good agreement with the measured data for sample currents below 1 A. The data plotted for currents above 1 A have been corrected for the eddy current loss by subtracting the eddy current-induced voltage drop indicated by the extrapolation of the fit to the data below 1 A.

Above 1 A conduction current, the loss goes as the cube of the current. However, the J_c for this sample, as determined by the 1 $\mu V/cm$ criterion is certainly no less than 70 A. Plugging a J_c corresponding to 70 A into the above formulae for the AC loss (3) or (4), results in the solid line on the graph, which is about an order of magnitude below the measured loss. A second very different type of sample was also measured with similar results that the measured

AC self field loss was about an order of magnitude greater than the loss calculated on the basis of the 1 $\mu V/cm$ J_c.

When the DC J_c is measured on a long (e.g., 1-metre) sample, a voltage drop is often detectable at a current as much as an order of magnitude below the current which produces 1 $\mu V/cm$. This suggests that for the present quality of HTSC conductors, the J_c relevant to AC self field loss may be as much as an order of magnitude less than the J_c typically measured according to the 1 $\mu V/cm$ criterion. Future experiments with a wider range of samples should improve understanding of this point.

CONCLUSION

An order of magnitude difference in the AC self field power dissipation has substantial implications for practical AC applications such as AC power cable. For example, Figure 6 shows the theoretical conductor power dissipation for a fixed current of 1500 A_{rms} as a function of operating current density (J_m) for various material critical current densities. Note that for a J_c of 20 kA/cm^2 the theoretical power dissipation is below 0.3 W/phase-m at a operating current density of 5000 A/cm^2, which corresponds to a conductor cross section of 0.43 cm or a 0.45 mm thick layer on a cryogen channel of 15 mm radius. A J_c of 20,000 A/cm^2 is already being achieved in appreciable lengths of flexible conductor. However, if the critical current density must be increased by an order of magnitude to achieve the power dissipations shown in Figure 6, this represents a significant technical challenge. As an HTSC power cable is among the simplest of all practical applications as a result of the relatively low operating current density required and the small operating magnetic field, the question of how to measure the J_c relevant to computation of AC self field loss is far from academic. Based on the work discussed above, the best guess would be the most sensitive possible measurement of initial voltage drop, which can be an order of magnitude below the current which produces a voltage drop of 1 $\mu V/cm$ for present conductor elements. Clearly further investigation of the AC self field power loss of HTSC conductors elements is warranted.

REFERENCES

1. E.W. Collings, K.R. Marken, M.D. Sumption, J.R. Clem, S.A. Boggs, and M.V. Parish. "AC Loss and Dynamic Resistance of a High T_c Strand Carrying A Direct Current in a Transverse Magnetic Field". Proceedings of the CEC/ICMC Conference, Huntsville, AL. June 1991.

2. W.J. Carr. "AC Loss and Macroscopic Theory of Superconductors". Gordon and Beach Science Publishers. New York, 1983. p. 76.

3. ibid., p. 70-73.

FABRICATION OF SUPERCONDUCTING TILES FOR MAGNETIC SHIELDING

N.X.Tan, A.J.Bourdillon, Y.Horan and H.Herman

Department of Materials Science & Engineering
State University of New York at Stony Brook
Stony Brook, NY 11794-2275, U.S.A.

ABSTRACT

Superconducting magnetic shields have potential weight advantages over conventional materials. Techniques have been developed for fabricating large area high T_c shields by precursor routes. Shielding properties in high fields depend on many materials features, including alignment, critical current densities and flux pinning. Diffusion texturing can be used to produce grain alignment with c-axis parallel to the sheet plane. The superconducting tile is made by bonding the ceramic precursor $SrCaCu_2$ to a metal substrate before texture growth with Pb-Bi-O. The bonding can be generated by sintering or by plasma spraying. A tile with transition temperature of 110 K and critical current density > 2000 A/cm^2 has been characterized.

1. INTRODUCTION

The following materials are used in various configurations for ac or dc magnetic shielding in applications such as MAGLEV transportation systems, magnetic resonance imaging, encephalography, etc. In the first of these systems weight is an important design consideration [1].

- Ferromagnetic shields or yokes (with some AC shielding) provided by high permeability materials [2] (e.g., Mumetal). Shields have a low magnetic reluctance in DC fields, but the reluctance increases in AC with increasing frequency. The required shield thickness is determined by the saturation field, B_s. The thickness, D, of shield needed to screen a field extended over height h, with average strength B is given by,

$$D = Bh/B_s. \tag{1}$$

Calculations show that 2 cm thick high-silicon sheet steel with B_s=1.2 T (12,000 Gauss) will shield DC fields of 200 Gauss to fields below 2 mT (20 Gauss) in a vehicle cabin [3]. The weight of such shielding is 160 kg m^{-2}.

- Eddy current shields for AC, typically of Aluminum [3]. Shielding occurs within a skin depth,
$$\delta = (\pi \sigma f \mu \mu_0)^{-1/2} \quad (2)$$
at frequency f, where σ and $\mu\mu_0$ are the conductivity and permeability of the aluminum, for which $\delta \sim 8.4 f^{-1/2}$ cm. The amplitude of electromagnetic waves decays with depth, x, inside the metal as $\exp(-x/\delta)$. Thus a field intensity with frequency 500 Hz is reduced by 100 in a thickness of 8 mm weighing 24 kg m^{-2}.

- Active coils to counteract external fields and controlled by sensors [3].

- superconducting shields, operating close to the temperature of liquid helium, e.g., at T<15 K for Nb_3Sn [3]. The thickness of material needed for perfect shielding depends on the penetration depth, l(T), which is proportional to temperature as $1-(T/T_c)^4$ [4]. The shields can be made as thin as 25×10^{-6} m, provided they are sufficiently homogeneous and that mechanical strength is provided by a substrate. The shields are efficient at screening both DC and AC. The chief weight factor involved in the use of conventional superconductor shielding lies in the cryogenic engineering.

- High T_c systems
Comparatively simple cryogenic insulation for the high T_c systems offer much greater design flexibility than is possible for conventional superconductors. This is particularly true if the shields dual as the outer temperature shield of the superconducting magnets. The critical transition temperature of $YBa_2Cu_3O_{7-x}$ [123], $Bi_2Sr_2Ca_2Cu_3O_{10}$ [2223] and $Tl_2Ba_2Ca_2Cu_3O_{10}$ [T2223] are 93, 110 and 125 K respectively. Published data show that high shielding factors observed [4,5] (at frequencies 0.1<f<1000 in an external field of B<0.1 mT (1 Gauss) and specimen temperature of 77 K) are justified by measured materials properties including the critical fields, H_{c1} [6] (fields of 100 Gauss, 10 mT, are shielded at 40 K) and H_{c2} [7], and critical current density, J_c [8,9]. Shielding occurs even when the applied field is greater than H_{c1}, as described by the Bean model [10]: there is a region of zero field inside the superconductor if the applied field,
$$B^* < \mu_0 J_c D/2, \quad (3)$$
where μ_0 is the permeability of free space, J_c is the critical current density of the specimen and d is its

thickness (see figure 1). The applicability of this model depends on sufficient flux pinning forces.

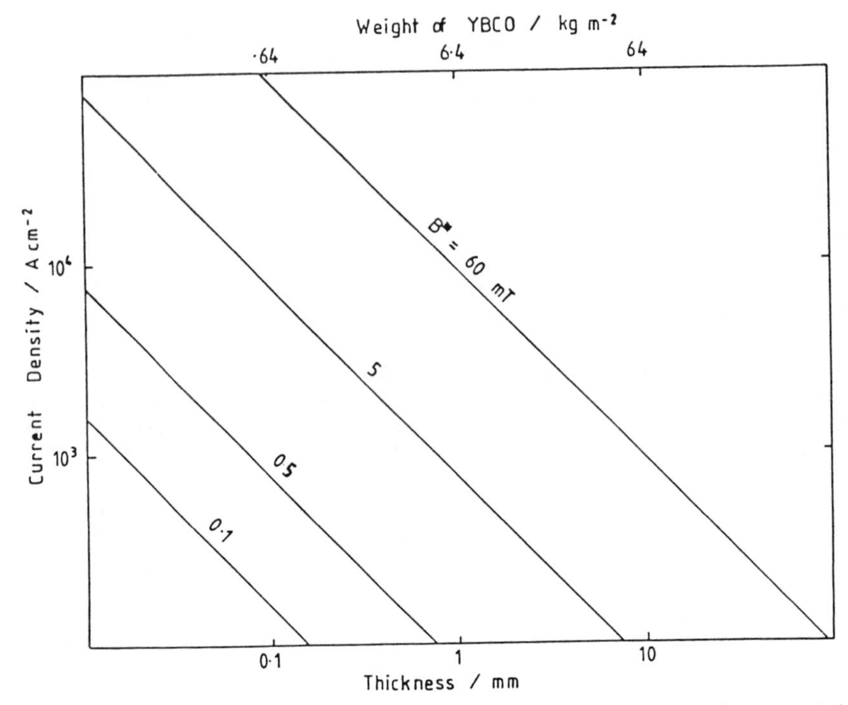

Figure 1. Relationship between current density, shield thickness and applied field, B^*, derived in equation 3. Also shown is the weight per m^2 of a shield made of $YBa_2Cu_3O_{7-x}$, dependent on thickness and density. The other superconductors described have similar densities.

2. FABRICATION OF HIGH-T_c TILES

Alignment is required in superconducting shields for the following three reasons: (1) to gain optimal shielding from anisotropic material against directional applied magnetic fields; (2) to shield, without excess weight, high magnetic field strengths requiring high critical current density in the superconducting material and (3) to maximize flux pinning through orientation effects.

We have previously reported a diffusion texture technique in the Bi-Sr-Ca-Cu-O system [11]. This texture technique involves liquid phase sintering of one Bi-Pb-O precursor (compounded from $3(Bi_2O_3).PbO$) and a second, novel, ceramic precursor, i.e. $SrCaCu_2O_x$ [0112], previously bonded to a metal substrate.

A feasibility study of tile fabrication was firstly carried out by the brush-on method as described below. In a second experiment, plasma spray deposition was used to produce the [0112] precursor. In plasma spraying a gas is ionised while it passes through a tungsten cathode and copper anode. Upon recombination, energy is released which results in a flame with a core temperature in the vicinity of 15,000°C. Steep temperature and velocity gradients in both the radial and axial directions exist within the flame. The particles are ejected after the gas exits the nozzle and, depending on the particles' trajectories, the powder melts or partially melts while being accelerated towards the substrate. Upon impact, the molten particles rapidly solidify to form lenticular splats. The splats impinge upon each other forming a layered coating as the gun traverses the substrate. Plasma spray processing is widely used to deposit both metallic and ceramic materials. Among the latter, high temperature superconductors have been successfully produced by APS (air plasma spraying) processing [12]. Furthermore as described here, large areas of homogeneous [0112] can be rapidly deposited by APS with a judicious choice of deposition parameters.

3. EXPERIMENTAL PROCEDURES

a) Brush-on method.

[0112] powder was pressed onto a Ni (99.99% pure) sheet (~ 10 μm thick, 5 X 5 cm^2). The composite was sintered at 985°C for 12 hours to form a firm bond between the metal substrate and the ceramic precursor. The specimen was naturally cooled down by switching off the furnace. Finally, a layer of Bi-Pb-O slurry mixed with methanol was brushed onto the substrate before sintering at 845°C for 100 h to promote the formation of [2223] phase [13].

b) Plasma Spray.

The substrate was grit-blasted and placed in an ultrasonic cleaner to enhance coating adhesion before spraying. A hand held Metro 3mB air plasma spray gun was used to deposit the [0112] powder onto Co-Fe alloy substrates (1 mm thick, 1 X 5 cm^2). The plasma parameters were 60 V, 500 A with Ar and H_2 plasma gases at flow rates of 755 SLPm and 2.5 SLPm respectively. The Ar carrier gas fed the powder into the flame with a flow rate of 4.5 SLPm, and the spray distance was about 75 mm.

After examination of the sprayed [0112] precursor, a layer of Bi-Pb-O mixture was brushed onto [0112], and finally sintered at 845°C for 72 hours.

4. RESULTS AND DISCUSSION

The microstructure of the deposited layer was studied on an ISI 30X Scanning Electron Microscope (SEM) equipped with a PGT energy-dispersive spectrometer (EDS). The superconducting transition temperature (T_c) and critical current density (J_c) of the specimens were measured resistively by the standard 4-probe method.

a) Brush on method.

Strong bonding was found with no visible cracks in the [0112] after the solid state reaction between [0112] and Ni sheet. The reaction of Bi-Pb-O with [0112] forms a superconducting BSCCO phase with T_c~110 K (figure 2). The J_c, measured at 77 K, is greater than 2000 A/cm^2, the measurement being limited by Joule heating at the resistive contacts.

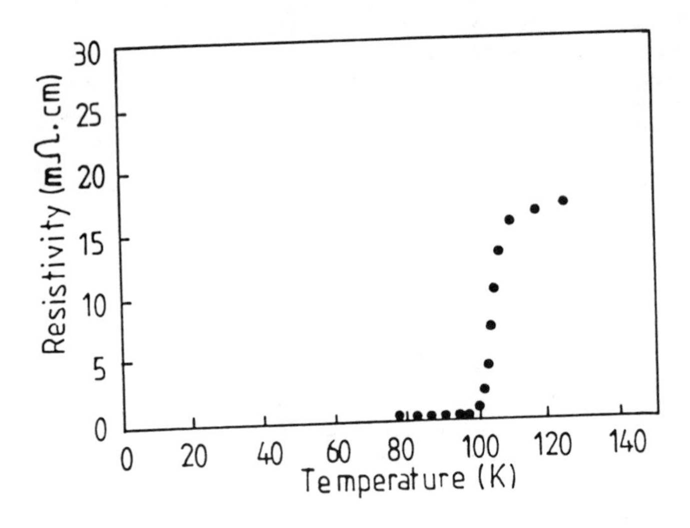

Figure 2. Temperature dependent resistivity of tile made by brush-on method after sintering for 100 h.

b) Plasma sprayed [0112] precursor.

Starting [0112] powder was prepared by repeated pressing, sintering and grinding. Several considerations determine optimum particle size: firstly if the particle size is below 10 μm, the fine particles within the powder clog the carrier feeding-tube so that inadequate amounts of powder are supplied to the flame. Secondly, fine particles lack sufficient momenta to enter the center of the plasma, where maximum heat transfer can be obtained. Instead, the fine particles travel through the outer turbulent gas-flow region. Thirdly oxide particles generally have relatively low thermal conductivities. The outer surfaces of larger particles can vaporize before heat is conducted to the inner core. Total melting does not then occur. On the other hand, finer particles may totally vaporize within the plasma flame. All of these considerations affect the deposition rate.

Thus a careful balance between powder size, composition and plasma parameters must be established to produce dense coating with the molten particles. The optimum [0112] particle size was found between 44 μm and 80 μm to produce a 0.5 mm thick coating. EDS analysis shows correct, homogeneous composition, and a typical splat-formed surface was observed (figure 3).

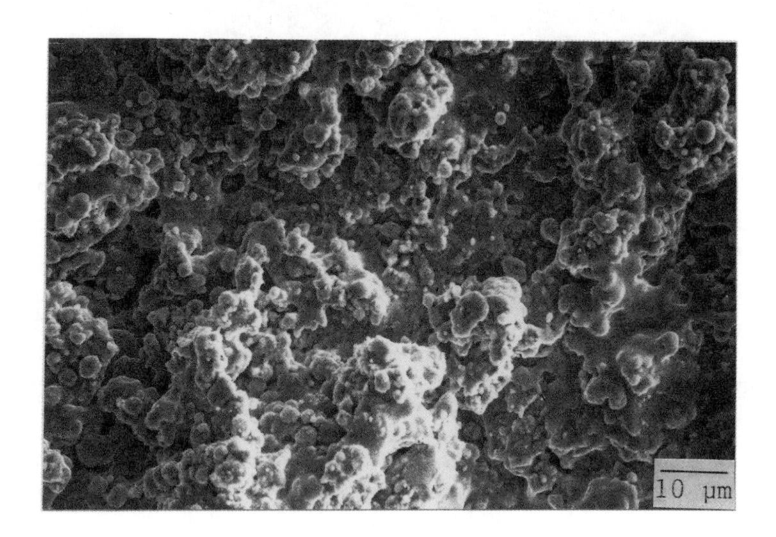

Figure 3. Secondary electron image showing the coated surface of [0112] precursor.

After the coated [0112] layer was reacted with the Pb-Bi-O compound, crystal alignment was observed in SEM (figure 4). The metal substrate, in this case a Co-Fe alloy, oxidized during the sintering reaction in air. By comparison with earlier results from the brush-on method, we found that Ni metal is the superior substrate for this application.

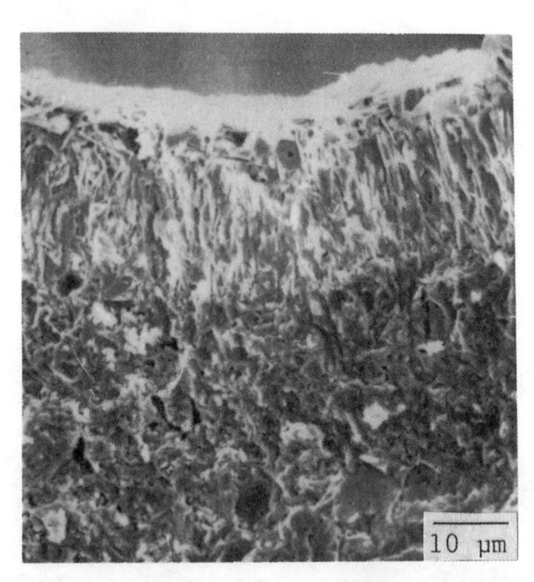

Figure 4. Secondary electron image showing alignment of superconducting BSCCO, after the coated [0112] precursor reacted with the Bi-Pb-O compound.

These experiments show that large areas of aligned superconducting sheets can be bonded to metal substrates by either brush-on or plasma-spray techniques through diffusion texture processing. The shields have great potential, since the planes with high J_c lie normal to the sheet plane. This is the orientation which, on the Bean model, gives greatest shielding in high fields, but further studies of flux pinning in directional magnetic fields are needed to confirm the applicability of the model.

Acknowledgement: The authors wish to thank Mr. G. Bancke for his help with plasma spraying. This work was carried out under grant number 90F009 from the New York State Institute for Superconductivity.

REFERENCES

1. A.J.Bourdillon, Proceedings of Workshop on MAGLEV, Brookhaven May 24-25, 1991, in press

2. V.O.Kelha, J.M.Pukki, R.S.Peltonen, A.J.Penttinen, R.J.Ilmoniemi and J.J.Heino, IEEE Trans. Magn. 18 260-270 (1982)

3. D.V.Gubser, S.A.Wolf, T.L.Francavilla, J.H.Classen and B.N.Das, IEEE Trans. Magn. 21, 320-323 (1985)

4. K.Hoshino, H.Ohta, E.Sudoh, K.Katoh, S.Yamazaki, H.Takahara and M.Aono, Jap. J. App. Phys. 29, L1435-L1438 (1990)

5. K.Shigematsu, H.Ohta, K.Oshino, H.Takayama, O.Yagishita, S.Yamazaki, H.Takahara and M.Aono, Jap. J. Appl. Phys. 28, L813-L815 (1989)

6. A.Umezawa, G.W.Crabtree, K.G.Vandervoort, U.Welp, W.K.Kwok and J.Z.Liu, Physica C, 162-164, 733-734 (1989)

7. U.Welp, W.K.Kwok, G.W.Crabtree, K.G.Vandervoort and J.Z.Liu, Physica C 162-164 735-736 (1989)

8. B.D.Biggs, M.N.Kunchur, J.J.Lin, S.J.Poon, T.R.Askew, R.B.Flippen, M.A.Subramanian, J.Gopalakrishnan and A.W.Sleight, Phys. Rev. B 39, 7309-7312 (1989).

9. B.D.Biggs, M.N.Kunchur, J.J.Lin, S.J.Poon, T.R.Askew, R.B.Flippen, M.A.Subramanian, J.Gopalakrishnan and A.W.Sleight, Phys. Rev. B 39, 7309-7312 (1989).

10. see e.g. D.R.Tilley and J.Tilley, Superfluidity and Superconductivity, Hilger 1986

11. N.X.Tan, A.J.Bourdillon and W.H.Tsai, submitted to Mod. Phys. Lett. B, (1991)

12. W.T.Elam, J.P.Kirkland, R.A.Neiser, E.F.Skelton, S.Sampath and H.Herman, Adv. Ceram. Mat. 2 411 (1987)

13. H.K.Liu, S.X.Dou, A.J.Bourdillon, M.Kviz, N.X.Tan and C.C.Sorrell, Phys. Rev. B 40 (1989) 5266

MAGNETIC RADIATION SHIELDING:
An Idea Whose Time Has Returned?

Geoffrey A. Landis
Sverdrup Technology, Inc.
NASA Lewis Research Center 302-1,
21000 Brookpark Rd., Cleveland, OH 44135

ABSTRACT

One solution to the problem of shielding crew from particulate radiation in space is to use active electromagnetic shielding. Practical types of shield include the magnetic shield, in which a strong magnetic field diverts charged particles from the crew region, and the magnetic/electrostatic plasma shield, in which an electrostatic field shields the crew from positively charged particles, while a magnetic field confines electrons from the space plasma to provide charge neutrality. Advances in technology include high-strength composite materials, high temperature superconductors, numerical computational solutions to particle transport in electromagnetic fields, and a technology base for construction and operation of large superconducting magnets. These advances make electromagnetic shielding a practical alternative for near-term future missions.

INTRODUCTION

A significant difficulty for a manned missions outside of the Earth's magnetosphere, including Mars missions, asteroid exploration, and space-based mining and manufacturing, is the hazard of crew exposure to particulate radiation. With the recent resurgence of interest in manned Mars missions, crew radiation shielding has again become an active problem for investigation [1].

Two types of radiation are particularly significant: solar flare protons, and high-energy galactic cosmic rays (GCR). Solar flare protons come in bursts, lasting a day or so, following an energetic solar event. The proton flux is omnidirectional: although the source of the radiation is the sun, the actual radiation comes from all directions, and hence the crew must be shielded in all directions, and not just in the direction of the sun. In the absence of shielding, a single large solar flare would likely be fatal to the crew, either immediately or as a result of cancers induced by the radiation dose. Cosmic rays are a continuous background consisting of extremely high energy heavy nuclei, and are also omnidirectional. While the GCR background is not immediately fatal, the integrated GCR dose over long (>1yr) missions will approach or exceed the recommended maximum allowable whole-body radiation dose, and also may result in other significant health problems to the crew.

On the Apollo missions, the approach to crew protection was simple: on notification of a large solar flare, the mission would be aborted to Earth. Since the missions were short, the cumulative fluence of galactic cosmic rays was not significant. This approach, however, is not possible for a Mars mission, where return to Earth times will be many months, not significantly shorter than the total mission duration; and would be unlikely for a space colony or manufacturing facility in Earth orbit, with the goal of continuous occupation.

The dangerous components in both solar flare and GCR radiation consists of positively charged particles. Neutral radiation (gammas, neutrons) are a negligible

component of the radiation ambient; negative particles (electrons), while present, can be easily shielded. The positive particles, however, are extremely penetrating, and require massive shields. In the case of GCR, a small amount of mass shielding has no benefit, or even negative benefit over no shielding at all, since the impact of GCR nuclei on a light shield can produce secondary radiation more intense than the original GCR.

ACTIVE ELECTROMAGNETIC SHIELDING

Since the particles involved are charged, an alternative solution to the problem of shielding is the use of active electromagnetic shields. The simplest such device is the magnetic dipole shield. The magnetic field of the Earth is a good example of a magnetic shield, and is responsible for the relatively benign radiation environment on Earth. A magnetic shield makes use of the fact that a charge particle's trajectory in a magnetic field is curved. As a particle enters the region of high magnetic field, its trajectory will curve away from the region to be protected. In essence, the principle is exactly the reverse of that involved in a magnetic bottle; in this case the intent is to trap the particles *outside* the region of interest, instead of inside. The advantages of a magnetic shield to crew safety and health are obvious.

An additional advantage of the magnetic shield is that no secondary radiation is produced by interaction of the shield with the incident radiation.

The concept of magnetic shielding using superconducting coils for space vehicles was first discussed by Levy [2] and analyzed in some detail in the 1960's and early 1970's [3-7], but was not pursued, primarily because there were no plans for long-duration manned flights which would benefit from such a shield. Magnetic radiation shielding received a brief resurgence of attention in the late 1970's, when the concept of large space colonies was introduced by O'Neill, as an alternative to simple mass shielding for the large populations envisioned [8-9].

A recent paper by Cocks [10] discusses many of the issues considered here, concluding hat the utility of magnetic shielding is highest for very large coil sizes, and suggesting that a large, deployable high-temperature superconducting coil may be usable to protect a Mars mission from solar flare protons.

The limit to the mass required to produce a magnetic field is set by the tensile strength of materials required to withstand the magnetic self-force on the conductors [11]. For the minimum structure, all the structural elements are in tension, and from the virial theorem, the mass required to withstand magnetic force can be estimated as [12]:

$$M = (\rho/S_o)\,(B^2V/2\mu_o) \qquad (1)$$

where ρ is the density of the structural material, S_o is the tensile strength, B the magnetic field, V the characteristic volume of the field, and μ_o the permeability of vacuum.

An alternative to the magnetic shielding is to use an electrostatic shield. Since both solar flare protons and heavy nuclei in GCR radiation are positively charged, it would be thought that shielding would simply require adding a positive charge to the object to be shielded sufficient to repel the particles. To shield against solar flare protons would require an electrostatic potential of on the order of 10^8 volts; shielding against cosmic ray nuclei as well would require as much as 10^{10} volts.

Unfortunately, the situation is complicated by the interplanetary plasma. Attempting to simply charge a vehicle in Earth orbit or in interplanetary space would result in electrons being attracted to the vehicle. This would discharge the vehicle very quickly. Levy and Janes [13] estimate that to maintain a charge of $2 \cdot 10^8$ volts would require a power of 10^7 kilowatts. Concentric spheres of opposing charge, proposed by Birch [7] to repel both positive and negative particles, would reduce the discharging, but maintaining the required voltage difference over a small distance is beyond the range of current technology.

A solution proposed by Levy and Janes [13-14] is a hybrid of the magnetic shield and the electrostatic shield, the "plasma shield." An electrostatic charge is applied to the vehicle (or habitat) shell to repel the positively charged radiation; a magnetic field then prevents the plasma electrons from discharging the vehicle. At large distances the shield is charge neutral, since the magnetically confined electrons exactly neutralize the charge on the shell (see figure 1). Since electrons are lighter than protons by a factor of 1860, the magnetic field required for the plasma shield is reduced over that for a simple magnetic shield by the same ratio, and hence the weight associated with the field generation and structure. This ratio is even more favorable for shielding from cosmic rays.

The required electron confinement time τ to maintain the charge, assuming a maximum allowable energy expenditure of 10 kW (for the 5m. radius torus assumed), is 100 minutes. With the assumed plasma density n of $2 \cdot 10^9$ e^-/cm^3, the confinement product $n\tau$ is 10^{13} cm^{-3}-sec. Magnetic containment systems for fusion applications have demonstrated $n\tau$ products in excess of 10^{14} cm^{-3}-sec at considerably higher temperature, so maintaining the charge is not unreasonable. They calculate that the system could likely work if the outgassing rate from the surfaces and the incident micrometeroid flux is low enough that the vehicle is not discharged by the ionized particles.

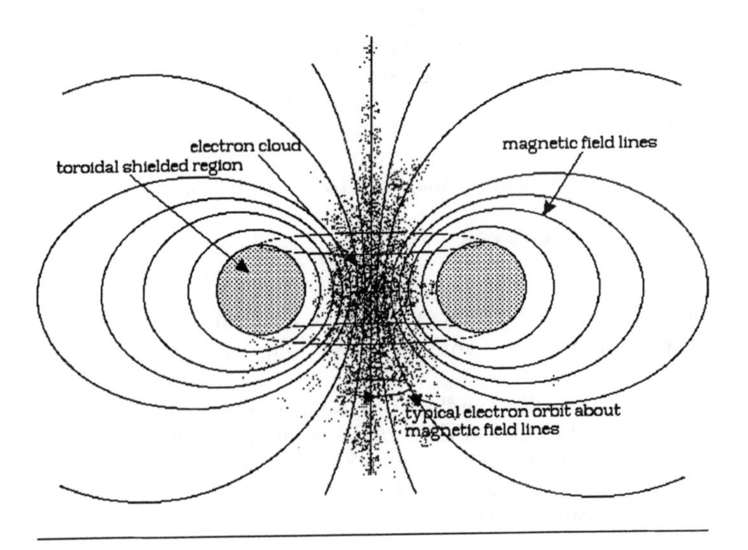

Figure 1. "Plasma" shielding of a spacecraft

French [15] notes that the plasma shield is made even more effective if it is used in combination with a passive mass shield, since the mass shield is most effective for low energy particles. The system is also discussed by Hannah [16], who notes that there are some advantages in confining the magnetic field (and hence also the electron cloud) to a small "core" region of the toroidal habitat.

RECENT ADVANCES

Four technical advances in recent years make magnetic shielding much closer to practicality than the early studies twenty years ago.

First, computers are now in universal use for solving particle trajectory problems by direct numerical integration. Previous solutions for the effective shielding produced by magnetic fields [2-9,17,18] relied on approximations, resulting in solutions in which the shield was completely effective for particles up to some cut-off energy and completely ineffective above this energy. In reality, the particle flux is still reduced, although not eliminated, for energies above the cut-off, and this reduction can be significant in the overall shielding effectiveness. Solving the problem of the shielding produced by any given magnetic field configuration for any energy spectrum is now straightforward, even for systems in which the magnetic field is of complicated geometry. This problem has been solved for other applications [19]. In terms of plasma shielding, the field of particle transport in a plasma, once nearly unknown, is now well understood from the work done in the fusion energy program.

Second, recent advances in high-temperature superconductors mean that it is reasonable to consider superconductors operating at 77°K or higher [12]. This is a range which will allow use of passive cooling, where the temperature is achieved by directly radiating excess heat to space [10,11], a considerable advantage over use of superconductors requiring liquid He temperatures.

Third, there is now a large body of experience in fabricating large superconducting magnets. Superconducting magnets are now standard technology on large particle colliders, such as the Fermilab "Tevatron" and the planned Superconducting Supercollider, as well as for large magnetic fusion experiments.

Finally, extremely high strength to weight composite materials are now available. Since the limit to magnetic field strength is produced by the tensile strength of materials required [12], composite materials with strength to weight ratios five times (Kevlar 49) to 7 times (PBO) higher than that of steel allow considerable weight reduction in the tension members.

These advances make magnetic shielding an extremely attractive option for long space missions.

REQUIRED TECHNOLOGY

The most important enabling technology is the ability to form high-temperature superconductors into wires. While magnetic shielding can be done using conventional (low temperature) superconductors, the payoff in simplicity and weight of higher temperatures is so great as to be mission-enabling. Clearly, advances in the critical current and the transition temperature also allow significant gains to be made as well.

A second required technology which must be demonstrated is the cooling of wires to the superconducting transition temperature using passive cooling, essentially shielding the wire from the sun and allowing it to radiate to deep space.

An as-yet unresolved question on the use of magnetic shields involves the question of whether the field region can be allowed to penetrate into the crew area. Almost all existing magnetic shield designs have been designed with the requirement that little or no magnetic field penetrate into the inhabited region, due to concern about the (unknown) hazardous effects of long-term exposure to magnetic fields. However, considerably simpler engineering designs can be made if some magnetic field is allowed to penetrate into the shielded region. While magnetic fields in general are not hazardous to humans, it is not known what effect high steady-state magnetic fields have.

CONCLUSIONS

Magnetic radiation shielding and magnetic/ electrostatic "plasma" shielding are concepts for shielding against the space radiation environment which have reached new levels of practicality due to advances in technology in the decades since they were first proposed.

Magnetic shielding is an idea whose time has returned.

ACKNOWLEDGEMENT

This paper was presented at the 10th biannual Princeton/SSI conference on Space manufacturing, May 15-19, 1991. The author would like to thank the AIP and the NY State Institute on Superconductivity for the opportunity to present it again to the audience interested in applications of superconductivity.

REFERENCES

1. B.C. Clark and L.W. Mason, The Radiation Showstopper to Mars Missions, *Case for Mars IV,* University of Colorado, June 4-8 1990.

2. R.H. Levy, Radiation Shielding of Space Vehicles by Means of Superconducting Coils, *ARS Journal, Vol. 31*:11, 1568-1570 (1961) .

3. R.E. Bernert and Z.J.J. Stekley, Magnetic Shielding Using Superconducting Coils, in A. Reetz, Jr., (Ed) *Second Symposium on Protection Against Radiations in Space*, NASA SP-71, Gatlinburg, TN, Oct. 1964, 199-209.

4. A.D. Prescott, E.W. Urban and R.D. Sheldon, Application of the Liouville Theorem to Magnetic Shielding Problems, in A. Reetz, Jr. (ed) *Second Symposium on Protection Against Radiations in Space*, NASA SP-71, Gatlinburg, TN, Oct. 1964, 189-198.

5. Lockheed Missiles and Space Co. (1964) Feasibility of a Magnetic Orbital Shielding System, Tech Report., Contract AF 04(695)-252, May 1964.

6. J.O. Helgesen and F.A. Spagnolo, The Motion of a Charged Particle in a Magnetic Field Due to a Finite Solenoid With Application to Solar Radiation Protection, AIAA Paper 66-512, 4th Aerospace Sciences Meeting, Los Angeles, CA, 1966.

7. G.V. Brown, Magnetic Radiation Shieding, Chapter 40, *High Magnetic Fields,* Kolm, Lax, Bitter, and Mills (eds.), 370-378, MIT Press, Cambridge, MA (1962).

8. E.C. Hannah, Meteoroid and Cosmic-Ray Protection, in: J. Grey (ed.) *Space Manufacturing Facilities*, proceedings of the Princeton/AIAA/ NASA Conference, May 7-9 1975, AIAA, 151-157 (1977).

9. P. Birch, Radiation Shields for Ships and Settlements, *J. Brit. Interplanetary Soc. Vol. 35*, 515-519 (1982).

10. F.H. Cocks, A Deployable High Temperature Superconducting Coil (DHTSC), *J. Brit Interplanetary Soc. Vol. 44,* 98-102 (1991).

11. R.H. Levy, Author's Reply to Willinski's comment on Radiation Shielding of Space Vehicles by Means of Superconducting Coils, *ARS Journal* **32**:5, 787 (1962).

12. D.J. Connolly, V.O. Heinen, P.R. Aron, J. Lazar and R.R. Romanofsky, Aerospace Applications of High Temperature Superconductivity, *41st Congress of the International Astronautical Federation,* paper IAF-90-054, Dresden, GDR, Oct. 6-12, 1990.

13. R.H. Levy and G.S. Janes, Plasma Radiation Shielding, *AIAA Journal* **2**:10, 1835-1838; also presented at *Second Symposium on Protection Against Radiations in Space*, Gatlinburg, TN, Oct. 1964, NASA SP-71, 211-215.

14. R.H. Levy and F.W. French, Plasma Radiation Shield: Concept and Applications to Space Vehicles, *J. Spacecraft* **5**:5, 570-577 (1968).

15. F.W. French, Solar Flare Radiation Protection Requirements for Passive and Active Shields, *J. Spacecraft* **7**:7, 794-800 (1970).

16. E.C. Hannah, Radiation Protection for Space Colonies, *J. British Interplanetary Soc.* **30**, 310-314 (1977).

17. A. Bhattacharajie and I. Michael, "Mass and Magnetic Dipole Shielding Against Electrons of the Artificial Radiation Belt, *AIAA Journal, Vol. 2* No. 12, 2198-2201 (1964).

18. S.W. Cash, Magnetic Space Shields, *Advances in Plasma Dynamics,* Anderson and Springer (eds.), 91-166, Northwestern University Press, Evanston, IL (1967).

19. D.G. Andrews and R.M. Zubrin, Use of Magnetic Sails for Advanced Exploration Missions, in: G. Landis (ed.) *Vision-21: Space Travel for the Next Millennium*, NASA CP-10059, 202-210, April 1990.

20. J.R. Hull, High Temperature Superconductors for Space Power Transmission Lines, *ASME Winter Annual Meeting Conf.,* San Francisco, CA, Dec. 10-15, 1989.

CRYOGENIC POWER CONVERSION: COMBINING HT SUPERCONDUCTORS and SEMICONDUCTORS

Otward Mueller
GE Corporate Research and Development
Schenectady, NY, 12301

ABSTRACT

The availability and use of high-temperature superconductors (HTS) will require and enforce completely new electronic systems concepts. One of many possible applications could and probably will be the field of ac/dc, dc/ac as well as RF power conversion at the multi-kilowatt level. Until HTS high frequency switches able to handle hundreds of volts and tens of amperes are invented and produced commercially existing semiconductor devices such as the power MOS field-effect transistor can be used advantageously in order to implement ultra-high efficiency circuits in combination with HTS components such as high Q inductors and capacitors. This marriage could result in a drastic size, weight and cost reduction for various suitable high power applications.

The key to high efficiency power conversion are so-called zero-voltage switching circuits known as single transistor Class E and half-bridge Class D amplifiers. This paper analyzes and discusses some relevant design criteria such as conversion efficiency etc. versus temperature down to 77K.

INTRODUCTION

Three independent developments and facts may in combination lead to an "age of cryogenics". One of these is the introduction of the first large-scale commercial application of (low-temperature) superconductivity at T=4K by the magnetic resonance imaging (MRI) industry. More than 4000 whole-body superconducting magnets have been installed worldwide in hospitals during the past ten years generating a cryogenic power base and demonstrating that cryogenics can be applied successfully, usefully and profitably on a large scale. If such whole-body magnets can be cooled, one certainly can do the same with suitable components of these and other systems.

The second development is, of course, the invention of high-temperature superconductors and the extensive research in this new field. Last but not least, one can mention the fact that most electronic properties of semiconductors and metals are drastically improving if cooled down to cryogenic temperatures. In addition, the thermal conductivity of these same semiconductors as well as that of insulating substrates such as berylliumoxide etc increases manyfold (factor 5-10) at low temperatures [1,6]. Thus, combining superconductors and semiconductors may lead to interesting new possibilities not only in computer technology but also in the field of power conversion providing substantial energy savings. [2-4]

Cryogenic ac/dc, dc/ac power conversion and radio frequency (RF) power generation at the multikilowatt levels can become feasible and economical only if ultra-high efficiency circuits (better than 99%) can be designed, or if at lower efficiencies the duty cycle of operation is relatively small. The feasibility of such circuits will be investigated in this article.

THE CRYOGENIC POWER MOSFET

It is well known that majority carrier field effect transistors are suitable for operation at low temperatures whereas the current gain of cooled bipolar transistors drops drastically [1]. Therefore key parameters such as on-resistance and switching times of commercially available power MOSFETs have been investigated and measured at liquid nitrogen temperature. The results have been reported elsewhere [5-10]. Figure 1 shows the typical on-resistance versus drain current curve of a IRFPG50 MOSFET (1000V, 6A, 2Ω, 180W) for room temperature and at 77K. One can see that at low drain currents the on-resistance of the cooled transistor is about a factor 15 times smaller than at 300K. With increasing drain current the improvement gets even better with factors of 30 and more. The measurements show that at room temperature thermal runaway occurs for the rated current of 6A whereas at T=77K twice the rated current (12A) can be obtained from the same chip in a rock-stable condition continuously. In Figure 2 the measured on-resistance of a 500 V, 11 A p-channel MOSFET IXTH11P50 is presented. Here again the improvement factor at low drain currents is about 15. It increases fast to about 30 at the rated drain current of 11A. In summary one can state that the on-resistance of high-voltage (400-1000V) MOSFETs is drastically decreased by cooling them down to liquid nitrogen temperature. In low voltage devices (100-300V) the improvement is less pronounced with factors of 4-10 [5,6]. Another advantage is the reduction of the reverse recovery time of the drain-source diode by cooling [7,10]. The ruggedness of a MOSFET is also increased due to the fact that the parasitic bipolar transistor inherent in every MOSFET device is becoming ineffective. A disadvantage of the cryogenic MOSFET (77K) is the slight reduction of the breakdown voltage by about 20-25% [10].

THE GAAS POWER MESFET

The on-resistance of a Fujitsu gallium-arsenide power MESFET FLM50MK (15 V, 1.8A) has also been measured in order to evaluate their usefulness for cryogenic operation. As shown in Figure 3 the improvement due to liquid nitrogen cooling is here relatively small in agreement with the fact that in low-voltage devices the channel on-resistance dominates.

THE HIGH-EFFICIENCY CLASS E POWER AMPLIFIER

High conversion efficiency is obtained in switchmode amplifiers which use resonance effects. A classical circuit is known as Class E amplifier invented by N.O. Sokal [11]. Figure 4 shows a circuit diagram. A computer design program for high efficiency RF power amplifier design, simulation and optimization called HEPA-PLUS is available [12] and necessary because all components must be just right if high efficiency is to be obtained. It has been used to carry out a paper design of a 1 kilowatt, 6.8 MHz high-efficiency RF power amplifier. The transistor used is the stripline n-channel enhancement MOSFET DE375-102N11 rated 1000V, 11A with an on-resistance of 1.2Ω. (Directed Energy, Inc., Fort Collins, CO, 303-493-1901). Switching times are 7ns or less due to the drastic reduction of the source lead inductances in the RF stripline package.

The design was done with the following parameters: On-resistance: 1Ω, (300K), turn-on switching time: 6ns, turn-off switching time 6ns (300K) and 3ns at low temperature (<200K), L1=1mH, C_{OSS}=175pF (210V), DC supply voltage: 210Vdc, load resistance R2=RL=20Ω. In order to see what inefficiency is generated by just the active element, the Q of all inductors and capacitors in the RF circuit were assumed to be Q=10,000. The gate losses are not included in the program. The "automatic preliminary design" with HEPA led to the following output circuit parameters: C1=69.4pF, L2=3.294μH and C2=202pF. The network loaded Q is 6.8. The computer results are as follows: The Class E amplifier should be able to deliver 1036W with an efficiency of 93.2% at T=300K.

The measurement results shown in Figures 1 and 2 demonstrate that by cooling n- or p-channel power MOSFETs down to T=77K one can achieve a reduction of the on-resistance by a factor 10-30 in high-voltage transistors. On the other hand data sheets of these devices show a relatively steep increase of the on-resistance with temperature above T=300K. (See data sheet of APT1001R1BN). The efficiency was now calculated with HEPA as a function of the on-resistance for the range 0.05 - 2Ω. This corresponds to a junction temperature range of 77K to about 400K. The result is shown in Figure 5 for the drain efficiency defined as RF output power divided by the dc input power. At a junction temperature of about 125°C (~400K) the efficiency has dropped sharply from 93.2% to 87.8%. At liquid nitrogen temperature (77K) it reaches under otherwise the same conditions more than 99%. (99.336% for R(on)=0.06Ω, 99.399% for 0.05Ω.) The inefficiency increases dramatically with the on-resistance and therefore with the junction temperature as shown in Figure 6. The calculated output power levels are 1153W, 1036W and 932W at 77K, 300K and 400K respectively.

In these calculations a quality factor Q=10,000 was assumed for the RF components C1, C2 and L2 (Figure 4) forming the zero voltage switching resonance circuit. The efficiency has also been computed as a function of Q of the series resonance circuit inductor L2 for room temperature (T=300K, R(on)=1Ω). In this case a more realistic value for the Q of the capacitor C2 of Q=3000 has been assumed. The results are plotted in Figure 7. One can see that the amplifier efficiency drops quite drastically from about 94% down to less than 88% if the Q of L2 drops from 10,000 to 100. For a Q=200 the efficiency is about 90%. One may note that for a 1 kilowatt amplifier the power dissipation at that number of 90% amounts to about 100W. For normal operation this requires a relatively big heatsink which gets hot pretty soon. Then the on-resistance goes up reducing the efficiency and increasing the power dissipation even more. In reality, circuits have to be designed also for operation at elevated ambient temperatures, for example 50°C, making everything worse. The use of an air cooling fan of at least 10W power consumption drops the system efficiency automatically by another percentage point and adds cost.

THE CLASS D AMPLIFIER

Another very useful topology for power conversion is the Class D or Half-Bridge amplifier shown in Figure 8. S.A. El-Hamamsy, GE-CRD, has developed a computer program for the zero voltage switching series resonant Class D amplifier [18]. Using two transistors APT6040HN with an on-resistance of R(on)=0.4Ω at T=300K the analysis provides for 500V operation an output power of 2000W at 6.8MHz with an theoretical efficiency of 94.1%. This includes the gate losses but not the losses of the series resonance circuit. A gate resistance of 0.59Ω was assumed. The efficiency mentioned could be obtained continuously only if the MOSFET junction temperature is kept at T=300K which is impossible with a power dissipation of about 120W, even with heatsinks and air blowers. The thermal resistance from junction to case is 0.5°C/W for that device. With a power dissipation of 60W per transistor the junction temperature rises by 30°C to 55°C even if the heatsink temperature could be kept at 25°C ambient temperature. But at 55°C the on-resistance has already increased by 30% which would reduce the efficiency even more generating a higher dissipation.

Assuming that the on-resistance drops by a factor 10 to 0.04Ω at T=77K the cryogenic operation of this Class D amplifier produces 2kW with an calculated efficiency of 99.3% at 450 V. The dissipation per transistor is now decreased by a factor of 8.4 from 60W to only 7W. As gate resistance a number of 0.59/4=0.148Ω was chosen for the analysis since a factor 4 reduction has been observed at 77K for this parameter.

COMBINING SEMICONDUCTORS WITH HT SUPERCONDUCTORS

The previous discussion leads to an obvious conclusion: The concept of ultra-high efficiency cryogenic power conversion with commercially available power MOSFETs or other devices such as IGBTs [13,14], MCTs [15] etc. would greatly be enhanced if the necessary passive components such as the resonance inductors (L2) and capacitors (C2) in Figures 4 and 8 could be implemented with high temperature superconductors reducing the losses of these necessary components by orders of magnitudes. Then, these components can also be cooled without contributing to a loss of cryogenic fluids or increasing the required power of a refrigerator. The challenge here lays in achieving high RF current densities in the HT superconductors. Thin-film HT superconducting inductors can be implemented in the form of spirules on existing 2 inch substrates [16]. Similar technology could be applied to implement ceramic high Q RF power capacitors with HTS thin film interleaved electrode plates. Ceramic capacitors have also been designed whose capacity increases by a factor 4-10 if cooled to cryogenic temperatures [17].

THE CRYOGENIC POWER DEVICE OF THE FUTURE

The measurements have shown that widely used plastic and hermetic packages such as the TO-247 (TO-3P), TO-254 etc. function very well if just dipped repeatedly into liquid nitrogen without breaking. The same is not true for some of the MOSFET power modules. It was also found that the so-called FREDFETs or HYPERFETs, designed for low drain-source diode reverse recovery time by metal doping of the drain region, stop functioning at 77K without being damaged. They recover perfectly if warmed up again. The reason for this is still unclear. Probably some carrier freeze-out occurs somewhere.

Of course, these commercially available power MOSFETs have in no way been designed for cryogenic operation. It is therefore herewith proposed that MOSFET manufacturers develop a specific "cryogenic power MOSFET" series, especially for the high-voltage range (400-1200V). Two approaches are feasible. First, existing MOSFET chips should be mounted differently in a way which is optimized for cryogenic application. In the case of insulated devices the beryllium-oxide substrate should be made much thicker than normally because its thermal conductivity is at T=77K higher than that of a copper metal plate [6] whereas it is just the opposite at room temperature. The problem with all insulated devices is that they cost many times more than non-isolated ones. This disadvantage disappears if liquid nitrogen is used as "heatsink" and insulator at the same time. Now, a chip can be mounted directly on a suitable metal plate such as for example beryllium with excellent thermal conductivity at T=77K. The latter in turn can be connected on an insulating plastic substrate thus providing low-cost isolation.

A second, probably better approach would consist in a radically new device design effort. History teaches that the first transistors have been made from germanium and not from silicon. For the last decades germanium lost out for two reasons: Its narrow band-gap limited the maximum junction temperature to about 100°C only. (Silicon: 200°C). Second, it was not possible to provide a stable surface layer as in silicon devices with SiO2 passivation. The first reason, of course, does not apply any more in the case of a cryogenic device. Actually, germanium has many advantages over silicon, especially at low temperatures. Its electron mobility is at -50°C about a factor two times higher than that in silicon for a doping level of N<10E+14/cubic-cm required for high voltage applications. Hole mobility of Ge is about three times higher under the same conditions than that of Si [20] opening up the possibility of good p-channel Ge MOSFETs. Many power electronic designers would welcome them.

The problem of passivation can be solved by using silicon-germanium heterostructures as has been done for example in the Hitachi Si-Ge HEMT device shown in Figure 9 [21]. It consists of a germanium substrate, a Si-Ge buffer layer and a thin germanium layer acting as a modulation doped high hole mobility area. The latter is covered by a 50%/50% germanium/silicon alloy region which is passivated with a silicon-oxide layer. Hitachi claims that this silicon-germanium high-electron mobility transistor (HEMT) has a hole mobility higher than that of comparable gallium-arsenide devices [21]. HEMTs, of course, work even better and faster at low temperatures. A look at the literature shows that recently a lot of work has been carried out on silicon-germanium transistors combining the best of two worlds [22-25]. One can conclude that these Si-Ge structures offering much higher electron and hole mobilities may be best suited for the realization of an optimized cryogenic power MOSFET. High mobility means lower on-resistance, therefore lower conduction losses and constitutes the real key to ultra-high efficiency (cryogenic) power conversion. Switching losses can more or less be minimized or eliminated with modern resonance techniques. A silicon-germanium high-voltage, high current MOSFET would certainly boost the concept of cryogenic power conversion in combination with HTS passive L/C resonators. Of course, some day, a HTS active three-terminal power device will be invented.

ADVANTAGES of CRYOGENIC POWER CONVERSION

The cryogenic power MOSFET combines the advantages of IGBTs, MCTs and MOSFETs providing low saturation voltages at high speed and at high frequencies. In addition current densities can be doubled or tripled. Bulky heatsinks, power consuming air blowers or water cooling systems are eliminated. Circuits could be made extremely small, light-weight and low-cost. New application areas for high-temperature superconductors are opened up.

APPLICATIONS OF CRYOGENIC POWER CONVERSION

Cryogenic power conversion (CPC) should find applications in those fields where cryogenics is already available such as in magnetic resonance imaging (MRI) machines and where high power levels (1-100kW) are required, for example in motor drive systems. Furthermore, this concept should be useful where space, weight and cost are of a premium such as in ships, airplanes, space vehicles etc. In all high-power pulsed systems such as radars etc. great advantages should be realizable. In addition CPC may even be necessary in order to implement certain HTS applications such as DC power transmission cables for dc/ac conversion etc.

SUMMARY and CONCLUSIONS

The marriage of high-temperature superconductors and high-power semiconductor devices, especially MOSFETs and perhaps also IGBTs, MCTs, GTOs, etc. should bring into existence many interesting new technological systems providing energy savings at reduced cost.

ACKNOWLEDGEMENTS

The author is grateful to Dr. W.A. Edelstein and Dr. S.A. El-Hamamsy for many helpful discussions on the subject of RF power conversion. The support of this project by Dr. P.B. Roemer, Dr. K.G. Vosburgh and GE-CRD management is greatly appreciated.

REFERENCES

1. R.K. Kirschman, Low Temperature Electronics, IEEE Press, NY, (1986).
2. R. Blanchard, Proceedings Powercon 10, D2,1-11, (1983).
3. B.W. Carsten, Proceedings, HFPC (High Frequency Power Conversion Conference, 140-150, (May 1988).
4. R. Severns, Powertechnics Magazine, 4, 32-34, (April 1988).
5. O. Mueller, Cryogenics, 29, 1009, (Oct.1989).
6. O. Mueller, Proc. 20th Int. Power Conversion Conf., Munich, pp.1-15, (June 1990)
7. O. Mueller, Cryogenics, 30, 1093, (Dec. 1990).
8. O. Mueller, Proc. RF EXPO East, (Oct. 1988), RF Design, 12, 29, (Jan. 1989).
9. O. Mueller, Proc. IEEE Workshop on Low Temp.Electr., 94-98, (Aug. 1989)
10. K.B. Hong, R.C. Jaeger, Cryogenics, 30, 1030, (1990)
11. N.O. Sokal, A.D. Sokal, US Patent 3,919,656, Nov. 11, 1975.
12. Design Automation, HEPA-PLUS Computer Program, Lexington, MA, (617-862-3769)

13. T.P. Chow, K.C. So, D. Lau, IEEE Electron Device Letters, 12, 498, Sept. 1991.
14. J.L. Hudgins, S. Menhart, W.M. Portnoy, Proc. EPE, Florence, Italy, (1991), 0-262.
15. C.S. Jezierski, V.A.K. Temple, Proc. IEEE Workshop on Low Temp. Semic. Electronics, (August 1989, 89.
16. Conductus, Inc, Data, 1991.
17. N. Kenny, private communication, (Kenny Associates, Inc. 919-872-2560).
18. S.A. El-Hamamsy, GE-CRD, private communication, 1991.
19. J.F. Kärner, H.W. Lorenzen, W. Rehm, Proc. EPE Firence, 2-500, (9/1991)
20. W.Gärtner, Transistors, Van Nostrand, (1960), 46.
21. C. Brown, Electronic Eng. Times, 4/22/1991, 37.
22. S. Verdonckt-Vandebroek, et.al. IEEE Electron Device Letter, 12, 447, (8/1991)
23. K. Ismail, B.S. Meyerson, P.J. Wang, Appl. Phys. Lett. 58, (19), 2117, (5/13/1991)
24. C.R. Selvakumar, B. Hecht, IEEE Electron Device Lett., 12, 444, (8/1991)
25. S.C. Jain, W. Haynes, Semicond. Sci. Technol. 6, 547, (1991)

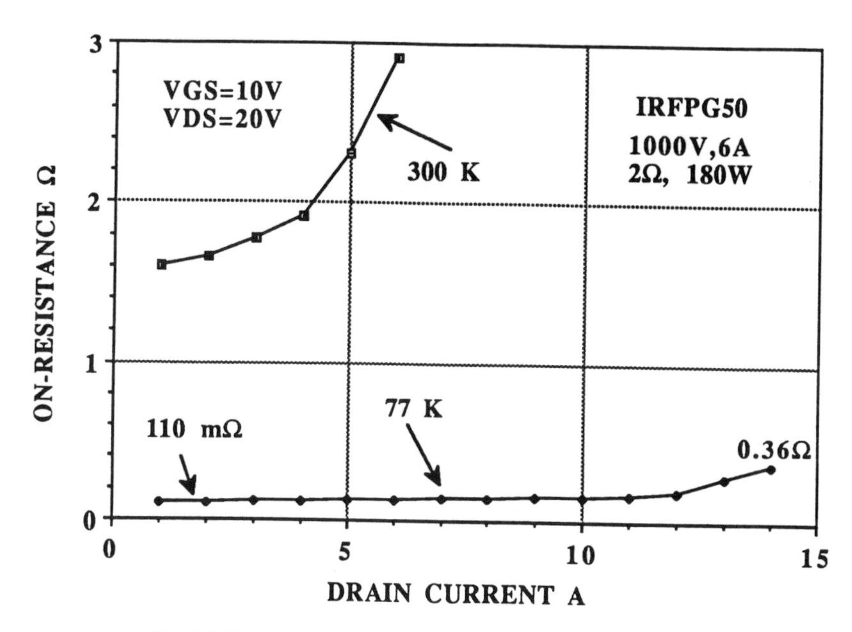

FIGURE 1: ON-RESISTANCE IRFPG50

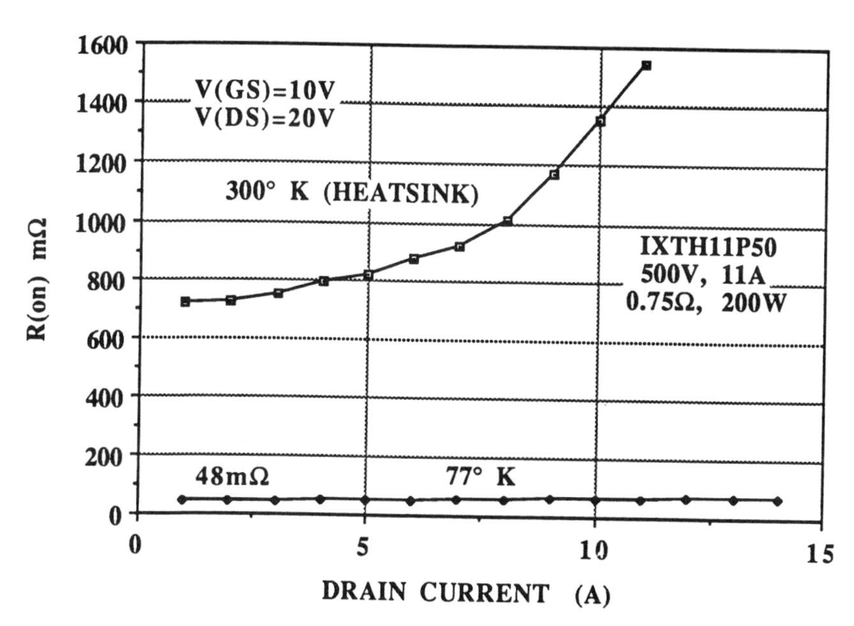

FIGURE 2: ON-RESISTANCE OF IXTH11P50

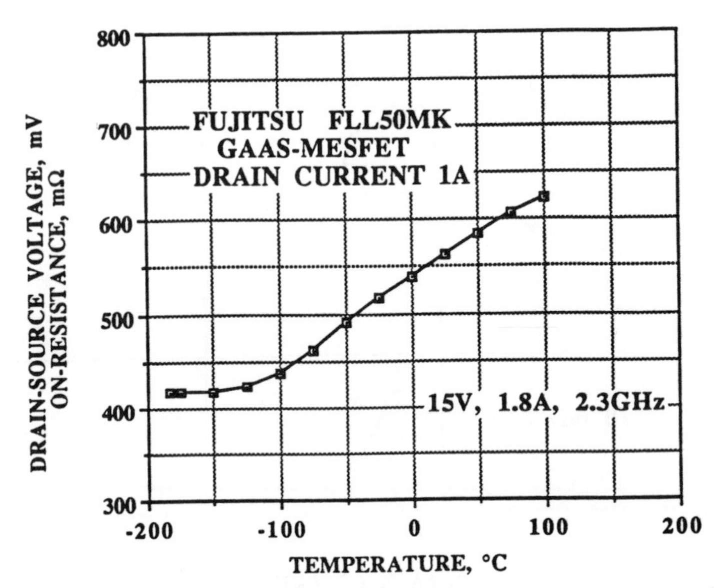

FLL50MK: TEMPERATURE DEPENDENCE of V(DS) and ON-RESISTANCE

FIGURE 3

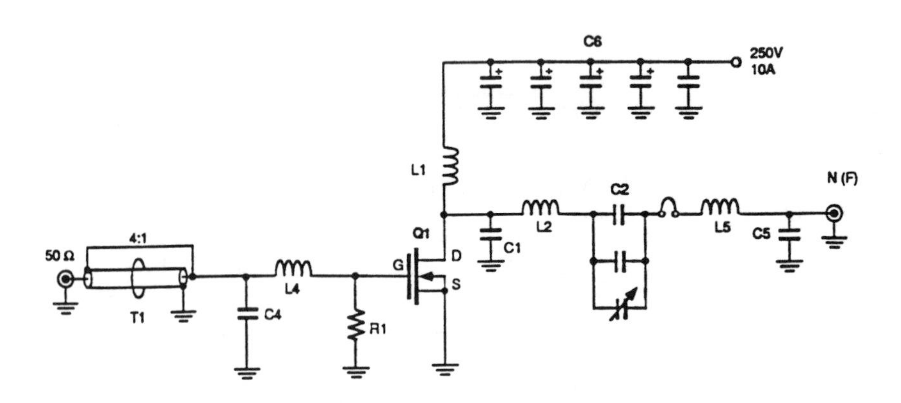

CLASS E AMPLIFIER

FIGURE 4

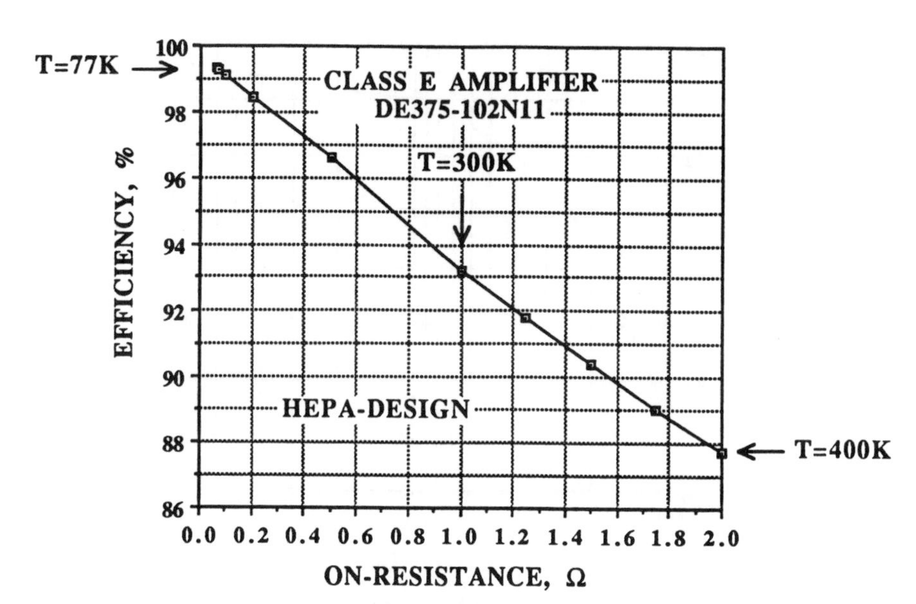

EFFICIENCY VERSUS ON-RESISTANCE
FIGURE 5

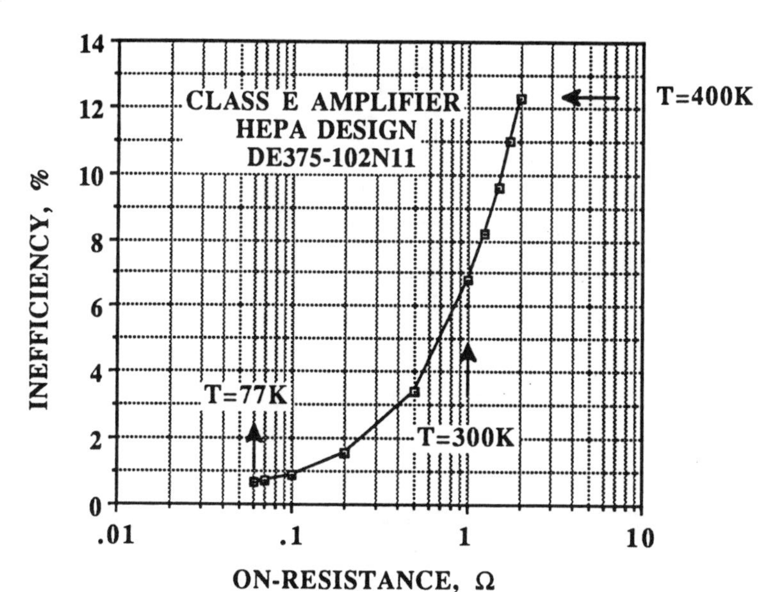

INEFFICIENCY VERSUS ON-RESISTANCE
FIGURE 6

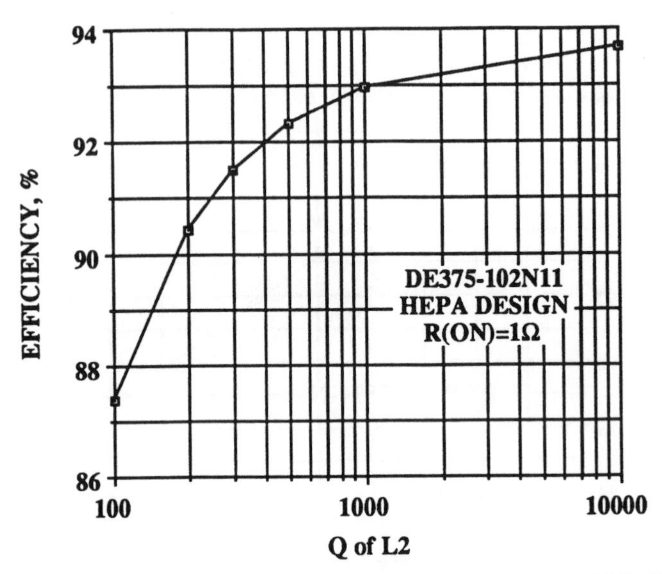

EFFICIENCY VERSUS Q OF RESONATOR INDUCTOR L2
FIGURE 7

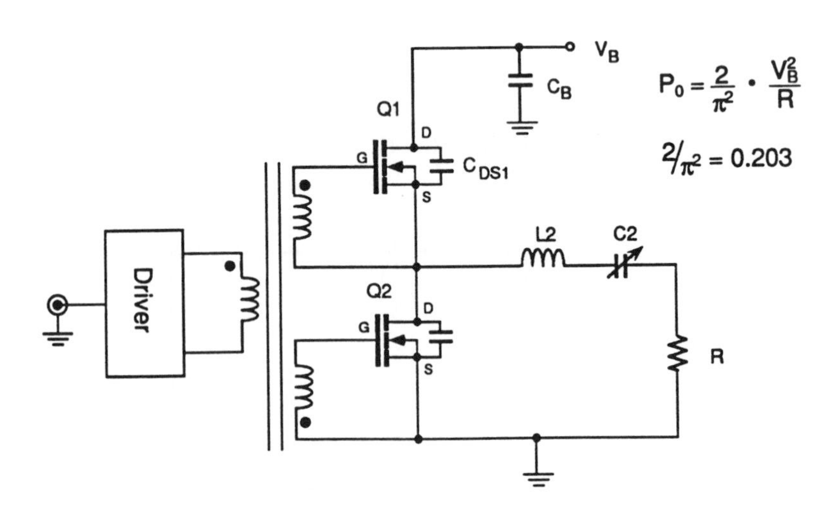

HIGH-EFFICIENCY HALF-BRIDGE CLASS-D RF AMPLIFIER
FIGURE 8

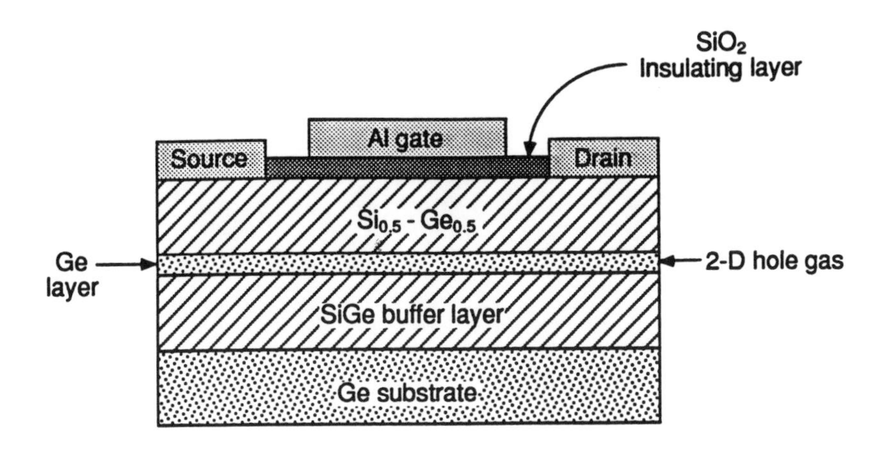

HITACHI'S SILICON/GERMANIUM HEMT
FIGURE 9

Author Index

G

H

I

J

K

L

M

N

O

P

Q

R

S

T

V

W

X

Y

Z

Subject Index
Conference on Superconductivity and Applications

AIP Conference Proceedings

		L.C. Number	ISBN
No. 236	Vacuum Design of Synchrotron Light Sources (Argonne, IL, 1990)	91-55527	0-88318-873-2
No. 237	Kent M. Terwilliger Memorial Symposium (Ann Arbor, MI, 1989)	91-55576	0-88318-788-4
No. 238	Capture Gamma-Ray Spectroscopy (Pacific Grove, CA, 1990)	91-57923	0-88318-830-9
No. 239	Advances in Biomolecular Simulations (Obernai, France, 1991)	91-58106	0-88318-940-2
No. 240	Joint Soviet-American Workshop on the Physics of Semiconductor Lasers (Leningrad, USSR, 1991)	91-58537	0-88318-936-4
No. 241	Scanned Probe Microscopy (Santa Barbara, CA, 1991)	91-76758	0-88318-816-3
No. 242	Strong, Weak, and Electromagnetic Interactions in Nuclei, Atoms, and Astrophysics: A Workshop in Honor of Stewart D. Bloom's Retirement (Livermore, CA, 1991)	91-76876	0-88318-943-7
No. 243	Intersections Between Particle and Nuclear Physics (Tucson, AZ, 1991)	91-77580	0-88318-950-X
No. 244	Radio Frequency Power in Plasmas (Charleston, SC, 1991)	91-77853	0-88318-937-2
No. 245	Basic Space Science (Bangalore, India, 1991)	91-78379	0-88318-951-8
No. 246	Space Nuclear Power Systems (Albuquerque, NM, 1992)	91-58793	1-56396-027-3 1-56396-026-5 (pbk.)
No. 247	Global Warming: Physics and Facts (Washington, DC, 1991)	91-78423	0-88318-932-1
No. 248	Computer-Aided Statistical Physics (Taipei, Taiwan, 1991)	91-78378	0-88318-942-9
No. 249	The Physics of Particle Accelerators (Upton, NY, 1989, 1990)	XX-XXXXX	0-88318-789-2
No. 250	Towards a Unified Picture of Nuclear Dynamics (Nikko, Japan, 1991)	92-70143	0-88318-951-8
No. 251	Superconductivity and its Applications (Buffalo, NY, 1991)	92-52726	1-56396-016-8
No. 252	Accelerator Instrumentation (Newport News, VA, 1991)	92-70356	0-88318-934-8
No. 253	High-Brightness Beams for Advanced Accelerator Applications (College Park, MD, 1991)	92-52705	0-88318-947-X
No. 254	Testing the AGN Paradigm (College Park, MD, 1991)	92-52780	1-56396-009-5
No. 255	Advanced Beam Dynamics Workshop on Effects of Errors in Accelerators, Their Diagnosis and Corrections (Corpus Christi, TX, 1991)	XX-XXXXX	1-56396-006-0